Birkhäuser

Control Engineering

More information about this series at http://www.springer.com/series/4988

Saša V. Raković • William S. Levine
Editors

Handbook of Model Predictive Control

Birkhäuser

Editors
Saša V. Raković
Independent Researcher
London, UK

William S. Levine
Department of Electrical and
Computer Engineering
University of Maryland
College Park, MD, USA

ISSN 2373-7719 ISSN 2373-7727 (electronic)
Control Engineering
ISBN 978-3-030-08466-0 ISBN 978-3-319-77489-3 (eBook)
https://doi.org/10.1007/978-3-319-77489-3

Mathematics Subject Classification (2010): 90C20, 93A30, 93B40, 93C10, 93C95, 93D15, 93E20

This book is published under the imprint Birkhäuser, www.birkhauser-science.com by the registered company Springer Nature Switzerland AG
The registered company address is: Gewerbestrasse 11, 6330 Cham, Switzerland

To my late parents Vojislav T. Raković and Zagorka O. Krunić. And, to Katarina and Teodor.

Preface

We created this handbook because we believe that model predictive control (MPC) is one of the most important recent developments in control theory and its applications. Our reasons for this belief are threefold.

1. MPC is an effective way to control a large and practical class of nonlinear multi-input multi-output (MIMO) systems.
2. Such systems are becoming more and more common and important as a result of improvements in sensors and communications coupled with engineer's desire to improve the overall system performance.
3. An important impediment to the use of MPC has been that it is computationally demanding. But computing has become ubiquitous, faster, and much less inexpensive.

The modern automobile, with over 100 computers, at least two communication networks, and its extensive collection of sensors, is a well-known system that exemplifies these points. A second example is the modern cell phone that includes—basically as throw-ins—GPS, an inertial measuring unit, a camera, and the ability to provide walking, driving, or public transit routes almost anywhere. Yet another example is that many PID controllers are now augmented by both anti-windup and self-tuning capabilities.

With this in mind, we have collected a set of articles that offer an introduction to the concepts, ideas, results, tools, and applications of MPC. It is our hope that anyone with an interest in some aspect of MPC will find, at least, useful information and suggestions for additional reading in this book. We also hope that this handbook will facilitate the further development of the theory and practice of MPC.

This handbook is really the work of the authors of the articles in it. We thank them. We also thank Ben Levitt, Editor, for his encouragement, advice, and support and Samuel DiBella, Assistant Editor, for his assistance with many of the details of publication.

We would also like to express our thanks to all the individuals who contributed to the articles in the book.

London, UK
College Park, MD, USA
March 1, 2018

Saša V. Raković
William S. Levine

Contents

Part I Theory

Part I
Theory

The Essentials of Model Predictive Control

William S. Levine

1 Introduction

Model Predictive Control (MPC) is a method of designing and implementing feedback control systems that, in many situations, perform better than those created by other methods. In addition, MPC provides an effective and general means to design control systems for a large and practically important set of multiple-input, multiple-output (MIMO) systems.

Recent technological advances have substantially decreased the costs and increased the capability of computers, sensors, and communications, making the benefit/cost ratio for computationally intensive control systems larger and larger. These advances have also greatly increased the need for an effective way to design controllers for complex multiple-input, multiple-output systems by making such systems more and more common. Simultaneously advances in the mathematics and computational algorithms for optimization have greatly improved the speed and reliability of the calculations required by MPC.

MPC requires a large amount of real-time computing but recent advances in computing hardware and software have also greatly reduced the cost and improved the speed and reliability of these computations.

This article is meant primarily as an introduction to MPC. By focusing on the historical background and on the simplest problem for which MPC is essential to the practical solution, it is hoped that it will help the reader understand the more sophisticated and complicated problems addressed in the rest of this handbook.

The following section provides the intuitive ideas and some of the history behind MPC. This is followed by a detailed development of MPC for linear systems with simple inequality constraints on their states and controls. The question of stability

W. S. Levine (✉)
Department of Electrical and Computer Engineering, University of Maryland, College Park, MD 20742, USA
e-mail: wsl@ece.umd.edu

S. V. Raković, W. S. Levine (eds.), *Handbook of Model Predictive Control*,
Control Engineering, https://doi.org/10.1007/978-3-319-77489-3_1

for linear time-invariant systems with simple inequality constraints on the states and controls is addressed next. The following section deals with the same collection of systems when they include exogenous inputs. The question of robustness is addressed next. This is followed by a somewhat detailed example of the use of MPC to elucidate the control of posture as it is done by humans. The article then ends with some brief concluding remarks.

2 Background

This section begins with an intuitive introduction to the basic ideas behind MPC. This is followed by a brief and biased summary of its history.

2.1 Intuition

Humans actually use an intuitive version of MPC in many aspects of their daily lives. It is the way that you catch a ball; it is a rational basis for investment. In both cases, you use a prediction of the future to inform your present decision. A particularly vivid example is in playing chess. A good player tries to predict all the possible moves as far into the future as possible before making a move. Unless it is very close to the end of the game, it is impossible to extend these predictions to the game's finish. Thus, the prediction horizon is finite and incomplete. The move the player chooses is the one that he or she believes is the best one (optimal) given the prediction. Once the opposing player has moved, the player repeats the prediction/optimization process based on the new situation. If the opponent has made a predicted move, the player has only to add one more step to his or her previous prediction. If the opponent's move was unexpected, the player adjusts his or her predictions appropriately. This iterative process is repeated until the end game, which has an explicitly computable solution, is reached.

There are several features of this method that should be emphasized. The first is that the prediction is only for a finite number of steps (moves). This is because the cost and complexity of prediction increases very rapidly with the number of steps. The second is that it is necessary to decide on a best move given the predictions. This decision requires a performance measure—a criterion that can be used to evaluate the possible moves. The choice of this performance measure is very difficult in chess. One has to trade off among positional advantage (strategy) and the benefit obtained by taking an opponent's piece (tactics). It is hard to judge the value of a positional advantage and to compare it to the advantage of having an extra piece or two. To some extent, the trade off is influenced by the quality of the opponent. A weak opponent is unlikely to be able to overcome the loss of a piece while a strong opponent may well have deliberately sacrificed a piece in return for a much better position.

The third important feature is that it can be very beneficial in chess to be able to prune the prediction tree. That is, because there are so many possible moves, and responses to those moves, it is very helpful to eliminate those sequences of moves that are obviously bad. It is obvious that the player who can accurately predict n moves ahead will generally beat a player who can only predict m $(m < n)$ moves ahead.

All of these aspects of chess repeat themselves in Model Predictive Control. The basic idea is to use a mathematical model of the system to be controlled to predict its behavior n time steps into the future (hence, model predictive). Of course, the prediction depends, as in chess, on the specific control used.

The second step is to choose the best (optimal) predicted future. This requires a performance measure, that is, a criterion for comparing the different possible futures. Usually, this criterion takes the form of a "cost" to be minimized. Once the best sequence of controls is chosen, the first element of the sequence is applied to the system. This is exactly analogous to making a move in chess. After this first control value is applied, the system responds, a new measurement is taken, and the problem is repeated, exactly as in chess. In contrast to chess, this sequence of steps is often repeated ad infinitum.

As in chess, it can be very difficult to choose a suitable performance measure. The most common choice is quadratic in both the control signal and the system states. The reasons for this will be explained shortly. An important issue in choosing the performance measure is robustness. If one thinks of nature as ones' opponent, the situation parallels that in chess. If the source of the disturbances is expected to be particularly clever and malicious, then a very robust controller is needed. If the disturbances are expected to be relatively benign, then a more aggressive but less robust controller would be superior. Again, this will be discussed at greater length.

2.2 History

There was great interest and excitement about control theory in the 1960s. Most of this centered around three areas of research.

1. The maximum principle—which produced necessary conditions for open-loop optimal controls under reasonably general assumptions.
2. Dynamic programming—which produced optimal feedback controls but required impossible amounts of computation for all but the simplest problems.
3. Lyapunov's method for determining stability—which gave sufficient conditions for stability when a suitable Lyapunov function could be found.

One of the most exciting results in controls during this period was the derivation of the linear quadratic regulator (LQR) and the linear quadratic Gaussian regulator (LQGR) theory [1] and [9]. Kalman, and others, using all three of the tools mentioned above, proved that the optimal controller for a linear time-invariant MIMO system was a linear time-invariant state feedback control. Optimality was with respect to a quadratic measure of performance. This was deemed to be reasonable because energy is, in many situations, quadratic. This linear feedback control was proven to have several very desirable properties. It was guaranteed to exist, to be unique, and to be asymptotically stable under very mild and reasonable assumptions. An explicit formula for the feedback gain was provided and it was relatively easy to compute.

Because understanding the LQR and LQGR is very important to understanding MPC, the theory is summarized here. The discrete-time LQR is first.

The discrete-time linear time-invariant dynamics are required to be

$$x(k+1) = Ax(k) + Bu(k), \tag{1}$$

where: $x(0) = \xi$ and $k = 0, 1, 2, \ldots$
A is an $n \times n$ real matrix and B is an $n \times m$ real matrix.
$x(k)$ is the state at time k and $u(k)$ is the control at time k.

The performance measure (to be minimized) is

$$J(x(o), u_{[0,\infty)}) = \frac{1}{2}\sum_{0}^{\infty}(x^T(k)Qx(k) + u^T(k)Ru(k)) \tag{2}$$

where:
$Q \geq 0$ is an $n \times n$ real symmetric matrix and $R > 0$ is an $m \times m$ symmetric real matrix.

There are two additional properties of the components of the problem (A, B, Q, and R) that are needed in order to guarantee that a solution exists, is unique, and asymptotically stabilizes the closed-loop system. These are that the system is stabilizable and detectable. These are refinements of the more commonly known properties of controllability and observability.

Definition: A linear system is stabilizable if and only if its uncontrollable part is asymptotically stable.

An example of a system that is not stabilizable.

$$x(k+1) = \begin{bmatrix} 2 & 0 \\ 0 & .3 \end{bmatrix} x(k) + \begin{bmatrix} 0 \\ 1 \end{bmatrix} u(k) \tag{3}$$

Note that

$$x_1(k+1) = 2x_1(k)$$

so $x_1(k) = 2^k x_1(0)$. Obviously, there is no controller that can stabilize such a system.

Definition: A linear system is detectable if and only if its unobservable part is asymptotically stable.

An example of a system that is not detectable.

$$x(k+1) = \begin{bmatrix} 2 & 0 \\ 0 & .3 \end{bmatrix} \tag{4}$$

Consider the observation (output)

$$y(k) = Hx(k) = \begin{bmatrix} 0 & 1 \end{bmatrix} x(k) = .3x_2(k)$$

As in the previous example,

$$x_1(k+1) = 2x_1(k)$$

so $x_1(k) = 2^k x_1(0)$

To understand the need for Detectability, note first that because Q is positive semi-definite and symmetric it can be factored as $Q = H^T H$. One can think of $x^T QX$ as $y^T y$ where $y = Hx$ is the "output" of the linear system (1). Thus, the performance measure assigns zero cost to a state that blows up. The optimal controller ignores this state—thereby allowing it to blow up.

Theorem 1. *Given the stabilizable and detectable LTI system* (1) *and the performance measure* (2)*, there exists (implied by stabilizability) a unique optimal feedback control*

$$u_{opt}(k) = -F_{opt} x(k) \tag{5}$$

where:

$$F_{opt} = (R + B^T PB)^{-1} B^T PA) \tag{6}$$

and P satisfies the Discrete-Time Algebraic Riccati Equation (ARE)

$$A^T (P - PB((R + B^T PB)^{-1} B^T P)A + Q - P = 0 \tag{7}$$

Furthermore, the closed-loop (optimal) system is asymptotically stable (implied by detectability).

Important Fact The solution, P, to the discrete-time algebraic Riccati equation (7) can be used to produce a Lyapunov Function $V(x) = \frac{1}{2} x^T Px$ that can then be used to prove that the closed-loop (optimal) system is asymptotically stable to the origin.

There is also a very nice solution to the finite-time version of this problem. In this case, one might as well allow the dynamics as well as Q and R to be time-dependent as this adds no real complication to either the proofs or the results.

Thus, the discrete-time linear time-varying dynamics are

$$x(k+1) = A(k)x(k) + B(k)u(k), \tag{8}$$

where:
$x(0) = \xi$ and $k = 0, 1, 2, \ldots, N+1$

$A(k)$ is an $n \times n$ real matrix and $B(k)$ is an $n \times m$ real matrix for all $k = 1,2,\dots,N$. $x(k)$ is the state at time k and $u(k)$ is the control at time k.

The performance measure (to be minimized) is

$$J(x(0),u_{[0,N)}) = \frac{1}{2}\sum_{0}^{N}(x^T(k)Q(k)x(k) + u^T(k)R(k)u(k)) + \frac{1}{2}x^T(N+1)Q_f x(N+1) \tag{9}$$

where $Q(k), Q_f \geq 0$ are $n \times n$ real symmetric matrices and $R(k) > 0$ is an $m \times m$ symmetric real matrix for all $k = 1,2,\dots,N$.

If N is larger than n, there are upper and lower bounds on the performance by stabilizability and detectability and these bounds are independent of N.

Theorem 2. *Given the stabilizable and detectable LTI system (1) and performance measure (9), there exists a unique optimal feedback control*

$$u_{opt}(k) = -F_{opt}(k)x(k) \tag{10}$$

where

$$F_{opt}(k) = (R(k) + B^T(k)P(k)B(k))^{-1}B^T(k)P(k)A(k) \tag{11}$$

and $P(k)$ satisfies the Discrete-Time Algebraic Riccati Equation

$$\begin{aligned} P(k) = {} & A^T(k)P(k+1) - P(k+1)B(k)((R(k) \\ & + B^T(k)P(k+1)B(k))^{-1}B^T(k)P(k+1))A(k) + Q(k) \end{aligned} \tag{12}$$

With the boundary condition at $k = N$ that $P(N+1) = Q_f$

Note that this implies that the optimal feedback gains are time dependent even if (A, B, C, and R) are constant. Note also that $P(0)$ is a function of N. Nonetheless, it is reasonable to expect that $\lim_{N\to\infty} P(0) = P$ provided that (A, B, C, and R) are constant. This is, in fact, true provided the system and performance criterion are stabilizable and detectable (Please see Proposition 3.1.1 in [2]).

One apparent drawback to these results is the need to feedback the entire state vector. This issue was addressed and a complete solution developed under two reasonable assumptions. The first step is to modify the model of the plant to include partial state feedback and perturbations to both the input and the output. The resulting model of the plant is shown in Figure 1.

In general, $y(k)$ is a p-vector and C is a $p \times n$ matrix. Additionally, the pair A, C is required to be detectable in order for the following results to produce a stable closed-loop system.

Given the basic model in Figure 1, the two most common assumptions are either that the two disturbance signals are independent of each other and individually White Gaussian Noise (WGN) or they are present but negligible. In either case the form of the full-order observer is the same, as shown in Figure 2.

It should be clear from comparing Figures 2 and 1 that the observer consists of two parts. The first part is a model of the plant to be observed. The second is

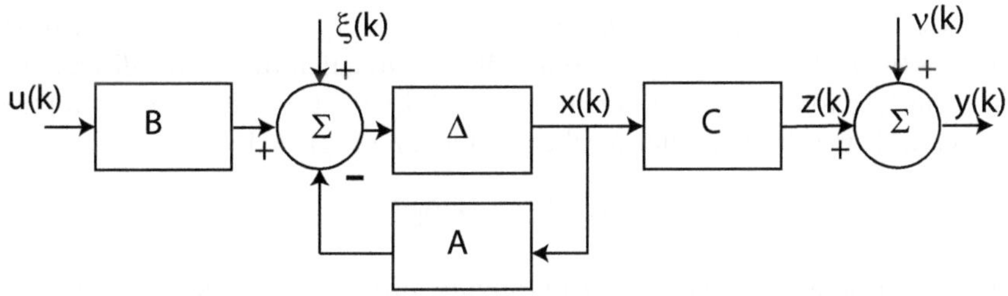

Fig. 1: An LTI plant with a control input, $u(k)$, a state disturbance input, $\xi(k)$, and an output disturbance input, $\nu(k)$.

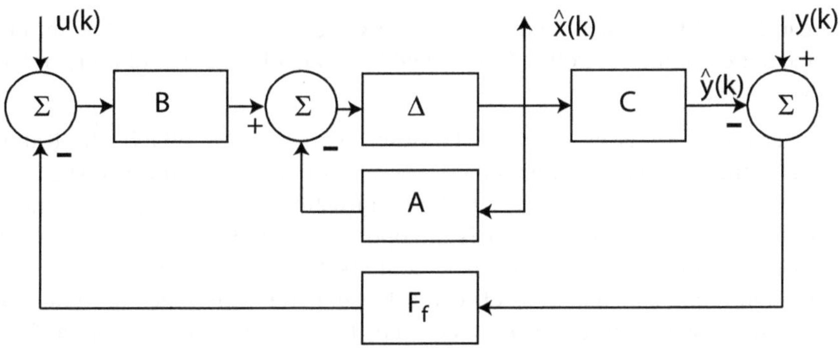

Fig. 2: A linear observer for the plant in Figure 1. The inputs are $y(k)$ and $u(k)$, both of which are known, and the output is $\hat{x}(k)$, an estimate of the state.

a feedback of the error between the predicted observation and the actual one. The intuitive idea is that the plant model will cause the observer to track the plant in the absence of errors in the initial conditions when the noise is negligible. When the noise is WGN, the plant model causes the observer to track the average (mean) state. The feedback of the observer error drives any deviations from the true value towards zero provided the closed-loop observer is exponentially stable.

The difference in the observers is in the choice of F_f. If the disturbances are assumed to be negligible, there is considerable freedom in the choice of F_f. The main requirement is that the resulting closed-loop system (the observer, not the plant) is exponentially stable. As the observer is implemented entirely in software, saturation is not a concern. Placing the observer poles further to the left in the complex plane speeds up the rate at which the estimated state converges to the true state. This is important because, initially, the state is not known. Placing the closed-loop poles of the observer so that the observer bandwidth is too large can be problematic because there is always some noise and it is best to filter it out to the extent possible.

When the disturbances are assumed to be WGN some additional assumptions are needed. Specifically, suppose that the expected values (means) of $\nu(k)$ and $\xi(k)$ are $E(\nu(k)) = 0$ and $E(\xi(k)) = 0$ for all k. Furthermore, assume that the covariance matrices satisfy $E(\xi(k)\xi^T(j)) = \Xi\delta(k-j), E(\nu(k)\nu^T(j)) = \Theta\delta(k-j)$—with $\Theta > 0$ and $\Xi \geq 0$—and $E(\nu(k)\xi^T(j)) = 0$ for all k, j. Note that $\delta(k)$ is the discrete time Dirac delta function so $\delta(k) = 0$ for all $k \neq 0$ and $\delta(0) = 1$. Under these assumptions, the observer is the Kalman filter for the plant and F_f is given by

$$F_f = A\hat{P}C^T(\Theta + C\hat{P}C^T)^{-1}, \tag{13}$$

where $\hat{P}$ is the solution to an ARE that is slightly different from the one for the LQR (please see Equation (7) for comparison).

$$\hat{P} = A(\hat{P} - \hat{P}C^T(\Theta + C\hat{P}C^T)^{-1}C\hat{P})A^T + \Xi \tag{14}$$

Regardless of the design method, the observer can be designed offline, prior to implementation of the controller. One should be aware of a number of issues related to these observers.

- Under the given assumptions—LTI system perturbed by WGN—the Kalman filter is the filter that minimizes the covariance of the error in the estimated state, $E((x(x) - \hat{x}(k))(x(x) - \hat{x}(k))^T)$. If the perturbations are random and White but not Gaussian, the Kalman filter is the optimal **linear** filter with respect to the same performance measure but there could be better nonlinear filters.
- One can think of Θ and Ξ as observer design parameters. They need not characterize the actual disturbances. One needs to be cautious about singular $\Xi x(k)$ because that can cause the Kalman filter to ignore some of its inputs.
- One might ask whether the observer and the state feedback controller are independent. In general, this would not be true. In the LTI case discussed here, it is true. A proof in the deterministic case (negligible noise) can be found in [6] starting on page 537. Please see [2] starting on page 197 for the stochastic case.

A good introduction to observers of all types can be found in [5]. There are many books and articles describing the Kalman filter and its variants. Many of the early papers are reprinted in [15]. A good introduction to the underlying theory is in [4].

Once you have random inputs to the plant the criterion by which you measure performance needs to account for this unpredictability. Intuitively, it makes sense to try to drive the average (mean) error to zero and minimize the error variance. Such a performance measure is

$$J(\Xi_0, u_{[0,\infty)}) = \frac{1}{2}E\{\lim_{N\to\infty}[\sum_0^N(x^T(k)\Xi x(k) + u^T(k)\Theta u(k))]\}. \tag{15}$$

Finding the control that minimizes this performance measure subject to the constraint described by the plant in Figure 1 results in the Linear Quadratic Gaussian Regulator (LQGR).

The LQGR is one of the triumphs of the state space version of control theory as it developed in the very late 1950s and early 1960s. It is a complete feedback solution to the problem of controlling an LTI system with additive WGN disturbances. It provides useful background for stochastic MPC and the previous few paragraphs

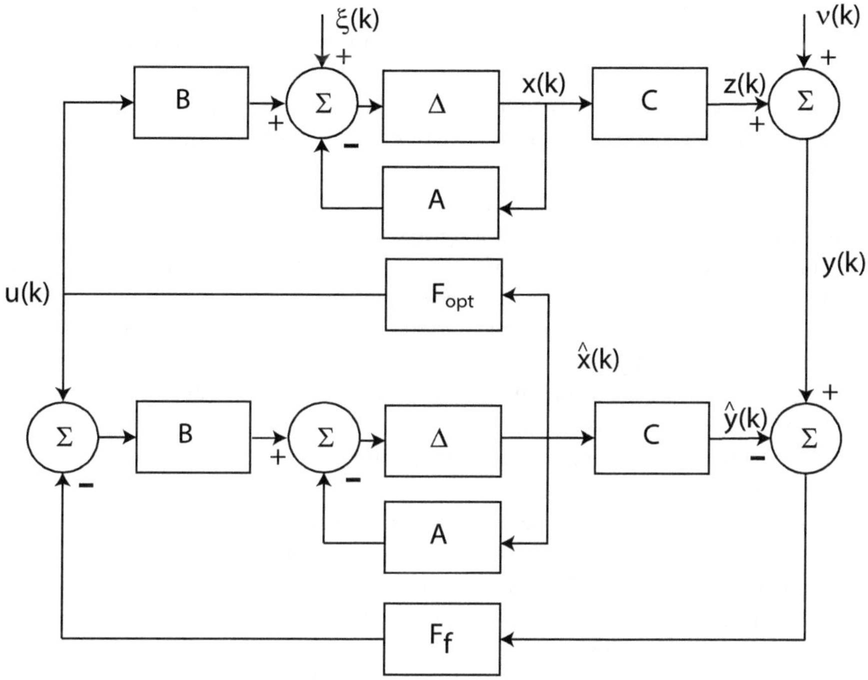

Fig. 3: The optimal control for a LTI plant perturbed by WGN, i.e., the LQGR

developed much of the theory. However, it was eventually shown to result in closed-loop controllers that were not robust. The basic LQGR result is as follows.

Under the previously given assumptions on the plant—including stabilizability of A, B and detectability ofA, H where $H^T H = \Xi$—and the additional assumptions that $x(0)$ is Gaussian, $E(x(0)) = \xi_0$, and $E((x(0)x^T(0)) = \Xi_0$, the optimal control is as shown in Figure 3.

Note that this is a regulator problem. The objective of the control is to drive the mean value of the state to zero. Also, there is a solution to the time-varying version of this problem. It combines the solution to the time-varying LQR with the time-varying version of the Kalman Filter.

Given that these results so thoroughly solve the control problem for linear-time invariant (LTI) systems, it was very surprising to many people that there were relatively few real engineering implementations of this controller. People speculated that there were at least three reasons.

1. Implementing the controller required *mn* gains and this might be prohibitively expensive.
2. Choosing Q and R could be very difficult.
3. The people who actually implemented controllers did not understand the theory well enough.

In fact, all of these conjectures were wrong. The most important reason for the lack of real applications was that the LQR completely ignored actuator saturation. In practice, this is a fundamental and inescapable limitation on most control systems, especially in the process industry. The effort to overcome this limitation on the theory—spearheaded by engineers involved in actually building controllers–led to MPC.

It was not the case that control engineers of all kinds did not know about actuator saturation. Theoreticians had long ago realized that actuator saturation was important and developed a theory of optimal control that could include the effects of saturation. This was the minimum principle of Pontryagin and his colleagues. Unfortunately, the minimum principle had three important flaws even under the best of circumstances. It only provided an open-loop optimal control; it was difficult to compute the solution; the optimal control problem had to cover only a finite time except in some very special cases.

Two approaches to solving the computational problem were considered, and their relative merits debated. The first was to formulate and solve the optimal control problem in continuous time. The continuous-time necessary conditions had to be discretized in time in order for them to be solved on the computer. The second was to discretize the original problem in time and then use linear, quadratic, convex, or nonlinear programming methods to solve the discrete time problem. It is also quite natural for digital controllers.

The open-loop optimal control depended on the initial condition and the future state and control vectors. In practice, feedback controls are far superior to even the best open-loop controls. Thus, it was important to find a way to obtain closed-loop controls even if that meant sacrificing the theoretical optimality.

A number of people recognized that a way to approximate the optimal open-loop controller by a closed-loop controller was to recompute the optimal control at each new discrete-time instant. In order to do this the future state and control had to be predicted out to the time at which the optimal control problem terminated. This prediction needed to be based on a mathematical model of the system to be controlled.

The potential advantages of such a closed-loop approximation to an optimal control were, and remain, obvious. However, optimality over a finite time interval does not, in and of itself, imply stability. This sets the stage for Model Predictive Control.

3 Basics of Model Predictive Control (MPC)

As indicate above, many people saw that a kind of feedback control could be constructed out of a sequence of open-loop optimal controls. Rather than give an abstract description of how this can be done, we describe the specific application of this idea as it is implemented in the simplest MPC algorithm.

Although one can describe a continuous-time theoretical version of MPC, the implementation of MPC is usually in discrete time. Thus, we give only the discrete-time version.

Given dynamics and performance measure identical to those for the infinite-time LQR (please see Equations (1) and (2)), add the common practical constraints

$$U_{min} \leq u_i(k) \leq U_{max} \quad \text{for all } i = 1,2,\ldots,M \text{ and } k = 0,1,2,\ldots,\infty \tag{16}$$

$$X_{min} \leq x_i(k) \leq X_{max} \quad \text{for all } i = 1,2,\ldots,N \text{ and } k = 0,1,2,\ldots,\infty \tag{17}$$

These constraints arise, for example, in a problem as simple as controlling the level of water in a tank with a hole in its bottom by pumping water into the tank. The water level cannot be less than zero or more than the height of the tank. The control is constrained by the maximum pumping capability, both into and out of the tank. Notice two things. First, there are constraints on the state in addition to those on the control. Second, the control constraints are impossible to violate but the state constraints must be enforced by the controller. It can be extremely important to insure these constraints are satisfied as terrible things can happen if they are not.

Once the constraints (Equations (16) and (17) are added to the LQR of Theorem 1, the results of the theorem are no longer true if the constraints are ever active, as they would be in most practical problems. In fact, it is impossible to determine or compute the optimal control exactly. It is possible to approximate the optimal control for this constrained LQR by several methods, including MPC.

To do this by MPC one first replaces the infinite-time limit in the performance measure by a finite-time, N. The resulting optimal control problem then consists of a set of linear equations with variables {x(1),x(2),...,x(N); u(0),u(1),...,u(N-1)}, a performance measure that is a quadratic function of those same variables, and a set of linear (really, affine) constraints on those same variables. This optimization problem starts at $k = 0$ where the initial state $x(0) = x$ is known. It is a quadratic programming (QP) problem. Under our assumptions below, the solution to such a QP always exists, is unique, and can be quickly and accurately computed.

Precisely, the problem is:

Minimize

$$J(x(0), u_{[0,N-1)}) = \frac{1}{2}\sum_{0}^{N-1}(x^T(k)Qx(k) + u^T(k)Ru(k)) + \frac{1}{2}x^T(N)Q_f x(N) \tag{18}$$

where:
$Q = H^T H \geq 0$ and $Q_f \geq 0$ are $n \times n$ real symmetric matrices and $R > 0$ is an $m \times m$ symmetric real matrix.

Subject to the constraints:

$$x(k+1) = Ax(k) + Bu(k), \tag{19}$$

and the additional constraints given by Equations (16) and (17). The requirements (assumptions) of stabilizability of A, B and detectability of A, H still apply.

The simplest MPC algorithm starts at $k = 0$ by solving the open-loop, finite-time, optimal control problem described above with known initial state $x(0) = x$. The result, as in chess, is a set of control inputs (moves in chess) that are optimal over the prediction interval (up to time $k = N$). Denote this sequence of controls by

$$\mathbf{u_{opt(x,N)}} = \{u_{opt(x,N)}(0)\ u_{opt(x,N)}(1)\ u_{opt(x,N)}(2)\ \ldots u_{opt(x,N)}(N-1)\} \tag{20}$$

The prediction includes a predicted optimal sequence of states

$$\mathbf{x_{opt(x,N)}} = \{x_{opt(x,N)}(0)\ x_{opt(x,N)}(1)\ x_{opt(x,N)}(2)\ \ldots x_{opt(x,N)}(N-1)\} \tag{21}$$

Note that the entire sequence is denoted by the bold font while the individual terms include the time and are not bold.

Having computed this sequence (it is temporarily assumed that this happens instantaneously) the first control value in the sequence, $u_{opt(x,N)}(0)$ is applied to the system which then produces the state $x(1)$. This result may or, more likely, may not be the predicted value which is $x_{opt}(1)$. This actual state is observed. It then becomes the initial value of the state in a new optimal control problem. This new problem is identical to the first one (already solved) except for the initial condition.

The simplest MPC algorithm then solves this new problem and computes an optimal sequence of optimal controls and predicted states starting at $k = 1$

$$\begin{aligned}\mathbf{u_{opt(x(1),N+1)}} = \{&u_{opt(x(1),N+1)}(1)\ u_{opt(x(1),N+1)}(2)\\ &u_{opt(x(1),N+1)}(3)\ \ldots\ u_{opt(x(1),N+1)}(N)\}\end{aligned} \tag{22}$$

$$\begin{aligned}\mathbf{x_{opt(x(1),N+1)}} = \{&x_{opt(x(1),N+1)}(1)\ x_{opt(x(1),N+1)}(2)\\ &x_{opt(x(1),N+1)}(3)\ \ldots x_{opt(x(1),N+1)}(N)\}\end{aligned} \tag{23}$$

Again, the first control value ($u(1) = u_{opt(x(1),N+1}(1)$) in the new sequence is applied to the system which again generates the next value for state vector, $x(2)$. This value is generally different from the one predicted so the next control value ($u(2)$) has to be computed. This is done by solving the same optimization problem with the new, known, initial condition $x(2)$.

This procedure is repeated ad infinitum and defines and describes the simplest version of MPC.

Note that the optimization problem is always the same except for the changing initial state. Thus, the control is a time-invariant function of the current state which we can denote by x. That is,

$$\kappa_N(x) = u_{opt(x,N)}(0) \tag{24}$$

1. Obviously, the computation of the optimal control cannot take zero time. But the computation has to be fast enough and reliable enough to produce a sufficiently accurate solution within a sample interval. In practice, the "optimal" control is implemented with a delay of no more than one sampling interval.

2. One also has to choose a suitable performance measure in order for the optimal control problem to produce a good sequence of control values.
3. While it is intuitively reasonable to believe that this procedure will produce a control sequence that is close to the truly optimal one (i.e., the solution to the infinite time optimal control problem), this is not guaranteed. Additional conditions to ensure that this is so are needed.
4. Without additional conditions, one cannot be certain that the sequence of actual states and controls will not violate the constraints.
5. There is not even a guarantee that this closed-loop system is stable. Again, conditions to guarantee asymptotic stability are needed.

In the following section, the conditions that guarantee stability and good performance of MPC controllers are developed.

4 Stability of MPC

There are three issues. First, the controlled system must not violate the constraints. This statement needs some qualification. Some constraints are "soft." That is, small violations of the constraint are tolerable. For example, suppose a tank would overflow if the level of liquid in it exceeded 5 meters. If the controller is designed with a constraint that the liquid level is less than 4 meters, small violations of the constraint are acceptable. Obviously, the same system has a hard level constraint at 5 meters.

There are also constraints that cannot possibly be violated because they are imposed by the hardware. An example is amplifier or motor saturation. These limit the performance of the closed-loop system but cannot be violated. For these constraints especially, but in general as well, controls and states that satisfy the constraints are described as feasible in the MPC literature.

Second, it is generally true that the controlled system must be stable.

The third and last issue is performance. An example might be the following. Once you have guaranteed feasibility and stability, you would like to minimize the cost of operating the closed-loop system.

The key to most results on the stability of MPC is a pair of results that were briefly mentioned immediately after Theorem 1. Specifically, the solution to the unconstrained LQG depends on the Algebraic Riccatti Equation (ARE). The unique positive definite solution to the ARE, denoted by P, can be used to create a Lyapunov function

$$V(x(k)) = \frac{1}{2}x^T(k)Px(k) \tag{25}$$

for the unconstrained LQG. This function $V(x(k))$ is also the performance of the unconstrained optimal control with initial condition $x(k)$, as measured by the performance measure (2).

This suggests that it would be very desirable to make the simplest MPC scheme outlined in the previous section be equivalent to the LQR for any problem for which the constraints are always inactive. Even when the constrains are active, you would like the simplest MPC to drive the state into a subset of the state space that has two properties:

1. The constraints are inactive;
2. It is a positive invariant set when the control is the optimal LQR feedback—when $u(k) = F_{opt}x(k)$.

This would, of course, guarantee exponential stability of the MPC controller.

Another property of the Lyapunov function (25) is very helpful in this program. Given any constant, c, the set of points (the level set) defined by

$$\{x : V(x) \leq c\} \tag{26}$$

is a positive invariant set for the LQR. This means that any initial condition inside that set remains in that set if the LQR feedback is the control and the constraints are never active.

The question is then, how can one construct the MPC problem so that the state is forced into a level set of the unconstrained LQR optimal controller and, once it is there, the MPC controller is identical to the optimal LQR controller. There are several ways to do this. These are nicely discussed in the paper by Mayne et al. [11]. One way is to change the MPC problem in two ways.

1. Change the terminal weighting from Q_f to P where P is the solution to the ARE, Equation (7). In this equation, A and B are determined by the plant but Q and R are design variables that can be chosen to modify and improve the MPC controller.
2. Augment the MPC performance measure by a terminal state target set. This takes the form,
$$x(N) \in X_f \tag{27}$$
where X_f is a feasible set, in the sense that it satisfies the state constraint of Equation (17) and has the additional property that the control constraint (16) is satisfied by $u_{opt}(k) = -F_{opt}x(k)$ (please see Equation (5)) for every $x \in X_f$. The obvious choice for X_f is the largest level set of the Lyapunov function defined by the P chosen above (please see Equation (26)) that satisfies these constraints.

The second change fundamentally alters the problem that must be solved in real time by the MPC controller. Adding the requirement that the state reach this set makes it certain that there will be initial states for which there is no solution to the basic MPC computation. That is, no control exists that satisfies the constraints and would drive the initial state to the target set in time N. Regardless of the choice of N there will be some initial states for which a solution exists. For these initial states it is possible to prove that the MPC controller is exponentially stable. Please see ([11] and/or [14]) for such proofs. In general, increasing N will enlarge the set of initial states for which the MPC controller results in stability.

It is logically impossible to prove mathematically that a physical system is stable because the mathematical model that is used in the proof is never exactly the same as the physical system. For this reason the sensitivity of the mathematical result to changes and other inaccuracies in the model is fundamentally important. This issue is discussed shortly.

5 Exogenous Inputs

So far, the only version of MPC that has been discussed is regulation. That is, a controller that takes a set of possible initial states, $x(0)$, to 0. Often, one wants the controlled system to track an external (exogenous) input. For example, in paper-making, the goal is to keep the thickness of the paper constant—but certainly not zero. Because the MPC controller is based on a prediction, it is necessary to supply the desired state to the controller as is shown in the block diagram in Figure 4.

Note that this completely changes the stability question. For LTI systems, exponential stability implies Bounded-Input Bounded Output (BIBO) stability. That is,

Definition 1. A system with input $u(k)$ and output $y(k)$ is BIBO stable if and only if any bounded input ($||u|| \leq M$) implies the output is bounded ($||y|| \leq C$). Note that the norm is arbitrary.

This is not true for nonlinear systems, not even for systems that are LTI except for saturation, the class of systems emphasized in this article.

In cases similar to paper-making, where the exogenous input is constant over long periods of time and is small enough, one can simply subtract the input from the actual state in the optimization problem that is internal to the MPC controller to obtain a regulator problem that is identical to the one we described previously. In this, and similar cases, there is a possible flaw in the simplest MPC. It is easy to show that the steady-state error in the response of the MPC-controlled system

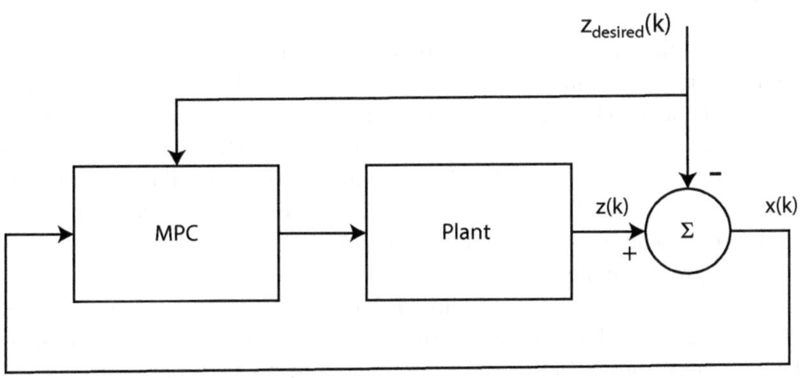

Fig. 4: A generic MPC-controlled system with an exogenous input that is available to the controller.

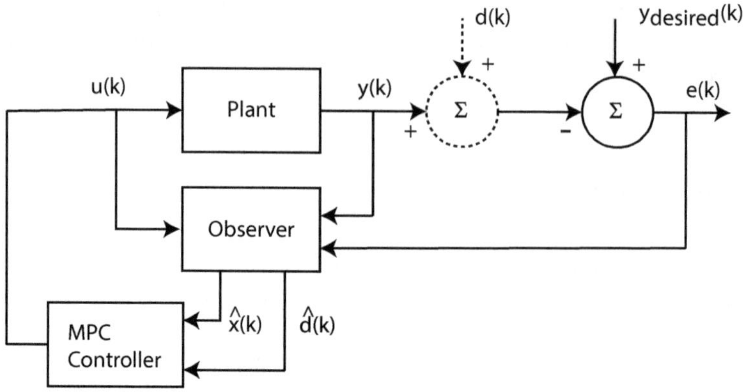

Fig. 5: An MPC controller augmented with an observer and a (hypothetical) disturbance so as to have zero steady-state error in response to a step input.

to a step input is generally not zero. This is a well-known and classic problem in feedback control. The classical solution is to add an integrator to the feedback loop. An MPC solution to this problem is outlined in Figure 5.

In this figure the desired response, $y_{desired} = y_c = constant$ for all $k \geq 0$ and is a scalar. The dotted line and circle are meant to indicate that the disturbance input $d(k)$ dos not really exist. Nonetheless, it adds a state to the plant model for which the controller is designed. The key to the set point tracking problem described here is the observer. An immediate application of the two main ideas underlying observer design—include a model of the plant and use error feedback to correct its errors—provides the solution to the set point tracking problem. One should augment the plant model in the observer to include a model for the hypothesized disturbance d.

$$\begin{bmatrix} \hat{x}(k+1) \\ \hat{d}(k+1) \end{bmatrix} \begin{bmatrix} A & 1 \\ 0 & 1 \end{bmatrix} \begin{bmatrix} \hat{x}(k) \\ \hat{d}(k) \end{bmatrix} + \begin{bmatrix} B \\ 0 \end{bmatrix} u(k) + \begin{bmatrix} F_f \\ F_d \end{bmatrix} \left(y(k) - \begin{bmatrix} C & 1 \end{bmatrix} \begin{bmatrix} \hat{x}(k) \\ \hat{d}(k) \end{bmatrix} \right) \tag{28}$$

To show that this leads to zero steady-state error in response to a step input one has to ignore any disturbances other than d as they would prevent the existence of a constant steady state. Then, assuming that there is a steady state, the last row of the observer equation satisfies $0 = F_d(y - y_s) - (C\hat{x} - y_s - \hat{d}(k))$. As long as F_d is not zero, this implies that $y = \hat{y} + \hat{d}$. The MPC controller is designed to make $\hat{y} + \hat{d} = y_s$.

A much more thorough discussion of MPC control to track a constant input can be found in [14] where it is called the set point tracking problem—standard terminology in the process control literature.

Even for the simple set point tracing problem for an LTI system with simple inequality constraints and a quadratic performance measure, too large an input signal could create infeasibility. In that case there would be no MPC solution.

The problem becomes very challenging when the input is less predictable, as it might be when a human operator is supplying it. An example of this might be in a so-called robot surgical system where a doctor explicitly controls the robot movement through a manipulandum. The extent to which the input is predictable over the desired MPC time horizon determines the usefulness of MPC for such problems.

6 Robustness

Precise analyses of robustness begins with precise descriptions of the ways in which the mathematical model can be different from the model for which we proved stability. There are two common ways in which this can happen. The first is that the measurement or estimate of the state that is fed back to the MPC controller can be corrupted in some way. The second is similar in its effect but different in its cause. Specifically, the model of the plant used in the MPC calculations is always an approximation to the real plant. The difference between the two can be viewed as a disturbance to the feedback. Figures 6 and 7 illustrate this.

It is also possible for the control signal to be perturbed, either by inaccuracies in its computation or by perturbations that occur in its implementation. These perturbations act on the input to the plant. An example would be the noise signal denoted by ξ in the LQGR.

The robustness problem is then to guarantee that the controlled system is (a) stable, (b) feasible. and (c) performs nearly as predicted despite whatever perturbations are present. As in the observer design problem, the solution depends on the assumed form of the perturbations. There are again two standard assumptions.

- The perturbations are WGN. This is somewhat optimistic. WGN will certainly excite every response mode of the system but it scatters its energy over all frequencies. It is also somewhat unrealistic because a physical perturbation must be bounded and Gaussian random variables are not.

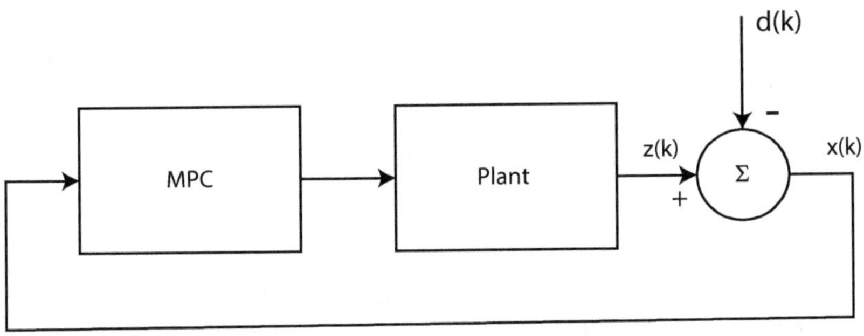

Fig. 6: A generic MPC-controlled system with an exogenous input that is not available to the controller and, thus, acts as a perturbation.

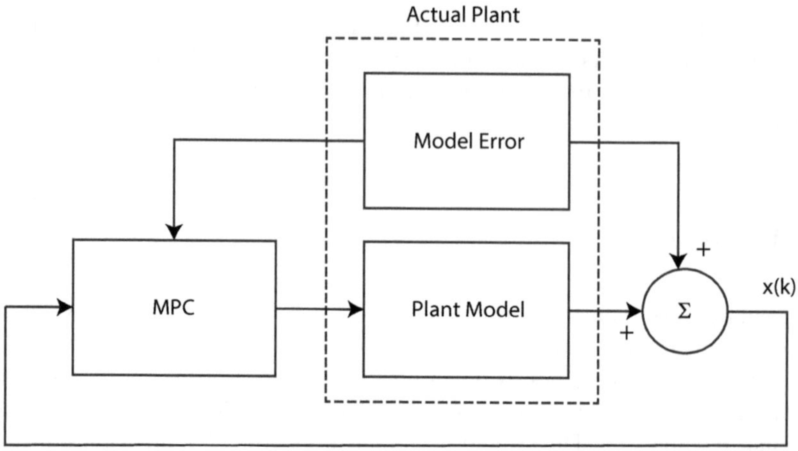

Fig. 7: A generic MPC-controlled system explicitly showing the error in the model used to compute the control.

- The perturbations are bounded but chosen by a malevolent nature to be the worst possible. This is somewhat pessimistic but there are situations where this is warranted.

It is known that the LQGR is not automatically robustly stable even though the LQR is [7]. Nonetheless, it has been successfully used in a variety of applications.

The worst-case assumption leads to H_∞ control in the LTI case when the performance measure is to minimize the response to the worst possible input (Please see [10] for a detailed introduction).

It is reassuring to know that the basic MPC problem described here (i.e., LTI systems with convex constraints and a quadratic performance measure) is known to be robustly asymptotically stable with respect to bounded state and measurement noise [14]. However, more general nonlinear MPC can be completely nonrobust. That is, arbitrarily small disturbances can cause instability [8].

Discussions of robustness in Model Predictive Control can be found in the articles [13] and [12].

7 Example

This example is meant to illustrate some of the ways that the simplest version of MPC can be used to solve interesting problems as well as some of the limitations of the approach. The problem is to describe how humans might regulate their upright standing posture. The full details of the problem and its solution by MPC are contained in [16]. Here the focus is on how MPC was used, how these ideas could be extended, and open questions about the solution.

7.1 Background

There are many aspects of the way humans regulate their upright standing posture that are not understood despite a great deal of experimental and theoretical research. If we consider only regulation in the sagittal plane (forward and backward but not side to side), experiments have shown that humans exhibit a continual back and forth sway. This indicates that the controller cannot be modeled as an LQR. Nonetheless, it is plausible to believe that the controller is optimal—or nearly so—with respect to some performance measure. Because the single overriding concern for living things during the entire history of life on earth has been getting enough food (there are local exceptions to this nowadays), it is reasonable to propose that the controller attempts to minimize the metabolic energy expended in maintaining posture. Of course, it is also important to maintain an upright posture and not to fall. Thus, one wants a performance measure that penalizes energy expenditure and promotes stability. The problem is important because falling is a major cause of injury, especially in the elderly. Understanding the controller might assist in developing methods to reduce the number of serious falls.

7.2 Dynamics

The dynamics of the human body are extremely complicated. It is necessary to develop a mathematical model that captures the essential dynamical features but is simple enough to understand and use in computations. It is also important to be able to either measure or estimate the parameters of the model. A common choice for the study of posture is to approximate the standing human by a multi-segment inverted pendulum as shown in Figure 8. Note that the foot does not have toes and is represented as a triangular solid. The actuators in this system are muscles. It is a gross oversimplification to represent them as ideal torque generators. This is done to simplify the control problem but it is a useful approximation for the following reason. The posture problem is mainly to minimize the effect of small perturbations. Thus, one loses very little in linearizing the dynamics. This linearization would be much more complicated if more realistic, nonlinear models of muscle were included. In addition, realistic muscle models would add many more parameters to the model and these would be difficult to determine accurately.

Experiments have shown that the human response to perturbations of their posture is different depending on the size of the perturbations. Small disturbances elicit a so-called "ankle strategy" whereby the knees and hips are locked and the only reaction is a rotation of the ankles. Larger disturbances are controlled by a "hip and ankle strategy." The knee is locked and the response is primarily bending at the hip and ankle. Still larger perturbations result in rotation at all three joints. However, in the interest of clarity and simplicity, it will be assumed that the knee joint is locked in this example. Thus, the model can be regarded as adequate for both small and intermediate amounts of perturbation.

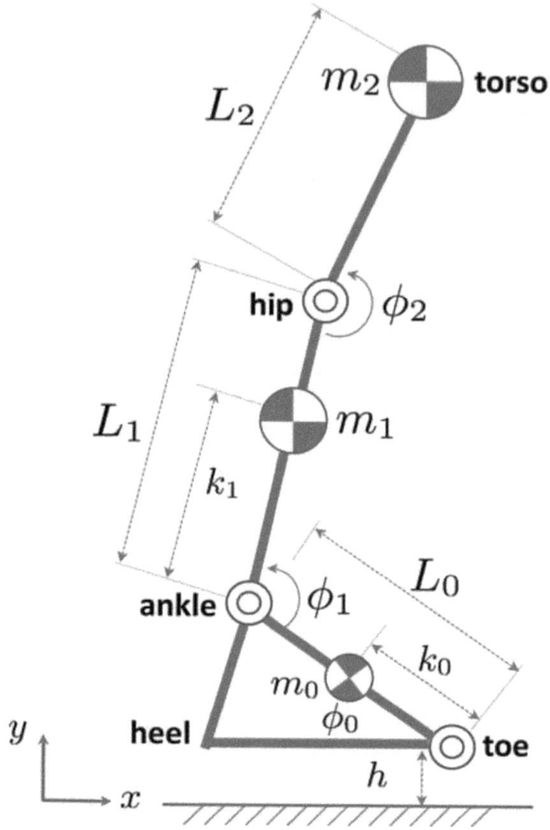

Fig. 8: A simplified model of a standing human as viewed from the side. Note that the knee is assumed to be locked and the arms are held still at the sides.

The mathematical model of this inverted pendulum system is nonlinear and includes sines, cosines, and squared velocities. Because the range of movements considered is small, linearization about a vertical (unstable) equilibrium state is reasonable. This posture is assumed to be with the joint angles $\phi_1 = \phi_2 = \pi$ and all the velocities and accelerations equal to zero. The result of linearization is a mathematical model that takes the form

$$\begin{bmatrix} Q_{11} & Q_{12} \\ Q_{21} & Q_{22} \end{bmatrix} \begin{bmatrix} \ddot{\phi}_1 \\ \ddot{\phi}_2 \end{bmatrix} + [\begin{bmatrix} R_1 \\ R_2 \end{bmatrix} = \begin{bmatrix} u_a \\ u_h \end{bmatrix} \tag{29}$$

where the linearized joint angles are denoted by ϕ_1 and ϕ_2 and u_a is the torque applied at the ankle and u_h is the torque applied at the hip.

We rewrite this in the standard state-space form as

$$\dot{x}(t) = Ax(t) + Bu(t) \tag{30}$$

where $x(t) = [\phi_1(t)\ \phi_2(t)\ \dot{\phi}_1(t)\ \dot{\phi}_2(t)]^T$ and $u(t) = [u_1(t)\ u_2(t)]^T$

7.3 Delay

It is known that the time delay between the onset of a postural perturbation and the response to that perturbation is significant. A detailed discussion of the sources of this delay can be found in [16]. It is helpful to separate the delay into two components, a delay in the feedback (sensing), which is denoted by τ_s and a delay in the control, represented by τ_c. Including these delays in the model results in

$$\dot{x}(t) = Ax(t) + Bu(t - \tau_c) \text{ with } y(t) = x(t - \tau_s) \tag{31}$$

where you will note that full state feedback is assumed. This is not unreasonable as there is a plethora of sensors, ranging from proprioceptors in the muscles and joints and tactile sensors in the feet to vision and the otoliths (accelerometers in the ears).

7.4 Performance Measure

A very good way to address the stability of posture is by means of the Center of Pressure (COP). The ground supports the human by creating a torque about the toes as well as both horizontal and vertical forces on the foot. As the foot does not move, the horizontal forces created by the human must be canceled by the horizontal ground force. Also, the downward vertical forces produced by the human must be canceled by the torque about the toe, τ_t, and the vertical force, f_v, of ground interaction. The COP replaces τ_t by locating f_v at the distance from the toe given by the COP as defined below.

$$l_{cop} = \frac{\tau_t}{f_v} \tag{32}$$

The horizontal position of the human's Center of Mass (CM) is another possible measure of stability but it does not account for the effects of velocity.

The COP is also linearized about the same nominal posture, resulting in a function that is linear in the perturbations of the states and controls.

The performance measure is thus assumed to be quartic or some higher even order in the linearized COP and quadratic in the linearized controls. Thus,

$$J(x(0), u_{[0,\infty)}) = \frac{1}{2}\int_0^\infty [q l_{cop}^4(t) + r_1 u_1^2(t) + r_2 u_2^2(t)]dt \tag{33}$$

7.5 Noise and Other Disturbances

There is a small amount of randomness in both neuronal sensing and activation. In addition to this, the experimental study of posture regulation generally involves deliberately inducing disturbances to the posture by horizontally moving the platform the subjects are standing on. Including this in the continuous-time model would involve stochastic differential equations and considerable mathematical technicalities. Instead, it is included in the discrete-time model where these complications are minimal. As solving the MPC problem requires discretization in time there is no additional loss of generality in doing this.

It is also assumed that the filtering of this noise can be separated from the design of the deterministic controller. This would certainly not be true if the nonlinearities were taken into account. But, the perturbations, including the random ones, are small and saturation does not occur.

7.6 Problem

The problem is then to choose the controls $u_a(k)$ and $u_h(k)$ for $0 \leq k < \infty$ to minimize the performance measure, Equation (33) subject to the constraints of Equation (31).

7.7 Solution

Because of the quartic weighting of the displacement of the COP, this is a good problem to solve by means of MPC. This is especially true because, once it is discretized in time, it will be a convex programming problem. Such problems always have a unique global solution if they have a solution at all. This problem certainly does have a solution.

The first step in applying MPC to this problem is to discretize it in time. This is a substantial simplification of the mathematics because of the delays and the random disturbances. In particular, the delays are described by so-called delay states. Thus, define the discretion interval to be δ and set $\delta = \tau_c/n$ where n is the number of time samples in a time interval of length τ_c. For simplicity it is assumed that $\tau_c = \tau_s$ but this is not necessary as long as $m\delta = \tau_s$ for some integer m.

Define the discrete-time state vector as

$$z(k) = [x^T(k-n),\, x^T(k-n+1),\, \ldots, x^T(k),\, u^T(k),\, x^T(k-n),\, \ldots,\, x^T(k-1)]^T \quad (34)$$

The discrete-time version of the problem is then

$$z(k+1) = A_d z(k) + B_d u(k) + \xi(k) \quad (35)$$

$$y(k) = C_d z(k) + \nu(k) \quad (36)$$

where:

$$z(k) = [z_1(k)\, z_2(k)\, z_3(k)\, \dots\, z_{16n_d+4(k)}(k)]^T \tag{37}$$

and $u(k) = [0\,0\,0\dots\, u_1(k)\, u_2(k)]^T$

The performance measure is also discretized in time and given a finite end time, as is required in order to apply MPC. It becomes

$$J(x(0), u_{[0,N_d-1]}) = \Sigma_{k=0}^{k=N_d}[ql_{cop}^4(k) + r_1 u_1^2(k) + r_2 u_2^2(k)] \tag{38}$$

7.8 Results

Detailed results are reported in [16]. Here, Figure 9 demonstrates that the results agree well with some experimental data. The Stabilogram Diffusion Function (SDF) was developed to better understand postural sway [3]. In these simulations the perturbations were WGN with variance = .005 and the delay was .125 milliseconds.

The analytical results also show that the ankle torque and movement are significantly larger than those at the hip for small perturbations. The reverse is true for large perturbations. In addition, the MPC-designed controller uses substantially less control energy than a similarly specified LQR-designed controller. This difference increases as the performance weighting on the COP increases from 4 to 6 to 8.

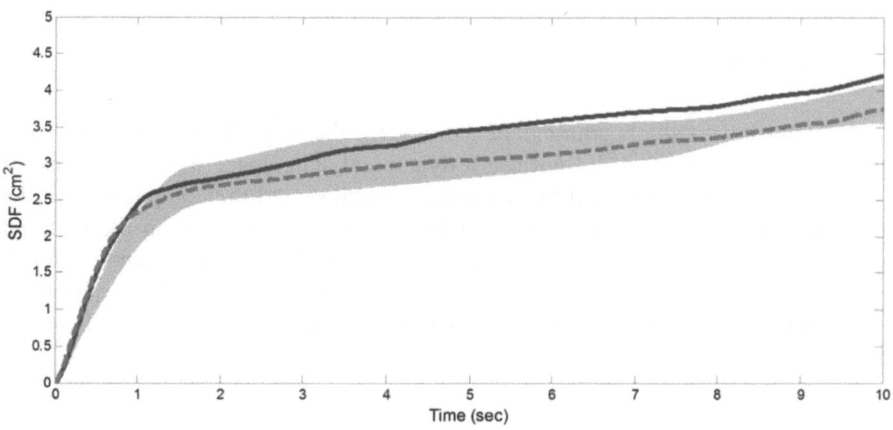

Fig. 9: Simulated SDF (dashed line) compared with experimental SDF (solid line). The dashed line is the mean of ten different simulated SDFs; light shading shows the range of ten different noise seeds with the same standard deviation.

7.9 Discussion

From the viewpoint of neuroscience, the question is whether the human brain and nervous system could implement a controller similar to the MPC-designed one. The controller produced by MPC is, ultimately, just a nonlinear controller that includes an observer. This could be learned or, to a degree, hardwired in the human.

From the viewpoint of the control engineer, there are a number of lessons.

- The problem of designing a controller for an LTI system that has convex state and actuator constraints can be formulated for performance measures that are convex in both control and states as in the example just discussed. Such problems can be solved quickly and with certainty. The result is a nonlinear controller with properties that can be very useful in some situations.
- Such a controller could use significantly less energy than a comparable linear controller, as in the example.
- Such a controller would be more aligned with specifications that require the system states to be in a range. Such a specification is much more realistic than an exact target.
- It should be noted that the closed-loop system in the example is not stable to the origin. It is believed, but not proven, that the state is confined to a neighborhood (but not an ε-neighborhood) of the origin provided that the perturbations are appropriately bounded. This is, again, the realistic requirement.

8 Conclusions

This chapter covers the beginnings of the modern theory of MPC. There are early applications of MPC that did not use a dynamical state space model. The use of a state space model facilitated the analysis of stability and the computation of the controls. Many of the applications of MPC require little more than the material covered here. More complex applications and theoretical problems require the much deeper knowledge covered in the rest of this handbook. However, the descriptions of the more profound results generally assume the knowledge this article is meant to convey.

References

1. Anderson, B.D.M., Moore, J.B.: Optimal Control: Linear Quadratic Methods. Prentice-Hall, Englewood Cliffs (1990)
2. Bertsekas, D.: Dynamic Programming and Optimal Control, vol. 1, 4th edn. Athena Scientific, Belmont (2017)
3. Collins, J.J., Luca, C.J.: The effects of visual input on open-loop and closed-loop postural control mechanisms. Exp. Brain Res. **103**, 151–163 (1995)

4. Davis, M.H.A.: Linear Estimation and Stochastic Control. Halstead Press, New York (1977)
5. Friedland, B.: Observers. In: Levine, W.S. (ed.) The Control Handbook. Control System Advanced Methods, 2nd edn. CRC Press, Boca Raton (2011)
6. Goodwin, G.C., Graebe, S.F., Salgado, M.E.: Control System Design. Prentice Hall, Upper Saddle River (2001)
7. Green, M., Limebeer, D.J.N.: Linear Robust Control. Prentice-Hall, Englewood Cliffs (1995)
8. Grimm, G., Messina, M.J., Tuna, S.E., Teel, A.R.: Examples when nonlinear model predictive control is nonrobust. Automatica **40**, 1729–1738 (2004)
9. Kwakernaak, H., Sivan, R.: Linear Optimal Control Systems. Wiley, New York (1972)
10. Lublin, L., Grocott, S., Athans, M.: In: Levine, W.S. (ed.) H_2 (LQG) and H_∞ Control. The Control Handbook. Control System Advanced Methods, 2nd edn. CRC Press, Boca Raton (2011)
11. Mayne, D.Q., Rawlings, J.B., Rao, C.V., Scokaert, P.O.M.: Constrained model predictive control: stability and optimality. Automatica **36**, 789–814 (2000)
12. Raković, S.V.: Robust control of constrained discrete time systems: characterization and implementation. Ph.D. thesis, Imperial College London, London, 2005
13. Raković, S.V.: Robust model predictive control. In: Baillieul, J., Samad, T. (eds.) Encyclopedia of Systems and Control, pp. 1225–1233. Springer, London (2015). (Also Available Online.)
14. Rawlings, J.B., Mayne, D.Q.: Model Predictive Control: Theory and Design. Nob Hill Publishing, Madison (2005)
15. Sorenson, H. (ed.) Kalman Filtering: Theory and Applications. IEEE Press, New York (1985)
16. Yao, L., Levine, W.S., Loeb, G.E.: A two-joint human posture control model with realistic neural delays. IEEE Trans. Neural Syst. Rehabil. Eng. **20**(5), 738–748 (2012)

Dynamic Programming, Optimal Control and Model Predictive Control

Lars Grüne

1 Introduction

Model Predictive Control (MPC), also known as Receding Horizon Control, is one of the most successful modern control techniques, both regarding its popularity in academics and its use in industrial applications [6, 11, 15, 28]. In MPC, the control input is synthesized via the repeated solution of finite horizon optimal control problems on overlapping horizons. Among the most fundamental properties to be investigated when analyzing MPC schemes are the stability and (approximate) optimality properties of the closed loop solutions generated by MPC. One interpretation of MPC is that an infinite horizon optimal control problem is split up into the repeated solution of auxiliary finite horizon problems [13].

Dynamic Programming (DP) is one of the fundamental mathematical techniques for dealing with optimal control problems [4, 5]. It provides a rule to split up a high (possibly infinite) dimensional optimization problem over a long (possibly infinite) time horizon into auxiliary optimization problems on shorter horizons, which are much easier to solve. While at a first glance this appears similar to the procedure just described for MPC, the approach is different, in the sense that in DP the exact information about the future of the optimal trajectories — by means of the corresponding optimal value function — is included in the auxiliary problem. Thus, it provides a characterization of the *exact* solution, at the expense that the auxiliary problems are typically difficult to formulate and the number of auxiliary problems becomes huge — the (in)famous "curse of dimensionality." In MPC, the future information is only approximated (for schemes with terminal conditions) or even completely disregarded (for schemes without terminal conditions). This makes the auxiliary problems easy to formulate and to solve and keeps the number of these

L. Grüne (✉)
Mathematical Institute, University of Bayreuth, 95440 Bayreuth, Germany
e-mail: lars.gruene@uni-bayreuth.de

S. V. Raković, W. S. Levine (eds.), *Handbook of Model Predictive Control*,
Control Engineering, https://doi.org/10.1007/978-3-319-77489-3_2

problems low, but now at the expense that it does *not* yield an exact optimal solution of the original problem anymore.

However, it may still be possible that the solution trajectories generated by MPC are stable and approximately optimal, and the key for proving such statements is to make sure that the neglected future information only slightly affects the solution. The present chapter presents a survey of a selection of results in this direction and in particular shows that ideas from dynamic programming are essential for this purpose. As we will show, dynamic programming methods can be used for estimating near optimal performance under suitable conditions on the future information (Proposition 6 and Theorem 15 are examples for such statements) but also for ensuring that the future information satisfies these conditions (as, e.g., in Proposition 8 or Lemma 14(ii)). Moreover, dynamic programming naturally provides ways to derive stability or convergence from optimality via Lyapunov functions arguments, as in Proposition 3.

The chapter is organized as follows. In Section 2 we describe the setting and the MPC algorithm we consider in this chapter. Section 3 collects the results from dynamic programming we will need in the sequel. Section 4 then presents results for stabilizing MPC, in which the stage cost penalizes the distance to a desired equilibrium. Both schemes with and without terminal conditions are discussed. Section 5 extends this analysis to MPC schemes with more general stage costs, which is usually referred to as economic MPC. Section 6 concludes the chapter.

2 Setting, Definitions and Notation

In this chapter we consider discrete time optimal control problems of the form

$$\text{Minimize } J_N(x_0,u) \text{ with respect to the control sequence } u, \tag{1}$$

where $N \in \mathbb{N}_\infty := \mathbb{N} \cup \{\infty\}$ and

$$J_N(x_0,u) = \sum_{k=0}^{N-1} \ell(x(k),u(k)),$$

subject to the dynamics and the initial condition

$$x(k+1) = f(x(k),u(k)), \qquad x(0) = x_0 \tag{2}$$

and the combined state and input constraints

$$(x(k),u(k)) \in \mathbb{Y}\ \forall = 0,\dots,N-1 \quad \text{and} \quad x(N) \in \mathbb{X} \tag{3}$$

for all $k \in \mathbb{N}$ for which the respective values are defined. Here $\mathbb{Y} \subset X \times U$ is the constraint set, X and U are the state and input value set, respectively, and $\mathbb{X} := \{x \in X \mid \exists u \in U \text{ with } (x,u) \in \mathbb{Y}\}$ is the state constraint set. The sets X and U are metric spaces with metrics $d_X(\cdot,\cdot)$ and $d_U(\cdot,\cdot)$. Because there is no danger of confusion we usually omit the indices X and U in the metrics. We denote the solution of (2) by $x_u(k,x_0)$. Moreover, for the distance of a point $x \in X$ to another point $y \in X$ we use the short notation $|x|_y := d(x,y)$.

For $x_0 \in \mathbb{X}$ and $N \in \mathbb{N}$ we define the set of admissible control sequences as

$$\mathbb{U}^N(x_0) := \{u \in U^N \,|\, (x_u(k,x_0),u(k)) \in \mathbb{Y} \;\forall k = 0,\dots,N-1 \text{ and } x_u(N,x_0) \in \mathbb{X}\}$$

and

$$\mathbb{U}^\infty(x_0) := \{u \in U^\infty \,|\, (x_u(k,x_0),u(k)) \in \mathbb{Y} \;\forall k \in \mathbb{N}\}$$

Since feasibility issues are not the topic of this chapter, we make the simplifying assumption that $\mathbb{U}^N(x_0) \neq \emptyset$ for all $x_0 \in \mathbb{X}$ and all $N \in \mathbb{N}_\infty$. If desired, this assumption can be avoided using the techniques from, e.g., [10], [15, Chapter 7], [21, Chapter 5], or [27].

Corresponding to the optimal control problem (1) we define the optimal value function

$$V_N(x_0) := \inf_{u \in \mathbb{U}^N(x_0)} J(x_0,u)$$

and we say that a control sequence $u_N^\star \in \mathbb{U}^N(x_0)$ is optimal for initial value $x_0 \in \mathbb{X}$ if $J(x_0,u_N^\star) = V_N(x_0)$ holds.

It is often desirable to solve optimal control problems with infinite horizon $N = \infty$, for instance because the control objective under consideration naturally leads to an infinite horizon problem (like stabilization or tracking problems) or because an optimal control is needed for an indefinite amount of time (as in many regulation problems). For such problems the optimal control is usually desired in feedback form, i.e., in the form $u_N^\star(k) = \mu(x(k))$ for a feedback map $\mu : \mathbb{X} \to \mathbb{U}$. Except for special cases like linear quadratic problems without constraints, computing infinite horizon optimal feedback laws is in general a very difficult task. On the other hand, very accurate approximations to optimal control sequences $u_N^\star$ for finite horizon problems, particularly with moderate N, can be computed easily and fast (sometimes within a few milliseconds), and often also reliably with state-of-the-art numerical optimization routines, even for problems in which the dynamics (2) are governed by partial differential equations. The following Receding Horizon or Model Predictive Control algorithm (henceforth abbreviated by MPC) is therefore an attractive alternative to solving an infinite horizon optimal control problem.

Algorithm 1 (Basic Model Predictive Control Algorithm)

(Step 0) *Fix a (finite) optimization horizon $N \in \mathbb{N}$ and set $k := 0$; let an initial value $x_{MPC}(0)$ be given*

(Step 1) *Compute an optimal control sequence $u_N^\star$ of Problem (1) for $x_0 = x_{MPC}(k)$*

(Step 2) *Define the MPC feedback law value $\mu_N(x_{MPC}(k)) := u_N^\star(0)$*

(Step 3) *Set $x_{MPC}(k+1) := f(x_{MPC}(k+1), \mu_N(x_{MPC}(k)))$, $k := k+1$ and go to (Step 1)*

We note that although derived from an open loop optimal control sequence $u_N^\star$, μ_N is indeed a map from $\mathbb{X}$ to U, however, it will in general not be given in the form of an explicit formula. Rather, given $x_{MPC}(k)$, the value $\mu_N(x_{MPC}(k))$ is obtained by solving the optimal control problem in Step 1 of Algorithm 1, which is usually done numerically.

In MPC, one often introduces additional terminal conditions, consisting of a terminal constraint set $\mathbb{X}_0 \subseteq \mathbb{X}$ and a terminal cost $F : \mathbb{X}_0 \to \mathbb{R}$. To this end, the optimization objective J_N is modified to

$$J_N^{tc}(x,u) = \sum_{k=0}^{N-1} \ell(x(k),u(k)) + F(x(N))$$

and the last constraint in (3) is tightened to

$$x(N) \in \mathbb{X}_0.$$

Moreover, we denote the corresponding space of admissible control sequences by

$$\mathbb{U}_0^N(x_0) := \{u \in \mathbb{U}^N(x_0) \,|\, x_u(N,x_0) \in \mathbb{X}_0\}$$

and the optimal value function by

$$V_N^{tc}(x_0) := \inf_{u \in \mathbb{U}_0^N(x_0)} J(x_0,u).$$

Observe that the problem without terminal conditions is obtained for $F \equiv 0$ and $\mathbb{X}_0 = \mathbb{X}$.

Again, a control $u_N^{tc\star} \in \mathbb{U}_0^N(x_0)$ is called optimal if $V_N^{tc}(x_0) = J_N^{tc}(x_0,u_N^{tc\star})$. Due to the terminal constraints it is in general not guaranteed that $\mathbb{U}_0^N(x_0) \neq \emptyset$ for all $x_0 \in \mathbb{X}$. We therefore define $\mathbb{X}_N := \{x_0 \in \mathbb{X} \,|\, \mathbb{U}_0^N(x_0) \neq \emptyset\}$. For MPC in which J_N^{tc} is minimized in Step 1 we denote the resulting feedback law by μ_N^{tc}. Note that μ_N^{tc} is defined on $\mathbb{X}_N$.

A priori, it is not clear, at all, whether the trajectory x_{MPC} generated by the MPC algorithm enjoys approximate optimality properties or qualitative properties like stability. In the remainder of this chapter, we will give conditions under which such properties can be guaranteed. In order to measure the optimality of the closed loop trajectory, we introduce its closed loop finite and infinite horizon values

$$J_K^{cl}(x,\mu_N) := \sum_{k=0}^{K-1} \ell(x_{MPC}(k),\mu_N(x_{MPC}(k)))$$

and

$$J_\infty^{cl}(x,\mu_N) := \limsup_{K\to\infty} J_K^{cl}(x_{MPC}(0),\mu_N)$$

where in both cases the initial value $x_{MPC}(0) = x$ is used.

3 Dynamic Programming

Dynamic programming is a name for a set of relations between optimal value functions and optimal trajectories at different time instants. In what follows we state those relations which are important for the remainder of this chapter. For their proofs we refer to [15, Chapters 3 and 4].

For the *finite horizon problem without terminal conditions* the following equations and statements hold for all $N \in \mathbb{N}$ and all $K \in \mathbb{N}$ with $K \leq N$ (using $V_0(x) \equiv 0$ in case $K = N$):

$$V_N(x) = \inf_{u \in \mathbb{U}^K(x)} \{J_K(x,u) + V_{N-K}(x_u(K,x))\} \tag{4}$$

If $u_N^\star \in \mathbb{U}^N(x)$ is an optimal control for initial value x and horizon N, then

$$V_N(x) = J_K(x,u_N^\star) + V_{N-K}(x_{u_N^\star}(K,x)) \tag{5}$$

and

the sequence $u_K := (u_N^\star(K),\dots,u_N^\star(N-1)) \in \mathbb{U}^{N-K}(x_{u_N^\star}(K,x))$
is an optimal control for initial value $x_{u_N^\star}(K,x)$ and horizon $N-K$. (6)

Moreover, for all $x \in \mathbb{X}$ the MPC feedback law μ_N satisfies

$$V_N(x) = \ell(x,\mu_N(x)) + V_{N-1}(f(x,\mu_N(x))). \tag{7}$$

For the *finite horizon problem with terminal conditions* the following holds for all $N \in \mathbb{N}$ and all $K \in \mathbb{N}$ with $K \leq N$ (using $V_0^{tc}(x) = F(x)$ in case $K = N$):

$$V_N^{tc}(x) = \inf_{u \in \mathbb{U}_{N-K}^K(x)} \{J_K(x,u) + V_{N-K}^{tc}(x_u(K,x))\}, \tag{8}$$

where $\mathbb{U}_{N-K}^K(x_0) := \{u \in \mathbb{U}^K(x_0) \,|\, x_u(N,x_0) \in \mathbb{X}_{N-K}\}$. If $u_N^{tc\star} \in \mathbb{U}_0^N(x)$ is an optimal control for initial value x and horizon N, then

$$V_N^{tc}(x) = J_K(x,u_N^{tc\star}) + V_{N-K}^{tc}(x_{u_N^{tc\star}}(K,x)) \tag{9}$$

and

the sequence $u_K^{tc} := (u_N^{tc\star}(K),\dots,u_N^{tc\star}(N-1)) \in \mathbb{U}^{N-K}(x_{u_N^{tc\star}}(K,x))$
is an optimal control for initial value $x_{u_N^{tc\star}}(K,x)$ and horizon $N-K$. (10)

Moreover, for all $x \in \mathbb{X}$ the MPC feedback law μ_N^{tc} satisfies

$$V_N^{tc}(x) = \ell(x,\mu_N^{tc}(x)) + V_{N-1}^{tc}(f(x,\mu_N^{tc}(x))). \tag{11}$$

Finally, for the *infinite horizon problem* the following equations and statements hold for all $K \in \mathbb{N}$:

$$V_\infty(x) = \inf_{u \in \mathbb{U}^K(x)} \{J_K(x,u) + V_\infty(x_u(K,x))\} \tag{12}$$

If $u^\star_\infty$ is an optimal control for initial value x and horizon N, then

$$V_\infty(x) = J_K(x,u^\star_\infty) + V_\infty(x_{u^\star_\infty}(K,x)) \tag{13}$$

and

$$\begin{array}{l}\text{the sequence } u_K := (u^\star_\infty(K), u^\star_\infty(K+1), \ldots) \in \mathbb{U}^\infty(x_{u^\star_\infty}(K,x)) \\ \text{is an optimal control for initial value } x_{u^\star_\infty}(K,x).\end{array} \tag{14}$$

The equations just stated can be used as the basis of numerical algorithms, see, e.g., [5, 17] and the references therein. Here, however, we rather use them as tools for the analysis of the performance of the MPC algorithm. Besides the equalities, above, which refer to the optimal trajectories, we will also need corresponding inequalities. These will be used in order to estimate J^{cl}_K and J^{cl}_∞ as shown in the following proposition.

Proposition 2 *Assume there is function $\varepsilon : \mathbb{X} \to \mathbb{R}$ such that the approximate dynamic programming inequality*

$$V_N(x) + \varepsilon(x) \geq \ell(x, \mu_N(x)) + V_N(f(x, \mu_N(x))) \tag{15}$$

holds for all $x \in \mathbb{X}$. Then for each MPC closed loop solution x_{MPC} and all $K \in \mathbb{N}$ the inequality

$$J^{cl}_K(x_{MPC}(0), \mu_N) \leq V_N(x_{MPC}(0)) - V_N(x_{MPC}(K)) + \sum_{k=0}^{K-1} \varepsilon_k \tag{16}$$

holds for $\varepsilon_k = \varepsilon(x_{MPC}(k))$. If, in addition, $\hat{\varepsilon} := \limsup_{K\to\infty} \sum_{k=0}^{K-1} \varepsilon_k < \infty$ and $\liminf_{K\to\infty} V_N(x_{MPC}(K)) \geq 0$ hold, then also

$$J^{cl}_\infty(x_{MPC}(0), \mu_N) \leq V_N(x_{MPC}(0)) + \hat{\varepsilon}$$

holds. The same statements are true when V_N and μ_N are replaced by their terminal conditioned counterparts V^{tc}_N and μ^{tc}_N, respectively.

Proof. Observing that $x_{MPC}(k+1) = f(x, \mu_N(x))$ for $x = x_{MPC}(k)$ and using (15) with this x we have

$$\begin{aligned} J^{cl}_K(x_{MPC}(0), \mu_N) &= \sum_{k=0}^{K-1} \ell(x_{MPC}(k), \mu_N(x_{MPC}(k))) \\ &\leq \sum_{k=0}^{K-1} [V_N(x_{MPC}(k)) - V_N(x_{MPC}(k+1)) + \varepsilon_k] \\ &= V_N(x_{MPC}(0)) - V_N(x_{MPC}(K)) + \sum_{k=0}^{K-1} \varepsilon_k, \end{aligned}$$

which shows the first claim. The second claim follows from the first by taking the upper limit for $K \to \infty$. The proof for the terminal conditioned case is identical. □

4 Stabilizing MPC

Using the dynamic programming results just stated, we will now derive estimates for J_∞^{cl} in the case of stabilizing MPC. Stabilizing MPC refers to the case in which the stage cost ℓ penalizes the distance to a desired equilibrium. More precisely, let $(x_*, u_*) \in \mathbb{Y}$ be an equilibrium, i.e., $f(x_*, u_*) = x_*$. Then throughout this section we assume that there is $\alpha_1 \in \mathscr{K}_\infty$ such that[1] ℓ satisfies

$$\ell(x_*, u_*) = 0 \quad \text{and} \quad \ell(x,u) \geq \alpha_1(|x|_{x_*}) \tag{17}$$

for all $x \in \mathbb{X}$. Moreover, for the terminal cost F we assume

$$F(x) \geq 0 \text{ for all } x \in \mathbb{X}_0. \tag{18}$$

We note that (18) trivially holds in case no terminal cost is used, i.e., if $F \equiv 0$.

The purpose of this choice of ℓ is to force the optimal trajectories — and thus hopefully also the MPC trajectories — to converge to x_*. The following proposition shows that this hope is justified under suitable conditions, where the approximate dynamic programming inequality (15) plays a pivotal role.

Proposition 3 *Let the assumptions of Proposition 2,* (17) *and* (18) *(in case of terminal conditions) hold with $\varepsilon(x) \leq \eta\alpha_1(|x|_{x_*})$ for all $x \in \mathbb{X}$ and some $\eta < 1$. Then $x_{MPC}(k) \to x_*$ as $k \to \infty$.*

Proof. We first observe that the assumptions imply $V_N(x) \geq 0$ or $V_N^{tc}(x) \geq 0$, respectively. We continue the proof for V_N, the proof for V_N^{tc} is identical. Assume $x_{MPC}(k) \not\to x_*$, i.e., there are $\delta > 0$ and a sequence $k_p \to \infty$ with $|x_{MPC}(k_p)|_{x_*} \geq \delta$ for all $p \in \mathbb{N}$. Then by induction over (15) with $x = x_{MPC}(k)$ we get

$$\begin{aligned} V_N(x_{MPC}(K)) &\leq V_N(x_{MPC}(0)) - \sum_{k=0}^{K-1} \big[\ell(x_{MPC}(k), \mu_N(x_{MPC}(k))) - \varepsilon(x_{MPC}(k))\big] \\ &\leq V_N(x_{MPC}(0)) - \sum_{k=0}^{K-1} (1-\eta)\alpha_1(|x_{MPC}(k)|_{x_*}) \\ &\leq V_N(x_{MPC}(0)) - \sum_{\substack{p\in\mathbb{N} \\ k_p \leq K-1}} (1-\eta)\alpha_1(|x_{MPC}(k_p)|_{x_*}) \\ &\leq V_N(x_{MPC}(0)) - \#\{p \in \mathbb{N} \,|\, k_p \leq K\}(1-\eta)\alpha_1(\delta). \end{aligned}$$

[1] The space $\mathscr{K}_\infty$ consists of all functions $\alpha : [0,\infty) \to [0,\infty)$ with $\alpha(0) = 0$ which are continuous, strictly increasing and unbounded.

Now as $K \to \infty$ the number $\#\{p \in \mathbb{N} \,|\, k_p \le K\}$ grows unboundedly, which implies that $V_N(x_{MPC}(K)) < 0$ for sufficiently large K which contradicts the non-negativity of V_N. □

We remark that under additional conditions (essentially appropriate upper bounds on V_N or V_N^{tc}, respectively), asymptotic stability of x_* can also be established, see, e.g., [15, Theorem 4.11] or [28, Theorem 2.22].

4.1 Terminal Conditions

In this section we use the terminal conditions in order to ensure that the approximate dynamic programming inequality (15) holds with $\varepsilon(x) \le 0$ and $V_N^{tc}(x) \ge 0$. Then Proposition 2 applies and yields $J_\infty^{cl}(x_{MPC}(0), \mu_N^{tc}) \le V_N^{tc}(x_{MPC}(0))$ while Proposition 3 implies $x_{MPC}(k) \to x_*$. The key for making this approach work is the following assumption.

Assumption 4 *For each $x \in \mathbb{X}$ there is $u_x \in U$ with $(x, u_x) \in \mathbb{Y}$, $f(x, u_x) \in \mathbb{X}$ and*

$$\ell(x, u_x) + F(f(x, u_x)) \le F(x).$$

While conditions like Assumption 4 were already developed in the 1990s, e.g., in [7, 8, 23], it was the paper [24] published in 2000 which established this condition as *the* standard assumption for stabilizing MPC with terminal conditions. The particular case $\mathbb{X} = \{x_*\}$ was investigated in detail already in the 1980s in the seminal paper [20].

Theorem 5. *Consider the MPC scheme with terminal conditions satisfying* (17), (18) *and Assumption 4. Then the inequality $J_\infty^{cl}(x, \mu_N^{tc}) \le V_N^{tc}(x)$ and the convergence $x_{MPC}(k) \to x_*$ for $k \to \infty$ hold for all $x \in \mathbb{X}_N$ and the closed loop solution $x_{MPC}(k)$ with $x_{MPC}(0) = x$.*

Proof. As explained before the theorem, it is sufficient to prove (15) with $\varepsilon(x) \le 0$ and $V_N^{tc}(x) \ge 0$; then Propositions 2 and 3 yield the assertions. The inequality$V_N^{tc}(x) \ge 0$ is immediate from (17) and (18). For proving (15) with $\varepsilon(x) \le 0$, using u_x from Assumption 4 with $x = x_u(N-1, x_0)$ we get

$$\begin{aligned}
V_{N-1}^{tc}(x_0) &= \inf_{u \in \mathbb{U}_0^{N-1}(x_0)} \sum_{k=0}^{N-2} \ell(x_u(k, x_0), u(k)) + F(x_u(N-1, x_0)) \\
&\ge \inf_{u \in \mathbb{U}_0^{N-1}(x_0)} \sum_{k=0}^{N-2} \ell(x_u(k, x_0), u(k)) + \ell(x, u_x) + F(f(x, u_x)) \\
&\ge \inf_{u \in \mathbb{U}_0^{N}(x_0)} \sum_{k=0}^{N-1} \ell(x_u(k, x_0), u(k)) + F(x_u(N, x_0)) \;=\; V_N^{tc}(x_0)
\end{aligned}$$

Inserting this inequality for $x_0 = f(x, \mu_N^{tc}(x))$ into (11) we obtain

$$V_N^{tc}(x) = \ell(x, \mu_N^{tc}(x)) + V_{N-1}^{tc}(f(x, \mu_N^{tc}(x))) \geq \ell(x, \mu_N^{tc}(x)) + V_N^{tc}(f(x, \mu_N^{tc}(x)))$$

and thus (15) with $\varepsilon \equiv 0$. □

A drawback of the inequality $J_\infty^{cl}(x, \mu_N^{tc}) \leq V_N^{tc}(x)$ is that it is in general quite difficult to give estimates for $V_N^{tc}(x)$. Under reasonable assumptions it can be shown that $V_N^{tc}(x) \to V_\infty(x)$ for $N \to \infty$ [15, Section 5.4]. This implies that the MPC solution is near optimal for the infinite horizon problem for N sufficiently large. However, it is in general difficult to make statements about the speed of convergence of $V_N^{tc}(x) \to V_\infty(x)$ as $N \to \infty$ and thus to estimate the length of the horizon N which is needed for a desired degree of suboptimality.

4.2 No Terminal Conditions

The decisive property induced by Assumption 4 and exploited in the proof of Theorem 5 is the fact that $V_{N-1}^{tc}(x_0) \geq V_N^{tc}(x_0)$. Without this inequality, (11) implies that (15) with $\varepsilon \equiv 0$ cannot in general be satisfied. Without terminal conditions and under the condition (17) it is, however, straightforward to see that the opposite inequality $V_{N-1}^{tc}(x_0) \leq V_N^{tc}(x_0)$ holds, where in most cases this inequality is strict. This means that without terminal conditions we need to work with positive ε. The following proposition, which was motivated by a similar "relaxed dynamic programming" inequality used in [22], introduces a variant of Proposition 2 which we will use for this purpose.

Proposition 6 *Assume there is a constant $\alpha \in (0,1]$ such that the relaxed dynamic programming inequality*

$$V_N(x) \geq \alpha \ell(x, \mu_N(x)) + V_N(f(x, \mu_N(x))) \tag{19}$$

holds for all $x \in \mathbb{X}$. Then for each MPC closed loop solution x_{MPC} and all $K \in \mathbb{N}$ the inequality

$$J_\infty^{cl}(x_{MPC}(0), \mu_N) \leq V_\infty(x_{MPC}(0))/\alpha$$

and, if additionally (17) *holds, the convergence $x_{MPC}(k) \to x_*$ for $k \to \infty$ hold.*

Proof. Applying Proposition 2 with $\varepsilon(x) = (1-\alpha)\ell(x, \mu_N(x))$ yields

$$\begin{aligned} J_K^{cl}(x_{MPC}(0), \mu_N) \leq\; & V_N(x_{MPC}(0)) - V_N(x_{MPC}(K)) \\ & + (1-\alpha) \underbrace{\sum_{k=0}^{K-1} \ell(x_{MPC}(k), \mu_N(x_{MPC}(k)))}_{=J_K^{cl}(x_{MPC}(0), \mu_N)}. \end{aligned}$$

Using $V_N \geq 0$ this implies $\alpha J_K^{cl}(x_{MPC}(0), \mu_N) \leq V_N(x_{MPC}(0))$ which implies the first assertion by letting $K \to \infty$ and dividing by α. The convergence $x_{MPC}(k) \to x_*$ follows from Proposition 3. □

A simple condition under which we can guarantee that (19) holds is given in the following assumption.

Assumption 7 *There are constants $\gamma_k > 0$, $k \in \mathbb{N}$ with $\sup_{k\in\mathbb{N}} \gamma_k < \infty$ and*

$$V_k(x) \leq \gamma_k \inf_{u\in U,(x,u)\in\mathbb{Y}} \ell(x,u).$$

A sufficient condition for Assumption 7 to hold is that ℓ is a polynomial satisfying (17) and the system can be controlled to x_* exponentially fast. However, via an appropriate choice of ℓ Assumption 7 can also be satisfied if the system is not exponentially controllable, see, e.g., [15, Example 6.7].

The following theorem, taken with modifications from [29], shows that Assumption 7 implies (19).

Proposition 8 *Consider the MPC scheme without terminal conditions satisfying Assumption 7. Then* (19) *holds with* $\alpha = 1 - (\gamma_2 - 1)(\gamma_N - 1)\prod_{k=0}^{N-1}\left(\frac{\gamma_k - 1}{\gamma_k}\right)$.

Proof. First note that for $x = x_*$ (19) always holds because all expressions vanish. For $x \neq x_*$, we consider the MPC solution $x_{MPC}(\cdot)$ with $x_{MPC}(0) = x$, abbreviate $\lambda_k = \ell(x_{u_N^\star}(k,x), u_N^\star(k))$ with $u_N^\star$ denoting the optimal control for initial value $x_0 = x$, and $\nu = V_N(f(x, \mu_N(x))) = V_N(x_{MPC}(1))$. Then (19) becomes

$$\sum_{k=0}^{N-1} \lambda_k - \nu \geq \alpha\lambda_0 \tag{20}$$

We prove the theorem by showing the inequality

$$\lambda_{N-1} \leq (\gamma_N - 1)\prod_{k=2}^{N-1}\left(\frac{\gamma_k - 1}{\gamma_k}\right)\lambda_0 \tag{21}$$

for all feasible $\lambda_0, \ldots, \lambda_{N-1}$. From this (20) follows since the dynamic programming equation (4) with $x = x_{MPC}(1)$ and $K = N-2$ implies

$$\nu \leq \sum_{n=1}^{N-2} \ell(x_{u_N^\star}(n,x), u_N^\star(n)) + V_2(x_{u_N^\star}(N-1,x)) \leq \sum_{n=1}^{N-2} \lambda_n + \gamma_2\lambda_{N-1}$$

and thus (21), $\gamma_2 \geq 1$ and $\lambda_0 = 1$ yield

$$\sum_{n=0}^{N-1} \lambda_n - \nu \geq \lambda_0 + (1-\gamma_2)\lambda_{N-1} \geq \lambda_0 - (\gamma_2 - 1)(\gamma_N - 1)\prod_{k=2}^{N-1}\left(\frac{\gamma_k - 1}{\gamma_k}\right)\lambda_0 = \alpha\lambda_0.$$

i.e., (20). In order to prove (21), we start by observing that since $u_K := (u_N^\star(K),\ldots, u_N^\star(N-1))$ is an optimal control for initial value $x_{u_N^\star}(K,x)$ and horizon $N-K$, we obtain $\sum_{k=p}^{N-1}\lambda_k = V_{N-p}(x_{u_N^\star}(p+1)) \le \gamma_{N-p}\lambda_p$, which implies

$$\sum_{k=p+1}^{N-1}\lambda_k \le (\gamma_{N-p}-1)\lambda_p \tag{22}$$

for $p = 0,\ldots,N-2$. From this we can conclude

$$\lambda_p + \sum_{k=p+1}^{N-1}\lambda_k \ge \frac{\sum_{k=p+1}^{N-1}\lambda_k}{\gamma_{N-p}-1} + \sum_{k=p+1}^{N-1}\lambda_k = \frac{\gamma_{N-p}}{\gamma_{N-p}-1}\sum_{k=p+1}^{N-1}\lambda_k.$$

Using this inequality inductively for $p = 1,\ldots,N-2$ yields

$$\sum_{k=1}^{N-1}\lambda_k \ge \prod_{k=1}^{N-2}\left(\frac{\gamma_{N-k}}{\gamma_{N-k}-1}\right)\lambda_{N-1} = \prod_{k=2}^{N-1}\left(\frac{\gamma_k}{\gamma_k-1}\right)\lambda_{N-1}.$$

Using (22) for $p = 0$ we then obtain

$$(\gamma_N-1)\lambda_0 \ge \sum_{k=1}^{N-1}\lambda_k \ge \prod_{k=2}^{N-1}\left(\frac{\gamma_k}{\gamma_k-1}\right)\lambda_{N-1}$$

which implies (21). □

This proposition immediately leads to the following theorem.

Theorem 9. *Consider the MPC scheme without terminal conditions satisfying Assumption 7. Then for all sufficiently large $N \in \mathbb{N}$ the inequality $J_\infty^{cl}(x,\mu_N) \le V_\infty(x)/\alpha$ and the convergence $x_{MPC}(k) \to x_*$ for $k \to \infty$ hold for all $x \in \mathbb{X}$ and the closed loop solution $x_{MPC}(k)$ with $x_{MPC}(0) = x$, with α from Proposition 8.*

Proof. Since $\gamma_\infty := \sup_{k\in\mathbb{N}}\gamma_k < \infty$ it follows that $(\gamma_k-1)/\gamma_k \le (\gamma_\infty-1)/\gamma_\infty < 1$ for all $k \in \mathbb{N}$, implying that α from Proposition 8 satisfies $\alpha \in (0,1]$ for sufficiently large N. For these N the assertion follows from Proposition 6. □

We note that α from Proposition 8 is not optimal. In [19] (see also [30] and [15, Chapter 6]) the optimal bound

$$\alpha = 1 - \frac{(\gamma_N-1)\prod_{k=2}^{N}(\gamma_k-1)}{\prod_{k=2}^{N}\gamma_k - \prod_{k=2}^{N}(\gamma_k-1)} \tag{23}$$

is derived, however, at the expense of a much more involved proof than that of Proposition 8. The difference between the two bounds can be illustrated if we assume $\gamma_k = \gamma$ for all $k \in \mathbb{N}$ and compute the minimal $N \in \mathbb{N}$ such that $\alpha > 0$ holds, i.e., the minimal N for which Theorem 9 ensures the convergence $x_{MPC}(k) \to x_*$. For α from Proposition 8 we obtain the condition $N > 2 + 2\ln(\gamma-1)/(\ln\gamma - \ln(\gamma-1))$

while for α from (23) we obtain $N > 2+\ln(\gamma-1)/(\ln\gamma-\ln(\gamma-1))$. The optimal α hence reduces the estimate for N roughly by a factor of 2.

The analysis can be extended to the situation in which α in (19) cannot be found for all $x \in \mathbb{X}$. In this case, one can proceed similarly as in the discussion after Theorem 15, below, in order to obtain practical asymptotic stability, i.e., inequality (34), on bounded subsets of $\mathbb{X}$.

5 Economic MPC

Economic MPC has become the common name for MPC schemes in which the stage cost ℓ does not penalize the distance to an equilibrium x_* which was determined a priori. Rather, ℓ models economic objectives, like high output, low energy consumption, etc. or a combination thereof.

For such general ℓ many of the arguments from the previous section do not work for several reasons. First, the cost J_N and thus the optimal value function V_N is not necessarily nonnegative, a fact which was exploited in several places in the proofs in the last section. Second, the infinite sum in the infinite horizon objective need not converge and thus it may not make sense to talk about infinite horizon performance. Finally, optimal trajectories need not stay close or converge to an equilibrium, again a fact that was used in various places in the last section.

A systems theoretic property which effectively serves as a remedy for all these difficulties is contained in the following definition.

Definition 10 (Strict Dissipativity and Dissipativity) *We say that an optimal control problem with stage cost ℓ is* strictly dissipative *at an equilibrium $(x^e, u^e) \in \mathbb{Y}$ if there exists a* storage function *$\lambda : \mathbb{X} \to \mathbb{R}$ bounded from below and satisfying $\lambda(x^e) = 0$, and a function $\rho \in \mathcal{K}_\infty$ such that for all $(x,u) \in \mathbb{Y}$ the inequality*

$$\ell(x,u) - \ell(x^e,u^e) + \lambda(x) - \lambda(f(x,u)) \geq \rho(|x|_{x^e}) \tag{24}$$

holds. We say that an optimal control problem with stage cost ℓ is dissipative *at (x^e,u^e) if the same conditions hold with $\rho \equiv 0$.*

We note that the assumption $\lambda(x^e) = 0$ can be made without loss of generality because adding a constant to λ does not invalidate (24).

The observation that strict dissipativity is the "right" property in order to analyze economic MPC schemes was first made by Diehl, Amrit, and Rawlings in [9], where strict duality, i.e., strict dissipativity with a linear storage function, was used. The extension to the nonlinear notion of strict dissipativity was then made by Angeli and Rawlings in [2]. Although recent studies show that for certain classes of systems this property can be further (slightly) relaxed (see [26]), here we work with this condition because it provides a mathematically elegant way for dealing with economic MPC.

Remark 11 *Strict dissipativity implies several important properties:*

(i) *The equilibrium $(x^e,u^e) \in \mathbb{Y}$ from Definition 10 is a strict optimal equilibrium in the sense that $\ell(x^e,u^e) < \ell(x,u)$ for all other admissible equilibria of f, i.e., all other $(x,u) \in \mathbb{Y}$ with $f(x,u) = x$. This follows immediately from* (24).

(ii) *The optimal equilibrium x^e has the turnpike property, i.e., the following holds: For each $\delta > 0$ there exists $\sigma_\delta \in \mathscr{L}$ such that*[2] *for all $N,P \in \mathbb{N}$, $x \in \mathbb{X}$ and $u \in \mathbb{U}^N(x)$ with $J_N^{uc}(x,u) \le N\ell(x^e,u^e)+\delta$, the set $\mathscr{Q}(x,u,P,N) := \{k \in \{0,\dots,N-1\} \mid |x_u(k,x)|_{x^e} \ge \sigma_\delta(P)\}$ has at most P elements. A proof of this fact can be found, e.g., in [15, Proposition 8.15]. The same property holds for all near optimal trajectories of the infinite horizon problem, provided it is well defined, cf. [15, Proposition 8.18].*

(iii) *If we define the modified or rotated cost $\tilde{\ell}(x,u) := \ell(x,u) - \ell(x^e,u^e) + \lambda(x) - \lambda(f(x,u))$, then this modified cost satisfies* (17), *i.e., the basic property we exploited in the previous section.*

The third property enables us to use the optimal control problem with modified cost $\tilde{\ell}$ as an auxiliary problem in our analysis. The way this auxiliary problem is used crucially depends on whether we use terminal conditions or not. We start with the case with terminal conditions. Throughout this section, we assume that all functions under consideration are continuous in x^e.

5.1 Terminal Conditions

For the economic MPC problem with terminal conditions we make exactly the same assumption on the terminal constraint set $\mathbb{X}_0$ and the terminal cost F as in the stabilizing case, i.e., we again use Assumption 4. We assume without loss of generality that $F(x^e) = 0$, which implies that F may attain negative values, because ℓ may be negative, too.

Now the main trick — taken from [1] — lies in the fact that we introduce an adapted terminal cost for the problem with the modified cost $\tilde{\ell}$. To this end, we define the terminal cost $\widetilde{F}(x) := F(x) + \lambda(x)$. We denote the cost functional for the modified problems without and with terminal conditions by $\widetilde{J}_N$ and $\widetilde{J}_N^{tc}$, respectively, and the corresponding optimal value functions by $\widetilde{V}_N$ and $\widetilde{V}_N^{tc}$. Then a straightforward computation reveals that

$$\widetilde{J}_N^{tc}(x,u) = J_N^{tc}(x,u) + \lambda(x) - N\ell(x^e,u^e), \tag{25}$$

which means that the original and the modified optimization objective only differ in terms which do not depend on u. Hence, the optimal trajectories corresponding to $\widetilde{V}_N^{tc}$ and V_N^{tc} coincide and the MPC scheme using the modified costs $\tilde{\ell}$ and $\widetilde{F}$ yields exactly the same closed loop trajectories as the scheme using ℓ and F.

[2] The space $\mathscr{L}$ contains all functions $\sigma : [0,\infty) \to [0,\infty)$ which are continuous and strictly decreasing with $\lim_{t\to\infty} \sigma(t) = 0$.

One easily sees that $\widetilde{F}$ and $\tilde{\ell}$ also satisfy Assumption 4, i.e., that for each $x \in \mathbb{X}_0$ there is $u_x \in U$ with $(x,u_x) \in \mathbb{Y}$, $f(x,u_x) \in \mathbb{X}_0$ and

$$\tilde{\ell}(x,u_x) + \widetilde{F}(f(x,u_x)) \leq \widetilde{F}(x) \tag{26}$$

if ℓ and F satisfy this property.

Moreover, if $\widetilde{F}$ is bounded on $\mathbb{X}_0$, then (26) implies $\widetilde{F}(x) \geq 0$ for all $x \in \mathbb{X}_0$. In order to see this, assume $F(x_0) < 0$ for some $x_0 \in \mathbb{X}_0$ and consider the control sequence defined by $u(k) = u_x$ with u_x from (26) for $x = x_u(k,x_0)$. Then, $\tilde{\ell} \geq 0$ implies $F(x_u(k,x)) \leq F(x_0) < 0$ for all $k \in \mathbb{N}$. Moreover, similar as in the proof of Theorem (3), the fact that $\tilde{\ell}$ satisfies (17) implies that $x_u(k,x_0) \to x^e$, because otherwise $F(x_u(k,x_0)) \to -\infty$ which contradicts the boundedness of F. But then continuity of F in x^e implies

$$F(x^e) = \lim_{k\to\infty} F(x_u(k,x_0)) \leq F(x_0) < 0$$

which contradicts $F(x^e) = 0$. Hence $\widetilde{F}(x) \geq 0$ follows for all $x \in \mathbb{X}_0$ (for a more detailed proof, see [15, Proof of Theorem 8.13]).

As a consequence, the problem with the modified costs $\tilde{\ell}$ and $\widetilde{F}$ satisfies all the properties we assumed for the results in Section 4.1. Hence, Theorem 5 applies and yields the convergence $x_{MPC}(k) \to x^e$ and the performance estimate

$$\widetilde{J}_\infty^{cl}(x,\mu_N^{tc}) \leq \widetilde{V}_N^{tc}(x).$$

As in the stabilizing case, under suitable conditions we obtain $\widetilde{V}_N^{tc}(x) \to \widetilde{V}_\infty(x)$ as $N \to \infty$. However, this only gives an estimate for the modified objective $\widetilde{J}_\infty^{cl}$ with stage cost $\tilde{\ell}$ but not for the original objective J_∞^{cl} with stage cost ℓ.

In order to obtain an estimate for J_∞^{cl}, one can proceed in two different ways: either one assumes $\ell(x^e,u^e) = 0$ (which can always be achieved by adding $\ell(x^e,u^e)$ to ℓ) and that the infinite horizon problem is well defined, which in particular means that $|V_\infty(x)|$ is finite. Then, from the definition of the problems, one sees that the relations

$$\widetilde{J}_\infty^{cl}(x,\mu_N^{tc}) = J_\infty^{cl}(x,\mu_N^{tc}) - \lim_{k\to\infty} \lambda(x_{MPC}(k))$$

and

$$\widetilde{V}_\infty(x) \leq V_\infty^{cl}(x) - \lim_{k\to\infty} \lambda(x_{u_\infty^\star}(k,x)) \quad \text{and} \quad V_\infty(x) \leq \widetilde{V}_\infty^{cl}(x) + \lim_{k\to\infty} \lambda(x_{\tilde{u}_\infty^\star}(k,x))$$

hold for $x_{MPC}(0) = x$ and $\tilde{u}_\infty^\star$ and $u_\infty^\star$ denoting the optimal controls corresponding to $\widetilde{V}_\infty(x)$ and $V_\infty(x)$, respectively.

Now strict dissipativity implies $x_{\tilde{u}_\infty^\star}(k,x) \to x^e$ and $x_{u_\infty^\star}(k,x) \to x^e$ as $k \to \infty$ (for details, see [15, Proposition 8.18]), moreover, we already know that $x_{MPC}(k) \to x^e$ as $k \to \infty$. Since $\lambda(x^e) = 0$ and λ is continuous in x^e this implies

$$J_\infty^{cl}(x,\mu_N^{tc}) \to V_\infty(x)$$

as $N \to \infty$, i.e., near optimal infinite horizon performance of the MPC closed loop for sufficiently large N.

The second way to obtain an estimate is to look at $J_K^{cl}(x, \mu_N^{tc})$, which avoids setting $\ell(x^e, u^e) = 0$ and making assumptions on $|V_\infty|$. However, while $x_{MPC}(k) \to x^e$, in the economic MPC context — even in the presence of strict dissipativity — the optimal trajectory $x_{u_N^\star}(k,x)$ will in general not end near x^e, see, e.g., the examples in [12, 13] or [15, Chapter 8]. Hence, comparing $J_K^{cl}(x, \mu_N^{tc})$ and $V_K(x)$ will in general not be meaningful. However, if for $x = x_{MPC}(0)$ we set $\delta(k) := |x_{MPC}(k)|_{x^e}$ and define the class of controls

$$\mathbb{U}^K_{\delta(K)}(x) := \{u \in \mathbb{U}^K(x) \,|\, |x_u(K,x)|_{x^e} \leq \delta(K)\} \tag{27}$$

then it makes sense to compare $J_K^{cl}(x, \mu_N^{tc})$ and $\inf_{u \in \mathbb{U}^K_{\delta(K)}(x)} J_K(x,u)$. More precisely, in [14] (see also [15, Section 8.4]) it was shown that there are error terms $\delta_1(N)$ and $\delta_2(K)$, converging to 0 as $N \to \infty$ or $K \to \infty$, respectively, such that the estimate

$$J_K^{cl}(x, \mu_N^{tc}) \leq \inf_{u \in \mathbb{U}^K_{\delta(K)}(x)} J_K(x,u) + \delta_1(N) + \delta_2(K) \tag{28}$$

holds. In other words, among all solutions steering x into the $\delta(K)$-neighborhood of the optimal equilibrium x^e, MPC yields the cheapest one up to error terms vanishing as K and N become large.

In summary, except for inequality (28) which requires additional arguments, by using terminal conditions the analysis of economic MPC schemes is not much more difficult than the analysis of stabilizing MPC schemes. However, in contrast to the stabilizing case, so far no systematic procedure for the construction of terminal costs and constraint sets satisfying (26) is known. Hence, it appears attractive to avoid the use of terminal conditions.

5.2 No Terminal Conditions

If we want to avoid the use of terminal conditions, the analysis becomes considerably more involved. The reason is that without terminal conditions the relation (25) changes to

$$\widetilde{J}_N(x,u) = J_N(x,u) + \lambda(x) - \lambda(x_u(N,x)) - N\ell(x^e,u^e). \tag{29}$$

This means that the difference between J_N and $\widetilde{J}_N$ now depends on u and consequently the optimal trajectories do no longer coincide. Moreover, the central property exploited in the proof of Proposition 8, whose counterpart in the setting of this section would be that $\lambda_{N-1} = \ell(x_{u_N^\star}(N-1,x), u_N^\star(N-1))$ is close to $\ell(x^e,u^e)$, is in general not true for economic MPC, not even for simple examples, see [12, 13] or [15, Chapter 8]. Hence, we cannot expect the arguments from the stabilizing case to work.

For these reasons, we have to use different arguments, which are combinations of arguments found in [12, 13, 18]. To this end we make the following assumptions.

Assumption 12 *(i) The optimal control problem is strictly dissipative in the sense of Definition 10.*

(ii) There exist functions γ_V, $\gamma_{\widetilde{V}}$, and $\gamma_\lambda \in \mathcal{K}_\infty$ as well as $\omega, \tilde{\omega} \in \mathcal{L}$ such that the following inequalities hold for all $x \in \mathbb{X}$ and all $N \in \mathbb{N}_\infty$:

$$\begin{aligned} (a)\ |V_N(x) - V_N(x^e)| &\leq \gamma_V(|x|_{x^e}) + \omega(N) \\ (b)\ |\widetilde{V}_N(x) - \widetilde{V}_N(x^e)| &\leq \gamma_{\widetilde{V}}(|x|_{x^e}) + \tilde{\omega}(N) \\ (c)\quad |\lambda(x) - \lambda(x^e)| &\leq \gamma_\lambda(|x|_{x^e}) \end{aligned}$$

Part (ii) of this assumption is a uniform continuity assumption in x^e. For the optimal value functions V_N and $\widetilde{V}_N$ it can, e.g., be guaranteed by local controllability around x^e, see [12, Theorem 6.4]. We note that this assumption together with the obvious inequality $V_N(x^e) \leq N\ell(x^e,u^e)$ and boundedness of $\mathbb{X}$ implies $V_N(x) \leq N\ell(x^e,u^e) + \delta$ with $\delta = \sup_{x\in\mathbb{X}} \gamma_V(|x|_{x^e}) + \omega(0)$. Hence, the optimal trajectories have the turnpike property according to Remark 11(ii).

For writing (in)equalities that hold up to an error term, we use the following convenient notation: for a sequence of functions $a_J : \mathbb{X} \to \mathbb{R}$, $J \in \mathbb{N}$, and another function $b : \mathbb{X} \to \mathbb{R}$ we write $a_J(x) \approx_J b(x)$ if $\lim_{J\to\infty} \sup_{x\in\mathbb{X}} a_J(x) - b(x) = 0$ and we write $a_J(x) \lesssim_J b(x)$ if $\limsup_{J\to\infty} \sup_{x\in\mathbb{X}} a_J(x) - b(x) \leq 0$. In words, $\approx_J$ means "$=$ up to terms which are independent of x and vanish as $J \to \infty$", and $\lesssim_J$ means the same for $\leq$.

With these assumptions and notation we can now prove the following relations. For simplicity of exposition in what follows we limit ourselves to a bounded state space $\mathbb{X}$. If this is not satisfied, the following considerations can be made for bounded subsets of $\mathbb{X}$. As we will see, dynamic programming arguments are ubiquitous in the following considerations.

Lemma 13 *Let $\mathbb{X}$ be bounded. Then under Assumptions 12 the following approximate equalities hold.*

$$\begin{aligned} (i)\quad V_N(x) &\approx_S J_M(x,u_N^\star) + V_{N-M}(x^e) \quad &&\text{for all } M \notin \mathcal{Q}(x,u_N^\star,P,N) \\ (ii)\ V_N(x^e) &\approx_S M\ell(x^e,u^e) + V_{N-M}(x^e) \quad &&\text{for all } M \notin \mathcal{Q}(x^e,u_N^{\star e},P,N) \\ (iii)\ \widetilde{V}_N(x) &\approx_N V_N(x) + \lambda(x) - V_N(x^e) \end{aligned}$$

Here $P \in \mathbb{N}$ is an arbitrary number, $S := \min\{P, N-M\}$, $u_N^\star$ is the control minimizing $J_N(x,u)$, $u_N^{\star e}$ is the control minimizing $J_N(x^e,u)$, and $\mathcal{Q}$ is the set from Remark 11(ii). Moreover, (i) and (ii) also apply to the optimal control problem with stage cost $\tilde{\ell}$.

Proof. (i) Observe that using the constant control $u \equiv u^e$ we can estimate $V_N(x^e) \leq J_N(x^e,u) = N\ell(x^e,u^e)$. Thus, using Assumption 12 we get $J_N(x,u_N^\star) \leq N\ell(x^e,u^e) +$

$\gamma_V(|x|_{x^e}) + \omega(N)$, hence the turnpike property from Remark 11(ii) applies to the optimal trajectory with $\delta = \gamma_V(|x|_{x^e}) + \omega(N)$. This in particular ensures $|x_{u_N^\star}(M,x)|_{x^e} \leq \sigma_\delta(P)$ for all $M \notin \mathcal{Q}(x, u_N^\star, P, N)$.

Now the dynamic programming equation (5) yields

$$V_N(x) = J_M(x, u_N^\star) + V_{N-M}(x_{u_N^\star}(M,x)).$$

Hence, (i) holds with remainder terms $R_1(x,M,N) = V_{N-M}(x_{u_N^\star}(M,x)) - V_{N-M}(x^e)$. For any $P \in \mathbb{N}$ and any $M \notin \mathcal{Q}(x, u_N^\star, P, N)$ we have $|R_1(x,M,N)| \leq \gamma_V(|x_{u_N^\star}(M,x)|_{x^e}) + \omega(N-M) \leq \gamma_V(\sigma_\delta(P)) + \omega(N-M)$ and thus (i).

(ii) From the dynamic programming equation (4) and $u \equiv u^e$ we obtain

$$V_N(x^e) \leq M\ell(x^e,u^e) + V_{N-M}(x^e).$$

On the other hand, from (5) we have

$$\begin{aligned} V_N(x^e) &= J_M(x, u_N^{\star e}) + V_{N-M}(x_{u_N^{\star e}}(M,x^e)) \\ &= \underbrace{\widetilde{J}_M(x, u_N^{\star e})}_{\geq 0} - \lambda(x^e) + \lambda(x_{u_N^{\star e}}(M,x^e)) + M\ell(x^e,u^e) + V_{N-M}(x_{u_N^{\star e}}(M,x^e)) \\ &\geq V_{N-M}(x^e) + M\ell(x^e,u^e) + \Big[V_{N-M}(x_{u_N^{\star e}}(M,x^e)) - V_{N-M}(x^e)\Big] \\ &\qquad\qquad + \Big[\lambda(x_{u_N^{\star e}}(M,x^e)) - \lambda(x^e)\Big] \end{aligned}$$

Now since V_{N-M} and λ satisfy Assumption 12(ii) and $x_{u_N^{\star e}}(M,x^e) \approx_P x^e$ for all $M \notin \mathcal{Q}(x^e, u_N^{\star e}, P, N)$, we can conclude that the differences in the squared brackets have values $\approx_S 0$ which shows the assertion.

(iii) Fix $x \in \mathbb{X}$ and let $u_N^\star$ and $\tilde{u}_N^\star \in \mathbb{U}^N(x)$ denote the optimal control minimizing $J_N(x,u)$ and $\widetilde{J}_N(x,u)$, respectively. We note that if the optimal control problem with cost ℓ is strictly dissipative then the problem with cost $\tilde{\ell}$ is strictly dissipative, too, with bounded storage function $\lambda \equiv 0$ and same $\rho \in \mathcal{K}_\infty$. Moreover, $V_N(x) \leq N\ell(x^e,u^e) + \gamma_V(|x|_{x^e}) + \omega(N)$ and $\widetilde{V}_N(x) \leq N\tilde{\ell}(x^e,u^e) + \gamma_{\widetilde{V}}(|x|_{x^e})$, since $V_N(x^e) \leq N\ell(x^e,u^e)$ and $\widetilde{V}_N(x^e) = 0$. Hence, the turnpike property from Remark 11(ii) applies to the optimal trajectories for both problems, yielding $\sigma_\delta \in \mathcal{L}$ and $\mathcal{Q}(x, u_N^\star, P, N)$ for $x_{u_N^\star}$ and $\tilde{\sigma}_{\tilde{\delta}}$ and $\widetilde{\mathcal{Q}}(x, \tilde{u}_N^\star, P, N)$ for $x_{\tilde{u}_N^\star}$. For all $M \notin \widetilde{\mathcal{Q}}(x, \tilde{u}_N^\star, P, N) \cup \mathcal{Q}(x^e, u_N^{\star e}, P, N)$ we can estimate

$$\begin{aligned} V_N(x) &\leq J_M(x, \tilde{u}_N^\star) + V_{N-M}(x_{\tilde{u}_N^\star}(M)) \\ &\leq J_M(x, \tilde{u}_N^\star) + V_{N-M}(x^e) + \gamma_V(\tilde{\sigma}_{\tilde{\delta}}(P)) + \omega(N-M) \\ &\leq \widetilde{J}_M(x, \tilde{u}_N^\star) - \lambda(x) + \lambda(x^e) + M\ell(x^e,u^e) + V_{N-M}(x^e) + \gamma_V(\tilde{\sigma}_{\tilde{\delta}}(P)) \\ &\qquad + \gamma_\lambda(\tilde{\sigma}_{\tilde{\delta}}(P)) + \omega(N-M) \\ &\lesssim_S \widetilde{V}_N(x) - \lambda(x) + V_N(x^e) \end{aligned}$$

for $S = \min\{P, N-M\}$, where we have applied the dynamic programming equation (4) in the first inequality, the turnpike property for $x_{\tilde{u}_N^\star}$ and Assumption 12 and (29) in the second and third inequality and (i) applied to $\widetilde{V}_N$, and (ii) applied to ℓ in the last step. Moreover, $\lambda(x^e) = 0$ and $\widetilde{V}_N(x^e) = 0$ were used.

By exchanging the two optimal control problems and using the same inequalities as above, we get

$$\widetilde{V}_N(x) \lesssim_S V_N(x) + \lambda(x) - V_N(x^e)$$

for all $M \notin \mathscr{Q}(x, u_N^\star, P, N) \cup \widetilde{\mathscr{Q}}(x^e, \tilde{u}_N^{\star e}, P, N)$. Together this implies

$$\widetilde{V}_N(x) \approx_S V_N(x) + \lambda(x) - V_N(x^e)$$

for all $M \notin \mathscr{Q}(x, u_N^\star, P, N) \cup \widetilde{\mathscr{Q}}(x, u_N^\star, P, N) \cup \mathscr{Q}(x, u_N^{\star e}, P, N) \cup \widetilde{\mathscr{Q}}(x^e, \tilde{u}_N^{\star e}, P, N)$ and $S = \min\{P, N-M\}$.

Now, choosing $P = \lfloor N/5 \rfloor$, the union of the four $\mathscr{Q}$-sets has at most $4N/5$ elements, hence there exists $M \le N/5$ for which this approximate inequality holds. This yields $S = \lfloor N/5 \rfloor$ and thus $\approx_S$ implies $\approx_N$, which shows (ii). □

We note that precise quantitative statements can be made for the error terms "hiding" in the $\approx_J$-notation. Essentially, these terms depend on the distance between the optimal trajectories to the optimal equilibrium in the turnpike property, as measured by the function σ_δ in Remark 11(ii), and by the functions from Assumption 12. For details we refer to [15, Chapter 8].

Now, as in the previous section we can proceed in two different ways. Again, the first way consists in assuming $\ell(x^e, u^e) = 0$ and the infinite horizon problem is well defined, implying that $|V_\infty(x)|$ is finite for all $x \in \mathbb{X}$. In this case, we can derive the following additional relations.

Lemma 14 *Let $\mathbb{X}$ be bounded, let Assumption 12 hold and assume $\ell(x^e, u^e) = 0$. Then the following approximate equalities hold.*

$$\begin{array}{llll} (i) & V_\infty(x) \approx_P J_M(x, u_\infty^\star) + V_\infty(x^e) & \text{for all } M \notin \mathscr{Q}(x, u_\infty^\star, P, \infty) \\ (ii) & J_M(x, u_\infty^\star) \approx_S J_M(x, u_N^\star) & \text{for all } M \notin \mathscr{Q}(x, u_N^\star, P, N) \\ & & \qquad \cup \mathscr{Q}(x, u_\infty^\star, P, \infty). \end{array}$$

Here $P \in \mathbb{N}$ is an arbitrary number, $S := \min\{P, N-M\}$ and $u_\infty^\star$ and $u_N^\star$ are the controls minimizing $J_\infty(x, u)$ and $J_N(x, u)$, respectively.

Proof. (i) The infinite horizon dynamic programming equation (13) yields

$$V_\infty(x) = J_M(x, u_\infty^\star) + V_\infty(x_{u_\infty^\star}(M, x)).$$

Hence, we obtain

$$V_\infty(x) = J_M(x, u_\infty^\star) + V_\infty(x^e) + \Big[V_\infty(x_{u_\infty^\star}(M, x)) - V_\infty(x^e)\Big].$$

From the turnpike property in Remark 11(ii) and Assumption 12 for $N = \infty$ we obtain that the term in square brackets is $\approx_P 0$ for all $M \notin \mathcal{Q}(x, u^\star_\infty, P, \infty)$, which shows (i).

(ii) The finite horizon dynamic programming equations (4) and (5) imply that $u = u^\star_N$ minimizes the expression $J_M(x,u) + V_{N-M}(x_u(M,x))$. Using the turnpike property and Assumption 12(ii) for V_N this yields

$$\begin{aligned} J_M(x,u^\star_N) + V_{N-M}(x^e) &\approx_S J_M(x,u^\star_N) + V_{N-M}(x_{u^\star_N}(M,x)) \\ &\le J_M(x,u^\star_\infty) + V_{N-M}(x_{u^\star_\infty}(M,x)) \approx_S J_M(x,u^\star_\infty) + V_{N-M}(x^e). \end{aligned}$$

for all $M \notin \mathcal{Q}(x, u^\star_N, P, N)$ and $S = \min\{P, N-M\}$.

Conversely, the infinite horizon dynamic programming equations (12) and (13) imply that $u^\star_\infty$ minimizes the expression $J_M(x,u^\star_\infty) + V_\infty(x_{u^\star_\infty}(M,x))$. Using the turnpike property and Assumption 12(ii) for V_∞ this yields

$$\begin{aligned} J_M(x,u^\star_\infty) + V_\infty(x^e) &\approx_P J_M(x,u^\star_\infty) + V_\infty(x_{u^\star_\infty}(M,x)) \\ &\le J_M(x,u^\star_N) + V_\infty(x_{u^\star_N}(M,x)) \approx_P J_M(x,u^\star_N) + V_\infty(x^e) \end{aligned}$$

for all $M \notin \mathcal{Q}(x, u^\star_\infty, P, \infty)$. Combining these two approximate inequalities then implies (ii). □

With these preparations we can state our first theorem on the performance of economic MPC without terminal conditions.

Theorem 15. *Consider the MPC scheme without terminal conditions satisfying Assumption 12 and let $\mathbb{X}$ be bounded. Then there is $\delta_1 \in \mathcal{L}$ such that for all $x \in \mathbb{X}$ the closed loop solution $x_{MPC}(k)$ generated by this scheme with $x_{MPC}(0) = x$ satisfies the inequality*

$$J^{cl}_K(x, \mu_N) + V_\infty(x_{MPC}(K)) \le V_\infty(x) + K\delta_1(N) \tag{30}$$

for all $K, N \in \mathbb{N}$.

Proof. We pick $x \in \mathbb{X}$ and abbreviate $x^+ := f(x, \mu_N(x))$. For the corresponding optimal control $u^\star_N$, the relation (6) yields that $u^\star_N(\cdot + 1)$ is an optimal control for initial value x^+ and horizon $N-1$. Hence, for each $M \in \{1, \ldots, N\}$ we obtain

$$\begin{aligned} \ell(x, \mu_N(x)) &= V_N(x) - V_{N-1}(x^+) = J_N(x, u^\star_N) - J_{N-1}(x^+, u^\star_N(\cdot+1)) \\ &= J_M(x, u^\star_N) - J_{M-1}(x^+, u^\star_N(\cdot+1)), \end{aligned}$$

where the last equality follows from the fact that the omitted terms in the sums defining $J_M(x,u^\star_N)$ and $J_{M-1}(x^+, u^\star_N(\cdot+1))$ coincide. Using Lemma 14(i) for N, x and M and for $N-1$, x^+ and $M-1$, respectively, yields

$$\begin{aligned} V_\infty(x) - V_\infty(x^+) &\approx_P J_M(x,u^\star_\infty) + V_\infty(x^e) - J_{M-1}(x^+, u^\star_\infty) - V_\infty(x^e) \\ &\approx_P J_M(x,u^\star_\infty) - J_{M-1}(x^+, u^\star_\infty). \end{aligned}$$

Putting the two (approximate) equations together and using Lemma 14(ii) yields

$$\ell(x,\mu_N(x)) \approx_S V_\infty(x) - V_\infty(x^+). \tag{31}$$

for all $M \in \{1,\dots,N\}$ satisfying $M \notin \mathcal{Q}(x,u_N^\star,P,N) \cup \mathcal{Q}(x,u_\infty^\star,P,\infty)$ and $M-1 \notin \mathcal{Q}(x^+,u_N^\star(\cdot+1),P,N-1) \cup \mathcal{Q}(x^+,u_\infty^\star(\cdot+1),P,\infty)$. Since each of the four $\mathcal{Q}$ sets contains at most P elements, their union contains at most $4P$ elements and hence if $N > 8P$ then there is at least one such M with $M \le N/2$.

Thus, choosing $P = \lfloor (N-1)/8 \rfloor$ yields the existence of $M \le N/2$ such that (31) holds with $S = \lfloor (N-1)/8 \rfloor$, implying that $\approx_S$ in (31) can be replaced by $\approx_N$. Hence, the error in (31) can be bounded by $\delta_1(N)$ for a function $\delta_1 \in \mathcal{L}$, yielding

$$\ell(x,\mu_N(x)) \le V_\infty(x) - V_\infty(x^+) + \delta_1(N). \tag{32}$$

Applying (32) for $x = x_{MPC}(k)$, $k = 0,\dots,K-1$, we can then conclude

$$\begin{aligned} J_K^{cl}(x,\mu_N) &= \sum_{k=0}^{K-1} \ell(x_{MPC}(k),\mu_N(x_{MPC}(k))) \\ &\le \sum_{k=0}^{K-1} \Big(V_\infty(x_{MPC}(k)) - V_\infty(x_{MPC}(k+1)) + \delta_1(N) \Big) \\ &\le V_\infty(x) - V_\infty(x_{MPC}(K)) + K\delta_1(N). \end{aligned}$$

This proves the claim. □

The interpretation of inequality (30) is as follows: If we concatenate the closed loop trajectory $(x_{MPC}(0),\dots,x_{MPC}(K))$ with the infinite horizon optimal trajectory emanating from $x_{MPC}(K)$, then the overall cost $J_K^{cl}(x,\mu_N) + V_\infty(x_{MPC}(K))$ is less than the optimal cost $V_\infty(x)$ plus the error term $K\delta_1(N)$. In other words, for large N the initial piece of the MPC closed loop trajectory is an initial piece of an approximately optimal infinite horizon trajectory.

With similar arguments as in the proofs of Lemmas 13 and 14 one can also prove the approximate equation

$$V_N(x) \approx_N V_{N-1}(x) + \ell(x^e,u^e).$$

Using this relation, Lemma 13(iii) and the dynamic programming equation (7), for $x^+ = f(x,\mu_N(x))$ we obtain

$$\begin{aligned} \widetilde{V}_N(x^+) &\approx_N V_N(x^+) + \lambda(x^+) - V_N(x^e) \\ &\approx_N V_{N-1}(x^+) + \ell(x^e,u^e) + \lambda(x^+) - V_N(x^e) \\ &= V_N(x) - \ell(x,\mu_N(x)) + \ell(x^e,u^e) + \lambda(x^+) - V_N(x^e) \\ &\approx_N \widetilde{V}_N(x) \underbrace{- \ell(x,\mu_N(x)) + \ell(x^e,u^e) + \lambda(x^+) - \lambda(x)}_{= -\tilde{\ell}(x,\mu_N(x))}. \end{aligned} \tag{33}$$

This implies that the modified optimal value function decays in each step, except for an error term which vanishes as $N \to \infty$. Since $\widetilde{V}_N(x) \geq \rho(|x|_{x^e})$ and $\tilde{\ell}(x, \mu_N(x)) \geq \rho(|x|_{x^e})$, from this we can conclude that as $k \to \infty$ the closed loop solution $x_{MPC}(k)$ converges to a neighborhood of x^e, which shrinks down to x^e for $N \to \infty$ (for a rigorous application of this argument, see [15, Section 8.6]). In fact, due to the upper bound on $\widetilde{V}_N$ induced by Assumption 12(ii), we can even conclude the existence of $\beta \in \mathscr{KL}$ and $\kappa \in \mathscr{L}$ such that for all $x \in \mathbb{X}$ the MPC closed loop solution $x_{MPC}(k)$ with $x_{MPC}(0) = x$ satisfies

$$|x_{MPC}(k)|_{x^e} \leq \max\{\beta(|x|_{x^e}, k),\ \kappa(N)\} \tag{34}$$

for all $N, k \in \mathbb{N}$, cf. [15, Theorem 8.33]. This means that the optimal equilibrium x^e is practically asymptotically stable for the MPC closed loop.

We note that already in very simple examples (see again [12, 13] or [15, Chapter 8]) convergence to the optimal equilibrium x^e will not hold for the MPC closed loop. Hence, in the absence of terminal conditions, practical asymptotic stability of x^e is in general the best one can obtain. This also explains the factor K before $\delta_1(N)$ in the estimate from Theorem 15. Since the trajectory always has a little distance to the optimal equilibrium, in each step we collect a small error and these errors sum up from 0 to $K-1$, resulting in the factor K in front of the error term. Note, however, that the fact that the trajectory stays near x^e prevents the solution from deteriorating as $k \to \infty$, even though the error term in (30) tends to infinite for large K.

Due to the fact that the closed loop solution converges to a neighborhood of x^e, it seems plausible that also without terminal conditions we can obtain a performance estimate for $J_K^{cl}(x, \mu_N)$ without reference to the infinite horizon problem, similar to (28). Our last theorem shows that this is indeed possible.

Theorem 16. *Consider the MPC scheme without terminal conditions satisfying Assumption 12 and let $\mathbb{X}$ be bounded. Then there are $\delta_1, \delta_2, \delta_3 \in \mathscr{L}$ such that for all $x \in \mathbb{X}$ the closed loop solution $x_{MPC}(k)$ generated by this scheme with $x_{MPC}(0) = x$ satisfies the inequality*

$$J_K^{cl}(x, \mu_N) \leq \inf_{u \in \mathbb{U}^K_{\delta(K)}(x)} J_K(x, u) + \delta_1(N) + K\delta_2(N) + \delta_3(K) \tag{35}$$

for all $K, N \in \mathbb{N}$, for $\mathbb{U}^K_{\delta(K)}(x)$ from (27) with $\delta(K) := |x_{MPC}(K)|_{x^e}$.

Proof. From (33) we obtain

$$\tilde{\ell}(x, \mu_N(x)) \approx_N \widetilde{V}_N(x) - \widetilde{V}_N(f(x, \mu_N(x))).$$

We denote the error in this approximate equation by $\delta_2(N)$. Summing $\tilde{\ell}(x, \mu_N(x))$ along the closed-loop trajectory then yields

$$\sum_{k=0}^{K-1} \tilde{\ell}(x_{MPC}(k), \mu_N(x_{MPC}(k))) \leq \widetilde{V}_N(x) - \widetilde{V}_N(x_{MPC}(K)) + K\delta_2(N). \tag{36}$$

Now the dynamic programming equation (4) and Assumption 12(ii) yield for all $K \in \{1,\dots,N\}$ and all $u \in \mathbb{U}^K_{\delta(K)}(x)$

$$\widetilde{J}_K(x,u) = \underbrace{\widetilde{J}_K(x,u) + \widetilde{V}_{N-K}(x_u(K,x))}_{\geq \widetilde{V}_N(x)} - \underbrace{\widetilde{V}_{N-K}(x_u(K,x))}_{\leq \gamma_{\widetilde{V}}(\delta(K))} \geq \widetilde{V}_N(x) - \gamma_{\widetilde{V}}(\delta(K)). \quad (37)$$

Due to the non-negativity of $\widetilde{\ell}$, for $K \geq N$ we get $\widetilde{J}_K(x,u) \geq \widetilde{V}_N(x)$ for all $u \in \mathbb{U}^K(x)$. Hence (37) holds for all $K \in \mathbb{N}$. Moreover, we have $\widetilde{V}_N(x) \geq 0$. Using (36), (37), (29) and the definition of δ_2, for all $u \in \mathbb{U}^K_{\delta(K)}(x)$ we obtain

$$\begin{aligned}
J^{cl}_K(x,\mu_N(x)) &= \sum_{k=0}^{K-1} \widetilde{\ell}(x_{MPC}(k),\mu_N(x_{MPC}(k))) - \lambda(x) + \lambda(x_{MPC}(K)) \\
&\leq \widetilde{V}_N(x) - \widetilde{V}_N(x_{MPC}(K)) + K\delta_2(N) - \lambda(x) + \lambda(x_{MPC}(K)) \\
&\leq \widetilde{J}_K(x,u) + \gamma_{\widetilde{V}}(\delta(K)) - \widetilde{V}_N(x_{MPC}(K)) + K\delta_2(N) - \lambda(x) + \lambda(x_{MPC}(K)) \\
&= J_K(x,u) + \gamma_{\widetilde{V}}(\delta(K)) - \widetilde{V}_N(x_{MPC}(K)) + K\delta_2(N) - \lambda(x_u(K,x)) + \lambda(x_{MPC}(K)) \\
&\leq J_K(x,u) + \gamma_{\widetilde{V}}(\delta(K)) + K\delta_2(N) + 2\gamma_\lambda(\delta(K)).
\end{aligned}$$

Now from (34) we obtain

$$\begin{aligned}
\gamma_{\widetilde{V}}(\delta(K)) + 2\gamma_\lambda(\delta(K)) &\leq \underbrace{\sup_{x\in\mathbb{X}} \gamma_{\widetilde{V}}(\beta(|x|_{x^e},K)) + 2\gamma_\lambda(\beta(|x|_{x^e},K))}_{=:\delta_3(K)} \\
&\quad + \underbrace{\gamma_{\widetilde{V}}(\kappa(N)) + 2\gamma_\lambda(\kappa(N))}_{=:\delta_1(N)}
\end{aligned}$$

which finishes the proof. □

The interpretation of this result is similar to that of (28): among all solutions steering x into the $\delta(K)$-neighborhood of the optimal equilibrium x^e, MPC yields the cheapest one up to error terms vanishing for large K and larger N.

We would like to note that the results from this section have been extended in various ways. For instance, in many examples it can be observed that the error terms $\delta_j(N)$ converge to 0 exponentially fast as $N \to \infty$, i.e., that they are of the form $\delta_j(N) = C\Theta^N$ for $C > 0$ and $\Theta \in (0,1)$. Conditions under which this can be rigorously proved can be found in [18]. Another extension concerns replacing the optimal equilibrium x^e by a periodic orbit. Corresponding results can be found, e.g., in [3, 25, 31]. Currently, one research focus is the extension of the results to arbitrary time varying problems, in which x^e is replaced by a general time varying trajectory with certain optimality properties. First results on this topic have appeared in [16].

6 Conclusions

We have presented a collection of results about the infinite horizon closed loop performance and stability of MPC closed loop trajectories, for both stabilizing and economic MPC and for schemes with and without terminal conditions. In the course of this analysis, we have shown that dynamic programming arguments are needed in a lot of different places and for various purposes. Dynamic programming thus forms an indispensable tool for understanding the behavior of MPC schemes.

References

1. Amrit, R., Rawlings, J.B., Angeli, D.: Economic optimization using model predictive control with a terminal cost. Annu. Rev. Control **35**, 178–186 (2011)
2. Angeli, D., Rawlings, J.B.: Receding horizon cost optimization and control for nonlinear plants. In: Proceedings of the 8th IFAC Symposium on Nonlinear Control Systems – NOLCOS 2010, pp. 1217–1223. Bologna, Italy (2010)
3. Angeli, D., Amrit, R., Rawlings, J.B.: On average performance and stability of economic model predictive control. IEEE Trans. Autom. Control **57**(7), 1615–1626 (2012)
4. Bellman, R.: Dynamic programming. Princeton Landmarks in Mathematics. Princeton University Press, Princeton (2010). Reprint of the 1957 edition
5. Bertsekas, D.P.: Dynamic Programming and Optimal Control, vols. 1 and 2. Athena Scientific, Belmont (1995)
6. Camacho, E.F., Bordons, C.: Model Predictive Control, 2nd edn. Springer, London (2004)
7. Chen, H., Allgöwer, F.: A quasi-infinite horizon nonlinear model predictive control scheme with guaranteed stability. Automatica **34**(10), 1205–1217 (1998)
8. de Nicolao, G., Magni, L., Scattolini, R.: Stabilizing receding-horizon control of nonlinear time-varying systems. IEEE Trans. Autom. Control **43**, 1030–1036 (1998)
9. Diehl, M., Amrit, R., Rawlings, J.B.: A Lyapunov function for economic optimizing model predictive control. IEEE Trans. Autom. Control **56**, 703–707 (2011)
10. Faulwasser, T., Bonvin, D.: On the design of economic NMPC based on approximate turnpike properties. In: Proceedings of the 54th IEEE Conference on Decision and Control — CDC 2015, pp. 4964–4970 (2015)
11. Forbes, M.G., Patwardhan, R.S., Hamadah, H., Gopaluni, R.B.: Model predictive control in industry: challenges and opportunities. In: Proceedings of the 9th IFAC Symposium on Advanced Control of Chemical Processes — ADCHEM 2015, *IFAC-PapersOnLine*, vol. 48, pp. 531–538. Whistler, Canada (2015)
12. Grüne, L.: Economic receding horizon control without terminal constraints. Automatica **49**(3), 725–734 (2013)
13. Grüne, L.: Approximation properties of receding horizon optimal control. Jahresber. DMV **118**(1), 3–37 (2016)
14. Grüne, L., Panin, A.: On non-averaged performance of economic MPC with terminal conditions. In: Proceedings of the 54th IEEE Conference on Decision and Control — CDC 2015, pp. 4332–4337. Osaka, Japan (2015)
15. Grüne, L., Pannek, J.: Nonlinear Model Predictive Control. Theory and Algorithms, 2nd edn. Springer, London (2017)
16. Grüne, L., Pirkelmann, S.: Closed-loop performance analysis for economic model predictive control of time-varying systems. In: Proceedings of the 56th IEEE Conference on Decision and Control — CDC 2017, pp. 5563–5569. Melbourne, Australia (2017)
17. Grüne, L., Semmler, W.: Using dynamic programming with adaptive grid scheme for optimal control problems in economics. J. Econ. Dyn. Control **28**, 2427–2456 (2004)

18. Grüne, L., Stieler, M.: Asymptotic stability and transient optimality of economic MPC without terminal conditions. J. Proc. Control **24**(8), 1187–1196 (2014)
19. Grüne, L., Pannek, J., Seehafer, M., Worthmann, K.: Analysis of unconstrained nonlinear MPC schemes with time varying control horizon. SIAM J. Control Optim. **48**, 4938–4962 (2010)
20. Keerthi, S.S., Gilbert, E.G.: Optimal infinite horizon feedback laws for a general class of constrained discrete-time systems: stability and moving horizon approximations. J. Optim. Theory Appl. **57**, 265–293 (1988)
21. Kerrigan, E.C.: Robust constraint satisfaction: invariant sets and predictive control. Ph.D. thesis, University of Cambridge (2000)
22. Lincoln, B., Rantzer, A.: Relaxing dynamic programming. IEEE Trans. Autom. Control **51**, 1249–1260 (2006)
23. Magni, L., Sepulchre, R.: Stability margins of nonlinear receding-horizon control via inverse optimality. Syst. Control Lett. **32**(4), 241–245 (1997)
24. Mayne, D.Q., Rawlings, J.B., Rao, C.V., Scokaert, P.O.M.: Constrained model predictive control: stability and optimality. Automatica **36**, 789–814 (2000)
25. Müller, M.A., Grüne, L.: Economic model predictive control without terminal constraints for optimal periodic behavior. Automatica **70**, 128–139 (2016)
26. Olanrewaju, O.I., Maciejowski, J.M.: Implications of dissipativity on stability of economic model predictive control—the indefinite linear quadratic case. Syst. Control Lett. **100**, 43–50 (2017)
27. Primbs, J.A., Nevistić, V.: Feasibility and stability of constrained finite receding horizon control. Automatica **36**(7), 965–971 (2000)
28. Rawlings, J.B., Mayne, D.Q.: Model Predictive Control: Theory and Design. Nob Hill Publishing, Madison (2009)
29. Tuna, S.E., Messina, M.J., Teel, A.R.: Shorter horizons for model predictive control. In: Proceedings of the 2006 American Control Conference, Minneapolis, pp. 863–868 (2006)
30. Worthmann, K.: Stability analysis of unconstrained receding horizon control schemes. Ph.D. thesis, Universität Bayreuth (2011)
31. Zanon, M., Grüne, L., Diehl, M.: Periodic optimal control, dissipativity and MPC. IEEE Trans. Autom. Control **62**(6), 2943–2949 (2017)

Set-Valued and Lyapunov Methods for MPC

Rafal Goebel and Saša V. Raković

1 Introduction

Model predictive control (MPC), sometimes referred to as the receding horizon control, is an optimization-based approach to stabilization of discrete-time control systems. It is well-known that infinite-horizon optimal control, with the Linear-Quadratic Regulator [1] as the fundamental example, can provide optimal controls that result in asymptotically stabilizing feedback [8]. MPC generates stabilizing feedback by using finite-horizon optimal control, which should be computationally accessible, and yet should preserve the stabilization properties of infinite-horizon problems. In fact, MPC is best summarized as a repetitive decision making process in which the underlying decision making takes the form of a finite horizon open loop optimal control. Because of its inherent ability to systematically handle constraints, guarantee stability, and optimize performance, MPC has attracted a great attention from both theoretical and practical control communities. MPC has been a very active research field that encapsulates a broad range of underlying conceptual and implementational issues [11, 12, 16], and that has seen a large number of real life implementations [13, 14]. A more detailed overview of the state of the affairs in MPC can be found in comprehensive survey papers [11, 12] and recent monographs [7, 16]. These references also provide a comprehensive overview of relevant literature.

This chapter provides a basic overview of MPC, starting from a common formulation in Section 2, and with the aim to demonstrate the utility of set-valued analysis

R. Goebel
Department of Mathematics and Statistics, Loyola University Chicago, 1032 W. Sheridan Road, Chicago, IL 60660, USA
e-mail: rgoebel1@luc.edu

S. V. Raković (✉)
Independent Researcher, London, UK
e-mail: sasa.v.rakovic@gmail.com

S. V. Raković, W. S. Levine (eds.), *Handbook of Model Predictive Control*,
Control Engineering, https://doi.org/10.1007/978-3-319-77489-3_3

in the study of structural properties of MPC. Set-valued analysis deals, among other things, with how sets — for example, sets of solutions to optimization problems — depend on parameters and how their structure behaves under operations that may include minimization. Relevant background in set-valued analysis is included in the chapter, following [17]; see also the monograph [2], or [15] for other applications to MPC. Here, set-valued and, more generally, variational analysis tools are particularly useful in the study of the finite-horizon optimal control problems but also considerations of robustness of the stabilization properties of MPC. Section 3 shows, by relying on general parametric optimization ideas, that MPC is well-posed and computationally applicable for a relatively large class of problems. Section 4 of this chapter outlines how Lyapunov techniques can be employed to provide *a priori* guarantees of invariance, stability and consistent improvement properties in MPC. The chapter is closed with Section 5 that outlines a further role of set-valued methods for analysis of MPC when applied to set-valued control systems.

2 Problem Statement and Assumptions

Consider a discrete-time control system

$$x^+ = f(x,u), \tag{1}$$

where $x \in \mathbb{R}^n$ represents the state, $u \in \mathbb{R}^m$ represents the control, and $f : \mathbb{R}^n \times \mathbb{R}^m \to \mathbb{R}^n$ is the state transition mapping. The state and the control may be subject to constraints. These are not mentioned explicitly here, but are explicit in the next subsection, where the constraints considered in the optimal control problem may include both the natural constraints on (1) and constraints introduced in the optimal control problem to induce desired properties of its solution. The goal is the design of a control feedback law that asymptotically stabilizes the origin for the closed loop system.

2.1 Open Loop Optimal Control Problem

The open loop optimal control problem to be solved is a discrete-time finite-horizon optimal control problem with the dynamics (1); subject to mixed (involving both x and u) stage constraints $(x,u) \in \mathbb{Y}$, where $\mathbb{Y} \subseteq \mathbb{R}^n \times \mathbb{R}^m$ is a set; subject to terminal state constraints $x \in \mathbb{X}_f$, where $\mathbb{X}_f \subseteq \mathbb{R}^n$ is a set; and with stage cost $\ell(x,u)$ and terminal cost $V_f(x)$, where $\ell : \mathbb{R}^n \times \mathbb{R}^m \to \mathbb{R}$ and $V_f : \mathbb{R}^n \to \mathbb{R}$ are functions.

The stage costs, the terminal cost and the terminal constraints, and some of the stage constraints — some, as there may be natural constraints on (1) already — might be design parameters (but they might be also specified as an integral part of the problem at hand). The rough idea is that the stage costs and constraints should be such that an infinite-horizon optimal control problem with them as the data should result in optimal feedback that is stabilizing. The terminal cost and constraints should be such that the properties of the finite-horizon optimal control problem approximate those of the infinite-horizon problem, and in particular such that the solution of the finite-horizon problem can be used for stabilization purposes.

Fix a positive time horizon $N \in \mathbb{N}$[1] — also a design parameter — and consider an initial condition $x \in \mathbb{R}^n$.

The **open loop optimal control** problem $\mathfrak{P}_N(x)$ to be solved is to minimize the cost

$$V_N(x, \mathbf{u}_{N-1})$$

over all feasible control sequences $\mathbf{u}_{N-1}$. The **decision variable** in $\mathfrak{P}_N(x)$ is the control sequence

$$\mathbf{u}_{N-1} := \{u_k\}_{k \in \mathbb{N}_{N-1}} = \{u_0, u_1, \ldots, u_{N-1}\}, \tag{2}$$

identified in what follows with a vector in $\mathbb{R}^{mN}$. The implementation of $\mathbf{u}_{N-1}$ results in the predicted sequence of controlled states $\mathbf{x}_N := \{x_k\}_{k \in \mathbb{N}_N}$, identified with a vector in $\mathbb{R}^{n(N+1)}$. The predicted states x_k and controls u_k are subject to

dynamical consistency constraints

$$x_0 = x \text{ and, } \forall k \in \mathbb{N}_{N-1}, \quad x_{k+1} = f(x_k, u_k); \tag{3}$$

stage constraints

$$\forall k \in \mathbb{N}_{N-1}, \quad (x_k, u_k) \in \mathbb{Y}; \tag{4}$$

and **terminal constraints**

$$x_N \in \mathbb{X}_f. \tag{5}$$

A **feasible control sequence** for $\mathfrak{P}_N(x)$ is a sequence (2) such that $\mathbf{u}_{N-1}$ and the resulting $\mathbf{x}_N$ satisfy (3), (4), and (5). Thus, the set of admissible[2] control sequences is

$$\mathbf{U}_N(x) := \{\mathbf{u}_{N-1} \in \mathbb{R}^{mN} \ : \ (3), (4) \text{ and } (5) \text{ hold}\}. \tag{6}$$

This defines a set-valued mapping $\mathbf{U}_N : \mathbb{R}^n \rightrightarrows \mathbb{R}^{mN}$. The set $\mathbb{X}_N$ of all initial conditions x for which there exists an admissible control sequence $\mathbf{u}_{N-1}$ is **the N-step controllability set**. In other words,

$$\mathbb{X}_N = \{x \in \mathbb{R}^n \ : \ \mathbf{U}_N(x) \neq \emptyset\}. \tag{7}$$

The **cost** to be minimized, over all admissible control sequences, is the sum of the associated stage costs $\ell(x_k, u_k)$, $k \in \mathbb{N}_{N-1}$ and the terminal cost $V_f(x_N)$, i.e., $V_N(x, \mathbf{u}_{N-1}) = \sum_{k=0}^{N-1} \ell(x_k, u_k) + V_f(x_N)$. For convenience, it is good to consider $V_N : \mathbb{R}^n \times \mathbb{R}^{mN} \to [0, \infty]$ defined as follows:

---

[1] $\mathbb{N}$ denotes the set of non-negative integers, and we use $\mathbb{N}_N := \{0, 1, \ldots, N-1, N\}$ for any $N \in \mathbb{N}$.

[2] Terms admissible control sequence/s are used interchangeably with feasible control sequence/s.

$$V_N(x,\mathbf{u}_{N-1}) := \begin{cases} \sum_{k=0}^{N-1} \ell(x_k,u_k) + V_f(x_N) & \text{if} \quad \mathbf{u}_{N-1} \in \mathbf{U}_N(x) \\ \infty & \text{if} \quad \mathbf{u}_{N-1} \notin \mathbf{U}_N(x). \end{cases} \tag{8}$$

The **optimal value** $V_N^0(x)$ for $\mathfrak{P}_N(x)$ or, when thought as a function dependent on x, the **value function** for $\mathfrak{P}_N$ is the function $V_N^0 : \mathbb{R}^n \to [0,\infty]$ given by

$$V_N^0(x) := \inf_{\mathbf{u}_{N-1}} V_N(x,\mathbf{u}_{N-1}). \tag{9}$$

Note that the constraint that $\mathbf{u}_{N-1}$ be admissible is implicitly present in (9), as, in view of (8), $V_N(x,\mathbf{u}_{N-1}) = \infty$ if the constraint is violated. An **open loop optimal control sequence** is any control sequence $\mathbf{u}_{N-1}$ at which the infimum in (9) is attained. The **set of optimal open loop control sequences** $\mathbf{u}_{N-1}^0(x)$ is then

$$\mathbf{u}_{N-1}^0(x) := \arg\min_{\mathbf{u}_{N-1}} V_N(x,\mathbf{u}_{N-1}), \tag{10}$$

and note that this defines a set-valued mapping $\mathbf{u}_{N-1}^0 : \mathbb{R}^n \rightrightarrows \mathbb{R}^{mN}$, which may have empty values — for example, $\mathbf{u}_{N-1}^0(x) = \emptyset$ if there is no admissible control sequence (i.e., when $x \notin \mathbb{X}_N$).

2.2 Closed Loop Dynamics

The closed loop dynamics for (1), resulting from iterative solutions to the open-loop optimal control problem $\mathfrak{P}_N(x)$ described in the previous section, is as follows:

- at a current state x of the system, one solves $\mathfrak{P}_N(x)$;
- assuming a solution exists, one selects an open loop optimal control sequence $\mathbf{u}_{N-1} \in \mathbf{u}_{N-1}^0(x)$ (uniqueness need not be guaranteed, even if existence is);
- one applies "the first" control value u_0, where $\mathbf{u}_{N-1} = \{u_0, u_1, \ldots, u_{N-1}\}$, to update the state according to $x^+ = f(x,u_0)$;

and the procedure is repeated.

2.3 Standing Assumptions

Unless otherwise mentioned, the following assumption is posed throughout the paper. The conditions in it are divided into three groups: regularity, which includes continuity of the functions and closedness of the sets in the data; growth conditions on the stage cost or on the stage constraint; and a condition requiring that 0 be a controlled equilibrium.

Assumption 1 *Regularity assumptions are:*

(a) *the state transition mapping* $f : \mathbb{R}^n \times \mathbb{R}^m \to \mathbb{R}^n$ *is continuous;*
(b) *the stage constraint* $\mathbb{Y} \subseteq \mathbb{R}^n \times \mathbb{R}^m$ *is closed;*
(c) *the terminal constraint* $\mathbb{X}_f \subseteq \mathbb{R}^n$ *is closed;*
(d) *the stage cost* $\ell : \mathbb{R}^n \times \mathbb{R}^m \to [0,\infty)$ *is continuous;*
(e) *the terminal cost* $V_f : \mathbb{R}^n \to [0,\infty)$ *is continuous.*

Coercivity/growth assumption is:

(f) *either* $\ell(x,u) \geq \psi(u)$ *for some radially unbounded* $\psi : \mathbb{R}^m \to [0,\infty)$, *or the set-valued mapping* $\mathbf{Y}_u : \mathbb{R}^n \rightrightarrows \mathbb{R}^m$, *given by*

$$\mathbf{Y}_u(x) := \{u \in \mathbb{R}^m \ : \ (x,u) \in \mathbb{Y}\},$$

is locally bounded.

"Properness" assumption is:

(g) $f(0,0) = 0$, $(0,0) \in \mathbb{Y}$, $0 \in \mathbb{X}_f$.

Coercivity assumption holds, for example, if

$$\ell(x,u) = x^T Qx + u^T Ru$$

for $Q = Q^T \geq 0$, $R = R^T > 0$, as then $\ell(x,u) \geq \psi(u) = u^T Ru$; or if

$$\mathbb{Y} = \{(x,u) \ : \ Cx + Du \in \mathbb{K}\}$$

for a compact $\mathbb{K}$ and invertible D, as then $\mathbf{Y}_u(x) = D^{-1}(\mathbb{K} - Cx)$ which is locally bounded (and continuous).

3 Properties of the Open Loop Optimal Control Problem

3.1 Set-Valued Analysis Background

The set-valued analysis background presented here follows [17]. Definitions are stated without precise pointers. A set-valued mapping $\mathbf{M}$ from $\mathbb{R}^n$ to $\mathbb{R}^m$, denoted $\mathbf{M} : \mathbb{R}^n \rightrightarrows \mathbb{R}^m$, is understood to associate to every $x \in \mathbb{R}^n$ a set $\mathbf{M}(x) \subseteq \mathbb{R}^m$. $\mathbf{M}$ is *outer semicontinuous* at $x \in \mathbb{R}^n$ if for every sequence $x_i \to x$, every convergent sequence $y_i \in \mathbf{M}(x_i)$, one has $\lim_{i\to\infty} y_i \in \mathbf{M}(x)$. An equivalent condition for outer semicontinuity of $\mathbf{M}$ at every $x \in \mathbb{R}^n$ is that the *graph* of $\mathbf{M}$, namely the set

$$\{(x,y) \in \mathbb{R}^n \times \mathbb{R}^m \ : \ y \in \mathbf{M}(x)\},$$

be closed. $\mathbf{M}$ is *continuous* at x if in addition to being outer semicontinuous at x, it is *inner semicontinuous* at x: for every $x_i \to x$ and every $y \in \mathbf{M}(x)$ there exist $y_i \in \mathbf{M}(x_i)$

for all large enough i so that $y_i \to y$. $\mathbf{M}$ is *locally bounded* at x if there exists a neighborhood $\mathbb{X}$ of x such that $\mathbf{M}(\mathbb{X}) := \bigcup_{x\in\mathbb{X}} \mathbf{M}(x)$ is bounded. If $\mathbf{M}$ is locally bounded at x and has closed values, then outer semicontinuity at x is equivalent to the condition that for every $\varepsilon > 0$ there exist $\delta > 0$ such that $\mathbf{M}(x') \subseteq \mathbf{M}(x) + \varepsilon\mathbb{B}$ for every $x' \in x + \delta\mathbb{B}$; see [17, Proposition 5.12]. Here, $\mathbb{B}$ is the closed unit ball in the Euclidean norm centered at 0; $x + \delta\mathbb{B}$ is the closed ball of radius δ centered at x; and, similarly, $\mathbf{M}(x) + \varepsilon\mathbb{B}$ is the closed ε-neighborhood around the set $\mathbf{M}(x)$. Similarly, under local boundedness, inner semicontinuity at x is equivalent to: for every $\varepsilon > 0$ there exist $\delta > 0$ such that $\mathbf{M}(x) \subseteq \mathbf{M}(x') + \varepsilon\mathbb{B}$ for every $x' \in x + \delta\mathbb{B}$. Consequently, if a locally bounded $\mathbf{M}$ has closed and nonempty values around x, it is continuous at x if and only if the Hausdorff distance between $\mathbf{M}(x')$ and $\mathbf{M}(x)$ tends to 0 as $x' \to x$; see [17, Corollary 5.21].

A set $\mathbb{X} \subseteq \mathbb{R}^n$ is said to be a polyhedral set if it can be expressed as the intersection of a finite family of closed half-spaces or hyperplanes, equivalently, can be specified by finitely many linear constraints, i.e., constraints $f_i(x) \leq 0$ or $f_i(x) = 0$ where f_i is affine. Image and pre-image of a polyhedral set under a linear mapping is polyhedral; see [17, Proposition 3.55]. A single-valued mapping $f : \mathbb{D} \to \mathbb{R}^m$ is *piecewise linear* if $\mathbb{D} \subseteq \mathbb{R}^n$ is a union of finitely many polyhedral sets, relative to each of which $f(x)$ is representable as $Ax + b$ for $A \in \mathbb{R}^{m\times n}$, $b \in \mathbb{R}^m$. A function $f : \mathbb{R}^n \to (-\infty, \infty]$ is piecewise linear if it is piecewise linear in the sense just defined, with $\mathbb{D}$ being the *effective domain* of f, namely, the set $\operatorname{dom} f := \{x \in \mathbb{R}^n \ : \ f(x) < \infty\}$. Similarly, a function $f : \mathbb{R}^n \to (-\infty, \infty]$ is *piecewise linear-quadratic* (PLQ) if $\operatorname{dom} f$ is a union of finitely many polyhedral sets, relative to each of which $f(x)$ is representable as $x^T Ax + b^T x + c$ for $A \in \mathbb{R}^{n\times n}$, $b \in \mathbb{R}^n$, and $c \in \mathbb{R}$. (The terms "piecewise linear" and "piecewise linear-quadratic" are utilized consistently with terminology of [17] instead of, perhaps more precise, terms "piecewise affine" and "piecewise affine-quadratic".) A PLQ function $f : \mathbb{R}^n \to (-\infty, \infty]$ has $\operatorname{dom} f$ closed and is continuous relative to $\operatorname{dom} f$, in the sense that $\lim_{i\to\infty} f(x_i) = f(x)$ whenever $x_i, x \in \operatorname{dom} f$ and $x_i \to x$, and thus lower semicontinuous[3] on $\mathbb{R}^n$; and if f is also convex then $\operatorname{dom} f$ is also polyhedral; see [17, Proposition 10.21]. A set-valued mapping $\mathbf{M} : \mathbb{R}^n \rightrightarrows \mathbb{R}^m$ is *piecewise polyhedral* if its graph is a union of finitely many polyhedral sets.

3.2 Parametric Optimization Background

Let $\phi : \mathbb{R}^n \times \mathbb{R}^m \to [0, \infty]$ be proper, in the sense that it is not ∞ everywhere; lower semicontinuous; and such that $\phi(x,u)$ is *level-bounded in u locally uniformly in x*: for each $\alpha \in \mathbb{R}$, $\overline{x} \in \mathbb{R}^n$ there exists a neighborhood $\mathbb{X}$ of $\overline{x}$ and a bounded set $\mathbb{U}$ such that $\{u \ : \ \phi(x,u) \leq \alpha\} \subseteq \mathbb{U}$ for every $x \in \mathbb{X}$. Let $p : \mathbb{R}^n \to [0, \infty]$ and $\mathbf{P} : \mathbb{R}^n \rightrightarrows \mathbb{R}^m$ be defined by

$$p(x) = \inf_{u\in\mathbb{R}^m} \phi(x,u), \quad \mathbf{P}(x) = \operatorname*{arg\,min}_{u\in\mathbb{R}^m} \phi(x,u).$$

[3] The function $f : \mathbb{R}^n \to [-\infty, \infty]$ is lower semicontinuous at $\overline{x} \in \mathbb{R}^n$ (on $\mathbb{R}^n$) if $f(\overline{x}) \leq \liminf_{x\to\overline{x}} f(x)$ ($f(\overline{x}) \leq \liminf_{x\to\overline{x}} f(x)$ for every $\overline{x} \in \mathbb{R}^n$).

It is illustrative to think about parametric optimization in terms of the *epigraph* of ϕ, namely the set

$$\operatorname{epi}\phi := \{(x,u,r) \in \mathbb{R}^n \times \mathbb{R}^m \times \mathbb{R} \;:\; \phi(x,u) \le r\},$$

and the epigraph of p, i.e., $\operatorname{epi} p := \{(x,r) \in \mathbb{R}^n \times \mathbb{R} \;:\; p(x) \le r\}$. Indeed, $\operatorname{epi} p$ is the projection $(x,u,r) \mapsto (x,r)$ of $\operatorname{epi}\phi$. Lower semicontinuity of ϕ is equivalent to $\operatorname{epi}\phi$ being closed. In general, a projection of a closed set need not be closed; for example project the epigraph of e^x onto the y-axis. The level-boundedness assumption addresses this issue. Furthermore, $\operatorname{epi}\phi$ is a convex set if and only if ϕ is a convex function, and since projections of convex sets are convex sets, this suggests that convexity of ϕ is inherited by p. Similarly, certain further piecewise structures of ϕ are also inherited.

General Regularity [17, Theorem 1.17] implies that p is proper, lower semicontinuous, and for each $x \in \operatorname{dom} p$, $\mathbf{P}(x)$ is nonempty and compact. [17, Theorem 7.41] implies that $\mathbf{P}$ is outer semicontinuous with respect to p-attentive convergence: if $x_i \to x \in \operatorname{dom} p$, $p(x_i) \to p(x)$, $u_i \in \mathbf{P}(x_i)$ and u_i converge, then $\lim_{i\to\infty} u_i \in \mathbf{P}(x)$, and $\mathbf{P}$ is locally bounded relative to any set on which p is bounded from above.

Continuity [17, Theorem 1.17] implies that p is continuous at every $\bar{x}$ for which there exists $\bar{u} \in \mathbf{P}(\bar{x})$ such that $x \mapsto \phi(x,\bar{u})$ is continuous at $\bar{x}$. Additionally, if ϕ is continuous relative to $\operatorname{dom}\phi$, then p is continuous at every $\bar{x}$ at which the set-valued mapping $x \mapsto \{u \;:\; (x,u) \in \operatorname{dom}\phi\}$ is continuous. This can be deduced from [17, Theorem 7.41(b)] but is easy to see directly. Indeed, under the assumptions above, let $\bar{u} \in \mathbf{P}(\bar{x})$ and consider $x_i \to \bar{x}$. By continuity of the mentioned set-valued mapping, there exist $u_i \to \bar{u}$ such that $(x_i,u_i) \in \operatorname{dom}\phi$. By continuity of ϕ relative to $\operatorname{dom}\phi$, $\phi(x_i,u_i) \to \phi(\bar{x},\bar{u})$. Thus

$$\limsup_{i\to\infty} p(x_i) \le \limsup_{i\to\infty} \phi(x_i,u_i) = \phi(\bar{x},\bar{u}) = p(\bar{x})$$

and p is upper semicontinuous[4] at $\bar{x}$. Together with previously established lower semicontinuity of p at every $\bar{x}$, this amounts to continuity at $\bar{x}$.

Convexity and Structure Suppose now that ϕ is additionally a convex function. Then p is a convex function and $\mathbf{P}$ has convex values, by [17, Proposition 2.22], while by [17, Theorem 7.41], $\mathbf{P}$ is locally bounded and outer semicontinuous on the interior of $\operatorname{dom} p$. One immediate consequence of convexity of p is the continuity of p on the interior of $\operatorname{dom} p$; see [17, Theorem 2.35]. Another consequence of convexity is preservation of further structure that ϕ may have. [17, Proposition 3.55] implies that if ϕ is convex and piecewise linear, then p is convex and piecewise linear. [17, Corollary 11.16 and Proposition 11.32] imply that if ϕ is convex and piecewise linear-quadratic, then p is convex and piecewise linear-quadratic and $\mathbf{P}(x)$ is polyhedral at each $x \in \operatorname{dom} p$. More can be shown, via an argument that underscores the value of analysis of mappings using their graphs. Convexity and PLQ structure of ϕ is equivalent to the subdifferential mapping $\partial\phi$ of ϕ, in the sense of convex

[4] The function $f : \mathbb{R}^n \to [-\infty,\infty]$ is upper semicontinuous at $\bar{x} \in \mathbb{R}^n$ (on $\mathbb{R}^n$), if $\limsup_{x\to\bar{x}} f(x) \le f(\bar{x})$ ($\limsup_{x\to\bar{x}} f(x) \le f(\bar{x})$ for every $\bar{x} \in \mathbb{R}^n$).

analysis,[5] being piecewise polyhedral [17, Proposition 12.30]. Then $\bar{u} \in \mathbf{P}(\bar{x})$ if and only if 0 is in the subdifferential of the convex function $u \mapsto \phi(\bar{x}, u)$ at $\bar{u}$. Equivalently, thanks to [17, Exercise 10.22], if and only if there exists $y \in \mathbb{R}^n$ such that $(y, 0) \in \partial\phi(\bar{x}, \bar{u})$. Thus, the graph of $\mathbf{P}$ is the projection $(x, u, y, z) \mapsto (x, u)$ of the intersection of the graph of $\partial\phi$ with $(\mathbb{R}^n \times \mathbb{R}^m \times \mathbb{R}^n \times \{0\})$. The graph of $\partial\phi$ is a union of finitely many polyhedral sets, hence the said intersection has this property, and thus so does the projection. Consequently, if ϕ is convex and PLQ, then $\mathbf{P}$ is piecewise polyhedral.

3.3 Existence and Structure of Optimal Solutions

Under Assumption 1, the N-step controllability set $\mathbb{X}_N$ is nonempty, as $0 \in \mathbb{X}_N$. Since $V_N(x, \mathbf{u}_{N-1})$ is nonnegative and finite for every admissible $\mathbf{u}_{N-1}$, $V_N^0(x)$ is finite for every $x \in \mathbb{X}_N$ and, by convention, infinite elsewhere. The set

$$\mathbb{U}_N := \left\{(x, \mathbf{u}_{N-1}) \in \mathbb{R}^n \times \mathbb{R}^{mN} \ : \ (3), (4) \text{ and } (5) \text{ hold}\right\}$$

is closed. Roughly, the limit of a sequence of feasible controls, for different initial conditions, is feasible for the limiting initial condition. This set is, in fact, the graph of the set-valued mapping $\mathbf{U}_N : \mathbb{R}^n \rightrightarrows \mathbb{R}^{mN}$ defined in (6), associating with each x the set of admissible control sequences. Consequently, $\mathbf{U}_N$ is outer semicontinuous. The cost to be minimized in $\mathfrak{P}_N(x)$, that is $V_N : \mathbb{R}^n \times \mathbb{R}^{mN} \to [0, \infty]$ given by (8), can be restated as

$$V_N(x, \mathbf{u}_{N-1}) := \begin{cases} \sum_{k=0}^{N-1} \ell(x_k, u_k) + V_f(x_N) & \text{if} \quad (x, \mathbf{u}_{N-1}) \in \mathbb{U}_N, \\ \infty & \text{if} \quad (x, \mathbf{u}_{N-1}) \notin \mathbb{U}_N. \end{cases} \tag{11}$$

Since $\sum_{k=0}^{N-1} \ell(x_k, u_k) + V_f(x_N)$ is continuous and $\mathbb{U}_N$ is closed, V_N is a lower semicontinuous function. It is also proper – it is finite at $x = 0$, since the zero sequence is an admissible control and has a finite cost. Finally, $V_N(x, \mathbf{u}_{N-1})$ is level bounded in $\mathbf{u}_{N-1}$, locally uniformly in x. In fact, if $\ell(x, u) \geq \psi(u)$ for some radially unbounded ψ then $V_N(x, \mathbf{u}_{N-1})$ is level bounded in $\mathbf{u}_{N-1}$ uniformly over all x (and independently of the properties of $\mathbf{U}_N$); while if $\mathbf{Y}_u$ is bounded in u locally uniformly in x, then $\mathbf{U}_N$ is locally bounded in x, which is sufficient for the needed property of V_N. Consequently, results in Section 3.2 yield:

Theorem 1. *$V_N^0 : \mathbb{R}^n \to [0, \infty]$ is proper and lower semicontinuous, $V_N^0(x)$ is finite if and only if $x \in \mathbb{X}_N$, and for every $x \in \mathbb{X}_N$, the set of minimizers $\mathbf{u}_{N-1}^0(x)$ is nonempty and compact.*

---

[5] For a convex $f : \mathbb{R}^n \to (-\infty, \infty]$, $\partial f(x) = \left\{y \in \mathbb{R}^n \ : \ \forall x' \in \mathbb{R}^n, \ f(x') \geq f(x) + y^T(x' - x)\right\}$, and if f is differentiable at $x \in \mathbb{R}^n$ then $\partial f(x)$ reduces to $\nabla f(x)$.

If $x \in \mathbb{X}_N$ and $\overline{\mathbf{u}}_{N-1}$ is such that $(\overline{x}, \overline{\mathbf{u}}_{N-1})$ is interior to $\mathbb{U}_N$, then $x \mapsto V_N(x, \overline{\mathbf{u}}_{N-1})$ is continuous at $\overline{x}$ and V_N^0 is continuous at $\overline{x}$. A different approach to continuity of V_N^0, since V_N is continuous relative to the set on which it is finite (i.e., $\mathbb{X}_N$), is through continuity of $\mathbf{U}_N$.

Theorem 2. *If $\overline{x} \in \mathbb{X}_N$ and $\mathbf{U}_N$ is continuous at $\overline{x}$, then V_N^0 is continuous at $\overline{x}$. If V_N^0 is continuous at $\overline{x}$, then $\mathbf{u}_{N-1}^0$ is outer semicontinuous and locally bounded at $\overline{x}$, and thus continuous if it is single-valued.*

Continuous dependence of $\mathbf{U}_N$ on x naturally occurs when the state transition mapping f is affine and the stage constraint set $\mathbb{Y}$ and the terminal constraint set $\mathbb{X}_f$ are convex. Then, the set $\mathbb{U}_N$ is convex and since $\mathbb{U}_N$ is the graph of $\mathbf{U}_N$, [17, Theorem 5.9] implies that $\mathbf{U}_N$ is continuous at every x in the interior of its domain. If, additionally, the stage and terminal costs are convex, the problem $\mathfrak{P}_N(x)$ is a convex optimization problem: the function V_N in (11) is convex. Indeed, convexity is preserved under summation and composition of a convex function with an affine mapping. This structure is preserved under the operation of minimization over $\mathbf{u}_{N-1}$.

Theorem 3. *Suppose that the state transition mapping f is affine; the stage constraint set $\mathbb{Y}$ and the terminal constraint set $\mathbb{X}_f$ are convex sets; and the stage cost ℓ and terminal cost V_f are convex functions. Then, $\mathbb{X}_N$ is convex, V_N^0 is convex, and hence continuous at every point in the interior of its domain $\mathbb{X}_N$, and $\mathbf{u}_{N-1}^0(x)$ is convex for every $x \in \mathbb{X}_N$.*

Under the assumptions of Theorem 3, which ensure convexity of V_N, further assumptions on the data ensure that V_N is also PLQ. The assumptions are that $\mathbb{Y}$ and $\mathbb{X}_f$ be polyhedral and ℓ and V_f be PLQ. Indeed, the PLQ structure of V_N follows as sums of convex PLQ functions and compositions of convex PLQ functions with affine mappings are PLQ; [17, Exercise 10.22]. The PLQ structure is also preserved under the operation of minimization over $\mathbf{u}_{N-1}$.

Theorem 4. *Suppose that, in addition to hypotheses of Theorem 3, $\mathbb{Y}$ and $\mathbb{X}_f$ are polyhedral. If ℓ and V_f are PLQ, then, in addition to properties asserted in Theorem 3, $\mathbb{X}_N$ is polyhedral, V_N^0 is PLQ, and $\mathbf{u}_{N-1}^0$ is piecewise polyhedral, while its values $\mathbf{u}_{N-1}^0(x)$ are polyhedral sets for all $x \in \mathbb{X}_N$, and in case of $\mathbf{u}_{N-1}^0$ being single-valued, $\mathbf{u}_{N-1}^0$ is piecewise linear. If ℓ and V_f are piecewise linear, then, furthermore, V_N^0 is piecewise linear.*

A special but important case that fits under the assumptions of Theorem 4 is the polyhedrally constrained linear-quadratic regulator, described in the remark below. For consequences of favorable structure, of the value function and of the minimizer in this case, for computation and implementation, see [3] and related literature.

Remark 1. Suppose that:

- $f(x,u) = Ax + Bu$ for appropriately-sized matrices A, B;
- $\mathbb{Y}$ and $\mathbb{X}_f$ are polyhedral;

– $\ell(x,u) = x^T Qx + u^T Ru$ with appropriately-sized symmetric matrices Q, R where Q is positive semidefinite and R is positive definite;
– $V_f(x) = x^T Px$ with appropriately sized positive semidefinite matrix P.

Then $\mathbb{U}_N$ is polyhedral; since ℓ and V_f are quadratic functions, V_N is quadratic on $\mathbb{U}_N$; and thus V_N is, in particular, PLQ (recall (11)). Furthermore, $\ell(x,u)$ is strictly convex in u for each fixed x, so V_N is strictly convex in $\mathbf{u}_{N-1}$ on $\mathbf{U}_N(x)$ for each fixed $x \in \mathbb{X}_N$. Thus, for each $x \in \mathbb{X}_N$ the minimization in (9) has a unique solution and so $\mathbf{u}_{N-1}$ is single-valued on $\mathbb{X}_N$. Theorem 4 and generic properties of convex PLQ functions imply that:

– $\mathbb{X}_N$ is nonempty and polyhedral;
– V_N^0 is a convex and PLQ function, continuous relative to $\mathbb{X}_N$;
– $\mathbf{u}_{N-1}^0$ is piecewise linear, continuous relative to $\mathbb{X}_N$.

4 Asymptotic Stability and Related Issues

The actual implementation of open loop optimal control problem $\mathfrak{P}_N(x)$ finds, while minimizing $V_N(x, \mathbf{u}_{N-1})$ over $\mathbf{u}_{N-1}$, the set $\mathbf{u}_N^0(x) := \{u_k^0(x)\}_{k \in \mathbb{N}_{N-1}}$ of open loop optimal control sequences. By Theorem 1, for every $x \in \mathbb{X}_N$, $\mathbf{u}_N^0(x)$ is nonempty and so open loop control sequences exist. Without assumptions guaranteeing the uniqueness of the open loop optimal control sequence at each x, this process leads to a set-valued feedback law

$$\kappa_N(x) = \{u_0^0 \ : \ \{u_0^0, u_1^0, \ldots, u_{N-1}^0\} \in \mathbf{u}_{N-1}^0(x)\} \tag{12}$$

As the MPC law κ_N is not necessarily a single-valued function, the resulting controlled MPC dynamics is most precisely modeled as:

$$x^+ \in \mathbf{F}_N(x), \quad \mathbf{F}_N(x) := \{f(x,u) \ : \ u \in \kappa_N(x)\}. \tag{13}$$

Section 3 stated basic properties of the optimization problem $\mathfrak{P}_N(x)$ under the standing assumptions. Further conditions on the data are required to ensure the following two properties:

- **Strong Positive Invariance (a.k.a. Recursive Feasibility)**, i.e., the property that given any $x \in \mathbb{X}_N$, we have that $(x,u) \in \mathbb{Y}$ for any $u \in \kappa_N(x)$, and that any $x^+ \in \mathbf{F}_N(x)$ remains in $\mathbb{X}_N$, i.e., $\mathbf{F}_N(x) \subseteq \mathbb{X}_N$, so that (13) can be iterated.

The strong positive invariance property ensures that complete solutions $\mathbf{x} := \{x_k\}_{k \in \mathbb{N}}$ to (13) exist from every initial point $x_0 \in \mathbb{X}_N$, and satisfy $x_k \in \mathbb{X}_N$ for every $k \in \mathbb{N}$. The corresponding realized control sequences $\mathbf{u} := \{u_k \in \kappa_N(x_k)\}_{k \in \mathbb{N}}$ are such that $(x_k, u_k) \in \mathbb{Y}$ for every $k \in \mathbb{N}$.

- **Strong Asymptotic Stability** of the origin for $x^+ \in \mathbf{F}_N(x)$ of (13) in the following sense:

– for every $\varepsilon > 0$ there exists $\delta > 0$ such that, for every solution $\mathbf{x} = \{x_k\}_{k\in\mathbb{N}}$ to $x^+ \in \mathbf{F}_N(x)$ of (13), if $x_0 \in \mathbb{X}_N$ satisfies $\|x_0\| < \delta$, then $\|x_k\| < \varepsilon$ for every $k \in \mathbb{N}$, and
– every solution $\mathbf{x} = \{x_k\}_{k\in\mathbb{N}}$ to $x^+ \in \mathbf{F}_N(x)$ of (13) with $x_0 \in \mathbb{X}_N$ converges to 0.

This asymptotic stability property is, usually, the true goal of MPC. In the current setting, where the dynamics (13) is set-valued, the term "strong" underlines that the two desired properties above hold for all solutions. This is in contrast with weak asymptotic stability, where the desired properties are expected to hold only for some solution. The term asymptotic stability is henceforth used interchangeably with strong asymptotic stability.

The set of conditions below stipulates that the terminal constraints set $\mathbb{X}_f$ and cost function V_f are a control invariant set and a local control Lyapunov-like function. The first condition leads to recursive feasibility, the second leads to the optimal value function V_N^0 being a Lyapunov-like function. Subject to further conditions in Section 4.3, V_N^0 is a Lyapunov function and strong asymptotic stability can be deduced.

Assumption 2 *For all $x \in \mathbb{X}_f$ there exists a $u \in \mathbb{R}^m$ such that*

$$(a) \quad (x,u) \in \mathbb{Y}, \text{ and } f(x,u) \in \mathbb{X}_f,$$

and

$$(b) \quad V_f(f(x,u)) \le V_f(x) - \ell(x,u).$$

4.1 Strong Positive Invariance (a.k.a. Recursive Feasibility)

The N-step controllability set $\mathbb{X}_N$ was defined in (7). It is clear that if $x \in \mathbb{X}_N$, then x^+ coming from (13) is in $\mathbb{X}_{N-1}$, but, in general, it need not be in $\mathbb{X}_N$.

Proposition 1. *Suppose that Assumption 2(a) holds. Then*

$$\forall x \in \mathbb{X}_N,\ \forall u \in \kappa_N(x), \quad (x,u) \in \mathbb{Y}, \text{ and } f(x,u) \in \mathbb{X}_N. \tag{14}$$

Proof. Take any $x \in \mathbb{X}_N$ and any $u \in \kappa_N(x)$. Then $u = u_0$ for some optimal, in particular admissible, $\mathbf{u}_{N-1} = \{u_0, u_1, \ldots, u_{N-1}\}$. Thus, $(x,u) \in \mathbb{Y}$. Let $\mathbf{x}_N := \{x_0, x_1, \ldots, x_N\}$ be the sequence of states resulting[6] from $\mathbf{u}_{N-1}$ and initial condition x. In particular, $x_N \in \mathbb{X}_f$. By Assumption 2(a) there exists a $u_N \in \mathbb{R}^m$ such that $(x_N, u_N) \in \mathbb{Y}$ and $x_{N+1} = f(x_N, u_N) \in \mathbb{X}_f$. Take any such u_N and corresponding $x_{N+1} = f(x_N, u_N)$. Consider

$$\mathbf{u}'_{N-1} = \{u_1, u_2, \ldots, u_N\}, \quad \mathbf{x}'_N = \{x_1, x_2, \ldots, x_{N+1}\}. \tag{15}$$

[6] We say that $\mathbf{x}_N$ results from $\mathbf{u}_{N-1}$ and x if, for each $k \in \mathbb{N}_{N-1}$, $x_{k+1} = f(x_k, u_k)$ with $x_0 = x$.

Notice that, by construction, $\mathbf{x}'_N$ results from $\mathbf{u}'_{N-1}$ and the initial condition $x' = f(x,u)$, and that, since $\mathbf{u}_{N-1}$ is admissible for the initial condition x, $\mathbf{u}'_{N-1}$ is admissible for the initial condition x'. Thus $f(x,u) \in \mathbb{X}_N$ and the proof is done.

Proposition 1 demonstrates that the control invariance of the terminal constraint set $\mathbb{X}_f$ is preserved through design of MPC, and it is reflected via strong positive invariance of the N-step controllability set $\mathbb{X}_N$. From the numerical point of view, the strong positive invariance property associated with the MPC law κ_N assures that its computation is not sensitive with respect to selections induced by a specific numerical algorithm used for solving the underlying open loop optimal control problem $\mathfrak{P}_N(x)$ (i.e., positive invariance of $\mathbb{X}_N$ is preserved with any selection $u(x) \in \kappa_N(x)$).

4.2 Strong Lyapunov Decrease (a.k.a. Cost Reduction)

It is a tradition in conventional MPC to utilize the value function V_N^0 as a Lyapunov function for the related closed loop dynamics. This is typically done by ensuring that the values of the value function V_N^0 decrease along the controlled state trajectories by the value of the stage cost functions ℓ. In other words, it is desirable to ensure that the Lyapunov decrease, as specified locally in Assumption 2(b), is maintained by MPC design. It turns out that this is indeed possible under Assumption 2.

Proposition 2. *Suppose that Assumption 2 holds. Then*

$$\forall x \in \mathbb{X}_N, \quad \forall u \in \kappa_N(x), \quad V_N^0(f(x,u)) \leq V_N^0(x) - \ell(x,u). \tag{16}$$

Proof. Take any $x \in \mathbb{X}_N$, any $u \in \kappa_N(x)$, and any optimal $\mathbf{u}_{N-1}$ such that $u = u_0$. By Assumption 2 there exists a $u_N \in \mathbb{R}^m$ such that $(x_N, u_N) \in \mathbb{Y}$ and $x_{N+1} = f(x_N, u_N) \in \mathbb{X}_f$, and $V_f(x_{N+1}) \leq V_f(x_N) - \ell(x_N, u_N)$. Take any such u_N and the corresponding $x_{N+1} = f(x_N, u_N)$. Define $\mathbf{u}'_{N-1}$ and $\mathbf{x}'_N$ as in (15). Let $x' = f(x,u)$. As in the proof of Proposition 1, $\mathbf{u}'_{N-1}$ is admissible for the initial condition x' and results in $\mathbf{x}'_N$. Furthermore,

$$\begin{aligned}
V_N(x', \mathbf{u}'_{N-1}) &= \sum_{k=1}^{N} \ell(x_k, u_k) + V_f(x_{N+1}) \\
&= \sum_{k=1}^{N-1} \ell(x_k, u_k) + \ell(x_N, u_N) + V_f(x_{N+1}) \\
&= -\ell(x_0, u_0) + \sum_{k=0}^{N-1} \ell(x_k, u_k) + \ell(x_N, u_N) + V_f(x_{N+1}) \\
&\leq -\ell(x_0, u_0) + \sum_{k=0}^{N-1} \ell(x_k, u_k) + V_f(x_N) \\
&= -\ell(x,u) + V_N^0(x),
\end{aligned}$$

where the last equality follows from optimality of $\mathbf{u}_{N-1}$. Since $V_N^0(x') \leq V_N(x', \mathbf{u}'_{N-1})$, one obtains

$$V_N^0(f(x,u)) \leq V_N^0(x) - \ell(x,u).$$

The proof is finished.

4.3 Strong Positive Invariance and Strong Asymptotic Stability

In view of Propositions 1 and 2, MPC law yields strong positive invariance of the N-step controllability set $\mathbb{X}_N$, and it also ensures the strong Lyapunov decrease. Assuming that $\mathscr{K}_\infty$-class functions[7] lower and upper bound the value function V_N^0 this means that the origin is strongly asymptotically stable for the MPC controlled dynamics $x^+ \in \mathbf{F}_N(x)$ with the region of attraction being equal to N-step controllability set $\mathbb{X}_N$. The existence of the related $\mathscr{K}_\infty$-class upper bound, under already displayed assumptions, reflects requirements for weak constrained controllability of considered problem setting. The developments of this section are summarized below.

Theorem 3. *Suppose that Assumption 2 holds and*

- *there exists a $\mathscr{K}_\infty$-class function α such that*

$$\forall (x,u) \in \mathbb{Y}, \quad \alpha(\|x\|) \leq \ell(x,u),$$

- *there exists a $\mathscr{K}_\infty$-class function β such that*

$$\forall x \in \mathbb{X}_N, \quad V_N^0(x) \leq \beta(\|x\|).$$

Then the N-step controllability set $\mathbb{X}_N$ is strongly positively invariant for the MPC controlled dynamics $x^+ \in \mathbf{F}_N(x)$. Furthermore, the origin is strongly asymptotically stable for the MPC controlled dynamics $x^+ \in \mathbf{F}_N(x)$ with the region of attraction being equal to the N-step controllability set $\mathbb{X}_N$.

Proof. Proposition 1 established strong positive invariance of the N-step controllability set $\mathbb{X}_N$, so that equivalently $\forall x \in \mathbb{X}_N$, $\forall u \in \kappa_N(x)$, $(x,u) \in \mathbb{Y}$, and

$$\forall x \in \mathbb{X}_N, \quad \mathbf{F}_N(x) \subseteq \mathbb{X}_N.$$

Proposition 2 established strong Lyapunov decrease property of the value function over the N-step controllability set $\mathbb{X}_N$: for all $x \in \mathbb{X}_N$, and all $u \in \kappa_N(x)$, $V_N^0(f(x,u)) \leq V_N^0(x) - \ell(x,u)$. Due to the assumed bound on ℓ, this decrease property becomes

$$\forall x \in \mathbb{X}_N, \quad \forall x^+ \in \mathbf{F}_N(x), \quad V_N^0(x^+) \leq V_N^0(x) - \alpha(\|x\|).$$

[7] A function $f : [0,\infty) \to [0,\infty)$ is called a $\mathscr{K}_\infty$-class function if it is continuous, strictly increasing, $f(0) = 0$, and $f(x) \to \infty$ as $x \to \infty$.

Since ℓ and V_f are nonnegative, $\ell(x,u) \leq V_N(x,\mathbf{u}_{N-1})$ for every $(x,u) \in \mathbb{Y}$ and so $\alpha(\|x\|) \leq V_N(x,\mathbf{u}_{N-1})$. This and the assumed bound on V_N^0 yield

$$\forall x \in \mathbb{X}_N, \quad \alpha(\|x\|) \leq V_N^0(x) \leq \beta(\|x\|),$$

so that the value function V_N^0 is a Lyapunov function verifying strong asymptotic stability of the origin for the MPC controlled dynamics $x^+ \in \mathbf{F}_N(x)$ with the region of attraction being equal to the N-step controllability set $\mathbb{X}_N$.

4.4 Set-Valued Approach to Robustness of Asymptotic Stability

Considering set-valued dynamics is convenient for the analysis of robustness of asymptotic stability, even if the nominal dynamics $x^+ \in \mathbf{F}_N(x)$ defined for $x \in \mathbb{X}_N$ is single-valued, i.e., reduces to $x^+ = \mathbf{F}_N(x)$. Indeed, if, for simplicity, $\kappa_N(x)$ is single-valued, then measurement error e and external perturbation d alter the nominal dynamics to

$$x^+ = f(x, \kappa_N(x+e)) + d,$$

and passing to set-valued dynamics can capture the behaviors resulting from all (small enough) e and d, and other perturbations. To that end, one considers dynamics which are "an enlargement" of the nominal dynamics. Let $\rho : \mathbb{R}^n \to [0,\infty)$ be a function. Define

$$\mathbb{X}_N^\rho := \{x \in \mathbb{R}^n \ : \ (x+\rho(x)\mathbb{B}) \cap \mathbb{X}_N \neq \emptyset\}, \text{ and}$$
$$\forall x \in \mathbb{X}_N^\rho, \quad \mathbf{F}_N^\rho(x) := \{v \in \mathbb{X}_N^\rho \ : \ v \in z+\rho(z)\mathbb{B},\ z \in \mathbf{F}_N(x+\rho(x)\mathbb{B})\}$$

and set $\mathbf{F}_N^\rho(x)$ to be empty for $x \notin \mathbb{X}_N^\rho$, just as $\mathbf{F}_N(x)$ is considered empty outside $\mathbb{X}_N$. If one thinks of $\rho(x)$ as a bound on the measurement error at x, then $\mathbb{X}_N^\rho$ includes all points that are within measurement error of $\mathbb{X}_N$. Similarly, then the enlarged dynamics $\mathbf{F}_N^\rho(x)$ considers all values z of $\mathbf{F}_N$ at points within the measurement error of x, and then allows for external perturbations of those z's with magnitude bounded by $\rho(z)$.

If $\mathbb{X}_N$ is closed, $\mathbf{F}_N$ is outer semicontinuous and locally bounded (as it is, in particular, if $\mathbf{F}_N$ is a continuous function on $\mathbb{X}_N$), and ρ is continuous, then $\mathbb{X}_N^\rho$ is closed and $\mathbf{F}_N^\rho$ is outer semicontinuous and locally bounded; see [17, Lemma 5.17]. Additionally, if $\mathbb{X}_N$ is strongly positively invariant for the nominal dynamics, then $\mathbf{F}_N^\rho(x)$ is nonempty at each $x \in \mathbb{X}_N^\rho$ and, since $v \in \mathbb{X}_N^\rho$ are considered, $\mathbb{X}_N^\rho$ is strongly positively invariant for the dynamics $x^+ \in \mathbf{F}_N^\rho(x)$. More importantly, the just mentioned regularity properties of $\mathbb{X}_N$ and $\mathbf{F}_N$ are sufficient to guarantee robustness of asymptotic stability, in the following sense:

Theorem 6. *Suppose that $\mathbb{X}_N$ is closed. Suppose that $\mathbf{F}_N$ is outer semicontinuous and locally bounded, as it automatically is if $\mathbf{F}_N$ is a continuous function on $\mathbb{X}_N$. Suppose that the origin is asymptotically stable for the nominal dynamics $x^+ \in$*

$\mathbf{F}_N(x)$ on $\mathbb{X}_N$, which entails strong positive invariance of $\mathbb{X}_N$ for $x^+ \in \mathbf{F}_N(x)$. Then there exists a continuous and positive definite $\rho : \mathbb{R}^n \to [0,\infty)$ such that the origin is strongly asymptotically stable for the enlarged dynamics

$$x^+ \in \mathbf{F}_N^{\rho}(x) \tag{17}$$

on $\mathbb{X}_N^{\rho}$, with the basin of attraction equal to $\mathbb{X}_N^{\rho}$.

The result is a special case of [5, Theorem 7.21]. In [5], a modified concept of asymptotic stability is considered; here, given the strong positive invariance of $\mathbb{X}_N^{\rho}$, that concept reduces to the standard (strong) asymptotic stability. In the result above, the magnitude of perturbations $\rho(x)$ decreases to 0 as the state x approaches the origin, and the result guarantees true asymptotic stability. Related results that allow for constant and positive magnitude of perturbations at all x usually conclude semiglobal practical stability; see, for example, [5, Lemma 7.20].

Section 3.3 provided some results on when outer semicontinuity of κ_N and thus of the closed loop dynamics $\mathbf{F}_N$ can be expected. These were usually tied to continuity of V_N^0. It turns out that continuity of V_N^0, if it is a Lyapunov function as described in the proof of Theorem 3, or continuity of any other Lyapunov function for the nominal dynamics is sufficient for robustness of asymptotic stability, even if $\mathbf{F}_N$ or $\mathbb{X}_N$ lack the regularity required by the result above. For an exposition of the relation between the regularity of Lyapunov functions and robustness, see the survey [9]. For multivalued dynamics in continuous time, the connection between smoothness of Lyapunov functions and robustness of asymptotic stability was made by [4]. For discrete time, equivalence between robustness of asymptotic stability and existence of *continuous* Lyapunov functions was shown in [10]. That such equivalences also hold when state constraints are present, subject to altering the asymptotic stability concept for the case where strong positive invariance may fail, is evidenced in [5].

For example when lack of continuous Lyapunov functions or lack of regularity of the feedback and the closed-loop dynamics leads to lack of robustness of asymptotic stability, and for further discussion, see [6], or Chapter 8 in [7].

4.5 Consistent Improvement

Increasing the horizon length N in the optimization problem $\mathfrak{P}_N(x)$ increases computational complexity. It is then desirable to know if it also improves the performance and applicability of MPC. Methods similar to those already employed in this section show that, under the same assumptions, this is indeed the case.

Let $\mathbf{u}_{N-1} = \{u_0, u_1, \ldots, u_{N-1}\}$ be an admissible control sequence for $\mathfrak{P}_N(x)$, i.e., $\mathbf{u}_{N-1} \in \mathbf{U}_N(x)$, and let $\mathbf{x}_N$ be the resulting trajectory. Then $x_N \in \mathbb{X}_f$, and under Assumption 2(a), there exists u_N so that $(x_N, u_N) \in \mathbb{Y}$ and $x_{N+1} := f(x_N, u_N) \in \mathbb{X}_f$. Then,

$$\mathbf{u}_N = \{u_0, u_1, \ldots, u_N\}$$

is an admissible control sequence for $\mathfrak{P}_{N+1}(x)$. Thus, admissible controls for the horizon N can be augmented to form admissible controls for the horizon $N+1$. In particular, for a given initial condition x, if there exists an admissible control for horizon N, then there exists one for the horizon $N+1$, and so $\mathbb{X}_f := \mathbb{X}_0 \subseteq \mathbb{X}_N \subseteq \mathbb{X}_{N+1}$. Under additional Assumption 2(b), the u_N above can be chosen so that $V_f(x_{N+1}) + \ell(x_N, u_N) \leq V_f(x_N)$ also holds. If $\mathbf{u}_{N-1}$ is optimal, one obtains

$$V_{N+1}^0(x) \leq \sum_{k=0}^{N-1} \ell(x_k, u_k) + \ell(x_N, u_N) + V_f(x_{N+1}) \leq \sum_{k=0}^{N-1} \ell(x_k, u_k) + V_f(x_N) = V_N^0(x).$$

In summary:

Proposition 3. *Under Assumption 2, for all $N \in \mathbb{N}$,*

$$\mathbb{X}_0 := \mathbb{X}_f \subseteq \mathbb{X}_N \subseteq \mathbb{X}_{N+1}, \quad \textit{and} \quad \forall x \in \mathbb{X}_{N+1}, \quad V_{N+1}^0(x) \leq V_N^0(x).$$

(Here, by convention, $V_0^0(x) = V_f(x)$ for all $x \in \mathbb{X}_f$, and $V_0^0(x) = \infty$ otherwise.)

5 Set-Valued Control Systems

Set-Valued System Discontinuous control systems and uncertain control systems are perhaps major classes of control systems that motivate the consideration of a generalized form of control systems, which are governed by

$$x^+ \in \mathbf{F}(x,u), \tag{18}$$

where, as before, $x \in \mathbb{R}^n$ and $u \in \mathbb{R}^m$ are the current state and control, respectively, while $x^+ \in \mathbb{R}^n$ is the successor state and $\mathbf{F} : \mathbb{R}^n \times \mathbb{R}^m \rightrightarrows \mathbb{R}^n$ is a set-valued mapping. These generalized control systems provide a convenient mathematical framework to treat discontinuous control systems, via their set-valued regularization, and single-valued systems with uncertainty (or a perturbation or opponent's action) where one wishes to account for every possible uncertain action. For a discontinuous control system, given by a discontinuous $f : \mathbb{R}^n \times \mathbb{R}^m \to \mathbb{R}^n$, following the ideas of Fillipov and Krasovskii for continuous-time systems, one may want to consider the control system given by the closure of f. The closure of f, denoted by $\mathbf{F} : \mathbb{R}^n \times \mathbb{R}^m \rightrightarrows \mathbb{R}^n$, is the set-valued mapping whose graph is the closure of the graph of f.[8] For a single-valued control system $x^+ = f(x,u,w)$ with uncertainty $w \in \mathbb{W}$, one may want to consider the set-valued control system $x^+ \in \mathbf{F}(x,u)$, $\mathbf{F}(x,u) := \{f(x,u,w) : w \in \mathbb{W}\}$. When discontinuous control systems are considered directly within MPC framework, the question of well-posedness becomes more delicate. Likewise, when the uncertain control systems are treated, it is necessary to consider robust variants of MPC as the use of open loop optimal control is not appropriate.

[8] In other words, for each $(x,u) \in \mathbb{R}^n \times \mathbb{R}^m$, the set $F(x,u)$ is the set of all limits $z = \lim_{i \to \infty} z_i$ with $z_i = f(x_i,u_i)$ and $(x_i,u_i) \to (x,u)$.

When set-valued control systems are used as models for underlying control systems, it is necessary to "define rules" determining the successor state at a given state and control. Depending on the nature of the corresponding process modeled as set-valued control system, the successor state might be chosen either by the decision maker or alternatively it might be produced as a result of the action by an external entity. The ramification of this observation is that it is necessary to differentiate between weak and strong formulations of MPC, as discussed briefly in what follows.

5.1 Weak Formulation of MPC

The weak formulation is concerned with set-valued control systems, for which the decision maker chooses, at a given current state x, both the current control u and successor state $x^+ \in \mathbf{F}(x,u)$. Thus, the optimization problem to be solved has, as decision variables, both the sequence of controls $\mathbf{u}_{N-1} := \{u_k\}_{k\in\mathbb{N}_{N-1}}$ and the sequence of states $\mathbf{x}_N := \{x_k\}_{k\in\mathbb{N}_N}$. For convenience, introduce $\mathbf{y}_N$ given by

$$\mathbf{y}_N := \{x_0, u_0, x_1, u_1, \ldots, x_{N-1}, u_{N-1}, x_N\}. \tag{19}$$

Within this setting, the **dynamical consistency constraints** take the following form

$$x_0 = x \text{ and, } \forall k \in \mathbb{N}_{N-1}, \quad x_{k+1} \in \mathbf{F}(x_k, u_k). \tag{20}$$

Then the set of admissible decision variables, given an initial condition x, is then

$$\mathbf{Y}_N(x) := \{\mathbf{y}_N \ : \ (20), (4) \text{ and } (5) \text{ hold}\}, \tag{21}$$

which defines a set-valued mapping $\mathbf{Y}_N : \mathbb{R}^n \rightrightarrows \mathbb{R}^{n(N+1)+mN}$. The cost to be minimized is the same cost as in Section 3, with the difference it now is a function of x and $\mathbf{y}_N$. (Since x is embedded in an admissible $\mathbf{y}_N$ through (20), one could omit the x dependence, but for connections to parametric optimization, it is reasonable to display it.)

$$V_N(x,\mathbf{y}_N) := \begin{cases} \sum_{k=0}^{N-1} \ell(x_k,u_k) + V_f(x_N) & \text{if} \quad \mathbf{y}_N \in \mathbf{Y}_N(x) \\ \infty & \text{if} \quad \mathbf{y}_N \notin \mathbf{Y}_N(x). \end{cases} \tag{22}$$

Thus, the optimization problem to be solved is

$$V_N^0(x) := \inf_{\mathbf{y}_N} V_N(x,\mathbf{y}_N), \tag{23}$$

and the set of optimal processes is

$$\mathbf{y}_N^0(x) := \arg\min_{\mathbf{y}_N} V_N(x,\mathbf{y}_N). \tag{24}$$

For the purposes of basic existence and lower semicontinuity results for this optimization problem, similar to what was done in Section 3 for $\mathfrak{P}_N(x)$, the assumption of continuity of the function f is replaced by outer semicontinuity of $\mathbf{F}$. This property of $\mathbf{F}$ arises naturally when discontinuous functions f are regularized, and can be equivalently expressed in terms of the graph of $\mathbf{F}$.

Assumption 3 *The graph $\mathbb{F}$ of the set-valued mapping* $\mathbf{F}$

$$\mathbb{F} := \{(x^+, x, u) \in \mathbb{R}^n \times \mathbb{R}^n \times \mathbb{R}^m \, : \, x^+ \in F(x,u)\}, \tag{25}$$

is a closed set such that $(0,0,0) \in \mathbb{F}$ *(i.e.,* $0 \in F(0,0)$*).*

When the condition (a) of the regularity assumptions, and $0 = f(0,0)$ part of the condition (g) of the "properness" assumptions in Assumption 1 are replaced by Assumption 3, the regularity of the related optimal solutions (i.e., the value function V_N^0 specified in (23) and the set of associated optimal processes specified in (24)) is guaranteed in the sense of Theorem 1. In particular, the value function $V_N^0 : \mathbb{R}^n \to [0,\infty]$ is proper and lower semicontinuous. Furthermore, $V_N^0(x)$ is finite if and only if $x \in \mathbb{X}_N$, and, for every $x \in \mathbb{X}_N$, the set of optimal processes $\mathbf{y}_N^0(x)$ is nonempty and compact. Here, the related N-step controllability set $\mathbb{X}_N$ is specified by

$$\mathbb{X}_N = \{x \in \mathbb{R}^n \, : \, \mathbf{Y}_N(x) \neq \emptyset\}.$$

The open loop optimal processes induce directly the MPC control law and controlled dynamics, both of which are well defined for every $x \in \mathbb{X}_N$. In general, these are set-valued feedback control law

$$\kappa_N(x) = \{u_0^0 \, : \, \{x_0^0, u_0^0, x_1^0, u_1^0 \ldots, x_{N-1}^0, u_{N-1}^0, x_N^0\} \in \mathbf{y}_N^0(x)\},$$

and related set-valued controlled dynamics

$$x^+ \in \mathbf{F}_N(x), \quad \mathbf{F}_N(x) = \{x_1^0 \, : \, \{x_0^0, u_0^0, x_1^0, u_1^0 \ldots, x_{N-1}^0, u_{N-1}^0, x_N^0\} \in \mathbf{y}_N^0(x)\}.$$

Clearly, for every $x \in \mathbb{X}_N$, and every $x^+ \in \mathbf{F}_N(x)$, there exists a $u \in \kappa_N(x)$ such that $x^+ \in \mathbf{F}(x,u)$. In fact, by construction, for every $x \in \mathbb{X}_N$, it holds that $\mathbf{F}_N(x) \subseteq \mathbf{F}(x, \kappa_N(x))$. It is worth noting that, at a given x, a simultaneous selection is made of a successor state and current control pair (x^+, u)

$$(x^+, u) \in \mathbf{C}_N(x) := \{(x_1^0, u_0^0) \, : \, \{x_0^0, u_0^0, x_1^0, u_1^0 \ldots, x_{N-1}^0, u_{N-1}^0, x_N^0\} \in \mathbf{y}_N^0(x)\}.$$

In order to guarantee the related invariance, stability, robustness, and consistent improvement properties, Assumption 2 needs to be modified as follows.

Assumption 4 *For all $x \in \mathbb{X}_f$ there exists a $u \in \mathbb{R}^m$ and $x^+ \in \mathbb{R}^n$ such that*

$$(a) \quad (x,u) \in \mathbb{Y}, \quad x^+ \in \mathbf{F}(x,u), \quad \textit{and } x^+ \in \mathbb{X}_f,$$

and

$$(b) \quad V_f(x^+) \le V_f(x) - \ell(x,u).$$

Under these hypotheses, the N-step controllability set $\mathbb{X}_N$ is strongly positively invariant for the MPC controlled dynamics, while the value function V_N^0 verifies the related Lyapunov decrease condition in the strong sense (relative to the MPC controlled dynamics), i.e.,

$$\forall x \in \mathbb{X}_N, \quad \forall (x^+, u) \in \mathbf{C}_N(x), \quad V_N^0(x^+) \le V_N^0(x) - \ell(x,u).$$

Consequently, under these modified hypotheses and with the lower and upper bounds on the value function V_N^0 as in Theorem 3, the conclusions of Theorem 3 remain valid. In particular, the N-step controllability set $\mathbb{X}_N$ is strongly positively invariant for the MPC controlled dynamics $x^+ \in \mathbf{F}_N(x)$, while the origin is strongly asymptotically stable for the MPC controlled dynamics $x^+ \in \mathbf{F}_N(x)$ with the region of attraction being equal to the N-step controllability set $\mathbb{X}_N$. The set-valued framing of robustness of asymptotic stability, as presented in Section 4.4, applies to this setting without change. Finally, within this setting, consistent improvement is also ensured, analogously to Proposition 3.

Remark 2. An alternative approach to the weak formulation may be, through some modeling tricks, to reduce this case to the standard setting discussed in previous sections. To that end, one can introduce auxiliary controls $v_0, v_1, \ldots, v_{N-1}$; add, to stage constraints, the constraint that, for each $k \in \mathbb{N}_{N-1}$, $v_k \in \mathbf{F}(x_k, u_k)$ which is equivalent to (v_k, x_k, u_k) being in the graph $\mathbb{F}$ of $\mathbf{F}$; and consider the trivial dynamics $x_{k+1} = v_k$. Such dynamics is continuous, and if Assumption 3 holds, the added stage constraints $(v_k, x_k, u_k) \in \mathbb{F}$ fit within our standing assumption since $\mathbb{F}$ is closed. How Assumption 4 and further stability considerations carry over is not pursued here.

5.2 Strong Formulation of MPC

The strong formulation is concerned with set-valued control systems, for which the decision maker chooses, at a given current state x, only the current control u, while the actual successor state x^+ is chosen by an external entity subject to no other rules than dynamical rule $x^+ \in \mathbf{F}(x,u)$. Clearly, in this setting, the state (and control) predictions become set-valued. Thus, it is necessary employ control policy Υ_{N-1}, which is a sequence $\Upsilon_{N-1} := \{\upsilon_k\}_{k \in \mathbb{N}_{N-1}}$ of, possibly set-valued, closed loop control functions υ_k with values $\upsilon_k(x_k)$. The use of control policy Υ_{N-1} allows for different control actions $u_k(x_k) = \upsilon_k(x_k)$ at different possible predicted states x_k. (The use of open loop control sequences restricts the control action $u_k(x_0)$ to be the same for all possible predicted states x_k.) Within this setting, the dynamical consistency constraints have to be taken in the strong sense, as expressed in terms of set-dynamics

$$\begin{aligned} \forall k \in \mathbb{N}_{N-1}, \quad & X_{k+1} := \{x_{k+1} \, : \, x_{k+1} \in \mathbf{F}(x_k, u_k), \; x_k \in X_k, \; u_k \in \upsilon_k(x_k)\} \\ \text{with} \quad & X_0 := \{x\}. \end{aligned} \tag{26}$$

Consequently, the stage constraints are required to be satisfied in the strong sense

$$\forall k \in \mathbb{N}_{N-1}, \quad \forall x_k \in X_k, \quad \forall u_k \in \upsilon_k(x_k), \quad (x_k, u_k) \in \mathbb{Y}, \tag{27}$$

and, likewise, the terminal constraints are required to be satisfied in the strong sense

$$\forall x_N \in X_N, \quad x_N \in \mathbb{X}_f. \tag{28}$$

Since the predicted states $x_k \in X_k$ and controls $u_k \in \upsilon(x_k)$ are set-valued, minimizing the cost function V_N of (8) also needs to be taken in an adequate sense. A frequently utilized approach is the worst-case design, i.e., minimizing, by selecting an admissible control policy Υ_{N-1}, the maximal value of the cost V_N over the possible predicated processes $\{x_0, u_0, x_1, u_1, \ldots, x_{N-1}, u_{N-1}, x_N\}$ formed from possible predicted states $x_k \in X_k$, controls $u_k \in \upsilon(x_k)$ and successor states $x_{k+1} \in \mathbf{F}(x_k, u_k)$. (Naturally, a number of alternatives to the worst-case design can also be considered.)

All in all, the strong formulation of MPC requires the utilization of closed loop optimal control (i.e., functional optimization), and thus further elaboration on this topic lies beyond the intended scope of this article.

Acknowledgement R. Goebel was partially supported by the Simons Foundation Grant 315326.

References

1. Anderson, B., Moore, J.: Optimal Control – Linear Quadratic Methods. Prentice-Hall, Upper Saddle River (1990)
2. Aubin, J.-P., Frankowska, H.: Set-Valued Analysis. Birkhauser, Boston (1990)
3. Bemporad, A., Morari, M., Dua, V., Pistikopoulos, E.: The explicit linear quadratic regulator for constrained systems. Automatica **38**(1), 3–20 (2002)
4. Clarke, F., Ledyaev, Y., Stern, R.: Asymptotic stability and smooth Lyapunov functions. J. Differ. Equ. **149**(1), 69–114 (1998)
5. Goebel, R., Sanfelice, R., Teel, A.: Hybrid Dynamical Systems: Modeling, Stability, and Robustness. Princeton University Press, Princeton (2012)
6. Grimm, G., Messina, M., Tuna, S., Teel, A.: Examples when nonlinear model predictive control is nonrobust. Automatica **40**, 1729–1738 (2004)
7. Grüne, L., Panek, J.: Nonlinear Model Predictive Control: Theory and Algorithms. Communications and Control Engineering Series. Springer, Cham (2017)
8. Keerthi, S.S., Gilbert, E.G.: Optimal, infinite horizon feedback laws for a general class of constrained discrete time systems: stability and moving–horizon approximations. J. Optim. Theory Appl. **57**, 265–293 (1988)
9. Kellett, C.M.: Classical converse theorems in Lyapunov's second method. Discrete Contin. Dyn. Syst. Ser. B **20**(8), 2333–2360 (2015)
10. Kellett, C., Teel, A.: Smooth Lyapunov functions and robustness of stability for difference inclusions. Syst. Control Lett. **52**, 395–405 (2004)
11. Mayne, D.Q.: Model predictive control: Recent developments and future promise. Automatica **50**, 2967–2986 (2014)
12. Mayne, D.Q., Rawlings, J.B., Rao, C.V., Scokaert, P.O.M.: Constrained model predictive control: stability and optimality. Automatica **36**, 789–814 (2000)

13. Qin, S.J., Badgwell, T.A.: A survey of industrial model predictive control technology. Control Eng. Pract. **11**, 733–764 (2003)
14. Qin, S.J., Badgwell, T.A.: Model-predictive control in practice. In: Baillieul, J., Samad, T. (eds.) Encyclopedia of Systems and Control. Springer, London (2015)
15. Raković, S.V.: Set theoretic methods in model predictive control. In: Nonlinear Model Predictive Control, pp. 41–54. Springer, Berlin (2009)
16. Rawlings, J.B., Mayne, D.Q.: Model Predictive Control: Theory and Design. Nob Hill Publishing, Madison (2009)
17. Rockafellar, R.T., Wets, R.J.-B.: Variational Analysis. A Series of Comprehensive Studies in Mathematics, vol. 317. Springer, Berlin (2009)

Stochastic Model Predictive Control

Ali Mesbah, Ilya V. Kolmanovsky, and Stefano Di Cairano

1 Introduction

Stochastic Model Predictive Control (SMPC) accounts for model uncertainties and disturbances based on their statistical description. SMPC is synergistic with the well-established fields of stochastic modeling, stochastic optimization, and estimation. In particular, SMPC benefits from availability of already established stochastic models in many domains, existing stochastic optimization techniques, and well-established stochastic estimation techniques. For instance, the effect of wind gusts on an aircraft can be modeled by stochastic von Kármán and Dryden's models [21] but no similar deterministic models appear to exist. Loads or failures in electrical power grids, prices of financial assets, weather (temperature, humidity, wind speed and directions), computational loads in data centers, demand for a product in marketing/supply chain management are frequently modeled stochastically thereby facilitating the application of the SMPC framework.

A comprehensive overview of various approaches and applications of SMPC has been given in the article [33]. Another overview article [27] in *Encyclopedia of Systems and Control* is focused on tube SMPC approaches. This chapter provides a tutorial exposition of several SMPC approaches.

A. Mesbah
University of California, Berkeley, CA, USA
e-mail: mesbah@berkeley.edu

I. V. Kolmanovsky (✉)
Department of Aerospace Engineering, University of Michigan, Ann Arbor, MI 48109, USA
e-mail: ilya@umich.edu

S. Di Cairano
Mitsubishi Electric Research Laboratories, Cambridge, MA, USA
e-mail: dicairano@ieee.org

S. V. Raković, W. S. Levine (eds.), *Handbook of Model Predictive Control*,
Control Engineering, https://doi.org/10.1007/978-3-319-77489-3_4

2 Stochastic Optimal Control and MPC with Chance Constraints

Consider a stochastic, discrete-time system,

$$x_{t+1} = f(x_t, u_t, w_t), \tag{1a}$$
$$y_t = h(x_t, v_t), \tag{1b}$$

where t is the time index; $x_t \in \mathbb{R}^{n_x}$, $u_t \in \mathbb{R}^{n_u}$, and $y_t \in \mathbb{R}^{n_y}$ are the system states, inputs, and outputs, respectively; $w_t \in \mathbb{R}^{n_w}$ denotes disturbances; $v_k \in \mathbb{R}^{n_v}$ denotes measurement noise; and functions $f : \mathbb{R}^{n_x} \times \mathbb{R}^{n_u} \times \mathbb{R}^{n_w} \to \mathbb{R}^{n_x}$ and $h : \mathbb{R}^{n_x} \times \mathbb{R}^{n_v} \to \mathbb{R}^{n_y}$ define system state and output equations, respectively. The uncertain initial state x_0 is described by the known probability distribution $\mathsf{P}[x_0]$. The independent and identically distributed random variables in the noise sequences $\{w_t\}$ and $\{v_t\}$ have known probability distributions $\mathsf{P}[w]$ and $\mathsf{P}[v]$, respectively.

The system (1) represents a Markov decision process, as the successor state x_{t+1} can be determined from the current state x_t and input u_t [29]. Let I_t denote the vector of system information that is causally available at time t,

$$I_t := [y_t, \dots, y_0, u_{t-1}, \dots, u_0],$$

with $I_0 := [y_0]$. The conditional probability of state x_t given I_t, i.e., $\mathsf{P}[x_t|I_t]$, can be computed via *recursive Bayesian estimation* [15]

$$\mathsf{P}[x_t|I_t] = \frac{\mathsf{P}[y_t|x_t]\mathsf{P}[x_t|I_{t-1}]}{\mathsf{P}[y_t|I_{t-1}]}, \tag{2a}$$
$$\mathsf{P}[x_{t+1}|I_t] = \int \mathsf{P}[x_{t+1}|x_t, u_t]\mathsf{P}[x_t|I_t]dx_t, \tag{2b}$$

with $\mathsf{P}[x_0|I_{-1}] := \mathsf{P}[x_0]$. We use $\mathbb{E}_{x_t}$ and $\mathbb{P}_{x_t}$ to denote, respectively, the expected value and the probability of an event with respect to the stochastic state x_t (with uncertainty $\mathsf{P}[x_t|I_t]$) as well as the random variables w_k and v_k for all $k > t$.

Let $N \in \mathbb{N}$ be the prediction horizon.[1] Consider an N-stage control sequence,

$$\boldsymbol{\Pi} := \{\pi_0, \pi_1, \dots, \pi_{N-1}\}, \tag{3}$$

where $\pi_k \in \mathbb{U} \subset \mathbb{R}^{n_u}$ and $\mathbb{U}$ is a nonempty measurable set for the inputs. Define the control cost function as

$$J_N(x_t, \boldsymbol{\Pi}) = \mathbb{E}_{x_t}\left[c_{t+N}(x_{t+N}) + \sum_{k=0}^{N-1} c_{t+k}(x_{t+k}, \pi_{t+k})\right], \tag{4}$$

where $c_{t+k} : \mathbb{R}^{n_x} \times \mathbb{U} \to [0, \infty)$ and $c_{t+N} : \mathbb{R}^{n_x} \to [0, \infty)$ denote the stage-wise cost incurred at the $(t+k)$th stage of control and at the terminal stage, respectively. Define

[1] For notational convenience, the control and prediction horizons are considered to be identical.

a *joint chance constraint* of the form

$$\mathbb{P}_{x_{t+k}}[g(x_{t+k}) \leq 0] \geq 1-\delta, \quad k=1,\ldots,N, \tag{5}$$

where $g : \mathbb{R}^{n_x} \to \mathbb{R}^{n_c}$ denotes state constraints, composed of $n_c > 1$ inequalities; and $\delta \in [0, 1)$ denotes the maximum allowed probability of state constraint violation. The chance constraint (5) is generally nonconvex and intractable [3, 7]; see [18, 36] for treatment of chance-constrained optimization.

Remark 1. When the probabilistic uncertainties in (1) are bounded, hard state constraints can be enforced by setting $\delta = 0$ in (5). This implies that $g(x_t) \leq 0$ must be satisfied for all realizations of system uncertainties.

Given the (uncertain) knowledge of the system state at sampling time t, i.e., $\mathsf{P}[x_t|I_t]$, the stochastic optimal control problem (OCP) for system (1) can be posed as

$$\min_{\boldsymbol{\Pi}} J_N(x_t, \boldsymbol{\Pi}) \tag{6a}$$

$$\text{s.t.: } x_{t+k+1|t} = f(x_{t+k|t}, \pi_{t+k|t}, w_{k|t}), \quad k=0,\ldots,N-1, \tag{6b}$$

$$\pi_{t+k|t} \in \mathbb{U}, \quad k=0,\ldots,N-1, \tag{6c}$$

$$\mathbb{P}_{x_{t+k|t}}[g(x_{t+k|t}) \leq 0] \geq 1-\delta, \quad k=1,\ldots,N, \tag{6d}$$

$$w_{k|t} \sim \mathsf{P}[w], \quad k=0,\ldots,N-1, \tag{6e}$$

$$x_{t|t} \sim \mathsf{P}[x_t|I_t], \tag{6f}$$

where $x_{t+k|t}$ and $u_{t+k|t}$ denote[2] the state and input computed at time $t+k$ based on the knowledge of $\mathsf{P}[x_t|I_t]$. Note that the chance constraint (6d) enforces the state constraints with respect to $\mathsf{P}[x_t|I_t]$ as well as the future noise sequence over the horizon N.

In theory, the stochastic OCP (6) can be solved offline using *Bellman's principle of optimality* [2]. The resulting optimal control policy $\boldsymbol{\Pi}^*$ can then be implemented in a receding-horizon manner by applying $u_t = \pi_t^*$ to the stochastic system (1) at every sampling time t that $\mathsf{P}[x_t|I_t]$ is estimated from (2). The principle of optimality requires that the optimal control cost at each control stage satisfy the Bellman equation for stochastic dynamic programming. To this end, the control input π_t at each stage t must be designed via a nested minimization of the expected sum of the current control cost and the optimal future control cost, which is computed based on the knowledge of the future state $\mathsf{P}[x_{t+1}|I_{t+1}]$ (e.g., see [34]). Although solving the Bellman equation will result in an optimal *closed-loop* control policy, it is well-known that stochastic dynamic programming suffers from the so-called *curse of dimensionality* for practically sized systems [6].

---

[2] Hereafter we use the common notation in predictive control to differentiate prediction time instances $t+k$ from time t at which predictions are made.

In recent years, a plethora of SMPC strategies have been presented that seek online solution of an approximate surrogate for the stochastic OCP (6) in a receding-horizon manner. Generally speaking, SMPC strategies neglect the effect of the control input π_t on the knowledge of the future state $\mathsf{P}[x_{t+1}|I_{t+1}]$ to avoid the formidable challenge of solving the Bellman equation [26, 33]. In the remainder of this chapter, three SMPC strategies are introduced for receding-horizon control of stochastic linear systems.

3 Scenario Tree-Based MPC

The scenario tree-based stochastic MPC approach was introduced in [4], and relies upon multi-stage stochastic programming [9]. The general idea behind tree-based stochastic MPC is to compute a closed-loop policy based on scenarios determined by predictions of the stochastic disturbance sequences. Due to causality, the predicted states and related control sequences result arranged in a tree structure. The first application of this methodology, in the context of robust min-max MPC was in [45]. In the stochastic context, each tree node is further associated with a probability of reaching it, based on the probability of the scenario to realize. Such a probability can be used to selectively trim parts of the tree that are unlikely to realize in order to restrict the number of nodes in the tree and reduce the computational effort. For other scenario-based approaches, see [10, 44] and references in [33].

In this section we modify the notations slightly, reserving subscripts to designate the nodes of the scenario tree, and using $x(t)$, $u(t)$, $w(t)$, etc., to denote the variables at the current time instant, t.

The system is modeled as a parameter varying discrete-time linear system, possibly with an additive disturbance,

$$x(t+1) = A(w(t))x(t) + B(w(t))u(t) + F(w(t)), \tag{7}$$

where $x(t) \in \mathbb{R}^{n_x}$ is the state, $u(t) \in \mathbb{R}^{n_u}$ is the input, and $w(t) \in \mathscr{W}$ is a scalar stochastic disturbance, which takes values in a finite set $\{\bar{w}_1, \ldots, \bar{w}_s\} \subset \mathbb{R}$. The state and input vectors in (7) are subject to the pointwise-in-time constraints of the form,

$$x(t) \in \mathscr{X},\ u(t) \in \mathscr{U},\ \forall t \in \mathbb{Z}_{0+}, \tag{8}$$

which must hold for all $t \geq 0$, where $\mathscr{X} \subseteq \mathbb{R}^{n_x}$, $\mathscr{U} \subseteq \mathbb{R}^{n_u}$, are polyhedral sets. The probability mass function $p(t)$ of w is assumed to be known or predictable at all times, that is, for all $t \in \mathbb{R}_{0+}$, $p_j(t) = \mathbb{P}[w(t) = \bar{w}_j]$, such that $p_j(t) \geq 0$, $\sum_{j=1}^{s} p_j(t) = 1$ is known, and it can be predicted for $\tau > t$ based only on the information known at time t. This includes the cases when $p(t)$ is constant, or varies in a pre-defined way, or when it is defined by a stochastic Markov process with state z, and $z(t+1) = f_M(z(t))$, $p(t) = p(z(t))$, where $z(t)$ is known at time t. In the latter case, the disturbance realization in $\mathcal{W}$ represents the combinations of disturbances on the system and the (discrete) transitions of the Markov process. The main restriction imposed by this assumption is that p cannot depend on the system state x, since the system evolution is affected by the input u, and hence $p(\tau)$, $\tau > t$, will not be predictable based only on data at time t.

3.1 Scenario-Tree Construction

Due to the presence of the stochastic disturbance, the MPC aims at minimizing the expected value of a given cost function, that is

$$\mathbb{E}[J_N(t)] = \mathbb{E}\left[\sum_{k=1}^{N} x(t+k|t)^\top Q x(t+k|t) + \sum_{k=0}^{N-1} u(t+k)^\top R u(t+k)\right], \tag{9}$$

where $N \in \mathbb{Z}_+$ is the prediction horizon, and $Q = Q^\top \geq 0$, $R = R^\top > 0$ are weight matrices of appropriate dimensions. Since $|\mathcal{W}|$ is finite, $p(t)$ is known, and $p(t+k)$ can be predicted based only on the information at time t, one can enumerate all the admissible realizations of the stochastic disturbance sequence along the finite horizon N, and their corresponding probabilities. The N-steps scenario $\omega_\ell^N \in \mathcal{W}^N$ is the ℓth realization of a sequence of N consecutive disturbance values, $\omega^N = [w(0), \ldots, w(N-1)]$, and its q steps prefix $\omega^{N,q}$ is the subsequence composed of only its first q elements $\omega^{N,q} = [w(0), \ldots, w(q-1)]$. Thus, one can optimize (9) by optimizing

$$\mathbb{E}[J_N(t)] = \sum_{\ell=1}^{s^N} J_N(t|\omega_\ell^N(t))\mathsf{P}[\omega_\ell^N(t)|z(t)],$$

with constraints that enforce causality, i.e., $u(t+k|\omega_j^N) = u(t+k|\omega_i^N)$ for all i, j such that $\omega_i^{N,k} = \omega_j^{N,k}$. However, the optimization problem obtained in this way is large, because it considers all disturbance sequences, even those that occur with arbitrarily small probability.

In the scenario tree-based MPC, (7), (8), and the predicted evolution of $p(t+k)$ are used to construct a variable horizon optimization problem where only the disturbance sequences that are more likely to realize are accounted for, and hence the size of the optimization problem is reduced. The scenario tree describes the most likely scenarios of future disturbance realizations, and is updated at every time step using newly available measurements of the state $x(t)$, and updated information to predict the disturbance probability $p(t+k)$. In order to explain the scenario tree-based approach, we introduce the following notations:

- $\mathcal{T} = \{\mathcal{N}_1, \mathcal{N}_2, \ldots, \mathcal{N}_n\}$: the set of the tree nodes. Nodes are indexed progressively as they are added to the tree, i.e., $\mathcal{N}_1$ is the root node and $\mathcal{N}_n$ is the last node added;
- $pre(\mathcal{N}) \in \mathcal{T}$: the predecessor of node $\mathcal{N}$;

- $succ(\mathcal{N}, w) \in \mathcal{T}$: the successor of node $\mathcal{N}$ for $w \in \mathcal{W}$;
- $\pi_{\mathcal{N}} \in [0,1]$: the probability of reaching $\mathcal{N}$ from $\mathcal{N}_1$;
- $x_{\mathcal{N}} \in \mathbb{R}^{n_x}$, $u_{\mathcal{N}} \in \mathbb{R}^{n_u}$, $w_{\mathcal{N}} \in \mathcal{W}$: the state, input, and disturbance value, respectively, associated with node $\mathcal{N}$, where $x_{\mathcal{N}_1} = x(t)$, and $w_{\mathcal{N}_1} = w(t)$;
- $\mathcal{C} = \{\mathcal{C}_1, \mathcal{C}_2, \ldots, \mathcal{C}_c\}$: the set of candidate nodes, i.e., $\mathcal{C} = \{\mathcal{N} \notin \mathcal{T} \mid \exists (i,j) : \mathcal{N} = succ(\mathcal{N}_i, w_j)\}$;
- $\mathcal{S} \subset \mathcal{T}$: the set of leaf nodes, with cardinality denoted by $n_{\text{leaf}} = |\mathcal{S}|$, i.e., $\mathcal{S} = \{\mathcal{N} \in \mathcal{T} \mid succ(\mathcal{N}, w_j) \notin \mathcal{T}, \forall j \in \{1, \ldots, s\}\}$.

Every path from the root node to a leaf node is a scenario in the tree and describes a disturbance realization that will be accounted for in the optimization problem. The procedure to construct the scenario tree is as follows.

Starting from the root node $\mathcal{N}_1$, which is associated with $w(t)$, we construct a list $\mathcal{C}$ of candidate nodes by considering all the possible s future values of the disturbance in $\mathcal{W}$ and their realization probabilities. Note that the probability of reaching a node can be computed by multiplying the conditional transition probabilities along the path leading to a given node from the root node. The candidate with maximum probability $\mathcal{C}_{i^*}$ is added to the tree and removed from $\mathcal{C}$. The procedure is repeated by adding at every step new candidates as children of the last node added to the tree, until the tree contains $n_{\max}$ nodes. The scenario-tree construction, summarized in Algorithm 1, expands the tree in the most likely direction, so that the paths with higher probability are extended longer in the future, because they may have larger impact on the overall performance. This leads to a tree with variable depth, where the paths from the root to the leaves may have different lengths and hence different prediction horizons, see Figure 1.

Algorithm 1: SMPC tree generation procedure

At any step k:
set $\mathcal{T} = \{\mathcal{N}_1\}$, $\pi_{\mathcal{N}_1} = 1$, $n = 1$, $c = s$;
set $\mathcal{C} = \bigcup_{j=1}^{s} \{succ(\mathcal{N}_1, w_j)\}$
while $n < n_{\max}$ **do**
  for all $i \in \{1, 2, \ldots, c\}$, **do**
    compute $\pi_{\mathcal{C}_i}$;
  end for
  set $i^* = \arg\max_{i \in \{1,2,\ldots,c\}} \pi_{\mathcal{C}_i}$;
  set $\mathcal{N}_{n+1} = \mathcal{C}_{i^*}$;
  set $\mathcal{T} = \mathcal{T} \cup \{\mathcal{N}_{n+1}\}$;
  set $\mathcal{C} = \bigcup_{j=1}^{s} \{succ(\mathcal{C}_{i^*}, w_j)\} \cup (\mathcal{C} \setminus \mathcal{C}_{i^*})$;
  set $c = c + s - 1$, $n = n + 1$;
end while

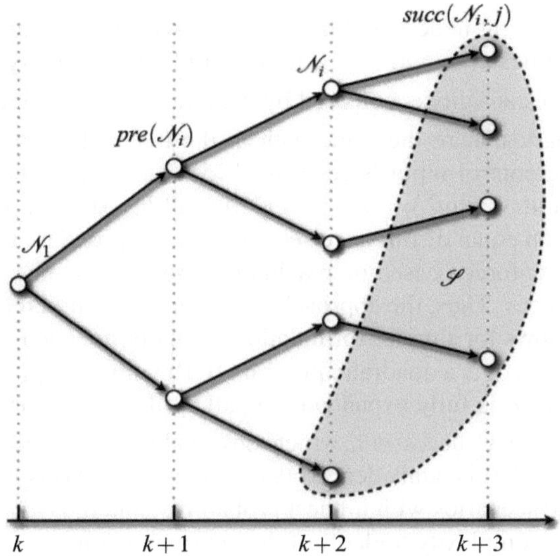

Fig. 1: Graphical representation of a multiple-horizon optimization tree. Some roof-to-leaves paths have length 2, others have length 3. Hence, different scenarios may have different prediction horizons.

3.2 Scenario-Tree Stochastic Optimization Problem

The scenario-tree constructed with Algorithm 1 is exploited to construct the MPC optimization problem at each time instant t. For the sake of simplifying the notations, in what follows we use x_i, u_i, y_i, w_i, π_i and $pre(i)$ to denote $x_{\mathcal{N}_i}$, $u_{\mathcal{N}_i}$, $y_{\mathcal{N}_i}$, $w_{\mathcal{N}_i}$, $\pi_{\mathcal{N}_i}$ and $pre(\mathcal{N}_i)$, respectively.

Given the maximum number of nodes, $n_{\max}$, at any time t, the scenario-tree based stochastic MPC performs the following operations: (i) it constructs the tree $\mathcal{T}(t, n_{\max})$ based on $w(t)$; (ii) it solves the following stochastic MPC based on $\mathcal{T}(t, n_{\max})$:

$$\min_{\mathbf{u}} \sum_{i \in \mathcal{T}(t,n_{\max}) \setminus \{\mathcal{N}_1\}} \pi_i x_i^\top Q x_i + \sum_{i \in \mathcal{T}(t,n_{\max}) \setminus \mathcal{S}} \pi_i u_i^\top R u_i \tag{10a}$$

$$\text{s.t. } x_1 = x(t), \tag{10b}$$

$$x_i = A_{pre(i)} x_{pre(i)} + B_{pre(i)} u_{pre(i)} + F_{pre(i)}, \; i \in \mathcal{T}(t, n_{\max}) \setminus \{\mathcal{N}_1\}, \tag{10c}$$

$$x_i \in \mathcal{X}, \; i \in \mathcal{T}(t, n_{\max}) \setminus \{\mathcal{N}_1\}, \tag{10d}$$

$$u_i \in \mathcal{U}, \; i \in \mathcal{T}(t, n_{\max}) \setminus \mathcal{S}, \tag{10e}$$

where $\mathbf{u} = \{u_i : \mathcal{N}_i \in \mathcal{T}(t, n_{\max}) \setminus \mathcal{S}\}$ is the multiple-horizon input sequence and where A_i, B_i, and F_i define the dynamics associated to node i and determined by the stochastic disturbance at node i, and hence by the scenario according to which i is reached; (iii) it applies $u(t) = u_1 = u_{\mathscr{N}_1}$ as a control input to the system (7).

In Problem (10), causality is enforced by the tree structure, since if $\omega_i^{N,k} = \omega_j^{N,k}$, i.e., the two scenarios share the same path in the tree at least until the kth level, and since a single control input is associated to each node in the tree, it follows automatically that $u(t+h|\omega_j^N) = u(t+h|\omega_i^N)$ for all $h = 0,\dots,k$. Indeed, causality is enforced based on equal disturbance sequences, as opposed to, for instance, [45], where causality is enforced based on reaching the same state, possibly by different disturbance sequences. Thus, the approach of (10) may result in redundant decision variables, but it allows for simpler formulation as an optimization problem.

In fact, Problem (10) is a quadratic program (QP) with $n_u(n_{\max} - n_{\text{leaf}})$ variables. If the scenario tree $\mathscr{T}$ is fully expanded, i.e., all the leaf nodes are at depth N and all parent nodes have s successors, which obviously requires $n_{\max} = \frac{s^{N+1}-1}{s-1}$, the objective function (10a) is equivalent to (9). Otherwise, (10a) is an approximation of (9) based on the scenarios with highest probability, and thus $n_{\max}$ determines the representativeness-complexity tradeoff of the approximation.

Based on Problem (10), the scenario-tree stochastic MPC results in a closed-loop prediction policy. Since there are multiple predictions, i.e., multiple tree nodes, of the future state values and to each a control action is associated, the input at any predicted step changes as a function of the disturbance realizations up to such step. Thus, the control action implicitly encodes feedback from the past disturbances.

3.3 Extensions and Applications

The Scenario-tree based MPC is a fairly general framework that allows for solving stochastic MPC problems with a precision that is related to the amount of available computational resources. Several extensions have been presented in the literature.

In terms of modeling, in [5] it is shown that $p(t)$ can be generated by several classes of stochastic processes, possibly in an approximate discretization, such as the generalized autoregressive conditional heteroskedasticity (GARCH), and Markov chains with transition matrix T, and emission matrix E, where z is the Markov chain state and

$$T_{ij} = \mathbb{P}[z(t+1) = z_j | z(t) = z_i], \tag{11a}$$
$$E_{ij} = \mathbb{P}[w(t) = \bar{w}_j | z(t) = z_i] = p_j(z_i). \tag{11b}$$

A simplified formulation of the Markov chain is the case where $z = w$, resulting in $\mathsf{P}[w(t+1)] = f_M(w(t))$ so that $T_{ij} = \mathbb{P}[w(t+1) = \bar{w}_j | w(t) = \bar{w}_i]$ and $E = I$.

For the case where $F(w) = 0$, for all $w \in \mathscr{W}$, and there are no (hard) constraints, uniform mean square exponential stability of the closed-loop system is demonstrated in [5] by designing offline a stochastic Lyapunov function satisfying

$\mathscr{V}(x) = x^{\mathrm{T}}Sx$, $\mathbb{E}[V(x(t+1))] - V(x(t)) \leq x(t)^{\mathrm{T}}Lx(t)$, where $S, L > 0$, which is then enforced as a constraint at the root node of the scenario tree $\mathscr{N}_1$ in Problem (10) by the quadratic constraint,

$$\sum_{i=1}^{S} p_i(t)(A_i x_1 + B_i u_1)^\top S(A_i x_1 + B_i u_1) \leq x_1^\top (S - L)x_1,$$

with $A_i = A(w^i)$, $B_i = B(w^i)$. For the constrained case, in [5] it is suggested to construct an invariant ellipsoid [25]

$$\mathscr{E} = \{x : x^\top Sx \leq \gamma\} \subset \mathscr{X},$$

and a linear controller $u = Kx$, such that $\mathscr{E}$ is robust positive invariant for the polytopic difference inclusion with vertices $[A(w), B(w)]$, $w \in \mathscr{W}$, controlled by $u = Kx$, and for all $x \in \mathscr{E}$, $Kx \in \mathscr{U}$. The invariant ellipsoid is exploited to construct another constraint to be added in Problem (10),

$$(A_i x_1 + B_i u_1)^\top S(A_i x_1 + B_i u_1) \leq \gamma, \ \forall i : \ p_i(t) > 0,$$

which guarantees recursive constraint satisfaction.

The scenario tree-based MPC is applied to energy management of a hybrid electric powertrain in [43]. In [16], also motivated by the application to the energy management of hybrid electric powertrains, the case where the stochastic disturbance representing driver actions is learned onboard during vehicle operation has been presented. In particular, in [16] the actions of the vehicle driver are modeled as a Markov chain with time-varying and initially unknown transition probabilities that are estimated with an iterative algorithm from the transition frequencies,

$$\mathsf{P}[w(t+1) = \bar{w}_j | w(t) = \bar{w}_i] = \frac{n_{ij}}{n_i}, \tag{12}$$

where n_{ij} is the number of transitions of w between values $\bar{w}_i$ and $\bar{w}_j$ and n_i is the number of time instants w takes the value $\bar{w}_i$ in the observed data. The so-estimated Markov chain is used to adapt the scenario tree construction in the stochastic MPC for optimizing the energy efficiency of the hybrid electric vehicle, and it is shown that with the learning of the Markov chain, the overall performance is very close to the one from an MPC with exact preview, on both synthetic and experimental data. Along these lines, [8] reports an application of the scenario tree-based stochastic MPC to adaptive cruise control where the Markov chain is used to model the actions of the vehicle which is ahead in traffic.

In terms of numerical algorithms, the scenario tree-based MPC results in QPs that are larger than those from nominal MPC, but have a special structure that can be exploited to reduce the computational effort. For instance, in [23] an algorithm based on the alternating direction method of multipliers (ADMM) was proposed, that exploits the structure and scales more favorably with the number of nodes in the tree than structure-ignoring algorithms, and allows for parallel implementation.

4 Polynomial Chaos-Based MPC

Polynomial chaos-based MPC strategies have been developed for receding-horizon control of stochastic linear [24, 38] and nonlinear systems [1, 17, 37] subject to probabilistic model uncertainty in initial conditions and parameters. The term *polynomial chaos* was introduced by Norbert Wiener in the seminal paper [47], in which a generalized harmonic analysis was applied to Brownian motion-like processes. The basic notion of polynomial chaos is to expand finite-variance random variables by an infinite series of Hermite polynomials, which are functions of a normally distributed input random variable [11]. The polynomial chaos framework has recently been generalized to non-Gaussian random variables by establishing the convergence properties of polynomials that are orthogonal with respect to possibly non-Gaussian input random variables [48]. The orthogonality of polynomials in generalized polynomial chaos (gPC) enables obtaining sample-free, closed-form expressions for propagation of high-order moments of states through the system dynamics. Alternatively, polynomial chaos expansions can be used as a surrogate for the system model for performing Monte Carlo simulations efficiently via algebraic operations in order to construct the probability distribution of states. This section uses gPC to present a sample-free formulation for SMPC of stochastic linear systems with probabilistic model uncertainty.

4.1 System Model, Constraints, and Control Input Parameterization

Consider a stochastic, linear system described by the prediction model

$$x_{t+k+1|t} = A(\theta)x_{t+k|t} + B(\theta)u_{t+k|t} + Dw_{t+k|t}, \tag{13}$$

where $\theta \in \mathbb{R}^{n_\theta}$ denotes the unknown system parameters that are modeled as (time-invariant) probabilistic uncertainties with probability distribution $\mathsf{P}[\theta]$; and the stochastic noise $w_{t+k|t}$ is a zero-mean Gaussian process with covariance Σ_w. The notation in (13) is as in (1). The probability distribution of parameters, $\mathsf{P}[\theta]$, quantifies our subjective belief in the unknown parameters, whereas the parameters are fixed in the true system.

A joint chance constraint of the form (5) is imposed, where the state constraint $g(x_{t+k|t}) \leq 0$ takes the form of a polytope

$$\mathbb{P}_{x_{t+k|t}}[\ C^\top x_{t+k|t} \leq d\ ] \geq 1-\delta, \quad k = 1,\ldots,N, \tag{14}$$

with $C \in \mathbb{R}^{n_x \times n_c}$ and $d \in \mathbb{R}^{n_c}$. The control cost function is defined as

$$J_N(x_t, \mathbf{U}) = \mathbb{E}_{x_t}\Big[\ \sum_{k=0}^{N-1} ||x_{t+k|t}||_Q^2 + ||u_{t+k|t}||_R^2\ \Big], \tag{15}$$

where Q and R are symmetric and positive definite weight matrices; and $\mathbf{U} := [u_{t|t}, \ldots, u_{t+N-1|t}]$ denotes the vector of control inputs over the prediction horizon. We choose to parameterize the control inputs $u_{t+k|t}$ as an affine function of state [20]

$$u_{t+k|t} = L_k x_{t+k|t} + m_k, \quad k = 1, \ldots, N, \tag{16}$$

where $L_k \in \mathbb{R}^{n_u \times n_x}$ and $m_k \in \mathbb{R}^{n_u}$ denote the feedback gains and control actions over the prediction horizon, respectively. The affine-state feedback parameterization (16) allows to account for the effect of state feedback over the prediction horizon. The underlying notion of (16) is that the system state will be known at the future time instants. Thus, the controller will have the state/disturbance information when designing the future control inputs over the prediction horizon.

Remark 2. It is generally impossible to guarantee satisfaction of the input bounds, i.e., $u_t(x_t) \in \mathbb{U}$, when the stochastic noise w_t is unbounded. To alleviate this shortcoming of (16), a *saturation function* can be incorporated into the affine feedback control policy to enable direct handling of hard input bounds in the presence of unbounded stochastic noise [22].

The key challenges in the above discussed setup for SMPC arise from: (i) propagation of the probabilistic model uncertainty and stochastic noise through the prediction model (13), and (ii) computational intractability of the joint chance constraint (14). We use a gPC-based uncertainty propagation method to obtain closed-form expressions for the mean and covariance of the predicted state as explicit functions of the control input. A moment-based surrogate is then presented for (14) in terms of the Mahalanobis distance [30], which is exact when the system state has a multivariate normal distribution.

4.2 Generalized Polynomial Chaos for Uncertainty Propagation

The gPC seeks to approximate a stochastic variable $\psi(\xi)$ in terms of a finite expansion of orthogonal polynomial basis functions

$$\psi(\xi) \approx \hat{\psi}(\xi) := \sum_{i=0}^{p} a_i \phi_i(\xi) = \mathbf{a}^\top \Lambda(\xi), \tag{17}$$

where $\mathbf{a} := [a_0, \ldots, a_p]^\top$ denotes the vector of expansion coefficients; $\Lambda(\xi) := [\phi_0(\xi), \ldots, \phi_p(\xi)]^\top$ denotes the vector of multivariate polynomial basis functions ϕ_i of maximum degree l with respect to the random variables $\xi \in \mathbb{R}^{n_\xi}$; and $p+1 = \frac{(n_\xi + l)!}{n_\xi! l!}$ denotes the total number of expansion terms. The basis functions belong to the Askey scheme of polynomials, which includes a set of orthogonal basis functions in the Hilbert space defined on the support of the random variables. Thus, the basis functions ϕ_i must be chosen in accordance with the probability distribution of the random variables ξ, as established in [48]. The orthogonality of the basis functions implies that $\langle \phi_i(\xi), \phi_j(\xi) \rangle = \langle \phi_i^2(\xi) \rangle \delta_{ij}$, where $\langle h(\xi), g(\xi) \rangle = \int_\Omega h(\xi) g(\xi) \mathsf{P}[\xi] d\xi$ denotes the inner product induced by $\mathsf{P}[\xi]$ and δ_{ij} denotes the Kronecker delta function. Hence, the expansion coefficients a_i in (17) can be obtained via

$$a_i = \frac{\langle \hat{\psi}(\xi), \phi_i(\xi) \rangle}{\langle \phi_i(\xi), \phi_i(\xi) \rangle},$$

which can be computed analytically for linear and polynomial systems [19].

For a particular realization of the stochastic system noise w in (13), the polynomial chaos expansions (PCEs) (17) can be used for efficient propagation of the model uncertainty θ through (13). Propagation of model uncertainty will yield the probability distribution of state conditioned on the noise realization, i.e., $\mathsf{P}[\hat{x}_{t+k|t}|w]$, which can then be integrated over all possible realizations of w to obtain the complete probability distribution of the (polynomial chaos-approximated) state

$$\mathsf{P}[\hat{x}_{t+k|t}] = \int_{-\infty}^{\infty} \mathsf{P}[\hat{x}_{t+k|t}|w] \mathsf{P}[w] dw. \tag{18}$$

When the distribution of stochastic noise, $\mathsf{P}[w]$, is Gaussian, the moments of the probability distribution $\mathsf{P}[\hat{x}_{t+k|t}]$ in (18) can be readily defined in terms of the coefficients of $\hat{x}_{t+k|t}$. To this end, we approximate each predicted state and control input as well as the unknown parameters in the system matrices $A(\theta)$ and $B(\theta)$ in (13) by PCEs of the form (17). Define $\tilde{x}_{i,t+k|t} = [a_{i_0,t+k|t}, \dots, a_{i_p,t+k|t}]^\top$ and $\tilde{u}_{i,t+k|t} = [b_{i_0,t+k|t}, \dots, b_{i_p,t+k|t}]^\top$ to denote the coefficients of PCEs for the ith predicted state and control input, respectively. The coefficients of PCEs for all states and control inputs can be concatenated into vectors $\tilde{\mathbf{x}}_{t+k|t} := [\tilde{x}_{1,t+k|t}^\top, \dots, \tilde{x}_{n_x,t+k|t}^\top]^\top \in \mathbb{R}^{n_x(p+1)}$ and $\tilde{\mathbf{u}}_{t+k|t} := [\tilde{u}_{1,t+k|t}^\top, \dots, \tilde{u}_{n_u,t+k|t}^\top]^\top \in \mathbb{R}^{n_u(p+1)}$, respectively. Using the Galerkin projection [19], the error in a gPC-based approximation of the prediction model (13), which arises from truncation in PCEs, can be projected onto the space of the multivariate basis functions $\{\phi_i\}_{i=0}^p$. This allows for expressing the prediction model (13) in terms of the coefficients of the PCEs for states and control inputs

$$\tilde{\mathbf{x}}_{t+k+1|t} = \mathbf{A}\tilde{\mathbf{x}}_{t+k|t} + \mathbf{B}\tilde{\mathbf{u}}_{t+k|t} + \mathbf{D}w_{t+k|t}, \tag{19}$$

where

$$\mathbf{A} = \sum_{i=0}^{p} A_i \otimes \Psi_i; \quad \mathbf{B} = \sum_{i=0}^{p} B_i \otimes \Psi_i; \quad \mathbf{D} = D \otimes e_{p+1};$$

$$\Psi_i := \begin{bmatrix} \sigma_{0i0} & \cdots & \sigma_{0ip} \\ \vdots & \ddots & \vdots \\ \sigma_{pi0} & \cdots & \sigma_{pip} \end{bmatrix};$$

A_i and B_i are the projections of $A(\theta)$ and $B(\theta)$ onto the ith basis function ϕ_i; $\sigma_{lmn} = \langle \phi_l, \phi_m, \phi_n \rangle / \langle \phi_l^2 \rangle$; and $e_a = [1, 0, \dots, 0]^\top \in \mathbb{R}^a$.

The orthogonality property of the multivariate polynomial basis functions can now be used to efficiently compute the moments of the conditional probability distribution $\mathsf{P}[\hat{x}_{t+k|t}|w]$ in terms of the coefficients $\tilde{\mathbf{x}}_{t+k|t}$. The conditional mean and variance of the ith predicted state are defined by

$$\mathbb{E}[\hat{x}_{i,t+k|t}|w] \approx \tilde{x}_{i_0,t+k|t}(w), \tag{20a}$$

$$\mathbb{E}\left[\hat{x}^2_{i,t+k|t}|w\right] \approx \sum_{j=0}^{p} \tilde{x}^2_{i_j,t+k|t}(w)\langle\phi_j^2\rangle. \tag{20b}$$

Similarly, the state feedback control law (16) is projected as

$$\tilde{\mathbf{u}}_{t+k|t} = \mathbf{L}_{t+k|t}\tilde{\mathbf{x}}_{t+k|t} + \mathbf{m}_{t+k|t}, \tag{21}$$

where $\mathbf{L}_{t+k|t} = L_{t+k|t} \otimes I_{p+1}$ and $\mathbf{m}_{t+k|t} = m_{t+k|t} \otimes e_{p+1}$. When w is a zero-mean Gaussian white noise with covariance Σ_w, $\tilde{\mathbf{x}}_{t+k|t}$ will be a Gaussian process with mean $\bar{\mathbf{x}}_{t+k|t}$ and covariance $\Gamma_{t+k|t}$ as defined by

$$\bar{\mathbf{x}}_{t+k+1|t} = (\mathbf{A} + \mathbf{B}\mathbf{L}_{t+k|t})\bar{\mathbf{x}}_{t+k|t} + \mathbf{B}\mathbf{m}_{t+k|t} \tag{22a}$$

$$\mathbf{\Sigma}_{t+k+1|t} = (\mathbf{A} + \mathbf{B}\mathbf{L}_{t+k|t})\mathbf{\Sigma}_{t+k|t}(\mathbf{A} + \mathbf{B}\mathbf{L}_{t+k|t})^\top + \mathbf{D}\mathbf{\Sigma}_w\mathbf{D}^\top. \tag{22b}$$

Note that $(\bar{\mathbf{x}}_{t|t}, \mathbf{\Sigma}_{t|t})$ is initialized based on the knowledge of state x_t. Using (20)–(22) and the law of iterated expectation, tractable expressions are derived for describing the mean and variance of each (polynomial chaos-approximated) state $\hat{x}_{i,t+k|t}$

$$\begin{aligned}\mathbb{E}[\hat{x}_{i,t+k|t}] &= \mathbb{E}\left[\mathbb{E}[\hat{x}_{i,t+k|t}|w]\right]\\ &\approx \mathbb{E}[\tilde{x}_{i_0,t+k|t}(w)] = \bar{x}_{i_0,t+k|t},\end{aligned} \tag{23}$$

and

$$\begin{aligned}\mathbb{E}[\hat{x}^2_{i,t+k|t}] &= \mathbb{E}\left[\mathbb{E}[\hat{x}^2_{i,t+k|t}|w]\right]\\ &\approx \mathbb{E}\Big[\sum_{j=0}^{p} \tilde{x}^2_{i_j,t+k|t}(w)\langle\phi_j^2\rangle\Big] = \sum_{j=0}^{p} \mathbb{E}[\tilde{x}^2_{i_j,t+k|t}]\langle\phi_j^2\rangle\\ &= \sum_{j=0}^{p} \left[\bar{x}^2_{i_j,t+k|t} + \mathbf{\Sigma}_{i_j i_j,t+k|t}\right]\langle\phi_j^2\rangle,\end{aligned} \tag{24}$$

respectively. It is important to note that the moments (23)–(24) can be expressed as explicit functions of the control inputs, i.e., the decision variables $\mathbf{L}$ and $\mathbf{m}$ in (22). The sample-free, closed-form expressions (23)–(24) for the moments of the predicted states are highly advantageous for gradient-based optimization methods since they avoid possible convergence problems associated with sampling.

4.3 Moment-Based Surrogate for Joint Chance Constraint

We look to replace the joint chance constraint (14) with a deterministic surrogate defined in terms of the mean and covariance of the predicted state $x_{t+k|t}$. Consider $x \sim \mathcal{N}(\bar{x}, \Sigma)$ as an n_x-dimensional multivariate Gaussian random vector and let $\mathcal{X} := \{\zeta : C^\top \zeta \leq d\}$. This allows for rewriting the joint chance constraint (14) as

$$\mathbb{P}(x \in \mathcal{X}) = \frac{1}{\sqrt{(2\pi)^{n_x}\det(\Sigma)}} \int_{\mathcal{X}} e^{-\frac{1}{2}(\zeta-\bar{x})^\top \boldsymbol{\Sigma}^{-1}(\zeta-\bar{x})} d\zeta \;\geq\; 1-\delta. \tag{25}$$

To obtain a relaxation for (25), define the ellipsoid $\mathcal{E}_r := \{\zeta : \zeta^\top \boldsymbol{\Sigma}^{-1} \zeta \leq r^2\}$ with radius r. Expression (25) is guaranteed to hold when

$$\bar{x} \oplus \mathcal{E}_r \subset \mathcal{X} \implies \mathbb{P}(x \in \mathcal{X}) > \mathbb{P}(x \in \bar{x} \oplus \mathcal{E}_r) = 1-\delta,$$

which indicates that the smallest radius of ellipsoid $\mathcal{E}_r$ must be chosen such that $\mathbb{P}(x \in \bar{x} \oplus \mathcal{E}_r) = 1-\delta$ [46]. Equivalently, $\mathbb{P}(x \in \bar{x} \oplus \mathcal{E}_r) = 1-\delta$ can be represented in terms of a chi-squared cumulative distribution function $F_{\chi_n^2}$ with n degrees of freedom

$$\mathbb{P}(x \in \bar{x} \oplus \mathcal{E}_r) = \mathbb{P}\big((x-\bar{x})^\top \boldsymbol{\Sigma}^{-1}(x-\bar{x}) \leq r^2\big) = F_{\chi_n^2}(r^2) = \frac{\gamma\left(\frac{n}{2}, \frac{r^2}{2}\right)}{\Gamma\left(\frac{n}{2}\right)}, \tag{26}$$

where γ is the lower incomplete Gamma function and Γ is the complete Gamma function. This implies that the radius r can be selected such that $F_{\chi_n^2}(r^2) = 1-\delta$ in order to guarantee that $\bar{x} \oplus \mathcal{E}_r \subset \mathcal{X}$. For the expression (26) to hold, the ellipsoid $\mathcal{E}_r$ must lie in the intersection of half-spaces $\mathcal{H}_j := \{\zeta : c_j^\top \zeta \leq d_j\}$, where $c_j \in \mathbb{R}^{n_x}$ is the jth column of C and $d_j \in \mathbb{R}$ is the jth element of d. We use the result of the following lemma to derive an expression for guaranteeing the inclusion of $\mathcal{E}_r$ in the half-spaces $\mathcal{H}_j$, which relies on the Mahalanobis distance $d_M(x) = \sqrt{(x-\bar{x})^\top \Sigma^{-1}(x-\bar{x})}$ [30].

Lemma 1. *The Mahalanobis distance to the hyperplane* $h^\top x = g$ *is given by* $d_M(x^\star) = \frac{(g-h^\top \bar{x})}{\sqrt{h^\top \Sigma h}}$, *where* $x^\star = \bar{x} + \frac{(g-h^\top \bar{x})}{\sqrt{h^\top \Sigma h}} \delta x$ *and* $\delta x = \frac{\Sigma h}{\sqrt{h^\top \Sigma h}}$.

Lemma 1 indicates that $x^\star$ is the "worst-case" vector at which the ellipsoid $\mathcal{E}_r$ with radius $d_M(x^\star)$ intersects the hyperplane, while δx is the direction along which $x^\star$ lies. Lemma 1 leads to the assertion that $\bar{x} \oplus \mathcal{E}_r \subset \mathcal{X}$ is equivalent to

$$\frac{(d_j - c_j^\top \bar{x})}{\sqrt{c_j^\top \Sigma c_j}} \geq r, \; j = 1, \ldots, n_c. \tag{27}$$

Expression (27) results in an exact moment-based surrogate for the joint chance constraint (14)

$$c_j^\top \bar{x}_{t+k|t} + r\sqrt{c_j^\top \Sigma_{t+k|t} c_j} \leq d_j,\ j = 1, \ldots, n_c, \tag{28}$$

where $\bar{x}_{t+k|t}$ and $\Sigma_{t+k|t}$ are, respectively, the mean and covariance of the predicted state $x_{t+k|t}$ in (13); and r must satisfy $F_{\chi^2_{n_c}}(r^2) = 1 - \delta$. The mean and covariance of the predicted state can be approximated in terms of the gPC-based moment expressions (23)–(24).

4.4 Sample-Free, Moment-Based SMPC Formulation

We now present a sample-free formulation for SMPC of system (13). Using the gPC-based prediction model (19) and the input parameterization (21), the control cost function (15) can be (approximately) rewritten as

$$J_N(x_t, \mathbf{L}^\mathbf{N}, \mathbf{m}^\mathbf{N}) = \mathbb{E}_{x_t}\Big[\ \sum_{k=0}^{N-1} \|\tilde{\mathbf{x}}_{t+k|t}\|^2_{\mathbf{Q}} + \|\mathbf{L}_{t+k|t}\tilde{\mathbf{x}}_{t+k|t} + \mathbf{m}_{t+k|t}\|^2_{\mathbf{R}}\ \Big],$$

where $\mathbf{Q} = Q \otimes W$; $\mathbf{R} = R \otimes W$; $W = \text{diag}(\langle\phi_0^2\rangle, \langle\phi_1^2\rangle, \ldots, \langle\phi_p^2\rangle)$; and $\mathbf{L}^\mathbf{N}$ and $\mathbf{m}^\mathbf{N}$ are the vectors of decision variables over the prediction horizon N. The sample-free SMPC algorithm involves solving the following OCP

$$\begin{aligned}
&\min_{\mathbf{L}^\mathbf{N}, \mathbf{m}^\mathbf{N}}\ J_N(x_t, \mathbf{L}^\mathbf{N}, \mathbf{m}^\mathbf{N}) && \\
&\text{s.t.: } \bar{\mathbf{x}}_{t+k+1|t} = (\mathbf{A} + \mathbf{B}\mathbf{L}_{t+k|t})\bar{\mathbf{x}}_{t+k|t} + \mathbf{B}\mathbf{m}_{t+k|t}, && k = 0, \ldots, N-1, \\
&\boldsymbol{\Sigma}_{t+k+1|t} = (\mathbf{A} + \mathbf{B}\mathbf{L}_{t+k|t})\boldsymbol{\Sigma}_{t+k|t}(\mathbf{A} + \mathbf{B}\mathbf{L}_{t+k|t})^\top + \mathbf{D}\boldsymbol{\Sigma}_w\mathbf{D}^\top, && k = 0, \ldots, N-1, \\
&c_j^\top \mathbb{E}[\hat{x}_{t+k|t}] + r\sqrt{c_j^\top(\mathbb{E}[\hat{x}_{t+k|t}\hat{x}_{t+k|t}^\top] - \mathbb{E}[\hat{x}_{t+k|t}]\mathbb{E}[\hat{x}_{t+k|t}]^\top)c_j} \leq d_j, && \forall j,\ k = 1, \ldots, N, \\
&\mathbf{L}_{t+k|t}\tilde{\mathbf{x}}_{t+k|t} + \mathbf{m}_{t+k|t} \in \mathbb{U}, && k = 0, \ldots, N-1.
\end{aligned} \tag{29}$$

The prediction model in the above OCP describes the evolution of the mean $\bar{\mathbf{x}}_{t+k|t}$ and covariance $\boldsymbol{\Sigma}_{t+k|t}$ of the coefficients of the PCEs of the states, $\tilde{\mathbf{x}}_{t+k|t}$, over the prediction horizon. The prediction model is initialized using the knowledge of the true state x_t at each sampling time t. The surrogate for the joint chance constraint is defined in terms of the mean and covariance of the polynomial chaos-approximated states $\hat{x}_{t+k|t}$, which are computed in terms of the mean $\bar{\mathbf{x}}_{t+k|t}$ and covariance $\boldsymbol{\Sigma}_{t+k|t}$ using the expressions (23)–(24).

4.5 Extensions

A limitation of gPC is the ability to handle random variables with arbitrary probability measures (e.g., parameter uncertainties with correlated or multimodal distributions obtained from Bayesian estimation). An alternative to gPC, termed arbitrary polynomial chaos (aPC), that can address this shortcoming has recently been presented [39]. The aPC allows for constructing a set of orthogonal polynomial basis in terms of the raw moments of the uncertainties using a multivariate generalization of the Gram-Schmidt process [35]. The aPC holds promise for devising efficient, sample-free algorithms for stochastic optimization and SMPC of systems with correlated probability uncertainty.

5 Stochastic Tube MPC

Tube MPC for control of deterministic systems under uncertainty has been developed in [31, 32, 40–42]. Stochastic tube MPC approaches have been proposed in [12–14, 28] and described in the book [26]. The approach in [28] exploits linear models affected by stochastic disturbances without assuming that disturbance values are normally distributed. This approach is further discussed in this section and illustrated with an example.

5.1 System Model, Disturbance Model and Constraints

The control design exploits a linear prediction model of the form,

$$x_{t+k+1|t} = Ax_{t+k|t} + Bu_{t+k|t} + Dw_{t+k|t}, \tag{30}$$

where $x_{t+k|t} \in \mathbb{R}^{n_x}$ and $u_{t+k|t} \in \mathbb{R}^{n_u}$ are the predicted state and control sequences k steps ahead starting from the current time t, and the elements of the disturbance sequence, $w_{t+k|t} \in \mathbb{R}^{n_w}$, are assumed to be zero mean, identically distributed, and independent for different k. Furthermore, these disturbance values are compactly-supported,

$$w_{t+k|t} \in \Pi = \{w : \ |w_i| \leq \alpha_i, \ i = 1, \cdots, n_w\}. \tag{31}$$

Probabilistic chance constraints are imposed as

$$\mathbb{P}_{x_t}[y_{t+k|t} = Cx_{t+k|t} \leq y_{max}] \geq 1 - \varepsilon, \tag{32}$$

where, to simplify the exposition, the case of the scalar output, $y \in \mathbb{R}^1$, is considered.

5.2 Tube MPC Design

The tube MPC controller is formed as a combination of the nominal state feedback and manipulatable input adjusted by the MPC controller,

$$u_{t+k|t} = Kx_{t+k|t} + g_{t+k|t}, \tag{33}$$

where the matrix $\Phi = (A+BK)$ is Schur and the sequence $g_{t+k|t}$, $k = 0, \cdots, N-1$, is optimized, with $g_{k|t} = 0$ for $k \geq N$. The receding horizon implementation involves using the first element of the optimized sequence, $g^*_{t|t}$, leading to a feedback law,

$$u_t = u_{MPC}(x_t) = Kx_t + g^*_{t|t}, \tag{34}$$

where x_t is the current state at the time instance t and u_t is the control input at t.

The linearity of the prediction model permits to decompose the predicted state based on the superposition principle as

$$x_{t+k|t} = z_{t+k|t} + e_{t+k|t}, \tag{35}$$

where $z_{t+k|t}$ is the state prediction based on the nominal system,

$$z_{t+k+1|t} = \Phi z_{t+k|t} + Bg_{t+k|t}, \tag{36}$$

and $e_{t+k|t}$ is the error induced by the disturbance, given by

$$e_{t+k+1|t} = \Phi e_{t+k|t} + Dw_{t+k|t}. \tag{37}$$

Tube MPC approaches generally proceed by steering the state of the nominal system (36) with tightened constraints to account for the contributions of the error system (37).

The constraint (32) imposed over the prediction horizon can now be re-stated as

$$C[\Phi^{k-1}B \;\Phi^{k-2}B \;\cdots\; B \;0 \cdots 0]\mathbb{G}_t + C\Phi^k z_t \leq y_{max} - \gamma_k, \; k = 1, 2, \cdots \tag{38}$$

where

$$\mathbb{P}\left[C[\Phi^{k-1}D \;\Phi^{k-2}D \;\cdots\; D \;0 \cdots 0]\mathbb{W}_t \leq \gamma_k\right] = 1 - \varepsilon, \tag{39}$$

and

$$\mathbb{G}_t = \begin{bmatrix} g_{t|t} \\ g_{t+1|t} \\ \vdots \\ g_{t+N-1|t} \end{bmatrix}, \quad \mathbb{W}_t = \begin{bmatrix} w_{t|t} \\ w_{t+1|t} \\ \vdots \\ w_{t+N-1|t} \end{bmatrix}.$$

The computation of γ_k in (39) requires constructing the distribution of

$$C(\Phi^{k-1}Dw_{t|t} + \cdots + Dw_{t+k-1|t}),$$

for $t = 0$ (since problem characteristics are time-invariant). This can be performed offline by numerical approximation of the convolution integrals or by random sampling methods. In the random sampling approach, N_w disturbance sequence scenarios are generated and the smallest number $\tilde{\gamma}_k$ is found such that $N^*/N_w \geq 1-\varepsilon$, where N^* is the number of sequences for which $C(\Phi^{k-1}Dw_{t|t} + \cdots + Dw_{t+k-1|t}) \leq \tilde{\gamma}_k$. As $N_w \to \infty$, we expect that $\tilde{\gamma}_k \to \gamma_k$.

In the case of $w_{k|t}$ being independent and identically distributed, the Chebyshev inequality can be used to replace γ_k with bounds based on

$$\gamma_k \leq \kappa\sqrt{CP_kC^{\mathrm{T}}},\ \kappa^2 = \frac{1-\varepsilon}{\varepsilon},\ P_{k+1} = \Phi P_k \Phi^{\mathrm{T}} + D\mathbb{E}[ww^{\mathrm{T}}]D^T,\ P_0 = 0,\ k = 0,1,2,\cdots. \tag{40}$$

To guarantee recursive feasibility, the constraint (38) is tightened to ensure that a feasible extension of $\mathbb{G}_t$ of the type "shift by 1 and pad by 0" exists at time $t+1$. This can be assured, as the disturbance takes values in a compact set, by the following "worst-case" constraints:

$$C[\Phi^{k-1}B\ \Phi^{k-2}B\ \cdots\ B\ 0\cdots 0]\mathbb{G}_t + C\Phi^k z_t \leq y_{max} - \beta_k,\ k = 1,2,\cdots, \tag{41}$$

$$\beta_k = \max\{\gamma_k, \gamma_{k-1} + a_{k-1}, \gamma_{k-2} + a_{k-2} + a_{k-1}, \cdots, \gamma_1 + a_1 + \cdots, a_{k-1}, 0\}, \tag{42}$$

$$a_k = \max_{w \in \Pi} C\Phi^k Dw.$$

The sequence $\{\beta_k\}$ is monotonically nondecreasing and upper bounded by a computable upper bound.

Minimizing the cost function defined for a given x_t by

$$J_N = \mathbb{E}_{x_t} \sum_{k=0}^{N-1} [x_{t+k|t}^{\top} Q x_{t+k|t} + u_{t+k|t}^{\top} R u_{t+k|t}] + \mathbb{E}[x_{t+N|t}^{\top} P x_{t+N|t}], \tag{43}$$

is replaced equivalently by minimizing

$$\tilde{J}_N = \sum_{k=0}^{N-1} [z_{t+k|t}^{\top} Q z_{t+k|t} + u_{t+k|t}^{\top} R u_{t+k|t}] + z_{t+N|t}^{\top} P z_{t+N|t},$$

which is a quadratic function of $\mathbb{G}_t$. Hence the following optimization problem is solved

$$\mathbb{G}_t^* = \arg\min_{\mathbb{G}_t} \tilde{J},$$

subject to

$$C[\Phi^{k-1}B\ \Phi^{k-2}B\ \cdots\ B\ 0\cdots 0]\mathbb{G}_t + C\Phi^k z_t \leq y_{max} - \beta_k,\ k = 1,2,\cdots,N,$$

$$z_{N|t} \in S,$$

where S is a finitely determined inner approximation of the maximum output admissible set defined by constraints,

$$\{z_N : C\Phi^j z_N \leq y_{max} - \beta_{N+j}, j = 0, 1, \cdots\}.$$

The terminal constraint set can be constructed as

$$S = \{z_N : C\Phi^j z_N \leq y_{max} - \beta_{N+j}, j = 1, \cdots, \hat{N},$$
$$C\Phi^l z_N \leq y_{max} - \bar{\beta},\ l = \hat{N}+1, \cdots, \hat{N}+n^*\},$$

where n^* must be sufficiently large, and

$$\bar{\beta} = \gamma_1 + \sum_{j=1}^{\bar{N}} a_j$$

for $\bar{N}$ sufficiently large.

5.3 Theoretical Guarantees

Theoretical guarantees for the closed-loop behavior under the tube SMPC law are available for the case when $\tilde{J}_N$ in (43) is modified to an infinite prediction horizon cost (the control horizon is still N),

$$\tilde{J}_N = \mathbb{E}_{x_t} \sum_{k=0}^{\infty} [x_{t+k|t}^\top Q x_{t+k|t} + u_{t+k|t}^\top R u_{t+k|t} - l_{ss}], \tag{44}$$

where

$$l_{ss} = \lim_{k \to \infty} \mathbb{E}_{x_t} (x_{t+k|t}^\top Q x_{t+k|t} + u_{t+k|t}^\top R u_{t+k|t}),$$

is the steady-state value of the stage cost under the control $u_{t+k|t} = K x_{t+k|t}$. This value can be computed as

$$l_{ss} = trace(\Theta(Q + K^\top R K)), \quad \Theta - \Phi \Theta \Phi^\top = D\mathbb{E}[ww^\top]D^\top.$$

Note that (44) is a quadratic function of $\mathbb{G}_t$ (the details of the computations of this function are given in Chapter 6 of [26]). Then, under suitable assumptions, given feasibility at the time instant $t = 0$, the problem remains feasible at all future time instants, and causes the closed loop system to satisfy the probabilistic constraint (39) and the quadratic stability condition,

$$\lim_{t \to \infty} \frac{1}{t} \sum_{k=0}^{t} \mathbb{E}_{x0} \left[x_k^\top Q x_k + u_k^\top R u_k \right] \leq l_{ss}.$$

5.4 Mass-Spring-Damper Example

We consider a mass-spring-damper example with the model and constraint given by

$$m\ddot{x}_1 + c\dot{x}_1 + kx_1 = w - u, \tag{45}$$

$$y = x_1 \leq y_{max}, \tag{46}$$

with $m = 1$, $c = 0.1$, $k = 7$, and $y_{max} = 1$. The model is converted to the form (30) by defining $x = [x_1,\ x_2]^{\mathrm{T}}$ and using the sampling period, $\Delta T = 0.1$ sec. The disturbance force samples, w_t, are assumed to be distributed according to the truncated Gaussian distribution, with zero mean, standard deviation of $\frac{1}{12}$ and truncation interval $[-0.2, 0.2]$.

Two methods to compute γ_k were considered, one based on random sampling and the other one based on Chebyshev's inequality. Over a 1000 simulated trajectories, a constraint violation rate metric was defined, as the maximum over t of the fraction of trajectories violating the constraints at the instant t. See Figure 2. The trajectories for $\varepsilon = 0.2$, corresponding to 80% confidence of constraint satisfaction, are shown in Figure 3.

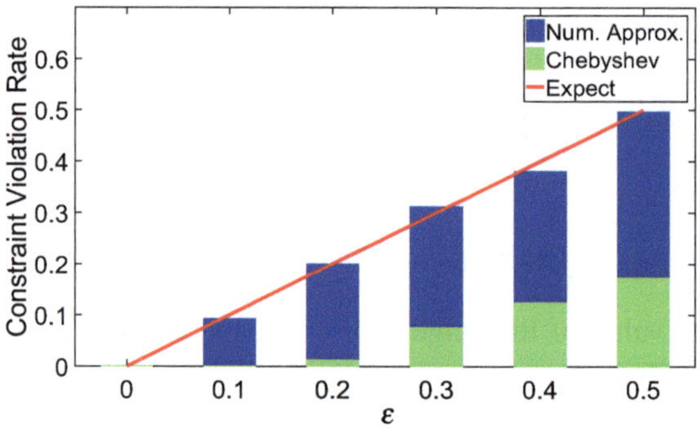

Fig. 2: Expected and estimated rate of constraint violations.

5.5 Extensions

The above approach utilizes the so-called polytopic stochastic tubes. Stochastic tubes with ellipsoidal cross-section to bound $e_{k|t}$ with probability at least $1 - \varepsilon$ can also be used. Polytopic tubes can be extended to handle both additive and multiplicative uncertainties. Typical assumptions involve

$$(A(q), B(q), w(q)) = (A^{(0)}, B^{(0)}, 0) + \sum_{j=1}^{m} (A^{(j)}, B^{(j)}, w^{j}) q^{(j)}$$

where $q^{(j)}$ are scalar random variables and $q_t = [q_t^{(1)}, q_t^{(2)}, \cdots, q_t^{(m)}]^\top$ are independent for different time instants, identically distributed and have known probability distribution.

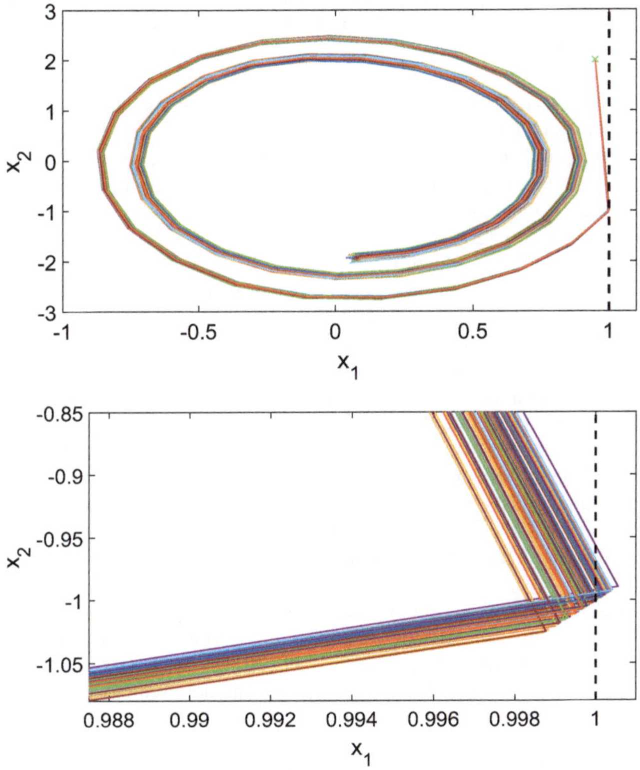

Fig. 3: Trajectories on x_1-x_2 plane for $\varepsilon = 0.2$, bottom plot: zoomed-in. Constraint is shown by the dashed vertical line.

Acknowledgement The second author would like to acknowledge Mr. Nan Li of the University of Michigan for the assistance and helpful comments.

References

1. Bavdekar, V., Mesbah, A.: Stochastic nonlinear model predictive control with joint chance constraints. In: Proceedings of the 10th IFAC Symposium on Nonlinear Control Systems, Monterey, pp. 276–281 (2016)
2. Bellman, R.E.: Dynamic Programming. Princeton University Press, New Jersey (1957)

3. Ben-Tal, A., Ghaoui, L.E., Nemirovski, A.: Robust Optimization. Princeton University Press, Princeton (2009)
4. Bernardini, D., Bemporad, A.: Scenario-based model predictive control of stochastic constrained linear systems. In: Proceedings of the 48th IEEE Conference on Decision and Control, Shanghai, pp. 6333–6338 (2009)
5. Bernardini, D., Bemporad, A.: Stabilizing model predictive control of stochastic constrained linear systems. IEEE Trans. Autom. Control **57**(6), 1468–1480 (2012)
6. Bertsekas, D.P.: Dynamic Programming and Optimal Control. Athena Scientific, Belmont (2000)
7. Bertsimas, D., Brown, D.B., Caramanis, C.: Theory and applications of robust optimization. SIAM Rev. **53**, 464–501 (2011)
8. Bichi, M., Ripaccioli, G., Di Cairano, S., Bernardini, D., Bemporad, A., Kolmanovsky, I.: Stochastic model predictive control with driver behavior learning for improved powertrain control. In: Proceedings of the 49th IEEE Conference on Decision and Control, pp. 6077–6082. IEEE, Piscataway (2010)
9. Birge, J., Louveaux, F.: Introduction to Stochastic Programming. Springer, New York (1997)
10. Calafiore, G.C., Fagiano, L.: Robust model predictive control via scenario optimization. IEEE Trans. Autom. Control **58**(1), 219–224 (2013)
11. Cameron, R.H., Martin, W.T.: The orthogonal development of non-linear functionals in series of fourier-hermite functionals. Ann. Math. **48**, 385–392 (1947)
12. Cannon, M., Kouvaritakis, B., Ng, D.: Probabilistic tubes in linear stochastic model predictive control. Syst. Control Lett. **58**(10), 747–753 (2009)
13. Cannon, M., Kouvaritakis, B., Wu, X.: Probabilistic constrained MPC for multiplicative and additive stochastic uncertainty. IEEE Trans. Autom. Control **54**(7), 1626–1632 (2009)
14. Cannon, M., Kouvaritakis, B., Rakovic, S.V., Cheng, Q.: Stochastic tubes in model predictive control with probabilistic constraints. IEEE Trans. Autom. Control **56**(1), 194–200 (2011)
15. Chen, Z.: Bayesian filtering: from Kalman filters to particle filters, and beyond. Statistics **182**, 1–69 (2003)
16. Di Cairano, S., Bernardini, D., Bemporad, A., Kolmanovsky, I.V.: Stochastic MPC with learning for driver-predictive vehicle control and its application to hev energy management. IEEE Trans. Control Syst. Technol. **22**(3), 1018–1031 (2014)
17. Fagiano, L., Khammash, M.: Nonlinear stochastic model predictive control via regularized polynomial chaos expansions. In: Proceedings of the 51st IEEE Conference on Decision and Control, Maui, pp. 142–147 (2012)
18. Geletu, A., Klöppel, M., Zhangi, H., Li, P.: Advances and applications of chance-constrained approaches to systems optimisation under uncertainty. Int. J. Syst. Sci. **44**, 1209–1232 (2013)
19. Ghanem, R., Spanos, P.: Stochastic finite elements – a spectral approach. Springer, New York (1991)
20. Goulart, P.J., Kerrigan, E.C., Maciejowski, J.M.: Optimization over state feedback policies for robust control with constraints. Automatica **42**, 523–533 (2006)
21. Hoblit, F.M.: Gust Loads on Aircraft: Concepts and Applications. AIAA, Reston (1988)
22. Hokayem, P., Cinquemani, E., Chatterjee, D., Ramponi, F., Lygeros, J.: Stochastic receding horizon control with output feedback and bounded controls. Automatica **48**, 77–88 (2012)
23. Kang, J., Raghunathan, A.U., Di Cairano, S.: Decomposition via ADMM for scenario-based model predictive control. In: American Control Conference (ACC), pp. 1246–1251. IEEE, Piscataway (2015)
24. Kim, K.K., Braatz, R.D.: Generalised polynomial chaos expansion approaches to approximate stochastic model predictive control. Int. J. Control **86**, 1324–1337 (2013)
25. Kothare, M.V., Balakrishnan, V., Morari, M.: Robust constrained model predictive control using linear matrix inequalities. Automatica **32**(10), 1361–1379 (1996)
26. Kouvaritakis, B., Cannon, M.: Model Predictive Control: Classical, Robust and Stochastic. Springer, Cham (2015)
27. Kouvaritakis, B., Cannon, M.: Stochastic model predictive control. Encyclopedia of Systems and Control, pp. 1350–1357. Springer, Berlin (2015)

28. Kouvaritakis, B., Cannon, M., Raković, S.V., Cheng, Q.: Explicit use of probabilistic distributions in linear predictive control. Automatica **46**(10), 1719–1724 (2010)
29. Kumar, P.R., Varaiya, P.: Stochastic Systems: Estimation, Identification, and Adaptive Control. SIAM, Philadelphia (2016)
30. Mahalanobis, P.C.: On the generalized distance in statistics. Proc. Natl. Inst. Sci. **2**, 49–55 (1936)
31. Mayne, D.Q., Seron, M.M., Raković, S.: Robust model predictive control of constrained linear systems with bounded disturbances. Automatica **41**(2), 219–224 (2005)
32. Mayne, D.Q., Raković, S., Findeisen, R., Allgöwer, F.: Robust output feedback model predictive control of constrained linear systems. Automatica **42**(7), 1217–1222 (2006)
33. Mesbah, A.: Stochastic model predictive control: an overview and perspectives for future research. IEEE Control Syst. **36**(6), 30–44 (2016)
34. Mesbah, A.: Stochastic model predictive control with active uncertainty learning: a survey on dual control. Annu. Rev. Control. **45**, 107–117 (2018)
35. Navarro, M., Witteveen, J., Blom, J.: Polynomial chaos expansion for general multivariate distributions with correlated variables (2014, Preprint). arXiv 1406.5483
36. Nemirovski, A., Shapiro, A.: Convex approximations of chance constrained programs. SIAM J. Optim. **17**, 969–996 (2006)
37. Paulson, J.A., Mesbah, A.: An efficient method for stochastic optimal control with joint chance constraints for nonlinear systems. Int. J. Robust Nonlinear Control (2017, https://onlinelibrary.wiley.com/doi/abs/10.1002/rnc.3999)
38. Paulson, J.A., Streif, S., Mesbah, A.: Stability for receding-horizon stochastic model predictive control. In: Proceedings of the American Control Conference, Chicago, pp. 937–943 (2015)
39. Paulson, J.A., Buehler, E.A., Mesbah, A.: Arbitrary polynomial chaos for uncertainty propagation of correlated random variables in dynamic systems. In: Proceedings of the IFAC World Congress, Toulouse, pp. 3607–3612 (2017)
40. Rakovic, S.V., Kouvaritakis, B., Cannon, M., Panos, C., Findeisen, R.: Parameterized tube model predictive control. IEEE Trans. Autom. Control **57**(11), 2746–2761 (2012)
41. Raković, S.V., Kouvaritakis, B., Findeisen, R., Cannon, M.: Homothetic tube model predictive control. Automatica **48**(8), 1631–1638 (2012)
42. Raković, S.V., Levine, W.S., Açıkmeşe, B.: Elastic tube model predictive control. In: American Control Conference (ACC), pp. 3594–3599. IEEE, Piscataway (2016)
43. Ripaccioli, G., Bernardini, D., Di Cairano, S., Bemporad, A., Kolmanovsky, I.: A stochastic model predictive control approach for series hybrid electric vehicle power management. In: American Control Conference, Baltimore, pp. 5844–5849 (2010)
44. Schildbach, G., Fagiano, L., Frei, C., Morari, M.: The scenario approach for stochastic model predictive control with bounds on closed-loop constraint violations. Automatica **50**(12), 3009–3018 (2014)
45. Scokaert, P.O., Mayne, D.: Min-max feedback model predictive control for constrained linear systems. IEEE Trans. Autom. Control **43**(8), 1136–1142 (1998)
46. Van Hessem, D.H., Scherer, C.W., Bosgra, O.H.: LMI-based closed-loop economic optimization of stochastic process operation under state and input constraints. In: Proceedings of the 40th IEEE Conference on Decision and Control, Orlando, pp. 4228–4233 (2001)
47. Wiener, N.: The homogeneous chaos. Am. J. Math. **60**, 897–936 (1938)
48. Xiu, D., Karniadakis, G.E.: The Wiener-Askey polynomial chaos for stochastic differential equations. SIAM J. Sci. Comput. **24**, 619–644 (2002)

Moving Horizon Estimation

Douglas A. Allan and James B. Rawlings

1 Introduction

Because model predictive control (MPC) uses a system model to predict and optimize system behavior, knowledge of the system's state is essential for good control performance. In practice, however, only noisy measurements are available. In the best case, they are measurements of the entire state, so only perturbations due to measurement noise need be considered in controller design. In a much more industrially relevant case, however, only a subset of system states are measured. Nevertheless, almost all MPC algorithms require a good estimate of the initial state. Thus reconstructing a system's state from noisy measurements is a vital control problem.

The most famous method of estimating a system's state from measurements is the Kalman filter [9]. For linear systems with Gaussian disturbances and noises, it is the statistically optimal state estimator, i.e., it maximizes the conditional density of the state given the measurements. Furthermore, because of its linear, recursive update formula, state estimates can be calculated quickly and efficiently. The Kalman filter is the first choice of state estimators when linear system models are used, and, by design, many nonlinear state estimators reduce to the Kalman filter when applied to linear systems.

In general, there have been two approaches to nonlinear state estimation. The first is to design a parallel dynamical system, termed an observer, that takes the original system's outputs as inputs and produces, as an output, an estimate of the original system's state. The design of such observers is an active area of research; a presentation of recent results on the subject is available in [17] and [16]. Two particular observers of note are the extended Kalman filter (EKF) and the unscented Kalman filter.

D. A. Allan · J. B. Rawlings (✉)
University of Wisconsin–Madison, 1415 Engineering Dr., Madison, WI 53706, USA
e-mail: dallan@wisc.edu; james.rawlings@wisc.edu

S. V. Raković, W. S. Levine (eds.), *Handbook of Model Predictive Control*,
Control Engineering, https://doi.org/10.1007/978-3-319-77489-3_5

The extended Kalman filter has two components to its state: an estimate of the system's actual state and an estimate of the covariance matrix of the estimated state. The estimated state is propagated forward in time using the nonlinear system equation. This estimate is corrected by linearizing the system model at the estimated state and applying the Kalman filter equations to the covariance matrix. This procedure is relatively easy to implement and produces a state estimate quickly, but is difficult to tune and is provably reliable only for systems that are nearly linear [8]. The unscented Kalman filter is an observer that takes an ensemble of states nearby the estimated state, runs the ensemble through the nonlinear system model, and uses the statistics of the ensemble for the state update [8]. This approach can work better than the EKF on systems with highly nonlinear dynamics [24] or measurements [25].

The second approach is to take a nonlinear process model and optimize to find the state trajectory that is "most likely" to have produced the observed measurements. Two types of disturbances are usually considered: state disturbances and measurement disturbances. State disturbances affect the system's state directly, and thus have effects that propagate forward in time. Measurement disturbances only appear to have affected the system's state and do not affect the system's evolution. These disturbances are usually decomposed into zero-mean components that do not persist, and persistent integrating disturbances. We do not consider estimating these integrating disturbances here; an interested reader can see [15] for disturbance estimation in the linear case and [23, Ch. 5] for the nonlinear case. For the zero-mean component of disturbances, a stage cost measures how "likely" each type of disturbance is to have occurred, with larger disturbances "less likely" than smaller disturbances. These naive notions of likelihood coincide with the statistical notion of choosing a state estimate with maximum likelihood for linear systems with zero-mean Gaussian noises when an appropriate quadratic stage cost is chosen [6]. A prior estimate is often used in order to sum up information not included in the optimization problem. So the optimization-based state estimator minimizes the sizes of noises and deviation from the prior estimate needed to explain the measurements of the system over a certain time horizon.

In order for a state estimator to be of use, it must faithfully reconstruct the system's state. Therefore, estimator *stability* is a vital property. Furthermore, the stability of the estimator must not be destroyed by the presence of state and measurement noise. Therefore estimator *robustness* is an equally vital property. Not all estimators are stable; if an unstable linear system has no outputs, the best estimate is created by the evolution of the state's prior estimate. If there is error in this prior estimate, the gap between the true state and the estimate increases without bound [23, Ch. 1]. Therefore, in order for a stable estimator to exist, the system must satisfy certain assumptions.

The most common assumptions are observability and detectability. A system is observable if its state can be reconstructed from a finite number of measurements, whereas a system is detectable if any two trajectories that produce the same measurements asymptotically approach one another. These properties are well-understood for linear systems, and one can determine if a system is observable or detectable by the satisfaction of certain matrix rank conditions [27, p. 271, 317]. Here, we consider nonlinear detectable systems. A popular notion of nonlinear detectability is incremental input/output-to-state stability (i-IOSS) [28]. This condition is necessary for a full-order (i.e., a state dimension equal to that of the system) observer to exist for a system, and has been used in many results on nonlinear optimization-based state estimation. The essential idea of i-IOSS is that the distance between any two trajectories of a system is bounded by the distance between the initial conditions, and the magnitude of the difference between the disturbances and the outputs of the system.

Optimization-based state estimation generally takes one of two forms: full information estimation (FIE) and moving horizon estimation (MHE). In FIE, all measurements are included in the optimization problem, while in MHE only a finite number of recent measurements are included. In general, FIE is computationally intractable, but it provides useful theoretical benchmarks for the performance of MHE. Furthermore, for unconstrained linear systems with a quadratic stage cost, it reduces to the Kalman filter. FIE was first shown to be a stable estimator in the case of observable nonlinear systems with no disturbances in [18], and it was shown to be a robustly stable and convergent estimator in the case of convergent disturbances in [23, 1st Ed.]. The proof of estimator convergence in the case of converging disturbances was subsequently streamlined in [20]. By adding an extra "max-term" to the objective function to penalize the largest stage cost, the authors in [7] proved the robust stability of FIE for bounded disturbances for a particular form of i-IOSS. This result was extended to additional forms of i-IOSS in [5]. Although these results with a max-term obtain robust stability in the case of bounded disturbances, they are *not* provably convergent when the disturbances converge to zero. An analysis that shows that FIE is robustly stable in the presence of bounded disturbances remains elusive.

Moving horizon estimation was first motivated for linear systems by the possibility of using constraints to improve the predictions of the Kalman filter by forbidding aphysical state and disturbance estimates in [10] and [14]. These results were extended to observable nonlinear systems with no prior weighting in [11]. The concept of the arrival cost for moving horizon estimation was introduced in [19]. The arrival cost is a function of the system's state that gives the cost of the full information problem such that it terminates at a given state value. It has the property that, when used as a prior weighting, MHE is equivalent to FIE. Although it cannot be found directly except in special cases such as the Kalman filter for linear systems, it is a useful point of comparison for the prior weighting. The authors of [19] found that MHE is stabilizing when the prior weighting is a lower bound of the arrival cost. This result was extended in [23, 1st. Ed.] to robust stability in the case of convergent disturbances.

Recently, there have been several new results demonstrating the robust stability for MHE. Inspired by the results in [7], in [12] it was shown that MHE with a max-term is stabilizing for certain i-IOSS systems so long as a sufficiently long horizon is used. This result was extended to MHE without a max-term in [13]. Furthermore, it was shown that MHE is convergent for convergent disturbances, with or without a max-term. In [4], it was shown for a more general class of systems that MHE with a max-term is stabilizing with a sufficiently long horizon if the functions dictating the robust stability of FIE using the same prior weighting and stage costs satisfy certain assumptions.

As we might expect in a field under active development, the assumptions that these works use are somewhat opaque, and the systems to which they apply are not fully understood. In particular, both results rely on exponential decay arguments and result in exponential convergence of the estimator. However, it is not completely clear how an estimator can be exponentially convergent without the system being exponentially detectable. Here, we present a version of the results in [13] generalized by using assumptions similar to those in [4]. Furthermore, we show, by arguments similar to the convergence proofs of MHE, that these systems indeed have a certain form of exponential detectability. The proof of MHE convergence has been streamlined by use of the maximization form, rather than the sum form, of i-IOSS. When the property of input-to-state stability (ISS) was originally introduced in [26], the author noted that taking the maximum of the $\mathscr{K}\mathscr{L}$ and $\mathscr{K}$ functions, rather than the sum, resulted in the same property, and the same is true of i-IOSS. We introduce some suggestive notation for pairwise maximization, inspired by the max-plus algebra, to assist in proofs when many elements are being maximized at once. The end result is shorter and simpler proofs than when using the standard sum approach.

Preliminaries The vector space of real numbers is denoted $\mathbb{R}^n$, and nonnegative scalar real numbers are denoted $\mathbb{R}_{\geq 0}$. The set of nonnegative integers is denoted $\mathbb{I}_{\geq 0}$, and sets of integers from i to j (inclusive) are denoted $\mathbb{I}_{i:j}$. We define $|\cdot|$ to be the Euclidean norm. We denote sequences $(d(0), d(1), d(2), \ldots) := \mathbf{d}$, and denote the supremum norm of a sequence $\sup_{k\geq 0} |d(k)| := \|\mathbf{d}\|$. Maximums of a sequence in certain intervals are denoted $\max_{k\in\mathbb{I}_{i:j}} |d(k)| := \|\mathbf{d}\|_{i:j}$.

A function $\alpha : \mathbb{R}_{\geq 0} \to \mathbb{R}_{\geq 0}$ is of class $\mathscr{K}$ if it is zero at the origin, continuous, and strictly increasing. It is said to be of class $\mathscr{K}_\infty$ if in addition $\lim_{s\to\infty} \alpha(s) = \infty$. A function $\phi : \mathbb{I}_{\geq 0} \to \mathbb{R}_{\geq 0}$ is of class $\mathscr{L}$ if it is nonincreasing and if $\lim_{k\to\infty} \phi(k) = 0$. A function $\beta : \mathbb{R}_{\geq 0} \times \mathbb{I}_{\geq 0} \to \mathbb{R}_{\geq 0}$ is of class $\mathscr{K}\mathscr{L}$ if for each fixed k the function $\beta(\cdot, k)$ is of class $\mathscr{K}$ and if for each fixed s the function $\beta(s, \cdot)$ is of class $\mathscr{L}$. The function $\beta(\cdot) \in \mathscr{K}\mathscr{L}$ is separable if there is some $\alpha(\cdot) \in \mathscr{K}$ and $\phi(\cdot) \in \mathscr{L}$ such that $\beta(s,k) = \alpha(s)\phi(k)$. The operator $\lfloor\cdot\rfloor$ denotes the floor function, the function that returns the greatest integer less than its argument, and the operator $\lceil\cdot\rceil$ denotes the ceiling function, the function that returns the least integer greater than its argument.

We wish to use the maximization form of i-IOSS because it is convenient for manipulating $\mathscr{K}$ functions, but do not wish for the notation to become burdensome. Inspired by the max-plus algebra, we denote pairwise maximization $\max\{a,b\} := a \oplus b$, in which $a, b \in \mathbb{R}$. This operation is associative, i.e., $(a \oplus b) \oplus c = a \oplus (b \oplus c) := a \oplus b \oplus c$ for $c \in \mathbb{R}$, and nondecreasing functions distribute across it, i.e., if $\sigma(s)$ is nondecreasing we have that $\sigma(a \oplus b) = \sigma(a) \oplus \sigma(b)$. In particular, we have that $(a \oplus b) + c = (a + c) \oplus (b + c)$, $c(a \oplus b) = ca \oplus cb$ for $c \geq 0$, and for $\sigma \in \mathscr{K}$ we have that $\sigma(a \oplus b) = \sigma(a) \oplus \sigma(b)$. Note that the operation $\oplus$ does *not* associate with addition, i.e., $a \oplus (b + c) \neq (a \oplus b) + c = (a + c) \oplus (b + c)$.

2 Systems of Interest

We consider general nonlinear discrete-time systems of the form

$$\begin{aligned} x^+ &= f(x,w) \\ y &= h(x) + v \end{aligned}$$

in which $x \in \mathbb{R}^n$ is the system state, $w \in \mathbb{R}^g$ is the process noise, $y \in \mathbb{R}^p$ is the process output, and $v \in \mathbb{R}^p$ is the measurement noise. For simplicity, we do not consider control inputs here. Under appropriate assumptions (e.g., requiring uniformity with respect to controls as in [23, Sec. 4.6]), these results extend to systems with control inputs.

One useful definition of detectability for nonlinear systems is incremental input/output-to-state stability (i-IOSS).

Definition 1 (i-IOSS). A system $x^+ = f(x,w)$ with a measurement function $y = h(x)$ is incrementally input/output-to-state stable (i-IOSS) if there exist $\beta(\cdot) \in \mathscr{KL}$ and $\gamma_w(\cdot), \gamma_v(\cdot) \in \mathscr{K}$ such that for any two initial conditions $x_1(0)$ and $x_2(0)$ and input sequences $\mathbf{w}_1$ and $\mathbf{w}_2$ we have that

$$\begin{aligned} |x_1(k) - x_2(k)| \leq \beta(|x_1(0) - x_2(0)|, k) &\oplus \gamma_w(\|\mathbf{w}_1 - \mathbf{w}_2\|_{0:k-1}) \\ &\oplus \gamma_v(\|\mathbf{y}_1 - \mathbf{y}_2\|_{0:k-1}) \end{aligned}$$

for all $k \geq 0$.

However, the recent results on the robust stability of MHE rely on a stronger assumption than i-IOSS. In [13], the authors assume that the $\mathscr{KL}$ function $\beta(s,k)$ admits an upper bound for times $k \geq 1$ of the form $Cs^a\phi(k)$ for some $C > 0$, $a \geq 1$, and $\phi \in \mathscr{L}$. Because at time $k = 0$ we have the trivial inequality $|x_1(0) - x_2(0)| \leq |x_1(0) - x_2(0)|$, any system that satisfies this assumption admits an upper bound of $(Cs^a + s)\phi(k)$ for all $k \geq 0$. We use a more general assumption from [4]. Because a stability link is derived from FIE in [4], the assumption there is about a $\mathscr{KL}$ function derived from the robust stability of FIE, but here we apply it to the $\mathscr{KL}$ function in the definition of i-IOSS.

Assumption 1 *The system $x^+ = f(x,w)$, $y = h(x)$ is i-IOSS and furthermore, for every $\bar{s} > 0$ there exists some $T \geq 0$ and $\eta \in (0,1)$ such that*

$$\beta(s,T) \leq \eta s$$

for all $s \leq \bar{s}$.

This assumption requires the $\mathscr{KL}$ function to eventually become a contraction map on all intervals from the origin to $\bar{s}$. In [4], a sufficient condition for this assumption is provided: that $\beta(s,k)$ admits an upper bound $\sigma(s)\phi(k)$ for $\sigma(\cdot) \in \mathscr{K}$ and $\phi(\cdot) \in \mathscr{L}$ such that $\sigma(\cdot)$ is Lipschitz continuous at the origin. This sufficient condition is a more general form of the assumption made in [13], and in fact permits a much stronger conclusion about the form of $\beta(\cdot)$. All systems that satisfy this sufficient condition are actually locally exponentially i-IOSS.

Definition 2 (Local Exponential i-IOSS). A system $x^+ = f(x,w)$ with a measurement function $y = h(x)$ is incrementally input/output-to-state stable (i-IOSS) if there exist $\gamma_w(\cdot), \gamma_v(\cdot) \in \mathscr{K}$ such that, for every $\delta > 0$ and for any two initial conditions $x_1(0)$ and $x_2(0)$ such that $|x_1(0) - x_2(0)| \leq \delta$ and input sequences $\mathbf{w}_1$ and $\mathbf{w}_2$, there exist $C(\delta) \geq 1$ and $\lambda(\delta) \in (0,1)$ such that

$$|x_1(k) - x_2(k)| \leq C(\delta)\,|x_1(0) - x_2(0)|\,\lambda(\delta)^k \oplus \gamma_w(\|\mathbf{w}_1 - \mathbf{w}_2\|_{0:k-1}) \\ \oplus \gamma_v(\|\mathbf{y}_1 - \mathbf{y}_2\|_{0:k-1})$$

for all $k \geq 0$.

Note that this exponential i-IOSS is "local" in the sense that it applies to all states that start sufficiently close to one another, rather than applying to all states in some set.

We first require a minor result on locally Lipschitz $\mathscr{K}$ function upper bounds of functions. This result is similar to that of Proposition 14 in [22].

Proposition 1 (Locally Lipschitz Upper Bound). *If a function $V : \mathscr{C} \to \mathbb{R}^n$, in which $\mathscr{C} \subseteq \mathbb{R}^m$ is closed, is Lipschitz continuous at a point x_0 and locally bounded, then there exists a locally Lipschitz function $\sigma(\cdot) \in \mathscr{K}$ such that $|V(x) - V(x_0)| \leq \sigma(|x - x_0|)$ for all $x \in \mathscr{C}$.*

The proof of this proposition is included in the appendix. Now we demonstrate that both Assumption 3 in [13] and the sufficient condition provided in [4] are equivalent to local exponential i-IOSS.

Proposition 2. *Suppose a system is i-IOSS with $\beta(\cdot) \in \mathscr{KL}$ and $\gamma_w(\cdot), \gamma_v(\cdot) \in \mathscr{K}$. Then the following statements are equivalent:*

1. *The system admits a $\mathscr{KL}$ function $\beta(s,k) = \sigma_x(s)\phi(k)$ for some $\sigma_x(\cdot) \in \mathscr{K}_\infty$ that is Lipschitz continuous at the origin and $\phi(\cdot) \in \mathscr{L}$.*
2. *For every $\bar{s} > 0$ there exists some $\eta \in (0,1)$ and some $T > 0$ such that $\beta(s,T) \leq \eta s$ for all $s \leq \bar{s}$. Furthermore, $\beta(s,0)$ is Lipschitz continuous at $s = 0$.*
3. *For every $\bar{s} > 0$, there exist $C > 0$ and $\lambda \in (0,1)$ such that for any two initial conditions $x_1(0)$ and $x_2(0)$ satisfying $|x_1(0) - x_2(0)| \leq \bar{s}$ and every two disturbance sequences $\mathbf{w}_1$ and $\mathbf{w}_2$, the system's state trajectories satisfy $|x_1(k) - x_2(k)| \leq C\,|x_1(0) - x_2(0)|\,\lambda^k \oplus \gamma_w(\|\mathbf{w}_1 - \mathbf{w}_2\|_{k-1}) \oplus \gamma_v(\|\mathbf{y}_1 - \mathbf{y}_2\|_{k-1})$.*
4. *The system admits a $\mathscr{KL}$ function $\beta(s,k) = \gamma_x(s)\lambda^k$ for some $\gamma_x(\cdot) \in \mathscr{K}_\infty$ that is Lipschitz continuous at the origin and $\lambda \in (0,1)$.*

The proof of this proposition is also provided in the appendix.

So all systems that satisfy the assumption made in [13] are locally exponentially i-IOSS. This result also explains why a result similar to robust exponential stability is given by Lemma 3 in [4], which uses the sufficient condition equivalent to Statement 2. Now, Assumption 1 is somewhat weaker. It is satisfied by systems that are "eventually" locally exponentially i-IOSS. Consider a system

$$\begin{aligned} x_1^+ &= \sqrt{x_2} \\ x_2^+ &= 0 \end{aligned}$$

This system is incrementally stable, and we have that

$$|\Delta x(k)| \leq \left(\sqrt{|\Delta x(0)|} + \Delta x(0)\right) (1/2)^{k-1}$$

for all $k \geq 0$, in which $\Delta x(k) := x_1(k) - x_2(k)$. Furthermore, we have that $|\Delta x(k)| = 0$ for all $k \geq 2$. Thus, when this system is augmented with the trivial measurement function $h(x) = 0$, it is i-IOSS and satisfies Assumption 1. However, because of the square-root term, it is not locally exponentially i-IOSS. Thus Assumption 1 is more general than local exponential i-IOSS. However, what this system shows is that an exponential bound can be used after a certain amount of time elapses in the system. We conjecture that all systems that satisfy Assumption 1 are this way.

Furthermore, by an argument similar to that used in the proof of Proposition 2, all systems that satisfy Assumption 1 admit a $\mathscr{KL}$ function of the form $\alpha(s)\lambda^k$ in their statements of i-IOSS, in which $\alpha(\cdot) \in \mathscr{K}_\infty$ but is not necessarily Lipschitz continuous at the origin. However, unlike in Proposition 2, not all systems that admit such a $\mathscr{KL}$ function satisfy Assumption 1. For example, consider the system

$$\begin{aligned} x_1^+ &= \sqrt{x_2} \\ x_2^+ &= (1/2)x_2 \end{aligned}$$

As in the previous system, this system is incrementally stable and admits the bound

$$|\Delta x(k)| \leq \left(\sqrt{|\Delta x(0)|} + |\Delta x(0)|\right) (1/2)^{k-1}$$

for all $k \geq 0$. However, because x_2 is not mapped to zero in finite time, the square-root term prevents any sort of bound of the form $|\Delta x(k)| \leq \eta s$ for all $s \leq \bar{s}$. Therefore Assumption 1 is not reducible to a $\mathscr{KL}$ bound of the form $\alpha(s)\lambda^k$.

Although there are major differences between how inputs and outputs affect the evolution of a system (namely, that one affects the evolution of the state and one does not), both process and measurement noises enter into the analysis of MHE in the same way. For simplicity, we define a combined disturbance

$$d := (w, v)$$

and note that if we define a function $\gamma_d(s) := \gamma_w(s) \oplus \gamma_v(s)$ we have the i-IOSS bound

$$|\Delta x(k)| \leq \beta(|\Delta x(0)|, k) \oplus \gamma_w(\|\Delta \mathbf{d}\|_{k-1})$$

for all $k \geq 0$, in which $\Delta \mathbf{d} := \mathbf{d}_1 - \mathbf{d}_2$.

3 MHE Setup

Now that the systems of interest have been characterized, we move to the design of the moving horizon estimator. MHE uses a nonlinear model to forecast a system's behavior over a certain horizon based on an initial estimate of the system's state (prior) and then optimizes to find the smallest disturbances to this forecast necessary to explain the system's measurements. This process is illustrated in Figure 1. The estimator's stage cost dictates which disturbances are deemed to be "more likely" than others. For example, a weighted least squares stage cost implies that many small disturbances are more likely than a few large ones, whereas an ℓ_1 stage cost implies that several small disturbances and one large disturbance are equally likely. The most popular stage cost is quadratic, because of its association with a Gaussian distribution of disturbances. The prior is taken into account with a prior weighting. Because the prior is usually not correct, it is important to be able to move it to produce a state trajectory that is more in line with the measurements produced. However, without any prior weighting, a system must be observable for an estimator to converge. Therefore careful choice of a prior weighting is necessary for good MHE performance.

We define the MHE objective function for some horizon length N to be

$$V_N(\chi,\boldsymbol{\omega},\boldsymbol{v},\bar{x}) := \rho V_p(\chi,\bar{x}) + \sum_{k=0}^{N-1} \ell(\omega(k),v(k)) \tag{1}$$

in which $V_p(\cdot)$ is the prior weighting and $\rho > 0$ is some constant that is chosen to achieve robust stability. We then define the optimal estimation problem

$$\begin{aligned} \min_{\chi,\boldsymbol{\omega},\boldsymbol{v}} \; & V_N(\chi(t-N),\boldsymbol{\omega},\boldsymbol{v},\bar{x}) \\ \text{s.t. } & \chi^+ = f(\chi,\omega) \\ & y = h(\chi)+v \end{aligned}$$

Note that when there are $k < N$ measurements, we use a sum from 0 to $k-1$, i.e., we define the MHE problem to be the full information problem. Note that in practice the final measurement, $y(N)$, should be included in the problem as well, but we do not include it for simplicity. All results derived for this prediction form of MHE, so-called because it estimates $x(N)$ without the measurement $y(N)$, can be extended to filtering MHE, the form of MHE that does include $y(N)$. State and state disturbance constraints can be included in MHE as well. For physical systems with variables such as absolute temperatures and concentrations, nonnegativity constraints do not just improve estimates by excluding nonsensical values, but are often necessary for the system model to be well-defined in the first place. If constraints are included in the formulation of MHE, the actual states and disturbances must always satisfy them, or else the estimator may not converge [23, Ch. 4].

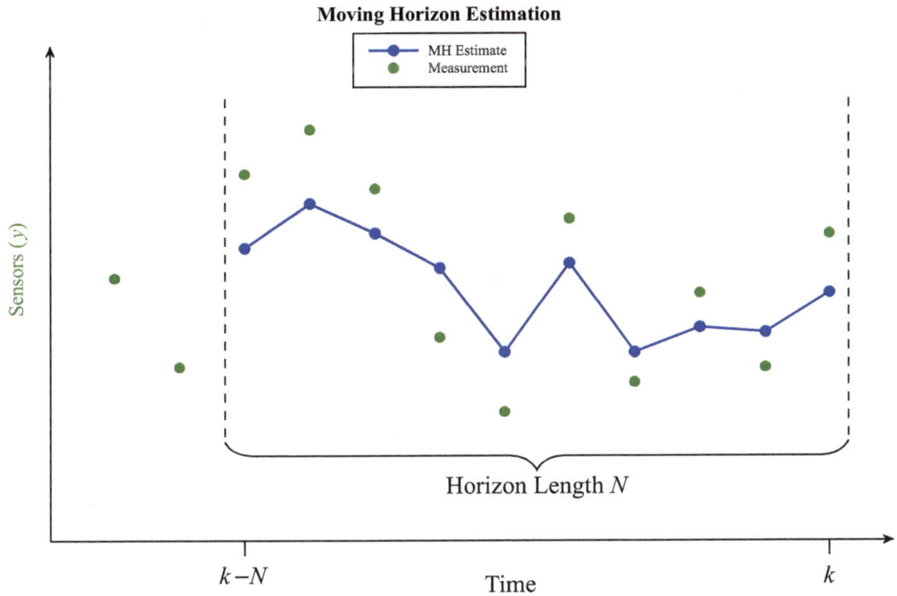

Fig. 1: Moving horizon estimation uses a system model and optimization to reconcile past measurements over a horizon of length N.

Next, the prior $\bar{x}$ needs to be chosen. We assume that there exists some initial estimate of the system's state $\bar{x}_0$ as part of the problem statement. But once the estimator has been online for at least N sampling times, this prior is obsolete. Ideally, the prior weighting would summarize all the previous measurements exactly, but except in the case of the Kalman filter's recursive update formula, such a prior weighting is computationally intractable. In order to include the most information in the moving horizon problem, then, MHE's estimate of the state at $k-N$, $\hat{x}(k-N)$, is used as a prior. Because the prior $\hat{x}(k-N)$ is estimated from measurements from $k-N-1$ to $k-2N$, the horizon is, in a way, extended to be twice as long. Furthermore, that MHE estimate uses a prior $\hat{x}(k-2N)$, so information is propagated from earlier problems to later ones.

However, the filtering update is not without problems. Because it uses a much earlier estimate, it can take a long time to recover from a bad initial prior. This recovery process can involve periodic behavior in the state estimate, as seen in [2] and [29]. As an alternative, the estimate of $x(k-N)$ at time $k-1$, $\hat{x}(k-N|k-1)$, can be used as a prior instead. Both strategies are compared in Figure 2. Because this prior is generated using measurements from $k-N-1$ to $k-1$, those measurements are "counted twice" in the optimization problem. For linear systems, this correlation can be corrected for by updating the prior weighting [23, Sec. 4.3.4]. For nonlinear systems, the prior is often somewhat arbitrary, so while this correlation should be kept in mind, there is not necessarily an easy way to correct for it. Despite its poorer the-

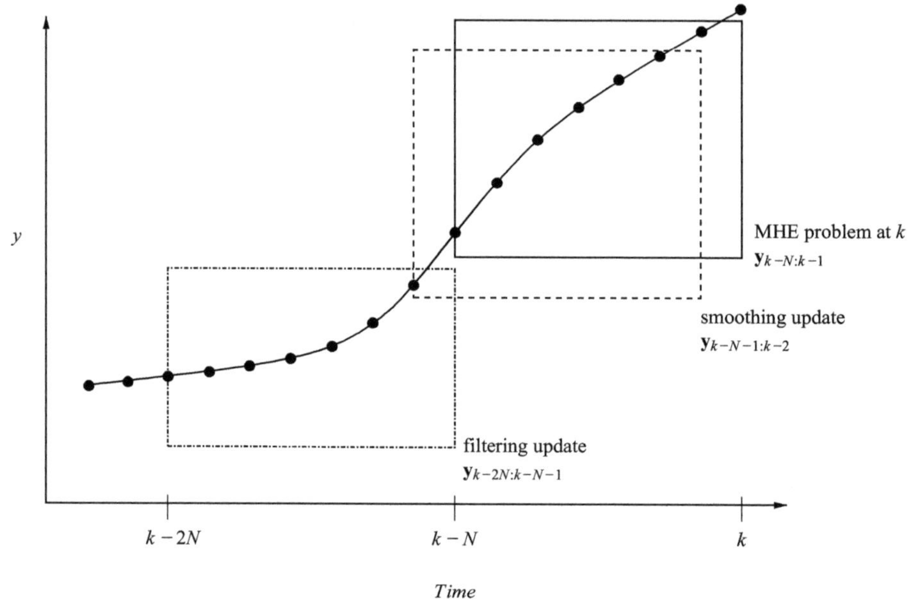

Fig. 2: The filtering prior is created using measurements from $k-2N$ to $k-N-1$, whereas the smoothing prior uses measurements from $k-N-1$ to $k-1$. As a result, measurements from $k-N$ to $k-1$ are counted twice in the MHE problem when the smoothing prior is used. (From [23, Ch. 4].)

oretical properties, it often gives much better performance than the filtering update in practice, as in [29]. Here, we use the filtering prior for its theoretical properties.

We next make several assumptions about the stage cost and prior weighting.

Assumption 2 (Continuity) *The functions $f(\cdot)$, $h(\cdot)$, $\ell(\cdot)$, and $V_p(\cdot)$ are continuous.*

Assumption 3 (Positive Definite Cost Function) *There exist $\underline{\gamma}_p(\cdot)$, $\overline{\gamma}_p(\cdot)$, $\underline{\gamma}_s(\cdot)$, $\overline{\gamma}_s(\cdot) \in \mathscr{K}_\infty$ such that*

$$\underline{\gamma}_p(|\chi-\overline{x}|) \leq V_p(\chi,\overline{x}) \leq \overline{\gamma}_p(|\chi-\overline{x}|)$$
$$\underline{\gamma}_s(|(\omega,\nu)|) \leq \ell(\omega,\nu) \leq \overline{\gamma}_s(|(\omega,\nu)|)$$

Assumption 4 (Lipschitz Continuity of Function Composition) *The function $\underline{\gamma}_p^{-1}(2\overline{\gamma}_p(s))$ is Lipschitz continuous at the origin.*

Assumption 5 (Contractivity of Function Composition) *For every $\overline{s}>0$ and $\eta \in (0,1)$, there exists a value of $\rho > 0$ such that*

$$\gamma_d(2\underline{\gamma}_s^{-1}(2\rho\overline{\gamma}_p(s))) \leq \eta s$$

for all $s \in [0,\overline{s}]$.

Assumptions 2 and 3 guarantee that the optimization problem is well posed and that, asymptotically, larger deviations from the prior and larger noises are penalized more than smaller deviations and noises. These assumptions are standard in the literature (e.g., [23, Ch.4]). The other assumptions are newer and require some discussion.

In [13], it is assumed (Assumptions 3 and 4) that $\underline{\gamma}_s(\cdot)$, $\overline{\gamma}_s(\cdot)$, $\underline{\gamma}_p(\cdot)$, and $\overline{\gamma}_p(\cdot)$ all have the form of Cs^α for some $C, \alpha > 0$, in which the upper and lower bounds of each function use the same power. Under those conditions, Assumption 4 holds. Furthermore, the combination of Assumptions 4 and 5 in [13] guarantees that the conditions of Assumption 5 here hold. The main assumption of Theorem 2 in [4] is essentially a combination of Assumptions 1 and 5.

In general, Assumptions 4 holds when the functions $\underline{\gamma}_p(\cdot)$ and $\overline{\gamma}_p(\cdot)$ admit lower and upper bounds, respectively, of the form Cs^α in a neighborhood of the origin, in which both power-law bounds have the same power. Similarly, if $\gamma_d(\cdot)$ is locally Lipschitz and $\underline{\gamma}_s(s) \geq Cs^\beta$, in which $\beta \leq \alpha$, in a neighborhood of the origin, then Assumption 5 holds.

Remark 1. The relationship between $\ell(\cdot)$ and $V_p(\cdot)$ required by Assumption 5 is the *opposite* of that between an MPC controller's stage cost and terminal cost. The controller requires a terminal cost that is as least as steep as that of the stage cost, so it can serve as a local control Lyapunov function. The estimator, by contrast, requires a prior weighting that is at most as steep as that of the stage cost. This assumption is similar to Condition C2 in [19], in which the prior weighting is assumed to be a lower bound to the estimator's arrival cost.

We use a definition of robust asymptotic stability similar to that provided in [23, Ch. 4].

Definition 3 (Robust Asymptotic Stability). A state estimator is robustly asymptotically stable if there exist $\delta_1, \delta_2 > 0$, $\beta_e(\cdot) \in \mathcal{KL}$, and $\gamma_e(\cdot) \in \mathcal{K}$ such that if $|x(0) - \overline{x}| \leq \delta_1$ and $\|\mathbf{d}\| \leq \delta_2$, then we have that

$$|\hat{x}(k) - x(k)| \leq \beta_e(|\overline{x} - x(0)|, k) \oplus \gamma_e(\|\mathbf{d}\|_{0:k-1})$$

for all $k \in \mathbb{I}_{\geq 0}$.

Remark 2 (Globality). Note that the authors of [13] and [4] call the properties they establish robust *global* asymptotic stability. While those properties are global in the sense that they apply regardless of where the initial condition $x(0)$ is, they are not global in the sense that the estimators are robustly stable regardless of the error in the prior estimate $|\overline{x} - x(0)|$. As we do here, those papers restrict the size of initial estimate error. Here, we prefer to reserve the term *global* for estimators that are global in both senses.

4 Main Results

In order to prove the robust stability of MHE, we first need a result that bounds the estimator error within the MHE problem in terms of the error in the prior and the size of the disturbances. This proposition is similar to Lemma 7 in [13]. Because both disturbances are combined into a single variable and the fact that maximization is used rather than addition, the proof is significantly streamlined, and we include it in the appendix.

Proposition 3. *For any MHE problem satisfying Assumptions 2 and 3 applied to an i-IOSS system, we have for all $k \leq N$ that*

$$\begin{aligned}|e(k)| \leq & \beta(2\underline{\gamma}_p^{-1}(2\overline{\gamma}_p(|e_p|)),k) \oplus \gamma_d(2\underline{\gamma}_s^{-1}(2\rho\overline{\gamma}_p(|e_p|))) \\ & \oplus \beta(2\underline{\gamma}_p^{-1}((2N/\rho)\overline{\gamma}_s(\|\mathbf{d}\|_{0:N-1})),k) \oplus \gamma_d(2\underline{\gamma}_s^{-1}(2N\overline{\gamma}_s(\|\mathbf{d}\|_{0:N-1})))\end{aligned}$$

in which $e_p := \overline{x} - x$ and $e(k) := x(k) - \hat{x}(k)$.

Finally, we present the result on robust asymptotic stability. First, we note that by choosing a long enough horizon N and small enough ρ in (1), we can obtain a contraction mapping for the error using the filtering prior, $|\hat{x}(k-T) - x(k-N)|$, using Proposition 3. Next, we apply this contraction mapping repeatedly to obtain exponential convergence in estimator error once the estimator's horizon is filled, i.e., once $k \geq N$. Finally, we bound the error from times 0 to $N-1$ when the initial prior is being used. These elements are combined into a statement of robust stability.

Theorem 1 (Robust Asymptotic Stability of MHE). *For every $\overline{s} > 0$, there exists a T and ρ such that if $N \geq T$ and $|e_p| \leq \overline{s}$, then there exist $\beta_e(\cdot) \in \mathcal{KL}$, $\gamma_e(\cdot) \in \mathcal{K}$, and $\delta > 0$ such that if $\|\mathbf{d}\| \leq \delta$, then*

$$|e(k)| \leq \beta_e(|e_p|,k) \oplus \gamma_e(\|\mathbf{d}\|)$$

for all $k \geq 0$. Furthermore, if $\lim_{k\to\infty} |d(k)| = 0$, then $\lim_{k\to\infty} |e(k)| = 0$.

Proof. Let $\tilde{s} := \beta(2\underline{\gamma}_p^{-1}(2\overline{\gamma}_p(\overline{s})),0)$. By Assumption 4 and Proposition 1 there exists some $L_p(\tilde{s}) > 0$ such that for all $s \in [0,\tilde{s}]$, we have that $2\underline{\gamma}_p^{-1}(2\overline{\gamma}_p(s)) \leq L_p s$. Let $\check{s} := L_p(\tilde{s})\tilde{s}$. By Assumption 1, for every $\eta \in (0,1)$ there exists $T \geq 0$ such that $\beta(s,T) \leq \eta s$ for all $s \in [0,\check{s}]$. Note that because $\underline{\gamma}_p^{-1}(2\overline{\gamma}_p(s)) \geq s$ and $\beta(s,0) \geq s$, we have that $\check{s} \geq \overline{s}$. Choose $\lambda \in (0,1)$. We thus have that there exists some $T \geq 0$ such that

$$\beta(2\underline{\gamma}_p^{-1}(2\overline{\gamma}_p(s)),T) \leq \beta(L_p(\tilde{s})s,T) \leq \eta L_p(\tilde{s})s \leq \lambda s$$

for all $s \leq \tilde{s}$.

By Assumption 5, we can fix ρ sufficiently small such that $\gamma_d(2\underline{\gamma}_s^{-1}(2\rho\overline{\gamma}_p(s))) \leq \lambda s$ for all $s \in [0,\tilde{s}]$. Suppose that $N \geq T$. Because we use a filtering prior (i.e., $\overline{x}(k) = \hat{x}(k-N)$ for all $k \geq N$), by Proposition 3 we have for all $k \geq 0$ that

$$\begin{aligned}|e(k+N)| \leq &\beta(2\underline{\gamma}_p^{-1}(2\overline{\gamma}_p(|e(k)|)),N) \oplus \gamma_d(2\underline{\gamma}_s^{-1}(2\rho\overline{\gamma}_p(|e(k)|)))\\
&\oplus \beta(2\underline{\gamma}_p^{-1}((2N/\rho)\overline{\gamma}_s(\|\mathbf{d}\|_{k:k+N-1})),N)\\
&\oplus \gamma_d(\underline{\gamma}_s^{-1}(2N\overline{\gamma}_s(\|\mathbf{d}\|_{k:k+N-1})))\\
\leq &\lambda\,|e(k)| \oplus \lambda\,|e(k)|\\
&\oplus \beta(2\underline{\gamma}_p^{-1}((2N/\rho)\overline{\gamma}_s(\|\mathbf{d}\|_{k:k+N-1})),0)\\
&\oplus \gamma_d(\underline{\gamma}_s^{-1}(2N\overline{\gamma}_s(\|\mathbf{d}\|_{k:k+N-1})))\\
= &\lambda\,|e(k)| \oplus \gamma_e(\|\mathbf{d}\|_{k:k+N-1}) \qquad (2)\end{aligned}$$

in which $\gamma_e(s) := \beta(2\underline{\gamma}_p^{-1}((2N/\rho)\overline{\gamma}_s(s)),0) \oplus \gamma_d(\underline{\gamma}_s^{-1}(2N\overline{\gamma}_s(s)))$, and note that $\gamma_e(\cdot) \in \mathscr{K}$.

Next, we require bounds on $|e(k)|$ for $k \in \mathbb{I}_{0:N-1}$. Because there are not enough measurements to fill the horizon yet, all of these MHE problems use $\overline{x}_0$ as their prior. Thus, by applying Proposition 3, we have that

$$\begin{aligned}|e(k)| \leq &\beta(2\underline{\gamma}_p^{-1}(2\overline{\gamma}_p(|e_p|)),0) \oplus \gamma_d(2\underline{\gamma}_s^{-1}(2\rho\overline{\gamma}_p(|e_p|)))\\
&\oplus \beta(2\underline{\gamma}_p^{-1}((2k/\rho)\overline{\gamma}_s(\|\mathbf{d}\|_{0:k-1})),0) \oplus \gamma_d(\underline{\gamma}_s^{-1}(2k\overline{\gamma}_s(\|\mathbf{d}\|_{0:k-1})))\\
\leq &\beta(2\underline{\gamma}_p^{-1}(2\overline{\gamma}_p(|e_p|)),0) \oplus \gamma_d(2\underline{\gamma}_s^{-1}(2\rho\overline{\gamma}_p(|e_p|)))\\
&\oplus \beta(2\underline{\gamma}_p^{-1}((2N/\rho)\overline{\gamma}_s(\|\mathbf{d}\|_{0:k-1})),0) \oplus \gamma_d(\underline{\gamma}_s^{-1}(2N\overline{\gamma}_s(\|\mathbf{d}\|_{0:k-1})))\\
= &\sigma_x(|e_p|) \oplus \gamma_e(\|\mathbf{d}\|_{0:k-1}) \qquad (3)\end{aligned}$$

in which $\sigma_x := \beta(2\underline{\gamma}_p^{-1}(2\overline{\gamma}_p(s)),0)$, and note both that $\sigma_x(\cdot) \in \mathscr{K}_\infty$ and that $\sigma_x(s) \geq s \geq \lambda s \geq \gamma_d(2\underline{\gamma}_s^{-1}(2\rho\overline{\gamma}_p(s)))$ for all $s \in [0,\tilde{s}]$. Because $\gamma_e(\cdot) \in \mathscr{K}$, there exists some $\delta > 0$ such that if $\|\mathbf{d}\| \leq \delta$, then $\gamma_e(\|\mathbf{d}\|) \leq \tilde{s}$.

Next, we prove by induction that

$$|e(k+jN)| \leq \lambda^j\sigma_x(|e_p|) \oplus \max_{i\in\mathbb{I}_{0:j-1}} \left(\lambda^{j-i-1}\gamma_e(\|\mathbf{d}\|_{iN:(i+1)N-1})\right) \oplus \gamma_e(\|\mathbf{d}\|_{jN:jN+k-1})$$

for all $j \geq 0$ and $k \in \mathbb{I}_{0:N-1}$. The base case is (3). Now we perform the inductive step.

Inductive Case Suppose that

$$|e(k+jN)| \leq \lambda^j\sigma_x(|e_p|) \oplus \max_{i\in\mathbb{I}_{0:j-1}} \left(\lambda^{j-i-1}\gamma_e(\|\mathbf{d}\|_{iN:(i+1)N-1})\right) \oplus \gamma_e(\|\mathbf{d}\|_{jN:jN+k-1})$$

for some $j \geq 0$. By applying (2), we have that

$$\begin{aligned}|e(k+(j+1)N)| \leq &\lambda\,|e(k+jN)| \oplus \gamma_e(\|\mathbf{d}\|_{k+jN:k+(j+1)N-1})\\
\leq &\lambda\Big(\lambda^j\sigma_x(|e_p|) \oplus \max_{i\in\mathbb{I}_{0:j-1}} \left(\lambda^{j-i-1}\gamma_e(\|\mathbf{d}\|_{iN:(i+1)N-1})\right)\end{aligned}$$

$$\begin{aligned}
&\oplus \gamma_e(\|\mathbf{d}\|_{jN:jN+k-1})\Big) \oplus \gamma_e(\|\mathbf{d}\|_{k+jN:k+(j+1)N-1}) \\
=&\lambda^{j+1}\sigma_x(|e_p|) \oplus \max_{i\in\mathbb{I}_{0:j-1}} \left(\lambda^{j-i}\gamma_e(\|\mathbf{d}\|_{iN:(i+1)N-1})\right) \\
&\oplus \lambda\gamma_e(\|\mathbf{d}\|_{jN:jN+k-1}) \oplus \gamma_e(\|\mathbf{d}\|_{k+jN:k+(j+1)N-1}) \\
\leq&\lambda^{j+1}\sigma_x(|e_p|) \oplus \max_{i\in\mathbb{I}_{0:j-1}} \left(\lambda^{j-i}\gamma_e(\|\mathbf{d}\|_{iN:(i+1)N-1})\right) \\
&\oplus \gamma_e(\|\mathbf{d}\|_{jN:(j+1)N-1}) \oplus \gamma_e(\|\mathbf{d}\|_{(j+1)N:k+(j+1)N-1}) \\
=&\lambda^{j+1}\sigma_x(|e_p|) \oplus \max_{i\in\mathbb{I}_{0:j}} \left(\lambda^{j-i}\gamma_e(\|\mathbf{d}\|_{iN:(i+1)N-1})\right) \\
&\oplus \gamma_e(\|\mathbf{d}\|_{(j+1)N:k+(j+1)N-1})
\end{aligned}$$

which is the required statement.

Note that an immediate consequence of this statement is that

$$|e(k+jN)| \leq \lambda^j\sigma_x(|e_p|) \oplus \gamma_e(\|\mathbf{d}\|_{0:jN+k-1})$$

for all $k \in \mathbb{I}_{0:N-1}$ and all $j \geq 0$. Let $\lambda^{\lfloor k/N \rfloor}\sigma_x(s) := \beta_e(s,k)$, and note that $\beta_e(\cdot) \in \mathcal{KL}$. Thus we have for all $k \geq 0$ that

$$|e(k)| \leq \beta_e(|e_p|,k) \oplus \gamma_e(\|\mathbf{d}\|_{0:k-1})$$

and thus MHE is robustly asymptotically stable.

Finally, we demonstrate that, if $\mathbf{d}$ converges to zero, then $|e(k)|$ converges to zero. Fix $\varepsilon > 0$. Because $\mathbf{d}$ converges to zero, there exists some K_1 such that $\gamma_e(\|\mathbf{d}\|_{k:\infty}) \leq \varepsilon$ for all $k \geq K_1$. Let $\bar{d} := \max_{k\in\mathbb{I}_{0:N\lceil K_1/N\rceil}} |d(k)|$. There exists some K_2 such that $\lambda^{\lfloor K_2/N\rfloor}\bar{d} \leq \varepsilon$. Finally, there exists some K_3 such that $\beta_e(|e_p|,K_3) \leq \varepsilon$. For all $Nj + k \geq (K_1+K_2+2N) \oplus K_3$, we have that

$$\begin{aligned}
|e(k+jN)| &\leq \lambda^j\sigma_x(|e_p|) \oplus \max_{i\in\mathbb{I}_{0:j-1}} \left(\lambda^{j-i-1}\gamma_e(\|\mathbf{d}\|_{iN:(i+1)N-1})\right) \oplus \gamma_e(\|\mathbf{d}\|_{jN:jN+k-1}) \\
&\leq \varepsilon \oplus \max_{i\in\mathbb{I}_{0:\lceil K_1/N\rceil}} \left(\lambda^{j-i-1}\gamma_e(\|\mathbf{d}\|_{iN:(i+1)N-1})\right) \oplus \gamma_e(\|\mathbf{d}\|_{N\lceil K_1/N\rceil:\infty}) \\
&\leq \varepsilon \oplus \bar{d}\lambda^{\lfloor (K_1+K_2+N)/N\rfloor - \lceil K_1/N\rceil - 1} \oplus \varepsilon
\end{aligned}$$

Because $\lfloor a+b \rfloor \leq \lfloor a \rfloor + \lfloor b \rfloor$ and $\lceil a \rceil \leq \lfloor a \rfloor + 1$, we have that

$$\begin{aligned}
|e(k+jN)| &\leq \varepsilon \oplus \bar{d}\lambda^{\lfloor K_1/N\rfloor + \lfloor K_2/N\rfloor + 1 - \lceil K_1/N\rceil - 1} \\
&\leq \varepsilon \oplus \bar{d}\lambda^{\lfloor K_1/N\rfloor + \lfloor K_2/N\rfloor + 1 - \lceil K_1/N\rceil} \\
&\leq \varepsilon \oplus \bar{d}\lambda^{\lceil K_1/N\rceil + \lfloor K_2/N\rfloor - \lceil K_1/N\rceil} \\
&\leq \varepsilon \oplus \varepsilon = \varepsilon
\end{aligned}$$

and thus $e(k)$ converges to zero.

This type of exponential decay argument was initially presented in [12] in the case of MHE with a maximum term in the objective function, and was extended to MHE without a maximum term in [13]. The form of the argument presented here more closely resembles that presented in [4] using a maximum term in its MHE for a broader class of systems than that in [12]. The argument here extends the results in [13] to a similar class of systems as in [4].

5 Numerical Example

We conclude with a numerical example illustrating the application of MHE, alongside the extended Kalman filter (EKF) and unscented Kalman filter (UKF), to a batch chemical reaction. This example is similar to the example in [3] in which the EKF fails, and was adapted from Example 4.41 in [23]. Consider the reversible dimerization of a species A into its dimer B catalyzed by an enzyme E

$$\begin{aligned} \mathrm{A} + \mathrm{E} &\rightleftharpoons \mathrm{AE} \\ 2\mathrm{AE} &\rightleftharpoons \mathrm{B} + 2\mathrm{E} \end{aligned}$$

For an isothermal batch reactor, the evolution of the concentrations of the species A, B, and the complex AE is given by the following differential equations

$$\begin{aligned} \frac{dc_{\mathrm{A}}}{dt} &= -k_1 c_{\mathrm{A}}(c_{\mathrm{t}} - c_{\mathrm{AE}}) + k_{-1} c_{\mathrm{AE}} \\ \frac{dc_{\mathrm{AE}}}{dt} &= k_1 c_{\mathrm{A}}(c_{\mathrm{t}} - c_{\mathrm{AE}}) - k_{-1} c_{\mathrm{AE}} - 2k_2 c_{\mathrm{AE}}^2 + 2k_{-2} c_{\mathrm{B}} \\ \frac{dc_{\mathrm{B}}}{dt} &= k_2 c_{\mathrm{AE}}^2 - k_{-2} c_{\mathrm{B}} \end{aligned}$$

in which c_i is the concentration of species i, c_{t} is the total concentration of enzyme initially in the reactor, k_i is the forward rate of reaction i, and k_{-i} is the reverse rate of reaction i. The parameters used were $k_1 = 0.05$, $k_{-1} = 0.005$, $k_2 = 0.02$, $k_{-2} = 0.001$, and $c_{\mathrm{t}} = 1$.

Suppose that the concentrations of these species cannot be measured individually, but rather the sum of c_{A}, c_{AE}, and c_{B} can be measured, i.e.,

$$h(c_{\mathrm{A}}, c_{\mathrm{AE}}, c_{\mathrm{B}}) = c_{\mathrm{A}} + c_{\mathrm{AE}} + c_{\mathrm{B}}$$

The code used to generate these figures was adapted from the code which generated Figures 4.5–4.7[1] in [23, Ch. 4]. We used CasADi[2] [1], an algorithmic (automatic) differentiation framework available for C++, Python, MATLAB, and Octave,

[1] Available at http://jbrwww.che.wisc.edu/home/jbraw/mpc/figures.html.

[2] Available at www.casadi.org.

and MPCTools,[3] a set of high-level functions for CasADi to streamline the formulation of optimal control and estimation problems, to discretize this set of ODEs. We used the fourth-order Runge-Kutta method with a sampling time of 0.5. We augmented this discrete-time model with an additive state disturbance, such that it had a form

$$x^+ = f(x) + w$$

When the system was simulated, each element of w was chosen by an independent normal distribution with zero mean and standard deviation 0.001. The measurement was disturbed with a normally distributed additive disturbance with zero mean and standard deviation 0.0076. For the MHE stage cost, we used

$$\ell(w,v) = \frac{1}{(0.001)^2}|w|^2 + \frac{1}{(0.0076)^2}|v|^2$$

i.e., a least squares objective with the inverse of the disturbance covariance matrices as weights, as in the recursive least squares formulation of the Kalman filter. These covariance matrices were also used for the EKF and UFK. A horizon of length 32 was used, and the final measurement was included.

For a prior weighting, we used

$$V_p(x,\bar{x}) = \frac{1}{4}|x-\bar{x}|^2$$

Nonnegativity constraints were enforced for all concentrations, while c_{AE} was constrained to be less than c_{t}. While simulating the system, if any disturbances caused the true concentrations to leave the feasible region, the concentrations were clipped to the nearest feasible point. We simulated two cases for the EKF and UKF, one case where the estimates were clipped to feasibility, and one in which they were not. The EKF's estimate was clipped immediately after its update formula. Because the UKF samples an ensemble of points, called sigma points, around the estimated state, both the estimate and the sigma points need to be clipped. The sigma points were rescaled to be feasible in x and weighted as in [30], run through the system model, then clipped again (due to w). The state estimators were all initialized with a prior of $\hat{x} = [0, 0.5, 3]^T$, while the actual system started with a state of $x = [0.8, 0.1, 0.0]^T$.

The results of the EKF and UKF are given in Figure 3 and the results of MHE are given in Figure 4. MHE converges to the true states much faster than the two other strategies, despite the bad prior, and all state estimates are physically realistic. Thus, in this case, MHE is a stable and reliable estimator.

---

[3] Available at https://bitbucket.org/rawlings-group/octave-mpctools.

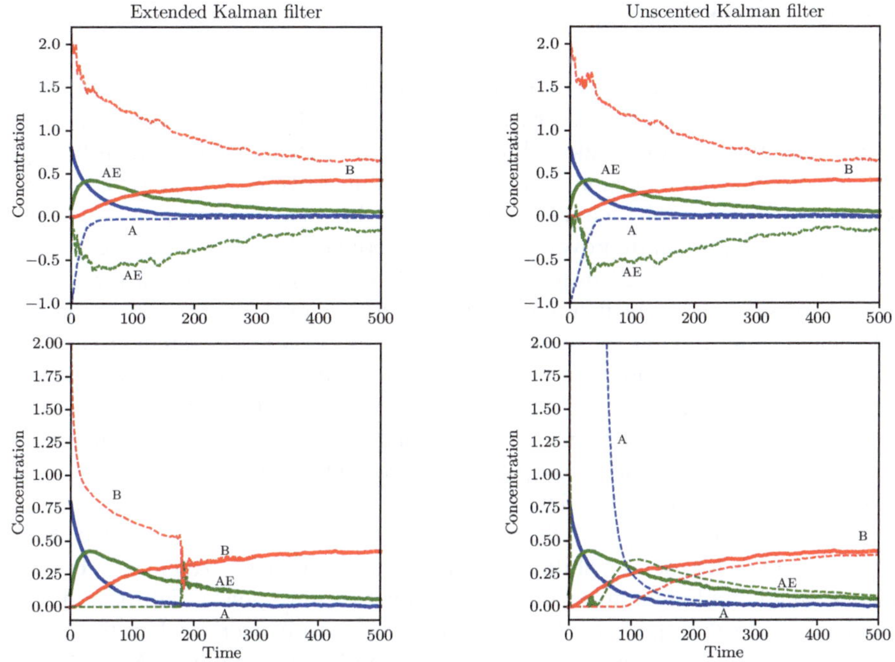

(a) When clipping is not used (top), the EKF predicts unphysical negative concentrations. When clipping is used (bottom), $\hat{c}_{\mathrm{B}}$ is initially very large while both $\hat{c}_{\mathrm{A}}$ and $\hat{c}_{\mathrm{AE}}$ are zero. At time 180, the estimates abruptly converge to the true states.

(b) When clipping is not used (top), the UKF also predicts negative concentrations. When clipping is used (bottom), the estimate $\hat{c}_{\mathrm{A}}$ grows to over 250, several orders of magnitude above its true value, before the estimator settles down by time 100 and converges.

Fig. 3: States (solid) and state estimates (dashed) for the EKF and UFK.

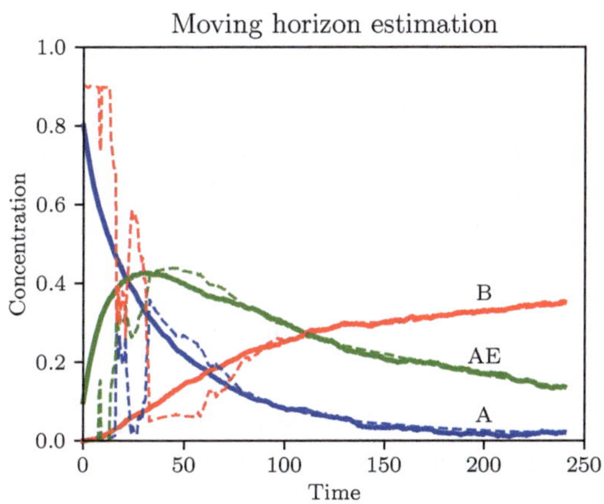

Fig. 4: States (solid) and state estimates (dashed) for MHE. Although $\hat{c}_{\mathrm{A}}$ initially follows c_{B} and $\hat{c}_{\mathrm{B}}$ initially follows c_{A}, MHE converges to the true states relatively quickly (note the changed time axis scale).

6 Conclusions

Here, we presented a version of the robust stability of MHE in the case of bounded disturbances in [13] generalized by assumptions from [4]. It may seem that these results settle the question of the stability and convergence of MHE for a fairly large class of systems, but many questions still remain unanswered. Probably the largest question for these results without max-terms in the objective function is whether or not a longer horizon improves estimator performance. Inspection of the $\mathscr{K}$ and $\mathscr{KL}$ function constructed in the proof of robust stability shows that these functions increase with the estimation horizon N. Although it is necessary to have a sufficiently long horizon for the estimator to work, it is not apparent from these results whether lengthening the horizon always results in improved estimator performance, or whether there is an optimal horizon length, after which estimator performance deteriorates. For MPC, it is known that a longer horizon always improves the controller's nominal closed-loop performance, and the main trade-off is between performance and computation time.

Addition of the max-term to the objective function does allow the $\mathscr{K}$ and $\mathscr{KL}$ function bounds for robust stability to be independent of the horizon length, but it is unclear whether a longer horizon impacts estimator convergence. Full-information estimation with the max-term has not been proven to converge when the disturbances converge, so it may be that the longer the horizon length in MHE with a max-term, the longer it takes for the state estimate to converge. For these reasons, despite the fact that MHE has been proven to be robustly stable without reference to the full-information problem, the full-information problem remains important. Understanding a problem with the longest horizon possible grants understanding of MHE problems with long horizons.

Appendix

Proof (Proposition 1). Without loss of generality, assume that $x_0 = 0$ and that $V(0) = 0$. Because $V(\cdot)$ is Lipschitz continuous at the origin, there exists some $\delta > 0$ and $L > 0$ such that if $|x| \leq \delta$ we have that $|V(x)| \leq L|x|$. Let $(\bar{s}(n))$ be a strictly increasing and unbounded sequence such that $\bar{s}(n) > \delta$ for all $n \in \mathbb{I}_{\geq 0}$. Define a sequence $(\tilde{M}(n))$ such that

$$\tilde{M}(n) := \sup_{x \in \mathscr{C}} |V(x)| \qquad \text{subject to } |x| \leq \bar{s}(n)$$

for all $n \in \mathbb{I}_{\geq 0}$. We have that $\tilde{M}(n)$ is finite for all $n \in \mathbb{I}_{\geq 0}$ because $V(\cdot)$ is locally bounded. Define another sequence $(M(n))$ such that

$$M(n) := \max(\tilde{M}(n), L\delta)$$

Note that $(M(n))$ is a nondecreasing sequence. Now define a piecewise linear function

$$\tilde{\alpha}(s) := \begin{cases} Ls & \text{if } s \in [0, \delta/2] \\ \frac{L\delta}{2} + (M(0) - L\delta/2)\frac{s-\delta/2}{\delta/2} & \text{if } s \in (\delta/2, \delta] \\ M(0) + (M(1) - M(0))\frac{s-\delta}{\bar{s}(0)-\delta} & \text{if } s \in (\delta, \bar{s}(0)] \\ M(n) + (M(n+1) - M(n))\frac{s-\bar{s}(n-1)}{\bar{s}(n)-\bar{s}(n-1)} & \text{if } s \in (\bar{s}(n-1), \bar{s}(n)] \end{cases}$$

The function $\tilde{\alpha}(s)$ is continuous, nondecreasing, and $\tilde{\alpha}(0) = 0$. Furthermore, because it is piecewise-linear, it is locally Lipschitz. Because $\tilde{\alpha}(\bar{s}(n-1)) = M(n)$, we also have that

$$|V(x)| \leq M(n) \leq \tilde{\alpha}(|x|) \quad \text{if } |x| \in (\bar{s}(n-1), \bar{s}(n)]$$

We have a similar bound for $|x| \in (\delta, \bar{s}(0)]$. Finally, from the Lipschitz bound, we have that $|V(x)| \leq \tilde{\alpha}(|x|)$ if $|x| \leq \delta$ and thus for all $x \in \mathscr{C}$. Finally, let $\alpha(s) := s + \tilde{\alpha}(s)$. We have that $\alpha(\cdot)$ is strictly increasing, continuous, zero at the origin, and asymptotically unbounded. Thus $\alpha(\cdot) \in \mathscr{K}_\infty$. Furthermore, $\alpha(\cdot)$ is piecewise-linear and thus locally Lipschitz, and is therefore the required bound.

Proof (Proposition 2). We prove this proposition by showing first that Statement 1 implies Statement 2, next that Statement 2 implies Statement 3, and then finally that Statement 3 implies Statement 4. Because Statement 4 is a restatement of Statement 1 with a particular form of the $\mathscr{L}$ function, the proof is then complete.

Proof (Statement 1 Implies Statement 2 (adapted from [1], Lemma 6)). Fix $\eta \in (0,1)$ and $\bar{s} > 0$. We seek to find T such that $\sigma_x(s)\phi(T) \leq \eta s$ for all $s \leq \bar{s}$. This condition is equivalent to $\phi(T) \leq \eta s/\sigma_x(s)$ for all $s \in (0, \bar{s}]$ and $\phi(T) \leq \eta \lim_{s \downarrow 0} s/\sigma_x(s)$. Because $\sigma_x(s)$ is Lipschitz continuous at zero, there exists some $L > 0$ and $\delta > 0$ such that $\sigma_x(s) \leq Ls$ for $0 \leq s \leq \delta$. Thus, we have that $s/\sigma_x(s) \geq s/(sL) = 1/L > 0$ for $s \leq \delta$. Furthermore, $s/\sigma_x(s) > 0$ and $s/\sigma_x(s)$ is continuous for all $s > 0$. Thus we have that $\inf_{s \leq \bar{s}} s/\sigma_x(s) := \zeta > 0$. Finally, we require T such that $\phi(T) \leq \eta\zeta \leq s/\sigma_x(s)$ for $s \leq \bar{s}$. Because both $\eta, \zeta > 0$ and $\phi(\cdot) \in \mathscr{L}$, there exists a T that fulfills this condition. Finally, because $\sigma_x(s)$ is Lipschitz at the origin, we have that $\beta(s,0)$ is Lipschitz at $s = 0$.

Remark 3. Because the only place where $s/\sigma_x(s)$ might equal zero is at $s = 0$, this proof also implies that if there exists a single $\bar{s} > 0$, $\eta \in (0,1)$, and $T > 0$ such that $\sigma_x(s)\phi(T) \leq \eta s$ for all $s \in [0, \bar{s}]$, then we can find such a T for every $\bar{s} > 0$.

Proof (Statement 2 Implies Statement 3). For brevity, we define $x_1(k) - x_2(k) := \Delta x(k)$, $\mathbf{w}_1 - \mathbf{w}_2 := \Delta \mathbf{w}$, and $\mathbf{y}_1 - \mathbf{y}_2 := \Delta \mathbf{y}$. Fix $\bar{s} > 0$ and choose T such that $\beta(s,T) \leq \eta s$ for all $s \in [0, \bar{s}]$. First, we prove by induction in n that

$$|\Delta x(k+nT)| \leq \eta^n |\Delta x(0)| \oplus \gamma_w(\|\Delta \mathbf{w}\|_{0:k+nT-1}) \oplus \gamma_y(\|\Delta \mathbf{y}\|_{0:k+nT-1}) \tag{4}$$

for all $k \geq 0$ and $n \geq 1$ if $\Delta x(0) \leq \bar{s}$.

Base Case Suppose that $\Delta x(0) \leq \bar{s}$. Then we have that

$$\begin{aligned}|\Delta x(k+T)| &\leq \beta(\Delta x(0), k+T) \oplus \gamma_w(\|\Delta \mathbf{w}\|_{0:k+T-1}) \oplus \gamma_y(\|\Delta \mathbf{y}\|_{0:k+T-1}) \\ &\leq \beta(\Delta x(0), T) \oplus \gamma_w(\|\Delta \mathbf{w}\|_{0:k+T-1}) \oplus \gamma_y(\|\Delta \mathbf{y}\|_{0:k+T-1}) \\ &\leq \eta\, |\Delta x(0)| \oplus \gamma_w(\|\Delta \mathbf{w}\|_{0:k+T-1}) \oplus \gamma_y(\|\Delta \mathbf{y}\|_{0:k+T-1})\end{aligned}$$

by the fact that $\beta(\cdot)$ is nonincreasing in its second argument and by Statement 2.

Inductive Case Suppose that (4) holds for some $n \in \mathbb{I}_{\geq 0}$. We have that

$$|\Delta x(k+nT)| \leq \eta^n |\Delta x(0)| \oplus \gamma_w(\|\Delta \mathbf{w}\|_{0:k+nT-1}) \oplus \gamma_y(\|\Delta \mathbf{y}\|_{0:k+nT-1})$$

for all $k \geq 0$. Suppose further that

$$\gamma_w(\|\Delta \mathbf{w}\|_{0:k+nT-1}) \oplus \gamma_y(\|\Delta \mathbf{y}\|_{0:k+nT-1}) > \bar{s}$$

We then apply the original i-IOSS bound to obtain

$$\begin{aligned}|\Delta x(k+(n+1)T)| \leq \beta(|\Delta x(0)|, k+(n+1)T) &\oplus \gamma_w(\|\Delta \mathbf{w}\|_{0:k+(n+1)T-1}) \\ &\oplus \gamma_y(\|\Delta \mathbf{y}\|_{0:k+(n+1)T-1})\end{aligned}$$

Furthermore, we have that

$$\beta(|\Delta x(0)|, k+nT) \leq \beta(|\Delta x(0)|, T) \leq \eta\, |\Delta x(0)| \leq \eta \bar{s} < \bar{s}$$

Thus we have that the $\mathcal{KL}$ function bound is unnecessary, and therefore we have that

$$\begin{aligned}|\Delta x(k+(n+1)T)| \leq &\gamma_w(\|\Delta \mathbf{w}\|_{0:k+(n+1)T-1}) \oplus \gamma_y(\|\Delta \mathbf{y}\|_{0:k+(n+1)T-1}) \\ \leq &\eta^{n+1} |\Delta x(0)| \oplus \gamma_w(\|\Delta \mathbf{w}\|_{0:k+(n+1)T-1}) \\ &\oplus \gamma_y(\|\Delta \mathbf{y}\|_{0:k+(n+1)T-1})\end{aligned}$$

trivially. Now suppose that

$$\gamma_w(\|\Delta \mathbf{w}\|_{0:k+nT-1}) \oplus \gamma_y(\|\Delta \mathbf{y}\|_{0:k+nT-1}) \leq \bar{s}$$

Because we have that $\eta^n |\Delta x(0)| \leq \bar{s}$, we have that $|\Delta x(k+nT)| \leq \bar{s}$ so we can apply the i-IOSS bound from $\Delta x(k+nT)$ to obtain

$$\begin{aligned}|\Delta x(k+(n+1)T)| \leq &\beta(|\Delta x(k+nT)|, T) \oplus \gamma_w(\|\Delta \mathbf{w}\|_{k+nT:k+(n+1)T-1}) \\ &\oplus \gamma_y(\|\Delta \mathbf{y}\|_{k+nT:k+(n+1)T-1}) \\ \leq &\eta\, |\Delta x(k+nT)| \oplus \gamma_w(\|\Delta \mathbf{w}\|_{k+nT:k+(n+1)T-1}) \\ &\oplus \gamma_y(\|\Delta \mathbf{y}\|_{k+nT:k+(n+1)T-1})\end{aligned}$$

$$\begin{aligned}&\leq \eta^{n+1}\left|\Delta x(0)\right| \oplus \eta\gamma_w(\|\boldsymbol{\Delta w}\|_{0:k+nT-1})\\&\quad\oplus \eta\gamma_y(\|\boldsymbol{\Delta y}\|_{0:k+nT-1}) \oplus \gamma_w(\|\boldsymbol{\Delta w}\|_{k+nT:k+(n+1)T-1})\\&\quad\oplus \gamma_y(\|\boldsymbol{\Delta y}\|_{k+nT:k+(n+1)T-1})\\&\leq \eta^{n+1}\left|\Delta x(0)\right| \oplus \gamma_w(\|\boldsymbol{\Delta w}\|_{0:k+(n+1)T-1})\\&\quad\oplus \gamma_y(\|\boldsymbol{\Delta y}\|_{0:k+(n+1)T-1})\end{aligned}$$

Thus we have that (4) for n implies (4) for $n+1$, completing the proof by induction.

Now we bound $|\Delta x(k)|$ for $k \in \mathbb{I}_{0:T-1}$. We have that

$$\begin{aligned}|\Delta x(k)| &\leq \beta(|\Delta x(0)|, k) \oplus \gamma_w(\|\boldsymbol{\Delta w}\|_{0:k-1}) \oplus \gamma_y(\|\boldsymbol{\Delta y}\|_{0:k-1})\\&\leq \beta(|\Delta x(0)|, 0) \oplus \gamma_w(\|\boldsymbol{\Delta w}\|_{0:k-1}) \oplus \gamma_y(\|\boldsymbol{\Delta y}\|_{0:k-1})\end{aligned}$$

because $\beta(\cdot)$ is nonincreasing in its second argument. Because $\beta(s,0)$ is Lipschitz at $s=0$, by Proposition 1 there exists some locally Lipschitz function $\bar{\alpha}(\cdot) \in \mathscr{K}_\infty$ such that $\beta(s,0) \leq \bar{\alpha}(s)$ for all $s \in \mathbb{R}_{\geq 0}$. Because $\bar{\alpha}(\cdot)$ is locally Lipschitz and $\beta(s,0) \geq s$, there exists some $K \geq 1$ such that for all $s \in [0, \bar{s}]$, we have that $\bar{\alpha}(s) \leq Ks$. Thus we have that

$$|\Delta x(k)| \leq K\left|\Delta x(0)\right| \oplus \gamma_w(\|\boldsymbol{\Delta w}\|_{0:k-1}) \oplus \gamma_y(\|\boldsymbol{\Delta y}\|_{0:k-1}) \tag{5}$$

for all $\Delta x(0)$ such that $|\Delta x(0)| \leq \bar{s}$. Because we have that (4) holds for $n \geq 1$ and $k \geq 0$ and that (5) holds for $n = 0$ and $k \geq 0$, we can combine these equations to write

$$|\Delta x(k+nT)| \leq K\eta^n \left|\Delta x(0)\right| \oplus \gamma_w(\|\boldsymbol{\Delta w}\|_{0:k+nT-1}) \oplus \gamma_y(\|\boldsymbol{\Delta y}\|_{0:k+nT-1})$$

for all Δx such that $|\Delta x| \leq \bar{s}$. We can then eliminate the index n using the floor function $\lfloor \cdot \rfloor$ to obtain the bound

$$|\Delta x(k)| \leq K\eta^{\lfloor k/T \rfloor} \left|\Delta x(0)\right| \oplus \gamma_w(\|\boldsymbol{\Delta w}\|_{0:k-1}) \oplus \gamma_y(\|\boldsymbol{\Delta y}\|_{0:k-1})$$

Note that $\eta^{\lfloor k/T \rfloor} \leq (1/\eta)\eta^{k/T}$, so we have that

$$|\Delta x(k)| \leq (K/\eta)\eta^{k/T} \left|\Delta x(0)\right| \oplus \gamma_w(\|\boldsymbol{\Delta w}\|_{0:k-1}) \oplus \gamma_y(\|\boldsymbol{\Delta y}\|_{0:k-1})$$

Finally, let $C := K/\eta$ and $\lambda := \eta^{1/T}$. We have that

$$|\Delta x(k)| \leq C\lambda^k \left|\Delta x(0)\right| \oplus \gamma_w(\|\boldsymbol{\Delta w}\|_{0:k-1}) \oplus \gamma_y(\|\boldsymbol{\Delta y}\|_{0:k-1})$$

for all $\Delta x(0)$ such that $|\Delta x(0)| \leq \bar{s}$.

Proof (Statement 3 Implies 4). We prove this statement in two steps. We first show that although both λ and C in Statement 3 depend on $\bar{s}$, the dependence of λ on $\bar{s}$ can be removed by increasing $C(\bar{s})$. We can remove this dependence because the term dependent on the initial conditions decays to be less than some smaller $\bar{s}$ within finite time. We then turn the function $C(\bar{s})s$ into a $\mathscr{K}_\infty$ function.

First, let $(\bar{s}(n))$ be a strictly increasing and unbounded sequence such that $\bar{s}(n) > 0$ for all $n \in \mathbb{I}_{1:\infty}$. By Statement 3, there exists sequences $(C(n))$ and $(\lambda(n))$ such that $C(n) \geq 1$ and $\lambda(n) \in (0,1)$ and that

$$|\Delta x(k)| \leq C(n)\lambda(n)^k |\Delta x(0)| \oplus \gamma_w(\|\Delta \mathbf{w}\|_{0:k-1}) \oplus \gamma_y(\|\Delta \mathbf{y}\|_{0:k-1})$$

for all $\Delta x(0)$ such that $|\Delta x(0)| \leq \bar{s}(n)$ for all $n \in \mathbb{I}_{1:\infty}$. Without loss of generality, assume that $(C(n))$ and that $(\lambda(n))$ are nondecreasing. Let $\underline{\lambda} := \lambda(1)$. We prove by induction that there exists a sequence $(\tilde{C}(n))$ such that

$$|\Delta x(k)| \leq \tilde{C}(n)\underline{\lambda}^k |\Delta x(0)| \oplus \gamma_w(\|\Delta \mathbf{w}\|_{0:k-1}) \oplus \gamma_y(\|\Delta \mathbf{y}\|_{0:k-1})$$

for all $\Delta x(0)$ such that $|\Delta x(0)| \leq \bar{s}(n)$ and all $n \in \mathbb{I}_{1:\infty}$. The base case is given by Statement 3 for $\bar{s}(1)$.

Inductive Case Suppose for some n we have that

$$|\Delta x(k)| \leq \tilde{C}(n)\underline{\lambda}^k |\Delta x(0)| \oplus \gamma_w(\|\Delta \mathbf{w}\|_{0:k-1}) \oplus \gamma_y(\|\Delta \mathbf{y}\|_{0:k-1})$$

for all $\Delta x(0)$ such that $|\Delta x(0)| \leq \bar{s}(n)$. We also have that

$$|\Delta x(k)| \leq C(n+1)\lambda(n+1)^k |\Delta x(0)| \oplus \gamma_w(\|\Delta \mathbf{w}\|_{0:k-1}) \oplus \gamma_y(\|\Delta \mathbf{y}\|_{0:k-1})$$

for all $\Delta x(0)$ such that $|\Delta x(0)| \leq \bar{s}(n+1)$. Let

$$N := \left\lceil \log_{\lambda(n+1)} \left(\frac{\bar{s}(n)}{\bar{s}(n+1)C(n+1)} \right) \right\rceil$$

in which $\lceil \cdot \rceil$ is the ceiling function. For all $\Delta x(0)$ such that $|\Delta x(0)| \leq \bar{s}(n+1)$, we have that $C(n+1)\lambda^N |\Delta x(0)| \leq \bar{s}(n)$. Suppose that $\gamma_w(\|\Delta \mathbf{w}\|_{0:k-1}) \oplus \gamma_y(\|\Delta \mathbf{y}\|_{0:k-1}) < C(n+1)\lambda^N |\Delta x(0)|$. Then we have that $|\Delta x(k)| \leq C(n+1)\lambda^N |\Delta x(0)|$ alone, and thus we can apply the bound for all $\Delta x(0) \leq \bar{s}(n)$ to obtain for all $k \geq N$ that

$$\begin{aligned}
|\Delta x(k)| \leq& \tilde{C}(n)\underline{\lambda}^{k-N} |\Delta x(N)| \oplus \gamma_w(\|\Delta \mathbf{w}\|_{N:k-1}) \oplus \gamma_y(\|\Delta \mathbf{y}\|_{N:k-1}) \\
\leq& \tilde{C}(n)\underline{\lambda}^{k-N} C(n+1)\lambda(n+1)^N |\Delta x(0)| \oplus \gamma_w(\|\Delta \mathbf{w}\|_{0:k-1}) \\
& \oplus \gamma_y(\|\Delta \mathbf{y}\|_{0:k-1}) \\
=& \frac{\tilde{C}(n)C(n+1)\lambda(n+1)^N}{\underline{\lambda}^N} \underline{\lambda}^k |\Delta x(0)| \oplus \gamma_w(\|\Delta \mathbf{w}\|_{0:k-1}) \\
& \oplus \gamma_y(\|\Delta \mathbf{y}\|_{0:k-1})
\end{aligned}$$

Define

$$\tilde{C}(n+1) := \tilde{C}(n)C(n+1) \left(\frac{\lambda(n+1)}{\underline{\lambda}} \right)^N$$

and we have the required bound. Now suppose that $\gamma_w(\|\Delta \mathbf{w}\|_{0:k-1}) \oplus \gamma_y(\|\Delta \mathbf{y}\|_{0:k-1}) \geq C(n+1)\lambda^N |\Delta x(0)|$. Then, because the $\mathcal{KL}$ function is nonincreasing, we have that

$$|\Delta x(k)| \leq \gamma_w(\|\Delta \mathbf{w}\|_{0:k-1}) \oplus \gamma_y(\|\Delta \mathbf{y}\|_{0:k-1})$$

$$\leq \tilde{C}(n+1)\underline{\lambda}^k |\Delta x(0)| \oplus \gamma_w(\|\Delta \mathbf{w}\|_{0:k-1}) \oplus \gamma_y(\|\Delta \mathbf{y}\|_{0:k-1})$$

which is the required bound. Both of these bounds apply for $k \geq N$. For $k < N$, note that

$$\begin{aligned}
C(n+1)\lambda(n+1)^k &= \frac{C(n+1)\lambda(n+1)^k}{\underline{\lambda}^k}\underline{\lambda}^k \\
&\leq \frac{C(n+1)\lambda(n+1)^N}{\underline{\lambda}^N}\underline{\lambda}^k \\
&\leq \tilde{C}(n+1)\underline{\lambda}^k
\end{aligned}$$

because $\underline{\lambda} \leq \lambda(n+1)$. Therefore we also have that

$$|\Delta x(k)| \leq \tilde{C}(n+1)\underline{\lambda}^k |\Delta x(0)| \oplus \gamma_w(\|\Delta \mathbf{w}\|_{0:k-1}) \oplus \gamma_y(\|\Delta \mathbf{y}\|_{0:k-1})$$

for $k < N$ and thus for all $k \in \mathbb{I}_{\geq 0}$. Thus the statement is proven for $n+1$, and the first part of the proof is complete.

Next, define

$$\tilde{\alpha}(s) := \begin{cases} C(1)s & \text{if } s \in [0, \overline{s}(1)] \\ C(n)s & \text{if } s \in (\overline{s}(n-1), \overline{s}(n)] \text{ for } n \geq 2 \end{cases}$$

Note that this function is Lipschitz continuous at the origin and locally bounded. Furthermore, note that by construction we have that

$$|\Delta x(k)| \leq \tilde{\alpha}(|\Delta x(0)|)\underline{\lambda}^k |\Delta x(0)| \oplus \gamma_w(\|\Delta \mathbf{w}\|_{0:k-1}) \oplus \gamma_y(\|\Delta \mathbf{y}\|_{0:k-1})$$

for all $\Delta x(0)$. By Proposition 1, there exists a locally Lipschitz function $\alpha(\cdot) \in \mathcal{K}_\infty$ such that $\tilde{\alpha}(s) \leq \alpha(s)$ for all $s \in \mathbb{R}_{\geq 0}$. Thus we have that

$$|\Delta x(k)| \leq \alpha(|\Delta x(0)|)\underline{\lambda}^k |\Delta x(0)| \oplus \gamma_w(\|\Delta \mathbf{w}\|_{0:k-1}) \oplus \gamma_y(\|\Delta \mathbf{y}\|_{0:k-1})$$

for all $\Delta x(0)$, and so the result is established.

Proof (Proposition 3). By Assumptions 2 and 3, the MHE problem has a solution $(\hat{x}(0), \widehat{\mathbf{d}})$. Denote the estimated state at time k *within* the MHE problem as $\hat{x}(k)$. By Assumption 3, we have that

$$\rho\underline{\gamma}_p(|\hat{e}_p|) \oplus \underline{\gamma}_s(\|\widehat{\mathbf{d}}\|_{0:N-1}) \leq \rho\underline{\gamma}_p(|\hat{e}_p|) + \underline{\gamma}_s(\|\widehat{\mathbf{d}}\|_{0:N-1}) \leq V_N(\hat{x}(0), \widehat{\mathbf{d}}, \overline{x})$$

in which $\hat{e}_p := \overline{x} - \hat{x}(0)$. Furthermore, by optimality, we have that

$$\begin{aligned}
V_N(\hat{x}(0), \widehat{\mathbf{d}}, \overline{x}) &\leq V_N(x(0), \mathbf{d}, \overline{x}) \\
&\leq \rho\overline{\gamma}_p(|e_p|) + N\overline{\gamma}_s(\|\mathbf{d}\|_{0:N-1}) \\
&\leq 2\rho\overline{\gamma}_p(|e_p|) \oplus 2N\overline{\gamma}_s(\|\mathbf{d}\|_{0:N-1})
\end{aligned}$$

Combining these bounds and rearranging, we obtain the following bounds

$$\begin{aligned}|\hat{e}_p| &\le \underline{\gamma}_p^{-1}(2\overline{\gamma}_p(|e_p|)\oplus(2N/\rho)\overline{\gamma}_s(\|\mathbf{d}\|_{0:N-1})) \\ &= \underline{\gamma}_p^{-1}(2\overline{\gamma}_p(|e_p|))\oplus\underline{\gamma}_p^{-1}((2N/\rho)\overline{\gamma}_s(\|\mathbf{d}\|_{0:N-1}))\end{aligned} \tag{6}$$

$$\begin{aligned}\|\widehat{\mathbf{d}}\| &\le \underline{\gamma}_s^{-1}(2\rho\overline{\gamma}_p(|e_p|)\oplus 2N\overline{\gamma}_s(\|\mathbf{d}\|_{0:N-1})) \\ &= \underline{\gamma}_s^{-1}(2\rho\overline{\gamma}_p(|e_p|))\oplus\underline{\gamma}_s^{-1}(2N\overline{\gamma}_s(\|\mathbf{d}\|_{0:N-1}))\end{aligned} \tag{7}$$

From the system's i-IOSS bound, we have that

$$\begin{aligned}|e(k)| &\le \beta(|e(0)|,k)\oplus\gamma_d(\|\mathbf{d}-\widehat{\mathbf{d}}\|_{0:N-1}) \\ &= \beta(|\hat{x}(0)-\overline{x}+\overline{x}-x(0)|,k)\oplus\gamma_d(\|\mathbf{d}-\widehat{\mathbf{d}}\|_{0:N-1}) \\ &\le \beta(|e_p|+|\hat{e}_p|,k)\oplus\gamma_d(\|\mathbf{d}\|_{0:N-1}+\|\widehat{\mathbf{d}}\|_{0:N-1}) \\ &\le \beta(2|e_p|\oplus 2|\hat{e}_p|,k)\oplus\gamma_d(2\|\mathbf{d}\|_{0:N-1}\oplus 2\|\widehat{\mathbf{d}}\|_{0:N-1})\end{aligned}$$

We next substitute (6) and (7) into this expression.

$$\begin{aligned}|e(k)| \le&\beta(2|e_p|\oplus 2\underline{\gamma}_p^{-1}(2\overline{\gamma}_p(|e_p|))\oplus 2\underline{\gamma}_p^{-1}((2N/\rho)\overline{\gamma}_s(\|\mathbf{d}\|_{0:N-1})),k) \\ &\oplus\gamma_d(2\|\mathbf{d}\|_{0:N-1}\oplus 2\underline{\gamma}_s^{-1}(2\rho\overline{\gamma}_p(|e_p|))\oplus 2\underline{\gamma}_s^{-1}(2N\overline{\gamma}_s(\|\mathbf{d}\|_{0:N-1}))) \\ =&\beta(2|e_p|,k)\oplus\beta(2\underline{\gamma}_p^{-1}(2\overline{\gamma}_p(|e_p|)),k)\oplus\gamma_d(2\underline{\gamma}_s^{-1}(2\rho\overline{\gamma}_p(|e_p|))) \\ &\oplus\beta(2\underline{\gamma}_p^{-1}((2N/\rho)\overline{\gamma}_s(\|\mathbf{d}\|_{0:N-1})),k)\oplus\gamma_d(2\|\mathbf{d}\|_{0:N-1}) \\ &\oplus\gamma_d(\underline{\gamma}_s^{-1}(2N\overline{\gamma}_s(\|\mathbf{d}\|_{0:N-1})))\end{aligned}$$

Note that because $\underline{\gamma}_p(s)\le\overline{\gamma}_p(s)\le 2\overline{\gamma}_p(s)$, we have that $s\le\underline{\gamma}_p^{-1}(2\overline{\gamma}_p(s))$. A similar argument follows for $\underline{\gamma}_s(\cdot)$ and $\overline{\gamma}_s(\cdot)$. Thus we have that the term $\beta(2|e_p|,k)\le\beta(2\underline{\gamma}_p^{-1}(2\overline{\gamma}_p(|e_p|)),k)$ and the term $\gamma_d(2\|\mathbf{d}\|_{0:N-1})\le\gamma_d(\underline{\gamma}_s^{-1}(2N\overline{\gamma}_s(\|\mathbf{d}\|_{0:N-1})))$, so we can eliminate them from the maximization. Thus we have

$$\begin{aligned}|e(k)| \le&\beta(2\underline{\gamma}_p^{-1}(2\overline{\gamma}_p(|e_p|)),k)\oplus\gamma_d(2\underline{\gamma}_s^{-1}(2\rho\overline{\gamma}_p(|e_p|))) \\ &\oplus\beta(2\underline{\gamma}_p^{-1}((2N/\rho)\overline{\gamma}_s(\|\mathbf{d}\|_{0:N-1})),k)\oplus\gamma_d(\underline{\gamma}_s^{-1}(2N\overline{\gamma}_s(\|\mathbf{d}\|_{0:N-1})))\end{aligned}$$

for all $k\ge 0$, which is the desired result.

References

1. Andersson, J.: A general-purpose software framework for dynamic optimization. Ph.D. thesis, Arenberg Doctoral School, KU Leuven (2013)

2. Findeisen, P.K.: Moving horizon state estimation of discrete time systems. Master's thesis, University of Wisconsin-Madison (1997)
3. Haseltine, E.L., Rawlings, J.B.: Critical evaluation of extended Kalman filtering and moving horizon estimation. Ind. Eng. Chem. Res. **44**(8), 2451–2460 (2005)
4. Hu, W.: Robust stability of optimization-based state estimation under bounded disturbances. ArXiv e-prints (2017)
5. Hu, W., Xie, L., You, K.: Optimization-based state estimation under bounded disturbances. In: 2015 54th IEEE Conference on Decision and Control CDC, pp. 6597–6602 (2015). https://doi.org/10.1109/CDC.2015.7403258
6. Jazwinski, A.H.: Stochastic Processes and Filtering Theory. Academic Press, New York (1970)
7. Ji, L., Rawlings, J.B., Hu, W., Wynn, A., Diehl, M.M.: Robust stability of moving horizon estimation under bounded disturbances. IEEE Trans. Autom. Control **61**(11), 3509–3514 (2016)
8. Julier, S.J., Uhlmann, J.K.: Unscented filtering and nonlinear estimation. Proc. IEEE **92**(3), 401–422 (2004)
9. Kalman, R.E.: A new approach to linear filtering and prediction problems. Trans. ASME J. Basic Eng. **82**(1), 35–45 (1960)
10. Meadows, E.S., Muske, K.R., Rawlings, J.B.: Constrained state estimation and discontinuous feedback in model predictive control. In: Proceedings of the 1993 European Control Conference, pp. 2308–2312. European Automatic Control Council (1993)
11. Michalska, H., Mayne, D.Q.: Moving horizon observers and observer-based control. IEEE Trans. Autom. Control **40**(6), 995–1006 (1995)
12. Müller, M.A.: Nonlinear moving horizon estimation for systems with bounded disturbances. In: 2016 American Control Conference ACC, pp. 883–888 (2016). https://doi.org/10.1109/ACC.2016.7525026
13. Müller, M.A.: Nonlinear moving horizon estimation in the presence of bounded disturbances. Automatica **79**, 306–314 (2017). https://doi.org/10.1016/j.automatica.2017.01.033. http://www.sciencedirect.com/science/article/pii/S0005109817300432
14. Muske, K.R., Rawlings, J.B., Lee, J.H.: Receding horizon recursive state estimation. In: Proceedings of the 1993 American Control Conference, pp. 900–904 (1993)
15. Pannocchia, G., Rawlings, J.B.: Disturbance models for offset-free MPC control. AIChE J. **49**(2), 426–437 (2003)
16. Rajamani, R.: Observers for nonlinear systems: part 2: an overview of the special issue. IEEE Control Syst. Mag. **37**(4), 30–32 (2017). https://doi.org/10.1109/MCS.2017.2696758
17. Rajamani, R.: Observers for nonlinear systems: introduction to part 1 of the special issue. IEEE Control Syst. Mag. **37**(3), 22–24 (2017). https://doi.org/10.1109/MCS.2017.2674400
18. Rao, C.V.: Moving horizon strategies for the constrained monitoring and control of nonlinear discrete-time systems. Ph.D. thesis, University of Wisconsin-Madison (2000)
19. Rao, C.V., Rawlings, J.B., Mayne, D.Q.: Constrained state estimation for nonlinear discrete-time systems: stability and moving horizon approximations. IEEE Trans. Autom. Control **48**(2), 246–258 (2003)
20. Rawlings, J.B., Ji, L.: Optimization-based state estimation: current status and some new results. J. Process Control **22**, 1439–1444 (2012)
21. Rawlings, J.B., Mayne, D.Q.: Model Predictive Control: Theory and Design, 576 p. Nob Hill Publishing, Madison, WI (2009). ISBN 978-0-9759377-0-9
22. Rawlings, J.B., Risbeck, M.J.: On the equivalence between statements with epsilon-delta and K-functions. Technical Report 2015–01, TWCCC Technical Report (2015). http://jbrwww.che.wisc.edu/tech-reports/twccc-2015-01.pdf
23. Rawlings, J.B., Mayne, D.Q., Diehl, M.M.: Model Predictive Control: Theory, Design, and Computation, 2nd edn., 770 p. Nob Hill Publishing, Madison, WI (2017). ISBN 978-0-9759377-3-0
24. Romanenko, A., Castro, J.A.A.M.: The unscented filter as an alternative to the EKF for nonlinear state estimation: a simulation case study. Comput. Chem. Eng. **28**(3), 347–355 (2004)

25. Romanenko, A., Santos, L.O., Afonso, P.A.F.N.A.: Unscented Kalman filtering of a simulated pH system. Ind. Eng. Chem. Res. **43**, 7531–7538 (2004)
26. Sontag, E.D.: Smooth stabilization implies coprime factorization. IEEE Trans. Autom. Control **34**(4), 435–443 (1989). https://doi.org/10.1109/9.28018
27. Sontag, E.D.: Mathematical Control Theory, 2nd edn. Springer, New York (1998)
28. Sontag, E.D., Wang, Y.: Output-to-state stability and detectability of nonlinear systems. Syst. Control Lett. **29**, 279–290 (1997)
29. Tenny, M.J., Rawlings, J.B.: Efficient moving horizon estimation and nonlinear model predictive control. In: Proceedings of the American Control Conference, pp. 4475–4480. Anchorage, Alaska (2002)
30. Vachhani, P., Narasimhan, S., Rengaswamy, R.: Robust and reliable estimation via unscented recursive nonlinear dynamic data reconciliation. J. Process Control **16**(10), 1075–1086 (2006)

Probing and Duality in Stochastic Model Predictive Control

Martin A. Sehr and Robert R. Bitmead

1 Introduction

In a general nonlinear setting, stochastic optimal control involves the propagation of the conditional probability density of the state given the input signal and output measurements. This density is known as the *information state* in control circles and as the *belief state* in artificial intelligence and robotics squares. The choice of control signal affects the information state so that state observability becomes control-dependent. Thus, the feedback control law needs to include aspects of probing in addition to, or more accurately in competition with, its function in regulation. This is called *duality* of the control. In the linear case, this connection is not problematic since the control signal simply translates or recenters the conditional density without other effect. But for nonlinear systems, this complication renders all but the simplest optimal control problems computationally intractable.

The usual recourse for receding horizon stochastic optimal control or stochastic MPC (SMPC) is to drop optimality and to use a more simply computed or approximated statistic from the conditional density, such as the conditional mean, and to move on from there. There have been a number of approaches, mostly hinging on replacement of the measured true state by a state estimate, which is computed via Kalman filtering [30, 39], moving-horizon estimator [36], tube-based minimax estimators [26], etc. These designs, often for linear systems, separate the estimator design from the control design. The control problem may be altered to accommodate the state estimation error by methods such as: constraint tightening [39], chance/probabilistic constraints [28], and so forth. We do *not* seek to provide a com-

M. A. Sehr
Siemens Corporate Technology, 1936 University Ave, Suite 320, Berkeley, CA 94704, USA
e-mail: martin.sehr@siemens.com

R. R. Bitmead (✉)
University of California, San Diego, 9500 Gilman Drive, La Jolla, CA 92093-0411, USA
e-mail: rbitmead@ucsd.edu

S. V. Raković, W. S. Levine (eds.), *Handbook of Model Predictive Control*,
Control Engineering, https://doi.org/10.1007/978-3-319-77489-3_6

prehensive survey of the myriad alternative approaches proposed for SMPC. For that, we recommend the numerous available references such as [13, 19, 25, 27].

In this chapter, we develop the connections and consequences of duality in SMPC. Our approach will begin with the analysis of stochastic observability and include three practical examples – from TCP/IP computer communications, from parameter estimation, and from cellular mobile communications – where duality plays an indispensable part of the control solution. That is, where control resources are dedicated to the diminution of plant state uncertainty. The aim is to draw attention to the presence of probing in these engineered solutions and to highlight the importance of duality in optimal stochastic control.

2 Stochastic Optimal Control and Duality

2.1 The State, the Information State, and the Bayesian Filter

Our formulation commences with the general nonlinear Markovian stochastic system

$$x_{t+1} = f(x_t, u_t, w_t), \;\; x_0, \tag{1}$$

$$y_t = h(x_t, v_t). \tag{2}$$

Here: x_t is the system state, u_t is the control input signal, y_t is the output measurement signal, w_t is the white process noise, v_t is the white measurement noise, and functions $f(\cdot,\cdot,\cdot)$ and $h(\cdot,\cdot)$ are presumed sufficiently well-behaved so that densities, joint, marginal and conditional, exist. Denote the signal measurements up to time t by

$$\mathbf{Z}^t = \{y_0, u_0, \ldots, u_{t-1}, y_t\}, \;\; \mathbf{Z}^0 = \{y_0\}. \tag{3}$$

Then the information state is defined to be the conditional probability density function (pdf)

$$\xi_t = \mathrm{pdf}(x_t | \mathbf{Z}^t). \tag{4}$$

The Markovian nature of (1–2) yields [20, 34] the immediate property that the information state is propagated by the Bayesian filter:

$$\xi_t = \frac{\mathrm{pdf}(y_t|x_t)\xi_{t|t-1}}{\int \mathrm{pdf}(y_t|x_t)\xi_{t|t-1}\,dx_t}, \qquad \xi_{0|-1} = \mathrm{pdf}(x_0), \tag{5}$$

$$\xi_{t+1|t} \triangleq \int \mathrm{pdf}(x_{t+1}|x_t)\xi_t\,dx_t. \tag{6}$$

The Bayesian filter consists of two familiar pieces:

(i) Measurement update (5) relies on: the measurement function $h(\cdot,\cdot)$ of (2), the density of v_t, and $\xi_{t|t-1}$. The denominator term serves to recover a normalized density.
(ii) Time update (6) propagates using: $f(\cdot,\cdot,\cdot)$, the control signal u_t, the densities of w_t, and ξ_t.

In the linear gaussian case, these comprise the Kalman filter equations. For modest state dimensions, the Bayesian filter can be computed either from density points over a grid in state-space or by using the Particle filter [7, 34].

2.2 Stochastic Optimal Control and the Information State

In stochastic optimal control, the state-space system (1–2) is accompanied by an objective function,

$$J_N(\xi_0,\mathbf{U}^{N-1}) \triangleq \mathbb{E}\left[\sum_{j=0}^{N-1} c(x_j,u_j)+c_N(x_N)\right], \tag{7}$$

to be minimized over feedback signals $\mathbf{U}^{N-1} = \{u_0,\dots,u_{N-1}\}$ causally computed from the measurements and previous controls. The Markovian property of the state equation further implies that the optimal control is a function solely of the information state [20],

$$u_j^{\text{optimal}} = \pi_j(\xi_j). \tag{8}$$

Here the sequence of functionals, $\{\pi_j(\cdot)\}$, is the sequence of optimal feedback policies. The optimal policies are found by solving the Stochastic Dynamic Programming Equation (SDPE),

$$V_t(\xi_t) = \inf_{\pi_t(\cdot)} \mathbb{E}\left[c(x_t,\pi_t(\xi_t))+V_{t+1}(\xi_{t+1})\right], \tag{9}$$

commencing from the terminal value

$$V_N(\xi_n) \triangleq \mathbb{E}[c_N(\xi_N)].$$

The expectations, $\mathbb{E}$, in these expressions are over the corresponding ξ_j densities and the future w_j and v_{j+1} densities.

An important and inherent feature of optimal control is that the computed value functions, $V_t(\xi_t)$ and particularly $V_0(\xi_0)$, inform us of the optimal controlled performance. Once approximations are introduced and optimality foregone, then the relationship between computed values and optimal values is compromised, although it is possible in some cases to relate achieved and computed values to optimal using monotonicity ideas [3].

2.3 Duality and the Source of Intractability

We note from time update (6) that the future information states depend explicitly on the control signals. This is captured in the SDPE calculation at time t through the appearance of ξ_{t+1} in the second term in (9). Further, since the SDPE is solved backwards in time from ξ_N, each calculation needs to propagate the dependence of this terminal information state on the intervening controls. In practice and if possible, the SDPE is solved backwards in time to yield, successively, value function $V_N(\cdot)$, then optimal feedback policy $\pi_{N-1}(\cdot)$, then value function V_{N-1}, optimal feedback policy $\pi_{N-2}(\cdot)$, etc. While, as in the deterministic case, the (famously cursed) dimensionality of the solution explodes with horizon N, the necessity of carrying forward the computation of the future information states adds a crippling burden, even in the simplest of optimal control problems.

An example in [15], and briefly reprised in Section 5.2 below, studies approximately optimal transmission power control in cellular mobile wireless communications. Four power values are considered with a stationary additive white gaussian noise channel, whose fade value may take one of four values. The known gaussian noise densities are sampled at twenty points and a horizon of $N = 5$ is taken. This approximate calculation of optimal controls, occupying perhaps a millisecond in real time, takes over thirty minutes on a high-performance desktop computer even at this level of coarseness.

3 Stochastic MPC and Deterministic MPC

In our variant of stochastic MPC, a horizon-N optimal control problem (7) is solved at time t from information state ξ_t. The solution of this horizon-N problem is, from (8),

$$u_t^{\mathrm{MPC}} = \pi_0(\xi_t). \tag{10}$$

As with deterministic MPC, this control (10) is applied, measurements taken, then the information state is updated to ξ_{t+1}, before the finite-horizon problem is resolved for $u_{t+1}^{\mathrm{MPC}} = \pi_0(\xi_{t+1})$. A bound on the infinite-horizon performance (with a discount factor) is established in [29] for this control in receding horizon relative to stochastic infinite-horizon-optimal control. Needless to say, this stochastic MPC preserves the dual aspects of the optimal control policy $\pi_0(\cdot)$ provided the horizon exceeds one. It also inherits the general computational intractability, even for modest horizons. Although some avenues to amelioration are explored in Section 6.

We highlight a central departure of stochastic optimal control from its deterministic counterpart. The solution generated via the SDPE (9) and the Bayesian filter (5)–(6) is either a sequence of optimal control policies $\{\pi_j(\cdot)\}$, $j = 0,\dots,N-1$, from (8) (at best) or the current value of the control signal $u_t = \pi_0(\xi_t)$ (at least). Since the Bayesian filter update depends explicitly on the measured output y_t, it is not possible to produce an a priori sequence of predicted information states, say $\{\xi_{t+j|t}\}$, for states more than one step ahead. By the same token, one cannot construct a sequence of future controls, say $\{u_{t+j|t}\}$, which might form part of a feasible but unused tail control sequence. In this fashion, stochastic optimal control and its receding horizon MPC variant deviate from the deterministic version. This complicates the establishment of recursive feasibility and asymptotic stability following from approaches such as those pioneered in [18]. Paraphrasing this paragraph, stochastic optimal control is inherently closed-loop and open-loop optimal control does not exhibit duality.

Linear quadratic Gaussian optimal control, and hence LQG-based MPC formulations such as [4], obey the separation principle; the optimal controller combines the optimal full-state feedback control with the state replaced by its least-squares optimal estimate based on input–output measurements. That is,

$$u_t^{\text{LQG}} = -K_t^{\text{LQ}} \hat{x}_t^{\text{LS}}.$$

Building on this approach, *certainty equivalent* MPC uses a state estimate in place of the actual state in the MPC solution. In linear quadratic problems, the conditional mean state estimate appears in the optimal output feedback control. Further, the quality (covariance) of this estimate is unaffected by the control value itself and so duality and excitation are unnecessary and absent. Although the closed-loop performance does depend on the covariance value.

Duality in stochastic optimal control is not an optional add-on but is inherent in the optimal solution and, true to appearances, is antagonistic to regulation performance, were the precise state known. One might be inclined to make an attempted end-run around the cost of duality with stabilization to the origin and argue that, since the controlled state should be small, there is no real need to probe it. But this belies the nature of the problem. If we truly know that the state is close to zero, then the information state will reflect this and the optimal control will adjust the excitation accordingly, i.e. not by much. However, if the quality of knowledge that the state is actually zero is poor, then excitation is needed to, sequentially and optimally, refine the state density and then apply the appropriate regulation action, which might indeed be close to zero. Of course, if one does not really care about the state provided that it is close to zero, one should alter the optimization criterion, not the solution.

4 Stochastic Reconstructibility and Its Dependence on Control

The role of probing and duality in stochastic optimal control lies in the dependence of the quality of the current state estimate on the choice of control signal. In linear systems, the archetypal such quality measure is the state estimate covariance and linearity dictates that there is no effect of the control on this covariance. Indeed, for linear systems the control signal's effect on the output is simply exactly computed and serves solely to translate the density of the output signal. By contrast for non-linear systems, there can be a strong dependence of estimate quality on the control signal. Further, there is no single measure of estimate quality such as conditional covariance, which might be easily attached to the information state arising from a control policy. Here we present two notions of estimate quality which will then be analyzed with regard to stochastic MPC.

4.1 Linear Regression and the Cramér-Rao Lower Bound

Consider the estimation of the fixed parameter vector $\theta_t \in \mathbb{R}^k$

$$\theta_{t+1} = \theta_t, \tag{11}$$

in the linear regression model with Gaussian noise,

$$y_t = \phi_t^T \theta_t + v_t, \; t = 0, 1, \ldots, M-1. \tag{12}$$

Here the k-vector regressor sequence $\{\phi_t\}$ is known and Gaussian noise $\{v_t\} \sim \mathcal{N}(0, \sigma^2 I)$ with σ^2 known. Rewrite this as M rows

$$Y = \Phi^T \theta + V.$$

Then we have the following characterization of identifiability, i.e. the uniqueness of the Maximum Likelihood estimate, and its corresponding goodness of fit.

Theorem 1 ([16]). *The Maximum Likelihood estimate,* $\hat{\theta}_M$, *of* θ *is*

$$\hat{\theta}_M = (\Phi\Phi^T)^{-1}\Phi Y = \left(\sum_{t=0}^{M-1} \phi_t \phi_t^T\right)^{-1} \sum_{t=0}^{M-1} \phi_t y_t,$$

and this achieves the Cramér-Rao lower bound on covariance

$$\mathbb{E}\left[(\theta - \hat{\theta}_M)(\theta - \hat{\theta}_M)^T\right] = \sigma^2 \mathscr{F}^{-1},$$

where

$$\mathscr{F} = \sum_{t=0}^{M-1} \phi_t \phi_t^T, \tag{13}$$

is the Fisher Information Matrix associated with this linear regression.

Regarding (11)–(12) as the state and measurement equations, this is a state estimation problem for constant state θ. One interprets the invertibility and conditioning of the Fisher Information Matrix, $\mathscr{F}$, of the regressor sequence, $\{\phi_t\}$, as central to the estimation quality via the Cramér-Rao bound. This is familiar from system identification [11] in terms of experiment design for parameter estimation, which is the context in which Fisher and others proposed these measures. Anderson and Johnson [2] extend these ideas from identification to adaptive control and illustrate that persistence of excitation of the input signal carries over to excitation of the regressor vector sequence subject to controllability conditions. This idea is taken further in the behavioral setting of linear systems by Willems *et al.* [38].

Theorem 2 ([38]). *Consider a system of McMillan degree n with input signal* $u_t \in \mathbb{R}^m$ *and output signal,* $y_t \in \mathbb{R}^p$, *for times* $t = 0, 1, \dots, M-1$. *Define the order-k regression vector*

$$\phi_t^k = \left[u_{t-1}^T \; u_{t-1}^T \; \cdots \; u_{t-k}^T \; y_{t-1}^T \; \cdots \; y_{t-k}^T\right]^T,$$

and corresponding order-ℓ input-only regression vector

$$\mathscr{U}_t^\ell = \left[u_{t-1}^T \; u_{t-2}^T \; \cdots \; u_{t-\ell}^T\right]^T.$$

Provided the parameter vector satisfies $\dim(\theta) = 2k \leq 2n$ *and the number of data samples* $M \geq 4k-1$,

$$\rho_1 I_{2k} > \sum_{j=2k}^{M} \mathscr{U}_j^{2k} \mathscr{U}_j^{2k^T} > \rho_2 I_{2k} > 0 \Rightarrow \rho_3 I_{2k} > \sum_{\ell=k}^{M} \phi_\ell^k \phi_\ell^{k^T} > \rho_4 I_{2k} > 0. \tag{14}$$

That is, Theorem 1 establishes the dependence of the parameter (state) estimate's variance on the input/output signal properties in regression problems via the Fisher Information Matrix $\mathscr{F}$. Theorem 2 shows that these excitation requirements can be transferred to the input/control signal alone. Further analysis is provided in [12, 14]. The province of [2] is adaptive control, where the parameters are continuously estimated – so the parameters' evolution is described by (11) driven by additive white noise (as in (20) below) – and rely on control excitation per (14) uniformly over time t. This is a nonlinear problem, since the regressor and the parameter vector, each of which is a part of the system state, are multiplied in (12). In the context of stochastic MPC with unknown and varying parameter θ_t, Genceli and Nikolau [10] and Marafioti *et al.* [24] present approaches to maintaining the excitation of the state-based regressor vectors in an MPC problem using constraints on the control signal similar to (14). This solution is imposed exogenously and is not part of the dual solution. The resultant *persistently exciting MPC* is a state-feedback with memory [24]. This is discussed in Section 5.3 shortly.

4.2 Conditional Entropy Measure of Reconstructibility

Reconstructibility is concerned with the capability to calculate precisely the current state value x_t from the current data Z^t defined in (3). Naturally, such precision is not feasible for a stochastic system such as (1). In [23], an alternative notion of stochastic reconstructibility is developed based on comparison between conditioning on the full input–output data, Z^t, and conditioning solely on the input data

$$U^t = \{u_0, \ldots, u_{t-1}\}, \quad U^{-1} = \emptyset. \tag{15}$$

Such an input-only state estimate might be generated by simulation of the system dynamics (1). Reconstructibility is linked to the state estimate quality improvement of $\xi_t = \text{pdf}(x_t|Z^t)$ versus $\varphi_t = \text{pdf}(x_t|U^t)$, which in turn attempts to quantify the benefit of including the output measurements with the input signals.

Effecting this comparison requires having a scalar measure of estimate quality. The authors of [23] use conditional entropy. The entropy of a random q-vector θ with $P(\theta) = \text{pdf}(\theta)$ is defined [6] as

$$H(\theta) = -\int_{\mathbb{R}^q} \ln P(\theta)\, dP(\theta).$$

Equations (1)–(2) may be used to propagate the joint, marginal and conditional densities: $P(x_t)$, $P(Z^t)$, $P(U^t)$, $P(x_t, Z^t)$, $P(x_t, U^t)$. From here, we define the conditional entropies

$$H(x_t|Z^t) = H(x_t, Z^t) - H(Z^t), \tag{16}$$

$$H(x_t|U^t) = H(x_t, U^t) - H(U^t). \tag{17}$$

Definition 1 ([23]). Stochastic system (1–2) with initial state density $\xi_{0|-1}$ is reconstructible if for any scalar measurable function $g(\cdot) : \mathbb{R}^n \to \mathbb{R}$, either $H(g(x_t)|U^t) = 0$ or $H(g(x_t)|Z^t) < H(g(x_t)|U^t)$.

Zero entropy events are deterministic. The separate consideration of zero entropy events is due to the inability to reduce further their entropy.

The difference in conditional entropies in Definition 1 serves as a measure of reconstructibility. For Gaussian systems, we have the following result for entropy.

Lemma 1 ([6]). *For Gaussian random n-vector $x \sim \mathcal{N}(\bar{x}, \Sigma)$,*

$$H(x) = \frac{1}{2} \ln\left((2\pi e)^n \det(\boldsymbol{\Sigma})\right).$$

When applied to linear systems driven by Gaussian noise, we arrive at the connection below, which result builds on the Kalman filter as the propagator of the (Gaussian) conditional state density.

Lemma 2 ([23]). *The linear system driven by Gaussian noises w_t, v_t possessing full-rank covariances and input u_t,*

$$x_{t+1} = Fx_t + Gu_t + w_t,$$

$$y_t = Hx_t + v_t,$$

is reconstructible according to Definition 1 if and only if

$$Ker\, F^n \supset Ker\, \mathcal{O},$$

where n is the dimension of x_t and $\mathcal{O}$ is the observability matrix of the pair $[F, H]$.

Evidently, the kernel test in Lemma 2 is the standard reconstructibility test for linear systems. Further, it depends in no way on the properties of the input sequence $\{u_t\}$, which is in contrast to the general nonlinear case.

In Section 5.1 below, we use this conditional entropy measure of reconstructibility to quantify the state estimation improvement achieved by specific control laws.

5 Three Examples of Dualized Stochastic Control

With these concepts of state reconstructibility, we now move on to consider three examples of "dualized" stochastic control. That is, artificially imposed control signal excitation injected to improve on-line state estimation even though at the expense of regulation performance. As the truly optimal dual solution is computationally intractable, these approaches are suboptimal and are included to provide examples of viable solutions to stochastic optimal control problems. There are many practical examples of such dualized control [1, 8, 9, 17] but at least the first two of the three examples below are revealing in their being implemented realized engineered solutions to stochastic control problems in which the state value must be estimated in order to achieve any control performance and the process of ensuring an adequate estimate runs counter to the regulation objective. That is, in Robotics parlance, a compromise between exploration and exploitation is selected. While evidently suboptimal, since the optimal solution is intractable, each of these approaches focuses on an achievable simplified compromise.

5.1 Internet Congestion Control in TCP/IP

Transmission Control Protocol/Internet Protocol (TCP/IP) is the familiar end-to-end network data communication standard. It operates with data packets sent into the network by the source computer addressed to the destination computer. Since the protocol is end-to-end, it does not rely on internal network signals from relay nodes and, accordingly, must manage network congestion without this information. Upon packet arrival at the destination, it responds to the source with a small acknowledgement (ACK) data packet indicating successful arrival of that specific packet. This latter packet is unacknowledged by the source.

Within the network, data packets are routed from node to node. Data arriving at a node is placed into a buffer (memory) and then released to a downstream node depending on the available link capacity. Congestion occurs when the capacity fails to accommodate the arriving data and the buffer overflows. Depending on the node logic, the arriving data can be dropped when the buffer overflows (Drop Tail operation) or randomly deleted when the buffer begins to approach overflow (Random Early Detection operation). In each case, the packet dropping causes a failed communication and, hence, absence of ACK, which indicates network congestion.

Network congestion is managed at the source computer using the Additive Increase Multiplicative Decrease (AIMD) window management control rule. Upon receipt of the ACK for a packet, the source increases the packet size by one and then transmits again. As ACKs arrive, the packet length increases linearly with the number of successfully transmitted packets. When an ACK fails to arrive, i.e. times out or arrives out of sequence, then the source responds by halving the packet size before retransmitting the presumed lost packet data. The packet send rate is illustrated in Figure 1 and exhibits the standard sawtooth pattern as the congested levels of transmission are broached [22]. Since the downstream capacity at each node is itself subject to stochastic packet arrivals, the capacity changes over time, which is depicted by the sharp drops in data send rates with AIMD.

From a control perspective, the additive increase of AIMD is destined to cause congestion and, thereby, to precipitate the halving of transmission rate with its attendant effect on overall data rates. This is immanent dual behavior – in order to facilitate control performance by revealing the capacity upper bound to transmission rate, it proves necessary to expose this rate and then suffer the control performance consequences. Yet, unless the capacity is revealed, it is not possible to transmit efficiently. In [23], the authors compute the *mutual information*

$$I(x;y) = H(x) - H(x|y),$$

for the initial downstream capacity limit given the received ACK sequence. They do this for the AIMD window control law and again for a constant transmission rate control law yielding

$$\begin{aligned} I_{\text{AIMD}}(c_0, \{\text{ACK}_k\}) &= H(c_0) - H(c_0|\{\text{ACK}_k\}) = 0.96951, \\ I_{\text{const}}(c_0, \{\text{ACK}_k\}) &= 0.52143. \end{aligned}$$

Evidently, AIMD significantly improves the reconstructibility of the bottleneck node capacity.

We do not imply that the presence of duality in the control law is evidence of optimality, only that this practical solution to capacity reconstructibility possesses this feature. Indeed, AIMD is devised as a congestion control approach rather than as a throughput maximizing technique. It is not immediately apparent what an optimization objective might be for such a problem.

5.2 Equalization in Cellular Wireless

In cellular mobile communications, transmission power control is critical for battery life and for interference management. The mobile station (MS) and the serving base station (BS) need to cooperate to manage transmission power in the face of the time-varying channel between MS and BS. The radio channel at cellular frequencies is well modeled by six-tap finite-impulse (FIR) response system to cap-

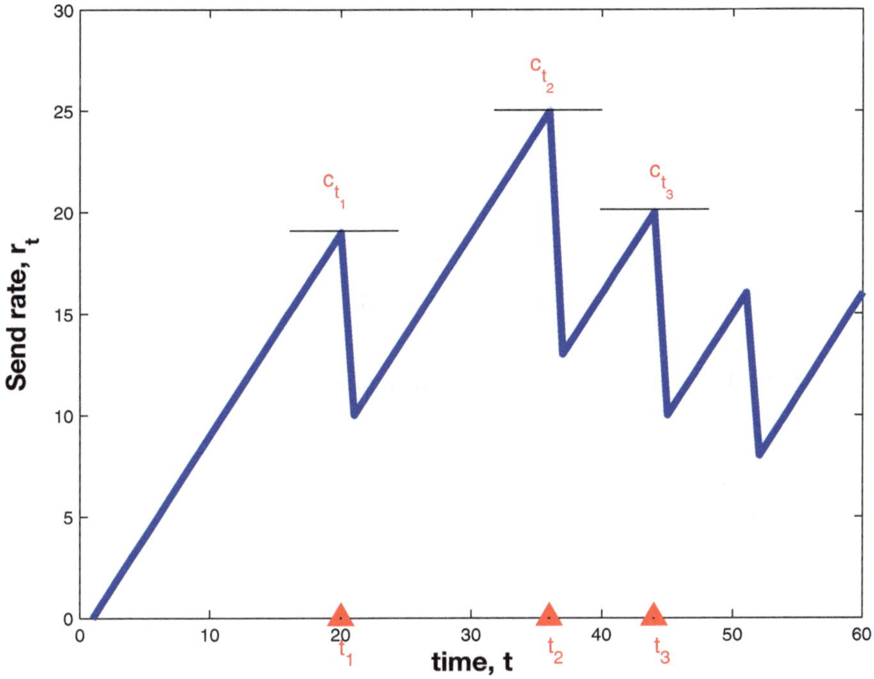

Fig. 1: Packet send rate for AIMD window control in TCP/IP. Duality is present through the enforced breaching of the capacity limits, $\{c_{t_j} : j = 1, 2, 3\}$, to make this performance-related state reconstructible.

ture both the signal fading properties and the inter-symbol interference introduced by reflections and path-length differences. Naturally, the two channels, MS-to-BS and BS-to-MS, which occupy different frequencies, need to be *equalized* to remove the inter-symbol interference and to maintain adequate signal-to-noise ratio at the receiver and thereby ensure reliable digital communication. The interference suppression objective tempers exceeding the reliable communication power.

To aid in fitting the FIR channel model, each transmitted packet contains a *midamble* training signal, which is known to both the transmitter and receiver and therefore is information-free. Figure 2 from Steele [35] depicts the presence of this training signal and quantifies the cost to the overall system data rate.

The training signal is a tangible artifact of duality; its presence diminishes the data rate yet its absence precludes power management and reliable communication without undue interference. The engineered cellular system clearly strikes a balance between these aspects without claims of optimality but demonstrated adequacy and robustness.

In the recent paper [15], Ha and Bitmead study a simplified variant of the cellular channel equalization problem in which solely the fade is present and an optimal power usage is sought. The transmission model is an additive white Gaussian noise

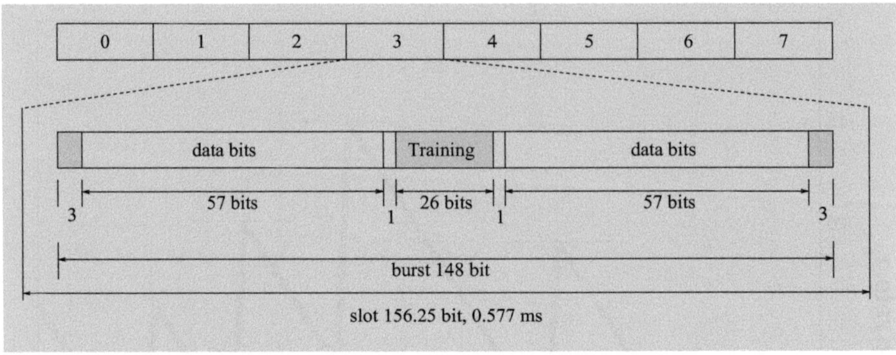

Fig. 2: GSM cellular packet data structure illustrating the 26-bit mid-amble training signal per 114 bits of message data (from [35]).

(AGWN) channel.

$$x_t = f u_{t-1}, \qquad x_0 = f,$$
$$y_t = x_t + v_t, \qquad t = 1,2,\ldots,N.$$

Parameter f is the constant but unknown channel fade. Input signal,

$$u_t = p_t a_t,$$

where a_t is a binary training signal symbol, known to MS and BS, and p_t is the square root of the transmission signal power known only to the transmitter. Channel noise, v_t, is zero-mean Gaussian white noise of known variance, σ_w^2. Signal y_t is the measured signal at the receiver. How should one choose the power sequence, $\{p_t : t = 1,\ldots,N\}$, to minimize the summed squared deviation from the optimal known f-value

$$p^{\star^2} = \frac{\gamma^\star \sigma_w^2}{f}, \tag{18}$$

along the horizon[1]? This now is a single-parameter nonlinear stochastic optimal control problem, which exhibits the beauty and horror of duality in its (approximate) solution. The information state of (4) is the conditional pdf of the fade f,

$$\xi_t = \text{pdf}(f|y^t).$$

Since the fade is presumed static, the time-update (6) has $\xi_{t+1|t} = \xi_t$, while the zero-mean, variance σ_w^2 Gaussian nature v_t yields

$$\text{pdf}(y_t|x_t) = \frac{1}{\sqrt{2\xi}\sigma_w} \exp\left(-\frac{1}{2\sigma_w^2}(y_t - x_t)^2\right),$$

[1] Here $\gamma^\star$ is the target signal-to-noise ratio for a given bit-error rate [21].

or,

$$\begin{aligned}\mathrm{pdf}(y_t|\hat{f}_t) &= \frac{1}{\sqrt{2\xi}\sigma_w}\exp\left(-\frac{1}{2\sigma_w^2}\left[v_t - u_{t-1}(f-\hat{f}_t)\right]^2\right). \\ &= \frac{1}{\sqrt{2\xi}\sigma_w}\exp\left(-\frac{1}{2\sigma_w^2}\left[v_t - p_{t-1}a_{t-1}(f-\hat{f}_t)\right]^2\right),\end{aligned} \tag{19}$$

where f is the true fade value. The role of the transmission power, p_{t-1}^2, becomes apparent in the measurement-update refinement of the density ξ_t via (5). Larger power causes the argument of the exponential in (19) to amplify the difference between the actual fade f and the argument of the conditional density, $\hat{f}_t$. That is, increased transmission power rapidly sharpens the conditional density of the fade value with each measurement y_t. This is depicted in Figure 3 below showing the passage from ξ_{t-1} to ξ_t for differing transmission power values.

Recall that the transmission power control problem is posed and solved at the transmitter and the selected power value is not communicated to the receiver, which separately computes and communicates the received signal-to-noise ratio. The information state ξ_t is computed at the receiver but forms part of the power control calculation at the transmitter via the SDPE (9). The overall aim is to achieve reliable communication, indicated by signal-to-noise ratio reaching or exceeding $\gamma^\star$ above, while minimizing the total message energy. The duality appears because correct estimation of fade f is required for reliability and depends on high-energy training with the objective of limiting the overall use of energy in the transmission. Here the selection of transmission power is the control signal which is optimized at the transmitter. The study in [15] sets up this nonlinear stochastic optimal control problem in a highly simplified situation of: four possible fade values, four corresponding signal power options, horizon-five optimization, and twenty-point approximation of Gaussian densities. The computational demand of this solution is prohibitive to its application in practice.

For real cellular communications systems, the duality is perhaps even more closely tied to the presence of the training signal itself. Clearly, were the channel fade and FIR model known precisely, the corresponding transmission power and equalizer would be clear. However, a significant fraction of available data-carrying capacity is devolved to the transmission of the information-free training signal to facilitate learning the channel model. In both the study of [15] and in the practical system, there is a crucial devotion of radio resources towards increasing the reconstructibility of the channel model state. This commitment both diminishes and enables reliable communication.

5.3 Experiment Design in Linear Regression for MPC

Section 4.1 introduced the idea of persistent input signal excitation and parameter identifiability via (14). Augmenting the system description (12) in Theorem 2 by

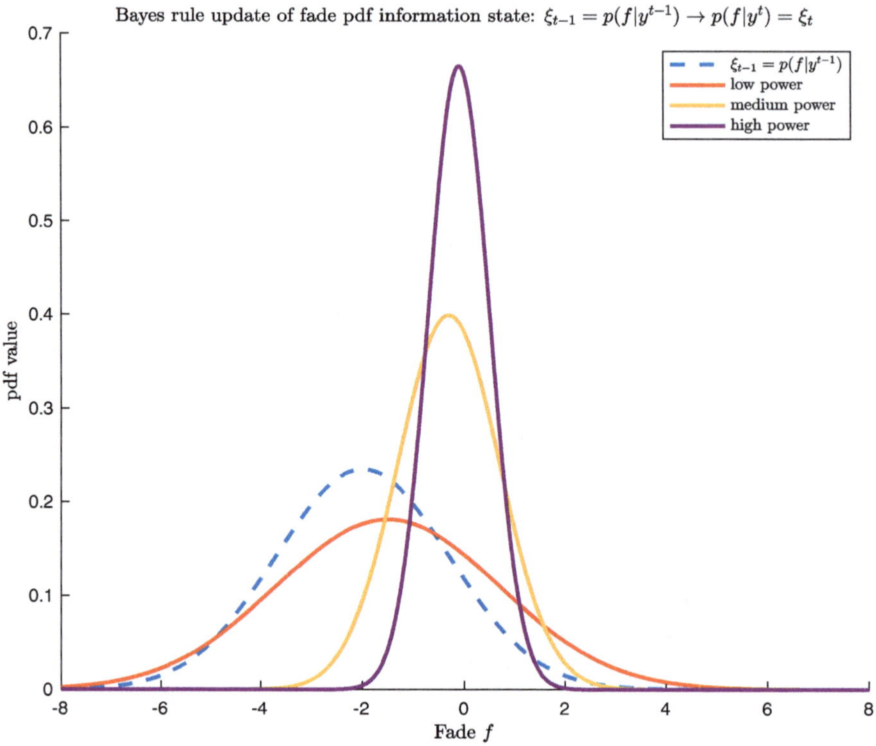

Fig. 3: A priori and a posteriori channel fade densities for differing transmission powers indicating the improvement in fade resolution for increased transmission energy.

the constant parameter state equation (11)

$$\theta_{t+1} = \theta_t + w_t, \tag{20}$$

it is immediate to equate persistent input excitation with uniform θ-state reconstructibility from the input–output data. Within the context of MPC, Marafioti *et al.* [24] study the inclusion of a control input persistent excitation condition (14) into the stationary constraint set associated with the determination of the receding-horizon optimal control at time t.

In stochastic MPC, a horizon-N optimal control problem (7) is solved at time t from information state ξ_t. Where the system state includes the parameter vector θ_t, ξ_t is the conditional joint state-parameter density. The solution of this horizon-N problem is given by (10),

$$u_t^{\text{MPC}} = \pi_0(\xi_t),$$

from (8). Since the raison d'être of MPC is the capacity for constraint handling, this optimal control problem will typically be a constrained optimization. To this set of constraints [24] adds an additional excitation constraint reflecting the left-hand side of (14) for $t \geq M-1$,

$$\rho_1 I_{2k} > \sum_{j=2k}^{M} \mathscr{U}^{2k}_{(t-M+1)+j} \mathscr{U}^{2k^T}_{(t-M+1)+j} > \rho_2 I_{2k} > 0, \tag{21}$$

where $2k$ is the dimension of the parameter vector, $\mathscr{U}^{\ell}_{j}$ is defined in Theorem 2 and M is now the excitation horizon. Where this additional constraint (21) is active, the resultant horizon-N MPC controller must necessarily be of diminished (finite-horizon N) computed performance $V_0(\xi_t)$ of (9) compared with the MPC controller without the excitation constraint. Thus, duality embodied by the excitation requirement negatively impacts performance.

Distinctions between stochastic MPC and deterministic MPC were explored in Section 3. Here we point to another intriguing feature of the persistent excitation requirement imbued by the receding horizon and dovetailing with the formulation of stochastic MPC discussed earlier. As with deterministic MPC, this stochastic MPC control (10) is: applied; measurements taken; information state updated to ξ_{t+1}; before the finite-horizon problem is re-solved for $u^{\text{MPC}}_{t+1} = \pi_0(\xi_{t+1})$. Along the MPC solution horizon $\{t+j : j = 0, \ldots, N-1\}$, constraint (21) applies solely to the first MPC control value u^{MPC}_{t} and asserts a constraint in terms of past control values $\{u^{\text{MPC}}_{t-M-2k+1}, \ldots, u^{\text{MPC}}_{t-1}\}$. It is pointless to apply an excitation condition along the future control horizon past the first value, since: it is only this first value which is computed in stochastic MPC; and, of the horizon-N solution, it is only this value which is actually applied. This results in a feedback controller with memory beyond just ξ_t, because the excitation constraint insists on carrying past control values. Formally, because of this requirement of the excitation constraint, one could augment the information state to include the precisely known prior control values and preserve the identity of the controller as separated.

6 Tractable Compromise Dualized Stochastic MPC Algorithms

The three examples of dualized stochastic control discussed above are provided to reinforce the requirement of control signal excitation in output feedback stochastic optimal control. The corresponding formulation of stochastic MPC in Section 3 illustrates the inheritance of computational intractability from stochastic optimal control in general for horizons beyond one. Recall that this intractability stems from the requirement to propagate the information state within the solution of the Stochastic Dynamic Programming Equation (9) to calculate the dual solution. If the excitation is otherwise managed, such as is done in the preceding Section 5.3, then more manageable formulations of MPC are possible at the expense of optimality. Indeed chapter "Stochastic Model Predictive Control" of this book explores a range of such approaches in stochastic MPC with an emphasis on linear systems, where duality requirements are diminished, relaxed, or more simply met via excitation.

We next explore in a preliminary fashion some tractable compromise MPC algorithms yielding a range of duality properties ranging from non-dual through to fully dual but approximate problems. The cost of moving away from the formal stochastic optimal control methodology is a disconnect from the optimal infinite-horizon performance. The following set of approaches is by no means complete, but serves our purpose of discussing duality in stochastic MPC before handing off to the more complete analysis of some of these algorithms in chapter "Stochastic Model Predictive Control". Most approaches considered in the literature fall into the following category of non-dual approaches to nonlinear stochastic MPC. We do note that performance estimates can always be computed for stochastic control solutions using simulation. What evanesces with approximations is the relationship to the optimal performance apart from providing a simulated upper bound to it.

6.1 Non-dual Approaches

6.1.1 Particle-Based Methods

One possible compromise algorithm inheriting the closed-loop structure of stochastic MPC relies on approximation of the information state, the conditional state density, via a particle filter [7, 34]. This approximation, which may be interpreted as a numerical implementation of the Bayesian filter, allows arbitrary degree of accuracy in terms of representing the information states at the cost of increasing the degree of computational complexity. The state conditional densities are propagated, not as complete functions on $\mathbb{R}^{n_x}$ as in the Bayesian filter, but as *particles* or samples from this conditional density. Samples of the state together with samples of the noise processes then can be iterated through the system dynamics to yield a sample of new state values as members of the predicted conditional density. The measurement update involves the resampling of this collection of samples using the Metropolis-Hastings Algorithm. The net result is that, at each stage of the Bayesian estimator, the state conditional density is described by a set of samples distributed according to that density. As the number of samples increases, so too does the accuracy in approximating the underlying densities. Needless to say, this procedure is amenable for moderate state dimensions only.

This numerically sampled information state can be combined naturally with the recent advances in Scenario MPC, described in chapter "Stochastic Model Predictive Control" Section 3, where stochastic state-feedback MPC problems are approximated by solving over a collection of sampled *scenarios*. In the output-feedback case considered here, these scenarios can naturally be initialized with the particles of the particle filter. Such approaches of combined particle propagation and scenario optimization have been explored in [5, 31]. The particle filter accounts for past data in its delivery of a set of samples distributed under the state conditional density, while the Scenario method then propagates randomly selected particles equipped with their own simulated realizations of the noise processes into the future.

Using combinations of particle filtering and simulation tools such as scenario methods allows tractable implementations of stochastic MPC on reasonable problem dimensions. However, while these approaches lead to computationally tractable solutions, they lose duality as a result of circumventing solution of the Stochastic Dynamic Programming Equation (9). Simulation of the process noise, as is performed when using scenario methods, results in evaluation of the MPC cost in an open-loop sense, without accounting for feedback through the measured output variables.

6.1.2 Certainty Equivalence Methods

The same non-dual, open-loop effects occur in certainty equivalent control, where the information state conditional density is replaced by a single dimension-n_x state estimate. This results in a very significant reduction in complexity of the problem and melds with linear quadratic Gaussian approaches to optimal output feedback control, where the solution separates into the optimal, conditional mean, state estimate coupled with the LQ-optimal linear state feedback gain. However, solving a full-state-feedback version of the Stochastic Dynamic Programming Equation (9) does not account for dependence of future information states on the control inputs and thereby does not allow for optimal probing through the control law. However, as mentioned previously in this chapter, both certainty equivalent control and the combined particle/scenario approach used to approximate stochastic MPC can be artificially augmented by imposing probing constraints in the optimization problem (see also Section 5.3). The resulting control signals will be of a probing nature, albeit not optimally so. This loss of optimality generally inhibits performance guarantees with respect to infinite-horizon stochastically optimally controlled closed-loop performance.

6.2 Dual Optimal POMDPs

An alternative approach to circumventing solution of the Stochastic Dynamic Programming Equation (9) with loss of duality is casting the stochastic MPC problem on a class of systems which permit computationally tractable solution of the stochastic control problem. One such class of systems is captured by Partially Observable Markov Decision Processes (POMDPs), which have been explored for use in stochastic MPC in [32, 33, 37]. POMDPs are output feedback control problems in which the system equation (1) preserves its Markov structure but operates over a finite state space. Likewise, the output process y_t of (2) also takes a finite set of possible values. That is, (1)–(2) describe a controlled Hidden Markov Model. The action space of control values determines the associated transition matrix of the Markov chain and it too is limited to be finite. The optimal control criterion function (7) is modified accordingly. Computationally, the exploration of the stochastic optimal output feedback control problem now takes place over a finite space, which if dimensions are managed well can be computed preserving finite-horizon optimality, and thereby duality, of the MPC solution.

There are two possible applications of this stochastic MPC on POMDPs. Firstly, the system dynamics may be naturally captured by a POMDP model, as is the case in a number of applications including decision problems in healthcare and robotics, where the finite state, observation, and action spaces associated with POMDPs are often natural to a given control problem. In this case, the resulting control inputs exhibit optimal probing and infinite-horizon performance results akin to those available for stochastic MPC. Secondly, one may approximate nonlinear dynamic models by POMDPs, resulting in a tradeoff between quality of the approximation and computational demand in solving the resulting stochastic MPC problem. In this latter case, optimal probing and infinite-horizon performance results hold with respect to the approximate POMDP dynamics. Varying the degree of approximation, e.g. by increasing the number of possible state values, enables approximation of the true dual optimal stochastic MPC law by the POMDP dual control, although it is not clear how the system approximation error extends to performance error bounds.

It is informative to mention the link between the particle-scenario approach and POMDP-MPC. In the former technique, the stochastic elements are managed by sampling, the propagation of samples, and the averaging of performance functions. In the POMDP approach, one might contemplate the state Markov chain as describing a fixed gridding of the state-space underpinning the approximate evolution of a continuous state. By extending this gridding into the control design, it is possible to include duality. In all of these approximate approaches, our experience is that the forward solution of the SDPE (9) – via gridding in POMDPs, sampling in scenarios, or full solution as in [15] – proves to be the bottleneck process in the solution compared with propagating the state conditional density or any of its approximants.

7 Conclusion

This chapter has broached the subject of duality in stochastic optimal control and, as a result, its presence in formulations of stochastic MPC based on stochastic optimal control in nonlinear systems. Our aim has been to highlight two central aspects: the reliance of optimality on probing to manage future information states along the horizon, and the attendant computational intractability. The benefit accrued is infinite-horizon stochastic MPC performance quantitatively comparable to truly stochastically optimal. We provide three examples of suboptimal control in which the probing requirements are necessarily included into the formulation of a workable engineering solution. We also briefly assess some approximations which address tractability. In chapter "Stochastic Model Predictive Control", further detailed formulation and analysis of stochastic MPC is explored, primarily in application to linear systems, where duality is frequently a non-issue or where adequate probing can be enforced simply through methods such as those described in

Section 5.3. These tractable approaches to stochastic MPC are very important and illuminating, since they handle practical uncertainty and robustness questions well. In a sense, the requirement to know densities and models so accurately as to be able to solve the stochastic optimal control problems is itself a show-stopper and approximate tractable methods are preferred. However, the practical examples of mobile wireless power control and of TCP/IP congestion control provide a touchstone for the importance of ensuring state reconstructibility through control and of the cost to performance due to excitation, which in turn is the price necessarily exacted for achieving any performance at all.

References

1. Allison, B., Ciarniello, J., Tessier, P., Dumont, G.: Dual adaptive control of chip refiner motor load. Automatica **31**(8), 1169–1184 (1995)
2. Anderson, B., Johnson, Jr., C.: Exponential convergence of adaptive identification and control algorithms. Automatica **18**(1), 1–13 (1982)
3. Bitmead, R., Gevers, M.: Riccati difference and differential equations: convergence, monotonicity and stability. In: Bittanti, S., Laub, A., Willems, J. (eds.) The Riccati Equation, Communications and Control Engineering, pp. 263–291. Springer, New York (1991)
4. Bitmead, R., Gevers, M., Wertz, V.: Adaptive Optimal Control: The Thinking Man's GPC. Series in Systems and Control Engineering. Prentice Hall, Englewood Cliffs (1990)
5. Blackmore, L., Ono, M., Bektassov, A., Williams, B.C.: A probabilistic particle-control approximation of chance-constrained stochastic predictive control. IEEE Trans. Robot. **26**(3), 502–517 (2010)
6. Cover, T., Thomas, J.: Elements of Information Theory. Wiley, New York (2006)
7. Doucet, A., de Freitas, J., Gordon, N.: Sequential Monte Carlo Methods in Practice. Springer, New York (2001)
8. Filatov, N., Unbehauen, H.: Adaptive Dual Control. Springer, New York (2004)
9. Filatov, N.M., Unbehauen, H.: Survey of adaptive dual control methods. IEE Proc. Control Theory Appl. **147**, 118–128 (2000)
10. Genceli, H., Nikolaou, M.: New approach to constrained predictive control with simultaneous model identification. AIChE J. **42**(10), 2857–2868 (1996)
11. Goodwin, G., Payne, R.: Dynamic System Identification: Experiment Design and Data Analysis. Academic, New York (1977)
12. Goodwin, C.G., Sin, K.S.: Adaptive Filtering Prediction and Control. Prentice-Hall, Englewood Cliffs (1984)
13. Goodwin, G.C., Kong, H., Mirzaeva, G., Seron, M.M.: Robust model predictive control: reflections and opportunities. J. Control Decis. **1**(2), 115–148 (2014)
14. Green, M., Moore, J.: Persistence of excitation in linear systems. Syst. Control Lett. **7**, 351–360 (1986)
15. Ha, M.H., Bitmead, R.: Optimal dual adaptive agile mobile wireless power control. Automatica **74**, 84–89 (2016)
16. Hayashi, F.: Econometrics. Princeton University Press, Princeton, NJ (2000)
17. Ismail, A., Dumont, G.: Dual adaptive control of paper coating. IEEE Trans. Control Syst. Technol. **11**(3), 289–309 (2003)
18. Keerthi, S., Gilbert, E.: Optimal, infinite horizon feedback laws for a general class of constrained discrete time systems: stability and moving horizon approximations. J. Optim. Theory Appl. **57**, 265–293 (1988)
19. Kouvaritakis, B., Cannon, M.: Model Predictive Control. Springer, Cham (2016)

20. Kumar, P.R., Varaiya, P.: Stochastic Systems: Estimation, Identification, and Adaptive Control. Prentice-Hall, Englewood Cliffs (1986)
21. Lee, E.A., Messerschmitt, D.G.: Digital Communication, 2nd edn. Kluwer Academic, Boston (1990)
22. Li, Y., Leith, D., Shorten, R.: Experimental evaluation of TCP protocols for high-speed networks. IEEE-ACM Trans. Netw. **15**, 1109–1122 (2007)
23. Liu, A., Bitmead, R.: Stochastic observability in network state estimation and control. Automatica **47**, 65–78 (2011)
24. Marafioti, G., Bitmead, R., Hovd, M.: Persistently exciting model predictive control. Int. J. Adapt. Control Signal Process. **28**, 536–552 (2014)
25. Mayne, D.Q.: Model predictive control: recent developments and future promise. Automatica **50**(12), 2967–2986 (2014)
26. Mayne, D.Q., Raković, S.V., Findeisen, R., Allgöwer, F.: Robust output feedback model predictive control of constrained linear systems: time varying case. Automatica **45**(9), 2082–2087 (2009)
27. Mesbah, A.: Stochastic model predictive control: an overview and perspectives for future research. IEEE Control System **36**(6), 30–44 (2016)
28. Schwarm, A.T., Nikolaou, M.: Chance-constrained model predictive control. AIChE J. **45**(8), 1743–1752 (1999)
29. Sehr, M., Bitmead, R.: Stochastic output feedback model predictive control. Automatica **94**, 315–323 (2018)
30. Sehr, M.A., Bitmead, R.R.: Sumptus cohiberi: the cost of constraints in MPC with state estimates. In: American Control Conference, pp. 901–906, Boston (2016)
31. Sehr, M.A., Bitmead, R.R.: Particle model predictive control: tractable stochastic nonlinear output-feedback MPC. In: Proceedings of IFAC World Congress, Toulouse (2017)
32. Sehr, M.A., Bitmead, R.R.: Tractable dual optimal stochastic model predictive control: an example in healthcare. In: Proceedings 1st IEEE Conference on Control Technology and Applications, Kohala Coast (2017)
33. Sehr, M.A., Bitmead, R.R.: Performance of model predictive control of POMDPs. In: Proceedings of 17th European Control Conference, Limassol Cyprus (2018)
34. Simon, D.: Optimal State Estimation: Kalman, H_∞, and Nonlinear Approaches. Wiley, Hoboken (2006)
35. Steele, R.: Mobile Radio Communications. IEEE Press, Piscataway (1992)
36. Sui, D., Feng, L., Hovd, M.: Robust output feedback model predictive control for linear systems via moving horizon estimation. In: American Control Conference, pp. 453–458, Seattle (2008)
37. Sunberg, Z., Chakravorty, S., Erwin, R.S.: Information space receding horizon control. IEEE Trans. Cybern. **43**(6), 2255–2260 (2013)
38. Willems, J., Rapisarda, P., Markovsky, I., De Moor, B.: A note on persistency of excitation. Syst. Control Lett. **54**(4), 325–329 (2005)
39. Yan, J., Bitmead, R.R.: Incorporating state estimation into model predictive control and its application to network traffic control. Automatica **41**(4), 595–604 (2005)

Economic Model Predictive Control: Some Design Tools and Analysis Techniques

David Angeli and Matthias A. Müller

1 Model-Based Control and Optimization

Designing a controller for a complex system is a process that entails multiple steps and decisions. Normally performance requirements, combined with economical and technological considerations, inform the selection of sensors and actuators. Even when such selection has been completed, and the scope of available control authority as well as information about the system's state have been identified, defining a detailed control strategy will normally entail a number of trade-offs and conflicting objectives. Given limited resources, possible disturbances and incomplete information, this is to be expected. In a realistic scenario, the task of settling such trade-offs may be daunting, unless appropriate mathematical tools are developed.

One of the most compelling techniques for approaching this problem is through model-based control. Rather than heuristically tuning knobs of a predefined control architecture, model-based control attempts, as a preliminary step, the formalization of a mathematical model underlying the evolution of all key variables involved in the system's dynamics. This may be done via first principles or, directly, by suitable identification techniques based on Input-Output data alone, or a combination of the two.

D. Angeli (✉)
Department of Electrical and Electronic Engineering, Imperial College London, Exhibition Road, SW7 2AZ London, UK

Dip. di Ingegneria dell'Informazione, Università di Firenze, Firenze, Italy
e-mail: d.angeli@imperial.ac.uk

M. A. Müller
Institute for Systems Theory and Automatic Control, University of Stuttgart, 70550 Stuttgart, Germany
e-mail: matthias.mueller@ist.uni-stuttgart.de

S. V. Raković, W. S. Levine (eds.), *Handbook of Model Predictive Control*,
Control Engineering, https://doi.org/10.1007/978-3-319-77489-3_7

A mathematical model allows one to translate the process of selecting and tuning a specific controller into an optimization problem. This is done provided performance requirements (often conflicting) can be quantified in terms of a single cost functional whose minimization is supposed to resolve such trade-offs in the best possible way. Designing cost functionals might itself be non-trivial as not all performance specifications have a clear "economic" interpretation. However, one can foresee a trial and error procedure in which cost functionals may be iteratively adapted and upgraded if the final control design is found to be underperforming in some respect. All other sorts of concerns, such as actuators and sensors limitations, safety and operational constraints, the allowed dynamics of the considered plant, maximal and minimal inflows and outflows, *etc.*, are instead represented as constraints of the optimization problem. The field of Optimal Control has developed in order to investigate mathematical tools for pursuing this type of approach. While physical insight might be useful in some respects, the mathematics is such that even an user with limited experience and understanding of the process to be controlled can find the optimal solution. Model Predictive Control, [36], on the other hand, has developed as an evolution of optimal control (introduced by practitioners) which is meant to allow a treatment of problems normally out of the range of classical optimal control applications, often due to the so-called "curse of dimensionality," [19]. Namely, every realistic problem, involving more than a few variables and constraints, can hardly be treated by means of the analytical set of tools available in optimal control. Model Predictive Control, in its simplest form, attempts to solve on-line a finite-horizon optimal control problem for a single given initial configuration, and deploys the optimal control action determined in this way according to a rolling-horizon strategy, viz. only implementing the first part of the optimal input sequence and reformulating, afterwards, a similar optimal control problem over a shifted time window.

When the process to be controlled is meant to spend most of the time in steady-state conditions, with all variables at equilibrium except for possible minor fluctuations due to process or sensor disturbances, it is tempting to separate profitability maximization (deciding at which point in state and input space it is best to operate the plant) from dynamical considerations, involving how to actually steer the plant's state towards the desired region and how to possibly stabilize it at some desirable equilibrium in the presence of disturbances that might lead it to drift away.

This hierarchical approach has, historically, developed in the control of chemical plants. Such systems, while often nonlinear and subject to hard constraints of different nature, have relatively slow process dynamics, so that one could afford to compute some solution to an optimal control problem within the inter-sample time interval, even with relatively modest computational power.

According to this paradigm, [25], a higher level device, called the Real Time Optimization (or RTO layer), is responsible for choosing set-points for all system's variables in order to maximize profit (in steady state) given the current operational constraints, market conditions (prices of raw materials, energy and products) or other factors which may affect profit. This is a static optimization problem in that the feasible set of considered input and states corresponds to equilibria which meet all operational constraints. It is in general a non-convex and non-linear (and for both reasons hard) optimization problem. Algorithms for its solutions might not have guaranteed convergence within any reasonable amount of time but, on the other hand, this is solved basically off-line, only when the operational constraints or, more likely, the market conditions have changed. Architectures with an even larger number of layers are often also envisioned, again as a result of tackling phenomena occurring at different time-scales.

The RTO forwards the computed set-points to an Advanced Control Layer, where an often linearized Model Predictive Control algorithm is responsible for solving on-line an optimal Tracking problem, namely, in the case of constant set-points, steering the systems' state towards the best possible equilibrium as quickly as possible. Notice that, during transient operations, when the Advanced Control Layer is acting and devising the appropriate control action, profitability concerns are no longer affecting the selection of the control variable. Indeed, while the predictive controller is still an optimization-based controller and operates on the basis of an underlying cost functional, the latter is devised in order to induce tracking towards the desired set-point, and need not bear any resemblance to the original profitability function maximized by the RTO.

This approach, which is widely adopted, has some important advantages:

- Computational: the hardest nonconvex programs are only solved off-line or on a much slower time-scale than Advanced Control Layer and do not involve dynamics, thus drastically reducing their size and complexity;
- Robustness: stability and robustness guarantees are easier to be tackled on a linearized model; moreover, nonlinear models, if not derived by first principles are often affected by uncertainty and may only be reliable near equilibrium or in small regions of state space so that envisioning a global optimization involving both the transient and the regime of operations may be unrealistic or even dangerous;

On the other hand, the hierarchical separation might not be ideal in situations where

- market conditions change frequently and on a time-scale which is comparable to the time-constants of the process dynamics;
- nonlinearity and non-convexity may result in complex optimal regimes of operations, viz. achievable only by keeping the system on a trajectory which is not an equilibrium state; in this respect, large profitability enhancements can sometimes occur, for specific models and within certain parameter ranges.

When the quality of the model at hand is deemed appropriate within a region much larger than the equilibrium manifold and if computational power is not a critical issue, it seems therefore appropriate to integrate these two layer into a single one that is responsible to optimize the dynamics (and not only the "static behavior") of the system on the ground of an optimal control problem deployed within a rolling-horizon approach and explicitly taking into account a profitability measure in its definition. This is the goal of Economic Model Predictive Control. While such an approach has a certain intuitive and practical appeal, it is only in recent years that a more systematic analysis of its implications regarding performance, stability and robustness (to name a few) has been attempted. This chapter will provide the reader a self-contained introduction to the main results in this relatively new area of research. Pointers to the relevant literature will be provided when space constraints do not allow an in depth discussion of the material presented.

2 Formulation of Economic Model Predictive Control

In its basic formulation Economic Model Predictive Control looks at deterministic plants governed by finite-dimensional difference equations of the following type:

$$x(t+1) = f(x(t), u(t)) \tag{1}$$

where $x(t) \in \mathbb{X} \subset \mathbb{R}^n$ is the state variable, and $u(t) \in \mathbb{U} \subset \mathbb{R}^m$ is the control variable. For the sake of simplicity, $\mathbb{X}$ and $\mathbb{U}$ are assumed to be compact sets. The function $f : \mathbb{X} \times \mathbb{U} \to \mathbb{X}$ is continuous, and ideally known without uncertainty.

In addition we assume that state and control variables be, in the light of operational and safety considerations, constrained to a certain compact set $\mathbb{Z} \subset \mathbb{X} \times \mathbb{U}$. We say that a solution $x(t)$ and corresponding control input $u(t)$ are feasible if the following is fulfilled:

$$(x(t), u(t)) \in \mathbb{Z} \qquad \forall t \in \mathbb{N}. \tag{2}$$

The goal of the control design is to maximize profit or, equivalently, minimize costs, both during transient and steady-state operation. The latter are quantified by means of a scalar valued function $\ell : \mathbb{X} \times \mathbb{U} \to \mathbb{R}$, with $\ell(x, u)$ representing the cost incurred for operating the plant at state x, subject to input u, throughout a sampling interval. In more general scenarios, of course, both f and ℓ might be time-dependent but, for the sake of simplicity, it is useful to consider situations in which costs and dynamics do not change significantly over the considered time window.

As customary in traditional MPC and Optimal Control, the stage cost is integrated over a (discrete) interval of length N, which is usually referred to as the prediction horizon. The rationale for this choice is to have a sufficiently long time window to assess (in a way which is not too short-sighted) the value of taking one particular course of action over another. Mathematically, the integrated cost is defined as below:

$$J_\ell(\mathbf{x}, \mathbf{u}) = \sum_{t=0}^{N-1} \ell(x(t), u(t)) + \psi_f(x(N)). \tag{3}$$

In the following we adopt the convention of denoting by bold fonts finite sequences of indexed variables: for instance, $\mathbf{x} := [x(0), x(1), \ldots, x(N)]$, $\mathbf{u} = [u(0), u(1), \ldots, u(N-1)]$ where the length of the sequence should be clear from the context. Notice that a final weighting function $\psi_f(\cdot)$ might or might not be present, depending on the considered formulation of EMPC. Its usefulness is intuitively understood considering that it could mitigate the effect of taking short-sighted actions by providing some bound to the best achievable cost incurred in operation over an infinite or very long horizon. More on this later, when the so-called terminal ingredients will be discussed in detail.

The main difference between tracking MPC and Economic MPC, at the definition level, is in the stage cost, $\ell(x,u)$. Typically, this is taken to be a positive definite quadratic form of state and input, in the former, while it may be an arbitrary continuous function in the latter case. For instance, $\ell(x,u) = x'Qx + u'Ru$ is a typical choice in tracking MPC. If a different equilibrium state-input pair is of interest, say (x_s, u_s), then a suitable expression is normally:

$$\ell(x,u) = (x - x_s)'Q(x - x_s) + (u - u_s)'R(u - u_s). \tag{4}$$

In other words, the stage-cost ℓ is designed in order to penalize deviations from an assigned set-point, rather than optimize the plant's profits. While non-quadratic and more general expressions could be used to such endeavor, the usual approach to induce convergence of the closed-loop system's trajectories towards the desired set-point is to take $\ell(x,u)$ positive definite with respect to the point (x_s, u_s). In other words:

$$0 = \ell(x_s, u_s) < \ell(x,u) \qquad \forall\, (x,u) \neq (x_s, u_s). \tag{5}$$

Inequality (5) need not hold for ℓ in Economic MPC set-ups, even if (x_s, u_s) is chosen to be the best feasible equilibrium, viz.

$$\ell(x_s, u_s) = \min_{\substack{(x,u) \in \mathbb{Z} \\ x = f(x,u)}} \ell(x,u). \tag{6}$$

This fact is emphasized in Figure 1.

We are now ready to formally define an Economic Model Predictive Control scheme. In particular, at each time t, and assuming exact knowledge of current state $x(t)$, the following optimization problem is solved:

$$\begin{array}{l} \min_{\mathbf{z},\mathbf{v}} J_\ell(\mathbf{z}, \mathbf{v}) \\ \text{subject to} \\ (z(k), v(k)) \in \mathbb{Z} \qquad k \in \{0, \ldots, N-1\} \\ z(k+1) = f(z(k), v(k)) \qquad k \in \{0, \ldots, N-1\} \\ z(N) \in \mathbb{X}_f \\ z(0) = x(t), \end{array} \tag{7}$$

where $\mathbb{X}_f$ is a compact set, whose properties will be later addressed when discussing the role of terminal ingredients in Economic MPC. Let $\mathbf{z}^\star, \mathbf{v}^\star$ denote any optimal

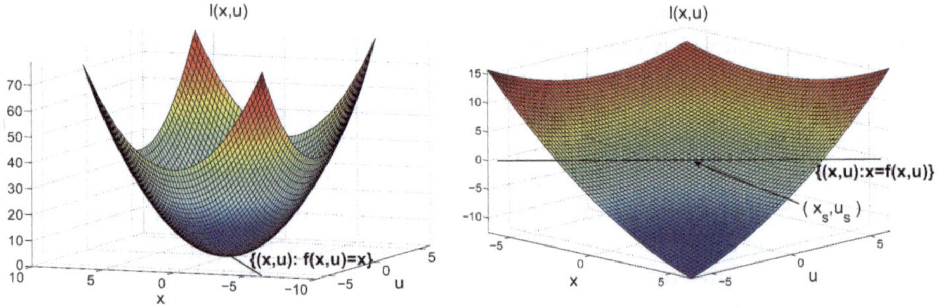

Fig. 1: Stage cost $\ell(x,u)$ for Tracking and Economic MPC

solution of problem (7) (this is non-unique in general), we define a state feedback by choosing the current control input $u(t)$ according to $u(t) = v^\star(0)$.

At the following sampling time the $\mathbf{v}^\star$ vector previously computed is discarded (though normally used as a warm start for the optimizer in a suitably shifted version of itself), and the problem is solved again from the resulting value of initial state $x(t+1)$. The iterative application of this procedure implicitly defines a state-feedback law,

$$u(t) = k(t,x(t)) := v^\star(0). \tag{8}$$

This is, in general, a time-varying, nonlinear, and possibly set-valued feedback. Time dependence, in particular, may arise as a result of time-varying terminal penalty functions or constraints, as well, of course, of time-varying costs and constraints which, for the sake of simplicity, we are not addressing.

Throughout this chapter, when we talk about the closed-loop system, we therefore understand the system evolving according to the following difference equation (or more precisely in the case of multiple optimal solutions, difference inclusion):

$$x(t+1) \in f(x(t),k(t,x(t))). \tag{9}$$

Given the current state $x(t)$, the feasible set $\mathscr{F}_t$ is defined as follows:

$$\mathscr{F}_t := \{x(t) \in \mathbb{X} : \exists\,(\mathbf{z},\mathbf{v}) \text{ fulfilling constraints in (7) }\}. \tag{10}$$

More sophisticated formulations of EMPC are possible to allow, for instance, average constraints (on infinite or finite windows) or take into account uncertainty in predictions. We will mention such variants in a later section.

3 Properties of Economic MPC

While solutions of the closed-loop system might not be uniquely defined, typically they share the same guaranteed stability, feasibility and performance properties. These are normally guaranteed through the adoption of suitable *terminal ingredients*. These are also commonly adopted in Tracking MPC and are usually terminal constraints (equality or set-membership) and terminal penalty functions. They have technical implications concerning several features of the closed-loop systems dynamics, such as *recursive feasibility*, *average asymptotic performance*, and *stability*. We discuss them separately throughout the following section.

3.1 Recursive Feasibility

Recursive feasibility is the property that for any initial condition $x(0) \in \mathscr{F}_0$ solutions of (9) fulfill for all $t \in \mathbb{N}$ $x(t) \in \mathscr{F}_t$. Indeed, due to the presence of hard constraints in (7), feasible solutions may fail to exist. The property of recursive feasibility then ensures that, provided the system's state is updated according to its nominal dynamics, feasibility is preserved at all times. This fact is normally shown by induction, by directly proving the following implication:

$$x(t) \in \mathscr{F}_t \Rightarrow x(t+1) = f(x(t), k(t, x(t))) \in \mathscr{F}_{t+1}. \tag{11}$$

There are several ways by which this can be guaranteed.

3.1.1 Terminal Equality Constraints

One way to ensure recursive feasibility is to constrain the final predicted state $z(t)$ to be equal to a predesigned feasible solution $x^\star(t)$. In particular, by endowing problem (7) with the following constraint:

$$z(N) = x^\star(t). \tag{12}$$

This can alternatively be formulated as $z(N) \in \mathbb{X}_f$ where $\mathbb{X}_f = \{x^\star(t)\}$. Notice that in such a case the penalty function ψ_f does not play any role in defining the optimal input sequence and may, without loss of generality, be taken as 0. A particularly simple choice of $x^\star(t)$ is when equilibrium solutions are considered. In particular one may take $x^\star(t) = x_s$, as defined in (6). This results in a time-invariant state-feedback law. Alternatively, another relevant situation arises when $x^\star(t)$ is periodic with some period T. In particular, then, $x^\star(t) = x^\star(t \mod T)$. The periodic solution can potentially be selected optimally, by solving the minimization problem shown below:

$$\begin{aligned}
&\min_{\mathbf{x},\mathbf{u}} \textstyle\sum_{t=0}^{T-1} \ell(x(t),u(t)) \\
&\text{subject to} \\
&x(1) = f(x(0),u(0)) \\
&\vdots \\
&x(T-1) = f(x(T-2),u(T-2)) \\
&x(0) = f(x(T-1),u(T-1)) \\
&(x(t),u(t)) \in \mathbb{Z} \qquad \forall t \in \{0,\ldots,T-1\}.
\end{aligned} \tag{13}$$

3.1.2 Terminal Set (or Terminal Inequality Constraint)

Another possibility is to enforce a terminal set-membership constraint

$$z(N) \in \mathbb{X}_f.$$

The closed set $\mathbb{X}_f$ is chosen to be a control invariant set. In particular, then:

$$\forall x \in \mathbb{X}_f,\ \exists u \in \mathbb{U} : (x,u) \in \mathbb{Z} \text{ and } f(x,u) \in \mathbb{X}_f.$$

Sometimes it may be useful to have an explicit expression for the control action u that keeps the state x in $\mathbb{X}_f$. We denote such a feedback policy by $\kappa_f(x)$. A special case of control invariant set is, for example, the equilibrium manifold. This choice corresponds to:

$$\mathbb{X}_f = \{x \in \mathbb{X} : \exists u : (x,u) \in \mathbb{Z} \text{ and } x = f(x,u)\}. \tag{14}$$

This was proposed in [24] in the context of tracking MPC and in [13] in the context of economic MPC and is normally referred to as *Generalized Terminal Equality Constraint*. The main advantage of such a generalized terminal equality constraint is that a possibly (much) larger feasible set is obtained than when using a fixed terminal equality constraint (or a region around some fixed equilibrium).

While, given the economic nature of the problem at hand, selection of $\mathbb{X}_f$ should seek to optimize the economic cost within $\mathbb{X}_f$ in some sense (for instance, on average when operating a viable solution), standard techniques for designing control invariant sets can be adopted as in classical MPC. For instance, the paper [1] provides simple linearization-based techniques for designing invariant ellipsoids (around preselected equilibrium states) and associated penalty functions. Other design techniques are of course possible, such as time-varying terminal sets, see [4]. Systematic numerical approaches to the design and selection of $\mathbb{X}_f$ in an economic and nonlinear context are, however, yet to be developed.

The basic idea behind most recursive feasibility proofs is that the optimal solution $\mathbf{z}_t^\star$, $\mathbf{v}_t^\star$ computed at time t can be used to generate a feasible solution at time $t+1$. In fact, in the absence of exogenous disturbances the following equality holds:

$$x(t+1) = f(x(t),k(t,x(t))) = f(z^\star(0),v^\star(0)) = z^\star(1)$$

In particular, then the sequence $[z^\star(1), \dots z^\star(N)]$, $[v^\star(1), \dots, v^\star(N-1)]$ may serve as the initial section of a solution feasible at time $t+1$. The last control move and terminal state can be obtained, in the case of terminal equality constraints simply by considering the sequences:

$$\tilde{\mathbf{z}} = [z^\star(1), \dots z^\star(N), x^\star(t+1)], \qquad \tilde{\mathbf{v}} = [v^\star(1), \dots, v^\star(N-1), u^\star(t)].$$

For the case of terminal set-membership constraints:

$$\tilde{\mathbf{z}} = [z^\star(1), \dots z^\star(N), f(z^\star(N), \kappa_f(z^\star(N)))], \qquad \tilde{\mathbf{v}} = [v^\star(1), \dots, v^\star(N-1), \kappa_f(z^\star(N))].$$

It is worth pointing out that the set of feasible initial conditions, $\mathscr{F}_0$, is, in general, dependent on both the terminal set and the length of the prediction horizon. We emphasize this dependence by denoting it $\mathscr{F}_0(N, \mathbb{X}_f)$. Having a larger feasible set at time 0 is of course of great practical interest. This can be achieved by considering the following monotonicity property:

$$N_1 \leq N_2 \text{ and } \mathbb{X}_f^1 \subseteq \mathbb{X}_f^2 \Rightarrow \mathscr{F}_0(N_1, \mathbb{X}_f^1) \subseteq \mathscr{F}_0(N_2, \mathbb{X}_f^2). \tag{15}$$

3.2 Asymptotic Average Cost

One way to assess the validity of an economic controller is to verify the long-run average cost incurred in closed-loop operation of the system. In particular, letting $x(t)$ be a solution of (9) and $u(t) \in k(t, x(t))$ the associated control input, we may define the average asymptotic cost as:

$$\bar{J} := \limsup_{T \to +\infty} \frac{\sum_{t=0}^{T-1} \ell(x(t), u(t))}{T}. \tag{16}$$

Many Economic MPC schemes come with a priori bounds on the possible values of $\bar{J}$ as a result of the chosen terminal ingredients.

3.2.1 Terminal Feasible Trajectory

Given a feasible trajectory $x^\star(t)$, $u^\star(t)$, that we adopt as a terminal constraint as in equation (12), we may let:

$$J^\star := \lim_{T \to +\infty} \frac{\sum_{t=0}^{T-1} \ell(x^\star(t), u^\star(t))}{T}. \tag{17}$$

Then, the following inequality holds:

$$\bar{J} \leq J^\star \tag{18}$$

as shown in Lemma 1 of [4]. In other words, the asymptotic average cost in closed-loop is never worse than the average cost of the feasible solution adopted as the terminal constraint. Both equality or strict inequality are possible, in fact. The simplest form of such bound is found when $x^\star(t)$ is a constant (i.e. the best feasible equilibrium) or periodic [3]. A general method to derive such bounds is based on feasibility at time $t+1$ of the shifted sequences $\tilde{\mathbf{z}}$ and $\tilde{\mathbf{v}}$. This property, as customary in traditional MPC, allows one to derive a useful dissipation inequality fulfilled by the cost-to-go function defined below

$$\begin{aligned} V_\ell(x) := \; & \min_{\mathbf{z},\mathbf{v}} J(\mathbf{z},\mathbf{v}) \\ & \text{subject to} \\ & (z(k),v(k)) \in \mathbb{Z} \qquad k \in \{0,\ldots,N-1\} \\ & z(k+1) = f(z(k),v(k)) \qquad k \in \{0,\ldots,N-1\} \\ & z(N) \in \mathbb{X}_f \\ & z(0) = x \end{aligned} \tag{19}$$

along solutions of the closed-loop system. In particular, exploiting suboptimality of the feasible shifted state and input sequences at time $t+1$, we see that

$$\begin{aligned} V_\ell(x(t+1)) \leq J_\ell(\tilde{\mathbf{z}},\tilde{\mathbf{v}}) &= J_\ell(\mathbf{z}^\star,\mathbf{v}^\star) - \ell(x(t),k(t,x(t))) + \ell(x^\star(t),u^\star(t)) \\ &= V_\ell(x(t)) - \ell(x(t),k(t,x(t))) + \ell(x^\star(t),u^\star(t)) \end{aligned} \tag{20}$$

Adding the previous inequality over a finite time interval implies:

$$\begin{aligned} V_\ell(x(T)) - V_\ell(x(0)) &= \sum_{t+0}^{T-1} V_\ell(x(t+1)) - V_\ell(x(t)) \\ &\leq \sum_{t=0}^{T-1} -\ell(x(t),k(t,x(t))) + \ell(x^\star(t),u^\star(t)). \end{aligned}$$

Dividing by T and taking lim infs letting T grow to infinity in both sides yields the desired inequality.

3.2.2 Terminal Penalty Function

When adopting a more general form of terminal constraint, such as the control-invariant set $\mathbb{X}_f$, a priori guaranteed bounds on the asymptotic closed loop average cost may still be possible. To this end, however, the terminal penalty function $\psi_f(\cdot)$ needs to fulfill a suitable inequality. This generalizes the idea of a control Lyapunov function, usually adopted for tracking MPC, to the context of Economic MPC. An appropriate condition, in the case of $\mathbb{X}_f$ being a control invariant neighborhood of an equilibrium (x_s,u_s) of interest, is the following:

$$\psi(f(x,\kappa_f(x))) \leq \psi(x) - \ell(x,\kappa_f(x)) + \ell(x_s,u_s), \tag{21}$$

a condition first proposed in [1]. In this case, suboptimality of the feasible shifted state and input sequences at time $t+1$ yields

$$\begin{aligned} V_\ell(x(t+1)) \le J_\ell(\tilde{\mathbf{z}},\tilde{\mathbf{v}}) &= J_\ell(\mathbf{z}^\star,\mathbf{v}^\star) - \ell(x(t),k(t,x(t))) + \ell(z^\star(N),\kappa_f(z^\star(N))) \\ &\quad + \psi_f(f(z^\star(N),\kappa_f(z^\star(N)))) - \psi_f(z^\star(N)) \\ &= V_\ell(x(t)) - \ell(x(t),k(t,x(t))) + \ell(z^\star(N),\kappa_f(z^\star(N))) \\ &\quad + \psi_f(f(z^\star(N),\kappa_f(z^\star(N)))) - \psi_f(z^\star(N)). \end{aligned} \tag{22}$$

Thanks to (21), the previous inequality can be further simplified into:

$$V_\ell(x(t+1)) \le V_\ell(x(t)) - \ell(x(t),k(t,x(t))) + \ell(x_s,u_s). \tag{23}$$

This is of the same form as (20) when the terminal feasible solution is constant. Hence, the following a priori performance bound can be guaranteed:

$$\bar{J} \le \ell(x_s,u_s). \tag{24}$$

3.2.3 Adaptive Terminal Weight

When Generalized Terminal Equality constraints are used, the following terminal weighting function is proposed:

$$\begin{aligned} \psi(x) := \; & \min_u \ell(x,u), \\ & \text{subject to} \\ & (x,u) \in \mathbb{Z} \\ & x = f(x,u) \end{aligned} \tag{25}$$

and $\psi_f(x)$ is then replaced by $\beta(t)\psi(x)$, where $\beta(t)$ is an adaptive coefficient, upgraded in order to tune the relative weight of the cost associated to the final equilibrium state. Under suitable upgrading rules for β, interesting asymptotic performance bounds are derived in [28]. In particular, the rationale behind the proposed update rules for the terminal weight is such that β is increased whenever the terminal steady-state of the optimal predicted trajectory is far away from the best reachable equilibrium (in N steps). To be more precise, define the best reachable steady-state cost (in N steps) from a given initial condition x as

$$\begin{aligned} \ell_{\min}(x) := \; & \min_{z,v} \ell(z,v), \\ & \text{subject to} \\ & (z,v) \in \mathbb{Z} \\ & z = f(z,v) \\ & x \in \mathscr{F}_0(N,\{z\}) \end{aligned}$$

A simple update rule can now be defined as

$$\beta(t+1) = \beta(t) + \alpha(\|\psi(z^\star(N)) - \ell_{\min}(x(t))\|). \tag{26}$$

This update rule ensures that $\beta \to \infty$ if the terminal predicted steady-state $z^\star(N)$ does not converge to the best reachable steady-state. This property (together with some technical conditions) can then be used to show that for the resulting closed-loop system, the terminal predicted steady-state cost $\psi(z^\star(N))$ can be upper bounded by the cost of the best steady-state which is *robustly* reachable from the ω-limit set of the closed loop [28]. Since as shown above in (18), the asymptotic average performance can be upper bounded by $J^\star$ as defined in (17), this result leads the conclusion that also the closed-loop asymptotic average performance can be upper bounded by the cost of the best steady-state which is *robustly* reachable from the ω-limit set of the closed loop. Besides the simple update rule (26), also more elaborate update rules have been proposed in [28], for which similar closed-loop performance statements can be derived. For example, one can allow resets of β (to some constant c) in order to avoid unnecessarily large terminal weights.

The above obtained closed-loop performance bounds for economic MPC with an adaptive terminal weight are rather of conceptual nature and not necessarily verifiable a priori, since they depend on the resulting ω-limit set of the closed loop. Improved bounds have been obtained in [29], for a setting where a generalized terminal region constraint instead of a generalized terminal equality constraint was used. This means that we use a terminal region around the whole equilibrium manifold instead of a terminal region around some (fixed) steady-state. In this case, it can be shown that if this generalized terminal region is designed properly, the terminal predicted steady-state cost $\psi(z^\star(N))$ (and hence also the closed-loop average performance) converges to a local minimum of the stage cost ℓ on the set of feasible steady-states. In case of linear systems with convex stage cost and constraints, convergence to the globally optimal steady-state cost can be shown, recovering the above result (24) when using a terminal equality constraint at the optimal equilibrium or a terminal region around this equilibrium.

3.3 Stability of Economic MPC

An advantage of designing terminal ingredients for tracking MPC is that the standard argument for recursive feasibility also allows a Lyapunov-based analysis of the controlled system's behavior. In fact, the cost to go function $V_\ell(x)$ is seen to fulfill a dissipation inequality along solutions of the closed-loop system and, under minor technical assumptions, can be used as a candidate Lyapunov function to carry out proofs of stability and convergence for the desired equilibrium state [22].

This is no longer the case for Economic Model Predictive Control. For instance, inequality (23) does not guarantee monotonicity with respect to time of $V_\ell(x(t))$, because it is not necessarily true that $\ell(x_s,u_s) \leq \ell(x(t),k(t,x(t)))$. In fact, the cost at the terminal equilibrium is not necessarily a global minimum of the stage-cost function, even if (x_s,u_s) might be the feasible state-input equilibrium pair of lowest cost.

As a result, even when initialized at x_s, the closed-loop system's solution might drift towards a different regime of operation: because of instability phenomena or because x_s is not an equilibrium in closed-loop. Bounds on asymptotic average cost, such as (24), tell us that when this happens it is in order to yield a regime of operation that is at least as economically rewarding as the best equilibrium. In general, however, stability of the underlying best equilibrium state x_s cannot be expected.

While convexity-based arguments were employed at first for the case of linear systems and convex cost functionals [37], only later Lyapunov-based insights were gained into stability of Economic MPC [9]. The major breakthrough in [9] was to introduce a rotated stage-cost function, defined as $\ell(x,u)+\bar{\lambda}^T[x-f(x-u)]$. Under suitable terminal equality constraints, it is seen that the extra-term in the cost-function does not affect the optimal solution of (7). A similar construction was later proposed in [3]. In particular, for any function $\lambda : \mathbb{X} \to \mathbb{R}$, one may define the associated rotated cost as:

$$L(x,u) = \ell(x,u) + \lambda(x) - \lambda(f(x,u)). \tag{27}$$

where the previous construction is thus a special case obtained for linear functions $\lambda(x) = \bar{\lambda}^T x$. Notice that, along any solution:

$$\sum_{t=0}^{N-1} L(x(t),u(t)) = \left(\sum_{t=0}^{N-1} \ell(x(t),u(t)) \right) + \lambda(x(0)) - \lambda(x(N)). \tag{28}$$

Hence, assuming, for instance, $z(N) = x^\star(t)$ is the terminal constraint, one may recognize that the optimal solution of (7) is not affected by having either ℓ or L as the stage-cost. This is of interest, for stability analysis, because choosing λ appropriately, the rotated stage cost might admit a global minimum at (x_s,u_s), namely:

$$L(x_s,u_s) = \min_{(x,u)\in\mathbb{Z}} L(x,u), \tag{29}$$

so that an Economic MPC scheme ends up being equivalent to a tracking MPC scheme with rotated stage cost. In fact, in such a case,

$$L(x_s,u_s) \le L(x,u) \qquad \forall (x,u) \in \mathbb{Z}. \tag{30}$$

Rearranging the different terms and exploiting the equation $x_s = f(x_s,u_s)$, the latter inequality reads:

$$\lambda(f(x,u)) - \lambda(x) \le \ell(x,u) - \ell(x_s,u_s), \tag{31}$$

and is therefore a *dissipativity* condition for system (1) with respect to the supply rate $s(x,u) = \ell(x,u) - \ell(x_s,u_s)$. For technical reasons, this condition needs to be slightly strengthened.

Definition 1. (Strict Dissipativity) System (1) is strictly dissipative with respect to the supply rate $s(x,u) := \ell(x,u) - \ell(x_s,u_s)$, if there exists a continuous function $\lambda : \mathbb{X} \to \mathbb{R}$ and a positive definite function ρ, such that:

$$\lambda(f(x,u)) - \lambda(x) \leq -\rho(|x-x_s|) + \ell(x,u) - \ell(x_s,u_s) \tag{32}$$

Notice that condition (32) implies that x_s is (strictly) the best feasible equilibrium and that the system is, in fact, suboptimally operated outside this equilibrium (in an asymptotic average sense), [2].

As anticipated, dissipativity plays a crucial role in establishing stability of Economic MPC. We sketch the basic argument below, for the case of a terminal equality constraint, $z(N) = x_s$. In fact, under strict dissipativity (32), we see that:

$$L(x_s,u_s) + \rho(|x-x_s|) \leq L(x,u), \qquad \forall (x,u) \in \mathbb{Z}. \tag{33}$$

Moreover, by virtue of (28), the optimal solution of

$$\begin{array}{l} \min_{\mathbf{z},\mathbf{v}} J_L(\mathbf{z},\mathbf{v}) \\ \text{subject to} \\ (z(k),v(k)) \in \mathbb{Z} \qquad k \in \{0,\dots,N-1\} \\ z(k+1) = f(z(k),v(k)) \qquad k \in \{0,\dots,N-1\} \\ z(N) = x_s \\ z(0) = x(t), \end{array} \tag{34}$$

where a rotated stage cost $L(x,u)$ is adopted, is the same as the one for (7). Letting $V_L(x)$ define the minimum of problem (34) and exploiting (33), we see that

$$V_L(x(t+1)) \leq V_L(x(t)) - L(x(t),k(t,x(t))) + L(x_s,u_s) \leq -\rho(|x(t)-x_s|). \tag{35}$$

Since $V_L(x) - V_L(x_s)$ is seen to be a positive definite function, condition (35) implies, by standard Lyapunov analysis, asymptotic stability of the equilibrium x_s within the feasibility region of problem (7).

When adopting a formulation with a terminal penalty function (rather than a terminal equality constraint), one may realize that:

$$\sum_{t=0}^{N-1} L(x(t),u(t)) + \Psi_f(x(N)) = \left(\sum_{t=0}^{N-1} \ell(x(t),u(t))\right) + \lambda(x(0)) + \Psi_f(x(N)) - \lambda(x(N))$$

$$= \left(\sum_{t=0}^{N-1} \ell(x(t),u(t))\right) + \psi_f(x(N)) + \lambda(x(0)).$$

provided the rotated terminal cost is defined as:

$$\Psi_f(x) = \psi_f(x) + \lambda(x). \tag{36}$$

The previous derivation shows that solution of (7) is not affected when L and Ψ_f replace ℓ and ψ_f, respectively, in the definition of the cost-functional. Thus, stability analysis can then be carried out under strict dissipativity, along the same lines as in the case of terminal equality constraints (see [1]).

Remarkably, dissipativity conditions are seen to be not only sufficient, but also necessary for optimal steady-state operation, in a suitable sense. Namely, as discussed above, under the dissipativity condition (31), the system is optimally operated at steady-state, and under the strict dissipativity condition (32) suboptimally operated outside the optimal equilibrium. This means that each other feasible state and input sequence pair yields a worse (strictly worse) asymptotic average performance than the optimal steady-state cost $\ell(x_s,u_s)$. One can show that the converse statement is also true under a certain controllability condition [33]. While the proof of sufficiency of dissipativity for optimal steady-state operation follows rather straightforwardly from the dissipation inequality and the definition of the latter property, the converse result is a bit more involved and can be shown by a contradiction argument. Namely, assuming that the system is not dissipative and using the controllability condition, one can construct a specific (periodic) state and input sequence which results in a lower average cost than the best equilibrium cost, contradicting optimal steady-state operation [33]. This converse result together with the above stability analysis allows the following interpretation. If steady-state operation is optimal, the system is dissipative with respect to the supply rate $s(x,u) := \ell(x,u) - \ell(x_s,u_s)$, which in turn (in its strict form) can be used to conclude that the closed loop converges to the optimal steady-state (x_s,u_s). This means that the closed loop "does the right thing", i.e., "finds" the optimal operating behavior.

It is worth mentioning that dissipativity conditions can not only be used to establish optimality of steady-state solutions over other more complex regimes of operation, but also exploited in order to find the best asymptotic average performance of the system. This may be defined as:

$$\begin{aligned} \ell^\star := \; & \inf_{x(.),u(\cdot)} \liminf_{T\to+\infty} \frac{\sum_{t=0}^{T-1} \ell(x(t),u(t))}{T} \\ & \text{subject to} \\ & (x(t),u(t)) \in \mathbb{Z} \\ & x(t+1) = f(x(t),u(t)) \end{aligned} \tag{37}$$

and happens to be related to dissipativity by the following equality:

$$\ell^\star := \sup\{\ell \in \mathbb{R} : \exists \lambda(\cdot) \text{ continuous } : \lambda(f(x,u)) \leq \lambda(x) + \ell(x,u) - \ell, \forall (x,u) \in \mathbb{Z}\}.$$

This equality was first derived for continuous time systems in [16] and later adapted to discrete-time systems in [21].

Finally, we remark that not only optimal steady-state operation can be characterized via a suitable dissipativity condition, but also the more general case where periodic operation is optimal [32, 40].

3.4 EMPC Without Terminal Ingredients

In economic MPC schemes without terminal ingredients, the terminal constraint $z(N) \in \mathbb{X}_f$ in (7) is omitted and the terminal cost in (3) is chosen to be zero. Using no terminal constraints may lead to algorithmic advantages (since the possibly restrictive terminal constraint is absent), and also the optimal steady-state, the optimal periodic orbit or some feasible terminal trajectory $x^\star(\cdot)$, $u^\star(\cdot)$ (depending on how the terminal region constraint is formulated, cf. Section 3.2) need not be known for implementing the economic MPC scheme. On the other hand, guaranteeing recursive feasibility is not as straightforward as in the case of suitably defined terminal ingredients, neither is the establishment of (a priori) closed-loop performance bounds that typically requires knowledge of the optimal operating behavior.

The main insight which is employed in economic MPC schemes without terminal constraints is the so-called turnpike property [11, 41]. This property means that open-loop optimal trajectories $\mathbf{z}^\star$ resulting from application of the optimal solution $\mathbf{v}^\star$ to Problem (7) spend most of the time in a neighborhood $\mathcal{N}$ of the optimal operating behavior (e.g., a neighborhood of the optimal steady-state x_s if steady-state operation is optimal, etc.). Importantly, the number of time instants where the optimal trajectory is outside of this neighborhood depends on the size of $\mathcal{N}$, but is independent of the prediction horizon N. In [17], it was shown that the same strict dissipativity condition as employed in Section 3.3, i.e., strict dissipativity with respect to the supply rate $s(x,u) = \ell(x,u) - \ell(x_s,u_s)$, together with suitable controllability conditions is sufficient for the turnpike property at the optimal steady-state x_s (compare also [15] for a continuous-time version of this result). In fact, also converse statements showing necessity of strict dissipativity for the turnpike behavior have recently been obtained [18]. The turnpike property at x_s can now be used to conclude practical asymptotic stability of the optimal steady-state for the resulting closed-loop system, using again V_L as a (practical) Lyapunov function, i.e., the optimal value function of the MPC problem using the rotated stage cost L, see [20] (compare also [14] for a continuous-time version of this result). In particular, it was shown that the size of the neighborhood of the optimal steady-state x_s into which the closed loop converges depends on the prediction horizon N and decreases to zero as $N \to \infty$. Interestingly, a candidate input sequence $\tilde{\mathbf{v}}$ for the next time instant $t+1$ is not necessarily constructed by appending some control value at the end (as is the case when using terminal constraints, see Section 3.1), but at some point where the optimal predicted trajectory is close to the optimal steady-state x_s as guaranteed by the turnpike property. A generalization of these results to the case where the optimal operating behavior is some general periodic orbit (instead of a steady-state) has been presented in [27].

4 EMPC with Constraints on Average

While classical Model Predictive Control always affords solutions that converge, at least nominally, towards the desired reference signal, Economic Model Predictive Control is not always stabilizing and may result in closed-loop behaviors that do not converge towards the best equilibrium or towards the underlying feasible solution adopted as a terminal ingredient. Under such circumstances it makes sense to want to guarantee, for the closed-loop behavior, specific additional constraints to be fulfilled on average, rather than pointwise in time. A typical example could be the outflow of a plant, which rather than constraining to be always greater or equal than a given flow rate, might be required to fulfill a similar inequality only in an average sense. Relaxing constraints in this way, if deemed suitable from an operational point of view, can in fact improve profitability margin of a plant simply because we are carrying out an optimization over a larger set of feasible solutions. Most EMPC control schemes can be endowed with constraints on average quantities. In particular, we may define the asymptotic average of a signal v as the following set:

$$\mathrm{Av}[v] = \left\{ \bar{v} : \exists \{T_n\}_{n=1}^{+\infty} : \lim_{n \to +\infty} T_n = +\infty \text{ and } \bar{v} = \lim_{n \to +\infty} \frac{\sum_{k=0}^{T_n - 1} v(k)}{T_n} \right\}. \tag{38}$$

For most signals v of interest, $\mathrm{Av}[v]$ is actually a singleton, corresponding to the asymptotic average of the signal. However, in general, a signal might have more than one asymptotic average when it keeps spending longer and longer time periods close to several values. Economic MPC with constraints on average allows to define an auxiliary output variable:

$$y(t) = h(x(t), u(t)) \tag{39}$$

where $h : \mathbb{X} \times \mathbb{U} \to \mathbb{R}^p$, and require that for the controlled system and given a *convex* set $\mathbb{Y}$ the following holds:

$$\mathrm{Av}[y] \in \mathbb{Y}.$$

To this end, the following definitions and augmented set-up were proposed in [3]:

$$\mathbb{Y}_{t+1} = \mathbb{Y}_t \oplus \mathbb{Y} \oplus \{-h(x(t), u(t))\}, \tag{40}$$

where $\oplus$ denotes set-sum. In particular $\mathbb{Y}_0$ can be initialized as an arbitrary compact convex set. The iteration leads to

$$\mathbb{Y}_t = t\mathbb{Y} \oplus \mathbb{Y}_0 \oplus \left\{ -\sum_{k=0}^{t-1} h(x(k), u(k)) \right\}.$$

Therefore the set $\mathbb{Y}_t$ grows in size with linear speed $\mathbb{Y}$/sampling time, in all allowed directions, while being shifted around by a quantity equal to the "integral" of h. Problem (7) can then be augmented as follows

$$
\begin{aligned}
&\min_{\mathbf{z},\mathbf{v}} J_\ell(\mathbf{z},\mathbf{v}) \\
&\text{subject to} \\
&(z(k),v(k)) \in \mathbb{Z} \qquad k \in \{0,\dots,N-1\} \\
&z(k+1) = f(z(k),v(k)) \qquad k \in \{0,\dots,N-1\} \\
&z(N) \in \mathbb{X}_f \\
&z(0) = x(t), \\
&\Sigma_{k=0}^{N-1} h(z(k),v(k)) \in \mathbb{Y}_t.
\end{aligned} \tag{41}
$$

Recursive feasibility, guaranteed satisfaction of all constraints in closed-loop and performance bounds can be derived for this EMPC scheme along similar steps as in the previous cases. When the set $\mathbb{Y}$ can be defined as a polyhedron, suitable relaxed notions of dissipativity on averagely constraints solutions are also proposed in [3]. These are used in [30] to derive asymptotic convergence results for Economic MPC with constraints on average. Another possibility suggested in [30] is to use suitably designed constraints on average in order to induce asymptotic convergence towards equilibria that would otherwise be unstable. Average constraints on finite time windows are also a possibility, and are proposed in [31].

5 Robust Economic Model Predictive Control

In real-world applications, the presence of disturbances and model uncertainties is typically unavoidable. Hence it is of paramount interest to obtain closed-loop performance and stability guarantees despite such disturbances. To this end, some of the techniques developed in the context of robust stabilizing (tracking) MPC can be helpful, in particular for addressing issues of robust feasibility, [23, 34, 35]. However, it turns out that just transferring, e.g., tube-based MPC approaches to an economic MPC context without suitable adaptations can lead to a suboptimal closed-loop performance. This observation can be illustrated by the following simple motivating example.

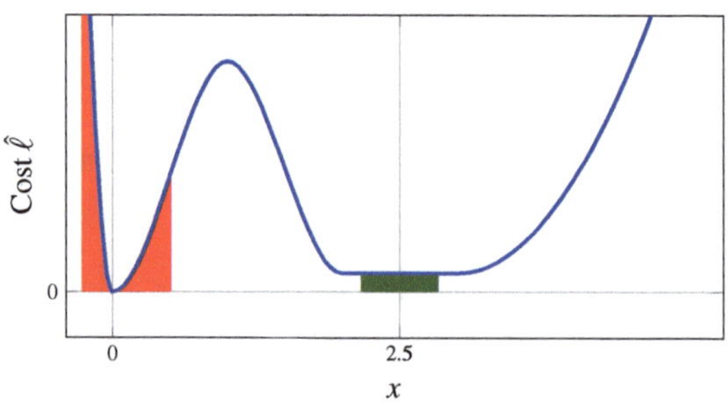

Fig. 2: Stage cost $\hat{\ell}$ in the motivating example of Section 5.

Consider the system $x(t+1) = x(t) + u(t)$ with stage cost $\ell(x,u) = \hat{\ell}(x)$ as shown in Figure 2. Clearly, the system is trivially optimally operated at the optimal steady-state $x = u = 0$, since the cost function $\hat{\ell}$ is positive definite (i.e., a tracking cost function). Now suppose that some additive disturbances are present, satisfying $w(t) \in \mathbb{W}$ for all $t \in \mathbb{N}$ and some compact set $\mathbb{W}$. If a standard tube-based MPC scheme such as the one in [26] is employed, one can show that the closed-loop system converges to a robust positively invariant (RPI) set Ω_1 centered at origin, which is exemplarily depicted in red in Figure 2. The size of this RPI set scales with the size of the disturbance set $\mathbb{W}$. Now if $\mathbb{W}$ is large enough, one can see that a better average performance might be obtained if the economic MPC scheme is such that the closed loop converges to an RPI set Ω_2 centered, e.g., at $x = 2.5$, which is exemplarily depicted in green in Figure 2. Namely, while $\hat{\ell}(2.5) > \hat{\ell}(0) = 0$, the average of $\hat{\ell}$ over Ω_2 is smaller than that of $\hat{\ell}$ over Ω_1.

Motivated by this fact, several different robust economic MPC schemes have been proposed in the literature where some knowledge about the present disturbances is incorporated into the cost function employed within the repeatedly solved optimization problem. Depending on the available knowledge of the disturbances, different closed-loop performance guarantees can then be derived. To this end, suppose that the real (disturbed) system is given by $x(t+1) = f(x(t), u(t), w(t))$ with $w(t) \in \mathbb{W}$ for all $t \in \mathbb{N}$ and some compact set $\mathbb{W}$, and denote the corresponding nominal system by $\xi(t+1) = f(\xi(t), v(t), 0)$. Furthermore, we assume that some robust control invariant set Ω has been determined such that if the error between the real and nominal state at time t is contained in Ω, the same is true at time $t+1$ independent of the disturbance $w(t) \in \mathbb{W}$ and the (nominal) input $v(t)$. This is typically achieved by using some prestabilization or additional (error) feedback, i.e., using $u = \phi(x, \xi, v)$ (in the linear case, e.g., $\phi(x, \xi, v) = K(x - \xi) + v$ can be chosen, in which case the computation of Ω reduces to determining an RPI set). As in standard (stabilizing) tube-based MPC, one can show that the real closed-loop system is contained in the set Ω around the nominal system state. In order to account for this fact within the repeatedly solved optimization problem, the following two cost functions have been proposed:

$$\ell^{\max}(\xi, v) := \max_{\omega \in \Omega} \ell(\xi + \omega, \phi(\xi + \omega, \xi, v)) \tag{42}$$

$$\ell^{\text{int}}(\xi, v) := \int_{\Omega} \ell(\xi + \omega, \phi(\xi + \omega, \xi, v)) \mathrm{d}\omega \tag{43}$$

Here, $\ell^{\max}$ is such that the worst case cost within the set Ω around the nominal state and input is considered, while in ℓ^{int} the average over all values in Ω is taken. Using these cost functions within a suitably defined tube-based MPC scheme based on the one in [26], the following closed-loop average performance bounds can be derived in a similar fashion as shown in Section 3.2 for the nominal case. For ℓ^{int} as defined in (43), it was shown in [5] that

$$\limsup_{T \to \infty} \frac{1}{T} \sum_{t=0}^{T-1} \ell^{\text{int}}(\xi(t), v(t)) \leq \ell^{\text{int}}(z_s, v_s),$$

where $\xi(\cdot)$ and $v(\cdot)$ are the resulting nominal state and input sequences and (z_s, v_s) is the optimal steady-state minimizing ℓ^{int}. Since the real closed-loop state is contained in the set Ω around the nominal closed-loop state, this can be interpreted as an average performance result for the real closed-loop system, averaged over all possible disturbances. For $\ell^{\max}$ as defined in (42), one can directly obtain an average performance bound for the real closed-loop system (independent of the realization of the disturbance), as shown in [6]:

$$\limsup_{T\to\infty} \frac{1}{T}\sum_{t=0}^{T-1} \ell(x(t), u(t)) \le \ell^{\max}(z_s, v_s).$$

Here, (z_s, v_s) is the optimal steady-state minimizing $\ell^{\max}$.

If additional information about the distribution of the disturbance is available, improved performance bounds can be obtained. This was shown in [7] for linear systems subject to additive disturbances. Here, one can directly minimize the expected value cost of predicted states and inputs, i.e., in (7) the following (prediction-time dependent) cost function is used:

$$\ell_k^{\text{int}}(z(k), v(k)) := \mathbb{E}\left\{\ell(x(t+k), u(t+k)) | x(t)\right\}.$$

Using a suitable tube-based MPC scheme based on [8], the following closed-loop average performance bound has been obtained in [7]:

$$\limsup_{T\to\infty} \frac{1}{T}\sum_{t=0}^{T-1} \mathbb{E}\left\{\ell(x(t), Kx(t) + u(t)) | x(0)\right\} \le \ell_\infty^{\text{int}}(z_s, v_s). \tag{44}$$

Here, ℓ_∞^{int} is defined by taking the expected value of the cost ℓ over the RPI set Ω according to the steady-state error distribution, and (z_s, v_s) is the optimal steady-state minimizing ℓ_∞^{int}. The bound (44) then says that the expected closed-loop average cost given the initial condition $x(0)$ is upper bounded by the best expected steady-state cost.

6 Conclusions

The chapter motivates and introduces Economic Model Predictive Control as a method to merge the Real Time Optimization layer and the Control Layer within a single optimization layer which is responsible of controlling the plant and optimizing its economic performance. The main relevant analytical tools and results are illustrated in a self-contained way and pointers to relevant in-depth literature on the topic are provided. In particular, we emphasized the role played by terminal ingredients in determining issues related to recursive feasibility, asymptotic performance and stability of nominal and robust Economic Model Predictive Control. The theory that emerged in the last few years has reached already a considerable maturity and is still developing with a number of issues of theoretical and practical relevance currently under investigation by the scientific control community. While we have made an effort to propose a coherent perspective on Economic Model Predictive Control we are aware that several recent developments did not find space in the present chapter. We would like to mention some of them:

- Formulation of tracking MPC schemes matching locally a given Economic Model Predictive Control, [39];
- Lyapunov-based Economic Model Predictive Control, [12];
- EMPC with generalized optimal regimes of operation, [10];
- Stochastic Economic Model Predictive Control, [38].

References

1. Amrit, R., Rawlings, J.B., Angeli, D.: Economic optimization using model predictive control with a terminal cost. Annu. Rev. Control **35**, 178–186 (2011)
2. Angeli, D., Rawlings, J.B.: Receding horizon cost optimization and control for nonlinear plants. In: 8th IFAC Symposium on Nonlinear Control Systems (NOLCOS), Bologna (2010)
3. Angeli, D., Amrit, R., Rawlings, J.B.: On average performance and stability of economic model predictive control. IEEE Trans. Autom. Control **57**(7), 1615–1626 (2012)
4. Angeli, D., Casavola, A., Tedesco, F.: Theoretical advances on economic model predictive control with time-varying costs. Annu. Rev. Control **41**, 218–224 (2016)
5. Bayer, F.A., Müller, M.A., Allgöwer, F.: Tube-based robust economic model predictive control. J. Process Control **24**(8), 1237–1246 (2014)
6. Bayer, F.A., Müller, M.A., Allgöwer, F.: Min-max economic model predictive control approaches with guaranteed performance. In: 55th IEEE Conference on Decision and Control (CDC), Las Vegas, pp. 3210–3215 (2016)
7. Bayer, F.A., Lorenzen, M., Müller, M.A., Allgöwer, F.: Robust economic model predictive control using stochastic information. Automatica **74**, 151–161 (2016)
8. Chisci, L., Rossiter, J.A., Zappa, G.: Systems with persistent disturbances: predictive control with restricted constraints. Automatica **37**(7), 1019–1028 (2001)
9. Diehl, M., Amrit, R., Rawlings, J.B.: A Lyapunov function for economic optimizing model predictive control. IEEE Trans. Autom. Control **56**(3), 703–707 (2011)
10. Dong, Z., Angeli, D.: A generalized approach to economic model predictive control with terminal penalty functions. In: IFAC World Congress, Toulouse (2017)
11. Dorfman, R., Samuelson, P.A., Solow, R.M.: Linear Programming and Economic Analysis. Dover Publications, New York (1987); Reprint of the 1958 original
12. Ellis, M., Liu, J., Christofides, P.D.: Economic Model Predictive Control: Theory, Formulations and Chemical Process Applications. Advances in Industrial Control. Springer, Berlin (2017)
13. Fagiano, L., Teel, A.: Generalized terminal state constraint for model predictive control. Automatica **49**(9), 2622–2631 (2013)
14. Faulwasser, T., Bonvin, D.: On the design of economic NMPC based on approximate turnpike properties. In: 54th IEEE Conference on Decision and Control (CDC), Osaka, pp. 4964–4970 (2015)

15. Faulwasser, T., Korda, M., Jones, C.N., Bonvin, D.: Turnpike and dissipativity properties in dynamic real-time optimization and economic MPC. In: 53rd IEEE Conference on Decision and Control (CDC), Los Angeles, pp. 2734–2739 (2014)
16. Finlay, L., Gaitsgory, V., Lebedev, I.: Duality in linear programming problems related to deterministic long run average problems of optimal control. SIAM J. Control Optim. **47**(4), 1667–1700 (2008)
17. Grüne, L.: Economic receding horizon control without terminal constraints. Automatica **493**, 725–734 (2013)
18. Grüne, L., Müller, M.A.: On the relation between strict dissipativity and turnpike properties. Syst. Control Lett. **90**, 45–53 (2016)
19. Grüne, L., Pannek, J.: Nonlinear Model Predictive Control: Theory and Algorithms. Communications and Control Engineering Series, 2nd edn. Springer, New York (2017)
20. Grüne, L., Stieler, M.: Asymptotic stability and transient optimality of economic MPC without terminal constraints. J. Process Control **24**(8), 1187–1196 (2014)
21. Gaitsgory, V., Parkinson, A., Shvartsman, I.: Linear programming formulations of deterministic infinite horizon optimal control problems in discrete time. arXiv:1702.00857 (2017)
22. Keerthi, S.S., Gilbert, E.G.: Optimal infinite-horizon feedback control laws for a general class of constrained discrete-time systems: stability and moving-horizon approximations. J. Optim. Theory Appl. **57**, 265–293 (1988)
23. Langson, W., Chryssochoos, I., Raković, S.V., Mayne, D.Q.: Robust model predictive control using tubes. Automatica **40**, 125–133 (2004)
24. Limon, D., Alvarado, I., Alamo, T., Camacho, E.F.: On the relation between strict dissipativity and turnpike properties. Automatica **44**(9), 2382–2387 (2008)
25. Marlin, T.E., Hrymak, A.N.: Real-time operations optimization of continuous processes. In: Kantor, J.C., Garca, C.E., Carnahan, B. (eds.) Chemical Process Control-V, pp. 156–164. CACHE, AIChE (1997)
26. Mayne, D.Q., Seron, M.M., Raković, S.V.: Robust model predictive control of constrained linear systems with bounded disturbances. Automatica **41**(2), 219–224 (2005)
27. Müller, M.A., Grüne, L.: Economic model predictive control without terminal constraints for optimal periodic behavior. Automatica **70**, 128–139 (2016)
28. Müller, M.A., Angeli, D., Allgöwer, F.: Economic model predictive control with self-tuning terminal cost. Eur. J. Control **19**(5), 408–416 (2013)
29. Müller, M.A., Angeli, D., Allgöwer, F.: On the performance of economic model predictive control with self-tuning terminal cost. J. Process Control **24**(8), 1179–1186 (2014)
30. Müller, M.A., Angeli, D., Allgöwer, F., Amrit, R., Rawlings, J.B.: Convergence in economic model predictive control with average constraints. Automatica **50**(12), 3100–3111 (2014)
31. Müller, M.A., Angeli, D., Allgöwer, F.: Transient average constraints in economic model predictive control. Automatica **50**(11), 2943–2950 (2014)
32. Müller, M.A., Grüne, L., Allgöwer, F.: On the role of dissipativity in economic model predictive control. In: 5th IFAC Conference on Nonlinear Model Predictive Control, Seville, pp. 110–116 (2015)
33. Müller, M.A., Angeli, D., Allgöwer, F.: On necessity and robustness of dissipativity in economic model predictive control. IEEE Trans. Autom. Control **60**(6), 1671–1676 (2015)
34. Raković, S.V., Kouvaritakis, B., Findeisen, R., Cannon, M.: Homothetic tube model predictive control. Automatica **48**, 1631–1638 (2012)
35. Rakovic, S.V., Kouvaritakis, B., Cannon, M., Panos, C., Findeisen, R.: Parameterized tube model predictive control. IEEE Trans. Autom. Control **57**, 2746–2761 (2012)
36. Rawlings,J.B., Mayne, D.Q.: Model Predictive Control: Theory and Design. Nob Hill Publishing, LLC, Madison (2009)
37. Rawlings, J.B., Bonne, D., Jörgensen, J.B., Venkat, A.N., Jörgensen, S.B.: Unreachable setpoints in model predictive control. IEEE Trans. Autom. Control **53**(9), 2209–2215 (2008)

38. Sopasakis, P., Herceg, D., Patrinos, P., Bemporad, A.: Stochastic economic model predictive control for Markovian switching systems. arXiv:1610.10014.
39. Zanon, M., Gros, S., Diehl, D.: A tracking MPC formulation that is locally equivalent to economic MPC. J. Process Control **45**, 30–42 (2016)
40. Zanon, M., Grüne, L., Diehl, M.: Periodic optimal control, dissipativity and MPC. IEEE Trans. Autom. Control **62**(6), 2943–2949 (2017)
41. Zaslavski, A.J.: Turnpike Properties in the Calculus of Variations and Optimal Control. Springer, New York (2006)

[illegible]

Nonlinear Predictive Control for Trajectory Tracking and Path Following: An Introduction and Perspective

Janine Matschek, Tobias Bäthge, Timm Faulwasser, and Rolf Findeisen

Setpoint stabilization subject to constraints is a common control task in many applications, spanning from temperature control of chemical processes to speed control of engines and drives. Nonlinear model predictive control has shown to be a valuable tool to stabilize such systems in an efficient and nearly optimal way. While many control problems can be reformulated as setpoint-stabilization problems, there are problems which are difficult to reformulate, or for which a reformulation leads to performance limitations. Examples are tracking problems, where the controlled variable should follow a known or unknown time-dependent reference signal, such as synchronization problems in electrical networks. In other cases, it is demanded to track a geometric curve as precisely as possible, while the timing where to be when is of secondary interest, which leads to so-called path-following problems. Furthermore, sometimes the main interest is to directly optimize a cost objective, while satisfying constraints, which is typically referred to as economic operation. The first part of this work provides an introduction to different control objectives, spanning from setpoint stabilization, trajectory tracking, path following, to economic operation. The second part outlines how these objectives can be achieved in the frame of nonlinear model predictive control. By presenting a theoretical analysis as well

---

J. Matschek
Laboratory for Systems Theory and Automatic Control, Otto-von-Guericke University Magdeburg, Magdeburg, Germany

T. Bäthge · R. Findeisen (✉)
Laboratory for Systems Theory and Automatic Control, Otto-von-Guericke University Magdeburg, Magdeburg, Germany

International Max Planck Research School for Advanced Methods in Process and Systems Engineering, Magdeburg, Germany

T. Faulwasser
Institute for Automation and Applied Informatics, Karlsruhe Institute of Technology, Karlsruhe, Germany
e-mail: timm.faulwasser@kit.edu

S. V. Raković, W. S. Levine (eds.), *Handbook of Model Predictive Control*,
Control Engineering, https://doi.org/10.1007/978-3-319-77489-3_8

as simulation studies, we underline the main differences and advantages of the approaches and objectives. We conclude with an outlook and perspective, providing insights into challenging open research problems.

1 Introduction and Motivation

The importance of control and automation in everyday life is increasing rapidly. Not only in industrial applications, like robotic manufacturing systems and chemical plants, but also in home-used devices ranging from basic kitchen equipment to entertainment gadgets, control is more widespread than ever.

Many control problems can be formulated as classical setpoint-stabilization problems, cf. Figure 1, top left.

Setpoint-stabilization approaches are best suited for systems that demand constant operation, e.g. operating a chemical plant at a given chemical equilibrium or stabilizing the operation of a gas turbine at a constant speed. Model Predictive Control (MPC) is well suited for such control tasks, especially if constraints are present [54]. Also, the theoretical aspects of MPC for setpoint stabilization are well-understood. There are many application examples of MPC for setpoint stabilization, such as the temperature control in a greenhouse [16], the control of liquid level and temperature in a continuous stirred tank reactor [63], or the control of the oxygen level in coke production [67].

Besides stabilization, additional control objectives are becoming increasingly important, such as operational performance and constraint satisfaction. However, the desired operating point of the system is often not fixed. Rather it may change frequently or continuously over time. Examples are the tracking of a reference orbit for satellites or the realization of a time-varying production recipe in pharmaceutical production systems. Following such time-varying references is typically denoted as *trajectory tracking* and the reference which the system should follow is parametrized by time, cf. Figure 1, top right. Note that different terminologies for trajectory-tracking problems can be found in the literature: if the reference signal is unknown, defined or generated by an exogenous system, terms like model following, servo control problem, or output regulation are used, see, e.g., [5] and [39]. We follow along classic lines of [7] and deliberately denote all these cases as trajectory-tracking problems.

By now, many results on predictive control for trajectory tracking exist, see, e.g., [21, 29, 42, 45, 47–49, 53]. Application examples range, e.g., from the control of UAVs [1, 41], mobile robots [43], to medical applications like artificial pancreas [38], just to mention a few.

Some tasks arising in applications are neither setpoint stabilization nor trajectory-tracking problems. Examples are automated driving or milling operations along predefined paths of a production machine. In these cases, the reference is a geometric path instead of a fixed time-dependent state or output trajectory, cf. Figure 1, bottom left. Furthermore, the *path-following* performance—i.e., staying on a track or road—

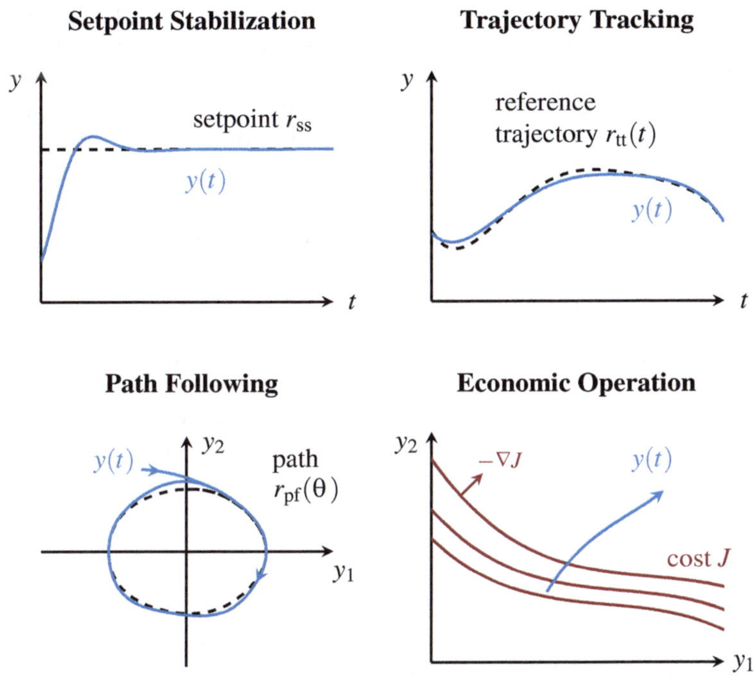

Fig. 1: Overview of different control objectives. Besides classical constant (setpoint stabilization) and time-dependent references (trajectory tracking), parametrized references (path following) or economic operation are of interest.

is of crucial interest. A series of nonlinear control approaches exist to tackle these problems, spanning from transversal feedback linearization [2, 8, 31, 59], backstepping [15, 64], to Lyapunov-based control strategies [1]. Model predictive control approaches allow to additionally consider performance requirements and constraints, cf. [22]. Applications of MPC for path-following problems span, for example, from robotic manipulators [26, 52], the control of an X-Y-table [44], up to the control of a tower crane [10].

Remark 1 (A Priori Unknown References). For certain applications, the complete reference that is to be tracked or followed might be available, e.g. due to an offline-planning procedure. This reference can then be exploited directly. In other cases, the reference is not known directly and must, for example, be recovered or cannot be used in a prediction of the system behavior. An example is the synchronization of satellite orbits, where the precise position of the other satellites is a priori unknown. We focus on the case that the reference is known. MPC approaches for unknown references can, for example, be found in [47–49].

Even more general, in some applications, *economically optimal operation* has highest priority. The values of the states or inputs are secondary, as long as constraints are satisfied. In such cases the controller should directly maximize an economic objective or benefit [23], see Figure 1, bottom right. Example applications are climate control for a building operated in a smart grid [37] or chemical processes with changing operating conditions [13].

In the following, we focus on the use of Nonlinear Model Predictive Control (NMPC) for setpoint tracking, trajectory tracking, path following, and economic

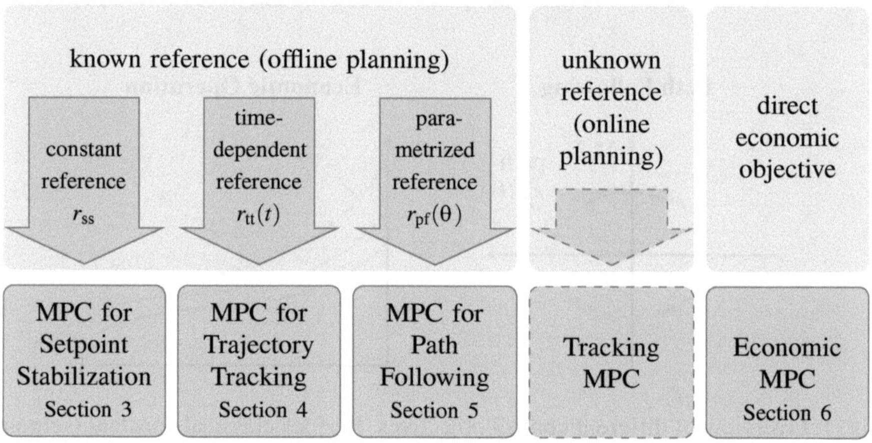

Fig. 2: The formulation of MPC explicitly relies on the specific task description. While MPC can use references that are either known or unknown a priori, it is also possible to directly consider an economic objective in the predictive control formulation (Economic MPC). In this chapter, we focus on MPC for trajectory tracking and path following for known references, considering offline planning. While in trajectory tracking the timing of the reference (i.e., the reference speed) is fixed, in path following the evolution of the reference is determined online by the controller.

operation. We outline the problem-specific objectives of the different problems in Section 2 and suitable NMPC formulations in Sections 3 to 6, see also Figure 2. Special emphasis is put on trajectory tracking and path following, also outlined by several simulation examples.

Besides an introduction to the solution of the different problems, we aim to underline that identifying the correct problem formulation for the different tasks is as important as selecting a suitable control approach.

2 Setpoint Stabilization, Trajectory Tracking, Path Following, and Economic Objectives

Often, a control task can be formulated and tackled from different directions. In this section, we introduce different control tasks and their respective objective. We distinguish those by the corresponding reference and its dependencies, e.g., on time, cf. Figure 1, while concentrating on the basic concepts behind the task definitions. Note that we do not focus on disturbance rejection or robust control problems. Instead, we refer to [60, 61] and the references therein.

We consider nonlinear, continuous-time, constrained systems of the form

$$\dot{x} = f(x,u), \quad x(0) = x_0 \tag{1a}$$

$$y = h(x) \tag{1b}$$

with the constrained system states $x \in \mathcal{X} \subseteq \mathbb{R}^{n_x}$, inputs $u \in \mathcal{U} \subseteq \mathbb{R}^{n_u}$, and outputs $y \subseteq \mathbb{R}^{n_y}$.[1] Note that system outputs do not necessarily correspond to the taken measurements. They rather refer to the variables of interest; i.e., they refer to the *controlled variables.*

The subsequently introduced control objectives focus on the specific task at hand, i.e. setpoint stabilization, trajectory tracking, path following, and purely economic desires. Performance objectives, such as minimizing energy, will be discussed in the predictive control formulations.

2.1 Setpoint Stabilization

Many control problems and methods consider the problem of regulation or stabilization of a setpoint, i.e. operating the system close to a setpoint or efficiently returning it to the setpoint is the main objective. This task is achieved by designing a controller for stabilizing the desired *stationary point/setpoint* r_{ss}, i.e. by minimizing the setpoint stabilization error

$$e_{ss}(t) = y(t) - r_{ss} \tag{2}$$

by steering the system output y to the setpoint r_{ss}.

If state constraints are present, it is important to verify that the reference point corresponds to feasible points in the state space:

[1] We focus on continuous-time systems described by differential equations. Extensions towards distributed parameter systems or discrete-time systems are subject to future work, respectively, easily possible.

Assumption 2.1. *The reference point r_{ss} is contained in the pointwise image of $\mathcal{X}$ under the output map $h(x)$: $r_{ss} \in h(\mathcal{X})$ where $h(\mathcal{X}) = \{y \in \mathbb{R}^{n_y} | \mathcal{X} \ni x \mapsto y = h(x)\}$.*

This leads to the following setpoint stabilization task:

Task 2.1 (Constrained Output Setpoint Stabilization) *Given system* (1) *and the reference setpoint r_{ss}, design a controller that achieves:*

i) Convergence to the Setpoint: The system output (1b) *converges to the setpoint r_{ss}, i.e.*

$$\lim_{t\to\infty} \|e_{ss}(t)\| = \lim_{t\to\infty} \|h(x(t)) - r_{ss}\| = 0.$$

ii) Constraint Satisfaction: The constraints on the states $x(t) \in \mathcal{X}$ and on the inputs $u(t) \in \mathcal{U}$ are satisfied for all times $t \geq 0$.

Typically, stabilizing a constant desired setpoint r_{ss} implies a corresponding steady state input u_{ss}, such that there exists $x_{ss} \in \mathcal{X}$ with $0 = f(x_{ss}, u_{ss})$ and $y_{ss} = h(x_{ss}) = r_{ss}$. Therefore, we introduce the "input error" variable $w_{ss}(t) = u(t) - u_{ss}$ that approaches zero as the output error $e_{ss}(t)$ goes to zero.

Note that it is assumed that the setpoint r_{ss} is explicitly given, see also Figure 2. Finding suitable setpoints is often based on static optimization, e.g. searching for a solution which minimizes a given cost function, usually subject to constraints.

2.2 Trajectory Tracking

Applications often require the tracking of changing setpoints or of time-dependent references, leading to *trajectory-tracking* control. Application examples are the dynamic operation of power plants, e.g. due to changing renewable stock source, and the tracking of cyclic reference signals in synchronization tasks, e.g. in electrical networks. Hereby, the time-varying reference can be given by the operator, generated by an offline planner, or it might be generated at run time by an online-planning algorithm.

Depending on whether the reference is a priori known or unknown (i.e., planned before/offline or at runtime/online), different control strategies exist, see Figure 2. In the case of a priori unknown references, the tracking problem is inherently uncertain. In particular, the prediction of the control error is lacking information about the future reference signal. In Section 4, we rely on a priori known reference signals, so that the prediction of the system error can be utilized to exploit the benefits of MPC.

The control task in trajectory tracking is to design a controller that achieves tracking of the time-dependent reference $r_{tt}(t)$, i.e. minimizing the tracking error $e_{tt}(t) = y(t) - r_{tt}(t)$. Note that this leads to an explicit dependency of the error dynamics on time and therefore to a time-varying control problem which needs particular attention in the controller design, see Section 4. Again, as we consider control of the outputs under state constraints, we need to define r_{tt} in such a way that no state constraints are violated:

Assumption 2.2. *The reference trajectory $r_{\text{tt}}(t)$ is contained in the pointwise image of $\mathcal{X}$ under the output map, i.e. $r_{\text{tt}}(t) \in h(\mathcal{X})$ for all t.*

Additionally, we assume that the input u_{tt} that is required to perfectly follow the reference is known a priori and can later on be used in the MPC scheme. Similar to the case of setpoint stabilization, tracking $r_{\text{tt}}(t)$ usually implies a corresponding input signal $u_{\text{tt}}(t)$, such that there exists $x_{\text{tt}}(t) \in \mathcal{X}$, defined for all $t \geq 0$, with $\dot{x}_{\text{tt}} = f(x_{\text{tt}}, u_{\text{tt}})$ and $y_{\text{tt}} = h(x_{\text{tt}}) = r_{\text{tt}}$. Hence, we introduce the "input error" $w_{\text{tt}}(t) = u(t) - u_{\text{tt}}(t)$ that goes to zero when $e_{\text{tt}}(t)$ converges to zero.

Task 2.2 (Constrained Output Trajectory Tracking) *Given system* (1) *and the reference trajectory $r_{\text{tt}}(t)$, design a controller that achieves:*

i) Convergence to the Trajectory: The system output (1b) *converges to the reference trajectory $r_{\text{tt}}(t)$, i.e.:*

$$\lim_{t\to\infty} \|e_{\text{tt}}(t)\| = \lim_{t\to\infty} \|h(x(t)) - r_{\text{tt}}(t)\| = 0.$$

ii) Constraint Satisfaction: The constraints on the states $x(t) \in \mathcal{X}$ and on the inputs $u(t) \in \mathcal{U}$ are satisfied for all times $t \geq 0$.

As the time dependency of the reference leads to a time-varying error and control objective, which needs special consideration in the controller design, trajectory-tracking formulations should only be used when this time dependency is inherent and relevant to the task. In tasks where, e.g., the following of a curve or manifold is the primary goal, without a predefined timing along the curve, path-following problems are an alternative.

2.3 Path Following

Many applications demand precise following of a given reference curve, rather than achieving a pre-defined speed. Considering the speed along a curve as a degree of freedom provides additional flexibility in the controller design. For example, in lane tracking for autonomous vehicles, human-robot co-operative tasks, deburring, or cutting, it is often not relevant *when* a state or position is reached, as long as it is reached within a close margin of error at some point in time.

This leads to so-called *path-following* problems: A geometric reference path should be followed as precisely as possible without an a priori fixed dependency on time. The reference path is often parametrized by a dimensionless path parameter θ, and the evolution of this parameter is decided by the controller.

We define the reference path as a parametrized regular curve in the output space

$$\mathcal{P} = \left\{ y \in \mathbb{R}^{n_y} \,|\, [\theta_{\text{start}}, \theta_{\text{end}}] \ni \theta \mapsto y = r_{\text{pf}}(\theta) \right\} \tag{3}$$

with the parametrization $r_{\text{pf}} : \mathbb{R} \to \mathbb{R}^{n_y}$. For constraint satisfaction we assume that:

Assumption 2.3. *The path $\mathcal{P}$ is contained in the pointwise image of the state constraints under the output map, i.e. $\mathcal{P} \subset h(\mathcal{X})$.*

Remark 2 (Connection to Manifold Stabilization and Reference Inputs). Exact path followability requires that the system can stay on the path once it started on it and that the error between the input and the reference-generating input should go to zero if started somewhere else. Those reference inputs can be obtained either in an analytical way (this is tractable, e.g., for input-affine systems with a well defined relative degree and for differentially flat systems) or by optimization, see, e.g., [18, 25]. However, it is important to note that convergence of

$$e_{\text{pf}}(t) = y(t) - r_{\text{pf}}$$

to zero implies (under mild technical assumptions) that the state x converges to a manifold in the state space [18, 57, 58]. Moreover, in many relevant applications, the reference input u_{pf} can be described as a function of θ and its time derivatives [18, 25].

Summarizing, the problem of path following can be stated as follows:

Task 2.3 (Constrained Output Path Following) *Given system* (1) *and the path $\mathcal{P}$* (3)*, design a controller that achieves:*

i) Path Convergence: The system output (1b) *converges to the set $\mathcal{P}$, i.e.:*

$$\lim_{t \to \infty} \|e_{\text{pf}}(t)\| = \lim_{t \to \infty} \|h(x(t)) - r_{\text{pf}}(\theta(t))\| = 0.$$

ii) Monotonous Forward Motion: The system moves along $\mathcal{P}$ in the direction of increasing values of θ, i.e. $\dot{\theta}(t) \geq 0$ and $\lim_{t \to \infty} \theta(t) = \theta_{\text{end}}$.

iii) Constraint Satisfaction: The constraints on the states $x(t) \in \mathcal{X}$ and on the inputs $u(t) \in \mathcal{U}$ are satisfied for all times $t \geq 0$.

Sub-task ii) can be relaxed, allowing backward motion. Furthermore, constraints on the desired velocities along the reference can be enforced, in the limit leading to trajectory-tracking formulations. As will be discussed later, it is important to note that path following, used to improve a tracking performance, will not necessarily lead to longer task execution times, cf. Section 5.2.1. In comparison to trajectory tracking, path following may achieve higher accuracy and smaller errors due to the direct utilization of an additional degree of freedom, i.e. the reference evolution.

Besides addressing path-following problems with predictive control, cf. Section 5, several other approaches to path-following control can be found in the literature. One large research field is path-following control using transversal feedback linearization, in which the task space is transformed into a transversal task direction (convergence to the geometric path or submanifold) and a tangential direction (moving along the path), done, e.g., in [2], [31], and [59]. In [1], Lyapunov-based control is combined with adaptive switching supervisory control to solve the path-following problem. Backstepping techniques are used, e.g., in [15] to obtain robust adaptive path following of underactuated ships, as well as in [64] for robust path following for a class of nonlinear systems.

2.4 Economic Objectives

Occasionally, no specific path, setpoint, or trajectory shall be tracked or is provided by a planner. Rather, the control input should be such that the system behaves optimally with respect to a given objective, while the achieved exact states and inputs are of secondary interest, cf. Figure 2. Examples are chemical plants where a maximum profit or product quality is desired, while the actual state evolution is of minor importance. Removing the separation between reference design/planning (e.g., calculating the optimal setpoint for a given cost) and controlling the system to reach this point can increase the flexibility and improve performance. Put differently, the controller can exploit all available degrees of freedom to improve performance. Predictive control is well suited to handle economic objectives, which is referred to as *Economic MPC*, see, e.g., [17, 23, 62] and the references therein.

3 A Brief Review of MPC for Setpoint Stabilization

We briefly recap MPC for the stabilization of constant references as it builds the foundation for the tracking of time-dependent references and path-following formulations. MPC allows for the consideration of nonlinear, coupled, and constrained systems, making it a good choice for the implementation of various control tasks [61].

We consider a sampled-data feedback perspective, see, e.g., [30, 32], as many real-world systems operate in continuous time. We assume that the system (1) fulfills the following assumptions:

Assumption 3.1. *The state constraint set $\mathcal{X}$ is closed, the input constraint set $\mathcal{U}$ is compact.*

Assumption 3.2. *The system dynamics* $f : \mathbb{R}^{n_x} \times \mathbb{R}^{n_u} \to \mathbb{R}^{n_x}$ *and the output map* $h : \mathbb{R}^{n_x} \to \mathbb{R}^{n_y}$ *are assumed to be sufficiently often continuously differentiable and locally Lipschitz for all* $(x,u)^\mathsf{T} \in \mathcal{X} \times \mathcal{U}$.

Assumption 3.3. *For any piecewise continuous input signal* $u(\cdot)$ *and for all initial points* $x_0 \in \mathcal{X}$, *system* (1) *admits a unique absolutely continuous solution.*[2]

In setpoint stabilization, we want to stabilize a constant reference point r_{ss}. Typically, in setpoint stabilization problems, the considered cost functional is

$$J_{\text{ss}}\left(x(t_k), \bar{u}_k(\cdot)\right) = \int_{t_k}^{t_k+T} L_{\text{ss}}\left(\bar{e}_{\text{ss}}(\tau), \bar{w}_{\text{ss}}(\tau)\right) \mathrm{d}\tau + E_{\text{ss}}\left(\bar{x}(t_k+T)\right). \tag{4}$$

Here, $L_{\text{ss}} : \mathbb{R}^{n_y} \times \mathbb{R}^{n_u} \to \mathbb{R}_0^+$ denotes the stage cost or cost function, $E_{\text{ss}} : \mathbb{R}^{n_x} \to \mathbb{R}_0^+$ is called terminal penalty, and T is the prediction horizon. We denote predicted signals with $\bar{\cdot}$, while the index k in $\bar{u}_k$ indicates that this is the optimal input based on the measurements available at sampling instant t_k. The signal to be optimized in the optimal control problem is the input $u_k(\cdot)$. In our case, the input is assumed to be piecewise continuous, i.e. $u_k(\cdot) \in \mathcal{PC}(\mathcal{U})$. Therefore, the optimal control problem that is solved at every sampling instant can be formulated as

$$\underset{\bar{u}_k(\cdot) \in \mathcal{PC}(\mathcal{U})}{\text{minimize}} \; J_{\text{ss}}\left(x(t_k), \bar{u}_k(\cdot)\right) \tag{5a}$$

subject to

$$\dot{\bar{x}}(\tau) = f(\bar{x}(\tau), \bar{u}_k(\tau)), \quad \bar{x}(t_k) = x(t_k) \tag{5b}$$
$$\bar{e}_{\text{ss}}(\tau) = h(\bar{x}(\tau)) - r_{\text{ss}} \tag{5c}$$
$$\bar{w}_{\text{ss}}(\tau) = \bar{u}_k(\tau) - u_{\text{ss}} \tag{5d}$$
$$\bar{x}(\tau) \in \mathcal{X}, \; \bar{u}_k(\tau) \in \mathcal{U} \tag{5e}$$
$$\bar{x}(t_k+T) \in \mathcal{E}_{\text{ss}} \subseteq \mathcal{X}, \tag{5f}$$

which have to hold for all $\tau \in [t_k, t_k+T]$. From the input signal $\bar{u}_k$, only the first part until the next sampling instant is used and the optimization is repeated in a receding-horizon fashion. Here, (5b) defines the system dynamics and (5c) can be regarded as the system output defining the stabilization error. It determines the difference between the system output and the constant reference r_{ss}. Additionally, (5d) defines the difference between the input and the reference-generating input value. In (5e), the state and input constraints are covered and (5f) restricts the state at the end of the prediction horizon to be inside the terminal region $\mathcal{E}_{\text{ss}}$.

Remark 3 (Structure of the Cost Functional). Since the reference (desired setpoint) is defined in the output space, the stage cost L_{ss} does merely penalize errors of

[2] The class of considered input functions could be readily extended to measurable controls. Here, for the sake of simplicity, we focus on the more application relevant setting of piecewise constant control signals.

the outputs of the system, which makes it semi-definite with respect to the states. However, the terminal cost E_{ss} and the terminal constraint set $\mathcal{E}_{ss}$ depend on the system states rather than the outputs.

3.1 Comments on Convergence and Stability

By now, stability of MPC for *discrete-time* formulations with constant references under *state feedback* is well-understood and can be guaranteed with an appropriate selection of terminal state constraints and end costs. For an overview of different approaches, see, e.g., [54], discussing terminal equality constraints, only a terminal cost, dual mode, i.e. no terminal cost, or both terminal cost and constraint set, among others. Additionally, one may also enforce stability without terminal constraints [33, 40, 46]. However, whenever L_{ss} penalizes *system outputs* instead of *states*, additional conditions are required, due to semi-definiteness of the stage cost with respect to the states. Stability guarantees can, for example, be obtained if additional detectability properties are satisfied. For example, [61] relies on input/output-to-state stability of the open-loop system, while [50] requires weak detectability of the considered system and the usage of a weak detector.

In *sampled-data* NMPC, achieving stability in the sense of convergence can, for example, be achieved by suitable choices of the stage cost L_{ss}, terminal cost E_{ss}, and terminal constraint set $\mathcal{E}_{ss}$ [28], which can be summarized as follows:

Assumption 3.4. *The stage cost* $L_{ss} : \mathbb{R}^{n_y} \times \mathbb{R}^{n_u} \to \mathbb{R}_0^+$ *is continuous,* $L_{ss}(0,0) = 0$ *and it is lower bounded by a class* $\mathcal{K}_\infty$ *function* α_1 *such that* $L_{ss}(e_{ss}, w_{ss}) \geq \alpha_1(\|e_{ss}\|)$ *for all* (e_{ss}, w_{ss}).

Assumption 3.5. *The terminal cost* $E_{ss} : \mathbb{R}^{n_x} \to \mathbb{R}_0^+$ *is positive semi-definite and continuously differentiable in* x.

Assumption 3.6. *The terminal constraint set* $\mathcal{E}_{ss} \subseteq \mathcal{X}$ *is compact.*

Assumption 3.7. *For all* $\tilde{x} \in \mathcal{E}_{ss}$ *and the considered sampling time* $\delta > 0$*, there exists an input* $u_{\mathcal{E}}(\cdot) \in \mathcal{PC}(\mathcal{U})$ *such that, for all* $\tau \in [0, \delta)$,

$$\frac{\partial E_{ss}}{\partial x} \cdot f(x(\tau), u_{\mathcal{E}}(\tau)) + L_{ss}(e_{ss}(\tau), w_{ss}(\tau)) \leq 0$$

and the closed-loop solution stays in the terminal region $x(\tau) = x(\tau, \tilde{x}|u_{\mathcal{E}}) \in \mathcal{E}_{ss}$*; i.e., the terminal region* $\mathcal{E}_{ss}$ *is control invariant.*

Provided that these assumptions hold, one can state the following theorem to address Task 2.1:

Theorem 1. *If the optimal control problem* (5) *is feasible for the initial time instant* t_0 *and its stage cost, terminal cost, and the terminal constraints satisfy Assumptions 3.4-3.7, then the optimal control problem* (5) *is recursively feasible and the error* $e_{ss}(t) = y(t) - r_{ss}$ *converges to zero under sampled-data NMPC.*

Note that while the stage cost depends on the outputs, the terminal cost and the terminal constraint depend on the states of the system; see also Remark 3. This choice is motivated by the fact that we need to upperbound the cost-to-go (which is state dependent) to show convergence. Convergence of the state instead of the output (error) is obtained, for example, by additionally assuming that the system is input/output-to-state stable and by replacing the lower bound in Assumption 3.4 with $L_{ss}(e_{ss}, w_{ss}) \geq \alpha_1(\|e_{ss}\|) + \alpha_1(\|w_{ss}\|)$. However, this might be of minor importance, since even without those additional assumptions, all states at the end of each prediction horizon are bounded in the compact set $\mathcal{E}_{ss}$ and therefore do not grow to infinity.

3.2 Setpoint Stabilization of a Lightweight Robot

We illustrate the different tasks and approaches by considering the control of a robotic manipulator, depicted in Figure 3. The robot's end-effector pose should follow a reference. For setpoint stabilization, this reference consists of a constant Cartesian position of the robot end effector. The KUKA lightweight robot [9] is modelled by the nonlinear dynamics

$$B(q)\ddot{q} = \mathcal{T},$$

with the inertia matrix B depending on the joint angles q, the angular accelerations $\ddot{q}$, and the motor torques $\mathcal{T}$ in each joint. For the sake of simplicity, we neglect friction and Coriolis effects and assume a complete compensation of torques originating from gravity.

With the transformation $\chi_1 = q$, $\chi_2 = \dot{q}$, and the input $\mu = \mathcal{T}$, the system dynamics become

$$\dot{\chi} = \begin{pmatrix} \dot{\chi}_1 \\ \dot{\chi}_2 \end{pmatrix} = \begin{pmatrix} \chi_2 \\ B^{-1}(\chi_1)\mu \end{pmatrix} \tag{6a}$$

$$\gamma_{ca} = h_{ca}(\chi_1). \tag{6b}$$

Fig. 3: Robot for manipulation tasks.

Here, the output function $h_{ca}(\chi_1)$ is the forward kinematics that map from the joint space into Cartesian space.

The performance of sampled-data setpoint stabilization NMPC, for a setpoint change of 10 cm in the Cartesian x-direction at $t = 2\,$s is shown in Figure 4. The instantaneous jump in the reference leads to a significant overshoot of the end-effector position, even when exploiting knowledge about it in the prediction. Avoiding such overshoot can be achieved in different ways. Firstly, one could tune the controller to be less aggressive, i.e. reduce the weights on the

Cartesian position errors. This would reduce the overshoot while, on the other hand side, it also reduces the position accuracy in the non-transient phases and increases the transition time. Secondly, one can reformulate the problem into the form of a time-dependent, sufficiently smooth trajectory. Doing so leads directly to trajectory-tracking formulations, as outlined in the following section.

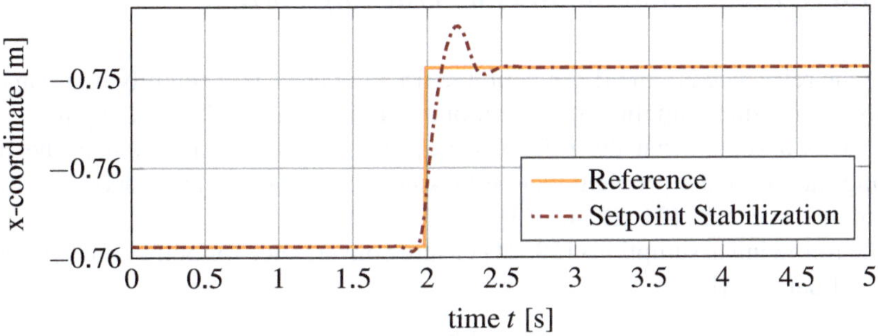

Fig. 4: Position of the end effector under NMPC.

4 Model Predictive Control for Trajectory Tracking

Trajectory tracking aims at following a timed reference, which is conceptually easy to integrate into the predictive control formulation, by modifying the considered cost functional to:

$$J_{\mathrm{tt}}\left(x(t_k),\bar{u}_k(\cdot)\right)=\int_{t_k}^{t_k+T} L_{\mathrm{tt}}\left(\bar{e}_{\mathrm{tt}}(\tau),\bar{w}_{\mathrm{tt}}(\tau)\right)\mathrm{d}\tau+E_{\mathrm{tt}}\left(\bar{x}(t_k+T)\right). \tag{7}$$

Here, $L_{\mathrm{tt}}:\mathbb{R}^{n_y}\times\mathbb{R}^{n_u}\to\mathbb{R}_0^+$ and $E_{\mathrm{tt}}:\mathbb{R}^{n_x}\to\mathbb{R}_0^+$ denote the cost function and the end penalty for the trajectory-tracking problem, respectively. The resulting optimal control problem, solved at the sampling instants, becomes

$$\underset{\bar{u}_k(\cdot)\in\mathcal{PC}(\mathcal{U})}{\operatorname{minimize}}\; J_{\mathrm{tt}}\left(x(t_k),\bar{u}_k(\cdot)\right) \tag{8a}$$

subject to

$$\dot{\bar{x}}(\tau)=f(\bar{x}(\tau),\bar{u}_k(\tau)),\quad \bar{x}(t_k)=x(t_k) \tag{8b}$$
$$\bar{e}_{\mathrm{tt}}(\tau)=h(\bar{x}(\tau))-r_{\mathrm{tt}}(\tau) \tag{8c}$$
$$\bar{w}_{\mathrm{tt}}(\tau)=\bar{u}_k(\tau)-u_{\mathrm{tt}}(\tau) \tag{8d}$$
$$\bar{x}(\tau)\in\mathcal{X},\ \bar{u}_k(\tau)\in\mathcal{U} \tag{8e}$$
$$\bar{x}(t_k+T)\in\mathcal{E}_{\mathrm{tt}}\subseteq\mathcal{X}, \tag{8f}$$

which have to hold for all $\tau \in [t_k, t_k + T]$. The main differences with respect to the setpoint stabilization problem (5) are that both the reference trajectory $r_{\mathrm{tt}}(t)$ and the input reference $u_{\mathrm{tt}}(t)$ are time-dependent.

4.1 Convergence and Stability of Tracking NMPC

Since the reference is time-dependent, the tracking error is inherently time-varying, which renders direct application of setpoint-stabilization results challenging.

As it is assumed that the reference trajectory is known in advance, the time-varying nature of the problem can be tackled by exploiting time-varying terminal constraints, see, e.g., [21, 29, 42, 53].

The following assumptions [18, 21] ensure convergence for output tracking using sampled-data NMPC:

Assumption 4.1. *The stage cost* $L_{\mathrm{tt}} : \mathbb{R}^{n_y} \times \mathbb{R}^{n_u} \to \mathbb{R}_0^+$ *is continuous,* $L_{\mathrm{tt}}(0,0) = 0$ *and is lower bounded by a class* $\mathcal{K}_\infty$ *function* α_1 *such that* $L_{\mathrm{tt}}(e_{\mathrm{tt}}, w_{\mathrm{tt}}) \geq \alpha_1(\|e_{\mathrm{tt}}\|)$ *for all* $(e_{\mathrm{tt}}, w_{\mathrm{tt}})$.

Assumption 4.2. *The terminal cost* $E_{\mathrm{tt}} : \mathbb{R}^{n_x} \to \mathbb{R}_0^+$ *is positive semi-definite and continuously differentiable in x.*

Assumption 4.3. *The terminal constraint set* $\mathcal{E}_{\mathrm{tt}} \subseteq \mathcal{X}$ *is compact and time-varying.*

Assumption 4.4. *For all* $\tilde{x} \in \mathcal{E}_{\mathrm{tt}}$ *and the considered sampling time* $\delta > 0$, *there exists an admissible input* $u_{\mathcal{E}}(\cdot) \in \mathcal{PC}(\mathcal{U})$ *such that, for all* $\tau \in [0, \delta)$,

$$\frac{\partial E_{\mathrm{tt}}}{\partial x} \cdot f(x(\tau), u_{\mathcal{E}}(\tau)) + L_{\mathrm{tt}}(e_{\mathrm{tt}}(\tau), w_{\mathrm{tt}}(\tau)) \leq 0$$

and the closed-loop solution fulfills $x(\tau) = x(\tau, \tilde{x}|u_{\mathcal{E}}) \in \mathcal{E}_{\mathrm{tt}}$*; i.e., the terminal region is control invariant.*

In comparison to setpoint stabilization, differences arise from the fact that the terminal constraint set $\mathcal{E}_{\mathrm{tt}}$ is now time-dependent (see Assumption 4.3) due to the inherently time-varying tracking error. For a detailed discussion on how to construct corresponding terminal regions, see [18, 21].

Similarly to Section 3, stability of the closed loop in the sense of convergence can be ensured to solve Task 2.2:

Theorem 2 ([18, 21]). *If the optimal control problem* (8) *is feasible for the initial time instant* t_0 *and the stage cost, the terminal cost and the terminal constraints satisfy Assumptions 4.1–4.4, then* (8) *is recursively feasible and the tracking error* e_{tt} *converges to zero under sampled-data NMPC.*

4.2 Trajectory-Tracking Control of a Lightweight Robot

Motivated by the simulation results for setpoint stabilization in Section 3.2, we formulate the control goal as a trajectory-tracking problem. The time-dependent reference $r_{tt}(t)$ is chosen to model a smooth transient between two constant values to make it comparable to the simulation of the setpoint stabilization. The NMPC formulation for trajectory tracking (8) uses this reference and the dynamic model of the robot (6). In both the setpoint stabilization and trajectory-tracking examples, the same cost function including the weights and same structure of the optimal control problem is used except from the reference definition. Figure 5 shows the performance of the closed loop for this setting. When comparing Figure 5 with Figure 4, the same time for the transition phase is adopted. As can be seen, the robot is able to follow the reference so that no overshooting occurs, which is crucial for safe performance. Nevertheless, tracking errors of up to 8 mm appear, as the robot is not able to follow the planned trajectory. Among other reasons, this originates from the fact that the offline planning of the reference did not take the actual abilities of the controlled system, resembled by the system dynamics and constraints, into account and planned a trajectory that is hard to follow given an a priori fixed timing of the reference. One way to overcome this problem is the use of path-following approaches, as outlined in the following section.

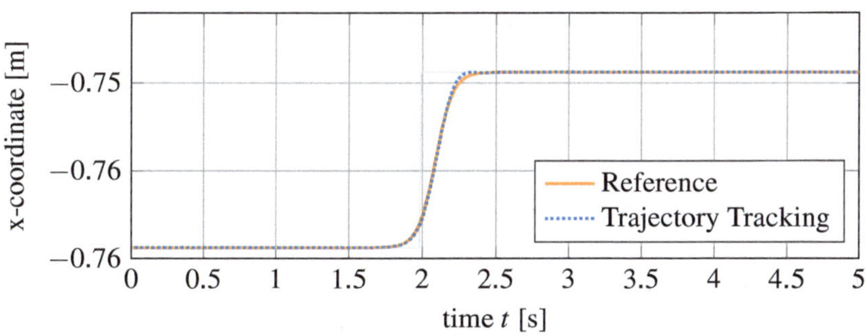

Fig. 5: Cartesian position of the end effector under NMPC with trajectory tracking.

5 Model Predictive Control for Path Following

As visible from the robot example in Section 4.2, it can be beneficial to consider the speed of reference evolution as a degree of freedom, in order to achieve optimal tracking performance. Such problems, i.e. following a curve or manifold as closely as possible, appear in many applications, spanning from autonomous driving, robotics, production systems, up to crystallization processes.

As outlined in Section 2.3, we consider that the reference path is known and given by a geometric curve, parametrized by a path parameter θ. The dependency of the path parameter θ on time can be considered as an additional degree of freedom. Often, however, influencing θ directly can lead to undesired effects, such as undesired jumps in the reference. To avoid these effects and for reasons of stability, adding additional filtering dynamics, i.e. "virtual" system dynamics can be advantageous [26, 52]. An example of simple virtual speed/filtering dynamics is the use of an integrator chain. To this end, the virtual system states $z_1, z_2, \ldots, z_\rho$, and the virtual system input v are introduced:

$$\begin{pmatrix} \dot{z}_1 \\ \dot{z}_2 \\ \vdots \\ \dot{z}_\rho \end{pmatrix} = \begin{pmatrix} 0 & 1 & 0 & \cdots & 0 \\ 0 & 0 & 1 & \cdots & 0 \\ \vdots & & \ddots & \ddots & 1 \\ 0 & 0 & 0 & \cdots & 0 \end{pmatrix} \begin{pmatrix} z_1 \\ z_2 \\ \vdots \\ z_\rho \end{pmatrix} + \begin{pmatrix} 0 \\ \vdots \\ 0 \\ 1 \end{pmatrix} v, \tag{9}$$

where $\theta = z_1$ is the virtual system output and $\rho \geq 1$ is some constant. A differential algebraic, respectively, geometric explanation of these dynamics can be found in [22, 58].

Depending on the task at hand, one can consider additional constraints on the virtual state z: limits on the path parameter to satisfy specific start and end points of the reference $\theta = z_1 \in [\theta_{\text{start}}, \theta_{\text{end}}]$, and limits on the reference speed by $\dot{\theta} = z_2 \geq 0$ to ensure forward motion along the reference. In particular, the states are restricted to a closed set, i.e. $z \in \mathcal{Z} \subset \mathbb{R}^\rho$, and the input v belongs to a compact set $v \in \mathcal{V}$, where both sets contain the origin in their interiors.

The predictive control formulation for path following is based on an expanded system state, adding the virtual states to the dynamics and introducing the virtual input v as additional input. In the cost functional, the deviation from the path is penalized suitably:

$$\begin{aligned} &J_{\text{pf}}\left(x(t_k), \bar{z}(t_k), \bar{u}_k(\cdot), \bar{v}_k(\cdot)\right) \\ &= \int_{t_k}^{t_k+T} L_{\text{pf}}\left(\bar{e}_{\text{pf}}(\tau), \bar{\theta}(\tau), \bar{u}(\tau), \bar{v}(\tau)\right) \mathrm{d}\tau + E_{\text{pf}}\left(\bar{x}(t_k+T), \bar{z}(t_k+T)\right). \end{aligned} \tag{10}$$

Note that the cost function $L_{\text{pf}} : \mathbb{R}^{n_y} \times \mathbb{R} \times \mathcal{U} \times \mathcal{V} \to \mathbb{R}_0^+$ and the end penalty $E_{\text{pf}} : \mathbb{R}^{n_x} \times \mathbb{R}^\rho \to \mathbb{R}_0^+$ now also include the virtual system states z, the input v, and the output θ.

The resulting optimal control problem, solved at the sampling instants, becomes

$$\underset{(\bar{u}(\cdot), \bar{v}(\cdot)) \in \mathcal{PC}(\mathcal{U} \times \mathcal{V})}{\text{minimize}} \quad J_{\text{pf}}\left(x(t_k), \bar{z}(t_k), \bar{u}_k(\cdot), \bar{v}_k(\cdot)\right) \tag{11a}$$

subject to

$$\dot{\bar{x}}(\tau) = f(\bar{x}(\tau), \bar{u}_k(\tau)), \quad \bar{x}(t_k) = x(t_k) \tag{11b}$$

$$\bar{e}_{\text{pf}}(\tau) = h(\bar{x}(\tau)) - r_{\text{pf}}(\bar{z}_1(\tau)) \tag{11c}$$

$$\dot{\bar{z}}(\tau) = l(\bar{z}(\tau), \bar{v}_k(\tau)), \ \bar{z}(t_k) = z(t_k) \tag{11d}$$

$$\bar{\theta}(\tau) = \bar{z}_1(\tau) \tag{11e}$$

$$\bar{x}(\tau) \in \mathcal{X}, \ \bar{u}_k(\tau) \in \mathcal{U} \tag{11f}$$

$$\bar{z}(\tau) \in \mathcal{Z},\ \bar{v}_k(\tau) \in \mathcal{V} \tag{11g}$$

$$(\bar{x}(t_k+T), \bar{z}(t_k+T))^\mathsf{T} \in \mathcal{E}_{\text{pf}} \subset \mathcal{X} \times \mathcal{Z}, \tag{11h}$$

which need to hold for all $\tau \in [t_k, t_k+T]$. Herein, (11b) are the system dynamics with the path-following error (11c) as output. Furthermore, the dynamics are augmented by the virtual system dynamics (11d) with the path parameter θ as its output (11e). Both the original and the virtual system inputs and states are constrained to their respective sets (11f)–(11g). Note that, in contrast to trajectory tracking and setpoint stabilization, the optimal control problem has as an additional input v, providing an additional degree of freedom for the path-following problem.

5.1 Convergence and Stability of Output Path-Following NMPC

As the controlled system now contains both the original dynamics plus the virtual path system, the following assumptions are required for concluding convergence:

Assumption 5.1. *The stage cost* $L_{pf}: \mathbb{R}^{n_y} \times \mathbb{R} \times \mathcal{U} \times \mathcal{V} \to \mathbb{R}_0^+$ *is continuous and is lower bounded by a class* $\mathcal{K}_\infty$ *function* α_1 *such that* $L_{\text{pf}}(e_{\text{pf}}, \theta, u, v) \geq \alpha_1(\|(e_{\text{pf}}, \theta - \theta_{\text{end}})^\mathsf{T}\|)$ *for all* $(e_{\text{pf}}, \theta, u, v)$.

Assumption 5.2. *The terminal cost* $E_{\text{pf}}: \mathbb{R}^{n_x} \times \mathbb{R}^{\rho} \to \mathbb{R}_0^+$ *is positive semi-definite and continuously differentiable in* x *and* z.

Assumption 5.3. *The terminal constraint set* $\mathcal{E}_{\text{pf}} \subseteq \mathcal{X} \times \mathcal{Z}$ *is compact.*

Assumption 5.4. *For all* $(\tilde{x}, \tilde{z})^\mathsf{T} \in \mathcal{E}_{\text{pf}}$ *and the considered sampling time* $\delta > 0$, *there exist inputs* $(u_{\mathcal{E}}, v_{\mathcal{E}})^\mathsf{T}(\cdot) \in \mathcal{PC}(\mathcal{U} \times \mathcal{V})$ *such that for all* $\tau \in [0, \delta)$

$$\left(\frac{\partial E_{\text{pf}}}{\partial x}, \frac{\partial E_{\text{pf}}}{\partial z}\right) \cdot \begin{pmatrix} f(x(\tau), u_{\mathcal{E}}(\tau)) \\ l(z(\tau), v_{\mathcal{E}}(\tau)) \end{pmatrix} + L_{\text{pf}}(e_{\text{pf}}(\tau), \theta(\tau), u_{\mathcal{E}}(\tau), v_{\mathcal{E}}(\tau)) \leq 0$$

and the closed-loop solution $x(\tau) = x(\tau, \tilde{x}|u_{\mathcal{E}})$ *and* $z(\tau) = z(\tau, \tilde{z}|v_{\mathcal{E}})$ *stay in* $\mathcal{E}_{\text{pf}}$*; i.e. the terminal region is control invariant.*

Considering that these assumptions hold, one can solve Task 2.3 and obtain the following convergence and stability results for predictive path following [18, 22]:

Theorem 3. *If the optimal control problem* (11) *is feasible for the initial time instant* t_0 *and the stage cost, the terminal cost, and the terminal constraints are chosen to fulfill Assumptions 5.1–5.4, then* (11) *is recursively feasible and the path-following error* e_{pf} *converges to zero under sampled-data NMPC.*

Proving Theorem 3 basically relies on the possibility of reformulating the path-following problem into the setpoint stabilization of an extended system, where the additional requirements of, e.g., the forward motion of the path are captured by the (extended) state constraints. For an extensive discussion and a complete convergence proof as well as insights into the computation of suitable terminal control laws and terminal regions to fulfill the stated assumptions, the reader is referred to [18, 22]. Stability proofs for path following for *discrete-time* systems and *state feedback* can be obtained in a straightforward manner from setpoint stabilization problems, following the classical ideas, as for example presented in [54].

5.2 Path-Following Control of a Lightweight Robot

We consider the control of the lightweight robot, see Section 3.2, to illustrate the concept of path following and to outline the main differences to trajectory tracking. To do so, we compare two trajectory-tracking approaches with different velocity profiles and a path-following implementation. Firstly, we show and compare the nominal behavior (no model-plant mismatch, no external disturbances) and secondly, we present the performance under an external, unknown disturbance.

The robot is supposed to follow a path which is given by a Lissajous figure in the y-z-plane, composed of cosine and sine terms, whereas the third dimension is composed of hyperbolic tangent functions, cf. Figure 6. The path is parametrized by time for the trajectory-tracking case and by the time-depending path parameter in the path-following case.

The virtual dynamics for the path evolution are chosen to be of second order:

$$\dot{\zeta} = \begin{pmatrix} \dot{\zeta}_1 \\ \dot{\zeta}_2 \end{pmatrix} = \begin{pmatrix} \zeta_2 \\ v \end{pmatrix} \tag{12a}$$

$$\theta = \zeta_1, \tag{12b}$$

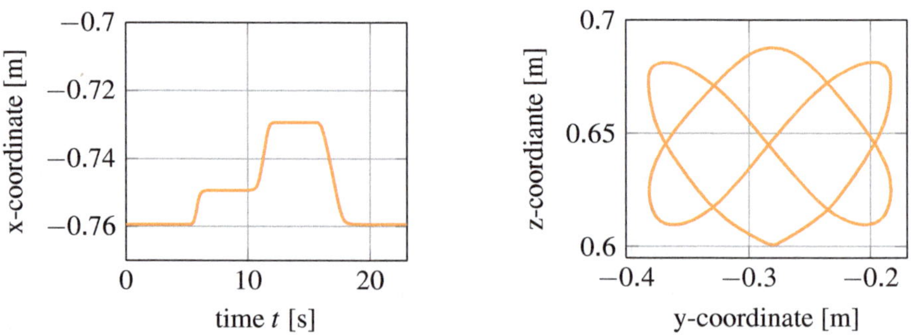

(a) Depth coordinate of the reference. (b) Lissajous figure in y-z-plane.

Fig. 6: Cartesian reference path for all shown simulations.

with the virtual system input v and the virtual states ζ. Note that, for trajectory tracking, the dynamics are given by (6) where the decision variables of the optimal control problem are the robot inputs (joint torques) $\mu_{tt} = \mathcal{T}$. In case of path following, one needs to consider the augmented system dynamics, given by (6) and (12).

The augmented input/decision variable $\mu_{pf} = (\mathcal{T}, v)^{\mathsf{T}}$ is composed of the joint torques and the virtual system input v that represents the additional degree of freedom in the path-following scheme.

5.2.1 Nominal Case

In robotics, large position errors are likely to occur due to high accelerations and acceleration changes (jerks). A naive way to compensate for this effect is to slow down the overall reference trajectory, as considered in Scenario 1.

Scenario 1 (trajectory tracking with a "low" velocity): The maximum velocity of the reference was set to 0.031 1/s so that the desired movement is completed after 40 s, see also Figure 7(a). In this case, the maximum absolute position errors are 3.99×10^{-2} mm, 1.33×10^{-2} mm, and 1.11×10^{-2} mm in x-, y-, and z-direction, respectively. Even though the errors are small, Scenario 1 will not be preferred in practice. Using a reduced overall reference speed, the transition times are increased, which often results in decreased economic profits. Gaining higher accuracy by slowing down the overall process is thus only a suboptimal option.

Scenario 2 (trajectory tracking with a "normal" velocity): We use a trajectory tracking with higher reference velocity than in Scenario 1. Allowing a higher maximum velocity of 0.1 1/s leads to an end time of 20 s instead of 40 s, see Figure 7(b) (dotted line). The maximum absolute tracking errors for Scenario 2 are 0.38 mm, 0.13 mm, and 0.12 mm in x-, y-, and z-direction, respectively. Comparing the position errors with Scenario 1, one sees that they differ in magnitude. This underlines

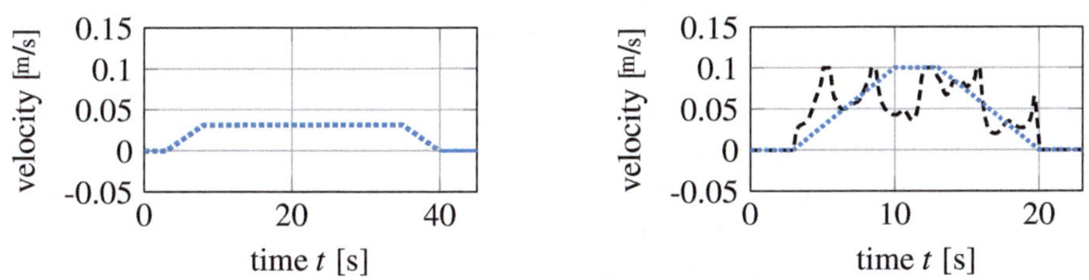

(a) Trajectory tracking (Scenario 1) with maximum reference speed of 0.031 1/s and runtime of 40 s.

(b) Trajectory tracking (Scenario 2, dotted), and path following (Scenario 3, dashed), both with the maximum reference speed of 0.1 1/s and run time of 20 s.

Fig. 7: Velocity profiles.

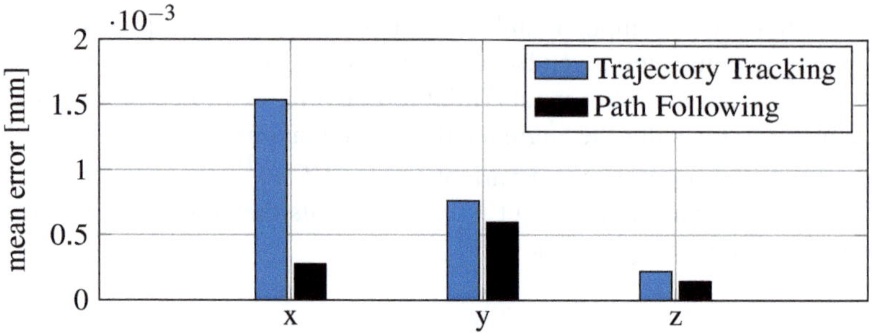

Fig. 8: Root mean squared error of Scenarios 2 (blue) and Scenario 3 (black) in all Cartesian directions. Unlike Scenario 1, Scenario 2 and 3 are comparable since they complete the task after the same time and use the same maximum velocity.

the sensitivity of the performance with respect to time. However, as before, slowing down the overall process is often not acceptable.

Scenario 3 (path following): For path following, no predefined timing for the reference is used. Rather, the controller adjusts the path evolution online based on the predicted position errors, cf. Figure 7(b). Hereby, a tradeoff between fast convergence to the endpoint of the path and high accuracy is found. As can be seen in Figure 7(b), both the second scenario (dotted) and the third scenario (dashed) require the same time for completion (20 s) and use the same maximum velocity of 0.1 $^1\!/_\mathrm{s}$.

For both scenarios, the mean errors are shown in Figure 8. As can be seen, position errors in all directions are smaller for path following, without slowdown of the overall process. Especially in the Cartesian x-direction, an improvement of 82 % was achieved.

The largest position error occurs around $t = 12\,\mathrm{s}$. At this time point, the reference velocity in the path-following scenario was decreased to improve the performance. The large errors at this time result from a rapid change of the reference acceleration in the Cartesian x-direction, as shown in Figure 9. Since both scenarios work with different timing laws, the plots of reference curves over time, depicted in solid line style, differ. In the zoomed-in region around $t = 12\,\mathrm{s}$, the path deviation in the trajectory tracking (dotted) from its reference can be seen, whereas the path following approach (dashed) follows the turn more precisely.

Note that the cost function weights and all other parameters for the trajectory tracking and the path following were chosen to be identical, i.e. the performance increase is a direct result of the problem formulation and not due to any tuning of cost functions.

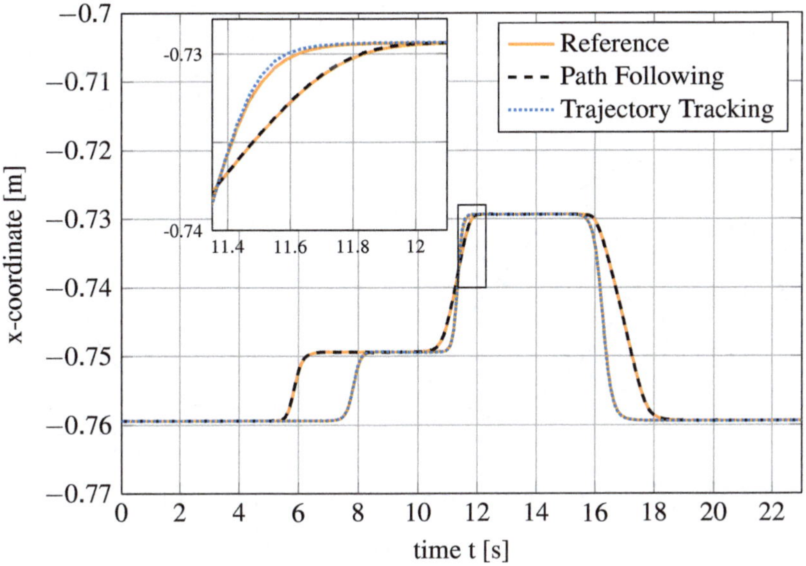

Fig. 9: Comparison of path following and trajectory tracking in the x-direction.

5.2.2 Disturbance Case

Next, we consider the closed-loop behaviors of the NMPC schemes under the influence of an external disturbance, appearing at $t = 12.5$ s: the end effector is displaced by 10 mm in y-direction and is locked there for 0.5 s. The considered disturbance is motivated by a collaborative working scenario, in which the robot could collide with or unexpectedly be stopped by another robot, human, or work piece. Similarly, unexpected material property changes, such as changes in solidity, can lead to the same effect. In the case of trajectory tracking, the reference is supposed to follow the predefined velocity profile under all circumstances. This leads to a larger error as, in the meantime, the reference continues to move on along the Lissajous curve with the predefined (non-adjustable) speed. Once the blockage is overcome, the robot resumes to the current reference position, resulting in a "shortcut." Thus, the reference is not tracked at all, see Figure 10, dotted line.

In contrast to this, the path-following approach is also able to adjust the reference speed to minimize the position errors, see also Figure 11. Right after the displacement occurs, the controller chooses to slow down the reference, so that the position error stays as small as possible and precise following is obtained, see also Figure 10.

As outlined, path following is able to generate velocity profiles for a geometric path that allow for higher position accuracy without sacrificing overall runtime. This includes the nice side effect of no need for an additional (iterative) path velocity planning. Secondly, in case of disturbances, which are of increasing importance in

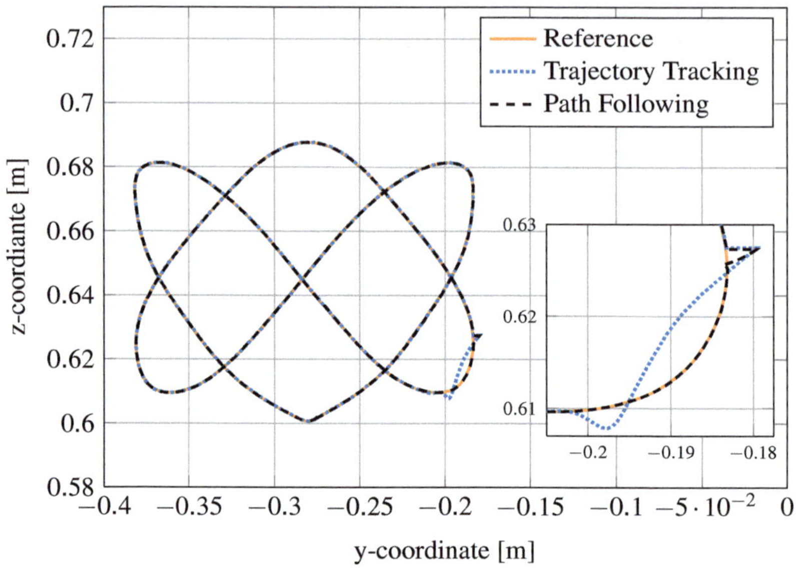

Fig. 10: Comparison of trajectory tracking and path following in the y-z-plane for the disturbance case.

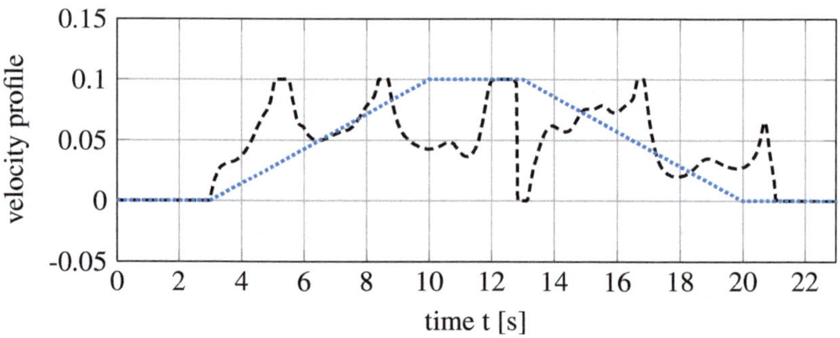

Fig. 11: Reference velocity profile of trajectory tracking (dotted) and path following (dashed) for the disturbance case.

future co-operative manufacturing systems, path following can utilize an additional degree of freedom in the control structure (the virtual system input) to achieve high performance.

5.3 Extensions of Path Following

The outlined approach for MPC for path-following problems can be expanded in several directions:

Path Following with Velocity Assignment Path following allows to pre-assign a velocity profile or corridor. Basically, the convergence of the path velocity (reference speed) to the desired velocity can be enforced, which can be expressed in a similar way as in Task 2.3, by adjusting sub-task ii):

Task 5.1 (Constrained Output Path Following with Velocity Assignment)
Given system (1) *and the path* $\mathcal{P}$ (3), *design a controller that achieves:*

i) Path Convergence: The system output $y = h(x)$ *converges to the set* $\mathcal{P}$, *i.e.:*

$$\lim_{t\to\infty} \|e_{\mathrm{pf}}(t)\| = \lim_{t\to\infty} \|h(x(t)) - r_{\mathrm{pf}}(\theta(t))\| = 0.$$

ii) Path Velocity Convergence: The path velocity $\dot{\theta}$ *converges to the desired velocity profile* $\dot{\theta}_{\mathrm{ref}}$, *i.e.* $\lim_{t\to\infty} \|\dot{\theta}(t) - \dot{\theta}_{\mathrm{ref}}(t)\| = 0$.

iii) Constraint Satisfaction: The constraints on the states $x(t) \in \mathcal{X}$ *and on the inputs* $u(t) \in \mathcal{U}$ *are satisfied for all times* $t \geq 0$.

One can easily adapt the optimal control problem (11) to achieve path following with velocity assignment; i.e., instead of penalizing the distance of the path parameter $\theta = z_1$ to its endpoint θ_{end} in the cost function L_{pf}, one considers the difference between its time derivative $\dot{\theta} = z_2$ and the desired velocity.

By choosing the weight on the velocity error high, the path following comes closer to a trajectory-tracking formulation. More details about path following with velocity assignment can, e.g., be found in [22].

Multi-Dimensional Path Following In some applications, additional degrees of freedom with respect to the path to be followed exist. An intuitive example is a car

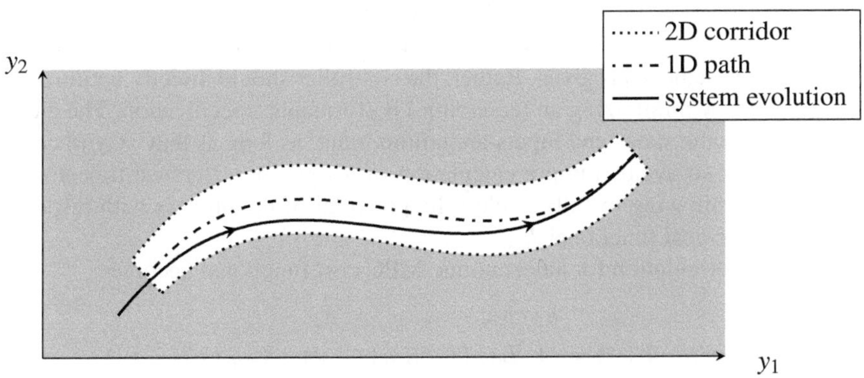

Fig. 12: Two-dimensional path following.

following a road, where an exact tracking of a curve is not desired. Rather, the car should stay on the lane, which could be wider than the car's dimensions. Similar cases occur for ships, airplanes, or unmanned aerial vehicles, which often are only required to stay in a certain corridor-path. They can and should utilize the additional degrees of freedom, for example to minimize time or energy consumption. We refer to those as multi-dimensional path following or general manifold stabilization problems. Hereby, the path $\mathcal{P}$, which in classical path following is a one-dimensional curve/manifold in the output space, is extended to be a higher-dimensional submanifold of the output space, cf. Figure 12. In this case, the extended path is characterized not only by one, but by multiple path parameters

$$\Theta = \left(\theta_1, \theta_2, \ldots, \theta_n\right)^{\mathsf{T}}, \tag{13}$$

with vector-valued constraints $\Theta_{\text{start}} \leq \Theta \leq \Theta_{\text{end}}$. Hereby, the first component of Θ corresponds to the path parameter in the previous sections, i.e. defining the evolution over time. The remaining parameters θ_i with $2 \leq i \leq n$ define spatial degrees of freedom. Their lower and upper bounds span a corridor around the original path, in which the controller is free to choose its actual path. For discussions and examples, see [24] and the recent results [11, 65].

Non-spatial Paths It is important to stress that the path does not need to be a "spatial curve." In principle, it can be defined with respect to any state variable or combinations/projections of such. For example, [52] considers the problem of force control in robotics, where the desired contact forces are parametrized by the path parameter. Similarly, one could consider temperatures in building control problems, concentrations in chemical reactors, etc. as path variables.

6 Economic MPC

We briefly discuss and introduce Economic MPC to set it in relation to the outlined setpoint stabilization, trajectory-tracking, and path-following formulations. In comparison to the aforementioned NMPC formulations, in Economic MPC no direct reference to be achieved is given. Rather, the controller should directly optimize system operation by minimizing an (economic) performance specification. The closed-loop values for the states and inputs are unimportant, as long as they stay in certain bounds. Doing so avoids the pre-calculation of a setpoint, reference trajectory, or path, possibly allowing to fully exploit the potential of the process with respect to some economic cost functional J_{eco}.

The basic formulation for an economic MPC cost functional becomes:

$$J_{\text{eco}}\left(x(t_k), \bar{u}_k(\cdot)\right) = \int_{t_k}^{t_k+T} L_{\text{eco}}\left(\bar{x}(\tau), \bar{u}_k(\tau)\right) \mathrm{d}\tau + E_{\text{eco}}\left(\bar{x}(t_k+T)\right), \tag{14}$$

while the optimal control problem is then the minimization of (14) subject to the system dynamics (1) and potentially state, input, and terminal constraints, $\bar{x} \in \mathcal{X}$, $\bar{u} \in \mathcal{U}$ and $\bar{x}(t_k+T) \in \mathcal{E}_{\text{eco}}$.

Note that no explicit reference appears. This allows to react to changing plant conditions (like external parameters or disturbances) without the explicit need of setpoint or reference calculations.

Other approaches use combined stage costs that result in a trade-off between stability and profitability, such as *Dual Objective MPC*, i.e. more than one (perhaps competitive) control goals should be achieved [3, 12, 51, 65]. Examples for such cases are, e.g., to find a tradeoff between observability and stabilizability of a system [12]. In general, a combination of economic and regulatory criteria can be formulated as

$$J_{\text{eco}}\left(x(t_k),\bar{u}_k(\cdot)\right) = \int_{t_k}^{t_k+T} a \cdot L_{\text{eco}}(\bar{x}(\tau),\bar{u}_k(\tau)) + (1-a) \cdot L_{\text{ss/tt/pf}}(\bar{e}(\tau),\bar{u}_k(\tau))\,\mathrm{d}\tau, \quad (15)$$

with a tuning parameter $0 \leq a \leq 1$ to define priorities in the tradeoff. For $a \to 1$, the economic cost would dominate, while for $a \to 0$, the stabilization/tracking cost would be the leading factor.

6.1 Convergence and Stability of Economic MPC

Recursive feasibility of Economic MPC can be shown in similar ways as with classical MPC approaches (e.g., setpoint stabilization), i.e. using terminal constraints [17, 62]. However, since optimal system operation might not be characterized by a steady state of the closed-loop system (e.g., it could also involve cyclic/periodic operation), stability or convergence cannot be expected in general [56]. In cases where the optimal operating point in fact is a steady state of the system—i.e., the system under considerations is said to be optimally operated at steady state [56]—, asymptotic stability of this steady state under Economic MPC can be established. Hereby, strict dissipativity of the underlying optimal control problem combined with suitable controllability properties is sufficient and (under suitable assumptions) necessary for the existence of such an optimal steady state [6, 56]. While [14] is using a terminal equality constraint to construct a Lyapunov function and show asymptotic stability, [4] replaces the terminal equality constraint by a terminal set and terminal cost. Furthermore, in [20, 34, 36], Economic MPC without terminal constraints and without any terminal conditions is discussed, respectively. It should be noted that [19, 20, 34, 36] exploit the fact that dissipativity of the optimal control problem implies the existence of a so-called turnpike property in the open-loop solutions to the optimal control problem. We remark that turnpike properties allow establishing recursive feasibility in Economic MPC without terminal constraints [19, 20, 23]. The close relations between dissipativity and turnpike properties have been investigated in detail for continuous-time optimal control problems [27] and discrete-time formulations [35]. However, there are cases in Economic MPC where the performance of the best reachable steady state can be improved by periodic orbits [55, 66]. For more details on Economic MPC, we refer to the recent overviews [17, 23].

7 Conclusions and Perspectives

Many control problems cannot be formulated as setpoint stabilization problems or the reformulation as such leads to a significantly decreased performance. In the first part of this chapter, we reviewed different types of control objectives and outlined the key differences. We especially focused on trajectory-tracking and path-following problems and set them into relation to setpoint stabilization and economic control objectives. As shown, it is very important to select and formulate the problem in the correct framework, otherwise the performance can be significantly deteriorated.

In the second part, we outlined how the desired control objectives can be tackled in the framework of nonlinear model predictive control, taking constraints into account. Based on a review of NMPC for setpoint stabilization, we outlined predictive control formulations for trajectory tracking and path following. Special focus was put on the comparison of NMPC for trajectory tracking and path following, to clarify the key differences with respect to the achievable performance, stability, and constraint satisfaction. The observations have been underlined considering the control of a lightweight robot subject to external disturbances. Finally, we outlined the main idea behind economic predictive control, which aims to directly optimize a desired economic objective without considering a reference.

It is important to note that we did not focus on robustness of trajectory-tracking or path-following predictive control. Furthermore, we limited the presentation to the case of known references, which can be directly exploited in the predictions.

While significant progress has been achieved in recent years, there are also many interesting and challenging problems for further development. Robust trajectory tracking and path following formulations and robustness analysis of nominal formulations are still in their infancy. As the direct consideration of uncertainty is important for many applications, such as automatic driving, there is a strong demand to develop suitable methods. Possible starting points can be MPC approaches based on robust invariant set considerations.

Moreover, the fusion of control and path planning becomes increasingly important in autonomous systems applications. Examples are robots that operate in a highly dynamic environment, as well as vehicles in dynamic traffic situations. Separating these tasks into a planning and a control level will lead to a decreased performance, as quick reactions to dynamic environment changes are often a necessity.

Furthermore, often references and paths are not known a priori, but might also not be free for optimization, as considered in economic predictive control. Examples are synchronization problems of robot swarms or power networks, as well as automatic driving subject to the uncertain behavior of other vehicles. This remedy might be overcome by fusing tracking and path following control with learning-based approaches.

Acknowledgements J. Matschek and R. Findeisen acknowledge support by the German Federal Ministry of Education and Research within the Forschungscampus STIMULATE under grant number 13GW0095A. T. Faulwasser acknowledges support from the Baden-Württemberg Stiftung under the Elite Programme for Postdocs.

References

1. Aguiar, A.P., Hespanha, J.P.: Trajectory-tracking and path-following of underactuated autonomous vehicles with parametric modeling uncertainty. IEEE Trans. Autom. Control **52**(8), 1362–1379 (2007)
2. Akhtar, A., Waslander, S.L., Nielsen, C.: Path following for a quadrotor using dynamic extension and transverse feedback linearization. In: Proceedings of the 51st IEEE Conference on Decision and Control (CDC), pp. 3551–3556 (2012)
3. Alessandretti, A., Aguiar, A., Jones, C.: On convergence and performance certification of a continuous-time economic model predictive control scheme with time-varying performance index. Automatica **68**, 305–313 (2016)
4. Amrit, R., Rawlings, J.B., Angeli, D.: Economic optimization using model predictive control with a terminal cost. Annu. Rev. Control **35**(2), 178–186 (2011)
5. Anderson, B., Moore, J.: Optimal Control - Linear Quadratic Methods. Information and System Science Series. Prentice Hall, Englewood Cliffs (1990)
6. Angeli, D., Amrit, R., Rawlings, J.B.: On average performance and stability of economic model predictive control. IEEE Trans. Autom. Control **57**(7), 1615–1626 (2012)
7. Athans, M., Falb, P.: Optimal Control - An Introduction to Theory and Its Applications. McGraw-Hill, New York (1966)
8. Banaszuk, A., Hauser, J.: Feedback linearization of transverse dynamics for periodic orbits. Syst. Control Lett. **26**(2), 95–105 (1995)
9. Bargsten, V., Zometa, P., Findeisen, R.: Modeling, parameter identification and model-based control of a lightweight robotic manipulator. In: Proceedings of the International Conference on Control Applications (CCA), pp. 134–139 (2013)
10. Böck, M., Kugi, A.: Real-time nonlinear model predictive path-following control of a laboratory tower crane. IEEE Trans. Control Syst. Technol. **22**(4), 1461–1473 (2014)
11. Böck, M., Kugi, A.: Constrained model predictive manifold stabilization based on transverse normal forms. Automatica **74**, 315–326 (2016)
12. Böhm, C., Findeisen, R., Allgöwer, F.: Avoidance of poorly observable trajectories: a predictive control perspective. IFAC Proc. Vol. **41**(2), 1952–1957 (2008)
13. Chen, X., Heidarinejad, M., Liu, J., Christofides, P.D.: Distributed economic MPC: application to a nonlinear chemical process network. J. Process Control **22**(4), 689–699 (2012)
14. Diehl, M., Amrit, R., Rawlings, J.B.: A Lyapunov function for economic optimizing model predictive control. IEEE Trans. Autom. Control **56**(3), 703–707 (2011)
15. Do, K.D., Jiang, Z.P., Pan, J.: Robust adaptive path following of underactuated ships. Automatica **40**(6), 929–944 (2004)
16. El Ghoumari, M., Tantau, H.J., Serrano, J.: Nonlinear constrained MPC: real-time implementation of greenhouse air temperature control. Comput. Electron. Agric. **49**(3), 345–356 (2005)
17. Ellis, M., Durand, H., Christofides, P.: A tutorial review of economic model predictive control methods. J. Process Control **24**(8), 1156–1178 (2014)
18. Faulwasser, T.: Optimization-Based Solutions to Constrained Trajectory-Tracking and Path-Following Problems. Number 3 in Contributions in Systems Theory and Automatic Control. Shaker Verlag, Herzogenrath, Otto-von-Guericke University Magdeburg (2013)
19. Faulwasser, T., Bonvin, D.: On the design of economic NMPC based on an exact turnpike property. In: Proceedings of the 9th IFAC Symposium on Advanced Control of Chemical Process (ADCHEM), pp. 525–530 (2015)

20. Faulwasser, T., Bonvin, D.: On the design of economic NMPC based on approximate turnpike properties. In: Proceedings of the 54th IEEE Conference on Decision and Control (CDC), pp. 4964–4970 (2015)
21. Faulwasser, T., Findeisen, R.: A predictive control approach to trajectory tracking problems via time-varying level sets of Lyapunov functions. In: Proceedings of the 50th IEEE Conference on Decision and Control and European Control Conference (CDC/ECC), pp. 3381–3386 (2011)
22. Faulwasser, T., Findeisen, R.: Nonlinear model predictive control for constrained output path following. IEEE Trans. Autom. Control **61**(4), 1026–1039 (2016)
23. Faulwasser, T., Grüne, L., Müller, M.A.: Economic nonlinear model predictive control. Found. Trends Syst. Control, **5**(1), 1–94 (2018)
24. Faulwasser, T., Kern, B., Findeisen, R.: Model predictive path-following for constrained nonlinear systems. In: Proceedings of the Joint 48th IEEE Conference on Decision and Control and 28th Chinese Control Conference (CDC/CCC), pp. 8642–8647 (2009)
25. Faulwasser, T., Hagenmeyer, V., Findeisen, R.: Constrained reachability and trajectory generation for flat systems. Automatica **50**(4), 1151–1159 (2014)
26. Faulwasser, T., Weber, T., Zometa, J.P., Findeisen, R.: Implementation of nonlinear model predictive path-following control for an industrial robot. IEEE Trans. Control Syst. Technol. **25**(4), 1505–1511 (2016)
27. Faulwasser, T., Korda, M., Jones, C.N., Bonvin, D.: On turnpike and dissipativity properties of continuous-time optimal control problems. Automatica **81**, 297–304 (2017)
28. Findeisen, R.: Nonlinear model predictive control: a sampled-data feedback perspective. Fortschr.-Ber. VDI, Reihe 8, Nr. 1087. VDI Verlag, Düsseldorf (2006)
29. Findeisen, R., Chen, H., Allgöwer, F.: Nonlinear predictive control for setpoint families. In: Proceedings of the IEEE American Control Conference (ACC), pp. 260–264 (2000)
30. Findeisen, R., Raff, T., Allgöwer, F.: Sampled-data nonlinear model predictive control for constrained continuous time systems. In: Tarbouriech, S., Garcia, G., Glattfelder, A.H. (eds.) Advanced Strategies in Control Systems with Input and Output Constraints. Lecture Notes in Control and Information Sciences, vol. 346, pp. 207–235. Springer, Berlin (2007)
31. Flixeder, S., Glück, T., Böck, M., Kugi, A.: Combined path following and compliance control with application to a biaxial gantry robot. In: Proceedings of the IEEE Conference on Control Applications (CCA), pp. 796–801 (2014)
32. Fontes, F.A.C.C.: A general framework to design stabilizing nonlinear model predictive controllers. Syst. Control Lett. **42**(2), 127–143 (2001)
33. Grüne, L.: Analysis and design of unconstrained nonlinear MPC schemes for finite and infinite dimensional systems. SIAM J. Control Optim. **48**(2), 1206–1228 (2009)
34. Grüne, L.: Economic receding horizon control without terminal constraints. Automatica **49**(3), 725–734 (2013)
35. Grüne, L., Müller, M.A.: On the relation between strict dissipativity and turnpike properties. Syst. Control Lett. **90**, 45–53 (2016)
36. Grüne, L., Stieler, M.: Asymptotic stability and transient optimality of economic MPC without terminal conditions. J. Process Control **24**(8), 1187–1196 (2014)
37. Halvgaard, R., Poulsen, N.K., Madsen, H., Jörgensen, J.B.: Economic model predictive control for building climate control in a smart grid. In: Proceedings of the IEEE Conference on Innovative Smart Grid Technologies, pp. 1–6 (2012)
38. Hovorka, R., Canonico, V., Chassin, L.J., Haueter, U., Massi-Benedetti, M., Federici, M.O., Pieber, T.R., Schaller, H.C., Schaupp, L., Vering, T., Wilinska, M.E.: Nonlinear model predictive control of glucose concentration in subjects with type 1 diabetes. Physiol. Meas. **25**(4), 905–920 (2004)
39. Isidori, A.: Nonlinear Control Systems, 3rd edn. Springer, Berlin (1995)
40. Jadbabaie, A., Hauser, J.: On the stability of receding horizon control with a general terminal cost. IEEE Trans. Autom. Control **50**(5), 674–678 (2005)

41. Kamel, M., Burri, M., Siegwart, R.: Linear vs nonlinear MPC for trajectory tracking applied to rotary wing micro aerial vehicles. In: Proceedings of 19th IFAC World Congress, pp. 3518–3524 (2017)
42. Kern, B., Böhm, C., Findeisen, R., Allgöwer, F.: Receding horizon control for linear periodic time-varying systems subject to input constraints. In: Magni, L., Raimondo, D.M., Allgöwer, F. (eds.) Nonlinear Model Predictive Control, pp. 109–117. Springer, Berlin (2009)
43. Kühne, F., Lages, W.F., da Silva Jr., J.M.G.: Mobile robot trajectory tracking using model predictive control. In: Proceedings of the 2nd IEEE Latin-American Robotics Symposium (2005)
44. Lam, D., Manzie, C., Good, M.: Application of model predictive contouring control to an X-Y table. In: Proceedings of the 18th IFAC World Congress, pp. 10325–10330 (2011)
45. Limon, D., Alamo, T.: Tracking model predictive control. In: Baillieul, J., Samad, T. (eds.) Encyclopedia of Systems and Control, pp. 1475–1484. Springer, Berlin (2015)
46. Limon, D., Alamo, T., Salas, F., Camacho, E.F.: On the stability of constrained MPC without terminal constraint. IEEE Trans. Autom. Control **51**(5), 832–836 (2006)
47. Limon, D., Alvarado, I., Alamo, T., Camacho, E.F.: MPC for tracking piecewise constant references for constrained linear systems. Automatica **44**(9), 2382–2387 (2008)
48. Limon, D., Pereira, M., de la Peña, D.M., Alamo, T., Jones, C.N., Zeilinger, M.N.: MPC for tracking periodic references. IEEE Trans. Autom. Control **61**(4), 1123–1128 (2016)
49. Maeder, U., Morari, M.: Offset-free reference tracking with model predictive control. Automatica **46**(9), 1469–1476 (2010)
50. Magni, L., De Nicolao, G., Scattolini, R.: Output feedback and tracking of nonlinear systems with model predictive control. Automatica **37**(10), 1601–1607 (2001)
51. Maree, J., Imsland, L.: Combined economic and regulatory predictive control. Automatica **69**, 342–347 (2016)
52. Matschek, J., Bethge, J., Zometa, P., Findeisen, R.: Force feedback and path following using predictive control: concept and application to a lightweight robot. In: Proceedings of 19th IFAC World Congress, pp. 10243–10248 (2017)
53. Mayne, D.Q., Michalska, H.: Receding horizon control of nonlinear systems. IEEE Trans. Autom. Control **35**(7), 814–824 (1990)
54. Mayne, D.Q., Rawlings, J.B., Rao, C.V., Scokaert, P.O.M.: Constrained model predictive control: stability and optimality. Automatica **36**(6), 789–814 (2000)
55. Müller, M.A., Grüne, L.: Economic model predictive control without terminal constraints: optimal periodic operation. In: Proceedings of the 54th IEEE Conference on Decision and Control (CDC), pp. 4946–4951 (2015)
56. Müller, M.A., Angeli, D., Allgöwer, F.: On necessity and robustness of dissipativity in economic model predictive control. IEEE Trans. Autom. Control **60**(6), 1671–1676 (2015)
57. Nielsen, C., Maggiore, M.: Output stabilization and maneuver regulation: a geometric approach. Syst. Control Lett. **55**, 418–427 (2006)
58. Nielsen, C., Maggiore, M.: On local transverse feedback linearization. SIAM J. Control Optim. **47**, 2227–2250 (2008)
59. Nielsen, C., Fulford, C., Maggiore, M.: Path following using transverse feedback linearization: application to a maglev positioning system. Automatica **46**(3), 585–590 (2010)
60. Raković, S.V.: Robust model-predictive control. In: Baillieul, J., Samad, T. (eds.) Encyclopedia of Systems and Control, pp. 1225–1233. Springer, Berlin (2015)
61. Rawlings, J.B., Mayne, D.Q., Diehl, M.M.: Model Predictive Control: Theory, Computation, and Design, 2nd edn. Nob Hill Publishing, Madison (2017)
62. Rawlings, J.B., Angeli, D., Bates, C.N.: Fundamentals of economic model predictive control. In: Proceedings of the 51st IEEE Conference on Decision and Control (CDC), pp. 3851–3861 (2012)
63. Santos, L.O., Afonso, P.A., Castro, J.A., Oliveira, N.M., Biegler, L.T.: On-line implementation of nonlinear MPC: an experimental case study. Control Eng. Pract. **9**(8), 847–857 (2001)

64. Skjetne, R., Fossen, T.I., Kokotović, P.V.: Robust output maneuvering for a class of nonlinear systems. Automatica **40**(3), 373–383 (2004)
65. van Duijkeren, N., Faulwasser, T., Pipeleers, G.: NMPC with economic objectives on target manifolds. In: Proceedings of the 56th IEEE Conference on Decision and Control (CDC), pp. 2519–2524 (2017)
66. Zanon, M., Grüne, L., Diehl, M.: Periodic optimal control, dissipativity and MPC. IEEE Trans. Autom. Control **62**(6), 2943–2949 (2017)
67. Zhang, R., Xue, A., Lu, R., Li, P., Gao, F.: Real-time implementation of improved state-space MPC for air supply in a coke furnace. IEEE Trans. Ind. Electron. **61**(7), 3532–3539 (2014)

Hybrid Model Predictive Control

Ricardo G. Sanfelice

1 Summary

Model Predictive Control (MPC) is a powerful method for the control of dynamical systems under constraints. Due to its computational nature, MPC is typically formulated in discrete time, though some continuous-time approaches are available. In addition, the literature features several MPC strategies that are labeled as hybrid, either due to features of the state of the system, its dynamics, or the control algorithm. The term *hybrid* in the context of MPC has been used to refer to systems that are to be controlled (or the control algorithm) with continuous-valued and discrete-valued state components; e.g., in the control of a thermostat system, a continuous-valued state component would represent temperature and a discrete-valued state component would represent whether the heating/cooling device is on or off. The term hybrid has also been used in the literature for systems with dynamics whose right-hand sides depend discontinuously on their state or on their input. In addition, the term hybrid has also been used to emphasize nonsmoothness in the control algorithm, for instance, when the algorithm switches between different control laws or when it is implemented using the sample-data control paradigm.

Due to the need for digitally implementable control algorithms, it is natural to consider dynamical models given in discrete time. In fact, the vast majority of the results in the literature of hybrid MPC fall into such a category. This article presents those strategies first. There are also a number of strategies that follow a sampled-data control approach. Rather than discretizing the system to control, such approaches incorporate into the mathematical models the continuous-time dynamics of the plant as well as the (periodic) discrete events at which computations occur. These type of strategies are presented after the ones for discrete-time systems. Strategies for systems with combined continuous and discrete dynamics in which the state variables

R. G. Sanfelice (✉)
University of California, Santa Cruz, 1156 High Street, Santa Cruz, CA, USA
e-mail: ricardo@ucsc.edu

S. V. Raković, W. S. Levine (eds.), *Handbook of Model Predictive Control*,
Control Engineering, https://doi.org/10.1007/978-3-319-77489-3_9

may flow and, at times exhibit jumps due to state or input conditions are scarce. As argued in Section 3, new advances in hybrid dynamical systems are needed to develop those strategies.

In light of the outlined state of the art, this chapter covers hybrid MPC results in each of the main three forms seen in the literature and is organized as follows:

1. Discrete-time MPC for systems modeled as discrete-time systems with discontinuous right-hand sides (Section 2.1);
2. Discrete-time MPC for systems modeled as discrete-time systems with a state that contains continuous and discrete-valued states (Section 2.2);
3. Discrete-time MPC for systems modeled as discrete-time systems using memory and logic variables (Section 2.3);
4. Continuous-discrete MPC for systems modeled as continuous-time systems, with piecewise continuous inputs (Section 2.4.1) and piecewise constant inputs (Section 2.4.2);
5. Continuous-discrete MPC for systems modeled as continuous-time systems with local static state-feedback controllers (Section 2.5);
6. Discrete-time MPC for systems modeled as continuous-time linear systems with impulses (Section 2.6).

Each of these approaches are summarized in an unifying framework to facilitate comparison. In particular, the core optimization problem to solve in each approach is formally stated. Computational solutions to each such problem can be obtained using algorithms to solve nonlinear optimization problems available literature; see, e.g., [20].

2 Hybrid Model Predictive Control

The MPC strategies presented in this section perform the following tasks:

- **Measure** the *current state* of the system to control;
- **Predict** for a finite amount of time – the so-called *prediction horizon* – the trajectories of the system to control from the current state and for a family of allowed input signals;
- **Select** an input signal that is a *minimizer* of a given cost functional, which potentially depends on the predicted trajectories and the input, and that satisfies a terminal constraint (if one is given);
- **Apply** the input signal for a finite amount of time – the so-called *control horizon*.

Most MPC algorithms perform these tasks repetitively in the order listed. The following sections provide details on these tasks for each of the strategies listed in Section 1. Regardless of the type of model used, and unless otherwise stated, the state and the input of the system to control are denoted as x and u, while the state and input constraints (if any) are denoted as $\mathcal{X}$ and $\mathcal{U}$, respectively. When the model of the system to control is of discrete-time type, the notation x^+ indicates the value of the state after a discrete-time step. Discrete time is denoted as

k, which takes values in $\mathbb{N} := \{0,1,2\dots\}$. For continuous-time models, the notation $\dot{x}$ denotes the derivative with respect to ordinary time. Ordinary time is denoted as t, and takes values from $\mathbb{R}_{\geq 0} := [0,\infty)$. The MPC strategies require solving an optimization problem using the current state of the system. Since the strategies presented in this article are stated for time-invariant systems, we treat the current state as an initial condition, and denote it as x_0. The prediction horizon in discrete time is denoted $N \in \mathbb{N}_{>0} := \{1,2,\dots\}$, while in continuous time is denoted by $T \in \mathbb{R}_{>0} := (0,\infty)$. Similarly, the control horizon in discrete time is denoted $N_c \in \mathbb{N}_{>0} := \{1,2,\dots\}$, while in continuous time is denoted by $T_c \in \mathbb{R}_{>0} := (0,\infty)$. We also define $\mathbb{N}_{<N} := \{0,1,2,\dots,N-1\}$ and $\mathbb{N}_{\leq N} := \{0,1,2,\dots,N\}$ for a given $N \in \mathbb{N}_{>0}$. Given a vector x, $|x|$ denotes its Euclidean norm and given $p \in [1,\infty]$, $|x|_p$ denotes its p-norm. Given $n \in \mathbb{N}_{>0}$, $\mathbb{R}^n$ denotes the Euclidean space of dimension n.

2.1 Discrete-Time MPC for Discrete-Time Systems with Discontinuous Right-Hand Sides

MPC for discrete-time systems that have piecewise-linear but discontinuous right-hand sides is studied in [23]. Under the name Piecewise Affine System (PWA), the systems considered in [23] take the form

$$x^+ = A_i x + B_i u + f_i \tag{1}$$

$$y = C_i x + D_i u \tag{2}$$

$$\text{subject to } x \in \Omega_i, u \in \mathcal{U}_i(x), i \in S \tag{3}$$

where $S := \{1,2,\dots,s\}$ with s finite, the sequence of constant matrices $\{(A_i,B_i,f_i,C_i,D_i)\}_{i\in S}$ has elements with appropriate dimensions, $\{\Omega_i\}_{i=1}^{s}$ is a collection of polyhedra such that

$$\bigcup_{i\in S} \Omega_i = \mathcal{X}$$

where $\mathcal{X}$ is the state space and

$$\text{int}(\Omega_i) \cap \text{int}(\Omega_j) = \emptyset \qquad \forall i \neq j, i, j \in S$$

where, for each $x \in \Omega_i$, $\mathcal{U}_i(x)$ is the set of allowed inputs. The subset of elements i in S for which $0 \in \overline{\Omega_i}$ is denoted as S_0, while all of the other elements in S define S_1. The origin of (1)–(3) is assumed to be an equilibrium state with $u = 0$, and the requirement $f_i = 0$ for all $i \in S_0$ is further imposed. It should be noted that in [23], this class of systems is referred to as hybrid, presumably due to the right-hand side being discontinuous – in fact, in general, the map

$$(x,u) \mapsto \{A_i x + B_i u + f_i \ : \ i \in S, x \in \Omega_i, u \in \mathcal{U}(x) \}$$

defined on $\bigcup_{i\in S} \bigcup_{x\in\Omega_i} (\{x\} \times \mathcal{U}(x))$ is discontinuous.

Given the current state x_0, a prediction horizon $N \in \mathbb{N}_{>0}$, a terminal constraint set $\mathcal{X}_f$, a stage cost $\mathcal{L}$, and a terminal cost $\mathcal{F}$, the problem of interest consists of minimizing the cost functional

$$\mathcal{J}(x,i,u) := \mathcal{F}(x(N)) + \sum_{k=0}^{N-1} \mathcal{L}(x(k),i(k),u(k))$$

whose argument is actually $k \mapsto (x(k),i(k),u(k))$, which is subject to the constrained dynamics in (1)–(3). Note that $k \mapsto x(k)$ is uniquely defined by x_0 and $k \mapsto (i(k),u(k))$. The initial state for the x component is such that $x(0) = x_0$ and the final value is restricted to $x(N) \in \mathcal{X}_f$. The argument $k \mapsto (x(k),i(k),u(k))$ of the functional is such that $x(k)$ is uniquely defined for each $k \in \mathbb{N}_{\leq N}$, while $(i(k),u(k))$ is uniquely defined for each $k \in \mathbb{N}_{<N}$.

The problem to solve at each discrete-time instant is as follows:

Problem 1. Given the current state x_0, a prediction horizon $N \in \mathbb{N}_{>0}$, a terminal constraint set $\mathcal{X}_f$, a stage cost $\mathcal{L}$, and a terminal cost $\mathcal{F}$

$$\begin{aligned}
&\min \mathcal{J}(x,i,u)\\
&\quad \text{subject to}\\
&\qquad x(0) = x_0\\
&\qquad x(N) \in \mathcal{X}_f\\
&\qquad x(k+1) = A_{i(k)}x(k) + B_{i(k)}u(k) + f_{i(k)} \qquad \forall k \in \mathbb{N}_{<N}\\
&\qquad \left.\begin{array}{l} y(k) = C_{i(k)}x(k) + D_{i(k)}u(k)\\ x(k) \in \Omega_{i(k)},\ u(k) \in \mathcal{U}_{i(k)}(x(k)),\ i(k) \in S \end{array}\right\} \quad \forall k \in \mathbb{N}_{\leq N}
\end{aligned}$$

A minimizer[1] $k \mapsto (x^*(k),i^*(k),u^*(k))$ defines the value of the cost functional $\mathcal{J}^*(x_0) = \mathcal{J}(x^*,i^*,u^*)$.

A typical choice of the functions $\mathcal{L}$ and $\mathcal{F}$ in the cost functional $\mathcal{J}$ is

$$\mathcal{L}(x,i,u) = |Q_i x|_p + |R_i u|_p, \qquad \mathcal{F}(x) = |Px|_p$$

for some $p \in [1,\infty]$, where $\{(Q_i,R_i)\}_{i\in S}$ and P are matrices of appropriate dimensions. When $p = 1$ or $p = \infty$, Problem 1 can be rewritten as a mixed integer linear program (MILP). When the stage and terminal costs are quadratic, Problem 1 can be rewritten as a mixed integer quadratic program (MIQP).

Key properties of Problem 1 were reported in [23], which due to space constraints are not included here. Under suitable assumptions, conditions guaranteeing recur-

[1] Or, equivalently, $k \mapsto (i^*(k),u^*(k))$, due to $k \mapsto x^*(k)$ being uniquely defined by x_0 and $k \mapsto (i^*(k),u^*(k))$.

sive feasibility and asymptotic stability of the origin are given in [23, Theorem III.2]. Properties of and techniques for the computation of the terminal cost and terminal constraint set are also given; see [23, Section IV and Section V]. The issue of existence of minimizers for Problem 1 requires careful treatment, in particular, due to the partitions of the state space introduced by the sets Ω_i. Furthermore, due to Problem 1 being a nonconvex nonlinear optimization problem, the authors of [23] suggest to use optimization solvers such as *fmincon* and *fminunc* in Matlab.

2.2 Discrete-Time MPC for Discrete-Time Systems with Mixed States

An MPC formulation for discrete-time systems to handle switching among different linear dynamics, on/off inputs, logic states and their transitions, as well as logic constraints on input and state variables is given in [5–8, 23]. The nominal models considered therein, which are called Mixed Logical Dynamical (MLD) systems, are discrete-time systems involving continuous-valued and discrete-valued states, inputs, and outputs, as well as constraints depending on the states, the inputs, and the outputs. These system models are given as

$$x^{+} = Ax + B_1 u + B_2 \delta + B_3 z + B_4 \tag{4}$$

$$y = Cx + D_1 u + D_2 \delta + D_3 z + D_4 \tag{5}$$

$$\text{subject to } E_2 \delta + E_3 z \leq E_1 u + E_4 x + E_5 \tag{6}$$

In most MLD models in the literature, the state vector x is partitioned as (x_c, x_ℓ), where $x_c \in \mathbb{R}^{n_c}$ are the continuous-valued components and $x_\ell \in \{0,1\}^{n_\ell}$ are the discrete-valued components of x. Similarly, the input u is partitioned as $(u_c, u_\ell) \in \mathbb{R}^{m_c} \times \{0,1\}^{m_\ell}$ and the output y as $(y_c, y_\ell) \in \mathbb{R}^{p_c} \times \{0,1\}^{p_\ell}$. The continuous-valued auxiliary variables $z \in \mathbb{R}^{r_c}$ and the discrete-valued auxiliary variables $\delta \in \{0,1\}^{r_\ell}$ are added to capture constraints, logic statements, and the such. The matrices A, $\{B_i\}_{i=1}^{3}$, B_4, C, $\{D_i\}_{i=1}^{3}$, D_4, and $\{E_i\}_{i=1}^{5}$ have suitable dimensions. Given the current state x_0, a prediction horizon $N \in \mathbb{N}_{>0}$, and a terminal constraint set $\mathcal{X}_f$, the problem of interest consists of minimizing the cost functional

$$\mathcal{J}(x, z, \delta, u)$$

whose argument is actually $k \mapsto (x(k), z(k), \delta(k), u(k))$, and is subject to the constrained dynamics in (4)–(6). The initial state for the x component is such that $x(0) = x_0$ and its final value is restricted to $x(N) \in \mathcal{X}_f$. In the literature, this class of dynamical systems is referred to as hybrid mainly due to having a discontinuous right-hand side and due to the states, inputs, and outputs having continuous-valued and discrete-valued components.

The problem to solve at each discrete-time instant is as follows:

Problem 2. Given the current state x_0, a prediction horizon $N \in \mathbb{N}_{>0}$, a terminal set $\mathcal{X}_f$, and a cost functional $\mathcal{J}$

$$\begin{aligned}
&\min \mathcal{J}(x,z,\delta,u)\\
&\quad \text{subject to}\\
&\qquad x(0) = x_0\\
&\qquad x(N) \in \mathcal{X}_f\\
&\qquad x(k+1) = Ax(k) + B_1u(k) + B_2\delta(k) + B_3z(k) + B_4 \quad \forall k \in \mathbb{N}_{<N}\\
&\qquad \left.\begin{array}{l} y(k) = Cx(k) + D_1u(k) + D_2\delta(k) + D_3z(k) + D_4\\ E_2\delta(k) + E_3z(k) \le E_1u(k) + E_4x(k) + E_5 \end{array}\right\} \quad \forall k \in \mathbb{N}_{\le N}
\end{aligned}$$

A minimizer $k \mapsto (x^*(k), z^*(k), \delta^*(k), u^*(k))$ defines the value of the cost functional $\mathcal{J}^*(x_0) = \mathcal{J}(x^*, z^*, \delta^*, u^*)$.

A particular choice of the cost functional $\mathcal{J}$ made in [5–8, 23] is

$$\mathcal{J}(x,z,\delta,u) = \sum_{k=0}^{N-1} \left(|Qx(k)|_p + |Ru(k)|_p + |Q_\delta \delta(k)|_p + |Q_z z(k)|_p\right) + |Px(N)|_p$$

for some $p \in [1,\infty]$. The term inside the sum is the stage cost, which, given matrices Q, Q_δ, and Q_z, involves the value of the current and predicted state x, input u, and auxiliary variables (δ, z) for $N-1$ steps in the future. The last term in $\mathcal{J}$ is the terminal cost.

Perhaps the most comprehensive reference about Problem 2 is Chapter 18 of the recent monograph [8]. Therein, the authors consider the same model (but with $B_4 = 0$ and $D_4 = D_5$) in Section 18.1. By picking the cost functional above, Problem 2 is formulated as a MIQP or MILP, according to the choice of p. A complete rewrite of Problem 2 including slack variables is given in (18.28) in the said reference. Mixed-integer optimization methods suitable to solve Problem 2 are also outlined. The chapter concludes with discussions on how to derive state feedback solutions via the batch approach and the recursive approach. This class of systems is referred to as hybrid due to the right-hand side being discontinuous and due to the states, inputs, and outputs having continuous-valued and discrete-valued components.

2.3 Discrete-Time MPC for Discrete-Time Systems Using Memory and Logic Variables

Variations of the basic MPC formulation, obtained by adding memory and logic states, for discrete-time systems of the following form is proposed in [42]:

$$x^+ = g(x,u) \tag{7}$$

$$\text{subject to } x \in \mathcal{X}, u \in \mathcal{U} \tag{8}$$

The set $\mathcal{X}$ defines the constraint on the state and $\mathcal{U}$ is the set of allowed inputs. Recall from chapter "The Essentials of Model Predictive Control" of this handbook (see also chapters "Dynamic Programming, Optimal Control and Model Predictive Control" and "Set-Valued and Lyapunov Methods for MPC") that the basic MPC formulation consists of minimizing the cost functional

$$\mathcal{J}(x,u) := \mathcal{F}(x(N)) + \sum_{k=0}^{N-1} \mathcal{L}(x(k),u(k))$$

where x is the current state, $N \in \mathbb{N}_{>0}$ is the prediction horizon, $\mathcal{L}$ is the stage cost, and $\mathcal{F}$ is the terminal cost. The function $k \mapsto x(k)$ in the cost functional $\mathcal{J}$ is the solution to (7)–(8) at time k, starting from the initial condition x_0 and under the influence of the input sequence $k \mapsto u(k)$. The two variations of this MPC formulation proposed in [42] are described next.

To incorporate memory in the selection of the input, define the buffer gain as $\mu > 1$, the memory horizon as $M \in \mathbb{N}_{>0}, M \leq N$, and the memory state as $\ell = (\ell_1, \ell_2, \ldots, \ell_M)$. The optimization problem in [42] involving the memory state ℓ that is to be solved at each discrete-time instant is as follows:

Problem 3. Given the current state x_0, a prediction horizon $N \in \mathbb{N}_{>0}$, a stage cost $\mathcal{L}$, a terminal cost $\mathcal{F}$, a buffer gain $\mu > 1$, a memory horizon $M \in \mathbb{N}_{>0}$ such that $M \leq N$, and the current memory state ℓ, solve the following problems:

Problem 3a:

$$\begin{aligned}
&\min \mathcal{J}(x,u)\\
&\quad \text{subject to}\\
&\qquad x(0) = x_0\\
&\qquad x(k+1) = g(x(k),u(k)) \qquad \forall k \in \mathbb{N}_{<N}\\
&\qquad u(k) \in \mathcal{U} \qquad \forall k \in \mathbb{N}_{\leq N}
\end{aligned}$$

Denote the solution to this problem as $k \mapsto (x^*(k), v^*(k))$ and define $V(x_0) = \mathcal{J}(x^*, v^*)$ as the associated value function.[2]

Problem 3b:

$$\begin{aligned}
&\min \mathcal{J}(x,u)\\
&\quad \text{subject to}
\end{aligned}$$

[2]Note that the only constraint on $v^*(N)$ is for it to belong to $\mathcal{U}$.

$$\begin{aligned} x(0) &= x_0 \\ x(k+1) &= g(x(k),u(k)) && \forall k \in \mathbb{N}_{<N} \\ u(k-1) &= \ell_k && \forall k \in \{1,2,\ldots,M\} \\ u(k) &\in \mathcal{U} && \forall k \in \mathbb{N}_{\leq N} \end{aligned}$$

Denote the solution to this problem as $k \mapsto (x^*(k), w^*(k))$ and define $W(x_0,\ell) = \mathcal{J}(x^*,w^*)$ as the associated value function.

After solving[3] Problem 3a and Problem 3b, update the memory state according to

$$\ell^+ = \begin{cases} (v^*(1), v^*(2), \ldots, v^*(M)) & \text{if } W(x_0,\ell) > \mu V(x_0) \\ (w^*(1), w^*(2), \ldots, w^*(M)) & \text{if } W(x_0,\ell) \leq \mu V(x_0) \end{cases}$$

and the minimizing control input $k \mapsto u^*(k)$ is[4]

$$u^* = \begin{cases} v^* & \text{if } W(x_0,\ell) > \mu V(x_0) \\ w^* & \text{if } W(x_0,\ell) \leq \mu V(x_0) \end{cases}$$

Problem 3a provides a solution to the standard MPC problem without memory states. The solution from this problem is used in Problem 3b, which uses the current value of the memory state ℓ as it enforces that the first M entries of u, namely, $(u(0), u(1), \ldots, u(M-1))$, are equal to ℓ. The selection of the control input is such that when the improvement provided by the solution to the standard MPC problem is significant when compared to the one with the memory states. The optimal control input u^* is given by v^* in Problem 3a when the improvement provided by the solution to that problem (namely, $k \mapsto (x^*(k), v^*(k))$) is "significantly better" – as characterized by the buffer gain $\mu > 1$ – than the improvement provided by the solution to the problem involving memory states (namely, $k \mapsto (x^*(k), w^*(k))$ in Problem 3b). More precisely, if the value function of the problem that does not use information about the previous solution (i.e., Problem 3a) is a factor $1/\mu \in (0,1)$ smaller than the value function of the problem solved using previous information (i.e., Problem 3b), namely,

$$V(x_0) < \frac{1}{\mu} W(x_0,\ell) \tag{9}$$

then the optimal solution comes from Problem 3a and the memory state is updated with the input component of the solution to that problem. Note that when $V(x_0) \geq$

[3]The solution component x^* in Problem 3a and in Problem 3b would be most likely different, but we use the same label due to not being part of the logic.

[4]The state component $k \mapsto x^*(k)$ associated to $k \mapsto u^*(k)$ is obtained by applying u^* to the system to control.

$\frac{1}{\mu}W(x_0,\ell)$ and the control horizon is equal to one, the input applied to the system to control would be ℓ_1 and that ℓ is subsequently updated to $(\ell_2,\ell_3,\dots,\ell_M,w^*(M))$.

To incorporate logic states in the selection of the input, define the buffer gain as $\mu > 1$, the logic state as q taking its value from $Q := \{1,2,\dots,\bar{q}\}$ where $\bar{q} \in \mathbb{N}_{>0}$, and, for each $q \in Q$, define the cost functional

$$\mathcal{J}_q(x,u) := \mathcal{F}_q(x(N)) + \sum_{k=0}^{N-1} \mathcal{L}_q(x(k),u(k))$$

where $\mathcal{L}_q$ is the stage cost and $\mathcal{F}_q$ the terminal cost associated with q. The proposed optimization problem involving a logic variable q to solve at each discrete-time instant is as follows:

Problem 4. Given the current state x_0, a prediction horizon $N \in \mathbb{N}_{>0}$, stage costs $\{\mathcal{L}_q\}_{q\in Q}$, terminal costs $\{\mathcal{F}_q\}_{q\in Q}$, and a buffer gain $\mu > 1$, solve the following problem for each $q \in Q$:

Problem 4-q:

$$\begin{aligned} &\min \mathcal{J}_q(x,u) \\ &\quad \text{subject to} \\ &\quad\quad x(0) = x_0 \\ &\quad\quad x(k+1) = g(x(k),u(k)) && \forall k \in \mathbb{N}_{<N} \\ &\quad\quad u(k) \in \mathcal{U} && \forall k \in \mathbb{N}_{\leq N} \end{aligned}$$

Denote the solution to this problem as $k \mapsto (x^{q*}(k), v^{q*}(k))$ and define $V_q(x_0) = \mathcal{J}_q(x^{q*}, v^{q*})$ as the associated value function.

After solving Problem 4-q for each $q \in Q$, pick

$$q^* \in \arg\min_{q\in Q} V_q(x_0)$$

update the logic state according to

$$q^+ = \begin{cases} q^* & \text{if } V_q(x_0) > \mu V_{q^*}(x_0) \\ q & \text{if } V_q(x_0) \leq \mu V_{q^*}(x_0) \end{cases}$$

and the minimizing control input $k \mapsto u^*(k)$ is

$$u^* = \begin{cases} v^{q^*} & \text{if } V_q(x_0) > \mu V_{q^*}(x_0) \\ v^q & \text{if } V_q(x_0) \leq \mu V_{q^*}(x_0) \end{cases}$$

The value functions $V_q(x_0)$ associated to each optimal solution obtained in Problem 4-q are compared when determining a new value of the logic variable. Such value of q is denoted as q^*, and is such that $V_{q^*}(x_0)$ is among those minimizers in $\{V_q(x_0)\}_{q \in Q}$ with "enough improvement" – as characterized by μ –when compared to the value function associated to the current value of q. In fact, according to Problem 4, a change on the value of the logic variable occurs when there exists $q^* \in Q$ such that

$$V_{q^*}(x_0) < \frac{1}{\mu} V_q(x_0) \tag{10}$$

Discussions in [42] indicate that the MPC strategies with memory states and logic variables guarantee robustness to small measurement noise. Such robustness is possible due to the hysteresis mechanism incorporated by conditions (9) and (10) in the strategies above. It is also likely that these MPC strategies confer robustness to other classes of perturbations, mainly due to the said hysteresis mechanism they implement, which, in particular, prevents the control law from chattering.

Several other variants of MPC for discrete-time systems are available in the literature. In particular, [44] and [45] propose a discrete-time MPC strategy that explicitly accounts for computation time and events.

2.4 *Periodic Continuous-Discrete MPC for Continuous-Time Systems*

In this section, we present model predictive control strategies for continuous-time systems that periodically recompute an input signal solving the optimization problem and apply it over a bounded horizon. Such MPC strategies appear in the literature under the name *continuous-discrete MPC*.

2.4.1 With Piecewise Continuous Inputs

MPC for continuous-time systems with input constraints is proposed in [10]. The class of systems is given by

$$\dot{x} = f(x,u) \qquad x \in \mathbb{R}^n,\ u \in \mathcal{U} \tag{11}$$

where $\mathcal{U}$ is the input constraint set. The right-hand side f is assumed to be twice continuously differentiable, to satisfy $f(0,0) = 0$, and such that it leads to unique solutions under the effect of piecewise right-continuous input signals. The input constraint set $\mathcal{U}$ is assumed to be compact, convex, and with the property that 0 belongs to the interior of $\mathcal{U}$.

Given the current state x_0, a prediction horizon $T > 0$, a terminal constraint set $\mathcal{X}_f$, a stage cost $\mathcal{L}$, and a terminal cost $\mathcal{F}$, the problem of interest consists of minimizing the cost functional

$$\mathcal{J}(x,u) := \mathcal{F}(x(T)) + \int_0^T \mathcal{L}(x(\tau),u(\tau))d\tau \tag{12}$$

whose argument is actually $t \mapsto (x(t), u(t))$ which is subject to the constrained dynamics in (11). The initial condition is such that $x(0) = x_0$, and the value of x after T seconds is restricted to $\mathcal{X}_f$. More precisely, the problem to solve every T seconds is as follows:

Problem 5. Given the current state x_0, a prediction horizon $T > 0$, a terminal constraint set $\mathcal{X}_f$, a stage cost $\mathcal{L}$, and a terminal cost $\mathcal{F}$

$$
\begin{aligned}
&\min \mathcal{J}(x,u) \\
&\quad \text{subject to} \\
&\quad\quad x(0) = x_0 \\
&\quad\quad x(T) \in \mathcal{X}_f \\
&\quad\quad \frac{d}{dt}x(t) = f(x(t),u(t)) \quad \forall t \in (0,T) \\
&\quad\quad u(t) \in \mathcal{U} \quad \forall t \in [0,T]
\end{aligned}
$$

A minimizer $t \mapsto (x^*(t), u^*(t))$ defines the value of the cost functional $\mathcal{J}^*(x_0) = \mathcal{J}(x^*, u^*)$.

In [10], the approach to solve this problem consists of picking $\mathcal{X}_f$ to be a neighborhood of the origin that is invariant in forward time for the closed-loop system resulting from using a (local) linear state-feedback law of the form Kx, and by picking $\mathcal{F}$ so that the terminal cost upper bounds the infinite horizon cost from $\mathcal{X}_f$. According to [10], the design of the set $\mathcal{X}_f$, the gain K, and the function $\mathcal{F}$ can be performed offline. Due to the value of the cost functional providing a bound to an infinite horizon cost problem, the authors refer to this strategy as *quasi-infinite horizon nonlinear MPC.*

The application of the stabilizing linear feedback law Kx to the system (11) generates a solution-input pair $t \mapsto (x(t), u(t))$ that satisfies the input and terminal constraints, for any initial condition $x_0 \in \mathcal{X}_f$. Therefore, the feasible set of initial conditions to Problem 5 includes $\mathcal{X}_f$. The actual moving horizon implementation of the MPC strategy in [10] would not use the (local) linear state-feedback law, but rather, guarantee feasibility. The moving horizon implementation would recursively apply the open-loop optimal control solution for $\delta < T$ seconds. The constant δ defines the sampling period for obtaining new measurements of the state of the plant. At such events, the optimal solution to the open-loop problem is recomputed and then the input to the plant is updated.

Note that forward/recursive feasibility of the closed-loop is guaranteed by the terminal constraint and the local feedback law Kx. This is because, as stated in [10], the MPC strategy can be thought of as a receding horizon implementation of the following switching control strategy:

- Over a finite horizon of length T, apply the optimal input obtained by solving Problem 5 so as to drive the state to the terminal set;

- Once the state is in the terminal set, switch the control input to the (local) linear state-feedback law so as to steer the state to the origin.

Recently, this strategy was extended in [1] to the case where the MPC law is recomputed at time instances that are not periodic.

2.4.2 With Piecewise Constant Inputs

A minimizing input $t \mapsto u^*(t)$ obtained from a solution to Problem 5 is a piecewise-continuous function defined on an interval of length equal to the prediction horizon T. Using a similar continuous-discrete MPC strategy, in [27], the class of inputs allowed is restricted to piecewise-constant functions and the strategy is of sample-and-hold type. More precisely, the input u satisfies the following:

($\star$) The input signal u is a piecewise-constant function with intervals of constant value of length δ seconds, within the control horizon $N_c\delta$, where $N_c \in \mathbb{N}_{>0}$ and $N_c\delta \leq T$.

In such a (zero-order) sample-and-hold approach, the input applied to the plant remains constant in between the sampling events. In [27], this mechanism is modeled by adding an extra state x_u to the system with the following dynamics:

$$\begin{aligned} \dot{x}_u &= 0 && \text{in between sampling events} \\ x_u^+ &= \kappa(x) && \text{at sampling events} \end{aligned}$$

*2inwhere κ denotes the function assigning the feedback at each event. Furthermore, the setting in [27] allows for state constraints $x \in \mathcal{X}$ in (11), where $\mathcal{X}$ is the state constraint set.

Given the current state x_0, a prediction horizon T, a sampling time $\delta \in (0,T]$, a control horizon $N_c\delta \leq T$, and a terminal constraint set $\mathcal{X}_f$, the problem formulated in [27] is that of minimizing (12) at every sampling time instant, where

$$\mathcal{F}(x) = x^\top Px, \qquad \mathcal{L}(x,u) = x^\top Qx + u^\top Ru \tag{13}$$

*2infor given matrices P, Q, and R of appropriate dimensions. The argument of (12) is actually $t \mapsto (x(t), u(t))$ with the input component being a piecewise constant function.

The problem to solve at each periodic sampling event occurring every δ seconds is as follows:

Problem 6. Given the current state x_0, a prediction horizon $T > 0$, a sampling time $\delta \in (0,T]$, a control horizon $N_c\delta \in (0,T]$, a terminal constraint set $\mathcal{X}_f$, and matrices P, Q, and R

$$\min \mathcal{J}(x,u)$$
$$\text{subject to}$$

$$x(0) = x_0$$
$$x(T) \in \mathcal{X}_f$$
$$\frac{d}{dt}x(t) = f(x(t),u(t)) \quad \forall t \in (0,T)$$
$$x(t) \in \mathcal{X}, \;\; u(t) \in \mathcal{U} \quad \forall t \in [0,T]$$
$$u \text{ satisfies } (\star)$$

A minimizer $t \mapsto (x^*(t), u^*(t))$ defines the value of the cost functional $\mathcal{J}^*(x_0) = \mathcal{J}(x^*, u^*)$.

A somewhat related problem that involves periodic continuous-discrete MPC for continuous-time systems with piecewise constant inputs was studied in [32]. In that reference, MPC is used to solve the problem of finding a sampled version of a continuous-time controller that leads to a trajectory of the resulting sample-data system that is as close as possible to the trajectory of the closed-loop system with the original continuous-time controller. To characterize closeness between them, the stage cost of the MPC problem in [32] penalizes the error between the two trajectories.

2.5 Periodic Continuous-Time MPC for Continuous-Time Systems Combined with Local Static State-Feedback Controllers

A strategy uniting two controllers for the asymptotic stabilization of the origin of continuous-time systems in affine control form is provided in [13]; see also [11, Chapter 5]. The family of continuous-time systems considered in [13] is given by

$$\dot{x} = f_1(x) + f_2(x)u \qquad x \in \mathbb{R}^n, \; u \in \mathcal{U} \tag{14}$$

where $\mathcal{U} = \{u \; : \; |u| \leq u_{\max} \}$ for some $u_{\max} \geq 0$ and $f_1(0) = 0$.

One of the controllers in the proposed strategy is a continuous-discrete MPC controller with piecewise-constant inputs and implemented with periodic sampling, similar to the strategy presented in Section 2.4.2. The stage cost used has the same form as in (13). In [13], this particular continuous-discrete MPC algorithm is designed so that, at each periodic sampling event, Problem 6 is solved with control horizon equal to the prediction horizon T and no state constraint set.

The second controller in the strategy in [13] consists of a family of finitely many locally stabilizing static state-feedback controllers $\{\kappa_1, \kappa_2, \ldots, \kappa_r\}$, $r \in \mathbb{N}_{>0}$, that are designed using a family of control Lyapunov functions $\{V_1, V_2, \ldots, V_r\}$, following the universal construction proposed in [24]. These individual control laws can be designed to satisfy the input constraint in (14). When the second controller is the one applied to (14), the particular element in the family that is actually used is such that x belongs to its basin of attraction, which in [13] is defined by a sublevel set of the control Lyapunov function associated with that controller.

The two controllers outlined above are combined via a strategy that uses the static state-feedback controllers as "fall-back" in the event that the continuous-discrete MPC controller is unable to achieve closed-loop stability, which could be the case when Problem 6 does not have a solution or does not terminate before δ seconds. The strategy proposed for combining them is as follows. The control system in (14) is treated as the switching system

$$\dot{x} = f_1(x) + f_2(x)u_\sigma \qquad x \in \mathbb{R}^n,\ u \in \mathcal{U}$$

where $t \mapsto \sigma(t) \in \{1,2\}$ is a switching signal that determines which controller is being used: $\sigma = 1$ indicates that $u = u_1$ with u_1 assigned by the MPC control law κ, and $\sigma = 2$ that $u = u_2$ with u_2 assigned by an element in the family of static state-feedback laws $\{\kappa_1, \kappa_2, \ldots, \kappa_r\}$. The particular selection of σ in [13] is

$$\sigma(t) = \begin{cases} 1 & \text{if } t \in [0, \bar{T}) \\ 2 & \text{if } t \in [\bar{T}, \infty) \end{cases} \tag{15}$$

where $\bar{T}$ is the smallest time such that

$$L_{f_1}V_i(x(\bar{T})) + L_{f_2}V_i(x(\bar{T}))\kappa(\bar{T}) \geq 0 \tag{16}$$

or the MPC algorithm fails to provide an output value, where $i \in K := \{1, 2, \ldots, r\}$ is such that $x(0)$ belongs to the basin of attraction induced by the static state-feedback controller κ_i. The idea behind the state-based triggering condition (16) is that since $x(0)$ is in the basin of attraction of a controller in the family $\{\kappa_1, \kappa_2, \ldots, \kappa_r\}$, then a solution guarantees a strict decrease of the control Lyapunov function associated with that controller.

The work in [13] also includes a switching strategy that is designed to enhance closed-loop performance. Also, an extension to these strategies for the case when the right-hand side of (14) includes additive uncertainties is proposed in [29]. See also [30].

2.6 *Periodic Discrete-Time MPC for Continuous-Time Linear Systems with Impulses*

MPC for linear time-invariant systems with impulses in the state, and with state and input constraints is proposed in [37]. The set of times at which impulses occur are predetermined and given by the sequence of times

$$\{t_k\}_{k \in \mathbb{N}}, \qquad t_k = k\delta$$

where $\delta > 0$ is the sampling (or, as defined in [37], the impulsive) period. The class of impulsive systems is given by

$$\dot{x}(t) = Ax(t) \qquad \forall t \in \mathbb{R}_{\geq 0},\ t \neq k\delta \tag{17}$$
$$x(t^+) = x(t) + Bu_k \qquad \forall t = k\delta \tag{18}$$

for each $k \in \mathbb{N}$, where $t \mapsto x(t)$ is a solution associated to $\{u_k\}_{k\in\mathbb{N}}$ and such that

$$x(t) \in \mathcal{X} \qquad \forall t \in \mathbb{R}_{\geq 0}, \qquad u_k \in \mathcal{U} \qquad \forall k \in \mathbb{N}$$

and $x(t^+)$ is the right limit of $x(t)$ at $t = k\delta$.

The MPC problem in [37] employs over approximation techniques to reduce the infinite number of constraints arising from the dynamics of (17)–(18) to a finite set of inequalities. For a given $\delta > 0$, the collection $\{A_i\}_{i=1}^K$ is introduced to define a polytopic over approximation for the flows of (17)–(18), namely, choose $\{A_i\}_{i=1}^K$ such that

$$\{\exp(At)\ :\ t \in [0,\delta]\ \} \subset \mathrm{co}\{A_i\}_{i=1}^K$$

To determine the stage cost $\mathcal{L}$, define the polytope

$$S(x,u) = \mathrm{co}\{A_i\}_{i=1}^K\,(x+Bu)$$

and, given a set $\mathcal{Z}$ and a terminal constraint set $\mathcal{X}_f \subset \mathcal{Z}$ that, for some feedback, is invariant for (17)–(18), define the input constraint

$$\mathcal{U}_f(x) = \left\{u \in \mathcal{U}\ :\ \exp(A\delta)(x+Bu) \in \mathcal{X}_f, S(x,u) \subset \mathcal{Z}\ \right\}$$

With these definitions, the stage cost $\mathcal{L}$ is given by the distance to the set

$$D = \left\{(x,u)\ :\ x \in \mathcal{X}_f, u \in \mathcal{U}_f(x)\ \right\}$$

which is the graph of $\mathcal{U}_f$ on $\mathcal{X}_f$.

Within the above setting, given the current state x_0, a prediction horizon $N \in \mathbb{N}_{>0}$, a terminal constraint set $\mathcal{X}_f$, and a set $\mathcal{Z}$, the problem formulated in [37] consists of minimizing the cost functional

$$\mathcal{J}(x,u) = \sum_{k=0}^{N-1} \mathcal{L}(x(\tau_k),u(\tau_k))$$

whose argument is $k \mapsto (x(\tau_k),u(\tau_k))$, where $x(\tau_k)$ is the evaluation at the N future impulse times τ_k of the solution to (17)–(18) from x_0 resulting from applying $u(\tau_k)$ at the impulse times, where for some $k_0 \in \mathbb{N}$,

$$\tau_k = t_{k+k_0}$$

and

$$u(\tau_k) = u_{k+k_0}$$

for each $k \in \{0,1,\ldots,N-1\}$. The constraints associated to the minimization problem are: (i) the polytope S remains within the state constraint set $\mathcal{X}$, and (ii) the value of the resulting solution reaches the terminal constraint set $\mathcal{X}_f$ at the end of the prediction horizon N. More precisely, the problem to solve at each periodic impulsive event is as follows:

Problem 7. Given the current state x_0, a prediction horizon $N \in \mathbb{N}_{>0}$, a terminal constraint set $\mathcal{X}_f$, a set $\mathcal{Z}$, and a stage cost $\mathcal{L}$

$$
\begin{aligned}
&\min \mathcal{J}(x,u) \\
&\quad \text{subject to} \\
&\quad\quad x(0) = x_0 \\
&\quad\quad x(\tau_N) \in \mathcal{X}_f \\
&\quad\quad \left.\begin{array}{ll}
\dot{x}(t) = Ax(t) & \forall t \in (0, N\delta), t \neq \tau_k \\
x(t^+) = x(t) + Bu(t) & \forall t = \tau_k \\
u(\tau_k) \in \mathcal{U} & \\
S(x(\tau_k), u(\tau_k)) \subset \mathcal{X} &
\end{array}\right\} \forall k \in \{0,1,\ldots,N-1\}
\end{aligned}
$$

A minimizer $k \mapsto (x^*(k), u^*(k))$ defines the value of the cost functional $\mathcal{J}^*(x_0) = \mathcal{J}(x^*, u^*)$.

In [37], instead of imposing the conditions involving the impulsive system in Problem 7 that are in the first two lines of the expressions within the brace, conditions on the solution x evaluated at each τ_k are imposed. Such a difference is possible due to the impulses occurring periodically and the continuous-time dynamics being linear. In fact, the values of the solution at the instants τ_k are given by the solution to the discrete-time system

$$x^+ = \exp(A\delta)(x + Bu)$$

from $x(0) = x_0$ and under the effect of the input equal to $u(\tau_k)$. The stability notion used therein only requires closeness and convergence of the values of the solution at the instants τ_k, which the authors refer to as a weak property. Following such a discretization approach, it is shown in [37] that Problem 7 can be formulated as a convex quadratic program (when $\mathcal{L}$ is convex). The MPC strategy in [37] combines features of impulsive systems and of sample-data systems, and is one of the MPC approaches found in the literature that is closest to hybrid dynamical systems, as introduced in the next section.

3 Towards MPC for Hybrid Dynamical Systems

Hybrid dynamical systems are systems with states that can evolve continuously (or flow) and, at times, have abrupt changes (or jump). Such systems may have state components that are continuous valued as well as components that are discrete valued, similar to the discrete-time systems described in Section 2.2. The conditions allowing continuous or discrete changes typically depend on the values of the state, the inputs, and outputs. The development of MPC strategies for such systems is in its infancy, possibly the most related strategy being the one described in Section 2.6 (even though it essentially replaces the flows by $\exp(A\delta)$ due to assuming periodic impulses occurring every δ seconds). On the other hand, research on methods to solve optimal control problems for hybrid dynamical systems has been quite active over the past few decades, and such developments could be exploited to develop MPC strategies for such systems. In particular, maximum principles of optimal control following Pontryagin's maximum principle [33] have been generated for systems with discontinuous right-hand side [38] and for certain classes of hybrid systems [15, 36, 39]. Shown to be useful in several applications [12, 39], these principles establish necessary conditions for optimality in terms of an adjoint function and a Hamiltonian satisfying the "classical" conditions along flow, in addition to matching conditions at jumps.

Numerous frameworks for modeling and analysis of hybrid systems have appeared in the literature. These include the work of Tavernini [40], Michel and Hu [31], Lygeros et al. [26], Aubin et al. [4], and van der Schaft and Schumacher [43], among others. In the framework of [16, 17] the continuous dynamics (or flows) of a hybrid dynamical system are modeled using differential inclusions while the discrete dynamics (or jumps) are captured by difference inclusions. Trajectories to a hybrid dynamical system conveniently use two parameters: an ordinary time parameter $t \in \mathbb{R}_{\geq 0}$, which is incremented continuously as flows occur, and a discrete time parameter $j \in \mathbb{N}$, which is incremented at unitary steps when jumps occur. The conditions determining whether a trajectory to a hybrid system should flow or jump are captured by subsets of the state space and input space. In simple terms, given an input $(t,j) \mapsto u(t,j)$, a trajectory $(t,j) \mapsto x(t,j)$ to a hybrid system satisfies, over intervals of flow,

$$\frac{d}{dt}x(t,j) \in F(x(t,j),u(t,j))$$

when

$$(x(t,j),u(t,j)) \in C$$

and, at jump times,

$$x(t,j+1) \in G(x(t,j),u(t,j))$$

when

$$(x(t,j),u(t,j)) \in D$$

The domain of a trajectory x is denoted $\operatorname{dom} x$, which is a *hybrid time domain* [17]. The above definition of trajectory (or solution) implicitly assumes that $\operatorname{dom} x = \operatorname{dom} u = \operatorname{dom}(x,u)$.

In this way, a hybrid dynamical system is defined by a set C, called the *flow set*, a set-valued map F, called the *flow map*, a set D, called the *jump set*, and a set-valued map G, called the *jump map*. Then, a hybrid system with state x and input u can be written in the compact form

$$\mathcal{H} \, : \, \begin{cases} \dot{x} \;\in F(x,u) & (x,u) \in C \\ x^{+} \in G(x,u) & (x,u) \in D \end{cases} \tag{19}$$

The objects defining the data of the hybrid system $\mathcal{H}$ are specified as $\mathcal{H} = (C,F,D,G)$. The state space for x is given by the Euclidean space $\mathbb{R}^n$ while the space for inputs u is given by the set $\mathcal{U}$. The set $C \subset \mathbb{R}^n \times \mathcal{U}$ defines the set of points in $\mathbb{R}^n \times \mathcal{U}$ in which flows are possible according to the differential inclusion defined by the flow map $F : C \rightrightarrows \mathbb{R}^n$. The set $D \subset \mathbb{R}^n \times \mathcal{U}$ defines the set of points in $\mathbb{R}^n \times \mathcal{U}$ from where jumps are possible according to the difference inclusion defined by the set-valued map $G : D \rightrightarrows \mathbb{R}^n$.

Given the current value of the state x_0, and the amount of flow time T and the number of jumps J to predict forward in time, which define a *hybrid prediction horizon* (T,J), an MPC strategy will need to compute trajectories of (19) over the window of hybrid time $[0,T] \times \{0,1,\ldots,J\}$ for all possibly allowed inputs. The fact that different inputs may be applied from the current state x_0 suggests that there may be multiple possible trajectories of (19) from such a point. While this feature is already present in the receding horizon approaches in [8, 9, 19, 27, 28, 34], the hybrid case further adds nonuniqueness due to the potential nonuniqueness of solutions to (19), in particular, due to overlaps between the flow and the jump sets. To deal with nonuniqueness, one would need a set-valued model for prediction that includes all possible predicted hybrid trajectories (and their associated inputs) from x_0 and over $[0,T] \times \{0,1,\ldots,J\}$.

An appropriate cost functional for an MPC strategy for (19), defined over the prediction horizon (T,J), may take the form

$$\begin{aligned} \mathcal{J}(x,u) := & \int_{t:(t,j)\in \operatorname{dom}(x,u), 0\leq t\leq T} \mathcal{L}_c(x(t,j),u(t,j))dt \\ & + \sum_{j:(t,j)\in \operatorname{dom}(x,u), 0<j\leq J} \mathcal{L}_d(x(t_j,j),u(t_j,j)) + \mathcal{F}(x(T,J)) \end{aligned} \tag{20}$$

where $t_1,t_2,\ldots,t_j,\ldots$ are the jump times of (x,u). The first two arguments of $\mathcal{J}$ correspond to a solution to (19) from $x_0 = x(0,0)$. The function $\mathcal{L}_c$ captures the stage cost of flowing and $\mathcal{L}_d$ captures the stage cost of jumping relative to desired subsets of the state space and the input space, respectively. The function $\mathcal{F}$ defines the terminal cost. The key challenge is in establishing conditions such that the value function, which at every point x_0 is given by

$$\mathcal{J}^{\star}(x_0) := \mathcal{J}(x_\star, u_\star)$$

with $(x_\star, u_\star)$ being minimizers of $\mathcal{J}$ from x_0, certifies the desired asymptotic stability property by guaranteeing that the stage cost approaches zero.

The goal of any MPC strategy for (19) would certainly be to minimize the cost functional $\mathcal{J}$ in (20) over the finite-time hybrid horizon $[0,T] \times \{0,1,\ldots,J\}$ defined by the hybrid prediction horizon (T,J). Given the current value of the state x_0, a potential form of this control law would be

$$\kappa_c(x_0) := u_\star \tag{21}$$

*2inwhere the choice of the function $u_\star$ is updated when a timer τ_c reaches the hybrid control horizon $N_c + T_c \leq T + J$, and the dynamics of τ_c are as follows:

$$\dot{\tau}_c = 1$$

*2inwhen $\tau_c \in [0, N_c + T_c]$, and

$$\tau_c^+ = \begin{cases} \tau_c + 1 & \text{when } (x,u) \in D, \tau_c < N_c + T_c \\ 0 & \text{when } (x,u) \notin D, \tau_c \geq N_c + T_c \\ \{\tau_c + 1, 0\} & \text{otherwise} \end{cases}$$

*2inwhen $(x,u) \notin D$ or $\tau_c \geq N_c + T_c$. These dynamics enforce that the timer increases during flows, so as to count ordinary time, and that at every jump of the hybrid dynamical system (19), the counter is incremented by one (this is in the first entry of difference equation for τ_c), while when the timer has counted at most $N_c + T_c$ seconds of flow and $N_c + T_c$ jumps, is reset to zero (this is the second entry in τ_c^+ – the last entry is when both events can occur). For the current value of the state x_0, the function $u_\star$ used for feedback could be chosen so that

$$u_\star \in \underset{u\,:\,(x,u)\in\mathcal{S}(x_0)\text{ subject to Problem } H}{\arg\min} \mathcal{J}(x,u) \tag{22}$$

which is then applied to the hybrid system over the hybrid horizon with length given by T seconds of flow and J jumps from the current time (t', j'). Above, $\mathcal{S}(x_0)$ denotes the set of state/input pairs (x,u) that satisfy the dynamics of $\mathcal{H}$ and also the conditions in the MPC strategy, which is denoted as Problem H and part of ongoing research efforts is to formally define it. An initial formulation appeared in [3]; see also [2].

It should be pointed out that, for purely continuous-time or discrete-time systems, it is not generally known if the controllers designed to satisfy the necessary conditions for optimality imposed by Pontryagin-like maximum principles or Bellman-like approaches confer a margin of robustness to perturbations of the closed loop. In fact, it is well known that discontinuous controllers obtained from solving optimal control laws may not be robust to small perturbations [21]; see also [35]. This difficulty motivates the generation of hybrid control strategies with prediction that guarantee optimality and robustness simultaneously.

For general nonlinear systems, continuity of the state-feedback law plays a key role in the establishment of robustness of the induced asymptotic stability property [25, 41]. Early results establishing that discontinuities in the feedback can lead to a closed-loop system with zero margin of robustness appeared in books by Filippov [14] and Krasovskii [22]; see also [21] for an insightful relationship between solution concepts to nonsmooth systems. Control laws (both open-loop and closed-loop) solving optimal control problems may not be continuous, which may indicate a lack of robustness when applied to the system to control. Such lack of robustness may also be present in receding horizon controllers. In particular, when the associated optimization problem involves state constraints or terminal constraints, and the optimization horizon is small, the asymptotic stability of the closed-loop system may have absolutely no robustness: arbitrarily small disturbances may keep the state away from the desired set [18]. On the bright side, results in [17] indicate that, for the case of no inputs, mild properties of the data of (19) lead to an upper semicontinuous dependence of the solutions with respect to initial conditions, which, in turn, guarantees that asymptotically stable compact sets for $\mathcal{H}$ (without inputs) are robust to small perturbations.

4 Further Reading

- Discrete-time MPC with hybrid flavor: [5–7, 23, 42];
- Continuous-discrete MPC with hybrid flavor: [10, 13, 27, 29, 32, 37];
- Hybrid dynamical systems: [1–3, 16, 17].
- Software tools for modeling and some MPC problems with hybrid flavor:
  - Multi-Parametric Toolbox (MPT) 3
    http://control.ee.ethz.ch/~mpt
  - The Hybrid Toolbox
    http://cse.lab.imtlucca.it/~bemporad/hybrid/toolbox

Acknowledgements This research has been partially supported by the National Science Foundation under CAREER Grant no. ECS-1450484, Grant no. ECS-1710621, and Grant no. CNS-1544396, by the Air Force Office of Scientific Research under Grant no. FA9550-16-1-0015, by the Air Force Research Laboratory under Grant no. FA9453-16-1-0053, and by CITRIS and the Banatao Institute at the University of California.

References

1. Altin, B., Sanfelice, R.G.: Model predictive control under intermittent measurements due to computational constraints: feasibility, stability, and robustness. In: American Control Conference, pp. 1418–1423 (2018)
2. Altin, B., Sanfelice, R.G.: Model predictive control under intermittent measurements due to computational constraints: feasibility, stability, and robustness (2017, Submitted)
3. Altin, B., Ojaghi, P., Sanfelice, R.G.: A model predictive control framework for hybrid systems. In: 6th IFAC Conference on Nonlinear Model Predictive Control (2018, to appear)
4. Aubin, J.-P., Lygeros, J., Quincampoix, M., Sastry, S.S., Seube, N.: Impulse differential inclusions: a viability approach to hybrid systems. IEEE Trans. Autom. Control **47**(1), 2–20 (2002)

5. Bemporad, A., Borrelli, F., Morari, M.: Optimal controllers for hybrid systems: stability and piecewise linear explicit form. In: Proceedings of the 39th IEEE Conference on Decision and Control, 2000, vol. 2, pp. 1810–1815. IEEE, Piscataway (2000)
6. Bemporad, A., Borrelli, F., Morari, M.: Piecewise linear optimal controllers for hybrid systems. In: Proceedings of the 2000 American Control Conference, vol. 2, pp. 1190–1194. IEEE, Piscataway (2000)
7. Bemporad, A., Heemels, W.M.H.P., De Schutter, B.: On hybrid systems and closed-loop MPC systems. IEEE Trans. Autom. Control **47**(5), 863–869 (2002)
8. Borrelli, F., Bemporad, A., Morari, M.: Predictive Control for Linear and Hybrid Systems. Cambridge University Press, Cambridge (2017)
9. Chen, H., Allgöwer, F.: A quasi-infinite horizon nonlinear model predictive control scheme with guaranteed stability. In: European Control Conference, pp. 1421–1426. IEEE, Piscataway (1997)
10. Chen, H., Allgöwer, F.: A quasi-infinite horizon nonlinear model predictive control scheme with guaranteed stability. Automatica **34**(10), 1205–1217 (1998)
11. Christofides, P.D., El-Farra, N.: Control of Nonlinear and Hybrid Process Systems: Designs for Uncertainty, Constraints and Time-Delays, vol. 324. Springer, New York (2005)
12. D'Apice, C., Garavello, M., Manzo, R., Piccoli, B.: Hybrid optimal control: case study of a car with gears. Int. J. Control **76**, 1272–1284 (2003)
13. El-Farra, N.H., Mhaskar, P., Christofides, P.D.: Hybrid predictive control of nonlinear systems: method and applications to chemical processes. Int. J. Robust Nonlinear Control **14**(2), 199–225 (2004)
14. Filippov, A.F.: Differential Equations with Discontinuous Right-Hand Sides. Kluwer, Dordrecht (1988)
15. Garavello, M., Piccoli, B.: Hybrid necessary principle. SIAM J. Control Optim. **43**(5), 1867–1887 (2005)
16. Goebel, R., Sanfelice, R.G., Teel, A.R.: Hybrid dynamical systems. IEEE Control. Syst. Mag. **29**(2), 28–93 (2009)
17. Goebel, R., Sanfelice, R.G., Teel, A.R.: Hybrid Dynamical Systems: Modeling, Stability, and Robustness. Princeton University Press, Princeton (2012)
18. Grimm, G., Messina, M.J., Tuna, S.E., Teel, A.R.: Examples when nonlinear model predictive control is nonrobust. Automatica **40**(10), 1729–1738 (2004)
19. Grimm, G., Messina, M.J., Tuna, S.E., Teel, A.R.: Nominally robust model predictive control with state constraints. IEEE Trans. Autom. Control **52**(10), 1856–1870 (2007)
20. Grüne, L., Pannek, J.: Nonlinear Model Predictive Control, pp. 45–69. Springer, Berlin (2017)
21. Hàjek, O.: Discontinuous differential equations I. J. Differ. Equ. **32**, 149–170 (1979)
22. Krasovskii, N.N., Subbotin, A.I.: Game-Theoretical Control Problems. Springer, Berlin (1988)
23. Lazar, M., Heemels, W.M.P.H., Bemporad, A.: Stabilizing model predictive control of hybrid systems. IEEE Trans. Autom. Control **51**(11), 1813–1818 (2006)
24. Lin, Y., Sontag, E.D.: A universal formula for stabilization with bounded controls. Syst. Control Lett. **16**(6), 393–397 (1991)
25. Lin, Y., Sontag, E.D., Wang, Y.: A smooth converse Lyapunov theorem for robust stability. SIAM J. Control Optim. **34**(1), 124–160 (1996)
26. Lygeros, J., Johansson, K.H., Simić, S.N., Zhang, J., Sastry, S.S.: Dynamical properties of hybrid automata. IEEE Trans. Autom. Control **48**(1), 2–17 (2003)
27. Magni, L., Scattolini, R.: Model predictive control of continuous-time nonlinear systems with piecewise constant control. IEEE Trans. Autom. Control **49**(6), 900–906 (2004)
28. Mayne, D.Q., Michalska, H.: Receding horizon control of nonlinear systems. IEEE Trans. Autom. Control **35**(7), 814–824 (1990)
29. Mhaskar, P., El-Farrah, N.H., Christofides, P.D.: Robust hybrid predictive control of nonlinear systems. Automatica **41**(2), 209–217 (2005)
30. Michalska, H., Mayne, D.Q.: Robust receding horizon control of constrained nonlinear systems. IEEE Trans. Autom. Control **38**(11), 1623–1633 (1993)

31. Michel, A.N., Hu, B.: Towards a stability theory of general hybrid dynamical systems. Automatica **35**(3), 371–384 (1999)
32. Nešić, D., Grüne, L.: A receding horizon control approach to sampled-data implementation of continuous-time controllers. Syst. Control Lett. **55**(8), 660–672 (2006)
33. Pontryagin, L.S., Boltyanskij, V.G., Gamkrelidze, R.V., Mishchenko, E.F.: The Mathematical Theory of Optimal Processes. Wiley, London (1962)
34. Rawlings, J.B., Mayne, D.Q.: Model Predictive Control: Theory and Design. Nob Hill Publishing, Madison (2009)
35. Sanfelice, R.G., Goebel, R., Teel, A.R.: Generalized solutions to hybrid dynamical systems. ESAIM Control Optim. Calculus Var. **14**(4), 699–724 (2008)
36. Shaikh, M.S., Caines, P.E.: On the hybrid optimal control problem: theory and algorithms. IEEE Trans. Autom. Control **52**, 1587–1603 (2007)
37. Sopasakis, P., Patrinos, P., Sarimveis, H., Bemporad, A.: Model predictive control for linear impulsive systems. IEEE Trans. Autom. Control **60**(8), 2277–2282 (2015)
38. Sussmann, H.J.: Some recent results on the maximum principle of optimal control theory. In: Systems and Control in the Twenty-First Century, pp. 351–372 (1997)
39. Sussmann, H.J.: A maximum principle for hybrid optimal control problems. In: Proceedings of 38th IEEE Conference on Decision and Control, pp. 425–430 (1999)
40. Tavernini, L.: Differential automata and their discrete simulators. Nonlinear Anal. Theory Methods Appl. **11**(6), 665–683 (1987)
41. Teel, A.R., Praly, L.: A smooth Lyapunov function from a class-$\mathcal{KL}$ estimate involving two positive semidefinite functions. ESAIM Control Optim. Calculus Var. **5**, 313–367 (2000)
42. Tuna, S.E., Sanfelice, R.G., Messina, M.J., Teel, A.R.: Hybrid MPC: open-minded but not easily swayed. In: Assessment and Future Directions of Nonlinear Model Predictive Control. Lecture Notes in Control and Information Sciences, vol. 358, pp. 17–34. Springer, Berlin (2007)
43. van der Schaft, A., Schumacher, H.: An Introduction to Hybrid Dynamical Systems. Lecture Notes in Control and Information Sciences. Springer, Berlin (2000)
44. Zhang, K., Sprinkle, J., Sanfelice, R.G.: Computationally-aware control of autonomous vehicles: a hybrid model predictive control approach. Auton. Robot. **39**, 503–517 (2015)
45. Zhang, K., Sprinkle, J., Sanfelice, R.G.: Computationally-aware switching criteria for hybrid model predictive control of cyber-physical systems. IEEE Trans. Autom. Sci. Eng. **13**, 479–490 (2016). https://doi.org/10.1109/TASE.2016.2523341

Model Predictive Control of Polynomial Systems

Eranda Harinath, Lucas C. Foguth, Joel A. Paulson, and Richard D. Braatz

1 Introduction

Model predictive control (MPC) is the most widely used approach for the advanced control of complex dynamical systems due to its ability to straightforwardly handle multivariable dynamics, incorporate input and state constraints, and trade-off between competing sets of objectives [1, 26]. Linear MPC refers to the family of MPC strategies in which linear models are used to predict the system dynamics subject to only linear constraints. Although most real industrial processes are inherently nonlinear, linear MPC has been well-studied for two main reasons: (i) a variety of highly reliable techniques and software are available for the identification of linear models and (ii) linear models yield good results when the plant is operating in the neighborhood of a specific operating point [3].

On the other hand, nonlinear model predictive control (NMPC) refers to MPC schemes involving a nonlinear objective function or nonlinear constraints, usually due to the use of nonlinear models. Demand for higher product quality, tighter specifications, and tougher environmental regulations necessitate high closed-loop performance over a wide range of operating conditions. This fact, combined with the inherent nonlinearity in most real systems (such as chemical, pharmaceutical, and biological systems), motivates the development of NMPC methods [1].

For discrete-time systems, linear MPC is directly formulated as a convex quadratic optimization (in case of quadratic objective function) which can be solved efficiently to the global minimum. A key advantage is that any local minimum detected is also a global minimum, and many efficient numerical algorithms exist that are guaranteed to converge to the global minimum. The inherent nonconvexity of general NMPC problems, however, makes them very expensive to solve to global optimality, which has limited their successful application to real problems [24].

E. Harinath · L. C. Foguth · J. A. Paulson · R. D. Braatz (✉)
Massachusetts Institute of Technology, 77 Massachusetts Avenue, Cambridge, MA 02139, USA
e-mail: eranda@mit.edu; lcfoguth@mit.edu; jpaulson@mit.edu; braatz@mit.edu

S. V. Raković, W. S. Levine (eds.), *Handbook of Model Predictive Control*,
Control Engineering, https://doi.org/10.1007/978-3-319-77489-3_10

Development of practical algorithms for NMPC that circumvent these issues is an active area of research (see, e.g., [1, 31] and the citations therein). This chapter describes an approach that considers a particular broad class of nonlinear dynamical systems and constraints described by means of polynomial functions [8, 24]. Many systems of industrial importance can be directly transformed into polynomial systems. Other nonlinear systems can be approximated as polynomial systems by expanding all nonlinear functions in terms of a Taylor series and truncating them to a finite number of terms. Furthermore, Taylor's theorem can be used to rigorously bound the approximation error [8].

For discrete-time polynomial systems, open-loop optimal control with a polynomial cost function is a polynomial optimization. Many different software packages (that use a suite of algorithms embedded with heuristics) are available for solving general polynomial optimizations. Furthermore, methods exist for reducing the original polynomial optimization to a semidefinite program (SDP) via the theory of moments [13] or the theory of nonnegative polynomials [14]. Using these methods, a global minimum of the original nonconvex polynomial MPC problem can be obtained by solving a series of convex SDPs. The fact that these algorithms converge fairly quickly in practice for a broad class of polynomial programs illustrates the vast potential of polynomial MPC.

The remainder of the chapter is organized as follows. In Section 2, the MPC problem for discrete-time polynomial systems is formulated as a polynomial optimization. In Section 3, the methods to solve these polynomial optimizations to global optimality are briefly discussed, and a promising sum-of-squares (SOS)-based method is reviewed. Section 4 reviews the methods for fast polynomial MPC, including methods to formulate the original MPC problem as a convex problem and methods for explicit polynomial MPC. Section 5 discusses methods by which nonlinear systems can be exactly written or approximated as polynomial systems. Finally, an outlook for future research is provided in Section 6.

Notation. Let $\mathbb{R}^n$ denote the set of real vectors with n elements, and $\mathbb{R}^{n\times m}$ denote the set of real matrices with n rows and m columns. For $x \in \mathbb{R}^n$, $\mathbb{R}[x]$ is the set of real polynomials, $\mathbb{R}[x]^p$ is the set of vectors of real polynomials of dimension p, and $\mathbb{R}[x]^{p\times q}$ is the set of matrices of real polynomials of dimension $p \times q$.

2 Model Predictive Control of Discrete-Time Polynomial Systems

Consider a discrete-time nonlinear dynamical system

$$x_{k+1} = f(x_k, u_k), \tag{1}$$

subject to state and input constraints of the form

$$x_k \in \mathbb{X}, \tag{2}$$

$$u_k \in \mathbb{U}, \tag{3}$$

where $k \geq 0$ is the discrete-time index, $x \in \mathbb{R}^n$ are the system states, and $u \in \mathbb{R}^m$ are the control inputs. The function $f : \mathbb{R}^n \times \mathbb{R}^m \to \mathbb{R}^n$ is a real polynomial function in x and u and has the origin as an equilibrium point, i.e., $f(0,0) = 0$. The set $\mathbb{X}$ is a closed subset of $\mathbb{R}^n$, and the set $\mathbb{U}$ is a compact subset of $\mathbb{R}^m$, both containing the origin. In their most general forms, $\mathbb{X}$ and $\mathbb{U}$ are semialgebraic sets given by the union of a finite number of real polynomial equations and inequalities in x and u, respectively.

Suppose that all initial states x_k are available at time instant k. Then, the open-loop optimal control problem to be solved at each time instant k is given by

$$\min_{\mathbf{u}_k} J_N(x_k, \mathbf{u}_k) \tag{4a}$$

$$\text{subject to } x_{k|k} = x_k, \tag{4b}$$

$$x_{i+1|k} = f(x_{i|k}, u_{i|k}), \tag{4c}$$

$$u_{i|k} \in \mathbb{U}, \tag{4d}$$

$$x_{i|k} \in \mathbb{X}, \tag{4e}$$

$$x_{k+N|k} \in X_f, \; i = k, \ldots, k+N-1, \tag{4f}$$

with the cost function

$$J_N(x_k, \mathbf{u}_k) := \sum_{i=k}^{k+N-1} l(x_{i|k}, u_{i|k}) + F(x_{k+N|k}), \tag{5}$$

where $\mathbf{u}_k := [u_{k|k}^\top, u_{k+1|k}^\top, \ldots, u_{k+N-1|k}^\top]^\top$ is the concatenated control input vector, $N > 0$ is the length of the control horizon, $x_{i|k}$ is the predicted state at time instant $i > k$ obtained by applying the input sequence $u_{k|k}, \ldots, u_{i-1|k}$ to the system (1) from initial states x_k, and X_f is the terminal constraint set.

The set X_f is assumed to be a semialgebraic subset of $\mathbb{R}^n$. Similarly to [24], the stage cost $l : \mathbb{R}^n \times \mathbb{R}^m \to \mathbb{R}$ is assumed to be a real polynomial function in x and u with $l(0,0) = 0$, and the terminal penalty $F : \mathbb{R}^n \to \mathbb{R}$ is assumed to be a real polynomial function in x with $F(0) = 0$. The optimal solution for the MPC problem (4) is denoted by

$$\mathbf{u}_k^\star := [u_{k|k}^{\star\top}, u_{k+1|k}^{\star\top}, \ldots, u_{k+N-1|k}^{\star\top}]^\top. \tag{6}$$

Under a receding horizon implementation, the first element of $\mathbf{u}_k^\star$, i.e.,

$$\kappa_N(x_k) := u_{k|k}^\star(x_k), \tag{7}$$

is applied to the system (1). The implicit MPC controller κ_N is a feedback control law as (4) is solved at each time instant k using the measured or estimated states x_k.

3 Polynomial Optimization Methods

This section summarizes global solution methods for the MPC optimization (4) that can be recast as a polynomial optimization, where the objective function and constraints are multivariate polynomials [24]. For simplicity of notation, consider a general polynomial optimization that can be described as

$$p_0^\star = \min_{x \in S} \; p_0(x), \tag{8}$$

where the closed semialgebraic set S is given by

$$S = \{x \in \mathbb{R}^n | p_i(x) \geq 0, i = 1, \cdots, k\}, \tag{9}$$

$p_0 : \mathbb{R}[x]^n \to \mathbb{R}$ is the polynomial objective function, $x \in \mathbb{R}^n$ are the decision variables, and p_i are polynomial constraints. The global optimum of the problem (8) is denoted by $p_0^\star$.

Although the general polynomial optimization (8) is nonconvex and NP-hard [19], the intrinsic properties of this class of optimization can be utilized to develop tailored algorithms to find the global solution. Development of polynomial programming algorithms has received much attention in the literature. As a result, polynomial programming solvers have been developed based on the solution of relaxed optimizations that can be solved in polynomial time, in which "relaxed" refers to an optimization whose solution provides a lower bound on the original optimization. A series of relaxed optimizations are solved, and their solutions are guaranteed to converge monotonically to the solution of the original polynomial optimization [9, 21].

The basic idea behind polynomial optimization algorithms is to derive convex relaxations of the original problem (8). The most common relaxation methods yield SDP or linear programs (LPs). Solving the relaxed problems provides a lower bound for the optimum $p_0^\star$. The theory of nonnegative polynomials [19] and the theory of moments [13] are the most popular relaxation techniques for polynomial optimizations and naturally give rise to SDP problems. The moment-based relaxation directly relaxes the primal problem (8) while the theory of nonnegative polynomials instead indirectly relaxes the primal problem by tightening the dual of (8) using SOS decompositions. Hierarchies of these SDP relaxation methods can be used to find lower bounds for (8) successively. In the high-level optimization tools SOSTOOLS [21] and Gloptipoly [9], the hierarchical SOS and moment-based SDP relaxations have been implemented, respectively.

Below is a description of an SOS-based relaxation method for solving the polynomial optimization (8).

3.1 Sum-of-Squares Decomposition

Checking whether a polynomial is nonnegative is an important task in algebraic geometry [19]. The existence of an SOS decomposition is a sufficient condition for nonnegativity of a polynomial and is defined below.

Definition 1 (SOS). A polynomial $h(x) \in \mathbb{R}[x]$ with degree $2d$, $d = 1,2,\ldots$, is called an *SOS polynomial* if there exists a finite number m of polynomials $h_i(x)$ such that

$$h(x) = \sum_{i=1}^{m} h_i^2(x). \tag{10}$$

Not every positive semidefinite polynomial can be written as an SOS polynomial, i.e.,

$$h(x) = \sum_i h_i^2(x) \in \sum[x] \subset N_+(x), \tag{11}$$

where $\sum[x]$ is the set of SOS polynomials and N_+ is the set of nonnegative polynomials in $\mathbb{R}[x]$. By employing a "Gram Matrix" technique, the necessary and sufficient conditions for the existence of an SOS decomposition for any given polynomial are presented in [5]. In [19], the Gram Matrix result is used to prove that the existence of an SOS decomposition can be checked by solving an SDP. This guaranteed certificate that can be checked in polynomial time for SOS polynomials leads to the development of SOS-based optimization algorithms (see [6, 19] for further discussions on SOS polynomials).

One of the earliest applications of SOS techniques is reported in [29], in which a method for finding the global lower bounds for polynomial functions is presented. Recently, the control literature has extensively used SOS-based techniques both for polynomial systems [19, 32] and non-polynomial systems [4, 18]. The next section summarizes a way in which SOS methods can be used to globally solve the polynomial optimization (8).

3.2 Dual Approach via SOS Decomposition

This section describes hierarchical SDP relaxation methods that successively find lower bounds for global polynomial optimizations [13, 19]. A dual problem formulation for (8) is presented by using the generalized Lagrangian function method as discussed in [12]. The generalized Lagrangian function for the polynomial optimization (8) is written as

$$L(x, s_1(x), \cdots, s_k(x)) = p_0(x) - \sum_{i=1}^{k} s_i(x) p_i(x), \tag{12}$$

where $s_i(x) \in \sum[x]$ with a finite degree can be interpreted as generalized Lagrangian multipliers [12] or generalized Karush-Kuhn-Tucker (KKT) multipliers [13].

The Lagrangian dual for (8) is given by

$$\lambda^\star = \max_{s_i(x)\in\sum[x]} \min_{x\in\mathbb{R}^n} L(x, s_1(x), \cdots, s_k(x)), \tag{13}$$

where $\lambda^\star \leq p_0^\star$. The Lagrangian dual (13) can be rewritten as

$$\begin{aligned} &\max\ t, && (14a)\\ &\text{subject to } L(x, s_1(x), \cdots, s_k(x)) - t \geq 0, && (14b)\\ &\forall\, x \in \mathbb{R}^n \text{ and } s_i(x) \in \textstyle\sum[x]. \end{aligned}$$

Replacing the nonnegativity condition (14b) with a stronger SOS polynomial s_0 gives

$$\begin{aligned} \alpha^\star &= \max\ t, && (15a)\\ &\text{subject to } L(x, s_1(x), \cdots, s_k(x)) - t = s_0, && (15b)\\ &x \in \mathbb{R}^n, \text{ and } s_i(x) \in \textstyle\sum[x],\ \forall\, i = 0, \ldots, k. \end{aligned}$$

Once the degree of the SOS polynomials s_i are fixed, such that

$$\max\{\deg(s_0), \deg(p_i s_i)\} \leq 2\beta, \tag{16}$$

where $\beta \geq \max\{\deg(p_0)/2, \deg(p_i)/2\}$ for all $i = 1, \ldots, k$, the dual problem (15) can be solved via an SDP [19]. The SOS relaxation gives a lower bound, i.e., $\alpha^\star \leq \lambda^\star \leq p_0^\star$.

In hierarchical SDP relaxation methods, the optimization (15) is solved sequentially, and each problem can be solved efficiently by using an interior-point SDP solver, e.g., SeDuMi [30]. Tools such as SOSTOOLS or YALMIP [49], which are implemented in MATLAB, can be used to parse the optimization problem.

Denote the optimum of the relaxed problem (15) at the jth iteration as $\alpha_j^\star$. By increasing the degree of the SOS polynomials s_i, the lower bound $\alpha_j^\star$ for the global optimization problem can be improved. However, this procedure increases the degrees of freedom in the relaxed SDP problem exponentially [24]. For hierarchical sequential SDP relaxation problems, it has been shown that

$$\alpha_j^\star \leq \alpha_{j+1}^\star \leq p_0^\star. \tag{17}$$

Furthermore, the relaxed solution has been shown to monotonically converge to the global optimum [13]. At a particular iteration, the global optimum can be detected via generalized KKT conditions as discussed in [13]. When the global optimum is reached, i.e., when $\alpha_j^\star = p_0^\star$, the constrained global optimization has a primal SDP formulation (corresponding to the dual of the relaxed dual problem). The optimal solution of this primal problem provides the global minimizer $x^\star$.

Remark 1. The dual optimization problem (15) can also be derived using Putinar's representation theorem [13, 23, 24], which utilizes the Positivstellensatz argument.

4 Fast Solution Methods for Polynomial MPC

As discussed in Section 2, the MPC optimization (4) is directly formulated as a polynomial optimization of the form (8). This problem can be solved to global optimality by iteratively solving a series of convex relaxations. However, the number of iterations, corresponding to the order of the SOS polynomials used in these relaxations, may be large for certain problems. Since the number of decision variables in the convex relaxation grows exponentially with the SOS polynomial order, each iteration becomes increasingly more expensive to solve. The computational effort required to solve the problem to global optimality is likely prohibitive for problems with a relatively large number of states, inputs, and/or horizon.

Methods have been developed to take advantage of sparsity within the convex relaxations of (8), which can provide a significant reduction in complexity [12]. This idea and others represent an active area of research in the field of polynomial optimization, which greatly benefits from the development of efficient SOS decomposition techniques and SDP solvers. The MPC problem adds additional structure and sparsity to the problem, which can be taken advantage of during implementation of the controller.

One interesting route, discussed below, is the direct formulation of the MPC problem as an SOS problem for a particular class of polynomial systems and constraints. This problem can then be solved to global optimality with a single convex SDP, which avoids the iterative approach discussed above. Alternatively, the original MPC problem (4) can be solved offline using parametric optimization (the so-called *explicit MPC*). Methods for solving (4) explicitly using algebraic geometry methods are also explored below.

4.1 Convex MPC for a Subclass of Polynomial Systems

One of the most studied subclasses of the more general polynomial systems (1) are the so-called input-affine polynomial systems, described by

$$x_{k+1} = f_a(x_k) + B(x_k)u_k, \tag{18}$$

where $f_a(x) \in \mathbb{R}[x]^n$ denotes a polynomial function with $f_a(0) = 0$, and $B(x) \in \mathbb{R}[x]^{n\times m}$ denotes the polynomial input matrix. Typically, the literature assumes that (18) can be written in the state-dependent representation [22]

$$x_{k+1} = A(x_k)Z(x_k) + B(x_k)u_k, \tag{19}$$

where $A(x_k) \in \mathbb{R}[x]^{n\times n}$ is the system polynomial matrix, and $Z(x_k) \in \mathbb{R}[x]^{n_n}$ is a polynomial vector. By exploiting the linear-like structure of (19), global stabilizing control [6], optimal control [11], and robust control [10] design techniques have been presented for continuous-time polynomial systems using state-dependent linear matrix inequalities (LMIs). These state-dependent LMIs are then represented directly with SOS decompositions. Convex constraints for the controller synthesis problem can also be included by parametrizing a Lyapunov function based on the structure of matrix $B(x)$. For details, see [22] and citations therein.

The current literature formulates the synthesis of feedback control laws by minimizing upper bounds of the infinite-horizon cost. These formulations are typically convex but suboptimal. Convex formulations are thus restricted to a particular class of polynomial systems even for optimal control. As such, MPC for these restricted systems is relatively unexplored. An exception is [16] which derives an MPC algorithm for input-affine polynomial systems by formulating the optimization directly as an SDP by parametrizing a Lyapunov function for the closed-loop system similarly as in [22].

4.2 Explicit MPC Using Algebraic Geometry Methods

As discussed in the introduction, limitations to online computations make the direct implementation of NMPC difficult or (in some cases) impossible. Even in the case of linear MPC, in which the optimization is a convex QP, the development of methods for quickly solving the MPC optimization online is still an active area of research. An interesting approach, introduced in [2], is to compute the control law offline by solving the optimization parametrically as a function of the initial states. The optimal control law is then implemented online as a lookup table, which greatly decreases the online computational cost of the controller.

When polynomial systems and constraints are considered, the standard MPC formulation requires the solution to a polynomial optimization. In contrast to the linear case, a closed-form expression for its solution is not guaranteed to exist (as it may involve implicit algebraic functions).

The parametric optimization of interest in this chapter is that of solving (4) for any value of the parameter x_k. Since the system is assumed to be time invariant, the time index k can be dropped, and (4) can be rewritten as

$$\min_{\mathbf{u}} J_N(\mathbf{u},x) \tag{20a}$$

$$\text{subject to } g(\mathbf{u},x) \leq 0, \tag{20b}$$

where $x \in \mathbb{R}^n$ are the initial states, $\mathbf{u} \in \mathbb{R}^{mN}$ are the decision variables representing the control inputs over the horizon, $J_N \in \mathbb{R}[x_1,\ldots,x_n,\mathbf{u}_1,\ldots,\mathbf{u}_{mN}]$ is the polynomial objective function, and $g \in \mathbb{R}[x_1,\ldots,x_n,\mathbf{u}_1,\ldots,\mathbf{u}_{mN}]^q$ is the polynomial vector representing all constraints in (4). The goal is to find a computational procedure for evaluating the maps

$$\mathbf{u}^\star(x) : \mathbb{R}^n \to \mathbb{R}^{mN}, \tag{21}$$

$$J_N^\star(x) : \mathbb{R}^n \to \mathbb{R}, \tag{22}$$

for any value of the parameter x in the region of interest. Methods for evaluating these maps are presented in [7] and [27]. These methods rely on techniques derived from the algebraic geometry literature and are summarized below.

In [7], cylindrical algebraic decomposition (CAD) is used to evaluate the map from the initial states to the corresponding optimizer (21) and optimal cost function (22). Given a finite set of polynomials in n variables, a CAD is a special partition of $\mathbb{R}^n$ into cells over which all polynomials have constant signs. The mathematical details behind CAD are illustrated by an example in [7].

In the first parametric optimization algorithm proposed in [7], the CAD corresponding to the polynomial expressions in the optimal control problem is constructed offline. Then, the online portion of the algorithm only consists of determining the cell in which the initial states x lie, finding the optimal cost $J_N^\star$ by exploring the cylinder above this cell, and determining the optimizer $\mathbf{u}^\star$ by lifting to the space of the decision variables. Thus, the algorithm only requires traversing a tree and solving univariate polynomial equations online. Readers interested in more details are referred to [7].

The second algorithm presented in [7] involves the parametric solution of the resulting KKT optimality conditions for (20), given by

$$\nabla_{\mathbf{u}} J(\mathbf{u},x) + \Sigma_{i=1}^{q} \mu_i \nabla_{\mathbf{u}} g_i(\mathbf{u},x) = 0, \tag{23a}$$

$$\mu_i g_i(\mathbf{u},x) = 0, \tag{23b}$$

$$\mu_i \geq 0, \tag{23c}$$

$$g(\mathbf{u},x) \leq 0, \tag{23d}$$

where g_i are the polynomial elements of the constraint vector g in (20b), and $\mu_i \in \mathbb{R}_+$ are the Lagrange multipliers for each $i = 1,\ldots,q$. The first two relations (23a) and (23b) form a square system of polynomial equations. The proposed algorithm involves solving this system of polynomial equations symbolically as a function of x. The idea is to use Gröbner bases to compute generalized companion matrices for a candidate optimizer (as a function of x) offline. Then, three steps must be performed online: (i) calculate all critical points of (23a) and (23b) using the eigenvalues of the companion matrices evaluated at a given x, (ii) eliminate any infeasible solutions by checking if $\mu_i \geq 0$ and $g(\mathbf{u},x) \leq 0$, and (iii) find the feasible candidate solution $\mathbf{u}^\star$ with the smallest objective function value $J_N^\star$.

Finally, a homotopy-based algorithm for evaluating the maps (21) and (22) is described in [27]. The basic idea is to solve the parametric optimization offline using homotopy continuation with a multihomogeneous start system. Online, the solution can then be constructed for different initial states $\tilde{x}$ using the known solutions for x when they are in the same family. Readers interested in more details are referred to [27].

The next section describes the use of Taylor series approximations to control general nonlinear systems by approximating them as polynomial systems. Also discussed is the use of Taylor's theorem to generate an uncertain polynomial system that allows for robust handling of the truncation error.

5 Taylor Series Approximations for Non-polynomial Systems

Different strategies for MPC can be used depending on the original description of the system. If the original system can be exactly written as a polynomial system, the above methods can be directly applied to nominal MPC. If not, the system can be approximated as a polynomial where the approximation error represents an additional source of uncertainty. This section discusses the use of Taylor's theorem to approximate non-polynomial systems as polynomial systems and subsequently provides rigorous bounds on the approximation error.

5.1 Taylor's Theorem

Consider a non-polynomial discrete-time system

$$x^+ = f(x), \tag{24}$$

where $x \in \mathbb{R}^n$ is the system state and x^+ is the successor state at the next time instant. When f is k times continuously differentiable, systems of this form can be approximated as polynomial systems by employing a kth-order Taylor series expansion. Polynomial control algorithms can then be applied directly to this approximation. Alternatively, the error associated with the approximation can be bounded using Taylor's theorem. The error can then be treated as an uncertainty, and the resulting uncertain system can be controlled by using robust control methods for polynomial systems. This section briefly summarizes Taylor's theorem for multivariate systems and demonstrates the utility of this theorem using a simple example. A discussion of robust control methods for polynomial systems is provided in Section 6.

To concisely state Taylor's theorem for higher dimensional systems, we begin by summarizing multi-index notation for an nth-order system. For the index $\alpha \in \mathbb{R}^n$, define the operations

$$|\alpha| = \alpha_1 + \cdots + \alpha_n, \tag{25a}$$

$$\alpha! = \alpha_1! \cdots \alpha_n!, \tag{25b}$$

$$x^\alpha = x_1^{\alpha_1} \cdots x_n^{\alpha_n}. \tag{25c}$$

Using this notation, the higher order partial derivatives of the multivariate system (24) can be written as

$$D^{\alpha} f_i = \frac{\partial^{|\alpha|} f_i}{\partial x_1^{\alpha_1} \cdots \partial x_n^{\alpha_n}}. \tag{26}$$

The kth-order Taylor series expansion of f_i, centered at a, is defined by

$$P_i(x;k,a) = \sum_{|\alpha| \leq k} \frac{D^{\alpha} f_i(a)}{\alpha!}(x-a)^{\alpha}, \quad i = 1,\ldots,n \tag{27}$$

when f_i is k times differentiable at the point a.

If f_i is $k+1$ times differentiable at the point a, the multivariate version of Taylor's theorem states that

$$f_i(x) = \sum_{|\alpha| \leq k} \frac{D^{\alpha} f_i(a)}{\alpha!}(x-a)^{\alpha} + \sum_{|\beta|=k+1} R_{i,\beta}(x)(x-a)^{\beta}, \tag{28}$$

where $R_{i,\beta}(x)$ are the remainder functions associated with $P_i(x;k,a)$.

Taylor's theorem is sometimes extended to state that, if x is in a compact set B, the remainder terms can be bounded using the inequality

$$|R_{i,\beta}(x)| \leq \frac{1}{\beta!} \max_{|\gamma|=|\beta|} \max_{y \in B} |D^{\gamma} f(y)|, \quad x \in B. \tag{29}$$

5.2 Example

The practical implementation of Taylor's theorem for bounding control trajectories is best illustrated by the use of an example. Consider the discrete-time nonlinear dynamical system

$$x_1^+ = f_1(x_1,x_2) = x_1 x_2^{0.7} + 0.3, \tag{30a}$$
$$x_2^+ = f_2(x_1,x_2) = \ln(x_2+1) + 0.05. \tag{30b}$$

Assume that, for the time period of interest, the states will lie within the region $0.3 \leq x_i \leq 0.7, i = 1,2$, which is denoted by Π (the validity of this assumption is confirmed later). Within this region Π, the system (30) can be (conservatively) represented as the uncertain polynomial system

$$x^+ = \begin{bmatrix} P_1(x;3,a) \\ P_2(x;3,a) \end{bmatrix} + \begin{bmatrix} \theta_1 \\ \theta_2 \end{bmatrix}, \tag{31}$$

where

$$\begin{aligned}P_i(x;3,a) =& f_i(a)+\frac{\partial f_i}{\partial x_1}(a)(x_1-a_1)+\frac{\partial f_i}{\partial x_2}(a)(x_2-a_2)+\frac{\partial^2 f_i}{\partial^2 x_1}(a)\frac{(x_1-a_1)^2}{2!}\\ &+\frac{\partial^2 f_i}{\partial x_1 \partial x_2}(a)(x_1-a_1)(x_2-a_2)+\frac{\partial^2 f_i}{\partial^2 x_2}(a)\frac{(x_2-a_2)^2}{2!}\\ &+\frac{\partial^3 f_i}{\partial x_1^3}(a)\frac{(x_1-a_1)^3}{3!}+\frac{\partial^3 f_i}{\partial^2 x_1 \partial x_2}(a)\frac{(x_1-a_1)^2(x_2-a_2)}{2!}\\ &+\frac{\partial^3 f_i}{\partial x_1 \partial^2 x_2}(a)\frac{(x_1-a_1)(x_2-a_2)^2}{2!}+\frac{\partial^3 f_i}{\partial^3 x_2}(a)\frac{(x_2-a_2)^3}{3!},\end{aligned}$$

the center of Π is $a = [a_1\ a_2]^\top = [0.5\ 0.5]^\top$, and the elements of the vector $\theta = [\theta_1\ \theta_2]^\top$ are bounded by

$$\min_{x\in\Pi}\{f_i(x) - P_i(x;3,a)\} \le \theta_i \le \max_{x\in\Pi}\{f_i(x) - P_i(x;3,a)\}. \tag{32}$$

These bounds on the uncertain parameters θ_i can be found using global optimization algorithms such as branch-and-bound methods. Although these bounds may be computationally expensive to compute, the computation is completely offline. Alternatively, relaxations can be used to compute the bounds on the θ_i at the expense of introducing additional conservatism. The system (31) is affine in the uncertain parameter θ, which can make robust controller algorithms easier to derive.

Alternatively, (29) can be applied directly to achieve an uncertain system of the form

$$x^+ = \begin{bmatrix} P_1(x;3,a)\\ P_2(x;3,a)\end{bmatrix} + \begin{bmatrix} \theta_1 \sum\limits_{|\beta|=4} \dfrac{(x-a)^\beta}{\beta!}\\ \theta_2 \sum\limits_{|\beta|=4} \dfrac{(x-a)^\beta}{\beta!}\end{bmatrix}, \tag{33}$$

where bounds on the parameters θ_i can be found by directly applying (29). In this case, the terms containing the uncertain parameters θ_i are dependent on the state. Although robust control algorithms are more difficult to derive with such state-dependent terms, the uncertain system (33) can be less conservative than the system (31).

The state trajectories for the original non-polynomial system (24) are shown in Figures 1 and 2, respectively, for both initial states equal to 0.5. Also shown in the figures are the trajectories of the approximate polynomial system and bounds on the trajectory of the true non-polynomial system (24) given by the uncertain systems (31) and (33). In this example, the state-dependent bounds obtained using system (33) are less conservative on average than the state-independent bounds obtained using system (31), but the upper bound for the state-independent uncertain system is slightly less conservative for much of the time.

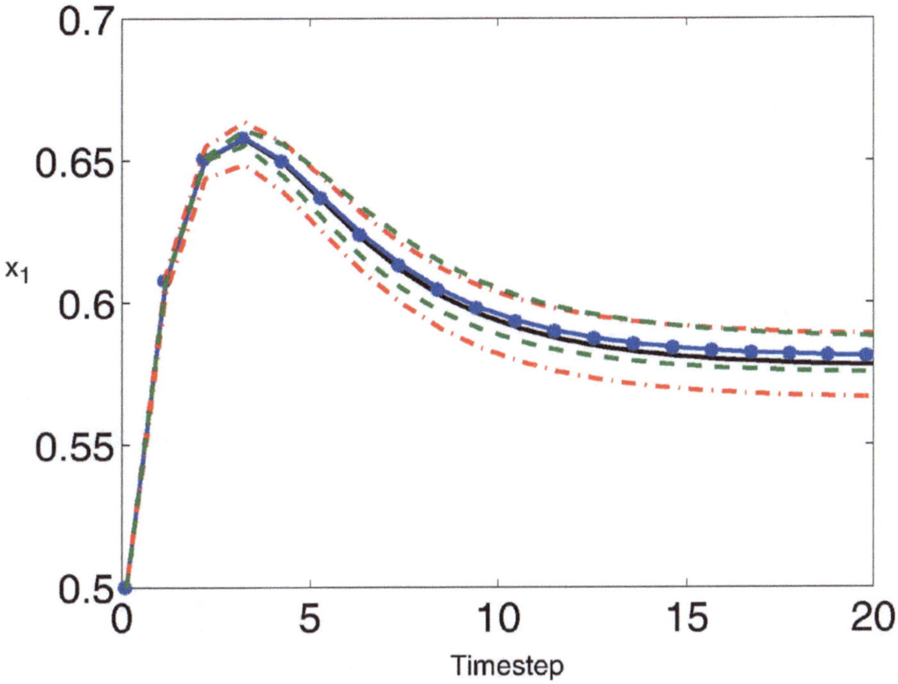

Fig. 1: The trajectory of the first state of the non-polynomial system (24) and of the approximate polynomial system are shown in black and dashed blue, respectively. State-independent and state-dependent bounds on the first state are shown in red and green, respectively.

6 Outlook for Future Research

While the field of deriving efficient methods for solving polynomial optimizations is relatively mature, many opportunities remain for contributing to the theory of MPC for polynomial systems.

One interesting direction involves developing tailored polynomial optimization algorithms for MPC. A current challenge with implementing MPC is that it may be impracticable to achieve the global solution of the nonconvex optimization within the required sampling time interval. However, as discussed in [17], solving for the global solution is not necessary for achieving asymptotic stability of the closed-loop system. Rather, only a feasible solution is needed to guarantee the existence of a Lyapunov function [28]. In the case of polynomial systems, although current polynomial optimization algorithms have the potential to find the global solution of polynomial programs within a few iterations (see, e.g., [13]), no direct way is available for extracting a feasible solution in a given iteration if the solver is unable to find the global solution within the required sampling time interval. It would be use-

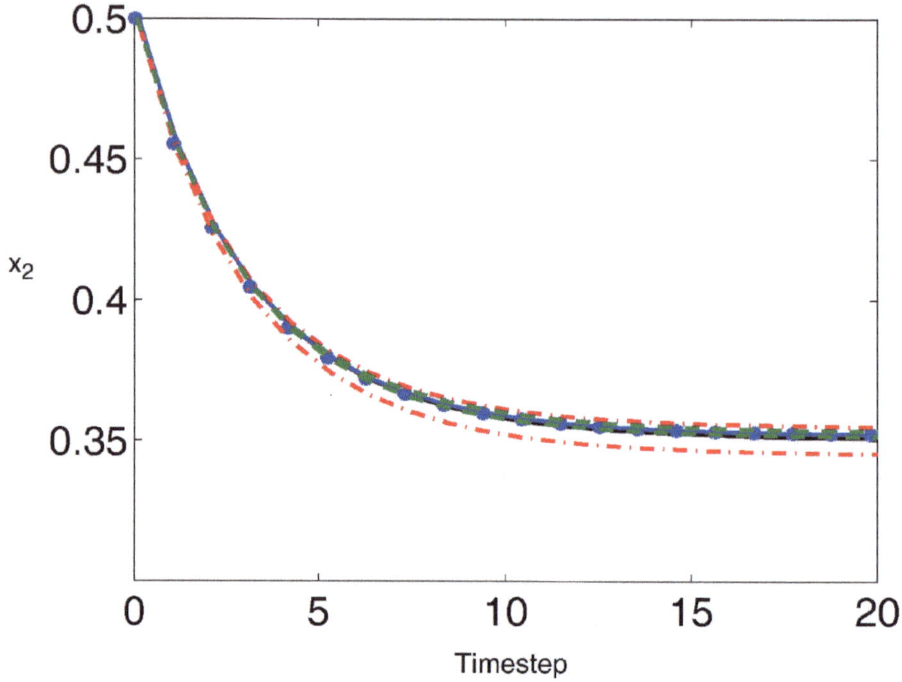

Fig. 2: The trajectory of the second state of the non-polynomial system (24) and of the approximate polynomial system are shown in black and dashed blue, respectively. State-independent and state-dependent bounds on the second state are shown in red and green, respectively.

ful if polynomial optimization solvers would provide a certificate for feasibility for MPC optimization problems within a sampling time interval. One strategy would be to first find a feasible solution within a specific time frame by exploiting inherent properties of the polynomial optimizations, and then try to solve for the global solution in the remaining time.

A method for reducing the online computational cost of polynomial MPC involves reducing the number of constraints in the polynomial optimization to be solved online. Zheng [33] proposed to enforce constraints on the first control move and relax constraints on subsequent control moves within the prediction horizon. The justification for the approach is that only the first calculated control move will actually be implemented. Although this method poses significant theoretical challenges (e.g., recursive feasibility), the approach could significantly reduce the online computational cost of NMPC algorithms.

Another opportunity for future research in MPC for polynomial systems involves the exploration of output-feedback MPC. The fact that the separation principle does not apply directly to nonlinear systems suggests that it may be beneficial to formulate the NMPC algorithm to simultaneously design the observer and controller. This concept has not been well-explored in the polynomial systems literature.

Another significant contribution to the field could be made by the application of robust MPC (RMPC) control techniques to polynomial systems. Uncertainty is unavoidable in real-world systems, and the approximation of a non-polynomial system as a polynomial system results in a bounded error term which can be treated as an uncertain parameter. Both inherent model uncertainty and approximation error motivate the application of RMPC techniques, which have been well-studied for linear systems [25], to polynomial systems.

Stochastic model predictive control (SMPC) has also been fairly well-studied for linear systems as a method to reduce the conservatism associated with set-based RMPC. The application of SMPC techniques to polynomial systems would be a valuable contribution. Many SMPC algorithms employ sampling techniques to propagate uncertainty through the dynamic system. In these cases, propagation of uncertainty through polynomial systems would not likely be much more difficult than propagation of uncertainty through linear systems. Other techniques employ methods such as polynomial chaos theory to propagate uncertainty [20]. These methods could also be extended to polynomial systems since both collocation and Galerkin projection can be applied to polynomial systems. A particular challenge for SMPC is the correct implementation of chance constraints in order to obtain meaningful guarantees for the closed-loop system (see [20] for details).

All of these directions for polynomial MPC have the potential to influence the field, both in terms of theoretical impact as well as practical implementation.

Acknowledgements Funding is acknowledged from the Novartis-MIT Center for Continuous Pharmaceutical Manufacturing.

References

1. Allgöwer, F., Findeisen, R., Nagy, Z.K.: Nonlinear model predictive control: from theory to application. J. Chin. Inst. Chem. Eng. **35**, 299–315 (2004)
2. Bemporad, A., Morari, M., Dua, V., Pistikopoulos, E.N.: The explicit linear quadratic regulator for constrained systems. Automatica **38**(1), 3–20 (2002)
3. Camacho, E.F., Bordons, C.: Nonlinear model predictive control: an introductory review. In: Findeisen, R., Allgöwer, F., Biegler, L. (eds.) Assessment and Future Directions of Nonlinear Model Predictive Control, pp. 1–16. Springer, Berlin (2007)
4. Chesi, G.: Estimating the domain of attraction for non-polynomial systems via LMI optimizations. Automatica **45**(6), 1536–1541 (2009)
5. Choi, M.-D., Lam, T.Y., Reznick, B.: Sums of squares of real polynomials. In: Proceedings of the Symposia in Pure Mathematics, vol. 58, pp. 103–126 (1995)
6. Ebenbauer, C., Allgöwer, F.: Analysis and design of polynomial control systems using dissipation inequalities and sum of squares. Comput. Chem. Eng. **30**(10), 1590–1602 (2006)
7. Fotiou, I.A., Rostalski, P., Parrilo, P.A., Morari, M.: Parametric optimization and optimal control using algebraic geometry methods. Int. J. Control **79**(11), 1340–1358 (2006)
8. Harinath, E., Foguth, L.C., Paulson, J.A., Braatz, R.D.: Nonlinear model predictive control using polynomial optimization methods. In: Proceedings of the American Control Conference, pp. 1–6 (2016)

9. Henrion, D., Lasserre, J.B.: GloptiPoly: global optimization over polynomials with Matlab and SeDuMi. ACM Trans. Math. Softw. **29**(2), 165–194 (2003)
10. Ichihara, H.: State feedback synthesis for polynomial systems with bounded disturbances. In: Proceedings of the 47th IEEE Conference on Decision and Control, pp. 2520–2525 (2008)
11. Ichihara, H.: Optimal control for polynomial systems using matrix sum of squares relaxations. IEEE Trans. Autom. Control **54**(5), 1048–1053 (2009)
12. Kim, S., Kojima, M., Waki, H.: Generalized Lagrangian duals and sums of squares relaxations of sparse polynomial optimization problems. SIAM J. Optim. **15**(3), 697–719 (2005)
13. Lasserre, J.B.: Global optimization with polynomials and the problem of moments. SIAM J. Optim. **11**(3), 796–817 (2001)
14. Lasserre, J.B.: Semidefinite programming vs. LP relaxations for polynomial programming. Math. Oper. Res. **27**(2), 347–360 (2002)
15. Löfberg, J.: YALMIP: a toolbox for modeling and optimization in MATLAB. In: Proceedings of the IEEE International Symposium on Computer Aided Control Systems Design, Taipei, Taiwan, pp. 284–289 (2004)
16. Maier, C., Böhm, C., Deroo, F., Allgöwer, F.: Predictive control for polynomial systems subject to constraints using sum of squares. In: Proceedings of the 49th IEEE Conference on Decision and Control, pp. 3433–3438 (2010)
17. Mayne, D.: Nonlinear model predictive control: challenges and opportunities. In: Allgöwer, F., Zheng, A. (eds.) Nonlinear Model Predictive Control, pp. 23–44. Springer, Berlin (2000)
18. Papachristodoulou, A., Prajna, S.: Analysis of non-polynomial systems using the sum of squares decomposition. In: Henrion, D., Garulli, A. (eds.) Positive Polynomials in Control, pp. 23–43. Springer, Berlin (2005)
19. Parrilo, P.A.: Structured semidefinite programs and semialgebraic geometry methods in robustness and optimization. PhD thesis, California Institute of Technology, Pasadena (2000)
20. Paulson, J.A., Harinath, E., Foguth, L.C., Braatz, R.D.: Nonlinear model predictive control of systems with probabilistic time-invariant uncertainties. In: Proceedings of the 5th IFAC Conference on Nonlinear Model Predictive Control, pp. 16–25 (2015)
21. Prajna, S., Papachristodoulou, A., Parrilo, P.A.: Introducing SOSTOOLS: a general purpose sum of squares programming solver. In: Proceedings of the 41st IEEE Conference on Decision and Control, pp. 741–746 (2002)
22. Prajna, S., Papachristodoulou, A., Wu, F.: Nonlinear control synthesis by sum of squares optimization: a Lyapunov-based approach. In: Proceedings of the 5th Asian Control Conference, pp. 157–165 (2004)
23. Putinar, M.: Positive polynomials on compact semi-algebraic sets. Indiana Univ. Math. J. **42**(3), 969–984 (1993)
24. Raff, T., Findeisen, R., Ebenbauer, C., Allgöwer, F.: Model predictive control for discrete time polynomial control systems: a convex approach. In: Proceedings of the 2nd IFAC Symposium on System, Structure and Control, pp. 158–163 (2004)
25. Raković, S.V.: Invention of prediction structures and categorization of robust MPC syntheses. In: Proceedings of the 4th IFAC Conference on Nonlinear Model Predictive Control, pp. 245–273 (2012)
26. Rawlings, J.B.: Tutorial overview of model predictive control. IEEE Control Syst. **20**(3), 38–52 (2000)
27. Rostalski, P., Fotiou, I.A., Bates, D.J., Beccuti, A.G., Morari, M.: Numerical algebraic geometry for optimal control applications. SIAM J. Optim. **21**(2), 417–437 (2011)
28. Scokaert, P.O., Mayne, D.Q., Rawlings, J.B.: Suboptimal model predictive control (feasibility implies stability). IEEE Trans. Autom. Control **44**(3), 648–654 (1999)
29. Shor, N.: Class of global minimum bounds of polynomial functions. Cybern. Syst. Anal. **23**(6), 731–734 (1987)

30. Sturm, J.F.: Using SeDuMi 1.02, a MATLAB toolbox for optimization over symmetric cones. Optim. Methods Softw. **11**(1–4), 625–653 (1999)
31. Tanartkit, P., Biegler, L.T.: A nested, simultaneous approach for dynamic optimizations – I. Comput. Chem. Eng. **20**(6–7), 735–741 (1996)
32. Tibken, B.: Estimation of the domain of attraction for polynomial systems via LMIs. In: Proceedings of the 39th IEEE Conference on Decision and Control, pp. 3860–3864 (2000)
33. Zheng, A.: A computationally efficient nonlinear MPC algorithm. In: Proceedings of the American Control Conference, pp. 1623–1627 (1997)

Distributed MPC for Large-Scale Systems

Marcello Farina and Riccardo Scattolini

1 Introduction and Motivations

In the past decades a number of popular methods have been developed for control of a multi-input and multi-output system $\mathscr{S}^o$, including optimal LQ control theory, pole-placement methods, $\mathscr{H}_2/\mathscr{H}_\infty$ control, and Model Predictive Control (MPC). These methods are intrinsically of centralized nature, i.e. the vector of control actions $\mathbf{u}$ is computed based on the knowledge of the whole state $\mathbf{x}$ or output $\mathbf{y}$ vectors. This allows to guarantee properties, in terms of stability, robustness, and performance. However, these methods display many limitations in case of large-scale [1] or complex [2] systems, i.e., systems with a large size, often characterized by the cooperation of many different parts (e.g., machines, reactors, robots, transportation systems), and possibly by uncertainties on the system components. Just to mention the main issues:

- the actuators and the transducers may be highly geographically distributed, and this may bring about *transmission* delays or failure issues.
- The control problem grows in size with the dimensionality of the system, and the related computational burden may, in turn, grow significantly. This is particularly true for methods, like MPC, where an optimization problem must be solved at any new sampling time. In turn, these scalability issues may induce large - and even inadmissible - computational delays, with serious limitations on the size of practically tractable problems.
- Single components or subsystems can be subject to structural changes, failures, and some may be removed, added, or replaced. Centralized control systems are normally non-robust and non-flexible with respect to such occurrences. This issue may have a big economic impact, since a new design and implementa-

M. Farina · R. Scattolini (✉)
Dipartimento di Elettronica, Informazione e Bioingegneria, Politecnico di Milano, via Ponzio 34/5, 20133 Milano, Italy
e-mail: marcello.farina@polimi.it; riccardo.scattolini@polimi.it

S. V. Raković, W. S. Levine (eds.), *Handbook of Model Predictive Control*, Control Engineering, https://doi.org/10.1007/978-3-319-77489-3_11

tion phase for the overall control system must be carried out, with significant expenses, e.g., due to the required downtime.

All the above reasons motivate the development of new methods for the design of more flexible control structures where the centralized controller is replaced by a set of M local regulators, possibly coordinated to recover to the maximum extent the performance guaranteed by a centralized solution. According to a terminology nowadays well accepted, we will define *decentralized control structures* those where the local regulators are fed by independent subsets of states and/or outputs and do not communicate with each other. As a middle-ground solution between centralized and decentralized control, we will denote by *distributed control structures* the schemes where the M local regulators are fed by not necessarily independent subsets of states and/or outputs and can exchange information to coordinate their actions. A pictorial representation of centralized, decentralized, and distributed control structures in the case $M = 2$ is reported in Figure 1.
The possibility to coordinate and negotiate the local control actions provided by distributed schemes and made possible thanks to suitable transmission networks fits very well with an optimization-based framework, where many ideas of game theory can be applied. Therefore, MPC is probably the best advanced control approach for the synthesis of distributed control algorithms and for this reason the aim of this chapter is to present in plain form the main ideas underlying some of the most popular Distributed MPC (DMPC) algorithms. Since nowadays many survey papers and books are dedicated to DMPC, this chapter is not aimed to provide an extensive review of all the available DMPC algorithms, for which the reader is referred to [3, 4], but rather to highlight the main features, properties, and requirements of the major classes of DMPC methods. With the goal to simplify the presentation some restrictive choices will be made: (i) the theoretical properties of the considered methods will not be examined in detail, and the reader will be referred to the relevant liter-

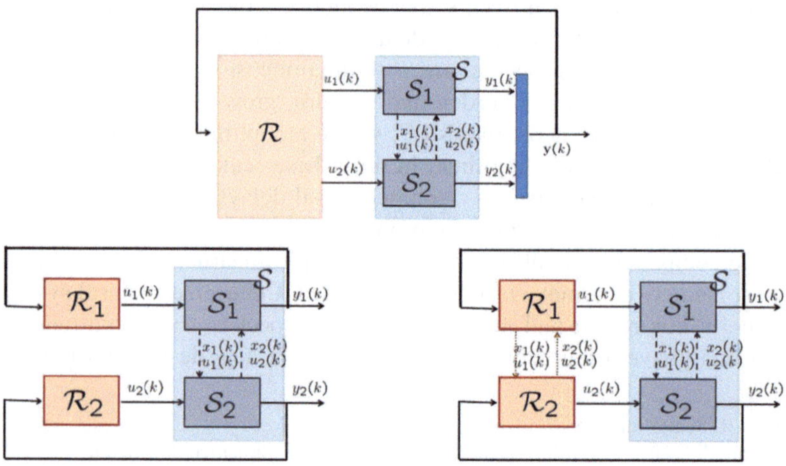

Fig. 1: Centralized (top panel), decentralized (bottom left panel), and distributed (bottom right panel) control structures.

ature; (ii) purely regulation problems will be considered; (iii) the system $\mathscr{S}^o$ to be controlled will be assumed to be described by a linear time-invariant discrete-time model. Some of these assumptions will be re-examined in the final section of the chapter.

2 Model and Control Problem Decomposition

As it is claimed in [1], "if such (large-scale) systems must be controlled,(...) they necessitate new ideas for decomposing and dividing the analysis and control problems of the overall system into rather independent subproblems." Under this viewpoint the decomposition of the dynamic large-scale system model and of the control problem is a preliminary - but key - step for the development of well-posed decentralized and distributed control procedures. In particular, it is definitely worth noting that the adopted model decomposition has a strong impact on the features, properties, and requirements of the control scheme which is designed based on it.

2.1 Model Decomposition

We assume that the large-scale system $\mathscr{S}^o$ is described by the following linear, discrete-time model

$$\begin{aligned} \mathbf{x}^o(k+1) &= \mathbf{A}^o\mathbf{x}^o(k) + \mathbf{B}^o\mathbf{u}(k) \\ \mathbf{y}(k) &= \mathbf{C}^o\mathbf{x}^o(k) \end{aligned} \tag{1}$$

where $\mathbf{x}^o \in R^n$, $\mathbf{u} \in R^m$, $\mathbf{y} \in R^p$ and, unless otherwise specified, the state $\mathbf{x}^o$ will be assumed to be measurable. The state, output, and/or input variables must satisfy constraints of the general type

$$\mathbf{x}^o \in \mathscr{X}^o \quad , \quad \mathbf{y} \in \mathscr{Y} \quad , \quad \mathbf{u} \in \mathscr{U} \tag{2}$$

where $\mathscr{X}^o$, $\mathscr{Y}$, and $\mathscr{U}$ are closed sets of proper dimensions containing the origin.

In order to design M decentralized or distributed MPC regulators guaranteeing the stability of the origin of the corresponding closed-loop system, the centralized model (1) must be decomposed into M small scale models $\mathscr{S}_i$, also denoted model partitions. The first problem is that of partitioning the input and output vectors $\mathbf{u}$ and $\mathbf{y}$, i.e., to identify, for each subsystem $\mathscr{S}_i$, a local input vector $u_i \in R^{m_i}$ and a local output vector $y_i \in R^{p_i}$ (commonly, it must hold that $\sum_{i=1}^M m_i = m$ and that $\sum_{i=1}^M p_i = p$). Indeed, a local input/output pair $(u_i(k), y_i(k))$ (also denoted *channel*) must be attributed to each subsystem on the basis of a specific partitioning criterion. Often this is based on physical insight; as an alternative, many methods have been developed and can be adopted to unveil the interactions of MIMO systems. Some of them are based on the analysis of the static and dynamic interactions among inputs and outputs see, e.g., [5, 6].

The state-space model, for each subsystem $\mathscr{S}_i$, is of the following general type.

$$
\begin{array}{ll}
x_i(k+1) & = A_{ii}x_i(k) + B_{ii}u_i(k) + \sum_{j\neq i}(A_{ij}x_j(k) + B_{ij}u_j(k)) \\
y_i(k) & = C_{ii}x_i(k) + \sum_{j\neq i}C_{ij}x_j(k)
\end{array}
\tag{3}
$$

where $i = 1,\dots,M$, $x_i \in R^{n_i}$. Many different methods can be used to obtain the models $\mathscr{S}_i$ from $\mathscr{S}^o$, each corresponding to a different model partition and to a different corresponding interconnection graph. A very rough classification can be made between *non-overlapping* and *overlapping* decompositions: in non-overlapping decompositions the aggregate state of the $\mathscr{S}_i$ model is still of order n, that is the one of the original system $\mathscr{S}^o$. On the contrary, in overlapping decompositions the overall state of the ensemble of models $\mathscr{S}_i$ is greater than the one of $\mathscr{S}^o$, i.e. $\sum_{i=1}^M n_i > n$. The characteristics of these two classes will be briefly analyzed in Sections 2.1.1 and 2.1.2.

A critical point regards which constraints should be enforced on the state, output, and/or input variables of $\mathscr{S}_i$, $i = 1,\dots,M$, to verify the centralized system constraints (2). Note that constraints involving the variables of more than one subsystems (the so-called *coupling* constraints) may be present, and even they may lie at the core of specific types of distributed and coordination control problems. For instance, when subsystems share a common, but limited, input resource some constraints of the type $\sum_{i=1}^M u_i \in \bar{\mathscr{U}}$ (for a suitably defined $\bar{\mathscr{U}}$) may be enforced. On the other hand, a number of power generation devices may be asked to produce a common, but bounded, output, leading to constraints of the type $\sum_{i=1}^M y_i \in \bar{\mathscr{Y}}$. Another similar example is the case of coordination of moving robots, where collision avoidance constraints are enforced: assuming that the robot - or the robot joints - positions are included in the outputs y_i and that two robots (i.e., $\mathscr{S}_1$ and $\mathscr{S}_2$) are involved, these constraints may be formulated as $(y_1, y_2) \in \mathscr{Y}_{12}$, for a suitably-defined set $\mathscr{Y}_{12}$. In some cases coupling constraints can be verified by enforcing a number of local constraints at the same time, e.g., when $u_i \in \mathscr{U}_i$ for all $i = 1,\dots,M$ involves $\sum_{i=1}^M u_i \in \bar{\mathscr{U}}$. However, this solution may be overly conservative and highly suboptimal in many contexts and should be discarded, at the price of including coupling (also said *complicating*) constraints in the - distributed - control problem formulation. Summing up, from now on we assume that constraints (2) allow to formulate two types of constraints on the model partition variables: local constraints, to be enforced for all $i = 1,\dots,M$

$$
x_i \in \mathscr{X}_i \quad , \quad y_i \in \mathscr{Y}_i \quad , \quad u_i \in \mathscr{U}_i \tag{4}
$$

and/or coupling constraints, which will be represented as follows for simplicity

$$
(x_1,\dots,x_M) \in \mathscr{X}_C \quad , \quad (y_1,\dots,y_M) \in \mathscr{Y}_C \quad , \quad (u_1,\dots,u_M) \in \mathscr{U}_C \tag{5}
$$

Note, in passing, that there may not be a trivial/direct correspondence between the state variable x_i of $\mathscr{S}_i$, $i = 1,\dots,M$ and the state $\mathbf{x}^o$ of $\mathscr{S}^o$. Therefore, especially - but not only - when overlapping decompositions are used, it may be critical and/or ambiguous to translate the constraint $\mathbf{x}^o \in \mathscr{X}^o$ into a number constraints on the local state variables x_i.

Subclasses of the representation (3) are called *input-decoupled* when $B_{ij} = 0$ for all i and $j \neq i$ or *state-decoupled* when $A_{ij} = 0$ for all i and $j \neq i$. In a wide class of control problems, for instance the coordination of independent vehicles with coupling constraints, the subsystems are *dynamically decoupled*, with $A_{ij} = 0$, $B_{ij} = 0$, $C_{ij} = 0$ for all i and $j \neq i$, but coupled through the state or control constraints (5).

Some final comments are in order. First, the model (1) is often computed as the discretization of a continuous time physical system made by the interconnection of subsystems. In this case, the corresponding matrices have a sparse structure which should be maintained after discretization. Unfortunately, if a Zero Order Hold transformation is used, this property is lost. To recover it, one should resort to the forward Euler (fE) discretization approach or to the approximate discretization method described in [7], specifically developed for distributed control.
Secondly, models of type (3) include the information on how subsystems have influence on each other. In other words, if (for $j \neq i$) $A_{ij} \neq 0$ and/or $B_{ij} \neq 0$, the state/input variables of subsystem $\mathscr{S}_j$ directly impact on the dynamics of subsystem $\mathscr{S}_i$. Consistently, we can define the set of neighbors (or parents) of subsystem $\mathscr{S}_i$ as $\mathscr{N}_i := \{j : \|A_{ij}\| + \|B_{ij}\| \neq 0\|\}$ and a corresponding direct interconnection graph, which highlights the system-wide dependencies between subsystems. Similar interconnection (possibly undirected) graphs can be drawn when coupling constraints are present, i.e., if a coupling constraint involves a set of subsystems, they should be considered as neighbors.

2.1.1 Non-overlapping Decompositions

Perhaps the most natural choice to decompose $\mathscr{S}^o$ is to partition the state $\mathbf{x}$ into M non-overlapping sub-vectors x_i with $\sum_{i=1}^{M} n_i = n$, so that, up to a suitable state variable permutation, $\mathbf{x}^o = (x_1, \ldots, x_M)$ and that the original system can be written as

$$\begin{bmatrix} x_1(k+1) \\ \vdots \\ x_M(k+1) \end{bmatrix} = \begin{bmatrix} A_{11} & \ldots & A_{1M} \\ \vdots & \ddots & \vdots \\ A_{M1} & \ldots & A_{MM} \end{bmatrix} \begin{bmatrix} x_1(k) \\ \vdots \\ x_M(k) \end{bmatrix} + \begin{bmatrix} B_{11} & \ldots & B_{M1} \\ \vdots & \ddots & \vdots \\ B_{M1} & \ldots & B_{MM} \end{bmatrix} \begin{bmatrix} u_1(k) \\ \vdots \\ u_M(k) \end{bmatrix}$$

$$\begin{bmatrix} y_1(k) \\ \vdots \\ y_M(k) \end{bmatrix} = \begin{bmatrix} C_{11} & \ldots & C_{1M} \\ \vdots & \ddots & \vdots \\ C_{M1} & \ldots & C_{MM} \end{bmatrix} \begin{bmatrix} x_1(k) \\ \vdots \\ x_M(k) \end{bmatrix}$$

It is intuitive that in the definition of the submodels (3) one should partition the original state so that the couplings among the subsystems are reduced as much as possible, i.e. A_{ij}, B_{ij}, C_{ij}, $i \neq j$ should be null or "small" (according to a proper norm or criterion) to the maximum possible extent. In the common practice, the state permutation/partition can be carried out based on the plant physical insight or based on available algebraic algorithms. For example, there exist graph-based methods for reordering the state, input, and output variables in order to highlight inherent structures, e.g., the presence of weakly interacting subsystems (see, e.g., the

ε-nested decomposition proposed in [2]) or cascaded configurations (e.g., the lower-block-triangular LBT decomposition discussed in [2]). Finally, methods have been proposed in the literature [2] for devising suitable changes of coordinates which lead to specific system decompositions (e.g., the *input and/or output decentralized forms*), at the price of a loss of physical insight.

2.1.2 Overlapping Decompositions

In some cases the coupling strength between different subsystems is relevant, which prevents the decomposition into disjoint subsystems to result into an effective control-oriented partition. An alternative to non-overlapping decompositions is to decompose the systems into subsystems which have some equations (and states) in common, i.e., carrying out a so-called *overlapping decomposition*. This may result in obtaining overlapping but weakly coupled subsystems. Overlapping decompositions have been widely studied in the past, mainly in the context of decentralized control, see, e.g., [2].
In the context of DMPC, a number of distributed control methods require the subsystem interconnections to be represented in the following *state-decoupled* form

$$\begin{cases} x_i(k+1) &= A_{ii}x_i(k) + \sum_{j=1}^{M} B_{ij}u_j(k) \\ y_i(k) &= C_i x_i(k) \end{cases} \tag{6}$$

However, it is rarely possible to obtain a representation of this type by simply applying a non-overlapping decomposition such as the one described in the previous paragraphs. Therefore, it is possible to adopt a *fully overlapping decomposition* where each subsystem is of full order. The simplest way to obtain this decomposition consists of replicating M times model (1), that is setting, for all $i, j = 1, \ldots, M$, $A_{ii} = \mathbf{A}^o$, $B_{ij} = \mathbf{B}^{j,o}$, $C_i = \mathbf{C}^{i,o}$, where $\mathbf{B}^{j,o}$ is the j-th block column of $\mathbf{B}^o$, while $\mathbf{C}^{j,o}$ is the j-th block row of $\mathbf{C}^o$; an alternative procedure is sketched in [8].
In general, overlapping decompositions are non-minimal, since the state dimension does not decrease (for each subsystem) under the application of the proposed partition. However, as better specified in the following, for subsystem $\mathscr{S}_i$, the local control variable is assumed to be u_i, while the u_j's, $j \neq i$ are considered as external signals, so that the number of variables to be optimized by the local DMPC algorithm is smaller than in the corresponding centralized problem.

2.2 Partition Properties and Control

The main properties of the model partitions which have a major impact on the resulting decentralized and distributed MPC-based schemes are the following.

- **Minimality of the state representation**. Minimality (in terms of dimension of the state/input/output variables for each subsystem) is required to limit the algorithm computational burden, since the number of involved variables generally directly impacts on the computational demands of the algorithm. Also, minimal representations demand minimal information to be stored by local memory units. This, besides requiring scalable memory load, allows for more flexibility of the resulting control systems. In fact, local model variations at subsystem level may require the re-design of local control systems only.

- **Minimality of the interconnection graph**. As discussed, to reduce the probability of network-induced issues (e.g., transmission delays, channel overloads, and package losses) the communication burden required by the adopted control architecture must be reduced as much as possible. To this end, one should reduce both (i) the number of communication links between subsystems (i.e., aiming to have a sparse supporting information transmission network), and (ii) the amount of communication that they should afford. While (ii) mostly depends on the type of the adopted control scheme (we defer the reader to Section 4 for a discussion on this point), (i) may strongly depend upon the approach taken for model partitioning, since the implementation of distributed schemes commonly requires to be supported by an underlying communication graph consistent with the subsystem interconnection graph.
- **Descriptive capabilities of the models**. Model partitioning may have a negative effect on the descriptive capabilities of the submodels. It would be desirable, in fact, that each local controller has the knowledge on how a control action, taken locally, can impact, not only on the local variables, but also on the variables of the surrounding subsystems. Similarly, a complete information on how the control action taken by other subsystems affects local state variables and outputs may be desired.

It is worth noting that the first and the second requirements are in general conflicting with the third one. Indeed, fully descriptive models, which are the ones that allow for full cooperation between local controllers, are often obtained using overlapping - often fully overlapping - partitions, which are the ones which typically lead to a non-minimal state representation and for which the interconnection graph is maximal.

2.3 MPC Problem Separability

To briefly introduce the optimization problems involved in the decentralized and distributed implementations summarized in this chapter, we first introduce the centralized (non-minimal) model which can be drawn by collecting together all the partitions (3). We define $\mathbf{x} = (x_1, \ldots, x_M)$, which corresponds to $\mathbf{x}^o$, up to a state permutation, only in case of non-overlapping decompositions. Consistently, we can describe the overall large-scale system dynamics with a model of the type

$$\begin{aligned} \mathbf{x}(k+1) &= \mathbf{A}\mathbf{x}(k) + \mathbf{B}\mathbf{u}(k) \\ \mathbf{y}(k) &= \mathbf{C}\mathbf{x}(k) \end{aligned} \tag{7}$$

Note that system (7) is nothing else than an *expansion* [2] of the corresponding contracted system (1), since their output free and forced motions must be indeed equal to each other. Starting with standard centralized MPC, the control action at each time instant k is obtained by solving a control problem of type

$$\min_{\mathbf{u}(k),\dots,\mathbf{u}(k+N-1)} J(k) \tag{8a}$$

subject to the transient constraints (7), (4), (5), for times $k,\dots,k+N-1$, and a terminal constraint of the type

$$\mathbf{x}(k+N) \in \mathscr{X}_f \tag{8b}$$

In (8a), $J(k)$ is the cost function, while $\mathscr{X}_f$ in (8b) is the terminal constraint set. As a common ground of the methods discussed in this chapter, problem (8) is required to be separable into a number of subproblems to be solved by local computing units. This first requires that the cost function to be minimized is formally separable, i.e.,

$$J(k) = \sum_{i=1}^{M} \rho_i J_i(k) \tag{9}$$

where $J_i(k)$ is a positive-definite (quadratic, for simplicity) function of local input and state variables of subsystem $\mathscr{S}_i$

$$J_i(k) = \sum_{j=0}^{N-1} [\|x_i(k+j)\|^2_{Q_i} + \|u_i(k+j)\|^2_{R_i}] + \|x_i(k+N)\|^2_{P_i} \tag{10}$$

and parameter $\rho_i > 0$, where $\sum_{i=1}^{M} \rho_i = 1$.
Secondly, we need to enforce the terminal constraint (8b) by imposing M local terminal constraints of type $x_i(k+N) \in \mathscr{X}_{f,i}$, and therefore $\mathscr{X}_f$ must be defined as the Cartesian product of M sets $\mathscr{X}_{f,i} \subseteq \mathbb{R}^{n_i}$.
However, separability is not the only requirement for the well-posedness of the problem, but also some assumptions - required for general-type MPC problems [8] - must be fulfilled by the "collective" terminal cost and terminal set, i.e., $V_f = \sum_{i=1}^{M} \rho_i \|x_i(k+N)\|^2_{P_i} = \|\mathbf{x}(k+N)\|^2_P$, with $P = \text{diag}(\rho_1 P_1, \dots, \rho_M P_M)$ and $\mathscr{X}_f$, respectively. More specifically V_f and $\mathscr{X}_f$ must also be a Lyapunov function and a positively invariant set, respectively, for the expanded system (7), controlled using a suitable (and possibly decentralized/distributed) auxiliary control law. The latter requirements are not easily compatible with the separability assumptions: in the sequel, for each of the described methods, we will discuss how they are guaranteed.

3 Decentralized MPC

A simple paradigmatic decentralized method is the one proposed in [9], for large-scale linear processes subject to local input saturations only. The global model (1), with stable matrix $\mathbf{A}^o$, is approximated by a number of (possibly overlapping) subsystems of type (3) with stable matrices A_{ii}, used for local predictions. The key feature of decentralized control is that there is no communication between the different local controllers, as shown in Figure 1. Therefore a modelling error is introduced by neglecting couplings, i.e., by setting $A_{ij} = 0$ and $B_{ij} = 0$ for all $j \neq i$, $i = 1,\ldots,M$. The proposed decentralized MPC algorithm requires that, at each time instant k, the following optimization problem is solved by each computational unit.

$$\min_{u_i(k),\ldots,u_i(k+N-1)} J_i(k) \tag{11}$$

subject to the dynamical model

$$x_i(k+1) = A_{ii}x_i(k) + B_{ii}u_i(k) \tag{12}$$

and the local input constraints $u_i(k) \in \mathscr{U}_i$, for times $k,\ldots,k+N-1$. Here no terminal constraint is required in view of the fact that no state constraints are imposed and that the local system matrices A_{ii} are stable. Indeed, the auxiliary control law is $u(k) = 0$ and, correspondingly, the separability of the cost function is obtained by selecting P_i in such a way that $A_{ii}^T P_i A_{ii} - P_i = -Q_i$. This, if we neglect the interconnections terms A_{ij} with $j \neq i$, between subsystems, makes V_f (as defined in the previous paragraph) a Lyapunov function for the overall - decentralized - system. The asymptotic stability properties of the control system are proved a posteriori if some inequality conditions are verified. In general, in order to achieve closed-loop stability as well as performance in the development of decentralized MPC algorithms, the interconnections (at least the terms A_{ij}) between different subsystems should be weak or the system should display peculiar structures (e.g., acyclic graphs).

A different approach consists of considering couplings as disturbances. For example, in [10], a decentralized MPC algorithm for nonlinear discrete time systems subject to decaying disturbances was presented. In the design of the decentralized MPC, the effects of interconnections between different subsystems are considered as perturbation terms whose magnitude depends on the norm of the system states. No information is exchanged between the local controllers and the stability of the closed-loop system relies on the inclusion of a contractive constraint in the formulation of each of the decentralized MPC problems. In [11], the stability of a decentralized MPC is analyzed from an input-to-state stability (ISS) point of view. For linear systems this approach consists of rewriting the subsystem $\mathscr{S}_i$ model (3) as

$$x_i(k+1) = A_{ii}x_i(k) + B_{ii}u_i(k) + v_i(k) \tag{13}$$

where $v_i(k) = \sum_{j\neq i}(A_{ij}x_j(k) + B_{ij}u_j(k))$ is regarded as an unknown, but bounded - if local state and input constraints (4) are enforced - disturbance which must be

compensated or rejected using an ad-hoc robust MPC scheme. Along this line, e.g., the algorithm in [12] has been proposed for linear systems, resorting to tube-based MPC (see [13, 14] and Chapter Robust Optimization for MPC). For more details, see also Chapter "Scalable MPC Design".
It is worth remarking that, in all the decentralized schemes described above, no information is required to be transmitted between local regulators, since only the information regarding the local state is used in the corresponding MPC optimization problem (11).

4 Distributed MPC

According to the taxonomy proposed in [3] and nowadays widely used, DMPC algorithms can be broadly classified as follows:

- *iterative* or *non-iterative*: in iterative algorithms information can be transmitted among the local regulators many times within the sampling time. This opens the way to the design of methods aimed at achieving a global consensus among the regulators on the actions to be taken within the sampling interval. On the contrary, in non-iterative algorithms information is transmitted only once in the sampling period, so that the regulators are required to possess some robustness properties to compensate for the reduced information available.
- *cooperating* and *non-cooperating*: in cooperating algorithms each local regulator tries to minimize a global cost function, so that Pareto - i.e., system-wide - optimal solutions can be computed, at least in principle. In non-cooperating algorithms each regulator minimizes its local cost function, with possible conflicting goals; in this case, Nash equilibria are to be expected.
- *fully connected* or *partially connected*: in fully connected algorithms information is transmitted and received from any local regulator to all the others. In partially connected methods the information is exchanged between any local regulator and a subset of the others. Although this is not a structural property, it can strongly influence the properties and the transmission load of the methods as well as their computational burden.

4.1 Cooperating DMPC

As a prototype algorithm for this class of DMPC methods, we make reference to the results reported in [8, 15]. The system to be controlled is assumed to be in the state decoupled form (6): this implies that the transition matrix of the expanded system (7) is block-diagonal, i.e., $\mathbf{A} = \text{diag}(A_{11}, \ldots, A_{MM})$. For simplicity we assume $\mathbf{A}$ to be asymptotically stable. This ensures the separability of the cost function $J(k)$ as in (9): in fact we can take $\mathbf{u}(k) = \mathbf{Kx} = 0$ (i.e., $\mathbf{K} = 0$) as a - decentralized - auxiliary control law for the expanded system and we can select P_i, $i = 1, \ldots, M$ in such a

way that $A_{ii}^T P_i A_{ii} - P_i = -Q_i$, $Q_i > 0$. This makes V_f a Lyapunov function for the system $\mathbf{x}(k+1) = \mathbf{A}\mathbf{x}(k)$, as required. Finally, in the present algorithm, only input constraints are enforced.
At time k all the local MPC control algorithms have knowledge of the overall state $\mathbf{x}(k)$ and of the full system dynamics, meaning that a fully connected communication network is required to support the transmission of the global state to all the local control stations.
The following iterative procedure is performed within the sampling period from time k and time $k+1$:

- at iteration (negotiation) step p, $p \geq 1$, each local controller in $\mathscr{S}_i$ has the information about the possible input sequences of the other subsystems, $u_j^{p-1}(k+l)$, for $l = 1, \ldots, N-1$ and $j \neq i$, to be broadcast thanks to the available fully connected communication network; the following (global) optimization problem is solved at a local level

$$\min_{u_i(k),\ldots,u_i(k+N-1)} J(k) \tag{14}$$

  subject to the expanded model (7) and, for $l = 0, \ldots, N-1$,

$$u_i(k+l) \in \mathscr{U}_i \tag{15}$$

$$u_j(k+l) = u_j^{p-1}(k+l), \forall j \neq i \tag{16}$$

- letting $u_i^o(k), \ldots, u_i^o(k+N-1)$ be the optimal solution, a convex combination between the solution at the previous iteration and the newly computed one is used, i.e.

$$u_i^p(k+l) = \alpha_i u_i^{p-1}(k+l) + (1-\alpha_i) u_i^o(k+l), l = 0, \ldots, N-1$$

  where $\alpha_i \in (0,1)$ and $\sum_{i=1}^M \alpha_i = 1$.
- if a termination condition, which can depend on the elapsed time within the sampling period or on an optimality test, is satisfied, the iterative procedure ends and the last computed value $u_i^p(k)$ is used as $u_i(k)$, otherwise a new iteration starts ($p \leftarrow p+1$)
- when a new measurement is received, ($k \leftarrow k+1$) the overall procedure is restarted.

The algorithm requires an initialization for $p = 0$, which can be obtained based on the optimal control sequence at the previous time $k-1$.
As shown in [8, 15] stability of the closed-loop system is guaranteed for any number of iterations performed within the sampling time; in addition, it can be proven that the computed solution converges to the one of the corresponding centralized control system as the number of iterations (p) increases. Finally, the method can be extended to cope with unstable open-loop systems, provided that a suitable zero terminal condition is included into the problem formulation, and with tracking problems [8].

Interestingly, in [8] it has been shown that, if each local controller adopts a non-cooperating (selfish) approach and minimizes only its own cost function J_i, convergence to a Nash equilibrium is achieved and no stability guarantees can be proven. In [16], a further step is done: more specifically, it is shown that it is possible to add, for each subsystem, a constraint related to the maximal satisfactory and sufficiently small (denoted *satisficing* in the papers) cost γ_i, i.e., $J_i(k) \leq \gamma_i$. According to [16], this variation allows to shift from a purely cooperating scheme (denoted also *categorical altruist* algorithm) to a scheme (denoted *situational altruist*) where local (*selfish*) constraints are introduced.

4.2 Non-cooperating Robustness-Based DMPC

The algorithm described in [17] is based on the idea that each subsystem i transmits to its neighbors its planned state reference trajectory $\tilde{x}_i(k+j)$, $j = 1,\ldots,N$, over the prediction horizon and guarantees that, for all $j \geq 0$, its actual trajectory lies in a "tube" centered in $\tilde{x}_i$, i.e. $x_i(k+j) \in \tilde{x}_i(k+j) \oplus \mathscr{E}_i$, where $\mathscr{E}_i$ is a compact set including the origin. Then, assuming here for simplicity that the system is input decoupled, Equation (3) can be written as

$$x_i(k+1) = A_{ii}x_i(k) + B_{ii}u_i(k) + \sum_j A_{ij}\tilde{x}_j(k) + w_i(k) \tag{17}$$

where $w_i(k) = \sum_j A_{ij}(x_j(k) - \tilde{x}_j(k)) \in \mathscr{W}_i$ is a bounded disturbance to be rejected using the tube-based MPC approach [13] (see also Chapter "Robust Optimization for MPC"), where $\mathscr{W}_i = \bigoplus_j A_{ij}\mathscr{E}_i$. The term $\sum_j A_{ij}\tilde{x}_j(k)$ is equivalent to a non-manipulable input, known in advance over the prediction horizon, to be properly compensated.

From (17), the i-th subsystem nominal model [13] is defined as

$$\hat{x}_i(k+1) = A_{ii}\hat{x}_i(k) + B_{ii}\hat{u}_i(k) + \sum_j A_{ij}\tilde{x}_j(k) \tag{18}$$

Letting $\mathbf{K}$ =diag$(K_1,\ldots,K_M)$ be a block-diagonal matrix such that both $\mathbf{A}+\mathbf{BK}$ and $A_{ii}+B_{ii}K_i$ are stable, the local control law is chosen as

$$u_i(k) = \hat{u}_i(k) + K_i(x_i(k) - \hat{x}_i(k)) \tag{19}$$

From (17) and (19), letting $z_i(k) = x_i(k) - \hat{x}_i(k)$, it holds that

$$z_i(k+1) = (A_{ii} + B_{ii}K_i)z_i(k) + w_i(k) \tag{20}$$

where $w_i \in \mathscr{W}_i$. Since $\mathscr{W}_i$ is bounded and $A_{ii}+B_{ii}K_i$ is stable, there exists a robust positively invariant set $\mathscr{Z}_i$ for (20) such that, for all $z_i(k) \in \mathscr{Z}_i$ and $w_i(k) \in \mathscr{W}_i$, one has $z_i(k+1) \in \mathscr{Z}_i$. According to the approach developed in [13], given $\mathscr{Z}_i$ and assuming that there exist neighborhoods of the origin $\Delta\mathscr{E}_i$ such that

$$\Delta\mathscr{E}_i \oplus \mathscr{Z}_i \subseteq \mathscr{E}_i \tag{21}$$

at any time instant k the i-th subsystem computes the value of $\hat{u}_i(k)$ in (19) as the solution to

$$\min_{\hat{x}_i(k),\hat{u}_i(k),\ldots,\hat{u}_i(k+N-1)} J_i(k) \tag{22a}$$

subject to (18) and the initial state constraint

$$x_i(k) - \hat{x}_i(k) \in \mathscr{Z}_i \tag{22b}$$

For $l = 0,\ldots,N-1$, to guarantee that the difference between x_i and $\tilde{x}_i$ is effectively limited as initially stated, we require that

$$\hat{x}_i(k+l) - \tilde{x}_i(k+l) \in \Delta\mathscr{E}_i \tag{22c}$$

Both local (4) and coupling (5) constraints can be imposed. This is done by requiring that, for $l = 0,\ldots,N-1$

$$\hat{x}_i(k+l) \in \hat{\mathscr{X}}_i \tag{22d}$$

$$(\tilde{x}_1(k+l),\ldots,\hat{x}_i(k+l)\ldots,\tilde{x}_M(k+l)) \in \hat{\mathscr{X}}_C \tag{22e}$$

This requires to suitably define the sets $\hat{\mathscr{X}}_i$ and $\hat{\mathscr{X}}_C$ as restricted ones, e.g., by setting $\hat{\mathscr{X}}_i \subseteq \mathscr{X}_i \ominus \mathscr{Z}_i$. Although state constraints only have been defined for simplicity, input constraints can be included similarly. Finally, the scheme calls for the definition of terminal constraints of the type

$$\hat{x}_i(k+N) \in \hat{\mathscr{X}}_{f,i} \tag{22f}$$

With the optimal solution at time k, it is also possible to compute the predicted value $\hat{x}_i(k+N)$, which is used to incrementally define the reference trajectory of the state to be used at the next time instant $k+1$, i.e. $\tilde{x}_i(k+N) = \hat{x}_i(k+N)$.

Condition (21) is a key condition for the well posedness of the present distributed control scheme. Despite its analysis goes beyond the scope of this chapter, it is worth remarking that it is equivalent to the so-called tube-based small gain condition for networks discussed in Chapter “Scalable MPC Design”.

Remarkably, each local control station uses only local state information (i.e., $x_i(k)$) and its neighbors' planned state trajectories $\tilde{x}_j(k)$. The latter is transmitted in a neighbor-to-neighbor fashion thanks to the available partially connected communication network.

A significant work has been devoted to the proper definition of separable terminal cost and constraint sets. In [17, 18] methods for a proper choice of the weights Q_i, R_i, P_i, of sets $\mathscr{E}_i$, and of the terminal set $\hat{\mathscr{X}}_{f,i}$ guaranteeing the well posedness and the stabilizing properties of the algorithm are proposed.

4.3 Distributed Control of Independent Systems

A prototype non-iterative algorithm for independent systems coupled through constraints is now described, inspired by the approach described in [19]. The system is assumed to be affected by an unknown, but bounded noise, so that the robust tube-based approach of [13] is used also in this case. The model of the i-th subsystem, $i = 1, \dots, M$, is described by

$$x_i(k+1) = A_{ii}x_i(k) + B_{ii}u_i(k) + d_i(k) \tag{23}$$

where $d_i(k) \in \mathscr{D}_i$ is a bounded disturbance. The M systems are subject to both local and coupling constraints (4) and (5), respectively.
For each system $i = 1, \dots, M$, the local nominal model

$$\hat{x}_i(k+1) = A_{ii}\hat{x}_i(k) + B_{ii}\hat{u}_i(k) \tag{24}$$

is defined and a stabilizing gain K_i is computed. Also in this case, the local stabilizing control law is given by

$$u_i(k) = \hat{u}_i(k) + K_i(x_i(k) - \hat{x}_i(k)) \tag{25}$$

and letting $z_i(k) = x_i(k) - \hat{x}_i(k)$, it holds that

$$z_i(k+1) = (A_{ii} + B_{ii}K_i)z_i(k) + d_i(k) \tag{26}$$

In view of the boundedness of the disturbance and the stability of $A_{ii} + B_{ii}K_i$, there exists a robust positively invariant set $\mathscr{Z}_i$ for (26) such that, for all $z_i(k) \in \mathscr{Z}_i$ and $d_i(k) \in \mathscr{D}_i$, one has $z_i(k+1) \in \mathscr{Z}_i$.
At any time instant k only one system, say the i-th one, is allowed to update its future plans by solving a suitable MPC problem, while all the others update their control variables according to the previously computed control sequence and the corresponding auxiliary law, i.e. their future nominal control moves computed at time k are, for all $j \neq i$

$$\begin{aligned} \hat{u}_j(k+l|k) &= \hat{u}_j(k+l|k-1)\,,\, l = 0, \dots, N-2 \\ \hat{u}_j(k+N-1|k) &= K_j\hat{x}_j(k+N-1|k-1) \end{aligned}$$

where $\hat{x}_j(k+N-1|k-1)$ is the evolution of the nominal state starting from $\hat{x}_i(k) = x_i(k)$ with the sequence $\hat{u}_j(k+l|k-1)$, $l = 0, \dots, N-2$. We also define $\hat{x}_j(k+N-1|k-1) = (A_{ii} + B_{ii}K_i)\hat{x}_j(k+N-1|k-1)$.
On the contrary, the i-th system computes the value of $\hat{u}_i(k)$ in (19) as the solution to

$$\min_{\hat{x}_i(k), \hat{u}_i(k), \dots, \hat{u}_i(k+N-1)} J_i(k) \tag{27a}$$

subject to (18) and the initial state constraint

$$x_i(k) - \hat{x}_i(k) \in \mathscr{Z}_i \tag{27b}$$

Both local (4) and coupling (5) constraints are forced by requiring that, for $l = 0,\dots,N-1$

$$\hat{x}_i(k+l) \in \hat{\mathscr{X}}_i \tag{27c}$$

$$(\hat{x}_1(k+l|k-1),\dots,\hat{x}_i(k+l)\dots,\hat{x}_M(k+l|k-1)) \in \hat{\mathscr{X}}_C \tag{27d}$$

where the sets $\hat{\mathscr{X}}_i$ and $\hat{\mathscr{X}}_C$ are properly restricted subsets of $\mathscr{X}_i$ and $\mathscr{X}_C$, e.g., by setting $\hat{\mathscr{X}}_i \subseteq \mathscr{X}_i \ominus \mathscr{Z}_i$. As in the previous algorithms, the scheme calls for the definition of terminal constraints of the type

$$\hat{x}_i(k+N) \in \hat{\mathscr{X}}_{f,i} \tag{27e}$$

Similarly to the non-cooperating robustness-based control scheme presented in Section 4.2, a partially connected communication network is required to support the transmission of the planned trajectories $\hat{x}_j(k+l|k-1)$ to each local control station from the (constraint-based) neighboring ones.
As described in [19, 20], the basic algorithm here described can be greatly enhanced to allow for more than one system updating its future plans with the optimization procedure at each time step. In addition, cooperation is achieved letting each system to minimize a global cost function and the communication requirements can be significantly reduced with respect to an all-to-all solution by exploiting the graph topology forced by the coupling constraints.
Similar schemes have been devised by other research teams, e.g., [21]. The paper [22] also extends this method to cope with economic-based cost functions.

4.4 Distributed Optimization

A different approach with respect to the distributed algorithms previously described consists of computing the optimal solution to the original centralized optimization problem as the iterative solution to smaller, more tractable, and independent ones. This idea, which is the basis of many popular decomposition methods, can be traced back to the early contributions, e.g., [23]. In the context of MPC, the reader is referred to the recent contributions [24–27].
A sketch of a simple version the popular dual decomposition approach, proposed in [24], applied to MPC is now described. Constraints of general type can easily be considered in the present framework, although for simplicity of presentation they will be neglected here. Consider the set of input decoupled systems

$$x_i(k+1) = A_{ii}x_i(k) + B_{ii}u_i(k) + \sum_{j\neq i} A_{ij}x_j(k) \tag{28}$$

and the following centralized problem

$$\min_{u(k),\dots,u(k+N-1)} J(k) \tag{29}$$

In view of the formal separability of the cost function $J(k)$, the coupling between the subproblems is due to the "coupling variables" $v_i = \sum_{j\neq i} A_{ij}x_j$ in (28). Now write Equation (28), similarly to (13), as

$$x_i(k+1) = A_{ii}x_i(k) + B_{ii}u_i(k) + v_i(k) \tag{30}$$

and, denoting by λ_i the Lagrange multipliers, consider the Lagrangian function

$$\mathscr{L}(k) = \sum_{i=1}^{M}[J_i(k) + \sum_{l=0}^{N-1} \lambda_i(k+l)(v_i(k+l) - \sum_{j\neq i} A_{ij}x_j(k+l))] \tag{31}$$

For the generic vector variable φ, let $\bar{\varphi}_i(k) = [\varphi_i^T(k), \dots, \varphi_i^T(k+N-1)]^T$ and $\bar{\varphi} = [\bar{\varphi}_1^T, \dots, \bar{\varphi}_M^T]^T$. Then, by relaxation of the coupling constraints, the optimization problem of Equation (29) can be stated as

$$\max_{\bar{\lambda}(k)} \min_{\bar{u}(k),\bar{v}(k)} \mathscr{L}(k) \tag{32}$$

or, equivalently

$$\max_{\bar{\lambda}(k)} \sum_{i=1}^{M} \tilde{J}_i(k) \tag{33}$$

where, letting $\bar{A}_{ji}$ be a block-diagonal matrix made by N blocks, all equal to A_{ji},

$$\tilde{J}_i(k) = \min_{\bar{u}_i(k),\bar{v}_i(k)} [J_i(k) + \bar{\lambda}_i^T(k)\bar{v}_i(k) - \sum_{j\neq i} \bar{\lambda}_j^T(k)\bar{A}_{ji}\bar{x}_i(k))] \tag{34}$$

The following two-step iterative procedure is then used at any time step to compute the optimal solution

1. for a fixed $\bar{\lambda}$, solve the set of M independent minimization problems given by Equation (34) with respect to $\bar{u}_i(k), \bar{v}_i(k)$;
2. given the collective values of $\bar{u}, \bar{v}$ computed at the previous step, solve the maximization problem given by (33) with respect to $\bar{\lambda}$. This problem can be solved in a distributed way using a gradient step, see [24].

To summarize, the described iterative algorithm must be supported by a partially connected communication network (for both steps 1 and 2, provided that the latter uses a distributed gradient step). More specifically, at each iteration, for step 1 it is required that, for all $i = 1,\dots,M$, the i-th local computing station receives the current values of $\lambda_j(k+l)$, $l = 1,\dots,N-1$ by the agents $j \neq i$ which have i as neighbor, i.e., such that $i \in \mathcal{N}_j$; on the other hand, for step 2, it is required that the i-th station receives the current values of $x_j(k+l)$ by its neighbors $j \in \mathcal{N}_i$.
It is well recognized that this kind of decomposition approaches are characterized by

slow convergence, due to the great number of iterations required to obtain a solution. However, to this regard, many efficient algorithms have been developed, see, for instance, [26]. In addition, the fundamental properties of recursive feasibility and stability are not a-priori guaranteed, and can be achieved by a proper definition of the optimization problem and of the constraints, see [27].
A final remark is due: as noted above and as apparent from (33), the separability of the cost function $J(k)$ is a major requirement also for this method, as well as - for constrained problems - the separability of the terminal constraint set. The previously discussed approaches can be used to this aim, including the one presented in [17, 18]. Also the recent work [28] is devoted to this problem and allows for scalable design. See Chapter "Scalable MPC Design" for more details.

5 Extensions and Applications

The DMPC algorithms discussed in the previous sections have been designed for linear, discrete-time, and time invariant systems, and the main approaches and ideas behind most of the nowadays available methods for the regulation problem have been described. However, the recent and tumultuous research activity in the field has produced a number of algorithms dealing with a large variety of systems and control problems. Among them, we here recall some of the most interesting research directions, with some references:

- DMPC algorithms have been developed for continuous-time and nonlinear systems in, e.g., [21, 29–31].
- The output feedback case has been studied in [32], while the tracking problem has been analyzed, e.g., in [8, 33]. An alternative approach, based on the particular class of MPC strategies called Command Governor methods, has been reported in [34] and in the papers referenced therein.
- DMPC for systems affected by stochastic noises has been considered in [35–37].
- Economic MPC (see Chapter "Economic Model Predictive Control: Some Design Tools and Analysis Techniques") has been extended to cope with a distributed implementation in [22, 38].
- The new and emerging field of coalitional control, which can be seen as an evolution of DMPC where the topology of the control structure can vary with time, has been treated in [39], where an up-to-date literature review is also reported.

Many applications domains of DMPC have been explored, although most of the reported research still makes reference to simulation studies, with only few real applications (mainly laboratory experiments). In view of its nature, DMPC fits very well with the control of large, spatially distributed systems, possibly interconnected through a network. For this reason, power networks, smart grids, and in general distributed energy management systems are the natural domains where DMPC can offer advantages with respect to a centralized solution, see, for instance, [40–44]. Another class of problems where DMPC can have a great potential impact concerns the management and control of irrigation canals, as discussed in [45]. The coordination of multi-vehicle systems with DMPC has been considered, e.g., in [46, 47]. Finally, applications in other fields are described in [48, 49].

6 Conclusions and Future Perspectives

The research activity in Distributed Model Predictive Control has been intense in the last decade and many methods are nowadays available, see, for instance, the many contributions reported in [4]. In parallel with the development of new algorithms, also the most significant fields of application of DMPC have been clarified. In our opinion, these include networked systems, like power, water, and traffic networks, the coordination of autonomous vehicles and flying systems (drones), the control of very large-scale, weakly coupled, systems. However, there is still a significant gap between research results and real-world applications since most of the DMPC algorithms have only been tested in simulations or with laboratory benchmarks. Among the most promising future research directions, we believe that the reconfigurability of DMPC will be a major topic. To this regard, plug-and-play and coalitional control strategies, possibly driven by external events, will be required to enhance the ability of DMPC to deal with many real control problems and to provide significant improvements with respect to the nowadays adopted control solutions.

References

1. Lunze, J.: Feedback Control of Large-Scale Systems. Series in Systems and Control Engineering, Prentice Hall International, Englewood Cliffs (1992)
2. Šiljak, D.D.: Decentralized control of complex systems. Courier Corporation, North Chelmsford (2011)
3. Scattolini, R.: Architectures for distributed and hierarchical model predictive control - a review. J. Process Control **19**, 723–731 (2009)
4. Maestre, J.M., Negenborn, R.R.: Distributed Model Preductive Control Made Easy. Springer, Berlin (2014)
5. Skogestad, S., Postlethwaite, I.: Multivariable Feedback Control: Analysis and Design. Wiley, London (2005)
6. Kariwala, V., Forbes, J.F., Meadows, E.S.: Block relative gain: properties and pairing rules. Ind. Eng. Chem. Res. **42**(20), 4564–4574 (2003)
7. Farina, M., Colaneri, P., Scattolini, R.: Block-wise discretization accounting for structural constraints. Automatica **49**(11), 3411–3417 (2013)
8. Rawlings, J.B., Mayne, D.Q.: Model Predictive Control: Theory and Design. Nob Hill Publishing, Madison (2009)
9. Alessio, A., Barcelli, D., Bemporad, A.: Decentralized model predictive control of dynamically-coupled linear systems. J. Process Control **21**(5), 705–714 (2011)
10. Magni, L., Scattolini, R.: Stabilizing decentralized model predictive control of nonlinear systems. Automatica **42**(7), 1231–1236 (2006)
11. Raimondo, D.M., Magni, L., Scattolini, R.: Decentralized MPC of nonlinear system: an input-to-state stability approach. Int. J. Robust Nonlinear Control **17**(5), 1651–1667 (2007)

12. Riverso, S., Farina, M., Ferrari-Trecate, G.: Plug-and-play decentralized model predictive control for linear systems. IEEE Trans. Autom. Control **58**(10), 2608–2614 (2013)
13. Mayne, D.Q., Seron, M.M., Raković, S.V.: Robust model predictive control of constrained linear systems with bounded disturbances. Automatica **41**(2), 219–224 (2005)
14. Raković, S.V., Levine, W.S., Açikmeşe, A.B.: Elastic tube model predictive control. In: Proceedings of the 2016 American Control Conference, pp. 3594–3599 (2016)
15. Rawlings, J.B., Stewart, B.T.: Coordinating multiple optimization-based controllers: new opportunities and challenges. J. Process Control **18**, 839–845 (2008)
16. Lopes de Lima, M., Camponogara, E., Limon, D., Muñoz de la Peña, D.: Distributed satisficing MPC. IEEE Trans. Control Syst. Technol. **23**(1), 305–312 (2015)
17. Farina, M., Scattolini, R.: Distributed predictive control: a non-cooperative algorithm with neighbor-to-neighbor communication for linear systems. Automatica **6**, 1088–1096 (2012)
18. Betti, G., Farina, M., Scattolini, R.: Realization issues, tuning, and testing of a distributed predictive control algorithm. J. Process Control **24**(4), 424–434 (2014)
19. Trodden, P., Richards, A.: Distributed model predictive control of linear systems with persistent disturbances. Int. J. Control **83**, 1653–1663 (2010)
20. Trodden, P., Richards, A.: Cooperative tube-based distributed MPC for linear uncertain systems coupled via constraints. In: Maestre, J.M., Negenborn, R.R. (eds.) Distributed Model Predictive Control Made Easy. Springer, Berlin (2014)
21. Müller, M.A., Reble, M., Allgöwer, F.: Cooperative control of dynamically decoupled systems via distributed model predictive control. Int. J. Robust Nonlinear Control **22**, 1376–1397 (2012)
22. Müller, M.A., Allgöwer, F.: Distributed economic MPC: a framework for cooperative control problems. In: Proceedings of IFAC World Congress, pp. 1029–1034 (2014)
23. Mesarovic, M.D., Macko, D., Takahara, Y.: Theory of Hierarchical, Multilevel, Systems. Academic Press, New York (1970)
24. Giselsson, P., Rantzer, A.: Distributed model predictive control with suboptimality and stability guarantees. In: Proceedings of the 49th Conference on Decision and Control, pp. 7272–7277 (2010)
25. Necoara, I., Nedelcu, V., Dumitrache, I.: Parallel and distributed optimization methods for estimation and control in networks. J. Process Control **21**, 756–766 (2011)
26. Giselsson, P., Doan, M.D., Keviczky, T., De Schutter, B., Rantzer, A.: Accelerated gradient methods and dual decomposition in distributed model predictive control. Automatica **49**, 829–833 (2013)
27. Giselsson, P., Rantzer, A.: On feasibility, stability and performance in distributed model predictive control. IEEE Trans. Autom. Control **59**, 1031–1036 (2014)
28. Conte, C., Jones, C.N., Morari, M., Zeilinger, M.N.: Distributed synthesis and stability of cooperative distributed model predictive control for linear systems. Automatica **69**, 117–125 (2016)
29. Farina, M., Betti, G., Scattolini, R.: Distributed predictive control of continuous-time systems. Syst. Control Lett. **74**, 32–40 (2014)
30. Dunbar, W.B.: Distributed receding horizon control of dynamically coupled nonlinear systems. IEEE Trans. Autom. Control **52**, 1249–1263 (2007)
31. Liu, J., Muñoz de la Peña, D., Christofides, P.D.: Distributed model predictive control of nonlinear process systems. AIChE J. **55**, 1171–1184 (2007)
32. Farina, M., Scattolini, R.: An output feedback distributed predictive control algorithm. In: Proceedings of the 50th IEEE Conference on Decision and Control, pp. 8139–8144 (2011)
33. Farina, M., Giulioni, L., Betti, G., Scattolini, R.: An approach to distributed predictive control for tracking - theory and applications. IEEE Trans. Control Syst. Technol. **22**(4), 1558–1566 (2014)
34. Casavola, A., Garone, E., Tedesco, F.: The distributed command governor approach in a nutshell. In: Maestre, J.M., Negenborn, R.R. (eds.) Distributed Model Preductive Control Made Easy. Springer, Berlin (2014)

35. Perizzato, A., Farina, M., Scattolini, R.: Stochastic distributed predictive control of independent systems with coupling constraints. In: IEEE Conference on Decision and Control, pp. 3228–3233 (2014)
36. Farina, M., Giulioni, L., Scattolini, R.: Distributed predictive control of stochastic linear systems with chance constraints. In: American Control Conference, pp. 20–25 (2016)
37. Dai, L., Xia, Y., Gao, Y., Cannon, M.: Distributed stochastic MPC of linear systems with additive uncertainty and coupled probabilistic constraints. IEEE Trans. Autom. Control **62**(7), 3474–3481 (2017)
38. Chen, X., Heidarinejad, M., Liu, J., Christofides, P.D.: Distributed economic MPC: application to a nonlinear chemical process network. J. Process Control **22**, 689–699 (2012)
39. Fele, F., Maestre, J.M., Camacho, E.F.: Coalitional control: cooperative game theory and control. IEEE Control Syst. Mag. **37**, 53–69 (2017)
40. Wang, D., Glavic, M., Wehenkel, L.: Comparison of centralized, distributed and hierarchical model predictive control schemes for electromechanical oscillations damping in large-scale power systems. Int. J. Electr. Power Energy Syst. **58**, 32–41 (2014)
41. Ma, M., Chen, H., Liu, X., Allgöwer, F.: Distributed model predictive load frequency control of multi-area interconnected power system. Int. J. Electr. Power Energy Syst. **62**, 289–298 (2014)
42. Larsen, G.K.H., van Foreest, N.D., Scherpen, J.M.A.: Distributed MPC applied to a network of households with micro-CHP and heat storage. IEEE Trans. Smart Grid **5**, 2106–2114 (2014)
43. del Real, A.J., Arce, A., Bordons, C.: An integrated framework for distributed model predictive control of large-scale power networks. IEEE Trans. Ind. Inf. **10**, 197–209 (2014)
44. Scherer, H.F., Pasamontes, M., Guzman, J.L., Alvarez, J.D., Camponogara, E., Normey-Rico, J.E.: Efficient building energy management using distributed model predictive control. J. Process Control **24**, 740–749 (2014)
45. Fele, F., Maestre, J.M., Hashemy, S.M., Muñoz de la Peña, D., Camacho, E.F.: Coalitional model predictive control of an irrigation canal. J. Process Control **24**, 314–325 (2014)
46. Zhu, M., Martinez, S.: On distributed constrained formation control in operator-vehicle adversarial networks. Automatica **49**, 3571–3582 (2013)
47. Farina, M., Perizzato, A., Scattolini, R.: Application of distributed predictive control to motion and coordination problems for unicycle autonomous robots. Robot. Auton. Syst. **72**, 248–260 (2015). Control **59**, 1439–1453
48. Kirubakaran, V., Radhakrishnan, T.K., Sivakumaran, N.: Distributed multiparametric model predictive control design for a quadruple tank process. Measurement **47**, 841–854 (2014)
49. Ferrara, A., Nai Oleari, A., Sacone, S., Siri, S.: Freeways as systems of systems: a distributed model predictive control scheme. IEEE Syst. J. **9**, 312–323 (2015)

Scalable MPC Design

Marcello Farina, Giancarlo Ferrari-Trecate, Colin Jones, Stefano Riverso, and Melanie Zeilinger

1 Introduction and Motivations

Nowadays, automation is rapidly evolving from the one-regulator-one-system setting to distributed control architectures for large interconnections of subsystems. Progresses in this direction are motivated by technologies such as CyberPhysical Systems (CPS), the Internet of Things, Industry 4.0, and the Industrial Internet [1–3], which are gaining wider and wider popularity in academia and industry. Fueled by progress in communication networks, MEMS, and distributed computing, these frameworks are closely related and hinge on the coupling of myriads of smart sensors and actuators for providing innovative services [3]. In fact, they are expected to impact a range of applications, including manufacturing, transportation, cooperative robotics, smart environments, green buildings, public utilities, and smart grids. In this vision, a central role is played by the concept of *flexibility* of CPSs. Ideally, the control technology should allow subsystems to join and leave a CPS with minimal supervision efforts.

M. Farina
Dipartimento di Elettronica, Informazione e Bioingegneria, Politecnico di Milano, Milan, Italy
e-mail: marcello.farina@polimi.it

G. Ferrari-Trecate (✉) · C. Jones
Automatic Control Laboratory, École Polytechnique Fédérale de Lausanne (EPFL), Lausanne, Switzerland
e-mail: giancarlo.ferraritrecate@epfl.ch; colin.jones@epfl.ch

S. Riverso
United Technologies Research Center, 4th Floor, Penrose Business Center, Penrose Wharf, Cork, Republic of Ireland
e-mail: riverss@utrc.utc.com

M. Zeilinger
Institute for Dynamic Systems and Control, ETH Zurich, Zurich, Switzerland
e-mail: mzeilinger@ethz.ch

S. V. Raković, W. S. Levine (eds.), *Handbook of Model Predictive Control*, Control Engineering, https://doi.org/10.1007/978-3-319-77489-3_12

This chapter describes modular control architectures based on MPC that can be easily updated when the CPS topology changes. In order to achieve such a level of flexibility, the design of local controllers should be *scalable*. Intuitively, this means that the complexity of the synthesis algorithm must be independent of the total number of subsystems. Moreover, the addition or removal of a subsystem should require, at most, the update of a limited number of controllers. In scalable design, the main challenge is how to preserve collective properties of interest, such as stability, after the plug-in and -out of subsystems. To this purpose, we will introduce the concept of Plug-and-Play (PnP) design [4], where the addition and removal of subsystems can be automatically denied if the control layer cannot be updated in a safe manner. The use of MPC for building local regulators is particularly interesting because it allows to easily handle multivariable subsystems and to guarantee the fulfilment of constraints on local variables.

The chapter is structured as follows. Scalable and PnP design are introduced in Section 2. These approaches will be described in general terms, as they are independent of the nature of local controllers. Section 3 is devoted to the mathematical tools needed for achieving scalability of MPC design in presence of constraints on variables of subsystems. The main PnP MPC schemes available in the literature [4–6] are reviewed in Section 4. We provide a tutorial description by adopting the simplest possible setting and by deferring generalizations and related approaches to Section 5. Section 6 discusses two applications of the proposed methods in the fields of power networks and smart grids. Concluding remarks, as well as directions of future research, are given in Section 7.

Notation. We use $a:b$ for the set of integers $\{a,a+1,\dots,b\}$. The column vector with s components $v_1,\dots,v_s$ is $\mathbf{v}=(v_1,\dots,v_s)$. The symbol $\oplus$ denotes the Minkowski sum, i.e. $A=B\oplus C$ means $A=\{a : a=b+c,\ b\in B,\ c\in C\}$. Moreover, $\bigoplus_{i=1}^{s} G_i = G_1\oplus\ldots\oplus G_s$. The symbol $\mathbf{1}_\alpha$ (resp. $\mathbf{0}_\alpha$) denotes a column vector with $\alpha\in\mathbb{N}$ elements all equal to 1 (resp. 0). The identity matrix of size n is $\mathbb{I}_n$. The pseudo-inverse of a matrix $A\in\mathbb{R}^{m\times n}$ is denoted with $A^\dagger$. A matrix $A\in\mathbb{R}^{n\times n}$ is asymptotically stable if all its eigenvalues λ verify $|\lambda|<1$.

Definition 1 (RPI and RCI sets). Consider the discrete-time system $x(t+1)=f(x(t),u(t),w(t))$, with state $x(t)\in\mathbb{R}^n$, input $u(t)\in\mathbb{U}\subseteq\mathbb{R}^m$, and disturbance $w(t)\in\mathbb{W}\subset\mathbb{R}^n$. The set $\mathbb{X}\subseteq\mathbb{R}^n$ is Robust Control Invariant (RCI) if $\forall x(t)\in\mathbb{X}$ there exists $u(t)\in\mathbb{U}$ such that $x(t+1)\in\mathbb{X}$, $\forall w(t)\in\mathbb{W}$. For the system $x(t+1)=f(x(t),w(t))$, the set $\mathbb{X}$ is Robust Positively Invariant (RPI) if $x(t)\in\mathbb{X}\Rightarrow x(t+1)\in\mathbb{X}$, $\forall w(t)\in\mathbb{W}$.

2 Scalable and Plug-and-Play Design

A CPS can be represented as a graph of interconnected subsystems $\Sigma_{[i]}$ $i\in\mathscr{M}=1:M$, see Figure 1. Edges represent physical or communication interactions involving the exchange of local variables, such as states and inputs.

The direction of coupling channels defines, for each subsystem $\Sigma_{[i]}$, a set of *parents (or influencing subsystems)* $\mathcal{N}_i$ and a set of *children (or influenced subsystems)* $\mathcal{S}_i$. As an example, in Figure 1 one has $\mathcal{N}_3 = \{2,4\}$ and $\mathcal{S}_3 = \{2\}$.

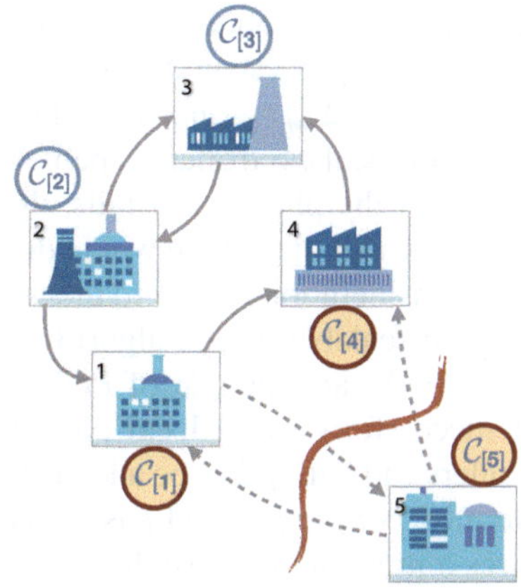

Fig. 1: CPS with a decentralized control architecture and plug-in of a subsystem. The original CPS is composed of subsystems $\Sigma_{[j]}$, $j \in 1:4$ (solid gray arrows are coupling channels). Plug-in of subsystem 5 in a parent based framework: plug-in tests and design of local controllers must be executed for $\Sigma_{[1]}$, $\Sigma_{[4]}$, and $\Sigma_{[5]}$. Dashed gray arrows are coupling channels that will be activated after the plug-in of $\Sigma_{[5]}$.

As also discussed in chapter "Distributed MPC for Large-Scale Systems", in decentralized and distributed control, the goal is to design local controllers $\mathcal{C}_{[i]}$, each associated to a subsystem $\Sigma_{[i]}$, $i \in \mathcal{M}$ for achieving desired collective behaviors, such as stability or the fulfilment of constraints on some variables.

In this chapter we focus on the problem of designing a single controller $\mathcal{C}_{[i]}$ and discuss the complexity of the synthesis algorithm as the system size grows. We use the maximal amount of information that one is allowed to use in control design as a proxy of complexity and distinguish between the following approaches. The design is *decentralized* if the synthesis of $\mathcal{C}_{[i]}$ is based on a model of $\Sigma_{[i]}$ only. This approach, introduced the 70s, has been considered in a series of papers, see [7–10] and the references therein. A less restrictive approach is *parent-based design*, where the synthesis of $\mathcal{C}_{[i]}$ exploits models $\Sigma_{[j]}$, $j \in \mathcal{N}_i$ (but *not* the associated controllers $\mathcal{C}_{[j]}$). One can extend decentralized and parent-based design by allowing for the use of l-th order parents, defined according to the coupling topology. We say that a control design algorithm is scalable if l is fixed and does not depend on M. For simplicity, in this chapter we set $l = 1$.

Parent-based design has two attractive features: (i) the information required for control synthesis flows as in the coupling graph, which is usually sparse, (ii) the plug-in of a subsystem (say $\Sigma_{[M+1]}$) requires to retune, at most, the controllers of its children. The latter point, illustrated in Figure 1, stems from the observation that the parents of $\Sigma_{[j]}$, $j \notin \{i\} \cup \mathcal{S}_{[i]}$ do not change if Σ_{M+1} is added, and hence no additional information is available for updating the controllers $\mathcal{C}_{[j]}$ in a parent-based fashion. Similar remarks apply to the unplugging of a subsystem.

The key problem in scalable design is how to guarantee collective properties in spite of the limited information available for local control synthesis. In some cases, this goal is impossible to achieve [9, 10] and this emphasizes the importance of providing conditions about feasibility of scalable synthesis.

A framework for addressing these issues is provided by PnP design, as defined in [4]. Assume subsystem Σ_{M+1} issues a plug-in request (i.e., Σ_5 in Figure 1). One would like that $\Sigma_{[M+1]}$ and its children execute a numerical test for checking the existence of local controllers capable of preserving key properties after the addition of $\Sigma_{[M+1]}$. If the plug-in test fails, the addition of $\Sigma_{[M+1]}$ is denied. Moreover, for preserving scalability of the whole design procedure, the test itself must be scalable. Similar considerations apply to the removal of a subsystem.

When a scalable design algorithm is complemented with a plug-in/out test, we call it PnP. As an example of PnP synthesis, in Figure 1, subsystems 1, 4, and 5 must successfully run a local plug-in test before re-designing the corresponding controllers and letting $\Sigma_{[5]}$ be added. The connection of $\Sigma_{[M+1]}$ at a given time instant $\bar{t}$ is termed *hot plug-in* and $\bar{t}$ is called the *plug-in time*. Sometimes, a hot plug-in requires a preparation phase for steering subsystems $\Sigma_{[j]}$, $j \in \mathscr{S}_i$ to a desired state. The attainability of these conditions must be also checked in the plug-in test.

The term PnP, borrowed from computer science, indicates the possibility of adding or removing subsystems in a safe way and with minimal effort. PnP is therefore naturally related to the concept of flexible CPSs that can be adapted over time in a seamless way. Besides flexibility, PnP design offers the following advantage. First, building and storing a global model of the system is not required, as local models must be transmitted only between parents and children. This allows the application of PnP control to systems with a very large number of subsystems. Second, as it will be shown in Sections 3 and 4, parents often transmit to children only partial information. This allows one to comply with privacy requirements, in terms of models, which arise in CPSs such as public utilities or smart grids where subsystems have different owners. Third, PnP design provides a framework for replacing components in industrial systems with minimal re-engineering and in a safe way.

In this chapter, we consider CPSs modeled as discrete-time Linear Time-Invariant (LTI) systems

$$\mathbf{x}^+ = \mathbf{A}\mathbf{x} + \mathbf{B}\mathbf{u}, \tag{1}$$

where $\mathbf{x} \in \mathbb{R}^n$ and $\mathbf{u} \in \mathbb{R}^m$ are the state and the input, respectively, at time t and $\mathbf{x}^+$ stands for $\mathbf{x}$ at time $t+1$. We will use the notation $\mathbf{x}(t)$, $\mathbf{u}(t)$ only when necessary. As more thoroughly discussed in chapter "Distributed MPC for Large-Scale Systems", the model (1) is decomposed in a number of interacting submodels. In this chapter we consider a non-overlapping decomposition and we partition the state into M vectors $x_{[i]} \in \mathbb{R}^{n_i}$, $i \in \mathscr{M}$ such that $\mathbf{x} = (x_{[1]}, \ldots, x_{[M]})$, and $n = \sum_{i \in \mathscr{M}} n_i$. Similarly, the input is partitioned into vectors $u_{[i]} \in \mathbb{R}^{m_i}$, $i \in \mathscr{M}$ such that $\mathbf{u} = (u_{[1]}, \ldots, u_{[M]})$ and $m = \sum_{i \in \mathscr{M}} m_i$.
The dynamics of the $i-th$ subsystem results to be described as follows:

$$\Sigma_{[i]}: \quad x_{[i]}^+ = A_{ii} x_{[i]} + B_i u_{[i]} + w_{[i]}, \quad w_{[i]} = \sum_{j=1, j \neq i}^{M} A_{ij} x_{[j]}, \tag{2}$$

where $A_{ij} \in \mathbb{R}^{n_i \times n_j}$, $i, j \in \mathscr{M}$ and $B_i \in \mathbb{R}^{n_i \times m_i}$. According to (2), the matrix $\mathbf{A}$ in (1) is formed by blocks A_{ij}, $i, j \in \mathscr{M}$ and, moreover, $\mathscr{N}_i = \{j : A_{ij} \neq 0, j \neq i\}$ and $\mathscr{S}_i = \{j : A_{ji} \neq 0, j \neq i\}$. From (2) one also obtains $\mathbf{B} = \text{diag}(B_1, \ldots, B_M)$. Finally, we assume that the states and the inputs of subsystems $\Sigma_{[i]}$, $i \in \mathscr{M}$ must fulfill local constraints $x_{[i]} \in \mathbb{X}_i$ and $u_{[i]} \in \mathbb{U}_i$. If we define the sets $\mathbb{X} = \prod_{i\in\mathscr{M}} \mathbb{X}_i$ and $\mathbb{U} = \prod_{i\in\mathscr{M}} \mathbb{U}_i$, then we obtain the following constraints for the collective system (1)

$$\mathbf{x} \in \mathbb{X}, \ \mathbf{u} \in \mathbb{U}. \tag{3}$$

Besides linearity of local dynamics and coupling terms $w_{[i]}$, the model (1)–(3) assumes no coupling through inputs or constraints. In spite of these simplifications, this setting will allow us to illustrate the main challenges of scalable and PnP control based on MPC. Relaxations of the above assumptions will be discussed in Section 5.

3 Concepts Enabling Scalable Design for Constrained Systems

The aim of this section is to present the main tools that have been used in the literature for achieving scalability of MPC design. Existing approaches consider decentralized or distributed MPC architectures, as defined in chapter "Distributed MPC for Large-Scale Systems", and aim at synthesizing local controllers $\mathscr{C}_{[i]}$ for providing collective stability and fulfillment of constraints (3) at all times. The key ingredient of local MPC design is the availability of structured state-feedback controllers guaranteeing the existence, for each subsystem, of an invariant set contained in the state constraints. For scalability of MPC design, these sets must be defined in a parent-based fashion. The approaches in Section 3.1 assume decentralized state-feedback controllers and invariant sets that do not change over time. Section 3.2, instead, considers a distributed state-feedback law and time-varying invariant sets that can be computed off-line (and updated on-line) using information only from parents. These approaches are complementary. The notions of invariance in Section 3.1 are tailored to the development of decentralized controllers. As such, they are simpler than those in Section 3.2, conceived for distributed architectures requiring a communication network. On the other hand, invariant sets in Section 3.1.2 exist only if a suitable small-coupling condition is verified. Instead, invariant sets in Section 3.2 might exist even for CPSs with tightly coupled subsystems.

3.1 Tube-Based Small-Gain Conditions for Networks

The MPC approach in [4, 5] is based on the idea of treating the coupling $w_{[i]}$ in (2) as a disturbance. For simplicity, in this section we neglect input constraints, i.e. $\mathbb{U} = \mathbb{R}^m$. Moreover, we assume that $\mathbb{X}_i$ are bounded sets containing the origin in their interior. In MPC, constraint satisfaction in presence of bounded disturbances can be addressed through the *tube* approach described in [11]. This requires to define

(i) the nominal (i.e., unperturbed) prediction model

$$\hat{\Sigma}_{[i]}: \quad \hat{x}^+_{[i]} = A_{ii}\hat{x}_{[i]} + B_i v_{[i]} \tag{4}$$

(ii) a local control law in the form

$$u_{[i]} = v_{[i]} + \kappa_i(x_{[i]} - \hat{x}_{[i]}), \tag{5}$$

which relates $u_{[i]}$ in (2) with $v_{[i]}$ in (4).

(iii) a state feedback function κ_i and a set $\mathbb{Z}_i$ including the origin in its interior such that $x_{[i]}(t)$ is confined in a tube of section $\mathbb{Z}_i$ centered in $\hat{x}_{[i]}(t)$,

$$x_{[i]}(0) \in \hat{x}_{[i]}(0) \oplus \mathbb{Z}_i \Rightarrow x_{[i]}(t) \in \hat{x}_{[i]}(t) \oplus \mathbb{Z}_i, \ \forall t \geq 0, \tag{6}$$

or, equivalently, such that the error $z_{[i]} = x_{[i]} - \hat{x}_{[i]}$ verifies

$$z_{[i]}(0) \in \mathbb{Z}_i \Rightarrow z_{[i]}(t) \in \mathbb{Z}_i, \ \forall t \geq 0, \tag{7}$$

Note that the set $\mathbb{Z}_i$ does not depend on $v_{[i]}$ because, from (2) and (4), the error dynamics is

$$z_{[i]}^{+} = A_{ii} z_{[i]} + B_i \kappa_i(z_{[i]}) + w_{[i]}. \tag{8}$$

Therefore, the design of the two control terms in (5) can be done separately, by

(a) choosing κ_i and the corresponding set $\mathbb{Z}_i$,

(b) designing a local MPC controller based on the *decoupled* model (4), so as to steer $\hat{x}_{[i]}$ as desired. From (6), the goal is to guarantee, at all times,

$$\hat{x}_{[i]}(t) \oplus \mathbb{Z}_i \subseteq \mathbb{X}_i, \tag{9}$$

so that constraints $x_{[i]} \in \mathbb{X}_i$ are fulfilled.

In order to achieve (9) for nonzero nominal states, the following strict inequality must be verified

$$\mathbb{Z}_i \subset \mathbb{X}_i \tag{10}$$

for all $i \in \mathscr{M}$. Moreover, (10) implies that there is a nonempty set $\hat{\mathbb{X}}_i$ such that

$$\hat{\mathbb{X}}_i \oplus \mathbb{Z}_i \subseteq \mathbb{X}_i \tag{11}$$

and then, if the local MPC controller guarantees $\hat{x}_{[i]} \in \hat{\mathbb{X}}_i$ at all times, constraints are fulfilled irrespective of the coupling.

Condition (10) is critical and it can be interpreted as a *tube-based small-gain condition for networks*. To clarify this recall that, from (7) and (8), $\mathbb{Z}_i$ is an RPI set for the disturbance set (i.e., the coupling set)

$$\mathbb{W}_i = \bigoplus_{j \in \mathscr{N}_i} A_{ij} \mathbb{X}_j. \tag{12}$$

Therefore $\mathbb{Z}_i$ is a function of $\mathbb{W}_i$, say $\mathbb{Z}_i = f_i^{IS}(\bigoplus_{j\in\mathcal{N}_i} A_{ij}\mathbb{X}_j)$. Accordingly, (10) becomes

$$f_i^{IS}(\bigoplus_{j\in\mathcal{N}_i} A_{ij}\mathbb{X}_j) \subset \mathbb{X}_i, \tag{13}$$

and since the inclusion must hold for all $i \in \mathcal{M}$, the network-wide nature of (13) becomes apparent. Indeed, f_i^{IS} are *gain functions* that represent the ability of each subsystem (8) to attenuate the effect of the bounded disturbance $w_{[i]}$ on $z_{[i]}$. In particular, the control function κ_i in (8) has the role to (and should be designed in order to) reduce as much as possible the impact of $w_{[i]}$ on $\mathbb{Z}_i$.

Two conditions are required for (13) to hold: (i) each local f_i^{IS} must be sufficiently small, and (ii) the coupling terms A_{ij} must be sufficiently small. However, the precise meaning of "sufficiently small" is largely application-dependent. Furthermore, the degree of coupling depends on how subsystems are defined. Often, subsystems are not identified a priori and there are several algorithms for computing decompositions where subsystems are weakly coupled [12]. Finally, the effect of coupling terms A_{ij} can be reduced (or even eliminated) by extending the control law (5) so as to embody coupling attenuation terms, as discussed in Section 5.

In Sections 3.1.1 and 3.1.2 we address how (13) can be rewritten in an analytical form; in particular, we will highlight that the main difference between the approaches in [4] and [5] lies in the criterion used for defining $\kappa_i(\cdot)$ and the set $\mathbb{Z}_i$.

3.1.1 Tube-Based Small-Gain Condition for Networks Using RPI Sets

The solution proposed in [4] requires $\kappa_i(\cdot)$ to be linear, i.e.

$$\kappa_i(x_{[i]} - \hat{x}_{[i]}) = K_i(x_{[i]} - \hat{x}_{[i]}), \tag{14}$$

where $K_i \in \mathbb{R}^{m_i \times n_i}$ is a gain matrix. This implies that (8) becomes

$$z_{[i]}^{+} = F_i z_{[i]} + w_{[i]}, \tag{15}$$

where $F_i = A_{ii} + B_i K_i$. If F_i is asymptotically stable, there is a nonempty minimal RPI (mRPI) set $\mathbb{Z}_i$ for (15) with disturbance $w_{[i]} \in \bigoplus_{j\in\mathcal{N}_i} A_{ij}\mathbb{X}_j$, given by [13]

$$\mathbb{Z}_i = \bigoplus_{k=0}^{\infty} F_i^k \bigoplus_{j\in\mathcal{N}_i} A_{ij}\mathbb{X}_j. \tag{16}$$

We assume, as in [4], that sets $\mathbb{X}_i$ are zonotopes, i.e., centrally symmetric convex polytopes described by $\mathbb{X}_i = \{\mathscr{F}_i x_{[i]} \leq \mathbf{1}_{\bar{r}_i}\}$ where the matrix $\mathscr{F}_i \in \mathbb{R}^{\bar{r}_i \times n_i}$ is given and $\mathrm{rank}(\mathscr{F}_i) = n_i$. In [4] it is shown that the small-gain condition (10) is verified if, for all $i \in \mathcal{M}$

$$\sum_{j \in \mathcal{N}_i} \sum_{k=0}^{\infty} ||\mathscr{F}_i F_i^k A_{ij} \mathscr{F}_j^\dagger||_\infty < 1. \tag{17}$$

The fulfillment of (17) also guarantees that $\mathbf{A} + \mathbf{BK}$ is asymptotically stable, where $\mathbf{K}$ =diag$(K_1, \ldots, K_M)$. This means that the local gains K_i define a decentralized controller stabilizing the CPS (1), when $v_{[i]} = 0$, $i \in \mathscr{M}$.
Note that (17) is a parent-based condition that can be checked locally by subsystem $\Sigma_{[i]}$. The computation of K_i under constraints (17) is therefore a nonlinear optimization problem that can be used as a plug-in test [4].

3.1.2 Tube-Based Small-Gain Condition for Networks Using RCI Sets

The method proposed in [14] computes the set $\mathbb{Z}_i$ before defining the feedback law κ_i that makes $\mathbb{Z}_i$ RPI. In other words, it first looks for an RCI set $\mathbb{Z}_i$ verifying (10) for the disturbance $w_{[i]} \in \mathbb{W}_i$. When constraints $\mathbb{X}_i$ are polytopes, this operation can be done following the procedure in [15]. In detail, by exploiting an appropriate parametrization of RCI sets, the computation of $\mathbb{Z}_i$ can be cast into a Linear Programming (LP) problem. The LP depends only on the polyhedron $\mathbb{W}_i$ and therefore, in view of (12), it provides a parent-based optimization problem that can be used as a plug-in test.

As shown in [15], for a given vector $z_{[i]}$, the function κ_i associated to $\mathbb{Z}_i$ can be evaluated by solving another LP problem. Differently from the computation of $\mathbb{Z}_i$, which is performed offline, this has to be done at each time instant for obtaining the control variable $u_{[i]}$ in (5).

Besides allowing the fulfillment local constraints, the feedback κ_i has another key feature: it guarantees that the origin of the CPS (1) is asymptotically stable when $\hat{x}_{[i]}$ and $v_{[i]}$ in (5) are null. This property, shown in [16, proof of Theorem 1, steps 2 and 3], exploits the fact that "disturbances" $w_{[i]}$ influencing the state dynamics are not exogenous variables (as in tube MPC, see [11]), but they depend on states of parent subsystems.

3.2 Distributed Invariance

If the coupling between subsystems is stronger, then robust control can no longer be used to compensate for system interactions, and the small-gain conditions outlined in the previous section will fail. In these cases, controllers are needed that utilize more information in their decision-making process via communication of the states of parent's subsystems. In this section, we outline how a sparse structured Lyapunov function can be designed, and then utilized to enforce a less-conservative form of distributed invariance, which comes at the cost of increased communication and design complexity.

The goal of this section is to define a structured invariant set, for which containment of a given local state can be determined using only information from the local subsystem and its parents. Such a set can then be employed as a terminal condition to ensure recursive feasibility and constraint satisfaction of the entire CPS, as outlined in Section 4.2.1. The primary challenge in this case is that invariance of a coupled collection of subsystems is a fundamentally global property.

We begin by assuming that a structured control law exists that stabilizes the unconstrained system.

Assumption 1 (Structured Linear Controller). There exists a linear control law $\mathbf{u} = \mathbf{Kx}$ such that the system $\mathbf{x}^+ = (\mathbf{A}+\mathbf{BK})\mathbf{x}$ is asymptotically stable and block K_{ij} is non-zero only if A_{ij} is non-zero.

We define the notation $x_{\mathscr{N}_i}$ to mean the vector containing only the states of the subsystems $\mathscr{N}_i$ (in an appropriate ordering), where we recall that $\mathscr{N}_i$ is the set of parents of the ith subsystem.

We begin by defining a structured Lyapunov function $V(x)$.

Theorem 1 (Adapted from [17]). If there exist, for all $i \in \mathscr{M}$, functions $V_i(x_{[i]})$ and $\gamma_i(x_{\mathscr{N}_i})$, as well as positive constants β_1, β_2 and β_3 such that for all $i \in \mathscr{M}$

$$\begin{aligned} \beta_1 \|x_{[i]}\|^2 \le V_i(x_{[i]}) \le \beta_2 \|x_{[i]}\|^2 \ , \\ V_i(x_{[i]}^+) - V_i(x_{[i]}) \le \gamma_i(x_{\mathscr{N}_i}) - \beta_3 \|x_{[i]}\|^2 \ , \end{aligned} \tag{18}$$

and the global condition

$$\sum_{i \in \mathscr{M}} \gamma_i(x_{\mathscr{N}_i}) \le 0 \tag{19}$$

is satisfied, then $V(x) = \sum_{i \in \mathscr{M}} V_i(x_{[i]})$ is a Lyapunov function for the system $\mathbf{x}^+ = (\mathbf{A}+\mathbf{BK})\mathbf{x}$.

Conditions (18) and (19) ensure that while some of the local terms $V_i(x_{[i]})$ may increase in any given time step, the global function V decreases at every time step. This property can be used to define time-varying local invariant sets, as shown in the following.

Based on the local function V_i, we define the parameterized level sets

$$X_i(\alpha_i) := \{x_{[i]} \mid V_i(x_{[i]}) \le \alpha_i\} \tag{20}$$

and the global set $\mathbf{X}(\boldsymbol{\alpha}) = \Pi_{i \in \mathscr{M}} X_i(\alpha_i)$, where $\boldsymbol{\alpha} = (\alpha_1, \ldots, \alpha_M)$. We evolve the parameter $\boldsymbol{\alpha}$ according to the dynamic system

$$\alpha_{[i]}^+ = \alpha_{[i]} + \gamma_i(x_{\mathscr{N}_i}) \qquad \forall i \in \mathscr{M} \tag{21}$$

and use the shorthand $\boldsymbol{\alpha}^+ = \boldsymbol{\alpha} + \boldsymbol{\gamma}(x)$ for (21). Note that we now have a dynamic set of sets $\mathbf{X}(\boldsymbol{\alpha})$, which evolve according to the state of the system.

The following Lemma follows from (21) and from Theorem 1.

Lemma 1. If $\mathbf{x} \in \mathbf{X}(\boldsymbol{\alpha})$, then $\mathbf{x}^+ = (\mathbf{A}+\mathbf{BK})\mathbf{x} \in \mathbf{X}(\boldsymbol{\alpha}^+) = \mathbf{X}(\boldsymbol{\alpha}+\boldsymbol{\gamma}(x))$.

Even if, as shown above in Lemma 1, the system states remain within the time-varying sets $X_i(\alpha_i)$, it is not obvious that the set $\mathbf{X}(\boldsymbol{\alpha})$ will still satisfy the system constraints: $\mathbf{X}(\boldsymbol{\alpha}) \subseteq \mathbb{X}$ and $\mathbf{KX}(\boldsymbol{\alpha}) \subseteq \mathbb{U}$. The following theorem provides conditions for such invariance.

Theorem 2 (Adapted from [18]). Let $\bar{\alpha} > 0$ be a scalar such that the level set $\bar{\mathbf{X}} := \{\mathbf{x} \,|\, V(\mathbf{x}) \leq \bar{\alpha}\}$ is a subset of $\mathbb{X}$ and $\mathbf{K}\bar{\mathbf{X}} \subset \mathbb{U}$. If $\sum \alpha_i = \bar{\alpha}$, $\alpha_i > 0$ and $\mathbf{X}(\boldsymbol{\alpha}) \subseteq \bar{\mathbf{X}}$, then

$$\mathbf{x}^+ = (\mathbf{A}+\mathbf{BK})\mathbf{x} \in \mathbf{X}(\boldsymbol{\alpha}^+) = \mathbf{X}(\boldsymbol{\alpha}+\boldsymbol{\gamma}(\mathbf{x}))$$
$$\mathbf{X}(\boldsymbol{\alpha}^+) \subset \bar{\mathbf{X}} \subset \mathbb{X}$$

We now have a system of sets $X_i(x)$ and functions $\gamma_i(x_{\mathcal{N}_i})$ that are based only on *local* and *parent* states. Specifically, containment of the state $\mathbf{x}$ in the set $\mathbf{X}(\boldsymbol{\alpha})$ can be tested locally using information stored at the local subsystem only. Evolution of the state dynamics $\boldsymbol{\alpha}$ can be done via a single communication of the states of the parent subsystems. In Section 4.2, we will demonstrate that these are the key properties required to develop a recursively feasible and stabilizing MPC scheme, and how these components can be adapted online due to changes in the system structure (plug-in/unplugging).

4 Scalable Design of MPC

In the following, we focus on MPC architectures that can be designed in a scalable way by exploiting the notions of invariance introduced in Section 3. More precisely, the decentralized MPC scheme in Section 4.1 will hinge on the results in Section 3.1, while the distributed MPC approach in Section 4.2 will leverage time-varying distributed invariance described in Section 3.2. In both cases, we will discuss how to perform local MPC design in a PnP fashion, so as to allow for the addition and removal of subsystems. While control design is here meant as an off-line task, in Section 4.2.3 we will illustrate how to enable the hot plug-in/out of subsystems without violating constraints on states and inputs. The algorithm will be presented as a complement to the distributed MPC scheme in Section 4.2. Nevertheless, it could be easily adapted to cope also with the decentralized MPC solution in Section 4.1.

4.1 PnP-MPC Based on Robustness Against Coupling

We describe on-line computations and off-line design steps required by the decentralized MPC schemes in [4, 5], which are based on the tube-based small-gain conditions for networks in Section 3.1. The goal of the local MPC regulator for system $\Sigma_{[i]}$

is to provide $v_{[i]}$ so as to guarantee (9). Following the approach in [11] the control signal $u_{[i]}$ in (5) is computed online as

$$u_{[i]}(t) = v_{[i]}(t|t) + \kappa_i(x_{[i]}(t) - \hat{x}_{[i]}(t|t)), \tag{22}$$

where $x_{[i]}(t)$ is the current local state and $v_{[i]}(t|t)$, $\hat{x}_{[i]}(t|t)$ are the optimal values of variables $v_{[i]}(0)$ and $\hat{x}_{[i]}(0)$, respectively, appearing in the MPC optimization problem

$$\min_{\substack{\hat{x}_{[i]}(0) \\ v_{[i]}(0:N_i-1)}} \sum_{k=0}^{N_i-1} \ell_i(\hat{x}_{[i]}(k), v_{[i]}(k)) + V_{f_i}(\hat{x}_{[i]}(N_i)) \tag{23a}$$

$$x_{[i]}(t) - \hat{x}_{[i]}(0) \in \mathbb{Z}_i \tag{23b}$$

$$\hat{x}_{[i]}(k+1) = A_{ii}\hat{x}_{[i]}(k) + B_i v_{[i]}(k) \qquad k \in 0:N_i-1 \tag{23c}$$

$$\hat{x}_{[i]}(k) \in \hat{\mathbb{X}}_i \qquad k \in 0:N_i-1 \tag{23d}$$

$$\hat{x}_{[i]}(N_i) \in \hat{\mathbb{X}}_{f_i} \tag{23e}$$

that must be solved at each time t. In (23), the integer $N_i > 0$ is the control horizon (that can be different for each subsystem), ℓ_i is the stage cost, V_{f_i} is the terminal cost, $\hat{\mathbb{X}}_{f_i}$ is the terminal set, and $\hat{\mathbb{X}}_i$ are state constraints for the dynamics (23c), which corresponds to the nominal system $\hat{\Sigma}_{[i]}$ in (4). Some remarks are due. First, (23) does not depend on states $x_{[j]}(t)$, $j \neq i$ and therefore the control law (22) is decentralized. Second, following the approach in [11], also the initial state of system (23c) is optimized at each time instant. Third, in terms of computational complexity, (23) is a standard tube MPC problem for a system of order n_i [11].

Next, we describe the *offline design problem* that has to be solved at location i in order to define all elements appearing in (22) and (23). As a first step, one has to check the existence of the RPI set $\mathbb{Z}_i = f_i^{IS}(\bigoplus_{j\in\mathcal{N}_i} A_{ij}\mathbb{X}_j)$ (and the associated control law $\kappa_i(\cdot)$) including the origin in its interior and verifying (10). The existence of $\mathbb{Z}_i$ and $\kappa_i(\cdot)$, as discussed in Section 3.1, requires the fulfillment of the tube-based small-gain condition for networks (13). Moreover, it provides a plug-in condition: if $\mathbb{Z}_i$ or $\kappa_i(\cdot)$ do not exist, the addition of subsystem $\Sigma_{[i]}$ must be denied. All these operations can be carried out by solving two different parent-based optimization problems, described in Sections 4.1.1 and 4.1.2.
Second, one has to derive the state constraints $\hat{\mathbb{X}}_i$ verifying (11). In Sections 4.1.1 and 4.1.2 we describe two algorithms, based on different parametrizations of $\hat{\mathbb{X}}_i$ and requiring an implicit representation of $\mathbb{Z}_i$ (Section 4.1.2) or not (Section 4.1.1).

Third, the remaining quantities in (23) must fulfill standard assumptions for guaranteeing recursive feasibility and convergence in a centralized MPC setting. Specifically, one must define an auxiliary control law $\kappa_i^{aux}(\hat{x}_{[i]})$ and a corresponding terminal set $\hat{\mathbb{X}}_{f_i}$ such that (i) $\hat{\mathbb{X}}_{f_i} \subseteq \hat{\mathbb{X}}_i$ is an invariant set for $\hat{x}_{[i]}^+ = A_{ii}\hat{x}_{[i]} + B_i\kappa_i^{aux}(\hat{x}_{[i]})$ and (ii) $\forall \hat{x}_{[i]} \in \hat{\mathbb{X}}_{f_i}$, $V_{f_i}(\hat{x}_{[i]}^+) - V_{f_i}(\hat{x}_{[i]}) \leq -\ell_i(\hat{x}_{[i]}, \kappa_i^{aux}(\hat{x}_{[i]}))$. We highlight that there are

several methods, discussed, e.g., in [19], for defining κ_i^{aux}, $\ell_i(\cdot)$, $V_{f_i}(\cdot)$ and $\hat{\mathbb{X}}_{f_i}$ verifying these assumptions. Importantly, these ingredients can be defined at a purely local level, requiring no information from neighboring systems.
The following statement, summarizing Theorem 1 in [4] and Theorem 9 in [5], characterizes stability and constraint satisfaction for the closed-loop collective system.

Theorem 1. *Assume that, for all subsystems, plug-in tests are passed and the local MPC controllers* (23) *are designed following one of the methods described above. Define the feasibility region for the MPC-i problem as*

$$\mathbb{X}_i^{\mathscr{F}} = \{s_{[i]} \in \mathbb{X}_i : \ (23) \textit{ is feasible for } x_{[i]}(t) = s_{[i]}\},$$

and the collective feasibility region as $\mathbb{X}^{\mathscr{F}} = \prod_{i\in\mathscr{M}} \mathbb{X}_i^{\mathscr{F}}$. *Then,*

(i) if $\mathbf{x}(0) \in \mathbb{X}^{\mathscr{F}}$, *i.e.* $x_{[i]}(0) \in \mathbb{X}_i^{\mathscr{F}}$ *for all* $i \in \mathscr{M}$, *the state constraints in* (3) *are fulfilled at all time instants;*
(ii) the origin of the closed-loop system is asymptotically stable and $\mathbb{X}^{\mathscr{F}}$ *is a region of attraction.*

Consider now the plug-in of a new subsystem $\Sigma_{[M+1]}$ at time $\bar{t}$. Theorem 1 implies that, for preserving stability and constraint satisfaction, one has to (i) run plug-in tests and design new controllers for system $\Sigma_{[M+1]}$ and its children and (ii) guarantee that $x_{[j]}(\bar{t}) \in \mathbb{X}_j^{\mathscr{F}}$ for $j \in \{M+1\} \cup \mathscr{S}_i$, where $\mathbb{X}_j^{\mathscr{F}}$ are the feasibility regions for the new MPC controllers. This provides an additional hot plug-in condition that might require a preparation phase before hot plug-in. Algorithms for this task are described in Section 4.2.3.

4.1.1 PnP-MPC Exploiting the Small-Gain Conditions for Networks Using RPI Sets

As recalled in Section 3.1.1 the PnP-MPC scheme in [4] assumes that κ_i is parametrized as in (14) and that $\mathbb{X}_i$ is a zonotope. The parent-based design phase relies on the following result [4, Proposition 1]: if there are matrices K_i such that $A_{ii} + B_iK_i$ is asymptotically stable and the small-gain condition (17) is verified, then one can always define suitable RPI sets $\mathbb{Z}_i$ and zonotopes

$$\hat{\mathbb{X}}_i = \{\hat{f}_{i,r}^T \hat{x}_{[i]} \leq \hat{l}_i, \forall r \in 1:\bar{\hat{r}}_i\} = \{\hat{x}_{[i]} = \hat{\Xi}_i \hat{d}_i,\ ||\hat{d}_i||_\infty \leq \hat{l}_i\} \tag{24}$$

In particular, there is a $\delta_i > 0$ such that $\mathbb{Z}_i$ can be chosen as a δ_i-outer approximation of the minimal RPI set, see [13]. We highlight that, in (24) the "shape" of $\hat{\mathbb{X}}_i$ (i.e., vectors $\hat{f}_{i,r}$ and matrix $\hat{\Xi}_i$) must be defined *a priori*, but the parameter $\hat{l}_i$, acting as a zooming factor, can be determined jointly with δ_i via simple scalar inequalities, see [4, Proposition 1] for details.
Based on this result, in [4] a nonlinear optimization problem is proposed for jointly computing K_i, $\delta_i > 0$, and $\hat{l}_i$, as well as for verifying (17). More specifically, in order

to guarantee asymptotic stability of the local dynamics $A_{ii}+B_iK_i$, the algorithm in [4] assumes that K_i is the LQ control gain associated to matrices $Q_i \succeq 0$ and $R_i \succ 0$, which are supplied by the user.

4.1.2 PnP-MPC Exploiting the Small-Gain Conditions for Networks Using RCI Sets

The parent-based design phase in [5] relies on a two-stage procedure, stemming from the considerations made in Section 3.1.2. The first step is to check for the existence of a suitable RCI $\mathbb{Z}_i$ by solving a parent-based LP (see Section 3.1.2). This is a plug-in test and, if passed, it provides as a by product points $\bar{z}^f_{[i]} \in \mathbb{R}^{n_i}$, $f \in 1:q_i$ whose convex hull is $\mathbb{Z}_i$. This implicit representation of $\mathbb{Z}_i$ allows one to derive LP problems both for evaluating the control law $\kappa_i(x_{[i]})$ and for computing the set $\hat{\mathbb{X}}_i$. As remarked in Section 3.1.2, the computation of $\kappa_i(x_{[i]})$ is required at each time instant for deriving $u_{[i]}$ in (22). Instead, the set $\hat{\mathbb{X}}_i$ must be computed once for all at the design stage.

4.2 PnP-MPC Based on Distributed Invariance

In this section we first note that the structured invariant sets and Lyapunov functions developed in Section 3.2 can be used to develop a recursively feasible MPC formulation, which can be deployed using distributed optimization. Following this, we focus on the problem of distributed design/synthesis and extend this to enable plug-and-play.

4.2.1 Implementation of Distributed MPC

The goal is to utilize standard distributed optimization tools to solve the MPC optimization problem online. To use such algorithms effectively, we require that the collective MPC problem can be written in the form $\min_{z_i \in Z_i} \sum f_i(z_i)$ s.t. $\sum L_i z_i = c$, where the matrices L_i are sparse and couple only adjacent subsystems.

By utilizing the structured Lyapunov function and the invariant sets developed in Section 3.2, we can pose a standard centralized MPC problem in exactly this desired form

$$V^\star(\bar{x}) = \underset{x,u}{\text{minimize}} \quad \sum_{i\in\mathcal{M}} J_i(x_{[i]},u_{[i]}) \tag{25a}$$

$$\text{subject to} \quad (x_{[i]},u_{[i]},w_{[i]}) \in \mathbb{C}_i(\bar{x}_{[i]},X_i(\alpha_i)) \quad \forall i \in \mathcal{M} \tag{25b}$$

$$w_{[i]} = \sum_{j=1,j\neq i}^{M} A_{ij}x_{[j]} \quad \forall i \in \mathcal{M} \tag{25c}$$

where the local value functions and constraint sets are defined as

$$J_i(x_{[i]}, u_{[i]}) = V_i(x_{[i]}(N)) + \sum_{k=0}^{N-1} l_i(x_i(k), u_i(k))$$

$$\mathbb{C}_i(\bar{x}_{[i]}, X_i(\alpha_i)) = \left\{ (x_{[i]}, u_{[i]}, w_{[i]}) \;\middle|\; \begin{array}{c} x_{[i]}(0) = \bar{x}_{[i]} \\ (x_{[i]}(k), u_{[i]}(k)) \in \mathbb{X}_i \times \mathbb{U}_i \\ x_{[i]}(N) \in X_i(\alpha_i) \\ x_{[i]}(k+1) = A_{ii}x_{[i]}(k) + B_i u_{[i]}(k) + w_{[i]}(k) \end{array} \right\}$$

where the current state of the system is $\bar{x}$, and the MPC control law for the ith subsystem is defined as $\kappa_{[i]}(x_{[i]}) := u_{[i]}(0)^\star$, where $u^\star$ is the optimizer of the above problem. In addition, at each time step, we update the level set vector α according to the dynamics (21).

Standard MPC theory tells us that if $V(\mathbf{x})$ is a Lyapunov function and $X_f \subset \mathbb{X}$ is an invariant set for an appropriately defined control law $\mathbf{u} = \mathbf{K}\mathbf{x}$, which is feasible (i.e., $\mathbf{K}\mathbf{x} \in \mathbb{U}$ for all states in $\mathbf{X}(\boldsymbol{\alpha})$), then the system $\mathbf{x}^+ = \mathbf{A}\mathbf{x} + \mathbf{B}\kappa(\mathbf{x})$ is asymptotically stable, satisfies system constraints and the feasible set of (25) is recursively feasible, where $\kappa(\mathbf{x})$ is the optimizer $\mathbf{u}^\star(0)$ of (25). (See [19] and [18] for full technical requirements and formal statements.)

The algorithm for evaluation of the control law is now:

1. Each system measures its local state $\bar{x}_{[i]}$.
2. Solve MPC problem (25) by distributed optimization, where subsystems iteratively communicate to reach consensus on the optimal trajectory[1]. Note that it is the specific sparsity structure developed for the terminal sets that enables this distributed optimization, and therefore a distributed control architecture.
3. Each subsystem applies the local control input $u_{[i]}(0)$ obtained in Step 2.
4. Each subsystem updates the local terminal set to $X_i(\alpha_i^+) = X_i(\alpha_i + \gamma_i(x_{\mathcal{N}_i}))$ according to (21).

4.2.2 Distributed Synthesis

Deployment of the distributed MPC scheme outlined above requires the computation of the elements $V_i(x)$ and $\gamma_i(x)$ of the Lyapunov function $\mathbf{V}(\mathbf{x})$ in a distributed and plug-and-play fashion.

We begin by assuming a quadratic structure for all required functions

$$\begin{aligned} V_i(x) &:= x^T P_i x, \qquad P_i \succeq 0 \\ \gamma_i(x) &:= x^T \Gamma_i x \\ l_i(x,u) &:= x^T Q_i x + u^T R_i u, \qquad Q_i \succeq 0,\ R_i \succ 0 \end{aligned}$$

[1] See, for example, [20] for an overview of distributed optimization methods.

where we note that Γ_i is not required to be positive definite. We restrict P_i, Q_i, and R_i to be nonzero only for the variables associated to the ith subsystem, $x_{[i]}$ and $u_{[i]}$, and the matrix Γ_i to be non-zero only for the variables associated to the ith subsystem and its parents $x_{\mathcal{N}_i}$.

Our goal is to develop an optimization problem that can be solved via distributed optimization through communication with only parent subsystems in order to synthesize the components P_i, Γ_i, and a structured control law $\mathbf{K}$, such that the closed-loop system satisfies the conditions of Theorem 2, which can now be stated as

$$(A_i + B_i K_i)^T P_i (A_i + B_i K_i) - P_i \preceq -(Q_i + K_i^T R_i K_i) + \Gamma_i \quad \forall i \in \mathcal{M} \tag{26}$$

$$\sum_{i \in \mathcal{M}} \Gamma_i = 0 \tag{27}$$

where K_i is the ith row of the linear control law $\mathbf{K}$, which is non-zero only for the variables associated to the ith subsystem and its parents $x_{\mathcal{N}_i}$.

We note that the conditions (26) are a standard LQR formulation for each subsystem independently, with a linear coupling constraint for the relaxation terms Γ_i. As a result, standard conversions can be used to transform (26) into a set of local convex LMI conditions, with a sparse linear coupling constraint (27), which can be solved using standard distributed optimization tools. In summary, when a subsystem wants to join the CPS, the plug-in test amounts to the solution of a parent-based LMI for guaranteeing that (26) and (27) will hold after plug-in. Moreover, the LMI can be solved through a distributed optimization scheme involving the subsystem and its parents only. See [18] for details.

4.2.3 Plug-and-Play and Hot-Transitions

The distributed MPC scheme presented in Section 4.2.1 guarantees recursive feasibility for a given network configuration. This section addresses recursive feasibility during a network change. If systems request to plug-in/-out during closed-loop operation, the parent-based plug-and-play principle involves the following main steps: A *redesign phase* updating the MPC problem components, i.e. structured terminal conditions, for children of dis-/connecting subsystems; and a *transition phase* computing a feasible reference to change safely between configurations.

The plug-in test in this case consists of feasibility of both steps, the existence of structured terminal conditions for the new network configuration and the existence of a feasible transition behavior. In the following, we present one realization of these two phases focusing on low-complexity calculations, more details can be found in [6]. Let $\mathcal{C}$ denote the set of children of all subsystems plugging-in or -out, respectively, which are collected in the set $\mathcal{P}$.

Redesign Phase:

1. Structured terminal cost: Compute $V_i(x_{[i]}), \gamma_i(x_{[N_i]})$ satisfying Theorem 1 for all $i \in \mathcal{C} \cup \mathcal{P}$, given $V_i(x_{[i]}), \gamma_i(x_{[N_i]})$ for all $i \in \mathcal{M} \setminus \mathcal{C}$.
2. Structured terminal constraint: Reallocate $\alpha_{[i]}$ such that $\sum \alpha_{[i]}{}_{i \in \mathcal{M}} = \bar{\alpha}$.

Note that because recursive feasibility is ensured by the transition phase, the set sizes $\alpha_{[i]}$ can be arbitrarily re-allocated for the new network configuration.

Transition Phase: If the PnP request happens during operation, the current state of the network may be an infeasible initial state for the modified MPC controller computed during the redesign phase for the modified network. This can be addressed by computing an intermediate reference steady-state $x^{ss}_{[i]}\ \forall i \in \mathcal{M}$ to be tracked, from which the modified MPC problem is feasible:

$$\text{minimize} \quad \sum_{i\in\mathscr{C}\cup\mathscr{P}} f_i(x_{[i]},\tilde{x}_{[i]},x^{ss}_{[i]}) \tag{28a}$$

$$\text{subject to} \quad (x_{[i]},u_{[i]},w_{[i]}) \in \mathbb{C}_i(\bar{x}_{[i]},\{x^{ss}_{[i]}\}) \quad \forall i \in \mathscr{C}\cup\mathscr{P} \tag{28b}$$

$$(\tilde{x}_{[i]},\tilde{u}_{[i]},\tilde{w}_{[i]}) \in \mathbb{C}_i(x^{ss}_{[i]},X^{\text{mod}}_i(\alpha_{[i]})) \quad \forall i \in \mathscr{C}\cup\mathscr{P} \tag{28c}$$

$$w_{[i]} = \sum_{j=1,j\neq i}^{M} A_{ij}x_{[j]}, \quad \tilde{w}_{[i]} = \sum_{j=1,j\neq i}^{M} A_{ij}\tilde{x}_{[j]} \quad \forall i \in \mathscr{C}\cup\mathscr{P} \tag{28d}$$

$$x^{ss}_{[i]} = \sum_{j=1}^{M} A_{ij}x^{ss}_{[j]} + B_i u^{ss}_{[i]}(k) \quad \forall i \in \mathscr{C}\cup\mathscr{P} \tag{28e}$$

where the cost functions $f_{[i]}$ can be chosen to realize a desired objective on the transition steady-state. Following similar arguments as in Sections 4.2.1 and 4.2.2 it can be shown that the synthesis in the redesign phase as well as the computation of the transition steady-state can be performed using standard distributed optimization tools. Note that we have limited the redesign and transition phase to subsystems plugging-in or -out as well as their children. This may however be restrictive and via distributed optimization, Step 2. of the redesign phase as well as the computation of the steady-state can similarly be performed for all systems $i \in \mathcal{M}$ to provide feasibility of the plug-in test for a larger set of network changes.

Given recursive feasibility ensured by the redesign together with the transition phase, constraint satisfaction and stability of the closed-loop system under the proposed plug-and-play procedure follows.

5 Generalizations and Related Approaches

The decentralized PnP-MPC regulators presented in Sections 4.1 and 4.2 have been extended along several directions. In [4] and [5] it is shown how to deal with input constraints and subsystems where some parameters change after the addition or removal of parent subsystems. An extension to distributed MPC architectures is presented in [14], where communication between controllers is used for counteracting coupling between subsystems. More precisely, by allowing the transmission of states from parents to children, one can add to $u_{[i]}$ in (5) the term $\sum_{j\in\mathscr{N}_i} K_{ij}x_{[j]}$ where the gains K_{ij} allow to shrink the coupling set $\mathbb{W}_i$ used in the design of the local MPC controllers. When subsystems are tightly interconnected, this approach can dramatically increase the chances to pass the plug-in tests described in Sections 3.1.1 and 3.1.2. PnP-MPC for subsystems affected by bounded additive disturbances is discussed in [21]. The key idea is to embed disturbances in the coupling terms that local tube-based MPC tries to counteract. For dealing with input-coupled subsystems, a similar approach can be taken, by treating terms $B_{ij}u_{[j]}$ as additional disturbances for subsystem $\Sigma_{[i]}$ [22].

The state-feedback controllers in Sections 4.1 and 4.2 are extended to the output-feedback scenario in [21]. The proposed solution relies on a distributed state-estimation scheme where each local estimator reconstructs the state of the corresponding subsystem only by using state estimates from parents. As shown in [21], in a bounded-error setting, the design of a local estimator can be done in a parent-based fashion and complemented with a plug-in test similar to (17). PnP state-estimation in a stochastic setting has been addressed in the recent paper [23], by exploiting an approach based on Kalman filtering.

Generalizations of PnP-MPC to subsystems with nonlinear dynamics have been proposed in [22], [24], and [25]. While [22] and [24] assume "matched" nonlinear terms that can be modified through the control input $u_{[i]}$, the control schemes in [26] apply to subsystems with general nonlinear dynamics and coupled through constraints. This requires the use of operators for bounding the magnitude of nonlinearities that can increase conservativity.

PnP-MPC has been also integrated with distributed fault detection for developing fault-tolerant control architectures. The main goal is to automatize the following operations without spoiling stability and constraint satisfaction for the whole CPS: (i) once a fault is detected, to unplug automatically the faulty subsystem for preventing fault propagation the CPS and (ii) once the issue has been solved, to plug in again the disconnected subsystem. Under the assumption that local states and coupling variables are measured, the PnP design of model-based local fault detectors has been presented in [24] for bounded disturbances and uncertainties. This approach has been extended to the case of stochastic uncertainties with known statistics in [27]. The case of stochastic uncertainties with unknown statistics and local output observations has been addressed in [28], relying on local Luenberger-like observers. Methods in [24] have been also complemented with a fault isolation logic for dealing with banks of faults [29].

A small number of approaches have been developed that generate distributed invariant sets, or invariant sets that require only communication between neighboring subsystems. While these approaches require a centralized synthesis step, they could be studied in a plug-and-play setting. Some ideas along this direction have been provided in [30]. In [31], distributed invariance conditions are developed for local sets evolving under a linear coupled dynamic system and in [32] an NMPC framework is developed that uses an implicit terminal invariance condition that can be encoded via distributed sets by using a cyclically changing horizon length, which thus avoids the requirement of explicitly incorporating a structured invariant set in the MPC formulation.

Most of the above PnP approaches to MPC and state estimation have been implemented in the PnPMPC toolbox for MatLab [33], which also offers a software framework for facilitating the modeling of large-scale systems and the computation of RPI and RCI sets.

6 Applications

In the recent years, scalable and PnP control design principles have been used to address control problems in a number of applications including district heating [34], HVAC systems [35], vapor compression systems [36], thermo-fluid processes [37], and microgrids [38–42].

In Section 6.1, we illustrate how the MPC approaches discussed in the previous sections can be used for the synthesis of secondary frequency controllers in power networks. In Section 6.2 we will show how methods for handling the hot plug-in of subsystems described in Section 4.2.3 can be exploited for managing the charging of electric vehicles.

6.1 Frequency Control in Power Networks

In this section, we describe an output-feedback PnP-MPC approach to the design of the Automatic Generation Control (AGC) layer for a Power Network System (PNS). For each area the main control objective is to regulate to zero the frequency deviation from a nominal value (e.g., 50 Hz in Europe), thus guaranteeing frequency regulation for the overall PNS. The proposed PNS is composed of 5 generation areas connected as in Figure 2: mathematical models, constraints on input, state, output, and disturbance variables, as well as control and simulation parameters, can be found in Chapter 9 and Appendix A in [22]. The design of each output-feedback controller follows the procedure described in Algorithm 9.2 in [22]. For each generation area a PnP distributed local state estimator is designed: stability and convergence of the overall state estimator can be guaranteed using similar arguments as in Section 3.1.1

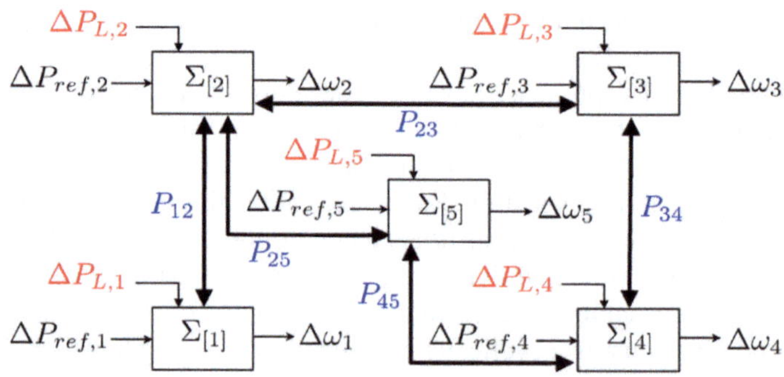

Fig. 2: Power network system. For each area $i \in 1:5$, $\Delta P_{L,i}$ is the load, $\Delta P_{ref,i}$ is the reference power setpoint, and $\Delta\omega_i$ is the frequency deviation from the nominal network value. Arrows represent tie-lines and P_{ij} is the transferred active power.

(see [22, Chapter 9]). Then, for each generation area, a PnP-MPC robust controller is designed as described in Section 5 and [22, Chapter 9].

In the following, we propose two scenarios: in Scenario 1 we consider $\Sigma_{[5]}$ disconnected from the PNS and then, in Scenario 2, we connect it by means of a plugging-in operation.
In Figure 3 we show the performance of the proposed output-feedback PnP-MPC architecture: frequency deviations in each area are kept close to zero despite persistent disturbances and power reference set-points change to compensate load steps in each area.[2]

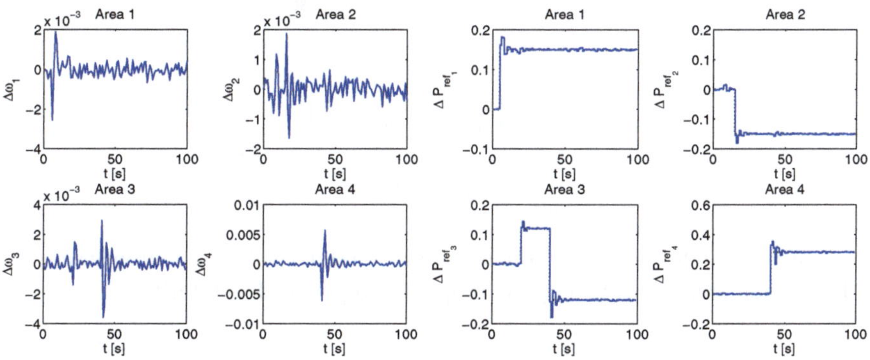

Fig. 3: Simulation Scenario 1: frequency deviation (left) and load reference set-point (right) in each area controlled by the proposed output-feedback PnP-MPC.

In Scenario 2, $\Sigma_{[5]}$ is connected to the PNS through areas $\Sigma_{[2]}$ and $\Sigma_{[4]}$: this plugging-in operation is completed by designing a new output-feedback PnP-MPC controller for $\Sigma_{[5]}$ and re-tuning controllers for areas $\Sigma_{[2]}$ and $\Sigma_{[4]}$. Therefore only the set of children of area 5 must re-tune their controllers, thus the plugging-in operation is not propagated in the network. In Figure 4 we show performance of the closed-loop system: thanks to the re-tuning of the output-feedback PnP-MPC controllers for $\Sigma_{[2]}$ and $\Sigma_{[4]}$, the plugging-in operation of $\Sigma_{[5]}$ does not compromise the overall stability of the PNS, thus frequency deviations in each area are kept close to zero. An example of unplugging operation is given in [22, Chapter 9]. The application of the distributed MPC schemes in Section 4.2 to the PNS model is described in [6].

It is worth noting that the design algorithms are completely scalable: even with a large number of generation areas the complexity of local control design scales with the number of parent subsystems only. Additionally, scalability is obtained with minor performance losses with respect to centralized MPC. Indeed, as shown for the state-feedback controller in [22, Chapter 6] the proposed control scheme obtains similar tracking performance compared to centralized MPC. On the other hand, the

[2] Tie-line powers are shown in Chapter 9 in [22].

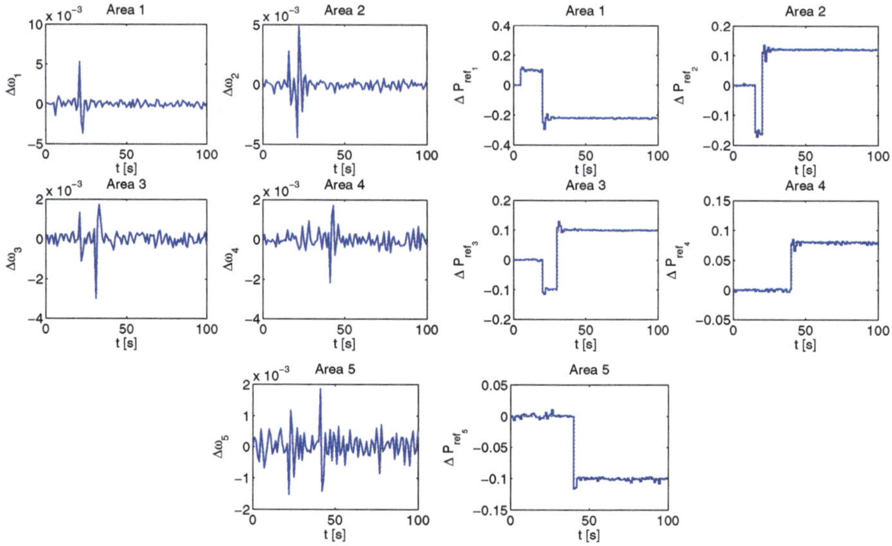

Fig. 4: Simulation Scenario 2: frequency deviation (left) and load reference set-point (right) in each area controlled by the proposed output-feedback PnP-MPC.

approach based on robustness against coupling helps in reducing the transfer of active power P_{ij} among areas during transients and at the state-steady. The PNS model, as well as the PnP MPC design algorithms used in this section, is available in the PnPMPC-toolbox for MatLab [33].

6.2 Electric Vehicle Charging in Smart Grids

The automatic management of electric vehicles (EVs) in the context of demand response schemes represents a new challenge for electricity grids due to their varying connectivity imposed by their users. This section discusses the use of PnP techniques for their optimal energy management in order to provide both voltage regulation at a local and short time scale, as well as load shaping services on a longer time scale. PnP MPC offers the capability to address this goal for varying connections of electric vehicles with user requirements in the form of deadlines, while ensuring satisfaction of critical grid constraints.

We consider a radial distribution network with traditional control elements in the form of capacitors, battery banks at select buses and three different types of loads: Fixed loads that cannot be controlled, shapeable EV loads with flexible power profile but requiring a fixed amount of energy in a fixed time period, and deferrable EV loads, which can be delayed but have a fixed power profile. It is assumed that the grid with only fixed loads is designed such that traditional control devices can maintain voltages within constraints. Additional loads, in particular electric vehicles, increase power demand and voltage drop and cannot be balanced purely with control devices. As shapeable loads can always be connected at zero power, the goal is to determine feasible connection times for deferrable loads, as well as optimal power profiles for shapeable loads to regulate voltage and minimize peak power.

The control scheme proposed in [43] addresses this problem by means of the hot plug-in strategy proposed in Section 4.2.3, ideally suited to address plug-in requests during closed-loop operation of other sub-systems. An MPC problem is first formulated for computing the optimal power profile for shapeable loads in the presence of connected deferrable loads. The optimal transition in case of additional deferrable loads requesting to connect is then obtained by solving a variant of the transition problem (28), returning the fastest connection time and the optimal profile for shapeable loads until connection. After connection, the MPC controller is again applied with the modified number of deferrable loads.

In the following case study, we consider a 55-bus distribution network (Southern California Edison) shown in Figure 5 (left), which was previously studied in [43, 44].

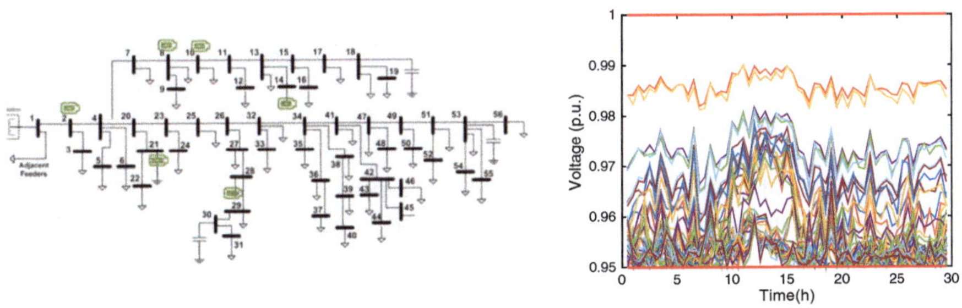

Fig. 5: 55-bus distribution test network with battery storage devices (left). Evolution of voltages at all buses during closed-loop and PnP operation (right).

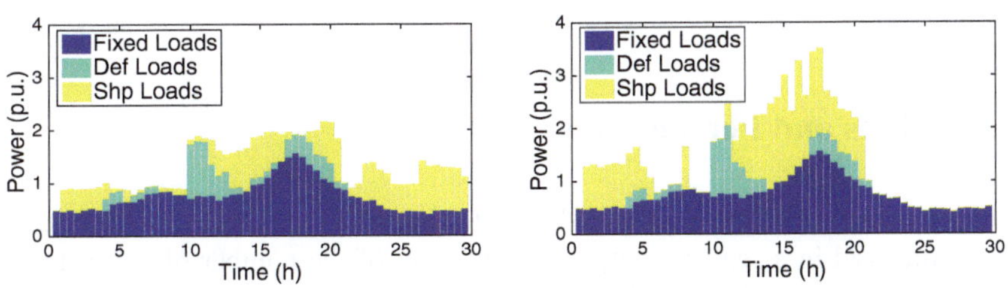

Fig. 6: Real power in the network for the controlled system with deferrable loads (left) and the uncontrolled system with deferrable loads (right).

Seven additional battery storage devices are connected as illustrated in Figure 5 (left). Fourteen shapeable EV loads connect over the simulation, between 1 and 2 vehicles at a time. Nine PnP requests with varying numbers of deferrable EV loads between 1 and 5 vehicles at a time are considered, out of which two requests have to be deferred. The details including the model, constraints, control and simulation parameters can be found in [43]. Figure 6 highlights the load shaping effect of the controller compared to the uncontrolled case. Figure 5 (right) shows the voltages at all 55 buses during closed-loop operation, including PnP requests and transition phases, demonstrating that constraints are satisfied at all times.

Figure 7 shows the overall cumulative deferrable and shapeable loads during a deferrable load request at 11h (bus 28), while four shapeable loads (bus 5, 6, 9, 19) and four deferrable loads (bus 4, 5, 16 and 17) are already connected and an additional shapeable load connects at 11h (bus 15).

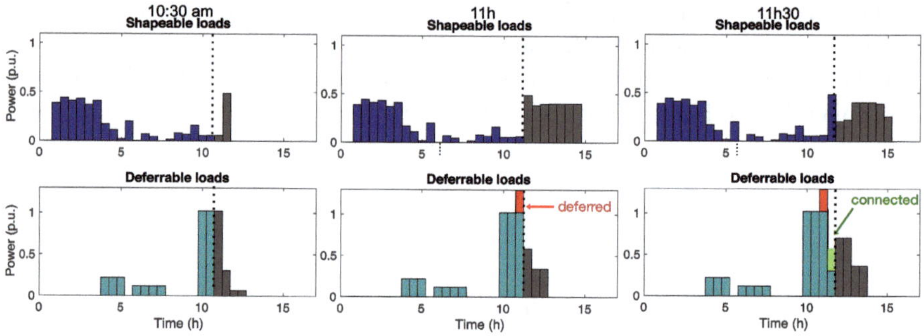

Fig. 7: Evolution of cumulative shapeable loads (top) and deferrable loads (middle): one deferrable load requests to plug-in at 11h and is delayed to connect at 11h30. Colored bars show implemented power, grey bars planned power.

The load is delayed to be connected at 11h30 to ensure feasibility of the grid. The total implemented and planned loads are shown before and after the delayed request. It can be seen that during the transition phase (11h-11h30), shapeable loads adapt their signal to accommodate the new load as quickly as possible while ensuring feasibility of the grid.

7 Conclusions and Perspectives

This chapter was devoted to a tutorial description of MPC solutions for CPSs that can be synthesized in a scalable way, independently of the number of subsystems and of their interconnection topology. We also reviewed the concept of PnP design and showed how to update existing controllers for performing plug-in/out of subsystems in a reliable and secure way. Until now, existing algorithms for scalable MPC design considered only "flat" decentralized or distributed control architectures. However, for managing large-scale CPS, hierarchical control structures might be more appropriate. This raises the issue of studying how the addition and removal of a subsystems can impact higher control layers, and how they can be updated in a scalable fashion.

The controllers reviewed in this chapter hinge on MPC formulations based on notions of invariance. Another interesting research topic would be to understand if scalable control design can be performed using alternative approaches to MPC that avoid the explicit use of invariant sets.

Acknowledgements The material in Section 6.2 is based on the work of Caroline Le Floch and we are grateful for making the simulation results available.

References

1. Atzori, L., Iera, A., Morabito, G.: The Internet of Things: a survey. Comput. Netw. **54**(15), 2787–2805 (2010). Available: http://dx.doi.org/10.1016/j.comnet.2010.05.010
2. Gubbi, J., Buyya, R., Marusic, S., Palaniswami, M.: Internet of Things (IoT): a vision, architectural elements, and future directions. Futur. Gener. Comput. Syst. **29**(7), 1645–1660 (2013). Available: http://dx.doi.org/10.1016/j.future.2013.01.010
3. Lee, E.A., Rabaey, J., Hartmann, B., Kubiatowicz, J., Pister, K., Simunic Rosing, T., Wawrzynek, J., Wessel, D., Sangiovanni-Vincentelli, A., Seshia, S.A., Blaauw, D., Dutta, P., Fu, K., Guestrin, C., Taskar, B., Jafari, R., Jones, D., Kumar, V., Mangharam, R., Pappas, G.J., Murray, R.M., Rowe, A.: The swarm at the edge of the cloud. IEEE Des. Test **31**(3), 8–20 (2014)
4. Riverso, S., Farina, M., Ferrari-Trecate, G.: Plug-and-play decentralized model predictive control for linear systems. IEEE Trans. Autom. Control **58**(10), 2608–2614 (2013)
5. Riverso, S., Farina, M., Ferrari-Trecate, G.: Plug-and-play model predictive control based on robust control invariant sets. Automatica **50**, 2179–2186 (2014)
6. Zeilinger, M.N., Pu, Y., Riverso, S., Ferrari-Trecate, G., Jones, C.N.: Plug and play distributed model predictive control based on distributed invariance and optimization. In: 52nd IEEE Conference on Decision and Control, pp. 5770–5776 (2013). Available: http://ieeexplore.ieee.org/lpdocs/epic03/wrapper.htm?arnumber=6760799
7. Hassan, M.F., Singh, M.G., Titli, A.: Near optimal decentralised control with a pre-specified degree of stability. Automatica **15**(4), 483–488 (1979)
8. Findeisen, W.: Decentralized and hierarchical control under consistency or disagreement of interests. Automatica **18**(6), 647–664 (1982)
9. Bakule, L., Lunze, J.: Decentralized design of feedback control for large-scale systems. Kybernetika **24**(8), 3–96 (1988)
10. Lunze, J.: Feedback Control of Large Scale Systems. Prentice Hall, Systems and Control Engineering, Upper Saddle River (1992)
11. Mayne, D.Q., Seron, M.M., Raković, S.V.: Robust model predictive control of constrained linear systems with bounded disturbances. Automatica **41**(2), 219–224 (2005)
12. Siljak, D.D.: Decentralized Control of Complex Systems. Academic Press, Boston (2011)
13. Raković, S.V., Kerrigan, E.C., Kouramas, K.I., Mayne, D.Q.: Invariant approximations of the minimal robust positively invariant set. IEEE Trans. Autom. Control **50**(3), 406–410 (2005)
14. Riverso, S., Ferrari-Trecate, G.: Plug-and-play distributed model predictive control with coupling attenuation. Optim. Control Appl. Methods. **36**(3), 292–305 (2015). https://doi.org/10.1002/oca.2142
15. Raković, S.V., Baric, M.: Parameterized robust control invariant sets for linear systems: theoretical advances and computational remarks. IEEE Trans. Autom. Control **55**(7), 1599–1614 (2010)

16. Riverso, S., Farina, M., Ferrari-Trecate, G.: Plug-and-play model predictive control based on robust control invariant sets. Università degli Studi di Pavia, Pavia, Italy, Tech. Rep., 2012. arXiv:1210.6927s
17. Jokić, A., Lazar, M.: On decentralized stabilization of discrete-time nonlinear systems. In: Proceedings of the American Control Conference, St. Louis, pp. 5777–5782 (2009)
18. Conte, C., Jones, C.N., Morari, M., Zeilinger, M.N.: Distributed synthesis and stability of cooperative distributed model predictive control for linear systems. Automatica **69**, 117–125 (2016). Available: http://www.sciencedirect.com/science/article/pii/S0005109816300413
19. Mayne, D., Rawlings, J., Rao, C., Scokaert, P.: Constrained model predictive control: stability and optimality. Automatica **36**, 789–814 (2000)
20. Boyd, S., Parikh, N., Chu, E., Peleato, B., Eckstein, J.: Distributed optimization and statistical learning via the alternating direction method of multipliers. Found. Trends Mach. Learn. **3**(1), 1–122 (2011). Available: http://dx.doi.org/10.1561/2200000016
21. Riverso, S., Farina, M., Ferrari-Trecate, G.: Plug-and-play state estimation and application to distributed output-feedback model predictive control. Eur. J. Control **25**, 17–26 (2015). https://doi.org/10.1016/j.ejcon.2015.04.001
22. Riverso, S.: Distributed and plug-and-play control for constrained systems. Ph.D. dissertation, Università degli Studi di Pavia, 2014. Available: http://sisdin.unipv.it/pnpmpc/phpinclude/papers/phd_thesis_riverso.pdf
23. Farina, M., Carli, R.: Partition-based distributed Kalman filter with plug and play features. IEEE Trans. Control Netw. Syst. **5**(1), 560–570 (2018). https://doi.org/10.1109/TCNS.2016.2633786
24. Riverso, S., Boem, F., Ferrari-Trecate, G., Parisini, T.: Plug-and-play fault detection and control reconfiguration for a class of nonlinear large-scale constrained systems. IEEE Trans. Autom. Control **61**(12), 3963–3978 (2016). https://doi.org/10.1109/TAC.2016.2535724
25. Lucia, S., Markus, K., Findeisen, R.: Contract-based predictive control of distributed systems with plug and play capabilities. IFAC-PapersOnLine **48**(23), 205–211 (2012). Available: http://dx.doi.org/10.1016/j.ifacol.2015.11.284
26. Lucia, S., Markus, K., Findeisen, R.: Contract-based predictive control of distributed systems with plug and play capabilities. IFAC-PapersOnLine **48**(23), 205–211 (2015). Available: http://dx.doi.org/10.1016/j.ifacol.2015.11.284
27. Boem, F., Riverso, S., Ferrari-Trecate, G., Parisini, T.: Stochastic fault detection in a plug-and-play scenario. In: Proceedings of the 54th IEEE Conference on Decision and Control, Osaka, 15–18 December 2015, pp. 3137–3142. https://doi.org/10.1109/CDC.2015.7402689
28. Boem, F., Carli, R., Farina, M., Ferrari-Trecate, G., Parisini, T.: Scalable monitoring of interconnected stochastic systems. In: Proceedings of the 55th IEEE Conference on Decision and Control, Las Vegas, 12–14 December 2016, pp. 1285–1290
29. Boem, F., Riverso, S., Ferrari-Trecate, G., Parisini, T.: A plug-and-play fault diagnosis approach for large-scale systems. In: IFAC 9th Safeprocess, Paris, 2–4 September 2015, pp. 601–606
30. Riverso, S., Rubini, D., Ferrari-Trecate, G.: Distributed bounded-error state estimation based on practical robust positive invariance. Int. J. Control **88**(11), 2277–2290 (2015)
31. Raković, S.V., Kern, B., Findeisen, R.: Practical set invariance for decentralized discrete time systems. In: 49th IEEE Conference on Decision and Control (CDC), December 2010, pp. 3283–3288
32. Kögel, M., Findeisen, R.: Stability of NMPC with cyclic horizons. IFAC Proc. Vol. **46**(23), 809–814 (2013). Available: http://www.sciencedirect.com/science/article/pii/S1474667016317591
33. Riverso, S., Battocchio, A., Ferrari-Trecate, G.: PnPMPC toolbox (2013). Available: http://sisdin.unipv.it/pnpmpc/pnpmpc.php
34. Knudsen, T., Trangbaek, K., Kallesøe, C.: Plug and play process control applied to a district heating system. In: 17th IFAC World Congress, vol. 5, pp. 325–330 (2008). Available: http://vbn.aau.dk/ws/files/56642547/IFAC08_DistHeatFinal.pdf

35. Hao, H., Lian, J., Kalsi, K., Stoustrup, J.: Distributed flexibility characterization and resource allocation for multi-zone commercial buildings in the smart grid. In: Proc. 54th IEEE Conference on Decision and Control, pp. 3161–3168 (2015). Available: https://www.scopus.com/inward/record.uri?eid=2-s2.0-84962013787&partnerID=40&md5=993b7319efddde0ed95c05d8abfd0119
36. Zhou, J., Burns, D.J., Danielson, C., Di Cairano, S.: A reconfigurable plug-and-play model predictive controller for multi-evaporator vapor compression systems. In: Proceedings of the American Control Conference, July 2016, vol. 2016, pp. 2358–2364
37. Bodenburg, S., Kraus, V., Lunze, J.: A design method for plug-and-play control of large-scale systems with a varying number of subsystems. In: Proceedings of the American Control Conference, pp. 5314–5321 (2016)
38. Riverso, S., Sarzo, F., Ferrari-Trecate, G.: Plug-and-play voltage and frequency control of islanded microgrids with meshed topology. IEEE Trans. Smart Grid **6**(3), 1176–1184 (2015). Available: http://ieeexplore.ieee.org/lpdocs/epic03/wrapper.htm?arnumber=6999972
39. Tucci, M., Riverso, S., Vasquez, J.C., Guerrero, J.M., Ferrari-Trecate, G.: A decentralized scalable approach to voltage control of DC islanded microgrids. IEEE Trans. Control Syst. Technol. **24**(6), 1965–1979 (2016)
40. Dorfler, F., Simpson-Porco, J.W., Bullo, F.: Plug-and-play control and optimization in microgrids. In: 53rd IEEE Conference on Decision and Control, pp. 211–216 (2014). Available: http://ieeexplore.ieee.org/lpdocs/epic03/wrapper.htm?arnumber=7039383
41. Ernane, A.A., Josep, M., Vasquez, J.C., Zhang, C., Member, S., Coelho, E.A.A., Guerrero, J.M., Juan, C.: Modular online uninterruptible power system plug 'n' play control and stability analysis. IEEE Trans. Ind. Electron. **63**(6), 3765–3776 (2016)
42. Sadabadi, M.S., Shafiee, Q., Karimi, A.: Plug-and-play voltage stabilization in inverter-interfaced microgrids via a robust control strategy. IEEE Trans. Control Syst. Technol. **25**(3), 781–791 (2017). https://doi.org/10.1109/TCST.2016.2583378
43. Le Floch, C., Bansal, S., Tomlin, C.J., Moura, S., Zeilinger, M.N.: Plug-and-play model predictive control for load shaping and voltage control in smart grids. IEEE Trans. Smart Grids (2017, to appear). https://doi.org/10.1109/TSG.2017.2655461
44. Farivar, M., Clarke, C.R., Low, S.H., Chandy, K.M.: Inverter VAR control for distribution systems with renewables. In: 2011 IEEE International Conference on Smart Grid Communications (SmartGridComm), pp. 457–462 (2011). https://doi.org/10.1109/SmartGridComm.2011.6102366

Part II
Computations

Efficient Convex Optimization for Linear MPC

Stephen J. Wright

1 Introduction

In linear model predictive control (linear MPC), the problem to be solved at each decision point has linear dynamics and a quadratic objective. This is a classic problem in optimization — quadratic programming (QP) — which is convex when (as is usually true) the quadratic objective is convex. It remains a convex QP even when linear constraints on the states and controls are allowed at each stage, or when only linear functions of the state can be observed. Moreover, this quadratic program has special structure that can be exploited by algorithms that solve it, particularly interior-point algorithms.

In deployment of linear MPC, unless there is an unanticipated upset, the quadratic program to be solved differs only slightly from one decision point to the next. The question arises of whether the solution at one decision point can be used to "warm-start" the algorithm at the next decision point. Interior-point methods can make only limited use of warm-start information. Active-set methods, which treat a subset of the inequality constraints (the *active set* or *working set*) as equality constraints at each iteration, are much better in this regard. Although the cost of solving the quadratic program from scratch is typically more expensive with the active-set approach than with interior-point, the cost of updating the solution at a decision point using solution information from the prior point is often minimal.

We start by outlining the most elementary control problem that arises in linear MPC — the LQR formulation — and interpret methods for solving it from both a control and optimization perspective. We then generalize this formulation to allow constraints on states and inputs, and show how interior-point methods and parametrized quadratic programming methods can be used to solve such models efficiently.

S. J. Wright (✉)
Computer Sciences Department, University of Wisconsin, 1210 W. Dayton Street, Madison, WI 53706, USA
e-mail: swright@cs.wisc.edu

S. V. Raković, W. S. Levine (eds.), *Handbook of Model Predictive Control*, Control Engineering, https://doi.org/10.1007/978-3-319-77489-3_13

2 Formulating and Solving LQR

We consider the following discrete-time finite-horizon LQR problem:

$$\min_{x,u} \frac{1}{2} \sum_{j=0}^{N-1} (x_j^T Q x_j + u_j^T R u_j + 2 x_j^T M u_j) + \frac{1}{2} x_N^T \tilde{Q} x_N \tag{1a}$$

$$\text{subject to} \quad x_{j+1} = A x_j + B u_j, \quad j = 0,1,\ldots,N-1, \quad (x_0 \text{ given}), \tag{1b}$$

where $x_j \in \mathbb{R}^n$, $j = 0,1,\ldots,N$ and $u_j \in \mathbb{R}^m$, $j = 0,1,\ldots,N-1$; and $x = (x_1, x_2, \ldots, x_N)$ and $u = (u_0, u_1, \ldots, u_{N-1})$. This is a convex quadratic program if and only if we have

$$\begin{bmatrix} Q & M \\ M^T & R \end{bmatrix} \succeq 0, \quad \tilde{Q} \succeq 0, \tag{2}$$

where the notation $C \succeq D$ indicates that $C - D$ is positive semidefinite. When the convexity condition holds, the solution of (1) can be found by solving the following optimality conditions (also known as the Karush-Kuhn-Tucker or KKT conditions) for some vector $p = (p_0, p_1, \ldots, p_{N-1})$ with $p_j \in \mathbb{R}^n$:

$$Q x_j + M u_j + A^T p_j - p_{j-1} = 0, \quad j = 1,2,\ldots,N-1, \tag{3a}$$

$$\tilde{Q} x_N - p_{N-1} = 0, \tag{3b}$$

$$R u_j + M^T x_j + B^T p_j = 0, \quad j = 0,1,\ldots,N-1, \tag{3c}$$

$$-x_{j+1} + A x_j + B u_j = 0, \quad j = 0,1,\ldots,N-1, \tag{3d}$$

for some given value of x_0. The costates p_j, $j = 0,1,\ldots,N-1$ can be thought of as Lagrange multipliers for the state equation (1b). Since (3) is simply a system of linear equations, we can obtain the solution using standard techniques of numerical linear algebra. This can be done in a particularly efficient manner, because when the variables and equations are ordered appropriately, the coefficient matrix of this linear system is banded. We order both equations and variables in a *stagewise* manner, to express (3) as follows:

$$\begin{bmatrix} R & B^T & & & & & & & & & & & & \\ B & 0 & -I & & & & & & & & & & & \\ & -I & Q & M & A^T & & & & & & & & & \\ & & M^T & R & B^T & & & & & & & & & \\ & & A & B & 0 & -I & & & & & & & & \\ & & & & -I & Q & M & A^T & & & & & & \\ & & & & & \ddots & \ddots & \ddots & \ddots & \ddots & & & & \\ & & & & & & A & B & 0 & -I & & & & \\ & & & & & & & & -I & Q & M & A^T & & \\ & & & & & & & & & M^T & R & B^T & & \\ & & & & & & & & & A & B & 0 & -I \\ & & & & & & & & & & & -I & \tilde{Q} \end{bmatrix} \begin{bmatrix} u_0 \\ p_0 \\ x_1 \\ u_1 \\ p_1 \\ \vdots \\ x_{N-1} \\ u_{N-1} \\ p_{N-1} \\ x_N \end{bmatrix} = \begin{bmatrix} -M^T x_0 \\ A x_0 \\ 0 \\ 0 \\ \vdots \\ 0 \\ 0 \end{bmatrix}. \tag{4}$$

This system is square, with dimension $N(2n+m)$ and bandwidth $2n+m-1$. (A matrix C is said to have bandwidth b if $C_{ij}=0$ whenever $|i-j|>b$.)

Since factorization of a square matrix of size q with bandwidth b requires $O(qb^2)$ operations, the cost of factoring the coefficient matrix in (4) is $O(N(m+n)^3)$, which is linear in the number of stages N. A careful implementation of the banded factorization can exploit the fact that the band is "narrower" in some places than others, and thus attain further savings. If we denote the coefficient matrix in (4) by C, the vector of unknowns by z, and the right-hand side by Ex_0, where E is the matrix $\left[-M\,|\,A^T\,|\,0\,|\,\ldots\,|\,0\right]^T$, an LU factorization of C with row partial pivoting has the form

$$PC = LU, \tag{5}$$

where P is a permutation matrix and L and U are lower- and upper-triangular factors whose bandwidth is a small multiple of $(m+n)$.[1] Using this factorization, we can write the system (4) as

$$PCz = LUz = PEx_0. \tag{6}$$

Thus we can obtain the solution z, for a given value of x_0, as follows:

- Calculate Ex_0;
- Apply permutations to obtain $P(Ex_0)$;
- Solve $Ly = PEx_0$ via forward-substitution to obtain y;
- Solve $Uz = y$ via back-substitution to obtain z.

Note that the LU factorization need not be recomputed for each x_0; only the four steps above need be performed. The two steps involving triangular substitution are the most computationally expensive; these require $O(N(m+n))$ operations.

The system (4) can alternatively be solved by a block-elimination technique that is equivalent to a well-known concept in control: the Riccati equation. We describe this approach below in the more general context of solving the banded linear system that arises at each iteration of an interior-point method.

3 Convex Quadratic Programming

Before introducing constraints into the LQR formulation (1), as happens in MPC subproblems, we introduce convex quadratic programs using general notation, and discuss their optimality conditions and the basic framework of primal-dual interior-point methods. We write the general problem as follows:

$$\min_w \tfrac{1}{2} w^T V w + c^T w \quad \text{subject to } Kw = b, \;\; Lw \le l, \tag{7}$$

[1] The LU factorization does not exploit the fact that the coefficient matrix is symmetric. The LDL^T factorization is commonly used for such matrices, but unfortunately the permutations required in this factorization tend to destroy the band structure, so it is not appropriate here.

where V is a positive semidefinite matrix. Solutions w of (7) are characterized completely by the following first-order optimality conditions (usually known as Karush-Kuhn-Tucker or KKT conditions): There are vectors λ and τ such that

$$Vw + c + K^T\lambda + L^T\tau = 0, \tag{8a}$$
$$Kw = b, \tag{8b}$$
$$0 \geq Lw - l \perp \tau \geq 0, \tag{8c}$$

where the notation $a \perp b$ means that $a^T b = 0$. Here, λ and τ are the Lagrange multipliers for the constraints $Kw = b$ and $Lw \leq l$, respectively.

Primal-dual interior-point methods for (7) are often motivated as path-following methods that follow a so-called *central path* to a solution of (8) (see [14]). To define the central path, we first rewrite (8) equivalently by introducing slack variables s for the inequality constraints:

$$Vw + c + K^T\lambda + L^T\tau = 0, \tag{9a}$$
$$Kw = b, \tag{9b}$$
$$Lw + s = l, \tag{9c}$$
$$0 \leq s \perp \tau \geq 0. \tag{9d}$$

The conditions on s and τ together imply that for each component of these vectors (s_i and τ_i) we have that both are nonnegative and at least one of the pair is zero. We can express these conditions by defining the diagonal matrices

$$S := \text{diag}\,(s_1, s_2, \ldots), \quad T := \text{diag}\,(\tau_1, \tau_2, \ldots),$$

and writing $s \geq 0$, $\tau \geq 0$, and $STe = 0$, where $e = (1, 1, \ldots)^T$. (Note that STe is the vector whose components are $s_1\tau_1, s_2\tau_2, \ldots$.) We can thus rewrite (9) as follows:

$$Vw + c + K^T\lambda + L^T\tau = 0, \tag{10a}$$
$$Kw = b, \tag{10b}$$
$$Lw + s = l, \tag{10c}$$
$$STe = 0, \tag{10d}$$
$$s \geq 0, \ \tau \geq 0. \tag{10e}$$

The central-path equations are obtained by replacing the right-hand side of (10d) by μe, for any $\mu > 0$, to obtain

$$Vw + c + K^T\lambda + L^T\tau = 0, \tag{11a}$$
$$Kw = b, \tag{11b}$$
$$Lw + s = l, \tag{11c}$$
$$STe = \mu e, \tag{11d}$$
$$s > 0, \ \tau > 0. \tag{11e}$$

It is known that (11) has a unique solution for each $\mu > 0$, provided that the original problem (7) has a solution.

Primal-dual interior-point methods generate iterates $(w^k, \lambda^k, \tau^k, s^k)$, $k = 0,1,2,\ldots$, that converge to a solution of (8). Strict positivity is maintained for all components of s^k and τ^k, for all k; that is, $s^k > 0$ and $\tau^k > 0$. At step k, a search direction is generated as a Newton-like step for the square nonlinear system of equations formed by the four equality conditions in (11), for some value of μ that is chosen by a (somewhat elaborate) adaptive scheme. These Newton equations are as follows:

$$\begin{bmatrix} V & K^T & L^T & 0 \\ K & 0 & 0 & 0 \\ L & 0 & 0 & I \\ 0 & 0 & S^k & T^k \end{bmatrix} \begin{bmatrix} \Delta w^k \\ \Delta \lambda^k \\ \Delta \tau^k \\ \Delta s^k \end{bmatrix} = \begin{bmatrix} r_w^k \\ r_\lambda^k \\ r_\tau^k \\ r_{st}^k \end{bmatrix}, \tag{12a}$$

$$\text{where} \begin{bmatrix} r_w^k \\ r_\lambda^k \\ r_\tau^k \\ r_{st}^k \end{bmatrix} = - \begin{bmatrix} Vw^k + c + K^T \lambda^l + L^T \tau^k \\ Kw^k - b \\ Lw^k + s^k - l \\ S^k T^k e - \mu_k e \end{bmatrix}, \tag{12b}$$

where $S^k = \operatorname{diag}(s_1^k, s_2^k, \ldots)$ and $T^k = \operatorname{diag}(t_1^k, t_2^k, \ldots)$. The step along this direction has the form

$$(w^{k+1}, \lambda^{k+1}, \tau^{k+1}, s^{k+1}) := (w^k, \lambda^k, \tau^k, s^k) + \alpha_k (\Delta w^k, \Delta \lambda^k, \Delta \tau^k, \Delta s^k),$$

where $\alpha_k > 0$ is chosen to maintain strict positivity on the s and τ vectors, that is, $\tau^{k+1} > 0$ and $s^{k+1} > 0$.

Some block elimination is usually applied to the system (12a), rather than factoring the matrix directly. By substituting out for Δs^k, we obtain

$$\begin{bmatrix} V & K^T & L^T \\ K & 0 & 0 \\ L & 0 & -(T^k)^{-1} S^k \end{bmatrix} \begin{bmatrix} \Delta w^k \\ \Delta \lambda^k \\ \Delta \tau^k \end{bmatrix} = \begin{bmatrix} r_w^k \\ r_\lambda^k \\ r_\tau^k - (T^k)^{-1} r_{st}^k \end{bmatrix}. \tag{13}$$

(Note that the matrix $(T^k)^{-1} S^k$ is positive diagonal.) We can further use the third row in (13) to eliminate $\Delta \tau^k$, to obtain the following block 2×2 system

$$\begin{bmatrix} (V + L^T (S^k)^{-1} T^k L) & K^T \\ K & 0 \end{bmatrix} \begin{bmatrix} \Delta w^k \\ \Delta \lambda^k \end{bmatrix} = \begin{bmatrix} r_w^k - L^T (S^k)^{-1} T^k (r_\tau^k - (T^k)^{-1} r_{st}^k) \\ r_\lambda^k \end{bmatrix}. \tag{14}$$

In practice, the forms (13) and (14) are most commonly used to obtain the search directions. We see below that the form (14) applied to a constrained version of (1) leads (with appropriate ordering of the variables) to a linear system with the same structure as (4).

4 Linear MPC Formulations and Interior-Point Implementation

We now describe an extension of (1), more common in applications of linear MPC, in which the inputs u_k and states x_k are subject to additional constraints at each stage $k = 1,2,\dots,N-1$. The final state x_N is often also constrained, with the goal of steering the state into a certain polyhedral set at the end of the time horizon, from which an unconstrained LQR strategy can be pursued from that point forward without fear of violating the stagewise constraints.

We start with the formulation of linear MPC, in its most natural form (involving both states and inputs at each time point) and in a condensed form in which all states beyond the initial state x_0 are eliminated from the problem. We then derive KKT conditions for the original formulation and show how the structure that is present in linear MPC allows primal-dual interior-point methods to be implemented efficiently.

4.1 Linear MPC Formulations

We define the linear MPC problem as follows:

$$\min_{x_1,\dots,x_N,u_0,\dots,u_{N-1}} \quad \frac{1}{2}\sum_{j=0}^{N-1}(x_j^T Q x_j + u_j^T R u_j + 2x_j^T M u_j) + \frac{1}{2}x_N^T \tilde{Q} x_N \tag{15a}$$

$$\text{subject to} \quad x_{j+1} = Ax_j + Bu_j, \quad j = 0,1,\dots,N-1, \quad (x_0 \text{ given}); \tag{15b}$$

$$Gu_j + Hx_j \le h, \quad j = 0,1,\dots,N-1; \tag{15c}$$

$$Fx_N \le f. \tag{15d}$$

The linear constraints (15c), (15d) define polyhedral regions; these could be bounds (upper and/or lower) on individual components of x_j and u_j, more complicated linear constraints that are separable in x_j and u_j, or mixed constraints that involve states and inputs together.

We obtain a condensed form of (15) by using the state constraints (15b) to eliminate $x_1, x_2, \dots, x_N$. We start by aggregating the variables and constraints in (15) into a representation that disguises the stagewise structure. We define

$$\bar{x} := \begin{bmatrix} x_0 \\ x_1 \\ x_2 \\ \vdots \\ x_N \end{bmatrix}, \quad \bar{u} := \begin{bmatrix} u_0 \\ u_1 \\ \vdots \\ u_{N-1} \end{bmatrix}, \quad \bar{h} := \begin{bmatrix} h \\ h \\ \vdots \\ h \\ f \end{bmatrix},$$

$$\bar{Q} := \begin{bmatrix} Q & & & & \\ & Q & & & \\ & & \ddots & & \\ & & & Q & \\ & & & & \tilde{Q} \end{bmatrix}, \quad \bar{R} := \begin{bmatrix} R & & & \\ & R & & \\ & & \ddots & \\ & & & R \end{bmatrix}, \quad \bar{M} := \begin{bmatrix} M & & & \\ & M & & \\ & & \ddots & \\ & & & M \\ 0 & 0 & 0 & 0 \end{bmatrix},$$

$$\bar{H} := \begin{bmatrix} H & & & & \\ & H & & & \\ & & \ddots & & \\ & & & H & \\ & & & & F \end{bmatrix}, \quad \bar{G} := \begin{bmatrix} G & & & \\ & G & & \\ & & \ddots & \\ & & & G \\ 0 & 0 & 0 & 0 \end{bmatrix},$$

and note that the objective (15a) can be written as

$$\tfrac{1}{2}\left[\bar{x}^T\bar{Q}\bar{x} + \bar{u}^T\bar{R}\bar{u} + 2\bar{x}^T\bar{M}\bar{u}\right], \tag{16}$$

while the constraints (15c) and (15d) can be written

$$\bar{G}\bar{u} + \bar{H}\bar{x} \le \bar{h}. \tag{17}$$

From the state equation (15b), we have

$$\bar{x} = \bar{A}x_0 + \bar{B}\bar{u}, \tag{18}$$

where

$$\bar{B} := \begin{bmatrix} 0 & 0 & 0 & 0 & 0 \\ B & & & & \\ AB & B & & & \\ A^2B & AB & B & & \\ \vdots & \vdots & & \ddots & \\ A^{N-1}B & A^{N-2}B & A^{N-3}B & \dots & B \end{bmatrix}, \quad \bar{A} := \begin{bmatrix} I \\ A \\ A^2 \\ A^3 \\ \vdots \\ A^N \end{bmatrix}. \tag{19}$$

By substituting (18) into (16) and (17), we obtain the following condensed form of (15):

$$\min_{\bar{u}} \tfrac{1}{2}\bar{u}^T(\bar{R} + \bar{B}^T\bar{Q}\bar{B} + \bar{B}^T\bar{M} + \bar{M}^T\bar{B})\bar{u} + x_0^T\bar{A}^T(\bar{Q}\bar{B} + \bar{M})\bar{u} \tag{20a}$$

$$\text{subject to } \left[\bar{G} + \bar{H}\bar{B}\right]\bar{u} \le h - \bar{H}\bar{A}x_0. \tag{20b}$$

We have omitted a quadratic term in x_0 from the objective in (20a), because it is independent of the variable $\bar{u}$ and thus does not affect the solution. That is, the problem defined by (16), (17), (18) is equivalent to the problem defined by (20) in that the solution of (20) is identical to the $\bar{u}$ component of the solution of (16), (17), (18).

4.2 KKT Conditions and Efficient Interior-Point Implementation

We can follow the methodology of Section 3 to write down the KKT optimality conditions. As in Section 2, we use p_j to denote the costates (or Lagrange multipliers for the state equation (15b)). We introduce λ_j, $j = 0, 1, \ldots, N-1$ as Lagrange multipliers for $Gu_j \geq g$, ζ_j, $j = 0, 1, \ldots, N-1$ as multipliers for the constraints $Hx_j \leq h$, and β as the vector of Lagrange multipliers for the constraint $Fx_N \leq f$. The KKT conditions, following the template (8), as follows:

$$Qx_j + Mu_j + A^T p_j - p_{j-1} + H^T \zeta_j = 0, \qquad j = 1, 2, \ldots, N-1, \tag{21a}$$

$$\tilde{Q}x_N + F^T \beta - p_{N-1} = 0, \tag{21b}$$

$$Ru_j + M^T x_j + B^T p_j + G^T \lambda_j = 0, \qquad j = 0, 1, \ldots, N-1, \tag{21c}$$

$$-x_{j+1} + Ax_j + Bu_j = 0, \qquad j = 0, 1, \ldots, N-1, \tag{21d}$$

$$0 \geq Gu_j + Hx_j - h \perp \lambda_j \geq 0, \quad j = 0, 1, \ldots, N-1, \tag{21e}$$

$$0 \geq Fx_N - f \perp \beta \geq 0. \tag{21f}$$

By introducing slack variables s_j^λ for the constraints $Gu_j + Hx_j \leq h$ and s^β for the constraint $Fx_N \leq f$, we obtain the following formula (cf. (9)):

$$Qx_j + Mu_j + A^T p_j - p_{j-1} + H^T \zeta_j = 0, \qquad j = 1, 2, \ldots, N-1, \tag{22a}$$

$$\tilde{Q}x_N + F^T \beta - p_{N-1} = 0, \tag{22b}$$

$$Ru_j + M^T x_j + B^T p_j + G^T \lambda_j = 0, \qquad j = 0, 1, \ldots, N-1, \tag{22c}$$

$$-x_{j+1} + Ax_j + Bu_j = 0, \qquad j = 0, 1, \ldots, N-1, \tag{22d}$$

$$Gu_j + Hx_j + s_j^\lambda = h, \qquad j = 0, 1, \ldots, N-1, \tag{22e}$$

$$Fx_N + s^\beta = f, \tag{22f}$$

$$0 \leq s_j^\lambda \perp \lambda_j \geq 0, \quad j = 0, 1, \ldots, N-1, \tag{22g}$$

$$0 \leq s^\beta \perp \beta \geq 0. \tag{22h}$$

By proceeding with the primal-dual interior-point approach, as described in Section 3, we solve the following system of equations to be solved at iteration k, obtained by specializing the form (14) to the structure of the MPC problem (see [13]):

$$
\begin{bmatrix}
R_0 & B^T \\
B & 0 & -I \\
& -I & Q_1 & M_1 & A^T \\
& & M_1^T & R_1 & B^T \\
& & A & B & 0 & -I \\
& & & & -I & Q_2 & M_2 & A^T \\
& & & & & \ddots & \ddots & \ddots & \ddots & \ddots \\
& & & & & & A & B & 0 & -I \\
& & & & & & & & -I & Q_{N-1} & M_{N-1} & A^T \\
& & & & & & & & & M_{N-1}^T & R_{N-1} & B^T \\
& & & & & & & & & A & B & 0 & -I \\
& & & & & & & & & & & -I & \tilde{Q}_N
\end{bmatrix}
\begin{bmatrix}
\Delta u_0 \\ \Delta p_0 \\ \Delta x_1 \\ \Delta u_1 \\ \Delta p_1 \\ \vdots \\ \Delta x_{N-1} \\ \Delta u_{N-1} \\ \Delta p_{N-1} \\ \Delta x_N
\end{bmatrix}
=
\begin{bmatrix}
\tilde{r}_0^u \\ \tilde{r}_0^p \\ \tilde{r}_1^x \\ \tilde{r}_1^u \\ \tilde{r}_1^p \\ \vdots \\ \tilde{r}_{N-1}^x \\ \tilde{r}_{N-1}^u \\ \tilde{r}_{N-1}^p \\ \tilde{r}_N^x
\end{bmatrix},
\tag{23}
$$

where

$$
\begin{aligned}
\begin{bmatrix} Q_j & M_j \\ M_j^T & R_j \end{bmatrix} &= \begin{bmatrix} Q & M \\ M^T & R \end{bmatrix} + \begin{bmatrix} G^T \\ H^T \end{bmatrix} (S_j^\lambda)^{-1} D_j^\lambda \begin{bmatrix} G & H \end{bmatrix}, \quad j = 1,2,\ldots,N-1, \\
R_0 &= R + G^T (S_0^\lambda)^{-1} D_0^\lambda G, \\
\tilde{Q}_N &= \tilde{Q} + F^T (S^\beta)^{-1} D^\beta F,
\end{aligned}
$$

and

$$
\begin{aligned}
S_j^\lambda &= \operatorname{diag}((s_j^\lambda)_1, (s_j^\lambda)_2, \ldots), \quad j = 0,1,\ldots,N-1, \\
S^\beta &= \operatorname{diag}((s^\beta)_1, (s^\beta)_2, \ldots), \\
D_j^\lambda &= \operatorname{diag}((\lambda_j)_1, (\lambda_j)_2, \ldots), \quad j = 0,1,\ldots,N-1, \\
D^\beta &= \operatorname{diag}(\beta_1, \beta_2, \ldots).
\end{aligned}
$$

We omit definitions of the terms in the right-hand side of (23); we refer to the general form of Section 3 for information on how to construct this vector. We note that the initial state x_0 appears linearly in $\tilde{r}_0^u$ and $\tilde{r}_0^p$.

This system can be solved using direct LU factorization of the coefficient matrix, as described in Section 2. But we describe here an alternative approach based on block-factorization of the matrix, which is essentially identical to solving a discrete-time, time-varying Riccati equation. We follow the derivation of [11, Section 3.3] to describe this technique. The key step is to use (23) to find matrices $\Pi_N, \Pi_{N-1}, \ldots, \Pi_1$ of size $n \times n$ and vectors $\pi_N, \pi_{N-1}, \ldots, \pi_1$ such that

$$
-p_{k-1} + \Pi_k \Delta x_k = \pi_k, \quad k = N, N-1, \ldots, 1. \tag{24}
$$

We find a recursive formula, working backwards from N. We see immediately from (4) that (24) is satisfied for $k = N$ by setting

$$
\Pi_N = \tilde{Q}_N, \quad \pi_N = \tilde{r}_N^x. \tag{25}
$$

Now supposing that the relationship (24) holds for some k, with known values of Π_k and π_k, we obtain formulas for Π_{k-1} and π_{k-1}. By combining (24) with three successive block rows from the system (23), we obtain the system

$$\begin{bmatrix} -I & Q_{k-1} & M_{k-1} & A^T & \\ & M_{k-1}^T & R_{k-1} & B^T & \\ & A & B & 0 & -I \\ & & & -I & \Pi_k \end{bmatrix} \begin{bmatrix} \Delta p_{k-2} \\ \Delta x_{k-1} \\ \Delta u_{k-1} \\ \Delta p_{k-1} \\ \Delta x_k \end{bmatrix} = \begin{bmatrix} \tilde{r}_{k-1}^x \\ \tilde{r}_{k-1}^u \\ \tilde{r}_{k-1}^p \\ \pi_k \end{bmatrix}. \tag{26}$$

By eliminating Δp_{k-1} and Δx_k, we obtain the reduced system

$$\begin{bmatrix} -I & Q_{k-1} + A^T \Pi_k A & A^T \Pi_k B + M_{k-1} \\ 0 & B^T \Pi_k A + M_{k-1}^T & R_{k-1} + B^T \Pi_k B \end{bmatrix} \begin{bmatrix} \Delta p_{k-2} \\ \Delta x_{k-1} \\ \Delta u_{k-1} \end{bmatrix} = \begin{bmatrix} \tilde{r}_{k-1}^x + A^T \Pi_k \tilde{r}_{k-1}^p + A^T \pi_k \\ \tilde{r}_{k-1}^u + B^T \Pi_k \tilde{r}_{k-1}^p + B^T \pi_k \end{bmatrix}. \tag{27}$$

Finally, by eliminating Δu_{k-1}, we obtain

$$-\Delta p_{k-2} + \Pi_{k-1} \Delta x_{k-1} = \pi_{k-1}, \tag{28}$$

where

$$\Pi_{k-1} = Q + A^T \Pi_k A - (A^T \Pi_k B + M)(R + B^T \Pi_k B)^{-1}(B^T \Pi_k A + M^T), \tag{29a}$$

$$\begin{aligned} \pi_{k-1} = {} & \tilde{r}_{k-1}^x + A^T \Pi_k \tilde{r}_{k-1}^p + A^T \pi_k \\ & - (A^T \Pi_k B + M)(R + B^T \Pi_k B)^{-1}(\tilde{r}_{k-1}^u + B^T \Pi_k \tilde{r}_{k-1}^p + B^T \pi_k). \end{aligned} \tag{29b}$$

A recursive argument based on symmetry of all Q_k and R_k reveals that all Π_k, $k = N, N-1, \ldots, 1$ are symmetric, and the matrix inversions in (29) can be performed whenever R is positive definite, a sufficient condition that is usually assumed in practice. The formula (29a) along with the initialization (25) is the *discrete-time, time-varying* Riccati equation. The more familiar algebraic Riccati equation is the limit of (29a) obtained by setting $\Pi_k = \Pi_{k-1} = \Pi$, and assuming that all Q_k, R_k, and M_k are identically equal to Q, R, and M, respectively. We obtain

$$\Pi = Q + A^T \Pi A - (A^T \Pi B + M)(R + B^T \Pi B)^{-1}(B^T \Pi A + M^T). \tag{30}$$

Having computed Π_k and π_k for $k = N, N-1, \ldots, 1$, we proceed to solve (23) as follows. By combining the first two rows of (23) with the formula (24) for $k = 1$, we obtain

$$\begin{bmatrix} R & B^T & \\ B & 0 & -I \\ & -I & \Pi_1 \end{bmatrix} \begin{bmatrix} \Delta u_0 \\ \Delta p_0 \\ \Delta x_1 \end{bmatrix} = \begin{bmatrix} \tilde{r}_0^u \\ \tilde{r}_0^p \\ \pi_1 \end{bmatrix}. \tag{31}$$

This square system (with $2n + m$ rows and columns) can be solved to find Δu_0, Δp_0, and Δx_1. We now use the second row of (27) along with the equations involving both Δx_k and Δx_{k-1} to obtain the following formulas:

$$\Delta u_{k-1} = (R + B^T \Pi_k B)^{-1} \left[\tilde{r}^u_{k-1} + B^T \Pi_k \tilde{r}^p_{k-1} + B^T \pi_k - (B^T \Pi_k A + M^T_{k-1}) \Delta x_{k-1} \right], \tag{32a}$$

$$\Delta x_k = A \Delta x_{k-1} + B \Delta u_{k-1} - \tilde{r}^p_{k-1}, \tag{32b}$$

which we iterate forward for $k = 2, 3, \ldots, N$ to obtain all inputs $\Delta u_0, \Delta u_1, \ldots, \Delta u_{N-1}$ and all states $\Delta x_1, \Delta x_2, \ldots, \Delta x_N$. The costates $\Delta p_1, \Delta p_2, \ldots, \Delta p_{N-1}$ can be recovered directly by substituting $\Delta x_2, \Delta x_3, \ldots, \Delta x_N$ into (28).

5 Parametrized Convex Quadratic Programming

The linear MPC is actually a convex QP that is parametrized by the current (initial) state x_0. In this section we formulate parametrized QP in general terms, write down optimality conditions, and describe a primal-dual active-set approach for finding its solution for some value of x_0, when the solution is already known for a (usually nearby) value of x_0. This approach leverages "warm-start" information from the current solution, and is often able to find the new solution quickly when the change to x_0 is small, as is often the case in the MPC context, unless an unmodeled upset occurs.

Omitting the equality constraints in (7) (which are not required for our application to linear MPC), we write the parametrized form as follows:

$$\min_w \tfrac{1}{2} w^T V w + c^T w - (J x_0)^T w \quad \text{subject to } Lw \le l + E x_0. \tag{33}$$

Note that this formulation corresponds exactly to the condensed form (20) of the linear MPC problem. We assume throughout this section that V is a *positive definite* matrix (so that this QP is strongly convex), L is a matrix of size $m_I \times p$, J and E are matrices, and x_0 is the parameter vector. Following (8), we can write the optimality conditions as follows:

$$Vw + c + L^T \tau = J x_0, \tag{34a}$$

$$0 \ge Lw - l - E x_0 \perp \tau \ge 0. \tag{34b}$$

Because of the complementarity conditions (34b), we can identify an *active set* $\mathscr{A} \subset \{1, 2, \ldots, m_I\}$ that indicates which of the m_I inequality constraints are satisfied *at equality*. That is, $i \in \mathscr{A}$ only if $L_{i\cdot} w = (l + E x_0)_i$, where $L_{i\cdot}$ denotes the ith row of L. It follows from this definition and (34b) that

$$L_{i\cdot} w < (l + E x_0)_i \text{ for all } i \notin \mathscr{A}; \quad \tau_{\mathscr{A}} \ge 0; \quad \tau_{\mathscr{A}^c} = 0; \tag{35}$$

where $\tau_{\mathscr{A}} = [\tau_i]_{i \in \mathscr{A}}$ and $\tau_{\mathscr{A}^c} = [\tau_i]_{i \notin \mathscr{A}}$. (Note that $\mathscr{A}^c$ denotes the complement of $\mathscr{A}$ in $\{1, 2, \ldots, m_I\}$.) We can substitute these definitions into (34) to obtain

$$Vw + c + L^T_{\mathscr{A}} \tau_{\mathscr{A}} = J x_0, \tag{36a}$$

$$L_{\mathscr{A}} w = (l + Ex_0)_{\mathscr{A}}, \tag{36b}$$
$$L_{\mathscr{A}^c} w \leq (l + Ex_0)_{\mathscr{A}^c}, \tag{36c}$$
$$\tau_{\mathscr{A}} \geq 0, \tag{36d}$$
$$\tau_{\mathscr{A}^c} = 0. \tag{36e}$$

At this point, we can make several interesting observations. Let us define the set $\mathscr{P}$ of points x_0 such that the feasible region for (33) is nonempty, that is,

$$\mathscr{P} := \{x_0 \,|\, Lw \leq l + Ex_0 \ \text{ for some } w\}.$$

First, because of the strong convexity of the quadratic objective, a solution of (36) is guaranteed to exist for all $x_0 \in \mathscr{P}$. Second, the set $\mathscr{P}$ is a polyhedron. This follows from the fact that $\mathscr{P}$ is the projection onto x_0 space of the polyhedral set

$$\{(w, x_0) \,|\, Lw \leq l + Ex_0\},$$

and the projection of a polyhedron onto a plane is itself polyhedral. Similar logic indicates that the subset of $\mathscr{P}$ that corresponds to a given active set $\mathscr{A}$ is also polyhedral. By fixing $\mathscr{A}$ in (36), we can note that the set

$$\mathscr{P}_{\mathscr{A}} := \{x_0 \,|\, (w, \tau, x_0) \text{ satisfies (36) for some } w, \tau\} \tag{37}$$

is the projection of the polyhedron defined by (36) onto x_0-space, so is itself polyhedral.

5.1 Enumeration

The last observation above suggests an enumeration approach, first proposed in [1, 2] and developed further by [9, 10, 12] and others. We describe this approach for the case in which the row submatrices $L_{\mathscr{A}}$ of L have full row rank, for all possible active sets $\mathscr{A}$ of interest. In this case, because V is positive definite, the vectors w and $\tau_{\mathscr{A}}$ are uniquely determined by the conditions (36a), (36b), that is,

$$\begin{bmatrix} w \\ \tau_{\mathscr{A}} \end{bmatrix} = \begin{bmatrix} V & L_{\mathscr{A}}^T \\ L_{\mathscr{A}} & 0 \end{bmatrix}^{-1} \begin{bmatrix} -c + Jx_0 \\ (l + Ex_0)_{\mathscr{A}} \end{bmatrix} = \begin{bmatrix} z_{w,\mathscr{A}} + Z_{w,\mathscr{A}} x_0 \\ z_{\tau,\mathscr{A}} + Z_{\tau,\mathscr{A}} x_0 \end{bmatrix}, \tag{38}$$

where

$$\begin{bmatrix} z_{w,\mathscr{A}} \\ z_{\tau,\mathscr{A}} \end{bmatrix} := \begin{bmatrix} V & L_{\mathscr{A}}^T \\ L_{\mathscr{A}} & 0 \end{bmatrix}^{-1} \begin{bmatrix} -c \\ l_{\mathscr{A}} \end{bmatrix}, \quad \begin{bmatrix} Z_{w,\mathscr{A}} \\ Z_{\tau,\mathscr{A}} \end{bmatrix} := \begin{bmatrix} V & L_{\mathscr{A}}^T \\ L_{\mathscr{A}} & 0 \end{bmatrix}^{-1} \begin{bmatrix} J \\ E_{\mathscr{A}} \end{bmatrix}, \tag{39}$$

where $E_{\mathscr{A}}$ is the row submatrix of E corresponding to $\mathscr{A}$. (Nonsingularity of the matrix that is inverted in (38) follows from positive definiteness of V and full row rank of $L_{\mathscr{A}}$.) We can substitute into (36c), (36d) to check the validity of this solution, that is,

$$L_{\mathscr{A}^c}(z_{w,\mathscr{A}} + Z_{w,\mathscr{A}} x_0) \le l_{\mathscr{A}^c} + E_{\mathscr{A}^c} x_0, \tag{40a}$$

$$z_{\tau,\mathscr{A}} + Z_{\tau,\mathscr{A}} x_0 \ge 0. \tag{40b}$$

In fact, these inequalities along with the definitions (39) provide another characterization of the set $\mathscr{P}_{\mathscr{A}}$ defined in (37): $\mathscr{P}_{\mathscr{A}}$ is exactly the set of vectors x_0 that satisfies the linear inequalities (40), which we can express as $Y_{\mathscr{A}} x_0 \le y_{\mathscr{A}}$, where

$$Y_{\mathscr{A}} := \begin{bmatrix} L_{\mathscr{A}^c} Z_{c,\mathscr{A}} - E_{\mathscr{A}^c} \\ -Z_{\tau,\mathscr{A}} \end{bmatrix}, \quad y_{\mathscr{A}} := \begin{bmatrix} l_{\mathscr{A}^c} - L_{\mathscr{A}^c} z_{w,\mathscr{A}} \\ z_{\tau,\mathscr{A}} \end{bmatrix}. \tag{41}$$

An enumeration approach stores the pairs $(Y_{\mathscr{A}}, y_{\mathscr{A}})$ for some or all of the $\mathscr{A}$ for which $\mathscr{P}_{\mathscr{A}}$ is nonempty. Then, when presented with a particular value of the parameter x_0, it identifies the set $\mathscr{A}$ for which $Y_{\mathscr{A}} x_0 \le y_{\mathscr{A}}$. The solution $(w, \tau_{\mathscr{A}})$ can then be recovered from (38), and we set $\tau_{\mathscr{A}^c} = 0$ to fill the remaining components of the Lagrange multiplier vector.

Enumeration approaches shift much of the work in calculating solutions of (33) offline. The pairs $(Y_{\mathscr{A}}, y_{\mathscr{A}})$ can be pre-computed for all $\mathscr{A}$ for which $\mathscr{P}_{\mathscr{A}}$ is nonempty. The online computation consists of testing the conditions $Y_{\mathscr{A}} x_0 \le y_{\mathscr{A}}$ for the given x_0. The order of testing can be crucial, as we want to identify the correct $\mathscr{A}$ for this value of x_0 as quickly as possible. (The approach in [9] maintains a table of the most frequently occurring instances of $\mathscr{A}$.) Full enumeration approaches become impractical quickly as the dimensions of the problem (and particularly the number of constraints m_I) increase. Partial enumeration stores only the pairs $(Y_{\mathscr{A}}, y_{\mathscr{A}})$ that have occurred most frequently and/or most recently during plant operation; when an x_0 is encountered that does not fall into any of the polyhedra currently stored, the solution can be computed from scratch, or some suboptimal backup strategy can be deployed. For plants of sufficiently high dimension, this approach too becomes impractical, but these enumeration approaches can be an appealing and practical way to implement linear MPC on small systems.

5.2 Active-Set Strategy

For problems that are too large for complete or partial enumeration to be practical, the conditions (36) can be used as the basis of an active-set strategy for solving (33). Active-set strategies make a series of estimates of the correct active set for (36), changing this estimate in a systematic way by a single index $i \in \{1, 2, \ldots, m_I\}$ (added or removed) at each iteration. This approach can be highly efficient in the context of linear MPC, when the parameter x_0 does not change greatly from one decision point to the next. If a solution of (33) is known for value of x_0 and we need to know a new solution for a nearby value, say x_0^{new}, it can often be found with just a few changes to the active set, which requires just a few steps of the algorithm. We give just an outline of the approach here; further details can be found in [3–5].

Before proceeding, we pay a little attention to the issue of *degeneracy*, which complicates significantly the implementation of active-set approaches. Degeneracy is present when there is ambiguity in the definition of the active set $\mathscr{A}$ for which (36) holds at a certain parameter x_0, or when the active constraint matrix $L_{\mathscr{A}}$ fails to have full row rank. Degeneracy of the former type occurs when there is some index $i \in \{1,2,\ldots,m_I\}$ such that both $(Lw - l - Ex_0)_i = 0$ and $\tau_i = 0$ for (w,τ) satisfying (34). Thus, for $\mathscr{A}$ satisfying (34), we may have either $i \in \mathscr{A}$ or $i \notin \mathscr{A}$; there is ambiguity about whether the constraint i is really "active." (Constraints with this property are sometimes called "weakly active.") Degeneracy of the latter type — rank-deficiency of $L_{\mathscr{A}}$ — leads to possible ambiguity in the definition of $\tau_{\mathscr{A}}$. Specifically, there may be multiple vectors $\tau_{\mathscr{A}}$ that satisfy conditions (36a) and (36d), that is,

$$L_{\mathscr{A}}^T \tau_{\mathscr{A}} = -Vw - c + Jx_0, \quad \tau_{\mathscr{A}} \geq 0. \tag{42}$$

For a particular $\mathscr{A}$ and a particular choice of (w,τ) satisfying (36), *both* types of degeneracy may be present. These degeneracies can be resolved by choosing $\tau_{\mathscr{A}}$ to be an extreme point of the polyhedron represented by (42), and removing from $\mathscr{A}$ those elements i for which $\tau_i = 0$.

We note that there is no ambiguity in the value of w for a given parameter x_0; the positive definiteness assumption on V precludes this possibility.

To account for the possibility of degeneracy, active-set algorithms introduce the concept of a *working set*. This is a subset $\mathscr{W}$ of $\{1,2,\ldots,m_I\}$ that is an estimate of the (possibly ambiguous) optimal active set $\mathscr{A}$ from (36), but with the additional property that the row constraint submatrix $L_{\mathscr{W}}$ has full rank. Small changes are made to the working set $\mathscr{W}$ at each step of the active-set method, to maintain the full-rank property but ultimately to converge to an optimal active set $\mathscr{A}$ for (36).

Let x_0 be the value of the parameter for which a primal-dual solution of (36) is known, for a certain active set $\mathscr{A}$. By doing some processing of (36), as discussed in the previous paragraph, we can identify a working set $\mathscr{W} \subset \mathscr{A}$ such that (36) holds when we replace $\mathscr{A}$ by $\mathscr{W}$ and, in addition, $L_{\mathscr{W}}$ has full row rank.

Suppose we wish to calculate the solution of (33) for a new value x_0^{new}, and define $\Delta x_0 := x_0^{\text{new}} - x_0$. We can determine the effect of replacing x_0 by x_0^{new} on the primal-dual solution components $(x, \tau_{\mathscr{W}})$ by applying (36a) and (36b) to account for the change in x_0 (similar to what we did in (38)):

$$\begin{bmatrix} \Delta w \\ \Delta \tau_{\mathscr{W}} \end{bmatrix} = \begin{bmatrix} V & L_{\mathscr{W}}^T \\ L_{\mathscr{W}} & 0 \end{bmatrix}^{-1} \begin{bmatrix} J\Delta x_0 \\ (E\Delta x_0)_{\mathscr{W}} \end{bmatrix}, \tag{43}$$

If we can take a full step along this perturbation vector without violating the other conditions (36c) and (36d), we are done! To check these conditions, we need to verify that

$$L_{\mathscr{W}^c}(w + \Delta w) \leq (l + Ex_0^{\text{new}})_{\mathscr{W}^c}, \quad \tau_{\mathscr{W}} + \Delta \tau_{\mathscr{W}} \geq 0. \tag{44}$$

If these conditions do not hold, it is necessary to make systematic changes to the active set $\mathscr{W}$. To start this process, we find the longest steplength that can be taken along $(\Delta w, \Delta \tau_{\mathscr{W}})$ while maintaining (36c) and (36d). That is, we seek the largest value of α in the range $(0,1]$ such that

$$L_{\mathscr{W}^c}(w + \alpha \Delta w) \leq (l + Ex_0 + \alpha E \Delta x_0)_{\mathscr{W}^c}, \quad \tau_{\mathscr{W}} + \alpha \Delta \tau_{\mathscr{W}} \geq 0. \tag{45}$$

We call this value $\alpha_{\max}$; it can be calculated explicitly from the following formulas:

$$i_P := \arg\min_{i\in\mathscr{W}^c} \frac{(l+Ex_0)_i - L_{i\cdot}w}{L_{i\cdot}\Delta w - (E\Delta x_0)_i}, \quad \alpha_{\max,P} := \frac{(l+Ex_0)_{i_P} - L_{i_P\cdot}w}{L_{i_P\cdot}\Delta w - (E\Delta x_0)_{i_P}}, \tag{46a}$$

$$i_D := \arg\min_{i\in\mathscr{W}} -\frac{\tau_i}{\Delta\tau_i}, \quad \alpha_{\max,D} := -\frac{\tau_{i_D}}{\Delta\tau_{i_D}}, \tag{46b}$$

$$\alpha_{\max} := \min(1, \alpha_{\max,P}, \alpha_{\max,D}). \tag{46c}$$

The constraint that "blocks" α at a value less than 1 — either i_P or i_D from (46) — motivates a change to the working set $\mathscr{W}$. If one of the Lagrange multipliers τ_i for $i \in \mathscr{W}$ goes to zero first, we remove this index from $\mathscr{W}$, allowing the corresponding constraint to move away from its constraint boundary at the next iteration. Alternatively, if one of the constraints i becomes active, we add this index to the working set $\mathscr{W}$ for the next iteration. We may need to do some postprocessing of the working set in the latter case, possibly removing some other element of the working set to maintain full rank of $L_{\mathscr{W}}$. In both of these cases, we update the values of x_0, w, and τ to reflect the step just taken, recalibrate the parameter perturbation vector Δx_0, and repeat the process. If there are no blocking constraints, and a full step can be taken, then we have recovered the solution and the active-set algorithm declares success and stops.

The active-set procedure is summarized as Algorithm 1. An example of the polyhedral decomposition of parameter space is shown in Figure 1. In this example, four steps of the active-set method (and three changes to the working set) are required to move from x_0 to the new parameter x_0^{new}.

Algorithm 1: Online Active Set Approach

Given current parameter x_0, new parameter x_0^{new}, current primal-dual solution (w,τ) and working set $\mathscr{W}$ obtained from the active set $\mathscr{A}$ satisfying (36) as described in the text;
Set $\alpha_{\max} = 0$;
while $\alpha_{\max} < 1$ **do**
  Set $\Delta x_0 := x_0^{\text{new}} - x_0$;
  Solve (43) for Δw and $\Delta\tau_{\mathscr{W}}$, and set $\Delta\tau_{\mathscr{W}^c} = 0$;
  Determine maximum steplength $\alpha_{\max}$ from (46);
  Set $\tilde{x}_0 \leftarrow x_0 + \alpha_{\max}\Delta x_0$, $\tilde{w} \leftarrow w + \alpha_{\max}\Delta w$, $\tilde{\tau} \leftarrow \tau + \alpha_{\max}\Delta\tau$;
  if $\alpha_{\max} = 1$ **then**
    Set $w^{\text{new}} \leftarrow \tilde{w}$, $\tau^{\text{new}} \leftarrow \tilde{\tau}$, $\mathscr{A} \leftarrow \mathscr{W}$ and STOP;
  else if $\alpha_{\max} = \alpha_{\max,D}$ **then**
    Remove dual blocking constraint i_D from working set: $\mathscr{W} \leftarrow \mathscr{W} \setminus \{i_D\}$;
  else if $\alpha_{\max} = \alpha_{\max,P}$ **then**
    Add primal blocking constraint i_P to the working set: $\mathscr{W} \leftarrow \mathscr{W} \cup \{i_P\}$, possibly removing some other element of $\mathscr{A}$ if necessary to maintain full rank of $L_{\mathscr{W}}$;
  end if
  Set $x_0 \leftarrow \tilde{x}_0$, $w \leftarrow \tilde{w}$, $\tau \leftarrow \tilde{\tau}$;
end while

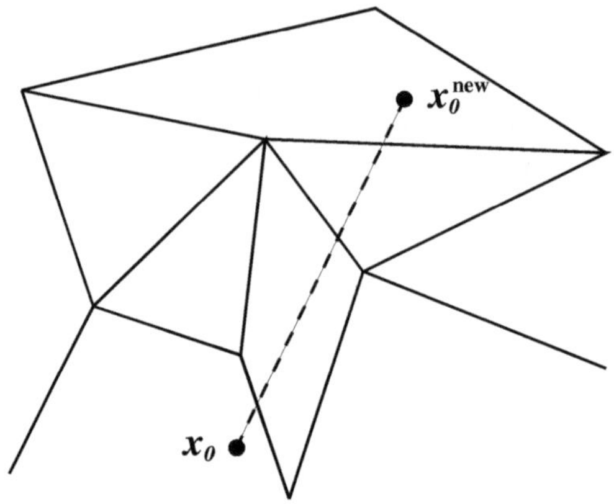

Fig. 1: Polyhedral decomposition of parameter space, showing path from x_0 to x_0^{new}

6 Software

Software packages are available online that facilitate efficient implementations of two of the approaches discussed in this chapter. The object-oriented code OOQP for structured convex quadratic programming [7, 8] can be customized to linear MPC problems; its $C++$ data structures and linear algebra modules can be tailored to problems of the form (15) and to solving systems of the form (23). The modeling framework YALMIP supports MPC; its web site shows several examples for setting up models and invoking underlying QP software (such as OOQP and general commercial solvers such as Gurobi).

The qpOASES solver [5, 6], which implements the approach described in Section 5.2, is available in an efficient and well-maintained implementation.

References

1. Bemporad, A., Borrelli, F., Morari, M.: Model predictive control based on linear programming—the explicit solution. IEEE Trans. Autom. Control **47**(12), 1974–1985 (2002)
2. Bemporad, A., Morari, M., Dua, V., Pistikopoulos, E.N.: The explicit linear quadratic regulator for constrained systems. Automatica **38**, 3–20 (2002)
3. Ferreau, H.J.: An online active set strategy for fast solution of parametric quadratic programs with applications to predictive engine control. Ph.D. thesis, Ruprecht-Karls-Universit at Heidelberg Fakult at fur Mathematik und Informatik (2006)
4. Ferreau, H.J., Bock, H.G., Diehl, M.: An online active set strategy to overcome the limitations of explicit MPC. Int. J. Robust Nonlinear Control **18**(8), 816–830 (2008)

5. Ferreau, H.J., Kirches, C., Potschka, A., Bock, H.G., Diehl, M.: qpOASES: a parametric active-set algorithm for quadratic programming. Math. Program. Comput. **6**(4), 327–363 (2014)
6. Ferreau, H.J., Potschka, A., Kirches, C.: qpOASES webpage. http://www.qpOASES.org/ (2007–2017)
7. Gertz, E.M., Wright, S.J.: OOQP. http://www.cs.wisc.edu/~swright/ooqp/
8. Gertz, E.M., Wright, S.J.: Object-oriented software for quadratic programming. ACM Trans. Math. Softw. **29**, 58–81 (2003)
9. Pannocchia, G., Rawlings, J.B., Wright, S.J.: Fast, large-scale model predictive control by partial enumeration. Automatica **43**, 852–860 (2007)
10. Pannocchia, G., Wright, S.J., Rawlings, J.B.: Partial enumeration MPC: robust stability results and application to an unstable CSTR. J. Process Control **21**, 1459–1466 (2011)
11. Rao, C.V., Wright, S.J., Rawlings, J.B.: Application of interior-point methods to model predictive control. J. Optim. Theory Appl. **99**, 723–757 (1998)
12. Tondel, P., Johansen, T.A., Bemporad, A.: Evaluation of piecewise affine control via binary search tree. Automatica **39**(5), 945–950 (2003)
13. Wright, S.J.: Applying new optimization algorithms to model predictive control. In: Kantor, J.C. (ed.) Chemical Process Control-V, AIChE Symposium Series, vol. 93, pp. 147–155. CACHE Publications, Austin (1997)
14. Wright, S.J.: Primal-Dual Interior-Point Methods. SIAM Publications, Philadelphia (1997)

Implicit Non-convex Model Predictive Control

Sebastien Gros

1 Introduction

In this chapter, we consider continuous non-convex MPC problems of the form:

$$\begin{aligned}
\mathrm{MPC}\,(\hat{\mathbf{x}}) \quad \min_{\mathbf{x},\mathbf{u}} \quad & T\,(\mathbf{x}(t_{\mathrm{f}})) + \textstyle\int_0^{t_{\mathrm{f}}} L\,(\mathbf{x}(.),\mathbf{u}(.))\,\mathrm{d}\tau && \text{(1a)}\\
\text{s.t.} \quad & \mathbf{x}(0) - \hat{\mathbf{x}} = 0 && \text{(1b)}\\
& \dot{\mathbf{x}}(t) = \mathbf{F}\,(\mathbf{x}(t),\mathbf{u}(t))\,, \qquad t \in [0,t_{\mathrm{f}}], && \text{(1c)}\\
& \mathbf{H}\,(\mathbf{x}(t),\mathbf{u}(t)) \leq 0, \qquad t \in [0,t_{\mathrm{f}}], && \text{(1d)}\\
& \mathbf{T}\,(\mathbf{x}(t_{\mathrm{f}})) \leq 0 && \text{(1e)}
\end{aligned}$$

where $\hat{\mathbf{x}} \in \mathbb{R}^n$ is the current estimation of the state of the physical system, function $\mathbf{F} : \mathbb{R}^n \times \mathbb{R}^m \mapsto \mathbb{R}^n$ is a model of the system dynamics, function $\mathbf{H} : \mathbb{R}^n \times \mathbb{R}^m \mapsto \mathbb{R}^c$ represents the physical limitations of the system, and functions $L : \mathbb{R}^n \times \mathbb{R}^m \mapsto \mathbb{R}$, $T : \mathbb{R}^n \mapsto \mathbb{R}$ are the Lagrange cost and terminal cost, respectively. Function $\mathbf{T} : \mathbb{R}^n \mapsto \mathbb{R}^{c_{\mathrm{T}}}$ imposes constraints on the terminal state of the MPC prediction, and is instrumental in most theories discussing the stability and recursive feasibility of the MPC scheme.

The continuous control law implicitly defined by (1) is the input $\mathbf{u}(0)$ obtained by solving (1) for a given state estimation $\hat{\mathbf{x}}$. In practice (1) cannot be solved continuously, and the control input delivered by (1) ought to be discretized in one form or another, see Section 4.

Non-convexity in Model Predictive Control can arise from any of the functions L, T being non-convex, the feasible set described by $\mathbf{H}$ or $\mathbf{T}$ being non-convex, or

S. Gros (✉)
Department of Electrical Engineering, Chalmers University of Technology, Gothenburg, Sweden
e-mail: grosse@chalmers.se

S. V. Raković, W. S. Levine (eds.), *Handbook of Model Predictive Control*,
Control Engineering, https://doi.org/10.1007/978-3-319-77489-3_14

from the model function $\mathbf{F}$ being nonlinear. The cost functions and feasible set are often (though not always) chosen convex by design, and the model nonlinearity is the most common source of non-convexity in practice.

From a computational point of view, problem (1) is approached via discretization techniques, whereby the continuous problem is transformed into a discrete one, holding a finite number of variables. The discretized MPC can then be written in a fairly generic form as a parametric Nonlinear Program (pNLP) [35]:

$$\mathrm{DMPC}\left(\hat{\mathbf{x}}_i\right): \quad \min_{\mathbf{w}} \quad \phi\left(\mathbf{w}, \hat{\mathbf{x}}_i\right) \tag{2a}$$

$$\text{s.t.} \quad \mathbf{g}\left(\mathbf{w}, \hat{\mathbf{x}}_i\right) = 0 \tag{2b}$$

$$\mathbf{h}\left(\mathbf{w}, \hat{\mathbf{x}}_i\right) \leq 0 \tag{2c}$$

where $\mathbf{w} \in \mathbb{R}^{\nu}$ is a vector of decision variables forming approximate representations of the continuous solution $\mathbf{x}(.), \mathbf{u}(.)$. The solution of the pNLP (2) cannot be computed continuously. In the following, we label Δt the sampling time at which the solution is updated, and $\hat{\mathbf{x}}_i := \hat{\mathbf{x}}(i \cdot \Delta t)$ the corresponding state estimations, which act as parameters in the pNLP. A crucial observation that is at the core of implicit real-time MPC techniques discussed in this chapter is that, assuming that the pNLP (2) is solved at a high frequency such that the successive discrete state estimations $\hat{\mathbf{x}}_i$, $i = 0, 1, \ldots$ are close to each other, then under some conditions the successive solutions $\mathbf{w}\left(\hat{\mathbf{x}}_i\right)$, $i = 0, 1, \ldots$ are also close to each other. This feature can be exploited to facilitate the computations of successive solution to pNLP (2), see Section 5.

The transformation from (1) to (2), often labelled *transcription* or more simply *discretization* gives rise to functions $\phi : \mathbb{R}^{\nu} \times \mathbb{R}^{n} \mapsto \mathbb{R}$, $\mathbf{g}\mathbb{R}^{\nu} \times \mathbb{R}^{n} \mapsto \mathbb{R}^{n_{\mathbf{g}}}$ and $\mathbf{h}\mathbb{R}^{\nu} \times \mathbb{R}^{n} \mapsto \mathbb{R}^{n_{\mathbf{h}}}$. The discretization will be briefly detailed in Section 4. We will consider here that functions ϕ, $\mathbf{g}$, and $\mathbf{h}$ are at least twice continuously differentiable. We will label $\mathbf{w}^{\star}\left(\hat{\mathbf{x}}_i\right)$ the optimal solution associated to (2). The non-convexity of the continuous problem (1) makes, in general, the NLP (2) non-convex.

In this chapter, we focus on tested numerical methods to solve pNLP (2) in real time, and detail the required conditions such that one can "safely" apply them. We will focus on methods aiming at computing numerical solutions to (2) that are close approximations of its exact solutions. We need to underline here that we will only be able to provide a summary of these methods here. Such methods are extensively relying on properties of pNLPs. In the next section, we briefly review those properties that will be important for us. Figure 1 offers a graphical overview of this chapter.

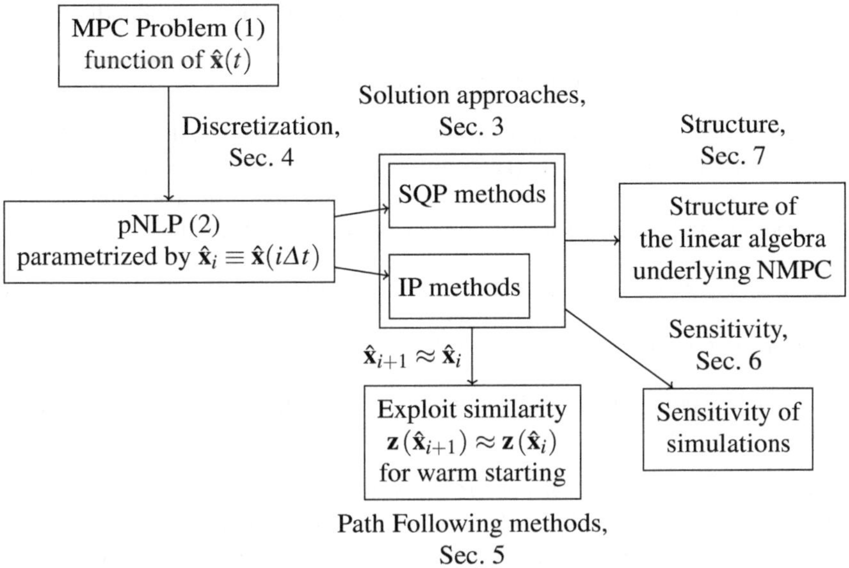

Fig. 1: Schematic of the material presented in this chapter

2 Parametric Nonlinear Programming

In order to develop solution approaches to MPC problems, we need to first point to some concepts underlying pNLPs. A natural question that arises in this context are the properties of function $\mathbf{w}^\star(\hat{\mathbf{x}}_i)$ implicitly defined by (2). In order to state the properties that will be helpful in the following, we need to briefly define some key concepts. We refer to, e.g., [44] for more details on the notions presented below.

Definition 1. the KKT conditions associated to problem (2) are defined as:

$$\nabla_{\mathbf{w}}\mathscr{L}(\mathbf{w},\lambda,\mu,\hat{\mathbf{x}}_i) = 0, \qquad \mu \geq 0 \tag{3a}$$

$$\mathbf{g}(\mathbf{w},\hat{\mathbf{x}}_i) = 0, \qquad \mathbf{h}(\mathbf{w},\hat{\mathbf{x}}_i) \leq 0 \tag{3b}$$

$$\mu^\top \mathbf{h}(\mathbf{w},\hat{\mathbf{x}}_i) = 0 \tag{3c}$$

where $\mathscr{L}$ is the Lagrange function of (2) defined as:

$$\mathscr{L}(\mathbf{w},\lambda,\mu,\hat{\mathbf{x}}_i) = \phi(\mathbf{w},\hat{\mathbf{x}}_i) + \lambda^\top \mathbf{g}(\mathbf{w},\hat{\mathbf{x}}_i) + \mu^\top \mathbf{h}(\mathbf{w},\hat{\mathbf{x}}_i) \tag{4}$$

and λ,μ are referred to as the multipliers or dual variables associated to (2). A point $\mathbf{w}$ is called a KKT point if there is a unique set of multipliers such that the triplet $(\mathbf{w},\lambda,\mu)$ satisfies the KKT conditions (3).

We observe that the primal-dual solution $\mathbf{z} :\equiv (\mathbf{w}^\star,\lambda^\star,\mu^\star)$ of (2) is an implicit function of the parameters $\hat{\mathbf{x}}_i$. We will use the notation $\mathbf{z}(\hat{\mathbf{x}}_i)$ in the following. Additionally, we will call the *active set* at a given solution $\mathbf{z}(\hat{\mathbf{x}}_i)$ the set of indices j of the inequality constraint function for which $\mathbf{h}_j = 0$, $\mu_j > 0$. The corresponding j^{th} constraint is then said (strictly) active. In a situation where $\mathbf{h}_j = 0$, $\mu_j = 0$ occurs, we label the corresponding constraint $\mathbf{h}_j$ as *weakly active*.

Definition 2. LICQ & SOSC: a feasible point $\mathbf{w}$ of (2) satisfies the Linear Independence Constraints Qualification (LICQ) iff the gradients of the equality and active inequality constraints are linearly independent. Moreover, such a point is said to satisfy the Strong Second Order Sufficient Conditions (SOSC) if

$$\mathbf{d}^\top \nabla^2_{\mathbf{w}} \mathscr{L}(\mathbf{w},\lambda,\mu,\hat{\mathbf{x}}_i)\mathbf{d} > 0, \qquad \forall\, \mathbf{d} \in C(\mathbf{w},\lambda,\mu,\hat{\mathbf{x}}_i),\ \mathbf{d} \neq 0, \tag{5}$$

where $C(\mathbf{w},\lambda,\mu)$ is the critical cone as defined in, e.g., [44, Chapter 12.5].

If $\mathbf{w}$ is a KKT point of (2) and satisfies the LICQ and SOSC, then it is a local minimizer for problem (2).

Theorem 1. *If problem* (2) *satisfies LICQ and strong SOSC at* $\mathbf{z}(\hat{\mathbf{x}}_i)$, *then the primal-dual parametric solution* $\mathbf{z}(\hat{\mathbf{x}}_i)$ *is (locally) continuous. Moreover, if the solution does not include weakly active constraints, then it is (locally) differentiable with respect to* $\hat{\mathbf{x}}_i$. *See, e.g., [44].*

In the following, the notion of *solution path* will be extensively used. This concept will be used in the following sense. We will assume that the state estimation $\hat{\mathbf{x}}(t)$ is continuous with respect to the physical time t. We will then label $\mathbf{z}(\hat{\mathbf{x}}(t))$ the solution path of the problem. Theorem (1) then lets us discuss the continuity and differentiability of $\mathbf{z}(\hat{\mathbf{x}}(t))$. It is important to observe here that the successive solutions $\mathbf{z}(\hat{\mathbf{x}}_i)$ for $i = 0,\dots$ are then points of the solution path, and that repeatedly solving pNLP (2) in real time for the successive state estimations $\hat{\mathbf{x}}_i$ is in fact an attempt at following the solution path $\mathbf{z}(\hat{\mathbf{x}}(t))$ as faithfully as possible.

The failure of Theorem (1) is illustrated via the simple following example

$$\min_{\mathbf{w}} \quad \tfrac{1}{2}\left(\mathbf{w}_1 - \hat{\mathbf{x}}_i\right)^2 + \mathbf{w}_2^2 \tag{6a}$$

$$\text{s.t.} \quad e^{-\mathbf{w}_1^2 - 2\mathbf{w}_2^2} - \tfrac{1}{2} \leq 0 \tag{6b}$$

whose solution path is displayed in Figure 2. One can observe that the determinant of the Hessian of the Lagrange function drops to zero (Figure 2, right graph) at a specific value of the (scalar) parameter $\hat{\mathbf{x}}_i$. A failure of SOSC ensues. At this specific point a bifurcation in the solution path occurs (Figure 2, left graph), i.e. the solution path splits in two individual branches.

3 Solution Approaches to Nonlinear Programming

Generic solution approaches to solve numerically a pNLP of the form (2) are iterative, and rely on the sufficient smoothness of the functions ϕ, $\mathbf{g}$, and $\mathbf{h}$ involved in the problem. It is important to understand here that the non-convexity of the pNLP (2) makes it in general impossible to guarantee that a solution computed by these methods is a global solution of the problem. Hence in practice, one relies instead on local solutions to (2). We ought to briefly discuss the two most popular solution approach to compute numerical solutions to pNLP (2).

3.1 SQP

Starting from an *initial guess* $\hat{\mathbf{z}}_i^0$ of the true solution $\mathbf{z}(\hat{\mathbf{x}}_i)$ of pNLP (2), *Successive Quadratic Programming* is based on iterating the parametric Quadratic Programs (pQPs):

$$\mathrm{pQP}(\hat{\mathbf{x}}_i): \quad \min_{\Delta \mathbf{w}} \quad \frac{1}{2}\Delta\mathbf{w}^\top \nabla^2_{\mathbf{w}}\mathscr{L}\Delta\mathbf{w} + \nabla_{\mathbf{w}}\phi^\top \Delta\mathbf{w}, \tag{7a}$$

$$\text{s.t.} \quad \mathbf{g} + \nabla_{\mathbf{w}}\mathbf{g}^\top \Delta\mathbf{w} = 0, \tag{7b}$$

$$\mathbf{h} + \nabla_{\mathbf{w}}\mathbf{h}^\top \Delta\mathbf{w} \leq 0, \tag{7c}$$

where $\nabla^2_{\mathbf{w}}\mathscr{L}$, $\nabla_{\mathbf{w}}\phi$, $\mathbf{g}$, $\nabla_{\mathbf{w}}\mathbf{g}$, $\mathbf{h}$, $\nabla_{\mathbf{w}}\mathbf{h}$ are evaluated at the current primal-dual guess $\hat{\mathbf{z}}_i^k$ (the superscript k denotes the iteration counter) and for the given set of parameter $\hat{\mathbf{x}}_i$, see, e.g., [11, 30, 45, 57]. The primal-dual guess is then updated according to

$$\hat{\mathbf{z}}_i^{k+1} \leftarrow \hat{\mathbf{z}}_i^k + \alpha\Delta\mathbf{z}, \tag{8}$$

where $\Delta\mathbf{z} = (\Delta\mathbf{w}, \lambda_{\mathrm{QP}}, \mu_{\mathrm{QP}})$ is the primal-dual solution of the pQP (7), and $\alpha \in]0,1]$ is the step-size.

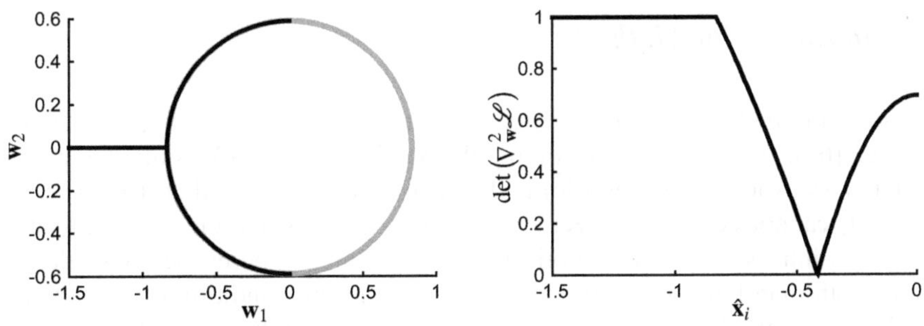

Fig. 2: Bifurcation in the pNLP (6) with $\mathbf{w} \in \mathbb{R}^2$ and $\hat{\mathbf{x}}_i \in \mathbb{R}$, resulting from the non-convex inequality constraint (6b) (grey ellipse, left graph), causing the solution path to divide in two branches (black curves, left graph). The Hessian of the Lagrange function becomes singular precisely where the bifurcation occurs (see right graph).

If LICQ and SOSC hold at the solution $\mathbf{z}(\hat{\mathbf{x}}_i)$ of pNLP (2), then one can show that the SQP iteration converges locally (i.e., for the initial guess $\hat{\mathbf{z}}_i^0$ being sufficiently close to the true solution $\mathbf{z}(\hat{\mathbf{x}}_i)$) for $\alpha = 1$. This local convergence occurs at a *quadratic rate*, i.e.

$$\left\|\hat{\mathbf{z}}_i^{k+1} - \mathbf{z}(\hat{\mathbf{x}}_i)\right\| \leq C \left\|\hat{\mathbf{z}}_i^k - \mathbf{z}(\hat{\mathbf{x}}_i)\right\|^2 \tag{9}$$

for some constant $C > 0$. This strong contraction rate occurs in a set $Q(\hat{\mathbf{x}}_i)$ of the primal-dual space, with $\mathbf{z}(\hat{\mathbf{x}}_i) \in Q(\hat{\mathbf{x}}_i)$. The set $Q(\hat{\mathbf{x}}_i)$ can be small and highly convoluted, even for simple problems. SQP methods often make use of reduced steps, by taking $\alpha < 1$. This allows the iteration to converge for a larger set of initial guesses $\hat{\mathbf{z}}_i^0$, but the quadratic contraction rate (9) can then be significantly degraded. See, e.g., [44] for more details.

The SQP iteration is typically carried out until the KKT conditions (3) associated to pNLP (2) are sufficiently close to being satisfied. It is important to underline here that the local convergence (9) of the SQP iteration is very strong: every SQP iteration k allows one to double the "number of accurate digits" in $\hat{\mathbf{z}}_i^{k+1}$ such that a few iterations are typically enough to reach machine precision. It is also crucial to observe that this strong convergence hinges on the initial guess being close enough to the true solution $\mathbf{z}(\hat{\mathbf{x}}_i)$, i.e. $\hat{\mathbf{z}}_i^0 \in Q(\hat{\mathbf{x}}_i)$.

It is also useful to point out here that the size of the set $Q(\hat{\mathbf{x}}_i)$ where a quadratic contraction occurs, as well as the magnitude of the constant C is directly related to how "close" the pNLP (2) is to a convex Quadratic Program (QP), i.e. how close the equality and inequality constraints (2b) and (2c) are to being linear and how close the cost function (2a) is to being quadratic. This notion of "closeness" of the pNLP to a QP is best formalized via quantifying how nonlinear the KKT conditions (3) are, i.e. via the Lipschitz constant of their Jacobian [44].

3.2 Interior-Point Methods

Interior-point methods have been extensively used in the context of optimal control, see, e.g., [63]. A difficulty with solving the KKT conditions (3) is that the complementarity slackness condition (3c) is disjunctive, in the sense that it imposes that $\mathbf{h}_i$ and μ_i cannot be both non-zero at the same time. Hence the manifold satisfying (3c) is intrinsically non-smooth. Primal-dual interior point methods relax this non-smooth manifold into a smooth one, by disturbing condition (3c) via a small positive constant τ [21, 22, 42, 46]. The relaxed KKT conditions read as:

$$\mathbf{r}_\tau(\mathbf{z}, \hat{\mathbf{x}}_i) = \begin{bmatrix} \nabla_{\mathbf{w}} \mathscr{L}(\mathbf{z}, \hat{\mathbf{x}}_i) \\ \mathbf{g}(\mathbf{w}, \hat{\mathbf{x}}_i) \\ \mu^\top \mathbf{h}(\mathbf{w}, \hat{\mathbf{x}}_i) + \tau \end{bmatrix} = 0, \tag{10}$$

where $\mathbf{h}(\mathbf{w},\hat{\mathbf{x}}_i) < 0$ and $\mu > 0$ must hold. For $\tau > 0$, a classic Newton method can then be applied to (10) to find solutions to $\mathbf{r}_\tau(\mathbf{z}_\tau(\hat{\mathbf{x}}_i),\hat{\mathbf{x}}_i) = 0$, via the following iteration starting from an initial guess $\hat{\mathbf{z}}^0_{i,\tau}$:

$$\Delta\mathbf{z} = -\left.\frac{\partial \mathbf{r}_\tau}{\partial \mathbf{z}}^{-1}\mathbf{r}_\tau\right|_{\hat{\mathbf{z}}^k_{i,\tau},\,\hat{\mathbf{x}}_i} \tag{11a}$$

$$\hat{\mathbf{z}}^{k+1}_{i,\tau} \leftarrow \hat{\mathbf{z}}^k_{i,\tau} + \alpha\Delta\mathbf{z} \tag{11b}$$

where $\alpha \in]0,1]$ is an adequately chosen step-size. The iteration (11) has similar convergence properties to the SQP iteration described above, locally converging to the relaxed solution $\mathbf{z}_\tau(\hat{\mathbf{x}}_i)$ at a quadratic rate when $\alpha = 1$ can be selected. However, $\mathbf{z}_\tau(\hat{\mathbf{x}}_i)$ differs from the true solution with an error $\|\mathbf{z}_\tau(\hat{\mathbf{x}}_i) - \mathbf{z}(\hat{\mathbf{x}}_i)\|$ that is in the order of τ, see, e.g., [63], hence it is desirable to solve (10) for τ small.

Because Newton methods are hindered when the Lipschitz constant associated to the Jacobian of the residual $\frac{\partial \mathbf{r}_\tau}{\partial \mathbf{z}}$ is high, a fairly large relaxation of the complementarity slackness condition (3c) makes it easier to solve (10), hence a reasonably large τ is desirable. This observation especially holds when the Newton iterations solving (10) need to perform a change of Active-Set, i.e. when it has to go through the (smoothened) "corner" in the complementarity slackness manifold. Primal-dual interior point methods resolve this trade-off by interweaving iterations of the form (11) with a careful reduction of parameter τ, until both τ and $\|\mathbf{r}_\tau\|$ are small [63]. We ought to underline here that the quadratic local convergence of the iteration (11) is degraded whenever τ is reduced, unless sufficiently small changes are made in τ and ad-hoc tools such as Mehrotra predictors [43] are deployed.

Similarly to SQP methods, primal-dual interior point methods must be provided with an initial guess $\hat{\mathbf{z}}^0_{i,\tau}$ to get the iteration (11) started. Unfortunately, since the iteration is typically started for a τ "large" so as to be able to negotiate possible changes of Active-Set efficiently, an initial guess close to the true solution, i.e. $\hat{\mathbf{z}}^0_{i,\tau} \approx \mathbf{z}(\hat{\mathbf{x}}_i)$, is not necessarily a good one because $\mathbf{z}_\tau(\hat{\mathbf{x}}_i)$ can differ significantly from $\mathbf{z}(\hat{\mathbf{x}}_i)$ for τ "large." This difficulty of "warm-starting" (i.e., forming good initial guesses for) interior point methods is well known, see, e.g., [56, 59], and poses some challenges in the context of real-time implicit MPC. Note that in the context of SQP methods, the pQP (7) is often itself solved using an Interior-Point method similar to what has been described in this section. In that case, the only nonlinearity in (10) stems from the relaxed complementarity slackness condition.

4 Discretization

We will now discuss the conversion of a continuous MPC scheme of the form (1) into the discrete pNLP (2). The former is infinite-dimensional as its solution is made of the continuous input profile $\mathbf{u}(t)$ and the corresponding continuous trajectory $\mathbf{x}(.)$, and must be approximated, in one way or another, by a finite set of variables that can be manipulated in the computer.

The discretization of (1) requires the discretization of both the continuous input profile $\mathbf{u}(.)$ and of the corresponding state trajectories $\mathbf{x}(.)$. The continuous input profile is commonly approximated via a piecewise-constant one, typically matching the discrete time grid $i \cdot \Delta t$ on which the solutions of the pNLP (2) are updated. On the prediction window $\tau \in [0, t_\mathrm{f}]$ used in the MPC scheme, the piecewise-constant input profile then reads as:

$$\mathbf{u}(\tau) = \mathbf{u}_k, \quad \tau \in [k\Delta t, (k+1)\Delta t), \quad \text{for} \quad k = 0, \dots, N-1 \tag{12}$$

The predicted state trajectories $\mathbf{x}(\tau)$ resulting from the initial conditions $\hat{\mathbf{x}}_i$ and an input profile $\mathbf{u}_{0,\dots,N-1}$ can then be obtained via a numerical simulation of the continuous dynamics (1c). Numerical simulations then hold an approximation of the continuous trajectory $\mathbf{x}(.)$ in the form of a finite number of "checkpoints" $\mathbf{x}_{k,j}$ corresponding to a (not necessarily uniform) time grid $\tau_{k,j}$, with $k = 0, \dots, N-1$, $j = 0, \dots, d-1$ and $\tau_{k,j} \in [t_k, t_{k+1}]$.

4.1 Single Shooting Methods

A straightforward approach to discretize the continuous MPC scheme (1) into its pNLP counterpart (2) can rely on adopting $\mathbf{w} = \{\mathbf{u}_0, \dots, \mathbf{u}_{N-1}\}$ cast in a vector format, as decision variables, and computing the discrete trajectories $\mathbf{x}_{i,j}$ as functions of the inputs and initial conditions $\hat{\mathbf{x}}_i$ via an ad-hoc computer code. For the sake of illustration let us consider the ineffective but simple explicit Euler approach, which would generate the discrete states $\mathbf{x}_{k,j}$ using the recursion:

$$\mathbf{x}_{0,0} = \hat{\mathbf{x}}_i, \qquad \mathbf{x}_{k+1,0} = \mathbf{x}_{k,d}, \tag{13a}$$

$$\mathbf{x}_{k,j+1} = \mathbf{x}_{k,j} + \left(\tau_{k,j+1} - \tau_{k,j}\right) \mathbf{F}\left(\mathbf{x}_{k,j}, \mathbf{u}_k\right), \qquad j = 0, \dots, d-1 \tag{13b}$$

for $k = 1, \dots, N-1$. Note that here $\tau_{k,d} = \tau_{k+1,0}$ would hold. The cost function (1a) can then be approximated using a quadrature rule such as:

$$\phi\left(\mathbf{w}, \hat{\mathbf{x}}_i\right) = T\left(\mathbf{x}_{N-1,d}\right) + \sum_{k=0}^{N-1} \sum_{j=0}^{d-1} \left(\tau_{k,j+1} - \tau_{k,j}\right) \frac{L\left(\mathbf{x}_{k,j+1}, \mathbf{u}_k\right) + L\left(\mathbf{x}_{k,j}, \mathbf{u}_k\right)}{2} \tag{14}$$

while the inequality constraints (1d)–(1e) can be enforced on the discrete time grid, i.e. function $\mathbf{h}$ gathers the constraints:

$$\mathbf{H}\left(\mathbf{x}_{k,j}, \mathbf{u}_k\right) \le 0, \qquad k = 0, \dots, N-1, \quad j = 0, \dots, d \tag{15}$$

and $\mathbf{T}\left(\mathbf{x}_{N-1,d}\right) \le 0$. Note that in this approach, the pNLP (2) does not hold an equality constraints function $\mathbf{g}$. This type of approach, labelled Single-Shooting, is not uncommon amongst practitioners of optimal control. While effective in some cases, Single-Shooting is known to be problematic for many optimal control and MPC problems, and alternative discretization schemes have been developed to address its shortcomings. The main issue with Single-Shooting is that the recursion (13) tends to "accumulate" the nonlinearity and (possible) instability of the continuous dynamics $\mathbf{F}$, see, e.g., [1]. As a result, even for mildly nonlinear and unstable continuous dynamics, the relationship from $\hat{\mathbf{x}}_i$, $\mathbf{u}_{0,\ldots,N-1}$ to the discrete state trajectory $\mathbf{x}_{k,j}$ can become extremely nonlinear and sensitive on a long prediction horizon t_f. This issue, in turn, has a detrimental impact on the convergence of iterative methods deployed on solving the resulting pNLP. Indeed, because of the potentially high nonlinearity of the simulations carried out in Single-Shooting, the region $Q(\hat{\mathbf{x}}_i)$ around the primal-dual solution $\mathbf{z}(\hat{\mathbf{x}}_i)$ where the quadratic convergence (9) occurs can become extremely small, hence requiring extremely good (i.e., very close to the true solution) initial guesses $\hat{\mathbf{z}}_i^0$. For these reasons, schemes tackling MPC numerically are typically based on alternative discretization methods.

4.2 Multiple Shooting Methods

Multiple Shooting methods address the problems described above by avoiding the deployment of long simulations in the discretization of the continuous MPC problem (1), see, e.g., [6–8]. The key idea behind Multiple-Shooting is to divide the time span $[0, t_\mathrm{f}]$ selected in the MPC scheme into smaller intervals, and perform the simulation separately on each on these intervals, starting from "artificial" initial conditions that we will label $\mathbf{x}_k$ in the following.

For the sake of simplicity, it is fairly common to perform this division using the time grid $t_{0,\ldots,N-1}$ selected for the discretization (12) of the input profile. Similarly to the remarks on (13), while high-performance simulation codes ought to be used for the simulations, let us illustrate the Multiple-Shooting approach here via the explicit Euler scheme. In that context, one would write the *simulation function* $\mathbf{f}(\mathbf{x}_k, \mathbf{u}_k) = \mathbf{x}_{k,d}$ where $\mathbf{x}_{k,d}$ is provided by the recursion:

$$\mathbf{x}_{k,0} = \mathbf{x}_k, \tag{16a}$$

$$\mathbf{x}_{k,j+1} = \mathbf{x}_{k,j} + \left(\tau_{k,j+1} - \tau_{k,j}\right) \mathbf{F}\left(\mathbf{x}_{k,j}, \mathbf{u}_k\right), \qquad j = 0, \ldots, d \tag{16b}$$

Note that the only difference between (16) and (13) is that in (13) the initial conditions $\mathbf{x}_{k,0}$ on each interval $[t_k, t_{k+1}]$ are inherited from the previous interval, while they are dictated by the variables $\mathbf{x}_k$ in (16). In Multiple-Shooting methods, the variables $\mathbf{x}_k$ with $k = 0, \ldots, N$ then become decision variables in the pNLP, and the continuity of the simulation between the time intervals becomes part of the equality constraint. The pNLP arising from a Multiple-Shooting approach to discretizating (1) is typically deployed as:

$$\min_{\mathbf{w}} \quad (14) \tag{17a}$$

$$\text{s.t.} \quad \mathbf{g}(\mathbf{w},\hat{\mathbf{x}}_{\mathbf{i}}) = \begin{bmatrix} \hat{\mathbf{x}}_i - \mathbf{x}_0 \\ \mathbf{f}(\mathbf{x}_0,\mathbf{u}_0) - \mathbf{x}_1 \\ \vdots \\ \mathbf{f}(\mathbf{x}_{N-1},\mathbf{u}_{N-1}) - \mathbf{x}_N \end{bmatrix} = 0 \tag{17b}$$

$$\mathbf{h}(\mathbf{w}) = \begin{bmatrix} \mathbf{h}(\mathbf{x}_0,\mathbf{u}_0) \\ \vdots \\ \mathbf{h}(\mathbf{x}_{N-1},\mathbf{u}_{N-1}) \\ \mathbf{T}(\mathbf{x}_N) \end{bmatrix} \leq 0 \tag{17c}$$

where the decision variables are $\mathbf{w} = \{\mathbf{x}_0,\mathbf{u}_0,\ldots\mathbf{x}_{N-1},\mathbf{u}_{N-1},\mathbf{x}_N\}$. The benefit of Multiple-Shooting stems from the integration functions $\mathbf{f}(\mathbf{x}_k,\mathbf{u}_k)$ becoming asymptotically linear as $\Delta t = t_{k+1} - t_k \to 0$, regardless of the integration scheme in use. Hence, the constraint function (17b) can be made in theory arbitrarily close to linear by dividing the prediction horizon $[0, t_f]$ using a sufficiently fine time grid $t_{0,\ldots,N-1}$ (i.e., a larger N). This is clearly limited in practice by the number of extra variables introduced in the pNLP (17), making it increasingly computationally heavy to tackle. As discussed in Section 3, decreasing the nonlinearity of the constraint functions in the pNLP tends to improve the behavior of the iterative methods (e.g., SQP or IP) deployed on solving the pNLP (in terms of speed and region of convergence).

We ought to underline again here that the numerical simulation methods (13) and (16) have been chosen for their simplicity in illustrating the discussion, but are in practice a poor choice of numerical integration method. More advanced integration methods, ranging from explicit Runge-Kutta schemes to implicit methods depending on the specific continuous dynamics at hand, prove to be more effective. A large suite of high-performance, efficient integration methods have been developed for the specific purpose of forming the constraint function $\mathbf{g}$ and its gradient $\nabla\mathbf{g}$ in (17b) at a minimum computational cost. Detailing these methods is out of the scope of this chapter, but we refer the reader to, e.g., [33, 48–50, 52] for detailed discussions on this question.

4.3 Direct Collocation Methods

Direct Collocation methods [4, 5, 62] take the Multiple-Shooting idea one step further. Multiple-Shooting introduces the intermediate discrete states $\mathbf{x}_k \equiv \mathbf{x}_{k,0}$ for $k = 0,\ldots,N-1$ as decision variables in the pNLP but "hides" the remaining discrete states $\mathbf{x}_{k,i}$ for $i = 1,\ldots,d-1$ from the pNLP, only reporting the functions $\mathbf{f}(\mathbf{x}_k,\mathbf{u}_k) = \mathbf{x}_{k,d}$ in the constraints. Hence the pNLP solver manipulates these hidden discrete states $\mathbf{x}_{k,i}$ "indirectly," via manipulating the intermediate initial conditions $\mathbf{x}_k = \mathbf{x}_{k,0}$ and the inputs $\mathbf{u}_{0,\ldots,N-1}$. In contrast, Direct Collocation methods introduce *all* discrete states $\mathbf{x}_{k,i}$ supporting the approximation of the continuous dynamics as decision variables in the pNLP, i.e. the set of decision variables becomes:

$$\mathbf{w} = \left\{\mathbf{x}_{0,0},\,\ldots,\,\mathbf{x}_{0,d},\,\mathbf{u}_0,\,\ldots,\,\mathbf{x}_{N-1,0},\,\ldots,\,\mathbf{x}_{N-1,d},\,\mathbf{u}_{N-1}\right\} \tag{18}$$

The constraint function $\mathbf{g}(\mathbf{w},\hat{\mathbf{x}}_\mathbf{i})$ then holds all the equations describing the relationship between successive states. For example, for the sake of illustration, in the case of a simulation based on a simple explicit Euler scheme, the Equation (13) would be all introduced in the equality constraints (2b) for $k = 0,\ldots,N-1$ and $j = 0,\ldots,d-1$.

In practice, the simulation of the dynamics ought to be supported by more efficient methods than the explicit Euler scheme used here in (13) for our illustration. In fact, Direct Collocation methods most often use specific high-order implicit Runge-Kutta schemes, see, e.g., [5]. It is useful to specify that all collocation methods yield constraint equations where the dynamics $\mathbf{F}\left(\mathbf{x}_{k,j},\mathbf{u}_k\right)$ appear linearly (as in (13)), i.e. the nonlinearity of the equality constraints in Direct Collocation methods comes from the nonlinearity of the continuous dynamics only, and are linear if the continuous dynamics are linear.

The inequality constraint function can then be imposed on every discrete state, i.e. function $\mathbf{h}(\mathbf{w})$ gathers the constraints $\mathbf{H}\left(\mathbf{x}_{k,j},\mathbf{u}_k\right) \leq 0$ for $k = 0,\ldots,N-1$ and $j = 0,\ldots,d$, and the terminal constraint $\mathbf{T}\left(\mathbf{x}_{N-1,d}\right) \leq 0$.

Because Direct Collocation methods perform the simulation (via enforcing the equations underlying the integration scheme) and the optimization together in the pNLP, they are often referred to as *simultaneous* optimal control methods. Multiple-Shooting methods arguably carry out a part of the simulation within the pNLP by enforcing continuity of the simulations, and are therefore often labelled as *semi-simultaneous* optimal control methods, in contrast to both single-shooting and direct collocation methods.

Similarly to Multiple-Shooting methods, Direct Collocation methods offer a way to treat the dynamics in the pNLP that reduces the nonlinearity of the simulations as the discrete time grid $\tau_{k,j}$ becomes finer. Direct Collocation methods have therefore the same benefit as Multiple-Shooting methods as they tend to yield regions where full-step iterative optimization methods have a quadratic convergence that are larger than in the Single-Shooting case. Because Direct Collocation methods "divide" the simulation into finer elements (time grid $\tau_{k,j}$ instead of t_k), it is often observed in practice that the resulting convergence of iterative methods is slightly better than using Multiple-Shooting methods, unless some specific corrections are deployed in the Multiple-Shooting scheme [54]. This benefit is obtained, however, at the price of having more decision variables in the resulting pNLP. We finally ought to specify that Direct Collocation methods share some similarities with pseudo-spectral methods, see, e.g., [28].

5 Predictors & Path-Following

Section 3 points to the benefit of having good initial guesses when deploying an iterative method to compute a numerical solution for pNLP (2). Theorem 1 discusses the continuity and differentiability of the solution path $\mathbf{z}(\hat{\mathbf{x}}(t))$, which can be exploited to form initial guesses in the context of real-time MPC.

In this section we will develop first-order, local predictors of the solution path at a given solution $\mathbf{z}(\hat{\mathbf{x}}_i)$, associated to the discrete time t_i. Such predictors can then form first-order approximation of the next solution $\mathbf{z}(\hat{\mathbf{x}}_{i+1})$ at the next discrete time t_{i+1}. This concept of predictors of the solution path plays a key role in many real-time NMPC methods. We detail them next.

The primal-dual solution to the pNLP (2), i.e. $\mathbf{z}(\hat{\mathbf{x}}_i)$ is implicitly described by (10) at $\tau = 0$, i.e. by $\mathbf{r}_0(\mathbf{z}, \hat{\mathbf{x}}_i) = 0$. The implicit function theorem (IFT) ensures that:

$$\left.\frac{\partial \mathbf{z}(\hat{\mathbf{x}}_i)}{\partial \hat{\mathbf{x}}_i}\right|_{\mathbf{z}(\hat{\mathbf{x}}_i),\hat{\mathbf{x}}_i} = -\left.\frac{\partial \mathbf{r}_0}{\partial \mathbf{z}}^{-1}\frac{\partial \mathbf{r}_0}{\partial \hat{\mathbf{x}}_i}\right|_{\mathbf{z}(\hat{\mathbf{x}}_i),\hat{\mathbf{x}}_i} \tag{19}$$

holds at all primal-dual solutions $\mathbf{z}(\hat{\mathbf{x}}_i)$ where $\frac{\partial \mathbf{r}_0}{\partial \mathbf{z}}$ is full rank, which is guaranteed if the pNLP fulfills both LICQ and SOSC and if no constraint is weakly active. The sensitivity of the solution $\mathbf{z}(\hat{\mathbf{x}}_i)$ of the pNLP described by (19) allows one to build tangential predictors of the pNLP solution path. Indeed, assuming one has computed a solution $\mathbf{z}(\hat{\mathbf{x}}_i)$ for a parameter value $\hat{\mathbf{x}}_i$, a first-order prediction for the next parameter value $\hat{\mathbf{x}}_{i+1}$ can be formed as:

$$\mathbf{z}(\hat{\mathbf{x}}_{i+1}) = \underbrace{\mathbf{z}(\hat{\mathbf{x}}_i) + \left.\frac{\partial \mathbf{z}(\hat{\mathbf{x}}_i)}{\partial \hat{\mathbf{x}}_i}\right|_{\mathbf{z}(\hat{\mathbf{x}}_i),\hat{\mathbf{x}}_i}(\hat{\mathbf{x}}_{i+1} - \hat{\mathbf{x}}_i)}_{1^{\text{st}}\text{-order predictor}} + \mathscr{O}\left(\|\hat{\mathbf{x}}_{i+1} - \hat{\mathbf{x}}_i\|^2\right), \tag{20}$$

as long as no constraint is weakly active.

Unfortunately, the first-order predictor (20) can be a poor predictor when a change of active set occurs near the solution $\mathbf{z}(\hat{\mathbf{x}}_i)$ at which the predictor is formed. At a change of active set, the continuity of the solution path entails that a constraint becomes weakly active. The solution path then typically loses its differentiability, as a change of active sets typically yields a "corner" in the solution path $\mathbf{z}(\hat{\mathbf{x}}(t))$, which is not captured by the first-order predictor (20). This effect can be observed in Figure 3.

Note that the first-order predictor described in (19)–(20) can be identically formed for the relaxed solution path $\mathbf{z}_\tau(\hat{\mathbf{x}}_i)$ with $\tau > 0$, replacing $\mathbf{r}_0$, $\mathbf{z}(\hat{\mathbf{x}}_i)$ by $\mathbf{r}_\tau$, $\mathbf{z}_\tau(\hat{\mathbf{x}}_i)$ in (19)–(20). See Figure 3 for an illustration. In that context, the "corners" in the solution path created by changes of active set are smoothed out into "sharp turns." The larger τ the smoother the turn is, making it easier to approximate using a first-order predictor of the form (20). Unfortunately, a large τ also increases the discrepancy between the relaxed solution path $\mathbf{z}_\tau(\hat{\mathbf{x}}(t))$ and the exact one $\mathbf{z}(\hat{\mathbf{x}}(t))$, rendering the relaxed solution less valid.

To address the problem of dealing with changes of active set in first-order predictors, one can turn to the QP predictor:

$$\begin{aligned}
\text{predQP}\left(\Delta\hat{\mathbf{x}}_i\right):\ \min_{\Delta\mathbf{w}} \quad & \frac{1}{2}\Delta\mathbf{w}^\top\nabla^2_{\mathbf{w}}\mathscr{L}\Delta\mathbf{w}+\nabla_{\mathbf{w}}\mathscr{L}^\top\Delta\mathbf{w}+\Delta\hat{\mathbf{x}}_i^\top\nabla_{\hat{\mathbf{x}}_i\mathbf{w}}\mathscr{L}\Delta\mathbf{w}, && \text{(21a)}\\
\text{s.t.} \quad & \mathbf{g}+\nabla_{\mathbf{w}}\mathbf{g}^\top\Delta\mathbf{w}+\nabla_{\hat{\mathbf{x}}_i}\mathbf{g}^\top\Delta\hat{\mathbf{x}}_i=0, && \text{(21b)}\\
& \mathbf{h}+\nabla_{\mathbf{w}}\mathbf{h}^\top\Delta\mathbf{w}+\nabla_{\hat{\mathbf{x}}_i}\mathbf{h}^\top\Delta\hat{\mathbf{x}}_i\leq 0, && \text{(21c)}
\end{aligned}$$

formed at a given point $\mathbf{z}\left(\hat{\mathbf{x}}_i\right)$ of the solution path, delivering the primal-dual update $\Delta\mathbf{z}=\left(\Delta\mathbf{w},\lambda_{\text{QP}},\mu_{\text{QP}}\right)$ parametrized by $\Delta\hat{\mathbf{x}}_i=\hat{\mathbf{x}}_{i+1}-\hat{\mathbf{x}}_i$, where $\lambda_{\text{QP}},\mu_{\text{QP}}$ are the dual variables delivered by the parametric QP (21). We can then write the predictor:

$$\mathbf{z}(\hat{\mathbf{x}}_{i+1})=\underbrace{\mathbf{z}(\hat{\mathbf{x}}_i)+\Delta\mathbf{z}\left(\hat{\mathbf{x}}_{i+1}-\hat{\mathbf{x}}_i\right)}_{\text{QP predictor}}+\mathscr{O}\left(\|\hat{\mathbf{x}}_{i+1}-\hat{\mathbf{x}}_i\|^2\right) \tag{22}$$

which is a (typically) non-smooth first-order predictor of the solution path, but valid regardless of changes of active set.

One can verify that if no constraint is weakly active at the solution $\mathbf{z}\left(\hat{\mathbf{x}}_i\right)$ where QP (21) is formed, then the predictor (22) is *locally* identical to the first-order predictor (20). However, the QP-based predictor (22) can anticipate (to a first-order accuracy) possible changes in the active set resulting from perturbations $\Delta\hat{\mathbf{x}}_i$. See Figure 4 for an illustration. One ought to observe that the anticipation of a coming change of active set is performed at the expense of solving QP (22), as opposed to using a simple matrix factorization in (19) and a matrix-vector multiplication in (20). The former is generally computationally considerably more expensive than the latter.

Note that $\nabla_{\mathbf{w}}\mathscr{L}$ can be replaced by $\nabla_{\mathbf{w}}\phi$ in (21a) without affecting the primal solution $\Delta\mathbf{w}$. However, this replacement changes the dual variables $\lambda_{\text{QP}},\mu_{\text{QP}}$ returned by the QP from being *steps* on $\mathbf{z}$ to being the *actual* new dual variables $\mathbf{z}$, such that the predictor formula (22) would have to be slightly modified for the dual variables.

5.1 Parametric Embedding

Consider the following pNLP parametrized by $\hat{\mathbf{x}}_{i+1}$:

$$\begin{aligned}
\text{EpNLP}\left(\hat{\mathbf{x}}_{i+1}\right):\quad \min_{\mathbf{w},\mathbf{p}} \quad & \phi\left(\mathbf{w},\mathbf{p}\right), && \text{(23a)}\\
\text{s.t.} \quad & \mathbf{g}\left(\mathbf{w},\mathbf{p}\right)=0, && \text{(23b)}\\
& \mathbf{h}\left(\mathbf{w},\mathbf{p}\right)\leq 0, && \text{(23c)}\\
& \mathbf{p}-\hat{\mathbf{x}}_{i+1}=0, && \text{(23d)}
\end{aligned}$$

where the decision variable $\mathbf{p}$ has been introduced and forced to match the parameter $\hat{\mathbf{x}}_{i+1}$ via constraint (23d). This procedure of introducing a "copy" of the pNLP

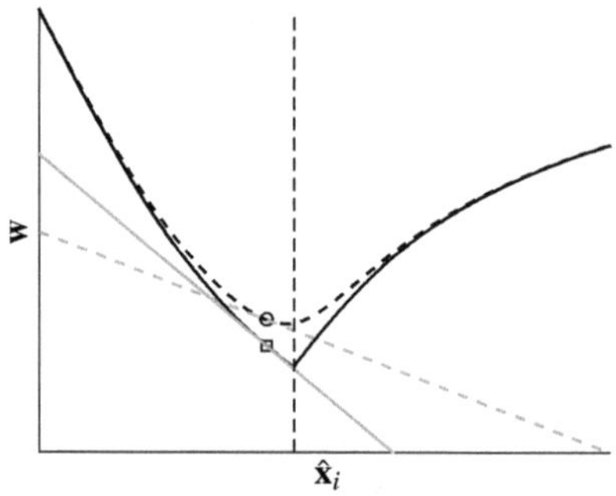

Fig. 3: Illustration of a solution path of a pNLP (solid black curve) and of the relaxed solution path ($\tau > 0$, dashed black curve). The tangential predictor associated to the solution path and its relaxed version yielded by (20) are depicted as grey lines (solid and dashed, resp.) associated to specific points $\mathbf{z}(\hat{\mathbf{x}}_i)$ (square) and $\mathbf{z}_\tau(\hat{\mathbf{x}}_i)$ (circle) of the solution paths. The pNLP underlying this solution path has a change of active set at the parameter value highlighted by the vertical dashed line. At this parameter value, the solution path $\mathbf{z}(\hat{\mathbf{x}}_i)$ is not differentiable, and is poorly approximated by the tangential predictor.

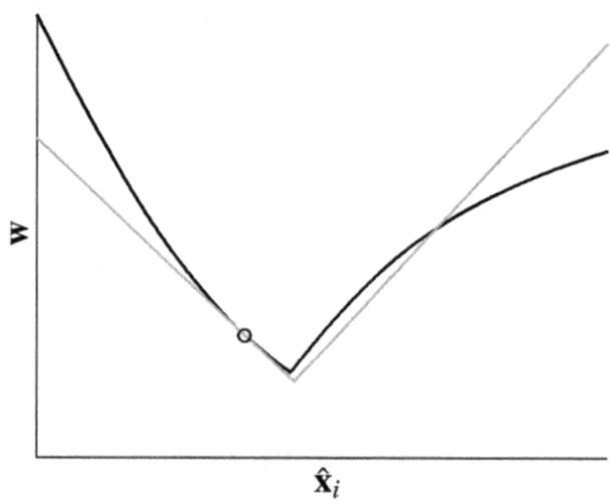

Fig. 4: Illustration of a solution path of a pNLP (solid black curve). The QP-based predictor resulting from (21)–(22) at the point outlined by a circle is represented by the solid grey lines. The QP-based predictor, while still being a first-order predictor, is capable of anticipating changes of active set, and "capturing" corners in the solution path.

parameters as decision variable is labelled parametric embedding. In the specific case where the embedded parameter is the vector of initial conditions $\hat{\mathbf{x}}_{i+1}$ arising in MPC, parametric embedding is also labelled *initial-value embedding*.

One can verify that for a primal-dual initial guess $\mathbf{z}(\hat{\mathbf{x}}_i)$ with matching initial value $\mathbf{p} = \hat{\mathbf{x}}_i$, a full (i.e., with $\alpha = 1$) step taken using the solution $\Delta\mathbf{z}$ delivered by (23) subjected to a parameter value $\hat{\mathbf{x}}_{i+1}$ matches the QP prediction (21)–(22). After the full step, the linear equality constraint (23d) ensures that $\mathbf{p} = \hat{\mathbf{x}}_{i+1}$. Subsequent steps holding $\hat{\mathbf{x}}_{i+1}$ as the parameter in use then yield classic SQP steps, often labelled *corrections* steps, which adjust the solution further until convergence is reached.

In the case the primal-dual initial guess provided to the SQP method deployed on (23) is not a primal-dual solution of (2) for the parameter value $\hat{\mathbf{x}}_i$, then the first SQP step deployed on (23) can be construed as a "mixture" of prediction (for the new parameter value $\hat{\mathbf{x}}_{i+1}$) and correction (for the inexact primal-dual guess).

Initial-value embedding is not fundamentally different than the QP predictor technique (21)–(22). Indeed, it can be verified that (23) delivers the same solution as (21), and the same subsequent steps (provided that $\Delta\hat{\mathbf{x}}_i$ is set to zero in (21) after the first step is taken). However, initial-value embedding offers a straightforward and convenient way of deploying a QP-based prediction-correction technique in the context of parametric Nonlinear-Programming.

5.2 Path Following Methods

The prediction techniques described so far in this section aim at forming efficient path-following methods, which in the context of real-time MPC aim at keeping up with the physical reality imposed by the system we aim at controlling. Indeed, since the state estimation $\hat{\mathbf{x}}(t)$ is changing all the time, the solution $\mathbf{z}(\hat{\mathbf{x}}(t))$ is continuously evolving along the solution path, following the "physical clock" represented by t. Path following in the context of real-time MPC then aims at delivering solutions $\hat{\mathbf{z}}_i$ that are always close to the current exact solution $\mathbf{z}(\hat{\mathbf{x}}(t))$.

Arguably the simplest approach to follow a solution path would rely on an algorithm of the form depicted in Figure 5, whereby upon receiving a new state estimation $\hat{\mathbf{x}}_{i+1}$, the solution estimated at time t_i, i.e. $\hat{\mathbf{z}}_i$, is updated to a new estimate $\hat{\mathbf{z}}_{i+1}$ matching $\hat{\mathbf{x}}_{i+1}$ by taking a (full) SQP step based on the available data $\hat{\mathbf{z}}_i$, $\hat{\mathbf{x}}_{i+1}$. Such a procedure does not make use of the predictor effects detailed earlier, and can be construed as a pure correction algorithm. It performs nonetheless effectively when $\hat{\mathbf{z}}_0 \approx \mathbf{z}(\hat{\mathbf{x}}_0)$ and if the successive state estimations $\hat{\mathbf{x}}_i$ and corresponding solutions $\mathbf{z}(\hat{\mathbf{x}}_i)$ are sufficiently close to each other, see Figure 7 left graph for an illustration. A caveat of this approach, though, lies in that the *preparation* phase whereby the QP problem (7) is formed (upper square block in Figure 5) requires that the state estimation $\hat{\mathbf{x}}_{i+1}$ is known. Since the preparation of the QP problem (7) requires computational time, it adds to the delay already imposed by solving the QP problem (7) between obtaining a new state estimation $\hat{\mathbf{x}}_{i+1}$ and delivering an updated solution $\hat{\mathbf{z}}_{i+1}$.

To circumvent this problem, prediction-correction techniques can help us. Indeed, since the prediction-correction QP (21) (or its "embedded version" (23)) is formed based on the data $\hat{\mathbf{z}}_i$, $\hat{\mathbf{x}}_i$ available at time t_i, it can be prepared *without knowledge* of $\hat{\mathbf{x}}_{i+1}$, and solved once $\hat{\mathbf{x}}_{i+1}$ is obtained. The procedure is illustrated in Figure 6. Because the prediction-correction QP (21) can be formed without knowing the new state estimation $\hat{\mathbf{x}}_{i+1}$, the delay between obtaining $\hat{\mathbf{x}}_{i+1}$ and delivering a solution update $\hat{\mathbf{z}}_{i+1}$ is limited to the computational time required to solve the prediction-correction QP (21) (or its "embedded version" (23)).

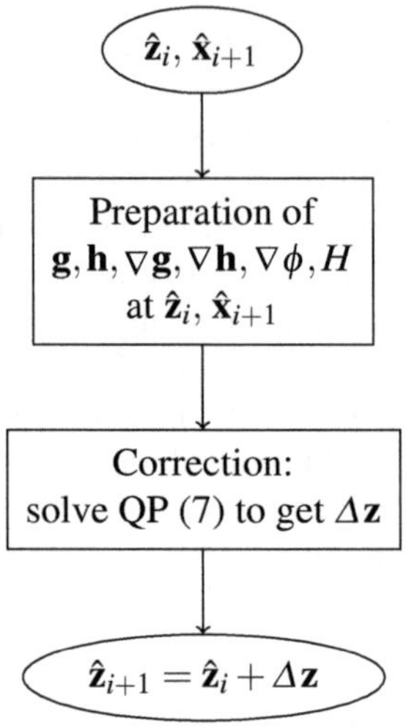

Fig. 5: Schematic of a (pure) Corrector path-following algorithm.

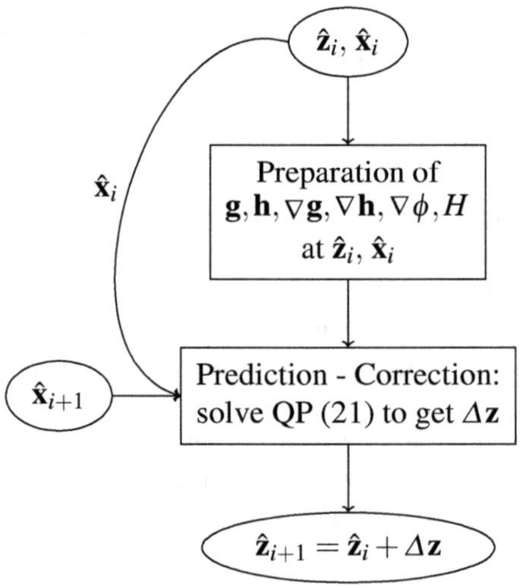

Fig. 6: Schematic of a Predictor-Corrector path-following algorithm.

For a sufficiently small difference between $\hat{\mathbf{x}}_i$ and $\hat{\mathbf{x}}_{i+1}$, and if the primal-dual guess $\hat{\mathbf{z}}_i$ is close to the exact solution $\mathbf{z}(\hat{\mathbf{x}}_i)$ for the parameter $\hat{\mathbf{x}}_i$, then the solution update $\hat{\mathbf{z}}_{i+1}$ delivered by Algorithm 6 is close to the exact new solution $\mathbf{z}(\hat{\mathbf{x}}_{i+1})$. The notion of "closeness" can be more formally described here via the bound:

$$\|\hat{\mathbf{z}}_{i+1} - \mathbf{z}(\hat{\mathbf{x}}_{i+1})\| \leq c \left\| \begin{matrix} \hat{\mathbf{x}}_{i+1} - \hat{\mathbf{x}}_i \\ \hat{\mathbf{z}}_i - \mathbf{z}(\hat{\mathbf{x}}_i) \end{matrix} \right\|^2 \tag{24}$$

which holds locally for some positive constants c. The proof of this bound is omitted here but follows from classical analysis on the contraction of the (exact) Newton iteration [44], and extends to the QP prediction-correction case. More detailed analysis of the contraction of predictor-corrector schemes can be, e.g., found in [15]. The intuition behind (24) can be construed fairly simply: the prediction-correction algorithm depicted in Figure 6 exploits all the first-order information available to compute the solution update, and therefore eliminates the first-order errors between $\hat{\mathbf{z}}_{i+1}$ and the true solution $\mathbf{z}(\hat{\mathbf{x}}_{i+1})$, such that the remaining error is of order two. The behavior of the prediction-correction algorithm is illustrated in Figure 7, center graph.

Prediction-correction techniques can also be deployed in the context of Interior-Point methods [68, 69], as detailed in Algorithm (PFIP). In this context, a linear predictor-corrector of the form (20) is used, adopting the adequate parameter τ in forming the sensitivities $\frac{\partial \mathbf{z}_\tau(\hat{\mathbf{x}}_i)}{\partial \hat{\mathbf{x}}_i}$ in (19). The resulting predictor-corrector step, however, does not in its simple form anticipate changes of active set, and tends to perform poorly when such a change has to be accounted for at a low value of τ. A natural trade-off then arises between having 1) a relatively large τ, allowing for an accurate following of the relaxed solution manifold $\mathbf{z}_\tau(\hat{\mathbf{x}}(t))$ but not being close to the exact manifold $\mathbf{z}(\hat{\mathbf{x}}(t))$, and 2) a small τ, allowing the relaxed manifold to be close to the exact one, i.e. $\mathbf{z}_\tau(\hat{\mathbf{x}}(t)) \approx \mathbf{z}(\hat{\mathbf{x}}(t))$, but making it harder to follow using predictor-corrector techniques. This behavior is illustrated in Figure 7, right graph.

In order to address this trade-off, a natural approach is to consider the path-following of a solution manifold for τ relatively large, and introduce in algorithm 3 extra iterations of the Newton method (11) while reducing τ in order to get closer to the exact solution manifold. The algorithm then ought to return both the predictor-corrector update (to be used in subsequent prediction-correction steps) and the refined solution based on the further iterations and a reduction of τ, see, e.g., [67].

Algorithm: Predictor-corrector IP path-following method (PFIP)

Input: $\hat{\mathbf{z}}_{i,\tau}$, $\hat{\mathbf{x}}_i$, $\hat{\mathbf{x}}_{i+1}$

Take full predictor-corrector step

$$\hat{\mathbf{z}}_{i+1,\tau} \leftarrow \hat{\mathbf{z}}_{i,\tau} - \left[\frac{\partial \mathbf{r}_\tau}{\partial \mathbf{z}}^{-1} \left(\mathbf{r}_\tau + \frac{\partial \mathbf{r}_\tau}{\partial \hat{\mathbf{x}}_i} (\hat{\mathbf{x}}_{i+1} - \hat{\mathbf{x}}_i) \right) \right]_{\hat{\mathbf{z}}_{i,\tau},\hat{\mathbf{x}}_i} \tag{25}$$

return $\hat{\mathbf{z}}_{i+1,\tau}$

5.3 *Real-Time Dilemma: Should We Converge the Solutions?*

The algorithm described in Figure 6 delivers at each discrete time t_i solution approximations $\hat{\mathbf{z}}_i$ that are meant to be close to the exact solutions $\mathbf{z}(\hat{\mathbf{x}}(t_i))$, but it does not

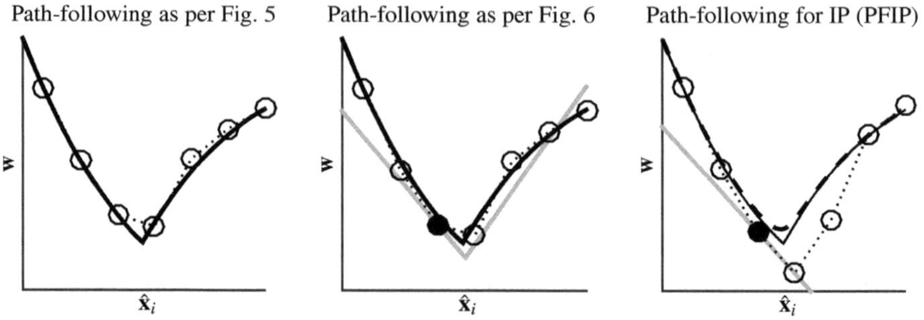

Fig. 7: Illustration of the corrector path-following algorithm depicted in Figure 5 (left graph) and of the predictor-corrector path-following algorithm depicted in Figure 6 (middle graph) of a pNLP solution path (solid black curves). The right graph depicts the behavior of the Interior-Point path-following algorithm (PFIP). The circles represent the solutions delivered by the algorithms, all starting with an exact solution at the leftmost side of the graphs. The predictors (QP-based or linear) built at the black circles are displayed as grey lines. The Interior-Point path-following algorithm uses a relaxation parameter $\tau = 2 \cdot 10^{-3}$.

consider iterating further these approximations to full convergence. Clearly, nothing prevents one to perform further iterations (whether they are SQP iterations or Newton iterations in the context of IP methods, possibly with a reduction of τ) in order to improve the accuracy of $\hat{\mathbf{z}}_i$ before delivering it as a control solution to the system. In this section, we discuss why doing so can be counterproductive.

In order to explain this statement in a simple way,let us assume that in order to deliver control solutions at each discrete time t_i, the algorithm described in Figure 6 is further iterated N times after the first solution estimate $\hat{\mathbf{z}}_i^0 = \hat{\mathbf{z}}_i$ obtained from the predictor-corrector. Let us assume that these further iterations (requiring one to reform and resolve the pQP) take a computational time of t_{iter}. The total time to perform these tasks is therefore $t_{\text{QP}} + N t_{\text{iter}}$, where t_{QP} is the time required to solve pQP (21).

By the time the N iterations have been performed, the system state will have moved to $\hat{\mathbf{x}}(t_i + t_{\text{QP}} + N t_{\text{iter}}) \neq \hat{\mathbf{x}}(t_i)$, which has a corresponding solution $\mathbf{w}(\hat{\mathbf{x}}(t_i + t_{\text{QP}} + N t_{\text{iter}})) \neq \mathbf{w}(\hat{\mathbf{x}}(t_i))$. Hence the larger N is, the more accurately one approximates the true solution $\mathbf{w}(\hat{\mathbf{x}}(t_i))$, but the worse the mismatch between that solution and the true solution $\mathbf{w}(\hat{\mathbf{x}}(t_i + t_{\text{QP}} + N t_{\text{iter}}))$ at that time. Because of the strong local contraction rate of the SQP iterations, very early in the iterations the mismatch dominates the solution inaccuracy. It is therefore recommended in practice to skip these further iterations altogether ($N = 0$), and rather to always base the computations on the most recent state estimation. This principle is graphically illustrated in Figure 8.

Clearly, one can improve this situation by accounting for the computational time delay, iterating on the MPC problem using a model-based prediction of $\hat{\mathbf{x}}(t + N t_{\text{iter}})$ obtained from the state estimation $\hat{\mathbf{x}}(t)$ instead of using $\hat{\mathbf{x}}(t)$ itself. However, be-

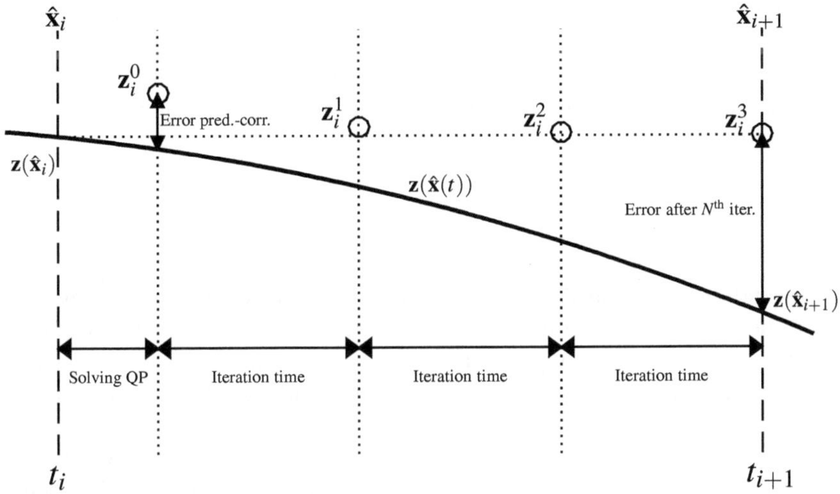

Fig. 8: Illustration of the real-time dilemma (see Section 5.3) for $N = 3$. Further iterations after the initial approximate solution $\mathbf{z}_i^0$obtained from prediction-correction yield a more accurate solution $\hat{\mathbf{z}}_i^N$, but also require computational time, during which $\hat{\mathbf{x}}(t)$ changes. If $\hat{\mathbf{z}}_i^0$ is close to the true solution, very early in the iterations $\hat{\mathbf{z}}_i^k \approx \mathbf{z}(\hat{\mathbf{x}}_i)$ will hold at close to machine precision, but $\mathbf{z}(\hat{\mathbf{x}}_i) \approx \mathbf{z}(\hat{\mathbf{x}}(t))$ will rapidly not hold anymore, such that after a few iterations, $\hat{\mathbf{z}}_i^k$ will become a poor approximation of $\mathbf{z}(\hat{\mathbf{x}}(t))$.

cause of disturbances and model error, the prediction is still corrupted by errors which grows with N and the trade-off is not fundamentally addressed.

5.4 *Shifting*

We need here to discuss a trivial yet crucial operation that needs to be introduced in schematic 6 in order to obtain a good path-following performance. This operation labelled *shifting* is computationally inexpensive, but allows one to modify $\hat{\mathbf{z}}_i$ into a guess for the next step $i+1$ that is as close as possible to the expected new solution $\mathbf{z}(\hat{\mathbf{x}}_{i+1})$. In order to best justify shifting, let us assume for the sake of the argument that the pNLP (2) is a discretization of an infinite-horizon MPC scheme, i.e. $t_f = \infty$, and let us assume that the model supporting the MPC scheme is perfect. The infinite-dimensional primal-dual solution $\mathbf{z}(\hat{\mathbf{x}}_i)$ would then include both the infinite discrete input profile $\mathbf{u}_{0,\ldots,\infty}$ and the corresponding discrete states $\mathbf{x}_{0,\ldots,\infty}$. If the model supporting the MPC scheme and the state estimation are perfect, then the next state of the physical system will match the prediction, i.e. $\hat{\mathbf{x}}_{i+1} = \mathbf{x}_1$. Moreover, if the prediction horizon is infinite, then the optimal control solution starting from $\hat{\mathbf{x}}_{i+1} = \mathbf{x}_1$ will be given by $\mathbf{u}_{1,\ldots,\infty}$ with the corresponding states $\mathbf{x}_{1,\ldots,\infty}$.

Using these observations, for the approximate primal-dual solution $\hat{\mathbf{z}}_i$ corresponding to the state estimation $\hat{\mathbf{x}}_i$ at physical time t_i, containing the discrete control input $\hat{\mathbf{u}}_{0,\ldots,N-1}$ and the corresponding discrete state trajectories $\hat{\mathbf{x}}_{0,\ldots,N}$, it is reasonable to update the guess for the next time instant t_{i+1} by *shifting* the primal-dual solution i.e. by performing:

$$\mathbf{u}_k \leftarrow \mathbf{u}_{k+1}, \qquad k = 0,\ldots,N-2 \tag{26a}$$

$$\mathbf{x}_k \leftarrow \mathbf{x}_{k+1}, \qquad k = 0,\ldots,N-1 \tag{26b}$$

in the ascending order of k. The shifting operation provides a very good primal initial guess if the prediction horizon is sufficiently long, and the model sufficiently good. In the case an exact Hessian $\nabla^2_{\mathbf{w}}\mathscr{L}$ is used, the same operation ought to be performed on the dual variables corresponding to the equality and inequality constraints in the pNLP.

One can observe that the operation (26) does not update the primal guess for the last input $\mathbf{u}_{N-1}$ and last state $\mathbf{x}_N$. Different approaches are in use here. It is actually common to not update these variables at all, and leave it to the predictor-corrector step to correct them. Alternatively, after the shifting procedure (26) is performed, one can update the last input $\mathbf{u}_{N-1}$ based on the LQR approximation of the MPC scheme at the current solution, based on the state $\mathbf{x}_{N-1} = \mathbf{x}_N$, and then update $\mathbf{x}_N$ using $\mathbf{x}_N \leftarrow \mathbf{f}(\mathbf{x}_{N-1}, \mathbf{u}_{N-1})$. In practice, and for a reasonably long discrete horizon N, it is often observed that the overall performance of the real-time MPC scheme is not much affected by the choice of strategy to generate $\mathbf{u}_{N-1}$ and $\mathbf{x}_N$.

Note that the shifting strategy described here applies to Multiple-Shooting methods, but it can be similarly deployed on Direct Collocation methods, by shifting together all the intermediate discrete states $\mathbf{x}_{k,j}$ belonging to the main time intervals k, i.e.:

$$\mathbf{x}_{k,j} \leftarrow \mathbf{x}_{k+1,j}, \qquad k = 0,\ldots,N-1, \quad j = 0,\ldots,d \tag{27}$$

5.5 *Convergence of Path-Following Methods*

The convergence of predictor-corrector path-following methods is formally analyzed in, e.g., [13–16, 69]. We will limit the discussion here to an informal reasoning. The difference between successive approximate solutions $\hat{\mathbf{z}}_i$ delivered by predictor-corrector path-following methods and the true solution $\mathbf{z}(\hat{\mathbf{x}}_i)$ arises from two sources: the error inherited from the previous solution $\hat{\mathbf{z}}_{i-1}$ used as an initial guess (with shifting) for time t_i and disturbances. The latter are actual physical disturbances, model errors and to a smaller extent the use of a finite horizon t_f (see discussion in Section 5.4). In the absence of disturbances, predictor-corrector path-following methods deployed on optimal control problems (and using shifting) inherit the quadratic local convergence rate of SQP techniques, and converge to the exact solution path $\mathbf{z}(\hat{\mathbf{x}}(t))$ very quickly, if started within the region $Q(\hat{\mathbf{x}}(t))$.

It is important to stress here that outside the region $Q(\hat{\mathbf{x}}(t))$, full prediction-correction steps may diverge, and reduced-steps may converge very slowly. A problem can then occur if the disturbances push the approximate solutions $\hat{\mathbf{z}}_i$ out of the region $Q(\hat{\mathbf{x}}_i)$. Indeed, if the disturbances "override" the strong contraction rate of the prediction-correction steps to the point of moving $\hat{\mathbf{z}}_i$ out of $Q(\hat{\mathbf{x}}_i)$, then the contraction rate is likely to degrade such that disturbances then are likely to move $\hat{\mathbf{z}}_{i+1}$ even further away from the solution path, and ultimately lead to a catastrophic failure of the predictor-corrector path-following algorithm. In order to prevent such failure, one ought to rely on $\hat{\mathbf{z}}_i \in Q(\hat{\mathbf{x}}_i)$ for all time instant t_i, see Figure 9 for an illustration. The problem then becomes one of setting up the algorithm so as to limit the impact of the disturbances on the predictor-corrector path-following algorithm.

In practice, this is achieved by making the sampling frequency of the MPC scheme Δt as small as possible. Indeed, provided that the dynamics (1c) have continuous solutions and assuming that the state estimation $\hat{x}(t)$ are noise-free, it is straightforward to verify that the smaller the sampling time Δt the "less time" the disturbance and model errors have to corrupt the guesses delivered via shifting. Indeed, to the limit $\Delta t \to 0$ the state estimation and solutions coincide, i.e. $\hat{\mathbf{x}}_{i+1} = \hat{\mathbf{x}}_i$ and $\hat{\mathbf{z}}_{i+1} = \hat{\mathbf{z}}_i$, such that disturbances lose their impact on the predictor-corrector path-following algorithm. The effect of a finite prediction horizon as a disturbance in the scheme also decreases as $\Delta t = t_f/N$ becomes smaller (for t_f held constant).

The reliability of implicit MPC schemes based predictor-corrector path-following methods therefore hinges on Δt being small. This observation is arguably at the core of the research effort aimed at delivering algorithms that achieve very high computational speeds.

6 Sensitivities & Hessian Approximation

A crucial aspect of the methods detailed so far is that deploying iterative methods to solve the pNLP arising from the discretization of the continuous MPC problem using simultaneous methods requires one to form the constraint function $\mathbf{g}$ and its gradient $\nabla\mathbf{g}$. In the context of Multiple-Shooting methods, the latter requires one to form the gradients $\nabla_{\mathbf{x}_k}\mathbf{f}(\mathbf{x}_k,\mathbf{u}_k)$ and $\nabla_{\mathbf{u}_k}\mathbf{f}(\mathbf{x}_k,\mathbf{u}_k)$ for $k = 0,\dots,N-1$. Additionally, forming the Hessian $\nabla^2_{\mathbf{w}}\mathscr{L}$, which, as detailed in (4), comprises the term:

$$\lambda^\top \mathbf{g}(\mathbf{w},\hat{\mathbf{x}}_i) = \lambda_0^\top(\hat{\mathbf{x}}_i - \mathbf{x}_0) + \sum_{k=0}^{N-1} \lambda_{k+1}^\top(\mathbf{f}(\mathbf{x}_k,\mathbf{u}_k) - \mathbf{x}_{k+1}), \tag{28}$$

where the vectors λ_k form a partition of the dual variables λ, requires forming the second-order directional sensitivities $\nabla^2_{\mathbf{w}}\left(\lambda_{k+1}^\top \mathbf{f}(\mathbf{x}_k,\mathbf{u}_k)\right)$ of the simulation functions for $k = 0,\dots,N-1$.

It ought to be underlined here that evaluating the simulation functions, $\mathbf{f}(\mathbf{x}_k,\mathbf{u}_k)$ their sensitivities and their second-order directional sensitivities can be computationally expensive, as it requires evaluating and differentiating the simulation codes used

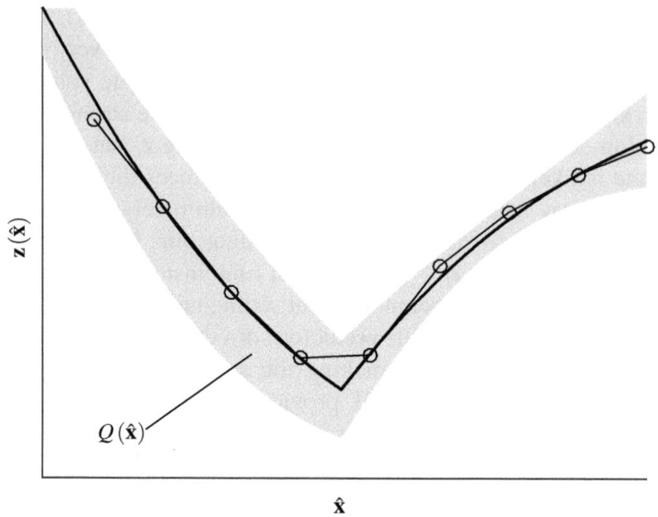

Fig. 9: Illustration of the region of quadratic convergence $Q(\hat{\mathbf{x}})$ around the true solution path $\mathbf{z}(\hat{\mathbf{x}})$ (in light grey), with a predictor-corrector path-following method deployed to a changing $\hat{\mathbf{x}}_i$ (circles).

to evaluate the functions $\mathbf{f}(\mathbf{x}_k, \mathbf{u}_k)$. These codes are typically recursive and therefore yield functions of high symbolic complexity, which are expensive to differentiate. However, the recursive structure of the simulation functions can be exploited for efficiency. Tools to perform the differentiation of the code evaluating the simulation functions $\mathbf{f}(\mathbf{x}_k, \mathbf{u}_k)$ are commonly referred to as Algorithmic Differentiation (AD) tools, and effective methods for differentiating simulation functions are available, see, e.g., [2, 3, 10, 19, 29, 34, 40, 52, 54, 55, 64].

While the simulation functions $\mathbf{f}(\mathbf{x}_k, \mathbf{u}_k)$ need to be evaluated at every step of iterative methods, it is not uncommon to deploy SQP or IP methods that do not evaluate the sensitivities at every step. It is, e.g., common in the context of MPC to skip the evaluation of the second-order sensitivities altogether, and replace the exact Hessian $\nabla^2_{\mathbf{w}}\mathscr{L}$ by an approximation [9, 12]. In the context of MPC, because least-squares cost functions are often used, the Gauss-Newton Hessian approximation is arguably the most popular one [44]. In that context, and assuming that the cost function of pNLP (2) can be put in the least-squares form $\phi(\mathbf{w}, \hat{\mathbf{x}}_i) = \frac{1}{2}\|\mathbf{R}(\mathbf{w}, \hat{\mathbf{x}}_i)\|^2$, the Gauss-Newton approximation reads as:

$$\nabla^2_{\mathbf{w}}\mathscr{L} \approx \nabla_{\mathbf{w}}\mathbf{R}(\mathbf{w}, \hat{\mathbf{x}}_i)\nabla_{\mathbf{w}}\mathbf{R}(\mathbf{w}, \hat{\mathbf{x}}_i)^\top \tag{29}$$

and is valid if the function $\mathbf{R}(\mathbf{w}, \hat{\mathbf{x}}_i)$ is either close to linear and/or close to zero at the solution of the pNLP, and if the functions $\mathbf{g}(\mathbf{w}, \hat{\mathbf{x}}_i)$ and $\mathbf{h}(\mathbf{w}, \hat{\mathbf{x}}_i)$ are close to linear. It is interesting to observe here that when using a Gauss-Newton Hessian approximation, one does not need to keep track of the dual part (λ, μ) of the solution in the path-following method, as it is then not used anywhere in forming the QP (21).

Alternative Hessian approximations can be used such as the iterative approximation BFGS [44], or even application-specific approximations, see, e.g., [31]. In any case, care should be taken to ensure that the sparsity structure present in the exact Hessian $\nabla^2_{\mathbf{w}}\mathscr{L}$ is not degraded in its approximation. Using inexact first-order sensitivities is also possible, see, e.g., [9, 23, 41, 65, 66], but must be performed carefully [47], or using a special treatment [51, 53].

It is interesting to note that when using a discretization based on Direct Collocation, the equality constraint function $\mathbf{g}(\mathbf{w}, \hat{\mathbf{x}}_i)$ becomes a list of linear combination of the nonlinear continuous dynamics $\mathbf{F}(.,.)$, such that computing the first and second-order sensitivities of the equality constraint function is fairly straightforward (though it can be computationally expensive).

7 Structures

Iterative methods such as SQP or IP ultimately rely on the factorization of (possibly) large matrices at every iteration, whether for solving the linear system (11) or for solving the QP (7). The computational complexity of matrix factorizations depends heavily on the sparsity and sparsity pattern of the matrix to be factorized. Hence, in this section we will briefly discuss the sparsity pattern of the matrices one needs to form QP (7), or in the Newton iteration (11). Indeed, while simultaneous methods such as Multiple-Shooting and Direct Collocation have strong benefits (discussed in Section 4), they also introduce a large number of decision variables in the pNLP (2) resulting from the discretization of (1), which in turn results in creating large matrices in the QP (7) or in the linear system (11). However, the matrices resulting from simultaneous methods also have a strong sparsity, and very specific sparsity patterns. Deploying real-time MPC using simultaneous methods hinges on exploiting these patterns effectively.

We display here the typical sparsity patterns arising from a specific optimal control problem with $\mathbf{x}(t) \in \mathbb{R}^2$, $\mathbf{u}(t) \in \mathbb{R}$, $N = 20$ and $d = 3$ for the three discretization methods detailed in Section 4, see Figures 10, 11 and 12 for Single-Shooting, Multiple-Shooting, and Direct Collocation, respectively. One can observe that while simultaneous optimization techniques yield larger matrices, these matrices are thinly banded, which can be exploited for achieving high computational speed in solving the QP (7) or iterating the linear system (11).

Note that it is fairly straightforward to verify here that the Lagrange function resulting from simultaneous methods is *separable*. That is, in, e.g., the Multiple-Shooting case, using (17) and (28) one can observe that

$$\frac{\partial^2 \mathscr{L}}{\partial \mathbf{x}_i \partial \mathbf{x}_j} = 0, \quad \frac{\partial^2 \mathscr{L}}{\partial \mathbf{x}_i \partial \mathbf{u}_j} = 0, \quad \frac{\partial^2 \mathscr{L}}{\partial \mathbf{u}_i \partial \mathbf{u}_j} = 0 \quad \text{for} \quad i \neq j \tag{30}$$

resulting in the block-diagonal structure one can observe in Figure 11, right graphs. Similar observations hold for Direct Collocation methods, yielding block-diagonal Hessians as in Figure 12, right.

Describing the techniques available to exploit these structures is beyond the scope of this chapter, we refer the reader to, e.g., [17, 18, 18–20, 23–27, 36–39, 58, 60, 60, 61, 61] for more details. It can be interesting to note that the structure present in the equation underlying the simulation of the dynamics can also be often exploited in several ways, see, e.g., [32, 50, 52–54].

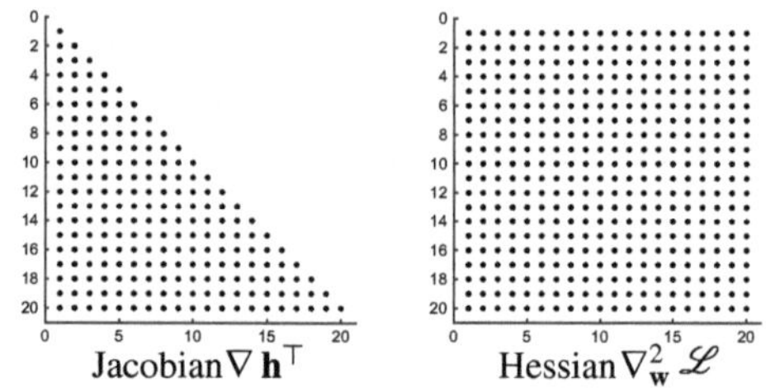

Fig. 10: Illustration of the sparsity pattern of the Jacobian of the inequality constraints and of the Hessian of the Lagrange function in the pNLP resulting from a discretization based on Single-Shooting, see Section 4.1. Note that Single-Shooting methods typically hold no equality constraints (beyond possibly initial value embedding).

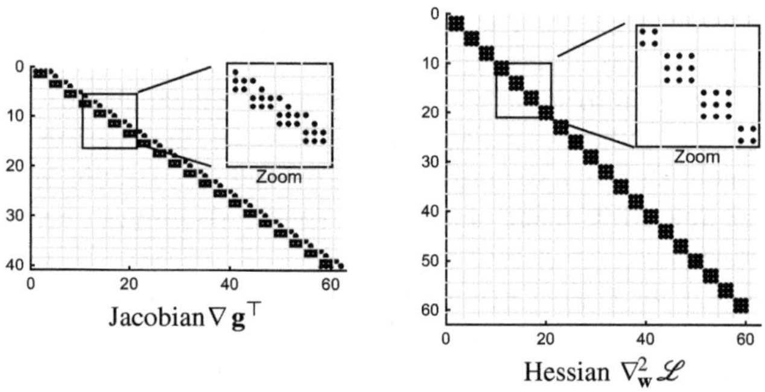

Fig. 11: Illustration of the sparsity pattern of the Jacobian of the inequality constraints and of the Hessian of the Lagrange function in the pNLP resulting from a discretization based on Multiple-Shooting, see Section 4.2. One can observe the banded structure of $\nabla \mathbf{g}$, and the block-diagonal structure of $\nabla^2_{\mathbf{w}} \mathscr{L}$. Note that in this context, the Jacobian of the inequality constraints is typically block-diagonal, and omitted here.

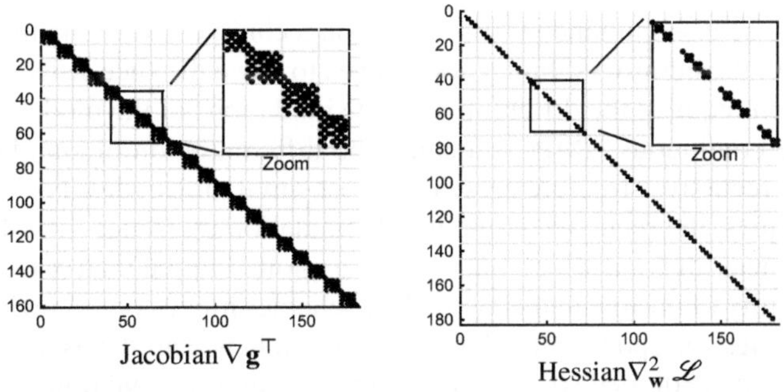

Fig. 12: Illustration of the sparsity pattern of the Jacobian of the inequality constraints and of the Hessian of the Lagrange function in the pNLP resulting from a discretization based on Direct Collocation, see Section 4.3. Note that in this context, the Jacobian of the inequality constraints is typically block-diagonal, and omitted here.

8 Summary

Let us summarize here the information provided in this chapter. The deployment of nonlinear MPC problems most often requires one to solve the underlying optimal control problem on-line (i.e., implicitly) and in real-time rather than explicitly and off-line. Unfortunately, nonlinear MPC problem usually yields non-convex optimal control problems, for which it is difficult to guarantee that the solutions computed are global.

In order to solve the continuous NMPC problem in the computer, a discretization of the problem is required. The discretization method adopted is often crucial for the performance of the overall real-time NMPC scheme. Simultaneous or semi-simultaneous discretization techniques are most often preferred, as they allow the existing iterative methods dedicated to solving the optimization problem numerically to converge faster and more reliably.

If the NMPC solution is updated at a high frequency, the successive solutions of the NMPC problem are typically not changing much between the discrete time instants where the solution is updated. The similarity between these solutions is exploited in real-time on-line NMPC schemes, using shifting and methods inherited from parametric Nonlinear Programming, which allow one at any discrete time to fully exploit the information obtained at the previous time instant.

Ultimately, all iterative solution approaches dedicated to solving the optimization problem underlying the discretized NMPC scheme rely heavily on matrix factorizations. Simultaneous optimal control methods yield optimal control problems that have a very distinctive structure, which translates into a linear algebra that has very distinctive sparsity patterns. In real-time NMPC, these patterns must be fully exploited when performing the matrix factorizations.

Finally, for semi-simultaneous discretization techniques, forming the linear algebra required in the iterative solution approaches to solve the successive optimization problems often requires the differentiation of simulations of the model dynamics involved in the continuous NMPC scheme. The differentiation of simulations ought to exploit the recursive structure present in most simulation codes. For fully simultaneous techniques, the recursive structure of simulation translates directly into structures in the linear algebra, which ought to be exploited in the matrix factorization.

The reliability of on-line real-time NMPC schemes hinges on achieving high sampling frequencies for updating the NMPC solution, hence all the aspects mentioned above ought to be exploited so as to minimize the computational time required to update the NMPC solution. It is then important to ensure that the solution path does not hold bifurcations (as illustrated in example (6)), and it is useful to keep in mind that "extremely" nonlinear continuous models remain challenging for on-line real-time NMPC schemes.

References

1. Albersmeyer, J., Diehl, M.: The lifted Newton method and its application in optimization. SIAM J. Optim. **20**(3), 1655–1684 (2010)
2. Albersmeyer, J.: Adjoint-based algorithms and numerical methods for sensitivity generation and optimization of large scale dynamic systems. PhD thesis, Ruprecht-Karls-Universitität Heidelberg (2010)
3. Andersson, J., Åkesson, J., Diehl, M.: CasADi – a symbolic package for automatic differentiation and optimal control. In: Forth, S., Hovland, P., Phipps, E., Utke, J., Walther, A. (eds.) Recent Advances in Algorithmic Differentiation, Lecture Notes in Computational Science and Engineering, pp. 297–307. Springer, Berlin (2012)
4. Biegler, L.T.: An overview of simultaneous strategies for dynamic optimization. Chem. Eng. Process. **46**, 1043–1053 (2007)
5. Biegler, L.T.: Nonlinear Programming. MOS-SIAM Series on Optimization. SIAM, Philadelphia (2010)
6. Bock, H.G., Plitt, K.J.: A multiple shooting algorithm for direct solution of optimal control problems. In: Proceedings of the IFAC World Congress, pp. 242–247. Pergamon Press, Oxford (1984)
7. Bock, H.G., Diehl, M., Leineweber, D.B., Schlöder, J.P.: Efficient direct multiple shooting in nonlinear model predictive control. In: Keil, F., Mackens, W., Voß, H., Werther, J. (eds.), Scientific Computing in Chemical Engineering II, vol. 2, pp. 218–227. Springer, Berlin (1999)
8. Bock, H.G., Diehl, M., Leineweber, D.B., Schlöder, J.P.: A direct multiple shooting method for real-time optimization of nonlinear DAE processes. In: Allgöwer, F., Zheng, A. (eds.) Nonlinear Predictive Control. Progress in Systems Theory, vol. 26, pp. 246–267. Birkhäuser, Basel (2000)
9. Bock, H.G., Diehl, M., Kostina, E.: SQP methods with inexact Jacobians for inequality constrained optimization. IWR-Preprint 04-XX, Universität Heidelberg, Heidelberg (2004)
10. Bücker, H.M., Petera, M., Vehreschild, A.: Chapter code optimization techniques in source transformations for interpreted languages. In: Advances in Automatic Differentiation, pp. 223–233. Springer, Berlin (2008)
11. Büskens, C., Maurer, H.: SQP-methods for solving optimal control problems with control and state constraints: adjoint variables, sensitivity analysis and real-time control. J. Comput. Appl. Math. **120**, 85–108 (2000)

12. Diehl, M., Bock, H.G., Schlöder, J.P.: Newton-type methods for the approximate solution of nonlinear programming problems in real-time. In: Di Pillo, G., Murli, A. (eds.) High Performance Algorithms and Software for Nonlinear Optimization, pp. 177–200. Kluwer Academic Publishers, Norwell (2002)
13. Diehl, M., Findeisen, R., Allgöwer, F., Bock, H.G., Schlöder, J.P.: Nominal stability of the real-time iteration scheme for nonlinear model predictive control. Technical Report #1910, IMA, University of Minnesota (2003)
14. Diehl, M., Findeisen, R., Allgöwer, F., Schlöder, J.P., Bock, H.G.: Stability of nonlinear model predictive control in the presence of errors due to numerical online optimization. In: Proceedings of the IEEE Conference on Decision and Control (CDC), Maui, pp. 1419–1424 (2003)
15. Diehl, M., Findeisen, R., Allgöwer, F., Bock, H.G., Schlöder, J.P.: Nominal stability of the real-time iteration scheme for nonlinear model predictive control. IEE Proc. Control Theory Appl. **152**(3), 296–308 (2005)
16. Diehl, M., Findeisen, R., Allgöwer, F.: A stabilizing real-time implementation of nonlinear model predictive control. In: Biegler, L., Ghattas, O., Heinkenschloss, M., Keyes, D., van Bloemen Waanders, B. (eds.) Real-Time and Online PDE-Constrained Optimization, pp. 23–52. SIAM, Philadelphia (2007)
17. Domahidi, A., Zgraggen, A., Zeilinger, M.N., Morari, M., Jones, C.N.: Efficient interior point methods for multistage problems arising in receding horizon control. In: Proceedings of the IEEE Conference on Decision and Control (CDC), Maui, December 2012, pp. 668–674
18. Ferreau, H.J., Bock, H.G., Diehl, M.: An online active set strategy for fast parametric quadratic programming in MPC applications. In: Proceedings of the IFAC Workshop on Nonlinear Model Predictive Control for Fast Systems, Grenoble, pp. 21–30 (2006). It is found on the official website of IFAC but without official version
19. Ferreau, H.J., Houska, B., Kraus, T., Diehl, M.: Numerical methods for embedded optimisation and their implementation within the ACADO toolkit. In: Mitkowski, W., Tadeusiewicz, R., Ligeza, A., Szymkat, M. (eds.) Proceedings of the 7th Conference Computer Methods and Systems, Krakow, November 2009, pp. 13–29. Oprogramowanie Naukowo-Techniczne
20. Ferreau, H.J., Kozma, A., Diehl, M.: A parallel active-set strategy to solve sparse parametric quadratic programs arising in MPC. In: Proceedings of the 4th IFAC Nonlinear Model Predictive Control Conference, Noordwijkerhout (2012)
21. Forsgren, A., Gill, P.E.: Primal-dual interior methods for nonconvex nonlinear programming. SIAM J. Optim. **8**(4), 1132–1152 (1998)
22. Forsgren, A., Gill, P.E., Wright, M.H.: Interior point methods for nonlinear optimization. SIAM Rev. **44**, 525–597 (2002)
23. Frasch, J.V., Wirsching, L., Sager, S., Bock, H.G.: Mixed-level iteration schemes for nonlinear model predictive control. In: Proceedings of the IFAC Conference on Nonlinear Model Predictive Control (2012)
24. Frasch, J.V., Vukov, M., Ferreau, H.J., Diehl, M.: A dual Newton strategy for the efficient solution of sparse quadratic programs arising in SQP-based nonlinear MPC (2013). Optimization Online 3972
25. Frison, G.: Algorithms and methods for fast model predictive control. PhD thesis, Technical University of Denmark (DTU) (2015)
26. Frison, G., Jørgensen, J.B.: A fast condensing method for solution of linear-quadratic control problems. In: Proceedings of the IEEE Conference on Decision and Control (CDC) (2013)
27. Frison, G., Sorensen, H.B., Dammann, B., Jørgensen, J.B.: High-performance small-scale solvers for linear model predictive control. In: Proceedings of the European Control Conference (ECC), June 2014, pp. 128–133
28. Garg, D., Patterson, M.A., Hager, W.W., Rao, A.V., Benson, D.A., Huntington, G.T.: A unified framework for the numerical solution of optimal control problems using pseudospectral methods. Automatica **46**(11), 1843–1851 (2010)
29. Gay, D.M.: Automatic differentiation of nonlinear AMPL models. In: In Automatic Differentiation of Algorithms: Theory, Implementation and Application, pp. 61–73. SIAM, Philadelphia (1991)

30. Gill, P., Murray, W., Saunders, M.: SNOPT: an SQP algorithm for large-scale constrained optimization. SIAM Rev. **47**(1), 99–131 (2005)
31. Gros, S., Quirynen, R., Diehl, M.: An improved real-time NMPC scheme for wind turbine control using spline-interpolated aerodynamic coefficients. In: Proceedings of the IEEE Conference on Decision and Control (CDC) (2014)
32. Gros, S., Quirynen, R., Schild, A., Diehl, M.: Implicit integrators for linear dynamics coupled to a nonlinear static feedback and application to wind turbine control. In: IFAC Conference (2017)
33. Hindmarsh, A.C., Brown, P.N., Grant, K.E., Lee, S.L., Serban, R., Shumaker, D.E., Woodward, C.S.: SUNDIALS: suite of nonlinear and differential/algebraic equation solvers. ACM Trans. Math. Softw. **31**, 363–396 (2005)
34. Houska, B., Ferreau, H.J., Diehl, M.: An auto-generated real-time iteration algorithm for nonlinear MPC in the microsecondrange. Automatica **47**(10), 2279–2285 (2011)
35. Jongen, H.Th., Weber, G.W.: On parametric nonlinear programming. Ann. Oper. Res. **27**, 253–283 (1990)
36. Kirches, C., Wirsching, L., Sager, S., Bock, H.G.: Efficient numerics for nonlinear model predictive control. In: Recent Advances in Optimization and Its Applications in Engineering, pp. 339–357. Springer, Berlin (2010)
37. Kirches, C., Wirsching, L., Bock, H.G., Schlöder, J.P.: Efficient direct multiple shooting for nonlinear model predictive control on long horizons. J. Process Control **22**(3), 540–550 (2012)
38. Kouzoupis, D., Ferreau, H.J., Peyrl, H., Diehl, M.: First-order methods in embedded nonlinear model predictive control. In: Proceedings of the European Control Conference (ECC) (2015)
39. Kouzoupis, D., Quirynen, R., Frasch, J.V., Diehl, M.: Block condensing for fast nonlinear MPC with the dual Newton strategy. In: Proceedings of the IFAC Conference on Nonlinear Model Predictive Control (NMPC) (2015)
40. Kvasnica, M., Rauova, I., Miroslav, F.: Automatic code generation for real-time implementation of model predictive control. In: Proceedings of the IEEE International Symposium on Computer-Aided Control System Design, Yokohama (2010)
41. Leibfritz, F., Sachs, E.W.: Inexact SQP interior point methods and large scale optimal control problems. SIAM J. Control Optim. **38**(1), 272–293 (2006)
42. Liu, X., Sun, J.: A robust primal-dual interior-point algorithm for nonlinear programs. SIAM J. Optim. **14**(4), 1163–1186 (2004)
43. Mehrotra, S.: On the implementation of a primal-dual interior point method. SIAM J. Optim. **2**(4), 575–601 (1992)
44. Nocedal, J., Wright, S.J.: Numerical Optimization. Springer Series in Operations Research and Financial Engineering, 2nd edn. Springer, Berlin (2006)
45. Pantoja, J.F.A.D., Mayne, D.Q.: Sequential quadratic programming algorithm for discrete optimal control problems with control inequality constraints. Int. J. Control **53**, 823–836 (1991)
46. Potra, F.A., Wright, S.J.: Interior-point methods. J. Comput. Appl. Math. **124**, 281–302 (2000)
47. Potschka, A.: A direct method for the numerical solution of optimization problems with time-periodic PDE constraints. PhD thesis, University of Heidelberg (2011)
48. Quirynen, R.: Automatic code generation of Implicit Runge-Kutta integrators with continuous output for fast embedded optimization. Master's thesis, KU Leuven, 2012
49. Quirynen, R., Vukov, M., Diehl, M.: Auto generation of implicit integrators for embedded NMPC with microsecond sampling times. In: Lazar, M., Allgöwer, F. (eds.) Proceedings of the 4th IFAC Nonlinear Model Predictive Control Conference, pp. 175–180 (2012)
50. Quirynen, R., Gros, S., Diehl, M.: Efficient NMPC for nonlinear models with linear subsystems. In: Proceedings of the IEEE Conference on Decision and Control (CDC), pp. 5101–5106 (2013)
51. Quirynen, R., Gros, S., Diehl, M.: Inexact Newton based lifted implicit integrators for fast nonlinear MPC. In: Proceedings of the IFAC Conference on Nonlinear Model Predictive Control (NMPC), pp. 32–38 (2015)

52. Quirynen, R., Gros, S., Diehl, M.: Lifted implicit integrators for direct optimal control. In: Proceedings of the IEEE Conference on Decision and Control (CDC) (2015)
53. Quirynen, R., Gros, S., Diehl, M.: Inexact newton-type optimization with iterated sensitivities. SIAM J. Optim. (2017)
54. Quirynen, R., Gros, S., Houska, B., Diehl, M.: Lifted collocation integrators for direct optimal control in ACADO toolkit. Math. Program. Comput. (2017)
55. Quirynen, R., Houska, B., Diehl, M.: Efficient symmetric hessian propagation for direct optimal control. J. Process Control (2017)
56. Rao, C.V., Wright, S.J., Rawlings, J.B.: Application of interior-point methods to model predictive control. J. Optim. Theory Appl. **99**, 723–757 (1998)
57. Schäfer, A.A.S., Bock, H.G., Schlöder, J.P., Leineweber, D.B.: An exact Hessian SQP method for ill-conditioned optimal control problems. IWR-Preprint 01-XX, Universität Heidelberg (2001)
58. Schmid, C., Biegler, L.T.: Quadratic programming methods for reduced Hessian SQP. Comput. Chem. Eng. **18**(9), 817–832 (1994)
59. Shahzad, A., Goulart, P.J.: A new hot-start interior-point method for model predictive control. In: Proceedings of the IFAC World Congress (2011)
60. Steinbach, M.C.: A structured interior point SQP method for nonlinear optimal control problems. In: Bulirsch, R., Kraft, D. (eds.) Computation Optimal Control, pp. 213–222. Birkhäuser, Basel (1994)
61. Steinbach, M.C.: Structured interior point SQP methods in optimal control. Z. Angew. Math. Mech. **76**(S3), 59–62 (1996)
62. von Stryk, O.: Numerical solution of optimal control problems by direct collocation. In: Optimal Control: Calculus of Variations, Optimal Control Theory and Numerical Methods, vol. 129. Bulirsch et al., 1993
63. Wächter, A.: An Interior Point Algorithm for Large-Scale Nonlinear Optimization with Applications in Process Engineering. PhD thesis, Carnegie Mellon University (2002)
64. Walther, A.: Automatic differentiation of explicit Runge-Kutta methods for optimal control. Comput. Optim. Appl. **36**(1), 83–108 (2006)
65. Wirsching, L.: An SQP Algorithm with inexact derivatives for a direct multiple shooting method for optimal control problems. Master's thesis, University of Heidelberg (2006)
66. Zanelli, A., Quirynen, R., Diehl, M.: An efficient inexact NMPC scheme with stability and feasibility guarantees. In: Proceedings of 10th IFAC Symposium on Nonlinear Control Systems, Monterey, August 2016
67. Zanelli, A., Quirynen, R., Jerez, J., Diehl, M.: A homotopy-based nonlinear interior-point method for NMPC. In: Proceedings of 20th IFAC World Congress, Toulouse, July 2017
68. Zavala, V.M., Biegler, L.T.: Nonlinear programming sensitivity for nonlinear state estimation and model predictive control. In: International Workshop on Assessment and Future Directions of Nonlinear Model Predictive Control (2008)
69. Zavala, V.M., Biegler, L.T.: The advanced step NMPC controller: optimality, stability and robustness. Automatica **45**, 86–93 (2009)

Convexification and Real-Time Optimization for MPC with Aerospace Applications

Yuanqi Mao, Daniel Dueri, Michael Szmuk, and Behçet Açıkmeşe

1 Introduction

Model Predictive Control (MPC) or Receding Horizon Control (RHC) is a form of control in which the control action at the current time is obtained by solving *online*, at each sampling instant, a finite horizon open-loop optimal control problem. This process yields a sequence of optimal control problems to be solved at each time instance. An important advantage of MPC is its ability to cope with hard constraints on controls and states, which are widely applicable in aerospace systems due to strict mission and resource constraints, and the need for control robustness (see [4, 46]). For a more detailed discussion on MPC, the reader is referred to [25, 38, 40].

The majority of the computational burden associated with MPC lies in solving finite horizon optimal control problems. Consequently, advances in optimal control theory are highly influential in the field of MPC. Optimal control theory has a rich history, and is an outgrowth of the classical calculus of variations. Centuries of work culminated in the 1950s and 1960s with the results of Pontryagin's Maximum Principle [45]. These results provide not only some analytical solutions, but more importantly powerful theoretical analysis tools.

In practice and in the absence of analytical solutions, optimal control problems must be solved numerically, and thus may be discretized and approximated as finite dimensional parameter optimization problems (i.e., the direct method). Therefore, the brunt of the work done in solving an MPC problem consists of solving such optimization problems. With recent advances in hardware and numerical optimization algorithms, the computational challenges of MPC have become more tractable. In particular, improvements in the efficiency of numerical convex optimization algorithms has promoted their proliferation in a wide variety of applications [12].

Y. Mao · D. Dueri · M. Szmuk · B. Açıkmeşe (✉)
University of Washington, Seattle, WA 98105, USA
e-mail: yqmao@uw.edu; dandueri@uw.edu; mszmuk@uw.edu; behcet@uw.edu

S. V. Raković, W. S. Levine (eds.), *Handbook of Model Predictive Control*,
Control Engineering, https://doi.org/10.1007/978-3-319-77489-3_15

Due to recent advances, the optimal solutions of a large class of convex optimization problems can be computed in polynomial time using either generic Second Order Cone Programming (SOCP) solvers [18] or customized solvers that take advantage of specific problem structures (see [19, 37]). Moreover, if a feasible solution exists, these algorithms are guaranteed to find the global optimal solution in a bounded number of iterations. Conversely, these algorithms provide a certificate of infeasibility if no feasible solution exists. As a result, these advances have enabled the use of MPC in a variety of contexts, including electronic automotive and aerospace systems. The interested reader is referred to [14, 53, 56] for more details on real-time MPC applications.

However, most real-world problems are inherently non-convex, and thus difficult to solve. Therefore, algorithms that solve non-convex optimization problems in real-time are needed for MPC. Most early attempts at solving non-convex optimal control problems (e.g., [13, 26]) first discretize the problem (see [30]) and then use general nonlinear programming solvers to obtain solutions. Same strategy is also employed in nonlinear MPC, however by taking advantage of specific problem structures (e.g., [17, 55]). Still, general nonlinear optimization methods have some challenges that are not easily overcome. First, there are few known time complexity bounds. For example, Interior Point Methods (IPMs) have polynomial time complexity when applied to convex problems [41], but require much more effort such as a Levenberg-Marquardt-type parameter search (see chapter 17 of [42]) when applied to non-convex problem, which is not particularly suitable for online MPC. Second, they can be sensitive to initial guesses, sometimes leading to divergence even when a feasible solution exists. To alleviate initial guess dependency, one may use, for instance, a line search framework in Sequential Quadratic Programming (SQP) (see chapter 18 of [42]), but this will again slow down the convergence. Besides, iterative methods like SQP often require additional steps, such as Broyden-Fletcher-Goldfarb-Shanno (BFGS) updates, to convexly approximate the Hessian for their subproblems. Such steps usually do not come cheap computationally [22]. In summary, these drawbacks usually render traditional non-convex optimization methods unsuitable for real-time and safety-critical applications, where computation speed and guaranteed convergence are of utmost importance.

In the meantime, the success of real-time convex optimization methods motivates the formulation of non-convex optimal control problems in a convex programming framework. We have seen a growing number of works exploring the idea of convexification for aerospace applications, such as rendezvous [29, 31], swarm formation flying [3, 24], planetary entry [32, 54], and obstacle avoidance [23]. While the results of the numerical experiments look promising, few theoretical proofs are provided.

Inspired by another aerospace application, the planetary soft landing of autonomous rockets [10], recent results have introduced a procedure known as *Lossless Convexification* [1, 2, 28]. It proves that a class of non-convex optimal control problems can be posed as equivalent convex optimization problems that recover the solution to the original problems. More recently, an iterative algorithm called *Successive Convexification* (`SCvx`) [21, 34, 35] has been introduced to solve problems with non-convex nonlinear dynamics and state-constraints, and it provides proofs of global convergence. In contrast to SQP based iterative methods, `SCvx` uses a first order model, which naturally convexifies the subproblem and hence does not require Hessian approximation steps. These two convexification techniques provide rigorous attempts to solve complex non-convex optimal control problems in real-time, and they form the main focus of this chapter.

2 Convexification

In this section, we cover basic convexification techniques for optimal control problems. Among these problems, convex ones have some of the most attractive properties, and are tied closely with their convex parameter optimization counterparts. [8] gives a rigorous treatment of this class of problems. Real-world optimal control problems, however, are generally non-convex, and oftentimes need to be convexified before they can be solved reliably and efficiently.

We first outline a typical finite horizon optimal control problem. To this end, we will use t to denote time, t_0 and t_f to denote the initial and (finite) final time. Note that t_0 and t_f can be free variables. We denote the system state as $x(t) : [t_0, t_f] \to \mathbb{R}^n$, and the control input as $u(t) : [t_0, t_f] \to \mathbb{R}^m$. The control input u is assumed to be at least Lebesgue integrable on $[t_0, t_f]$. More specifically, we assume that u is measurable and essentially bounded, i.e. bounded almost everywhere (a.e.), on $[t_0, t_f]$: $u \in L_\infty[t_0, t_f]^m$, with the ∞-norm defined as

$$\|u\|_\infty := \operatorname*{ess\,sup}_{t \in [t_0, t_f]} \|u(t)\|,$$

where $\|\cdot\|$ is the Euclidean vector norm on $\mathbb{R}^m$, and ess sup represents the essential supremum. The system dynamics, with the system state x and the control input u, are described as

$$\dot{x}(t) = f(x(t), u(t), t) \qquad a.e. \quad t_0 \le t \le t_f,$$

where $f : \mathbb{R}^n \times \mathbb{R}^m \times \mathbb{R} \to \mathbb{R}^n$ is the control-state mapping. f is Fréchet differentiable with respect to all arguments, and is nonlinear in general. Due to the differentiability of f and the Lebesgue integrability of u, x is continuous on $[t_0, t_f]$. Hence, $x \in W_{1,\infty}[t_0, t_f]^n$, where $W_{1,\infty}[t_0, t_f]^n$ is the space of absolutely continuous functions on $[t_0, t_f]$ with measurable and essentially bounded (first order) time derivatives. The 1-norm of this space is defined by

$$\|x\|_{1,\infty} := \max\{\|x\|_\infty, \|\dot{x}\|_\infty\}.$$

With these two norms, it can be shown that both $L_\infty[t_0, t_f]^m$ and $W_{1,\infty}[t_0, t_f]^n$ are Banach spaces (see [33] for more details). In addition to dynamics, most real-world problems include control and state constraints that must hold for (almost) all $t \in [t_0, t_f]$. For simplicity, we assume these constraints are time invariant. That is, we assume that $u(t) \in \mathscr{U}$ and $x(t) \in \mathscr{X}$. In general, $\mathscr{U} \subseteq \mathbb{R}^m$ and $\mathscr{X} \subseteq \mathbb{R}^n$ are assumed to be non-empty non-convex sets. Lastly, the problem includes an objective (or cost) functional, that is to be minimized. The objective functional is generally assumed to be convex, since non-convex objectives can simply be reformulated as non-convex constraints (i.e., moving the non-convexity from the cost to the constraints).

We are now ready to present the general formulation of a non-convex constrained optimal control problem.

Problem 1 (Original Non-convex Problem). Determine a control function $u^* \in L_\infty[t_0,t_f]^m$, and a state trajectory $x^* \in W_{1,\infty}[t_0,t_f]^n$, which minimize the convex functional

$$J(x,u) := \varphi(x(t_f),t_f) + \int_{t_0}^{t_f} L(x(t),u(t),t)\,dt, \tag{1a}$$

subject to the constraints

$$\dot{x}(t) = f(x(t),u(t),t) \qquad a.e. \quad t_0 \leq t \leq t_f, \tag{1b}$$

$$u(t) \in \mathscr{U} \qquad a.e. \quad t_0 \leq t \leq t_f, \tag{1c}$$

$$x(t) \in \mathscr{X} \qquad t_0 \leq t \leq t_f, \tag{1d}$$

where $\varphi : \mathbb{R}^n \times \mathbb{R} \to \mathbb{R}$ is the terminal cost, $L : \mathbb{R}^n \times \mathbb{R}^m \times \mathbb{R} \to \mathbb{R}$ is the running cost, and both are convex and Fréchet differentiable.

Different sources of non-convexity can be observed in Problem 1. In particular, if the dynamics are nonlinear, then (1b) represents a non-convex constraint. Similarly, a non-convex $\mathscr{U}$ renders (1c) into a non-convex constraint, and a non-convex $\mathscr{X}$ renders (1d) non-convex. These sources of non-convexity can be addressed using the following convexification techniques:

- Lossless convexification.
- Successive convexification.
- Successive convexification with state constraints.

Each of the aforementioned methods is introduced to handle one source of non-convexity in Problem 1. The methods can be applied independently, or simultaneously with proper precautions.

2.1 Lossless Convexification of Control Constraints

The first technique we consider is called *lossless convexification*. Its primary purpose is to handle non-convex control constraints, which appear in a wide range of engineering applications. This method introduces a relaxation to the non-convex control constraint through the use of a slack variable. Lossless convexification guarantees that if a feasible solution exists, the optimal solution of the relaxed optimal control problem is also the optimal solution to the original non-convex problem. This property is called the *equivalence*.

Lossless convexification was introduced in [2] for a fuel optimal planetary landing problem. Though there are a number of state constraints in this problem, the equivalence proof was carried out by assuming that the state constraints could only be active on a set with measure zero. This is a strong assumption since it cannot be verified before solving the problem. A more general result was obtained in [1]. Instead of focusing on a specific landing problem, lossless convexification was proven for a more general class of optimal control problems with convex cost, linear dynamics, convex state constraints, and a special class of non-convex control constraints. This paper represents the first instance where the equivalence property of lossless convexification was related to properties of the dynamical system. It was shown that, under a few other minor assumptions, the equivalence property holds if the linear system is controllable. This is a very powerful result since most systems of interest are designed to be controllable. The theory presented in this section is primarily based on this paper.

The theory of lossless convexification of control constraints was extended to nonlinear systems by [11]. Here, the authors identified the dependence of the equivalence property on gradient matrices having full rank along the optimal state and control trajectories – a condition that is difficult to verify a-priori (nevertheless, some special cases were treated rigorously). Since no convexification was applied to the nonlinear dynamics, Mixed Integer Linear Programming (MILP) is required to solve the optimal control problem, which quickly becomes computationally intractable.

Attention returned to planetary landing in [5], where the authors focused on an additional thrust pointing constraint. In this paper, the authors developed a geometric insight that establishes a connection with *normal systems* (i.e., systems where the Hamiltonian is maximized at the extreme points of a projection of the relaxed set of feasible controls). However, a small perturbation to the problem was introduced in order to complete the proof, thus rendering the results less rigorous in terms of the treatment of active state constraints and the handling of thrust pointing constraints. Both of these were addressed in [28], which contains the most general lossless convexification results. It states that the *equivalence* holds whenever the state space is a strongly controllable subspace, which extended the controllability concept used by [1], and recovered their result as a special case. The work also naturally handled pointing constraints in a rigorous manner and answered the question of when lossless convexification can be achieved without having to perturb the problem.

2.1.1 Theory

The theory of lossless convexification is largely based on the result in [1], and also includes enhancements from [28]. The system we consider has convex cost, linear dynamics, and convex state constraints, which makes the control constraints the single source of non-convexity. The corresponding optimal control problem can be formulated as follows.

Problem 2 (Linear Dynamics, Non-convex Control Constraints). Determine a control function $u^* \in L_\infty[t_0,t_f]^m$, and a state trajectory $x^* \in W_{1,\infty}[t_0,t_f]^n$, which minimize the functional

$$J(x,u) := h_0(t_0,t_f,x(t_0),x(t_f)) + k\int_{t_0}^{t_f} g_0(u(t))\,dt, \tag{2a}$$

subject to the constraints

$$\dot{x}(t) = A(t)x(t) + B(t)u(t) + E(t)w(t) \qquad a.e. \quad t_0 \le t \le t_f, \tag{2b}$$

$$u(t) \in \mathscr{U}, \quad x(t) \in \mathscr{X} \qquad a.e. \quad t_0 \le t \le t_f, \tag{2c}$$

$$\big(t_0,t_f,x(t_0),x(t_f)\big) \in \mathscr{E}, \tag{2d}$$

where $h_0 : \mathbb{R}\times\mathbb{R}\times\mathbb{R}^n\times\mathbb{R}^n \to \mathbb{R}$ is the end cost, $g_0 : \mathbb{R}^m \to \mathbb{R}$ is the running cost, and both are convex and Fréchet differentiable. $k \ge 0$ is a scalar, $A : \mathbb{R}_+ \to \mathbb{R}^{n\times n}$, $B : \mathbb{R}_+ \to \mathbb{R}^{n\times m}$, and $E : \mathbb{R}_+ \to \mathbb{R}^{n\times p}$ are piecewise analytic functions of time t. $w(t) \in \mathbb{R}^p$ is a known exogenous input. $\mathscr{X} \subseteq \mathbb{R}^n$ is the convex set of feasible states, $\mathscr{U} \subseteq \mathbb{R}^m$ is the set of feasible control inputs, and $\mathscr{E} \subset \mathbb{R}^{2n+2}$ is the set of feasible boundary conditions.

Here $\mathscr{U}$ represents a class of non-convex sets that satisfy

$$\mathscr{U} = \mathscr{U}_1 \setminus \mathscr{U}_2, \quad \mathscr{U}_2 = \bigcap_{i=1}^{q} \mathscr{U}_{2,i} \subset \mathscr{U}_1, \tag{3}$$

where $\mathscr{U}_1$ and $\mathscr{U}_2$ are, respectively, compact convex and open convex sets with

$$\mathscr{U}_{2,i} = \{u \in \mathbb{R}^m : \; g_i(u) < 1\}, \quad i = 1,\dots,q,$$

where g_i, $i = 1,\dots,q$, are convex functions that are bounded on $\mathscr{U}_1$, that is, there exists some $\bar{g} \in \mathbb{R}$ such that $g_i(u) \le \bar{g}$, $\forall u \in \mathscr{U}_1$, $i = 1,\dots,q$. Note that $\mathscr{U}_2 \cap \partial\mathscr{U}_1$ is empty, where $\partial\mathscr{U}_1$ represents the set of extremal points of $\mathscr{U}_1$. This follows from the fact that $\mathscr{U}_2 \subset \mathscr{U}_1$ and that $\mathscr{U}_2 \cap \partial\mathscr{U}_2$ is empty.

The main difficulty in the convexification of Problem 2 is the non-convex control constraints defined by the set $\mathscr{U}$ (shown in Figure 1a). As an example, Figure 1b gives the thrust bound constraints, $\rho_1 \le \|T_c(t)\| \le \rho_2$, which are quite broadly applicable in the realm of optimal control.

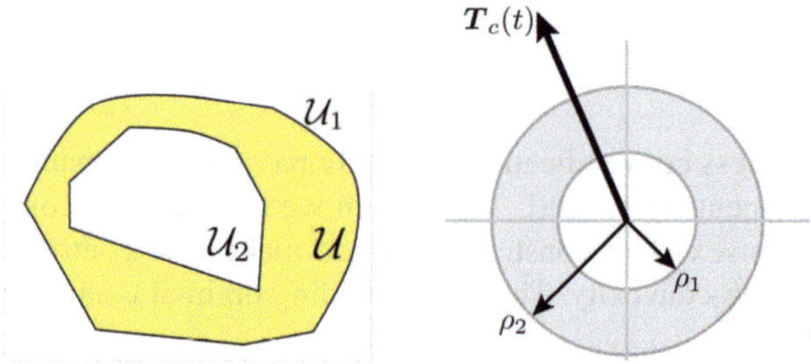

(a) General $\mathscr{U}$, shaded **(b)** Example, thrust bound constraints

Fig. 1: Non-convex set of feasible control inputs.

To address this difficulty, we will use lossless convexification, and the concept of which is best captured by geometric insight. Using planetary soft landing as an example, we have $\mathscr{U} = \{u \in \mathbb{R}^2 : 1 \le \|u\| \le \rho\}$, which is a two-dimensional non-convex annulus. However, we can lift it to a convex cone by introducing a third dimension σ, and extending the annulus in this direction (see Figure 2). We denote the resulting set as $\mathscr{V}$, and in this case $\mathscr{V} = \{(u,\sigma) \in \mathbb{R}^3 : \sigma \ge 1,\ \|u\| \le \min(\rho,\sigma)\}$. Clearly, $\mathscr{V}$ is in a higher dimensional space and $\mathscr{U} \subset \mathscr{V}$, i.e. we are solving a relaxed

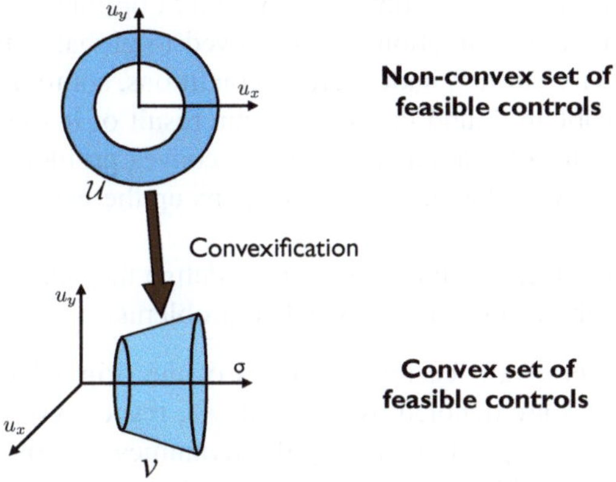

Fig. 2: Convexification of the Control Magnitude Constraint for Planetary Soft Landing. The annulus represents the actual non-convex control constraints $\mathscr{U}$ in (u_x, u_y) space, which is lifted to a convex cone $\mathscr{V}$ in (u_x, u_y, σ) space.

problem. The set $\mathscr{V}$ also contains control inputs that are not feasible for Problem 2. Hence, it is not trivial to establish the *equivalence* of solutions between these two sets. However, by carefully designing the convex relaxation, it can be shown that under certain conditions, an optimal solution of the relaxed problem will be feasible and optimal for the original non-convex problem (Problem 2).

Problem 3 (Convex Equivalence of Problem 2). Determine a control function $u^* \in L_\infty[t_0,t_f]^m$, and a state trajectory $x^* \in W_{1,\infty}[t_0,t_f]^n$, which minimize the functional

$$J(x,u) := h_0(t_0,t_f,x(t_0),x(t_f)) + k\xi(t_f), \tag{4a}$$

subject to the constraints

$$\dot{x}(t) = A(t)x(t) + B(t)u(t) + E(t)w(t) \qquad a.e. \quad t_0 \le t \le t_f, \tag{4b}$$

$$\dot{\xi}(t) = \sigma(t) \qquad a.e. \quad t_0 \le t \le t_f, \tag{4c}$$

$$(u(t),\sigma(t)) \in \mathscr{V}, \quad x(t) \in \mathscr{X} \qquad a.e. \quad t_0 \le t \le t_f, \tag{4d}$$

$$\big(t_0,t_f,x(t_0),x(t_f),\xi(t_0),\xi(t_f)\big) \in \tilde{\mathscr{E}}, \tag{4e}$$

where $\sigma(t) \in \mathbb{R}$ is a slack control variable, $\xi(t) \in \mathbb{R}$ is the corresponding state variable, and $\tilde{\mathscr{E}} \subset \mathbb{R}^{2n+4}$ is the set of relaxed feasible boundary conditions. Here

$\mathscr{V} \subseteq \mathbb{R}^{m+1}$ is the set of relaxed feasible control inputs,

$$\mathscr{V} = \left\{(u,\sigma) \in \mathbb{R}^{m+1} : \sigma \geq 1 \text{ and } u \in \mathscr{U}_1 \cap \mathscr{V}_2(\sigma)\right\} \tag{5}$$

with $\mathscr{V}_2(\sigma) = \bigcap_{i=i_o}^{q} \mathscr{V}_{2,i}(\sigma)$ where

$$\mathscr{V}_{2,i}(s) := \{u \in \mathbb{R}^m : g_i(u) \leq s\},\ i_o = \begin{cases} 0 & \text{for } k > 0 \\ 1 & \text{for } k = 0 \end{cases}.$$

This $\mathscr{V}$ set is a generalization of the convex cone in Figure 2, and it can easily be shown to be convex. Therefore, Problem 3 is a convex optimal control problem, and its finite dimensional approximation can be solved to global optimality in polynomial time, see [12, 41]. Further, under certain conditions, solutions of Problem 3 are also solutions of Problem 2, and that is the main result of lossless convexification, because it enables us to solve a much harder non-convex problem by solving its convex relaxation in a *lossless* fashion. It really opens up the gate to real-time optimal control.

To concisely explain the result, first we may define the sets of feasible and optimal solutions for both original and convexified problems.

Definition 1. The sets of all feasible solutions of the original Problem 2 and the convexified Problem 3 are denoted by $\mathscr{F}_O$ and $\mathscr{F}_R$, respectively. $(t_0,t_f,x,u) \in \mathscr{F}_O$ and $(t_0,t_f,x,\xi,u,\sigma) \in \mathscr{F}_R$ if they satisfy the dynamics and the state and control constraints of the two problems, respectively, *a.e.* $[t_0,t_f]$. $\mathscr{F}_O^*$ and $\mathscr{F}_R^*$ represent the respective sets of optimal solutions, with optimal costs J_O^* and J_R^*.

Next are the two generally required conditions in lossless convexification. In many cases it is straightforward to verify these conditions. A detailed discussion can be found in [1].

Condition 1. The pair $\{A(\cdot),B(\cdot)\}$ is *controllable* and the set of feasible controls $\mathscr{U}$ satisfies $\mathscr{U}_2^\dagger = \{\mathbf{0}\}$, where $\mathscr{U}^\dagger := \left\{v \in \mathbb{R}^m : \exists c \in \mathbb{R} \text{ s.t. } v^T u = c\ \forall u \in \mathscr{U}\right\}$.

Condition 2. $(t_0,t_f,x,\xi,u,\sigma) \in \mathscr{F}_R^*$ and

$$\left(-k\sigma(t_0) - \frac{\partial h_0}{\partial t_0},\ k\sigma(t_f) - \frac{\partial h_0}{\partial t_f},\ \frac{\partial h_0}{\partial x(t_0)},\ \frac{\partial h_0}{\partial x(t_f)}\right)$$

is not orthogonal to $\mathscr{E}$, where $\mathscr{E}$ is given by (2d).

The next theorem presents a fundamental result that establishes conditions under which the optimal solution of the convexified problem is indeed feasible for the original non-convex problem.

Theorem 1. *Suppose that Condition 1 holds. If (t_0,t_f,x,ξ,u,σ) satisfies Condition 2 and additionally*

$$x(t) \in \operatorname{int}\mathscr{X} \qquad \forall t \in [t_0,t_f], \tag{6}$$

then $(t_0,t_f,x,u) \in \mathscr{F}_O$.

The proof of this result is non-trivial and it uses Pontryagin's Maximum Principle (see in [9, 45]), which can be found in [1].

Besides this general result concerning feasibility, [1] also addresses optimality for several different cases often seen in aerospace optimal control applications. For instance, for the class of problems that has an integral cost on the controls, which is applicable to many minimum fuel planetary soft landing applications, we have

$$\mathscr{U} = \{u \in \mathbb{R}^m : 1 \le g_0(u) \le \rho\}, \tag{7}$$

and the relaxed convex set given by (5) is

$$\mathscr{V} = \left\{(u,\sigma) \in \mathbb{R}^{m+1} : \sigma \ge 1,\ g_0(u) \le \min(\rho,\sigma)\right\}. \tag{8}$$

Then the following theorem gives the conditions for an optimal control of the convexified problem to also define an optimal solution for the original problem, see [1] for a proof.

Theorem 2. *Suppose that $\mathscr{U}$ satisfies (7), $k>0$, and Condition 1 is satisfied for the original Problem 2. If (t_0,t_f,x,ξ,u,σ) satisfies Condition 2 and additionally the condition in* (6)*, then $(t_0,t_f,x,u) \in \mathscr{F}_O^*$.*

Although lossless convexification results presented in Theorems 1 and 2 provide a theoretical tool to tackle non-convexities, they have their limitations, too. First, they do not handle the thrust pointing constraints, or more generally, constraints of the form $Cu(t) \le d$. The pointing constraints ensure that the translational maneuver does not require the spacecraft to be oriented outside of a desired pointing cone, which usually results in a reduction of performance [5]. Another imperfection would be the additional interior condition in (6), which is more or less restrictive and not particularly easy to verify beforehand. These two shortcomings were addressed in [28] by introducing the notion of the *friend* of a linear system and the strongly controllable subspace. All of the conditions in that paper can be checked a priori, and are satisfied for many applications since strong controllability is often designed into the systems.

2.1.2 Application

Throughout the section, we have mentioned the planetary soft landing problem as an interesting example, which is gaining renewed interest due to emergence of reusable rockets [10], of the lossless convexification technique. The objective is to search for the thrust (control) profile T_c and an accompanying translational state trajectory $(r,\dot{r})$ that guide a lander from an initial position r_0 and velocity $\dot{r}_0$ to rest at the prescribed target location on the planet while minimizing the fuel consumption. The problem considers planets with a constant rotation rate (angular velocity), a uniform gravity field, and negligible aerodynamic forces during the powered-descent phase of landing. When the target point is unreachable from a given initial state, a precision landing problem (or *minimum landing error* problem) is considered instead, with the objective to first find the closest reachable surface location to the target and second to obtain the minimum fuel state trajectory to that closest point. The setup of this problem is shown in Figure 3.

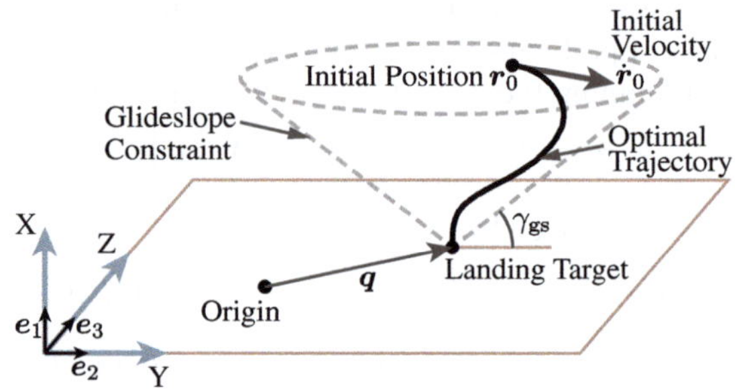

Fig. 3: The setup of minimum landing error powered descent guidance problem. The glide-slope constraint requires the spacecraft to remain in a cone defined by the minimum slope angle γ.

In this problem there are several state and control constraints. The main state constraints are the glide-slope constraint on the position vector and an upper bound constraint on the velocity vector magnitude. The glide-slope constraint is imposed to ensure that the lander stays at a safe distance from the ground until it reaches its target. The upper bound on velocity is needed to avoid supersonic velocities for planets with atmosphere, where the control thrusters can become unreliable. Both of these constraints are convex and they fit well to the lossless convexification framework.

The control constraints, however, are challenging since they define a non-convex set of feasible controls. We have three control constraints (see Figure 4): Given any maneuver time (time-of-flight) t_f, for all $t \in [0, t_f]$,

- Convex upper bound on thrust, $\|T_c(t)\| \leq \rho_2$.
- Non-convex lower bound on thrust, $\|T_c(t)\| \geq \rho_1 > 0$.

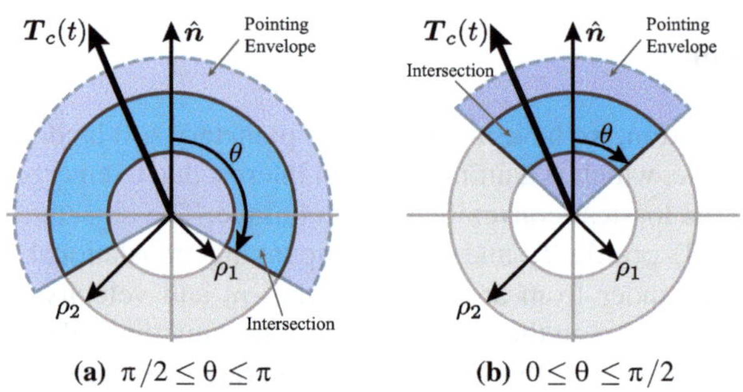

Fig. 4: Planar representation of thrust bounds and thrust pointing limits constraints.

- Thrust pointing constraint $\hat{n}^T T_c(t)/\|T_c(t)\| \geq \cos\theta$ where $\|\hat{n}\| = 1$ is a unit vector and $0 \leq \theta \leq \pi$ is the maximum allowable angle of deviation from the direction given by $\hat{n}$, which is convex when $\theta \leq \pi/2$ and non-convex when $\theta > \pi/2$.

The details of the modeling and the parameters of the numerical simulation can be found in [5]. Here our focus is to demonstrate the effectiveness of the lossless convexification by showing some simulation results. Three numerical experiments were performed for various pointing-constraints: i) unconstrained; ii) 90° constraint; iii) 45° constraint. The results are overlaid in Figure 5.

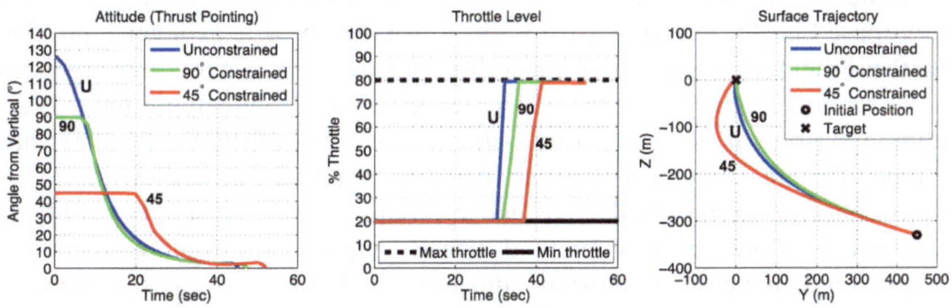

Fig. 5: Simulation results with three different pointing constraints: Unconstrained (U), $\theta = 90°$ (90) and $\theta = 45°$ (45). Thrust pointing and magnitude constraints are satisfied for the optimal solution, as seen in the first two plots. The last plot is the position trajectory of each solution.

The pointing angle is relative to local vertical, which aligns the pointing cone $\hat{n}$ vector along the coordinate frame X axis. The attitude pointing plot indicates that the solution of the relaxed problem ensures the satisfaction of the pointing constraints for the original problem. The throttle level plot shows that the thrust bounds are satisfied. These two plots indicate that the solution of the convexified relaxed problem remains valid for the original problem. The last plot of Figure 5 overlays the position trajectories. The 45° case overshoots the target along the Y axis to satisfy the pointing constraint. Interestingly, the 90° constrained path takes a more direct route.

2.2 Successive Convexification

Compared with lossless convexification, Successive Convexification (`SCvx`) is a recent development, and the main reference for this subject is [34]. `SCvx` is an iterative algorithm designed to tackle the non-convexity resulting from nonlinear dynamics, and the extended versions of `SCvx` can also handle certain types of non-convex state constraints [35] and inter-sample constraints [21].

The basic idea is to successively linearize the dynamical equations, and solve a sequence of convex subproblems, specifically Second Order Cone Programmings (SOCPs). While similar ideas in both finite dimensional optimization problems (e.g., [27, 43]) and optimal control problems (e.g., [39, 47]) have long been tried, few convergence results were reported. The `SCvx` algorithm, on the other hand, presents a systematic procedure. More importantly, through a continuous-time convergence analysis, we guarantee that the `SCvx` algorithm will converge, and the solution it converges to will at least recover local optimality for the original problem. To facilitate convergence, *virtual control* and *trust regions* are incorporated into our algorithm. The former acts like an exact penalty function (see [39, 57]), but with additional controllability features. The latter is similar to standard trust-region-type updating rules as in [16], but the distinction lies in that we solve each convex subproblem to full optimality.

Note that similar ideas based on Sequential Convex Programming (SCP) have been proposed in, for example, [6, 15, 49] to handle obstacle avoidance problems. While these methods usually perform well in practice, few general convergence results are reported. Hence here the convergence result accompanying the `SCvx` algorithm marks one of the first rigorous attempts on this subject. On the other hand, SQP type of methods along with a line search (or trust region) framework do have rigorous convergence results derived under certain conditions (see, e.g., [44]). As aforementioned, however, these methods usually require more computational effort each iteration than the `SCvx` algorithm due to the additional Hessian approximation steps, and because of the nature of line search (or in the trust region case, the fact that each subproblem is only approximately solved), the overall rate of convergence will also be slower than `SCvx`. Based on these two observations, we consider the `Scvx` algorithm to be more suitable for real-time applications.

2.2.1 Theory

Since `SCvx` focuses on nonlinear dynamics, the original problem we are solving will be Problem 1, with $\mathscr{U}$ and $\mathscr{X}$ assumed to be convex or at least already convexified. Also, to eliminate potential non-convexity as a result of the free final time, we assume $t_0 = 0$ and $t_f = T$ are fixed. Note that we only take this measure for theoretical simplicity, because in practice there are a couple of ways to get around this issue (see, e.g., [50]). Now, the only source of non-convexity lies in the nonlinear dynamics (1b). Since it is an equality constraint, an obvious way to convexify it is linearization by using its first order Taylor series approximation. The solution to the convexified problem, however, won't necessarily be the same as its non-convex counterpart. To recover optimality, we need to come up with an algorithm that can find a solution, which satisfies at least the first order optimality condition of the original problem. A natural approach would be executing this linearization successively, i.e. at k^{th} succession, we linearize the dynamics about the control and the corresponding trajectory computed in the $(k-1)^{th}$ succession. This procedure is repeated until convergence, which essentially forms the basic idea behind the `SCvx` algorithm.

Linearization

Assume the $(i-1)^{th}$ succession gives us a solution $(x^{i-1}(t), u^{i-1}(t))$. Let $A(t), B(t)$, and $C(t)$ be the partial derivative of $f(x^{i-1}(t), u^{i-1}(t), t)$ with respect to x, u, and t respectively, and $d(t) = x(t) - x^{i-1}(t)$, and $w(t) = u(t) - u^{i-1}(t)$, then the first order Taylor expansion about that solution will be

$$\dot{d}(t) = A(t)d(t) + B(t)w(t) + C(t) + H.O.T.. \tag{9}$$

This is a linear system with respect to $d(t)$ and $w(t)$, which are our new states and control, respectively. The linearization procedure gets us the benefit of convexity, but it also introduces two new issues, namely *artificial infeasibility* and *approximation error*. We will address them in the following two subsections.

Virtual Control

`SCvx` method can sometimes generate infeasible problems, even if the original nonlinear problem itself is feasible. That is the *artificial infeasibility* introduced by the linearization. In such scenarios, the undesirable infeasibility obstructs the iteration process and prevents convergence. The most evident example of this arises when the problem is linearized about an unrealistically short time horizon, i.e. the final time T is too small. In such a case, one can intuitively see that there is no feasible control input that can satisfy the prescribed dynamics and constraints. To prevent this artificial infeasibility, we introduce an additional control input $v(t)$, called *virtual control*, to the linear dynamics (9) (without the higher order terms):

$$\dot{d}(t) = A(t)d(t) + B(t)w(t) + E(t)v(t) + C(t), \tag{10}$$

where $E(t)$ can be chosen based on $A(t)$ such that the pair $(A(t), E(t))$ is controllable. Then, since $v(t)$ is unconstrained, any state in the feasible region can be reachable in finite time. This is why this virtual control can eliminate the artificial infeasibility. For example, on autonomous vehicles the *virtual control* can be understood as a synthetic acceleration that acts on the vehicle, which can drive the vehicle virtually anywhere in the feasible area. Since we want to resort to this *virtual control* as needed, it will be heavily penalized via an additional term $\lambda\gamma(Ev)$ in the cost, where λ is the penalty weight, and $\gamma(\cdot)$ is the penalty function, defined by

$$\gamma(\cdot) := \operatorname*{ess\,sup}_{t\in[0,T]} \|\cdot(t)\|_1,$$

where $\|\cdot\|_1$ is the L_1 norm on $\mathbb{R}^n$. For example, $\|x(t)\|_1 = \Sigma_{i=1}^{n} |x(t)|$. Thus we have

$$\gamma(Ev) := \operatorname*{ess\,sup}_{t\in[0,T]} \|E(t)v(t)\|_1.$$

Now the penalized cost after linearization will be defined as

$$L(d,w) := J(x,u) + \lambda\gamma(Ev), \tag{11}$$

while the penalized cost before linearizing can be formulated in a similar fashion:

$$P(x,u) := J(x,u) + \lambda\gamma(\dot{x} - f), \tag{12}$$

where $J(x,u)$ is the original cost functional defined in (1a).

Trust Regions

Another concern is that the SCvx algorithm can potentially render the problem unbounded. A simple example will be linearizing the cost $y_1(x) = 0.5x^2$ at $x = 1$ to get $y_2(x) = x - 0.5$. Now if going left is a feasible direction, then the linearized problem could potentially be unbounded while the nonlinear problem will definitely find its minimum at $x = 0$. The reason is: when large deviation is allowed and occurs, the linear approximation sometimes fails to capture the distinction made by nonlinearity, for instance $y_1(x)$ attains its stationary point at $x = 0$, while $y_2(x)$ certainly does not.

To mitigate this risk, we ensure that the linearized trajectory does not deviate significantly from the nominal one obtained in the previous succession, via a *trust region* on our new control input,

$$\|w\|_\infty \leq \Delta, \tag{13}$$

and thus our new state will be restricted as well due to the dynamic equations. The rationale is that we only trust the linear approximation in the trust region. Figure 6 shows a typical convergence process of this trust-region type algorithm in solving a 2-D problem. The algorithm can start from virtually anywhere, and manages to converge to a feasible point. Note that the figure also demonstrates *virtual control*, as the trajectory deviates from the constraint in the first few successions.

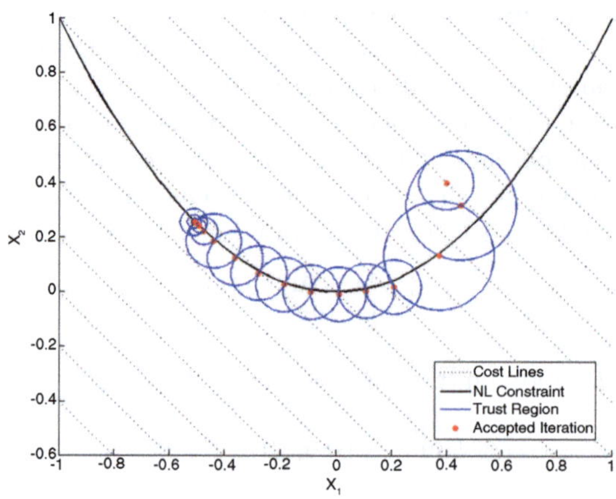

Fig. 6: Typical convergence process with *virtual control* and *trust regions*.

The SCvx Algorithm

The final problem formulation and the SCvx algorithm can now be presented. Considering the virtual control and the trust regions, a convex optimal control subproblem is solved at the kth succession:

Problem 4 (Convex Subproblem). Determine $w^* \in L_\infty[0,T]^m$, and $d^* \in W_{1,\infty}[0,T]^n$, which minimize $L(d,w)$ in (11), subject to constraints (10), (13), as well as

$$u^k(t) + w(t) \in U,$$
$$x^k(t) + d(t) \in X.$$

With this convex subproblem in hand, SCvx algorithm is given by Algorithm 1.

Algorithm 1: Successive Convexification (SCvx)

Input Select initial state x^1 and control u^1 s.t. $x^1 \in X$ and $u^1 \in U$. Initialize trust region radius with positive Δ^1 and lower bound Δ_l. Select positive penalty weight λ, and parameters $0 < \rho_0 < \rho_1 < \rho_2 < 1$ and $\alpha > 1$.

Step 1 At each succession k, solve Problem 4 at (x^k, u^k, Δ^k) to get an optimal solution (d^k, w^k).

Step 2 Compute the *actual* change in the penalized cost (12):

$$\Delta J(x^k, u^k) := \Delta J^k = J(x^k, u^k) - J(x^k + d^k, u^k + w^k),$$

and the *predicted* change by linear approximation:

$$\Delta L(d^k, w^k) := \Delta L^k = J(x^k, u^k) - L(d^k, w^k).$$

if $\Delta L^k = 0$ **then**
  Stop, and **return** (x^k, u^k);
else
  Compute the ratio $\rho^k = \Delta J^k / \Delta L^k$.
end if

Step 3
if $\rho^k < \rho_0$ **then**
  Reject this step, contract the trust region radius, i.e. $\Delta_k \leftarrow \Delta_k/\alpha$ and go back to **Step 1**;
else
  Accept this step, i.e. $x^{k+1} \leftarrow x^k + d^k$, $u^{k+1} \leftarrow u^k + w^k$, and update Δ^{k+1} by

$$\Delta^{k+1} = \begin{cases} \Delta^k/\alpha, & \text{if } \rho^k < \rho_1; \\ \Delta^k, & \text{if } \rho_1 \le \rho^k < \rho_2; \\ \alpha\Delta^k, & \text{if } \rho_2 \le \rho^k. \end{cases}$$

end if
$\Delta^{k+1} \leftarrow \max\{\Delta^{k+1}, \Delta_l\}$, $k \leftarrow k+1$, and go to **Step 1**.

This algorithm is of trust region type, and follows standard trust region radius update rules with some modifications. One important distinction lies in the subproblem to be solved at each succession. Since the subproblem (usually QP that may not be convex) is relatively expensive to solve, conventional trust region algorithms perform a line search along the Cauchy arc to achieve a "sufficient" cost reduction [16]. In the `SCvx` algorithms, however, a full convex optimization problem is solved to speed up the overall process. As a result, the number of successions can be significantly less by achieving more cost reduction at each succession. Thanks to the algorithm customization techniques [19, 20], we are able to solve each convex subproblem fast enough to outweigh the negative impact of solving to full optimality.

In Step 2, the ratio ρ^k is used as a metric for the quality of linear approximations. A desirable scenario is when ΔJ^k agrees with ΔL^k, i.e. ρ^k is close to 1. Hence $\rho^k \geq \rho_2$ means our linear approximation predicts the cost reduction well, then we may choose to enlarge the trust region in Step 3, i.e., we put more faith in our approximation. Otherwise, we may keep the trust region unchanged, or contract its radius if needed. The most unwanted situation is when ρ^k is negative, or close to zero. The current step will be rejected in this case, and one has to contract the trust region and re-optimize at (x^k, u^k).

Convergence Analysis

One can find the full details of the convergence analysis in [34], which made use of exact penalty function, separating hyperplane theorem, generalized derivatives and asymptotic analysis, etc. For simplicity, here we will only state the final result without proofs.

Theorem 3. *If the* `SCvx` *algorithm (Algorithm 1)generates an infinite sequence* $\{x^k\}$*, then* $\{x^k\}$ *has limit points, and if any limit point* $\bar{x}$ *is feasible for Problem 1, then it is a local optimum of Problem 1.*

2.2.2 Application

Throughout this chapter, we have alluded to a translational 3-degree-of-freedom (DoF) planetary landing problem. When aerodynamic forces are neglected (e.g., for Mars applications), the vehicle's equations of motion can be expressed as

$$\begin{aligned}\dot{m} &= -\alpha\Gamma(t),\\ \dot{r}(t) &= v(t),\\ \dot{v}(t) &= \frac{1}{m(t)}T_c(t) + g,\end{aligned}$$

where m, r, and v are the mass, position, and velocity of the vehicle, respectively. We use T_c to denote the commanded thrust, and Γ to denote the slack control variable used to upper bound the norm of T_c (as outlined in Problem 3). g denotes the local gravitational acceleration, which we assume is constant, and α is a positive constant that relates the commanded thrust magnitude to the rate of mass consumption.

Since the mass depletion rate is a function of the control variable, m becomes a state variable, and the dynamics are rendered nonlinear in the variables m, r, v, Γ, and T_c. Fortunately, using the variable substitution introduced in [2], we can resolve this issue by defining the following variables.

$$z(t) \triangleq \log m(t)\,, \qquad u(t) \triangleq \frac{T_c(t)}{m(t)}\,, \qquad \sigma(t) \triangleq \frac{\Gamma(t)}{m(t)}.$$

Thus, we are able to express the dynamics in linear form as

$$\begin{aligned}
\dot{z}(t) &= -\alpha\sigma(t),\\
\dot{r}(t) &= v(t),\\
\dot{v}(t) &= u(t) + g.
\end{aligned}$$

The dynamics are now linear, the problem can be solved via *lossless convexification*, and the final mass and thrust profiles can be obtained by applying the variable substitution backwards to solve for m and T_c.

Now, consider a more complicated problem: modeling the rocket/rover using 6-DoF equations of motion. The equations of motion are now given by

$$\begin{aligned}
\dot{m}(t) &= -\alpha\|T_B(t)\|_2,\\
\dot{r}_I(t) &= v_I(t),\\
\dot{v}_I(t) &= \frac{1}{m(t)}C\big(q_{B\leftarrow I}(t)\big)T_B(t) + g_I,\\
\dot{q}_{B\leftarrow I}(t) &= \frac{1}{2}\Omega(\omega_B(t))q_{B\leftarrow I}(t),\\
J\dot{\omega}_B &= [l\times]T_B(t) - [\omega_B(t)\times]J\omega_B(t),
\end{aligned}$$

where subscript I and B are used to denote the inertial and body frame, respectively. We use $q_{B\leftarrow I}$ to denote a unit quaternion rotating from inertial frame to body frame, and $C\big(q_{B\leftarrow I}(t)\big)$ to denote the corresponding direction cosine matrix. ω is the angular velocity, l is the constant position vector of the engine gimbal point, and J represents the moment of inertia.

Due to the introduction of quaternion kinematics and rigid-body dynamics, we are represented with a set of tightly coupled nonlinear equations. This problem is highly non-convex, and thus we can no longer apply the results of lossless convexification. Instead, we turn to *successive convexification* to handle the non-convexities. In addition to the artificial infeasibility and unboundedness problems discussed in Section 2.2.1, the successive convexification method (as well as other sequential optimization methods) often suffers from high-frequency oscillations in the control solution (e.g., see [32]). To circumvent this, the control variables can be augmented to the state, and their rate can be controlled instead. This has the effect of decoupling state and control constraints, resulting in smoother control profiles.

Below we present a landing trajectory in Figure 7, generated by a successive convexification algorithm for a 6-DoF rocket equipped with a single gimbaled engine. The parameters of this problem were chosen to accentuate the non-minimum phase behavior that appears when using a 6-DoF model. Note that such behavior is not observed when a 3-DoF model is used instead. In this example, the solution was obtained in seven iterations. The discrepancy between the dots and the solid curves in the early iterations is due to linearization errors and the use of virtual controls. For a more detailed treatment, the reader is referred to [51].

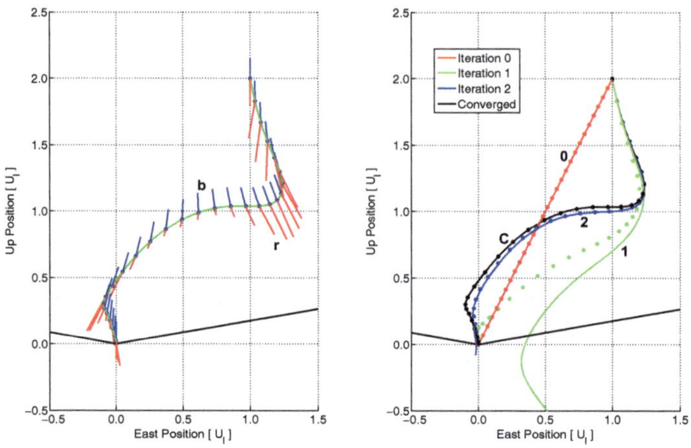

Fig. 7: On the left, the planar 6-DoF trajectory is shown. The solid points indicate time discretization points, the blue (b) and red (r) lines represent the vehicle body axis and thrust vector at each instance in time, and the solid green line represents the path of the vehicle. On the right, the red line (0) represents the trajectory used to initialize the process, the green (1) and the blue (2) represent the first two iterations, and the black (C) represents the converged solution (7th iteration). The dots represent the results of the parameter optimization problem, whereas the solid curves represent the trajectories propagated using the generated control input.

3 Real-Time Computation

So far, we have shown that convexification is a powerful tool for solving real-world, constrained optimal control problems. Now, we focus on methods for solving the resulting convex optimal control problems in real-time using onboard embedded computers. These methods have direct application to iterative techniques such as successive convexification.

The first step towards solving constrained optimal control problems in real-time is to discretize the continuous time system dynamics and constraints, which is typically accomplished by selecting a finite sequence of points in time and utilizing either a zero or first order hold control discretization (see [2] for more details on the discretization). The method used to discretize the continuous time problem should be chosen with care, as the selection will have a large impact on the real-world performance of the control system. The discretized optimal control problem is then converted into the canonical form of optimization problems in (14).

$$\text{minimize} \quad c^T x \quad \text{subject to}: \quad Ax = b, \quad x \in \mathscr{K}, \tag{14}$$

where $x \in \mathbb{R}^n$ is the solution variable, $c \in \mathbb{R}^n$ is a linear functional that maps the solution variable to a cost, $A \in \mathbb{R}^{p\times n}$ relates solution variables to constraint equations with $b \in \mathbb{R}^p$ on the right-hand side, and $\mathscr{K} \subset \mathbb{R}^n$ is the domain of the solution variable and is formed by the Cartesian product of convex cones. A straightforward, but tedious process produces the A matrix along with the b and c vectors, and there are freely available tools online that perform this task (e.g., CVX and YALMIP).

Once in canonical form, a variety of Interior Point Method solvers can be used to solve convex optimization problems to global optimality with polynomial time complexity and convergence guarantees (SDPT3, SeDuMi, ECOS, Bsocp, Gurobi, CPLEX). In the case of successive convexification, these same solvers are used to solve the convex sub-problems for each iteration.

While computationally efficient, convex solvers may not always perform at the level needed to solve optimal control problems onboard in real-time. Moreover, many applications require solving *similar* optimization problems over and over again, only varying some problem parameters (such as initial conditions and final conditions). Customized solvers for specific problem classes were developed for these reasons (see [19, 36]). We begin by noting that optimal control problems naturally translate into sparse optimization problems, that is, the A matrix is typically over 90% sparse (i.e., only 10% of the elements are non-zero). In order to classify problems according to their sparsity structure, we formally define a problem class.

Definition 2. Given $A_0 \in \mathbb{R}^{p\times n}$, $b_0 \in \mathbb{R}^p$, $c_0 \in \mathbb{R}^n$, and $\mathscr{K}_0 \subset \mathbb{R}^n$, a problem class, $\mathscr{P}$, is defined as:

$$\mathscr{P} = \{A \in \mathbb{R}^{p\times n}, b \in \mathbb{R}^p, c \in \mathbb{R}^n, \mathscr{K} \subset \mathbb{R}^n : \text{str}(A) \leq \text{str}(A_0), \text{str}(b) \leq \text{str}(b_0), \\ \text{str}(c) \leq \text{str}(c_0), \mathscr{K} = \mathscr{K}_0\},$$

where the $\leq$ operator denotes element-wise inequality, and str maps any non-zero element to a 1 and leaves 0 elements undisturbed, thereby forming the sparsity structure of its input. Thus, $(A,b,c,\mathscr{K}) \in \mathscr{P}$ if any zero element in $(A_0, b_0.c_0, \mathscr{K}_0)$ is also a zero element in $(A,b,c,\mathscr{K})$. Consequently, a problem class can be interpreted as an upper bound on the sparsity structure of (A,b,c). The framework for generating a customized solver for a specific problem class is shown in Figure 8.

Embedded systems typically operate in environments with stringent time requirements and limited memory. With this in mind, suppose that a given embedded system solves a set of problems $\mathscr{P}_0 \subseteq \mathscr{P}$. Then, the problem class and an upper bound on its size is known at compile time. Thus, the exact amount of memory that is necessary to solve $\mathscr{P}$ is statically allocated, removing the need for dynamic memory allocation altogether. This property is highly valued in the development of software for safety-critical systems because it eliminates memory leaks and more importantly, reduces the complexity of flight software verification. Furthermore, the customized solver is free of the logical operations introduced by sparse algorithms. This is accomplished by keeping track of non-zero element interactions and generating code that handles sparse operations directly. That is, once the interaction between two non-zero elements has been determined, one line of C code is generated to directly and correctly handle the interaction without any logic. This customized sparse framework supports IPM algorithms, and customized solvers include an IPM tailored to the problem class.

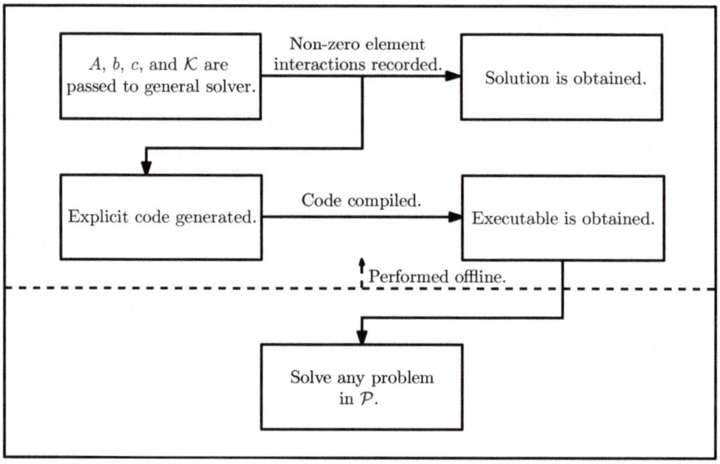

Fig. 8: Flow diagram for explicit code generation.

For comparison, the planetary landing, also known as powered descent guidance problem (see [2]) was solved using a variety of solvers, and their performance is presented in Figure 9a. ECOS and Bsocp (in-house solver) are among the fastest of the tested generic solvers, so Figure 9b focuses on the performance of ECOS, Bsocp, and customized Bsocp solvers. We note that 700 solution variables were sufficient for planning a divert trajectory onboard a rocket-powered vertical take-off and landing vehicle during real-world tests in 2012–2013 (see [48] and [20]). For problems of this size, customized solvers provide at least an order of magnitude improvement in computation times.

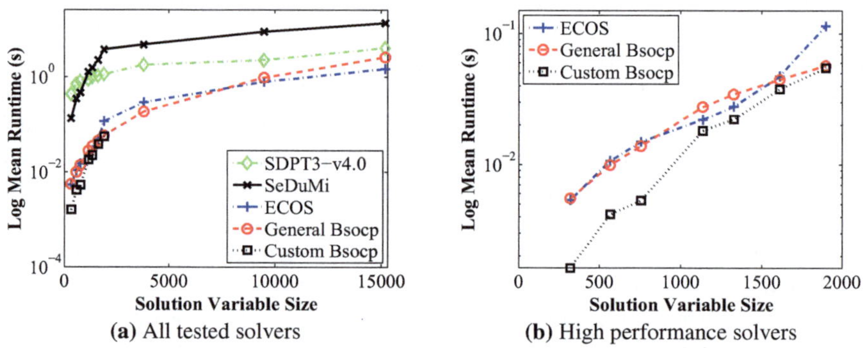

Fig. 9: Runtime Benchmarks for a broad range of solvers and problem sizes.

Recently, both convexification methods are also tested in real-world experiments for quadrotor motion planning problems including real-time obstacle avoidance. Timing statistics from the experiments are reported in [52], and they again demonstrate the real-time capability of these convexification technologies. Interested reader can find videos of these experiments in the YouTube channel [7].

4 Concluding Remarks

To summarize, this chapter gives a general picture of convexification methods for MPC and how they can be implemented in real-time computation. In Section 2, we introduce two convexification methods. Namely, the *lossless convexification* and the *successive convexification*. The former addresses a class of non-convex control constraints, while the latter aims at non-convexities raised by nonlinear dynamics and certain types of non-convex state constraints. They can be used independently or in combination, depending on the type of problems one tries to solve. In Section 3, we present methods for solving convex optimal control problems in real-time using onboard embedded computers. Moreover, we give the general idea behind the implementation of customized solvers, where sparsity in the problem structure is explored. Throughout this chapter, the planetary soft landing problem is used as an example to demonstrate how convexification works in real-world settings.

Though there are proposed solutions to nonlinear MPC (e.g. [17, 55]), so far most MPC methods focus on linear control systems, due to the lack of efficient and reliable solutions to online nonlinear programming. Therefore, convexification techniques could potentially have major impact on the progress of MPC research. They not only offer methods or algorithms that can be implemented in real-time, but also provide theoretical guarantees (i.e. proofs of equivalence or convergence), which are extremely valuable for aerospace applications. Thus, we may conclude this chapter by claiming that the convexification technologies, such as what we covered here, have the clear potential to make MPC research applicable to current control systems, as well as enabling new applications, especially in the emerging area of autonomous systems.

Acknowledgements We would like to thank David S. Bayard, John M. Carson, and Daniel P. Scharf of JPL, Lars Blackmore of SpaceX, John Hauser of University of Colorado, and Eric Feron of Georgia Institute of Technology for insightful discussions in this area. This research was supported in part by the Office of Naval Research Grant No. N00014-16-1-2318 and by the National Science Foundation Grants No. CMMI-1613235 and CNS-1619729.

References

1. Açıkmeşe, B., Blackmore, L.: Lossless convexification of a class of optimal control problems with non-convex control constraints. Automatica **47**(2), 341–347 (2011)
2. Açıkmeşe, B., Ploen, S.R.: Convex programming approach to powered descent guidance for Mars landing. AIAA J. Guid. Control Dyn. **30**(5), 1353–1366 (2007)
3. Açıkmese, B., Scharf, D.P., Murray, E.A., Hadaegh, F.Y.: A convex guidance algorithm for formation reconfiguration. In: Proceedings of the AIAA Guidance, Navigation, and Control Conference and Exhibit (2006)
4. Açikmese, B., Carson, J.M., Bayard, D.S.: A robust model predictive control algorithm for incrementally conic uncertain/nonlinear systems. Int. J. Robust Nonlinear Control **21**(5), 563–590 (2011)
5. Açıkmeşe, B., Carson, J., Blackmore, L.: Lossless convexification of non-convex control bound and pointing constraints of the soft landing optimal control problem. IEEE Trans. Control Syst. Technol. **21**(6), 2104–2113 (2013)
6. Augugliaro, F., Schoellig, A.P., D'Andrea, R.: Generation of collision-free trajectories for a quadrocopter fleet: a sequential convex programming approach. In: 2012 IEEE/RSJ International Conference on Intelligent Robots and Systems, pp. 1917–1922. IEEE, Piscataway (2012)
7. Autonomous Control Labortory (ACL): YouTube channel (2017). https://www.youtube.com/channel/UCZwV0cPCR3QeGn4dSfXxkKw
8. Azhmyakov, V., Raisch, J.: Convex control systems and convex optimal control problems with constraints. IEEE Trans. Autom. Control **53**(4), 993–998 (2008)
9. Berkovitz, L.D.: Optimal Control Theory. Springer, Berlin (1974)
10. Blackmore, L.: Autonomous precision landing of space rockets. Bridge Natl. Acad. Eng. **46**(4), 15–20 (2016)
11. Blackmore, L., Açıkmeşe, B., Carson, J.M.: Lossless convexfication of control constraints for a class of nonlinear optimal control problems. Syst. Control Lett. **61**(4), 863–871 (2012)
12. Boyd, S., Vandenberghe, L.: Convex Optimization. Cambridge University Press, Cambridge (2004)
13. Buskens, C., Maurer, H.: SQP-methods for solving optimal control problems with control and state constraints: adjoint variables, sensitivity analysis, and real-time control. J. Comput. Appl. Math. **120**, 85–108 (2000)
14. Canale, M., Fagiano, L., Milanese, M.: Set membership approximation theory for fast implementation of model predictive control laws. Automatica **45**(1), 45–54 (2009)
15. Chen, Y., Cutler, M., How, J.P.: Decoupled multiagent path planning via incremental sequential convex programming. In: 2015 IEEE International Conference on Robotics and Automation (ICRA), pp 5954–5961. IEEE, Piscataway (2015)
16. Conn, A.R., Gould, N.I., Toint, P.L.: Trust Region Methods, vol 1. SIAM, Philadelphia (2000)
17. Diehl, M., Bock, H.G., Schlöder, J.P., Findeisen, R., Nagy, Z., Allgöwer, F.: Real-time optimization and nonlinear model predictive control of processes governed by differential-algebraic equations. J. Process Control **12**(4), 577–585 (2002)
18. Domahidi, A., Chu, E., Boyd, S.: ECOS: an SOCP solver for embedded systems. In: Proceedings European Control Conference (2013)
19. Dueri, D., Zhang, J., Açikmese, B.: Automated custom code generation for embedded, real-time second order cone programming. In: 19th IFAC World Congress, pp. 1605–1612 (2014)
20. Dueri, D., Açıkmeşe, B., Scharf, D.P., Harris, M.W.: Customized real-time interior-point methods for onboard powered-descent guidance. J. Guid. Control Dyn. **40**, 197–212 (2017)
21. Dueri, D., Mao, Y., Mian, Z., Ding, J., Açıkmeşe, B.: Trajectory optimization with inter-sample obstacle avoidance via successive convexification. In: IEEE 56th Conference on Decision and Control (CDC) (2017)
22. Fletcher, R.: Practical Methods of Optimization: Vol. 2: Constrained Optimization. Wiley, New York (1981)

23. Franzè, G., Lucia, W.: The obstacle avoidance motion planning problem for autonomous vehicles: a low-demanding receding horizon control scheme. Syst. Control Lett. **77**, 1–10 (2015)
24. Frazzoli, E., Mao, Z.H., Oh, J.H., Feron, E.: Resolution of conflicts involving many aircraft via semidefinite programming. J. Guid. Control Dyn. **24**(1), 79–86 (2001)
25. Garcia, C., Morari, M.: Model predictive control: theory and practice — a survey. Automatica **25**(3), 335–348 (1989)
26. Gerdts, M.: A nonsmooth Newton's method for control-state constrained optimal control problems. Math. Comput. Simul. **79**, 925–936 (2008)
27. Griffith, R.E., Stewart, R.: A nonlinear programming technique for the optimization of continuous processing systems. Manag. Sci. **7**(4), 379–392 (1961)
28. Harris, M.W., Açıkmeşe, B.: Lossless convexification of non-convex optimal control problems for state constrained linear systems. Automatica **50**(9), 2304–2311 (2014)
29. Harris, M.W., Açıkmeşe, B.: Minimum time rendezvous of multiple spacecraft using differential drag. J. Guid. Control Dyn. **37**, 365–373 (2014)
30. Hull D (1997) Conversion of optimal control problems into parameter optimization problems. J. Guid. Control Dyn. **20**(1), 57–60
31. Liu, X., Lu, P.: Solving nonconvex optimal control problems by convex optimization. J. Guid. Control Dyn. **37**(3), 750–765 (2014)
32. Liu, X., Shen, Z., Lu, P.: Entry trajectory optimization by second-order cone programming. J. Guid. Control Dyn. **39**(2), 227–241 (2015)
33. Machielsen, K.C.P.: Numerical solution of optimal control problems with state constraints by sequential quadratic programming in function space. Technische Universiteit Eindhoven (1987)
34. Mao, Y., Szmuk, M., Açıkmeşe, B.: Successive convexification of non-convex optimal control problems and its convergence properties. In: 2016 IEEE 55th Conference on Decision and Control (CDC), pp. 3636–3641 (2016)
35. Mao, Y., Dueri, D., Szmuk, M., Açıkmeşe, B.: Successive convexification of non-convex optimal control problems with state constraints. IFAC-PapersOnLine **50**(1), 4063–4069 (2017)
36. Mattingley, J., Boyd, S.: Automatic code generation for real-time convex optimization. In: Eldar, Y., Palomar, D. (eds.) Convex Optimization in Signal Processing and Communications. Cambridge University Press, Cambridge (2010)
37. Mattingley, J., Boyd, S.: Cvxgen: a code generator for embedded convex optimization. Optim. Eng. **13**(1), 1–27 (2012)
38. Mayne, D.Q.: Model predictive control: recent developments and future promise. Automatica **50**(12), 2967–2986 (2014)
39. Mayne, D.Q., Polak, E.: An exact penalty function algorithm for control problems with state and control constraints. IEEE Trans. Autom. Control **32**(5), 380–387 (1987)
40. Mayne, D., Rawlings, J., Rao, C., Scokaert, P.: Constrained model predictive control: stability and optimality. Automatica **36**(6), 789–814 (2000)
41. Nesterov, Y., Nemirovskii, A.: Interior-Point Polynomial Algorithms in Convex Programming. Society for Industrial and Applied Mathematics, Philadelphia (1994)
42. Nocedal, J., Wright, S.J.: Numerical Optimization. Springer, Berlin (2006)
43. Palacios-Gomez, F., Lasdon, L., Engquist, M.: Nonlinear optimization by successive linear programming. Manag. Sci. **28**(10), 1106–1120 (1982)
44. Polak, E.: Optimization: Algorithms and Consistent Approximations, vol 124. Springer, Berlin (2012)
45. Pontryagin, L.S.: Mathematical Theory of Optimal Processes. CRC Press, Boca Raton (1987)
46. Richards, A., How, J.P.: Robust variable horizon model predictive control for vehicle maneuvering. Int. J. Robust Nonlinear Control **16**(7), 333–351 (2006)
47. Rosen, J.B.: Iterative solution of nonlinear optimal control problems. SIAM J. Control **4**(1), 223–244 (1966)
48. Scharf, D.P., Açıkmeşe, B., Dueri, D., Benito, J., Casoliva, J.: Implementation and experimental demonstration of onboard powered-descent guidance. J. Guid. Control Dyn. pp. 213–229 (2016)

49. Schulman, J., Duan, Y., Ho, J., Lee, A., Awwal, I., Bradlow, H., Pan, J., Patil, S., Goldberg, K., Abbeel, P.: Motion planning with sequential convex optimization and convex collision checking. Int. J. Robot. Res. **33**(9), 1251–1270 (2014)
50. Szmuk, M., Açıkmeşe, B., Berning, A.W.: Successive convexification for fuel-optimal powered landing with aerodynamic drag and non-convex constraints. In: AIAA Guidance, Navigation, and Control Conference, p 0378 (2016)
51. Szmuk, M., Eren, U., Açıkmeşe, B.: Successive convexification for mars 6-dof powered descent landing guidance. In: AIAA Guidance, Navigation, and Control Conference, p. 1500 (2017)
52. Szmuk, M., Pascucci, C.A., Dueri, D., Açıkmeşe, B.: Convexification and real-time on- board optimization for agile quad-rotor maneuvering and obstacle avoidance. In: IEEE/RSJ International Conference on Intelligent Robots and Systems (IROS) (2017)
53. Wang, Y., Boyd, S.: Fast model predictive control using online optimization. IEEE Trans. Control Syst. Technol. **18**(2), 267–278 (2010)
54. Wang, Z., Grant, M.J.: Constrained trajectory optimization for planetary entry via sequential convex programming. J. Guid. Control Dyn. **40**(10), 2603–2615 (2017)
55. Zavala, V.M., Biegler, L.T.: The advanced-step nmpc controller: optimality, stability and robustness. Automatica **45**(1), 86–93 (2009)
56. Zeilinger, M.N., Raimondo, D.M., Domahidi, A., Morari, M., Jones, C.N.: On real-time robust model predictive control. Automatica **50**(3), 683–694 (2014)
57. Zhang, J., Kim, N.H., Lasdon, L.: An improved successive linear programming algorithm. Manag. Sci. **31**(10), 1312–1331 (1985)

Explicit (Offline) Optimization for MPC

Nikolaos A. Diangelakis, Richard Oberdieck, and Efstratios N. Pistikopoulos

1 Introduction

MPC has become the accepted standard for complex constrained multivariable control problems in the process industries [15]. Starting from the current state, an open-loop optimal control problem is solved over a finite horizon. The computation is repeated at the next time step, considering the new state and over a shifted horizon. This moving horizon policy relies on a discrete-time linear/non-linear dynamic model, respects all input and output constraints, and optimizes a performance index. In this chapter, we discuss (i) the reformulation of the linear MPC problem with quadratic performance index into a (mixed integer) quadratic programming problem, free of equality constraints, via successive substitution of state and output variables, over the finite horizon, (ii) the explicit/multi-parametric solution of the (mixed integer) quadratic programming problem as a function of parameters and (iii) the theoretical properties of the solution.

1.1 From State-Space Models to Multi-Parametric Programming

As an example, consider the problem of regulating to the origin the discrete-time, linear, time invariant system:

N. A. Diangelakis · E. N. Pistikopoulos (✉)
Texas A&M University and Texas A&M Energy Institute, College Station, TX, USA
e-mail: nikos@tamu.edu; stratos@tamu.edu

R. Oberdieck
Texas A&M University and Texas A&M Energy Institute, College Station, TX, USA

Ørsted A/S, Kraftværksvej 53, Skærbæk, 7000 Fredericia, Denmark
e-mail: ricob@orsted.dk

S. V. Raković, W. S. Levine (eds.), *Handbook of Model Predictive Control*,
Control Engineering, https://doi.org/10.1007/978-3-319-77489-3_16

$$\begin{cases} x_{k+1} = Ax_k + Bu_k \\ y_k = Cx_k, \end{cases} \tag{1}$$

under the following constraints:

$$x_{\min} \leq x_k \leq x_{\max},\ y_{\min} \leq y_k \leq y_{\max},\ u_{\min} \leq u_k \leq u_{\max}. \tag{2}$$

The MPC problem corresponding to regulating a discrete-time, linear, time invariant system to its origin is called a linear quadratic regulator and is presented in Equation (3) for the case of a quadratic performance index.

$$\begin{aligned} \underset{u}{\text{minimize}} \quad & x_N^T P x_N + \sum_{k=1}^{N-1} x_k^T Q x_k + \sum_{k=0}^{M-1} u_k^T R u_k \\ \text{subject to} \quad & \begin{cases} x_{k+1} = Ax_k + Bu_k \\ y_k = Cx_k \end{cases} \\ & x_{\min} \leq x_k \leq x_{\max},\ \forall k \in [0, N] \\ & y_{\min} \leq y_k \leq y_{\max},\ \forall k \in [0, N] \\ & u_{\min} \leq u_k \leq u_{\max},\ \forall k \in [0, M-1] \\ & u_k = u_M,\ \forall k \in [M+1, N-1] \end{aligned} \tag{3}$$

where N is the prediction horizon, $M \leq N$ the control horizon, $Q \succeq 0$ the weight for the states, $R \succeq 0$ the weight for the inputs, P the final state weight calculated as discussed in [15], $x_{\min}$ and $x_{\max}$ the lower and upper bound on x_k, respectively, $u_{\min}$ and $u_{\max}$ the lower and upper bound on u_k, respectively and $y_{\min}$ and $y_{\max}$ the lower and upper bound on y_k, respectively.

Remark 1. It is equally possible to use a linear performance index such as the $1/\infty$-norm.

Based on Equation (1) every state vector for which $k \geq 1$ can be reformulated as follows:

$$\begin{aligned} x_k = & A^k x_0 + \sum_{k \in [1,M],\ i=0}^{k-1} A^{k-i-1} B u_i + \\ & + \sum_{k \in [M+1,N],\ j=M}^{k} A^{N-j-1} B u_{M-1},\ \forall k \in [1, N]. \end{aligned} \tag{4}$$

Equation (4) is linear in x_0 and u_i, $\forall i \in [0, M-1]$. A simple example of its application is given in Table 1 for $N = 6$, $M = 4$.

Table 1: State substitution based on Equation (1)

Prediction horizon: 6, Control horizon: 4	
$k=1$	$x_1 = Ax_0 + Bu_0$
$k=2$	$x_2 = A^2x_0 + ABu_0 + Bu_1$
$k=3$	$x_3 = A^3x_0 + A^2Bu_0 + ABu_1 + Bu_2$
$k=4$	$x_4 = A^4x_0 + A^3Bu_0 + A^2Bu_1 + ABu_2 + Bu_3$
$k=5$	$x_5 = A^5x_0 + A^4Bu_0 + A^3Bu_1 + A^2Bu_2 + (AB+B)\,u_3$
$k=6$	$x_6 = A^6x_0 + A^5Bu_0 + A^4Bu_1 + A^3Bu_2 + \left(A^2B + AB + B\right)u_3$

The states of the system can therefore be expressed as a linear function of the form of Equation (5).

$$x_k = A_r x_0 + \sum_{i=0}^{M} B_{r,i} u_i, \ \forall i \in [0, M-1], \tag{5}$$

where A_r and $B_{r,i}$ are fixed matrices corresponding to the linear reformulation based on the original discrete-time, linear, time invariant system. Note that the states of the system are expressed as a linear function of the initial state values x_0 and the control variables u_i, $\forall i \in [0, M-1]$.

Equivalently, the outputs of the system can be expressed as a linear function by substitution (Equation 6):

$$y_k = C\left[A_r x_0 + \sum_{i=0}^{M} B_{r,i} u_i\right], \ \forall i \in [0, M-1]. \tag{6}$$

By substituting Equations (5) and (6) into the original MPC formulation of Equation (3) we get linear inequality constraints based on the upper and lower bounds and three quadratic terms corresponding to (i) the quadratic term of the control variables, (ii) the bilinear term between the control variables and the initial states, and (iii) the quadratic term of the initial states. In order for this to be more comprehensive, we consider a simple example based on Equations (3), (4), and (6) for a prediction horizon of 2 and a control horizon of 1.

After rearranging the terms, the resulting quadratic programming problem is presented in Equation (7):

$$
\begin{aligned}
\underset{u_0}{\text{minimize}} \quad & u_0^T \underbrace{\left[[AB+B]^T P [AB+B] + B^T Q B + R \right]}_{H} u_0 + \\
& + u_0^T \underbrace{\left[[AB+B]^T \left[P + P^T \right] AA + B^T \left[Q + Q^T \right] A \right]}_{Z} x_0 + \\
& + x_0^T \underbrace{\left[[AA]^T PAA + A^T QA \right]}_{\hat{M}} x_0 \\
\text{subject to} \quad & \underbrace{\begin{bmatrix} B \\ -B \\ AB+B \\ -AB-B \\ CB \\ -CB \\ \mathbf{I} \\ -\mathbf{I} \end{bmatrix}}_{G} u_0 \leq \underbrace{\begin{bmatrix} x_{\max} \\ -x_{\min} \\ x_{\max} \\ -x_{\min} \\ y_{\max} \\ -y_{\min} \\ u_{\max} \\ -u_{\min} \end{bmatrix}}_{W} + \underbrace{\begin{bmatrix} -A \\ A \\ -AA \\ AA \\ -CA \\ CA \\ \mathbf{0} \\ \mathbf{0} \end{bmatrix}}_{S} x_0 \\
& \underbrace{\begin{bmatrix} \mathbf{I} \\ -\mathbf{I} \\ C \\ -C \end{bmatrix}}_{CR_A} x_0 \leq \underbrace{\begin{bmatrix} x_{\max} \\ -x_{\min} \\ y_{\max} \\ -y_{\min} \end{bmatrix}}_{CR_b}
\end{aligned} \tag{7}
$$

where H is the quadratic term for the control variables, Z the bilinear term between the control variables and initial states, $\hat{M}$ the quadratic term for the initial states, $\mathbf{I}$ the identity matrix and $\mathbf{0}$ a zero matrix, all of appropriate dimensions. Note that the term $x_0^T \hat{M} x_0$ does not affect the outcome of the optimization problem as it is a positive number that cannot be affected via the optimization procedure, therefore it will hereon be omitted.

Remark 2. In the cases where the performance index of the MPC problem (Equation 3) is linear via a $1/\infty$-norm formulation, it can be equivalently shown that:

- The objective function of the optimization problem remains linear. The two terms that result from the reformulation correspond to (i) the linear term of the control variables and (ii) the linear term of the initial states.
- The reformulation of the constraints of the problem remains the same. A set of constraints may be added to preserve the properties of the infinite norm.
- The corresponding optimization problem becomes a linear programming problem under linear constraints.

Hereon for simplicity, the real control variables will be presented in the vector form $u = [u_0^T, u_1^T, \ldots, u_{M-1}^T]^T$ and x will denote the state values at $T = 0$. For completion, a linear term $c^T u$ will be taken into account in the objective function (i.e., Equation 8)

$$\begin{aligned} &\underset{u}{\text{minimize}}\ u^T H u + u^T Z x + c^T u \\ &\text{subject to}\ Gu \leq W + Sx \\ &\qquad\qquad CR_A x \leq CR_b. \end{aligned} \tag{8}$$

Problem 8 corresponds to a convex [39] multi-parametric quadratic programming (mp-QP) problem where the parameters are the initial states x and the control variables u are the optimization variables.

1.2 When Discrete Elements Occur

In many applications, there are situations where the state-space system in Equation (1) is not sufficient to describe the dynamics of the system to a sufficient degree of accuracy. Reasons for this may be the presence of discrete actuators or states, decision variables or nonlinearities in the system response. One way to deal with these challenges is the introduction of binary variables, i.e. variables which only adhere to the values 0 and 1. Based on the introduction of these variables, it is possible to (i) model any bounded integer variable, (ii) model switches and binary decisions as well as (iii) partition a nonlinear system into smaller regions, where linearized versions of the system response can be combined together (piecewise linearizations).

Reviewing the literature and advancements of modelling with binary variables goes beyond the scope of this chapter. For further reference, the interested reader is directed to the following references which provide some key results in this area: [13, 35, 61].

One of the key advantages of using binary variables to approximate nonlinearities is thereby that the linear nature of the constraints can be preserved. Thus, the resulting MPC problems are mixed-integer linear or quadratic programming (MILP and MIQP, respectively) problems, depending on the performance index chosen. Thus, consequently to the continuous case, if the dependence on the initial state is considered explicitly, then this results in a multi-parametric MILP or MIQP (mp-MILP and mp-MIQP, respectively).

2 Multi-Parametric Linear and Quadratic Programming: An Overview

Consider the following mp-QP problem:

$$\begin{aligned} z(x) = &\underset{u}{\text{minimize}} \quad (Hu + Zx + c)^T u \\ &\text{subject to}\ Gu \leq W + Sx \\ &\qquad\qquad u \in \mathbb{R}^n \\ &\qquad\qquad x \in X := \{x \in \mathbb{R}^q \,|CR_A x \leq CR_b\} \end{aligned} \tag{9}$$

with $H \in \mathbb{R}^{n\times n} \succ 0$, $Z \in \mathbb{R}^{n\times q}$, $c \in \mathbb{R}^n$, $G \in \mathbb{R}^{m\times n}$, $W \in \mathbb{R}^m$, $S \in \mathbb{R}^{m\times q}$, $CR_A \in \mathbb{R}^{r\times q}$, $CR_b \in \mathbb{R}^r$ and X is compact.

Remark 3. The properties discussed below are also valid for mp-LP problems of the form:

$$\begin{aligned} z(x) = \underset{u}{\text{minimize}} \quad & c^T u \\ \text{subject to } & Gu \leq W + Sx \\ & u \in \mathbb{R}^n \\ & x \in X := \{x \in \mathbb{R}^q \,|CR_A x \leq CR_b\} \end{aligned} \tag{10}$$

Note however that due to the positive semi-definite nature of problem (10),[1] this might lead to dual degeneracy, as discussed in Section 2.2.

Remark 4. In order to facilitate readability, throughout this chapter equality constraints will be omitted in the problem formulations of multi-parametric programming problems as they can be understood as inequality constraints which have to be active in the entire parameter space (i.e., they are always part of the active set).

As sets defined by halfspaces are closely related to multi-parametric programming, we define the notion of a polytope as follows:

Definition 1. The set $\mathscr{P}$ is called an n-dimensional polytope if and only if it satisfies:

$$\mathscr{P} := \left\{x \in \mathbb{R}^n \,|a_i^T x \leq b_i, i = 1, \ldots, m\right\}, \tag{11}$$

where m is finite.

2.1 Theoretical Properties

The key question when considering problem (9) is how to obtain the parametric solution $u(x)$ and $\lambda(x)$, where λ denote the Lagrangian multipliers.[2] In the open literature, two ways have been presented:

Post-optimal sensitivity analysis: Consider problem (9), let $f(u,x)$ and $g_i(u,x) \leq 0$ denote the objective function and the i-th constraint, respectively and let x be fixed to x_0, such that the inequalities $CR_A x_0 \leq CR_b$ are satisfied. Then the resulting quadratic programming (QP) problem can be solved using the Karush-Kuhn-Tucker (KKT) conditions, which are given by:

[1] Problem (10) can be viewed as a special case of problem (9) with $Q = 0_{n\times n}$ and $H = 0_{n\times q}$, which is inherently positive semi-definite.

[2] For an introduction into the concept of Lagrangian multipliers and duality in general, the reader is referred to the excellent textbook by C. A. Floudas [28].

$$\nabla_u \mathscr{L} = \nabla_u f(u, x_0) + \sum_{i=1}^{m} \lambda_i \nabla_u g_i(u, x_0) = 0 \tag{12a}$$

$$g_i(u, x_0) \leq 0, \lambda_i \geq 0, \forall i = 1, \ldots, m \tag{12b}$$

$$\lambda_i g_i(u, x_0) = 0, \forall i = 1, \ldots, m, \tag{12c}$$

where the optimal solution is given by the optimizer u_0 and the Lagragian multipliers $\lambda_0 = [\lambda_1, \lambda_2, \ldots, \lambda_m]^T$. This consideration leads to the main theorem on post-optimal sensitivity analysis:

Theorem 1 (Basic Sensitivity Theorem [27]). Let x_0 be a vector of parameter values and (u_0, λ_0) the solution derived from the KKT conditions in Equation (12), where λ_0 is non-negative and u_0 is feasible. Also assume that: (i) strict complementary slackness (SCS) holds; (ii) the binding constraint gradients are linearly independent (LICQ: Linear Independence Constraint Qualification); and (iii) the second-order sufficiency conditions (SOSC) hold [11, 28, 45, 46]. Then, in the neighborhood of x_0, there exists a unique, once differentiable function $[u(x), \lambda(x)]$ satisfying Equation (12) with $[u(x_0), \lambda(x_0)] = (u_0, \lambda_0)$, where $u(x)$ is a unique isolated minimizer for problem (9) and

$$\left(\frac{du(x_0)}{dx}, \frac{d\lambda(x_0)}{dx}\right) = -\left(M_0^{-1}\right) N_0, \tag{13}$$

where

$$M_0 = \begin{pmatrix} \nabla_{uu}^2 \mathscr{L} & \nabla_u g_1 & \cdots & \nabla_u g_m \\ -\lambda_1 \nabla_u^T g_1 & -g_1 & & \\ \vdots & & \ddots & \\ -\lambda_m \nabla_u^T g_m & & & -g_m \end{pmatrix} \tag{14a}$$

$$N_0 = \left(\nabla_{x,u}^2 \mathscr{L}, -\lambda_1 \nabla_x^T g_1, \ldots, -\lambda_m \nabla_x^T g_m\right)^T \tag{14b}$$

$$\mathscr{L} = f(u, x) + \sum_{i=1}^{m} \lambda_i g_i(u, x). \tag{14c}$$

As a result of Theorem 1 the parametric solutions $u(x)$ and $\lambda(x)$ are affine functions of x around x_0.

Parametric solution of the KKT conditions: Consider problem (9) and Equation (12) without fixing x to x_0. Additionally, let k be a candidate active set, then the corresponding KKT conditions are given as[3]:

$$\begin{aligned} \nabla_u \mathscr{L}(u, \lambda, x) &= \nabla_u \left((Hu + Zx + c)^T u\right) + \nabla_u \left(\sum_{i \in k} \lambda_i (G_i u - W_i - S_i x)\right) \\ &= Hu + Zx + c + G_k^T \lambda_k = 0 \end{aligned} \tag{15a}$$

[3] Assuming no degeneracy, in the case of mp-LP problems, the cardinality of the active set k is $\text{card}(k) = n$ and thus the parametric solution is directly given as $u(\theta) = G_k^{-1}(W_k + S_k x)$.

$$G_k u - W_k - S_k x = 0. \tag{15b}$$

Thus, Equation (15a) is reformulated such that

$$u = -H^{-1}\left(Zx + c + G_k^T \lambda_k\right). \tag{16}$$

Note that H is invertible since it is positive definite. The substitution of Equation (16) into Equation (15b) results in:

$$\begin{aligned} -G_k H^{-1}\left(Zx + c + G_k^T \lambda_k\right) - W_k - S_k x = 0 \\ \Rightarrow \lambda_k(x) = -\left(G_k H^{-1} G_k^T\right)^{-1}\left(W_k + S_k x + G_k H^{-1}(Zx + c)\right), \end{aligned} \tag{17}$$

which can be substituted into Equation (16) to obtain the full parametric solution.

Once the parametric solution has been obtained, the set over which it is valid is defined by feasibility and optimality requirements:

$$Gu(x) \leq W + Sx \qquad \text{(Feasibility of } u(x)\text{)} \tag{18a}$$

$$\lambda(x) \geq 0 \qquad \text{(Optimality of } u(x)\text{)} \tag{18b}$$

$$CR_A x \leq CR_b \qquad \text{(Feasibility of } x\text{)} \tag{18c}$$

For mp-LP and mp-QP problems, Equation (18) denotes a set of linear inequalities, and thus the critical region where a parametric solution is optimal is a polytope. Since this analysis is valid for any feasible point x_0, the main properties of mp-LP and mp-QP solutions are given as follows:

Definition 2. A function $u(x) : X \to \mathbb{R}^n$, where $X \in \mathbb{R}^q$ is a polytope, is called piecewise affine if it is possible to partition X into non-overlapping polytopes, called critical regions, CR_i and

$$u(x) = K^i x + r^i, \; \forall x \in CR_i. \tag{19}$$

Remark 5. The definition of piecewise quadratic is analogous.

Theorem 2 (Properties of mp-QP solution [15, 24])**.** Consider the mp-QP problem (9). Then the set of feasible parameters $X_f \subseteq X$ is polytopic, the optimizer $u(x) : X_f \mapsto \mathbb{R}^n$ is continuous and piecewise affine, and the optimal objective function $z(x) : X_f \mapsto \mathbb{R}$ is continuous, and piecewise quadratic.

Remark 6. The case of mp-LP problems is more complex due to the positive semidefiniteness of the objective function. This may lead to issues with degeneracies (see Section 2.2) which require the use of special algorithms for their consideration. However, note that Theorem 2 still holds and an optimal objective function $z(x) : X_f \mapsto \mathbb{R}$ can be found which is continuous, convex, and piecewise affine [31].

*Remark 7 (*Active set representation*).* Each critical region in an mp-LP or mp-QP problem is uniquely defined by the optimal active set associated with it, and the solution of problem (9) can be represented as the set of all optimal active sets.

2.1.1 Literature Review

Parametric[4] programming was first considered in 1952: for the linear case, Orchard-Hays considered in the unpublished M.S. thesis the case of switching the optimal basis as a function of a varying parameter [30], while Markowitz considered the quadratic case conceptually in his famous "Portfolio selection" paper [48]. Since then, many researchers have discovered new properties of parametric and multi-parametric programming problems: in 1972, Gal and Nedoma showed that the solution to an mp-LP problem is continuous, its optimal objective function is convex and the optimal active sets form a connected-graph [32]. In 2002 Dua *et al.* proved that the solution to an mp-QP problem is continuous and its optimal objective function is conditionally convex[5] [24], while Tøndel *et al.* showed a year later that under certain conditions the Lagrange multipliers are continuous and it is possible to infer the optimal active sets of adjacent critical regions [73]. Lastly, in 2017 it was shown that the optimal active set of an mp-QP problem is also given by a connected graph[6] [55].

2.2 Degeneracy

One of the most important issues encountered in linear and quadratic programming is degeneracy. However, since the solution to a strictly convex QP is guaranteed to be unique, some types of degeneracy do not occur in QP and consequentially in mp-QP problems. Thus, for completion consider a standard mp-LP problem, where degeneracy generally refers to the situation where the active set for a specific LP problem (e.g., problem (10) with $x = 0$) cannot be identified uniquely.[7] Commonly, the two types of degeneracy encountered are primal and dual degeneracy (see Figure 1):

Primal degeneracy: In this case, the vertex of the optimal solution of the LP is overdefined, i.e. there exist multiple sets $k_1 \neq k_2 \neq \ldots \neq k_{tot}$ such that:

$$u_{k_1} = u_{k_2} = \ldots = u_{k_{tot}}, \tag{20}$$

where $u_k = G_k^{-1} W_k$.

---

[4] In general, the term "parametric" refers to the case where a single parameter is considered, while "multi-parametric" suggests the presence of multiple parameters.

[5] It is convex, if the following z-transformation is applied $u = z - \frac{1}{2} H^{-1} Z x$, based on the nomenclature of problem (9).

[6] In an excellent technical note from 2002, Baotić actually already commented on the connected graph nature of the problem, however without providing a formal proof or further discussion on the topic [7].

[7] This does not consider problems arising from scaling, round-off computational errors or the presence of identical constraints in the problem formulation.

By inspection of Figure 1a, it is clear that primal degeneracy is caused by the presence of constraints which only coincide with the feasible space, but do not intersect it. Thus, if any of these constraints would be chosen to be part of the active set of the corresponding parametric solution, this results in a lower-dimensional critical region,[8] and only one active set k exists for which a full-dimensional critical region results, and it is constituted by those constraints which intersect with the feasible space.

Remark 8. Constraints which coincide but do not intersect with the feasible space are also referred to as weakly redundant constraints.

Dual degeneracy: If there exists more than one point u having the same optimal objective function value z, then the optimal solution is not unique. Thus, there exist multiple sets $k_1 \neq k_2 \neq \ldots \neq k_{tot}$ with $u_{k_1} \neq u_{k_2} \neq \ldots \neq u_{k_{tot}}$ such that:

$$z_{k_1} = z_{k_2} = \ldots = z_{k_{tot}}, \tag{21}$$

where $z_k = c^T u_k$.

In general, the effect of primal degeneracy within the solution procedure of mp-LP problems is manageable, since it can be detected by substituting u_k into the constraints and if necessary solving one LP problem for each constraint. However, dual degeneracy is more challenging as the different active sets might result in full-dimensional, but potentially overlapping, critical regions. In particular since the optimal solutions u_k differ, the presence of dual degeneracy might eliminate the continuous nature of the optimizer described in Theorem 2. However, three approaches have been proposed to generate continuous optimizers as well as non-overlapping critical regions [44, 56]. The most promising one is thereby the application of lexicographic perturbation techniques, which is based on the idea that the problem of dual-degeneracy only arises because of the specific numerical structure of the objective function and the constraints [44]. In order to overcome the degeneracy, the right-hand side of the constraints and the objective function are symbolically perturbed in order to obtain a single, continuous optimizer for the solution of the mp-LP problem. Note that the problem is not actually perturbed, but only the result of a proposed perturbation is analyzed and enables the formulation of a continuous optimizer.

2.2.1 Literature Review

Degeneracy is a crucial topic for mp-LP and mp-QP problems, as they result in the problem to be not well-behaved anymore, which may lead to overlapping regions,

[8] Consider Figure 1a: if the constraint which only coincides at the single point with the feasible space is chosen as part of the active set, the corresponding parametric solution will only be valid in that point.

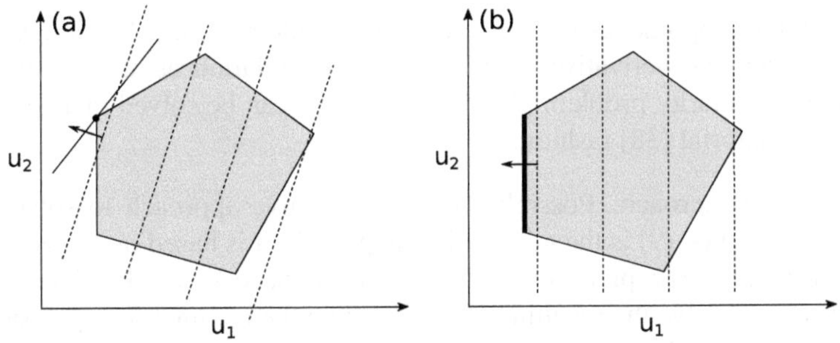

Fig. 1: Primal and dual degeneracy in linear programming. In (a), primal degeneracy occurs since there are three constraints which are active at the solution, while in (b) dual degeneracy occurs since there is more than one point (u_1, u_2) which features the optimal objective function value.

unexplored regions, and failures of the solution algorithms. Thus, it has been discussed by several researchers in depth: in 1972, Gal and Nedoma showed that for mp-LP problems, dual degeneracy leads to a disconnected graph and more than one graph needs to be obtained from the algorithm [32].[9] Other theoretical discussions by Rockafellar and Klatte and co-workers considered the conditions under which a continuous optimizer would be obtained, including minimum norm selection and Steiner point selection [62, 63]. However, despite these theoretical developments dual degeneracy was not taken into account in the first mp-LP algorithms [14, 17]. These were only considered later, when Spjøtvold *et al.* suggested to reformulate mp-LP problems locally into mp-QP problems to obtain a non-overlapping and complete solution. [66]. However, the most comprehensive and important treatment of degeneracy was published in 2007, when Jones *et al.* discussed the use of lexicographic perturbation, which enables the direct solution of the mp-LP problem in the face of dual degeneracy by observing the effect of a small perturbation on the problem [44].

2.3 Solution Algorithms for mp-LP and mp-QP Problems

Based on Theorem 2 and Remark 7, it is possible to consider the solution to problem (9) either as a set of non-overlapping polytopes which cover the feasible parameter space X_f or as a set of optimal active sets, which generate the critical regions based on the parametric solution $u(x), \lambda(x)$. This has given rise to three distinct types of solution approaches: a geometrical approach, a combinatorial approach, and a connected-graph approach for mp-LP problems.

[9] A similar approach was presented in 2006 by Olaru and Dumur [56].

Remark 9. Other approaches for the solution of problem (9) involve vertex enumeration [51], graphical derivatives [59], or the reformulation as a multi-parametric linear complementarity problem [20, 42, 47], which can be solved in a geometrical [37] or combinatorial [38] fashion.

The geometrical approach: Possibly the most intuitive approach to solve mp-QP problems of type (9) is the geometrical approach. It is based on the geometrical consideration and exploration of the parameter space X. The key idea is to fix a point $x_0 \in X$, solve the resulting QP, and obtain the parametric expressions $u(x)$ and $\lambda(x)$ alongside the corresponding critical region CR. Then, a new, feasible point $x_1 \notin CR$ is fixed and the same procedure is repeated until the entire parameter space has been explored. The different contributions differ in the way the parameter space is explored: in [15, 24], the constraints of the critical region are reversed, yielding a set of new polytopes which are considered separately. As this introduces a large number of artificial cuts [73], the step-sized approach has gained importance, as it calculates a point on the facet of each critical region and steps away from it orthogonally (see Figure 2) [7, 12].

However the geometrical approach presented in [7, 12] is only guaranteed to provide the full parametric map if the so-called facet-to-facet property is fulfilled [68]:

Definition 3 (Facet-to-facet property)**.** Let CR_1 and CR_2 be two full-dimensional disjoint critical regions. Then the facet-to-facet property is said to hold if $F = CR_1 \cap CR_2$ is a facet of both CR_1 and CR_2.

Additionally, researchers have proposed techniques to infer the active set of the adjacent critical region:

Theorem 3 (Active set of adjacent region [73])**.** Consider the active set of a full-dimensional critical region CR_0 in minimal representation, $k = \{i_1, i_2, \ldots, i_k\}$. Additionally, let CR_i be a full-dimensional neighboring critical region to CR_0 and assume that the linear independent constraint qualification holds on their common facet $F = CR_0 \cap H$, where H is the separating hyperplane. Moreover, assume that there are no constraints which are weakly active at the optimizer $u(x)$ for all $x \in CR_0$. Then:

Type I: If H is given by $G_{i_{k+1}} u(x) = W_{i_{k+1}} + S_{i_{k+1}} x$, then the optimal active set in CR_i is $\{i_1, \ldots, i_k, i_{k+1}\}$.

Type II: If H is given by $\lambda_{i_k}(x) = 0$, then the optimal active set in CR_i is $\{i_1, \ldots, i_{k-1}\}$.

Consequently, the following corollary is stated:

Corollary 1 (Facet-to-facet conditionality [68])**.** The facet-to-facet property holds between CR_0 and CR_i, if the conditions of Theorem 3 are fulfilled.

Remark 10. In mp-LP problems, it can be shown that the facet-to-facet property is inherently violated [55]. Therefore, it is tendentially not advisable to use an algorithm based on the facet-to-facet property for mp-LP problems.

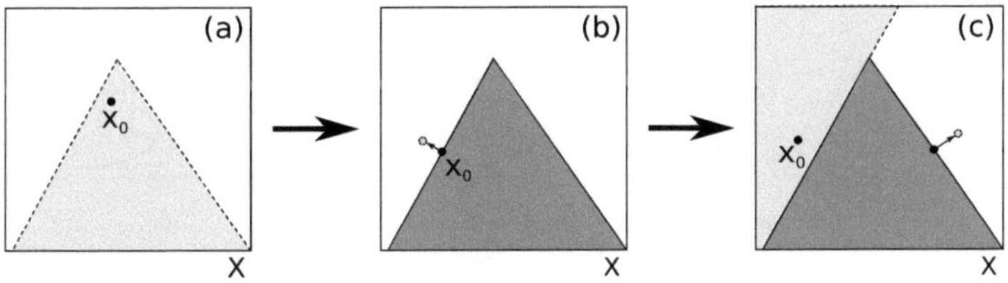

Fig. 2: A graphical representation of the geometrical solution procedure of exploring the parameter space based on the step-size approach. Starting from an initial point $x_0 \in X$, in (a) the first critical region CR_0 is calculated (shown with dashed lines). In (b), a facet of CR_0 is identified and a step orthogonal to that facet is taken to identify a new point $x_0 \notin CR_0$, while in (c) the new critical region associated with x_0 is identified, and the remaining facet from CR_0 is identified combined with the orthogonal step from it to identify a new point.

The combinatorial approach: As stated in Remark 7, every critical region is uniquely defined by the corresponding optimal active set. Thus, a combinatorial approach has been suggested, which considers the fact that the possible number of active sets is finite, and thus can be exhaustively enumerated. In order to make this approach computationally tractable, the following pruning criterion is stated:

Lemma 1 (Pruning of active sets [34]). Let k be an infeasible candidate active set, i.e.

$$\left\{ (u,x) \;\middle|\; \begin{array}{c} G_k u = W_k + S_k x \\ G_j u \leq W_j + S_j x, \;\; \forall j \notin k \\ x \in X \end{array} \right\} = \emptyset. \tag{22}$$

Then any set $k' \supset k$ is also infeasible and may be pruned.[10]

Thus, the following branch-and-bound approach has been presented [34] (see Figure 3):

Step 1: Generate a tree consisting of all possible active sets.
Step 2: Select the candidate active set with the lowest cardinality of the active set and check for feasibility. If it is infeasible, prune that node and all its child nodes.
Step 3: Obtain the parametric solution of the selected node accordingly and check whether the resulting region is non-empty.
Step 4: If there are nodes to explore, go to Step 2. Otherwise terminate.

This approach has been shown to be particularly efficient when symmetry is present [25, 26], and has been extended to include Theorem 3 to only consider active sets with a matching cardinality [2, 3].

[10] In other words: if k is infeasible, so is its powerset.

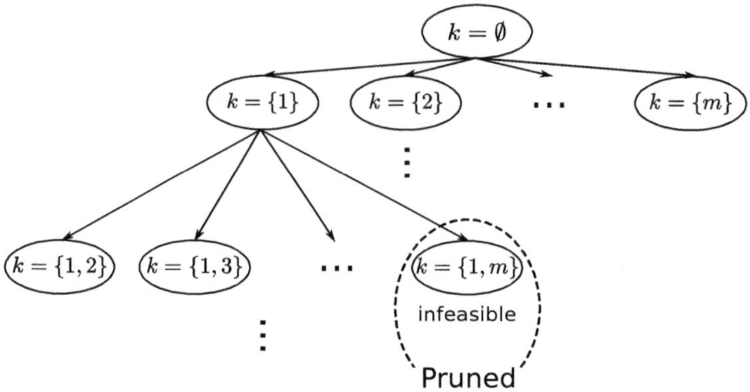

Fig. 3: A graphical representation of the combinatorial approach for the solution of mp-QP problems. All candidate active sets are exhaustively enumerated based on their cardinality. The computational tractability arises from the ability to discard active sets if infeasibility is detected for a candidate active set which is a subset of the currently considered candidate.

The connected graph approach [32, 55]: Since the parametric solution of a critical region can be obtained solely based on the active set k (see Equation (16)), the combinatorial approach is a simple and robust solution approach to problem (9), as it does not feature the limitations of the geometrical approach such as the necessity to consider facet-to-facet properties and step-size determination. However, even when considering the pruning criteria stated in Lemma 1, only a small percentage of the considered active sets result in a full-dimensional critical region. Thus, the key to a more efficient algorithm is to decrease the number of candidate active sets. In order to achieve this, the results on connected graphs from mp-LP problems [32] are extended to the mp-QP case:

Definition 4 (mp-QP Graph)**.** Let each optimal active set k of an mp-QP problem be a node in $\mathscr{S}$. Then the nodes k_1 and k_2 are connected if (a) there exists $x^* \in X_f$ such that k_1 and k_2 are both optimal active sets and (b) the conditions of Theorem (3) are fulfilled on the facet or it is possible to pass from k_1 to k_2 by one step of the dual simplex algorithm. The resulting graph G is fully defined by the nodes $\mathscr{S}$ as well as all connections Γ, i.e. $G = (\mathscr{S}, \Gamma)$

Corollary 2 (Connected graph for the mp-QP solution)**.** Consider the solution to an mp-QP problem and let $x_1, x_2 \in X_f$ be two arbitrary feasible parameters and $k_1 \in \mathscr{S}$ be given such that $x_1 \in CR_1$. Then there exists a path $\{k_1, \ldots, k_j\}$ in the mp-QP graph $G = (\mathscr{S}, \Gamma)$ such that $x_2 \in CR_j$.

3 Multi-Parametric Mixed-Integer Linear and Quadratic Programming: An Overview

Consider the following multi-parametric mixed-integer quadratic programming (mp-MIQP) problem

$$\begin{aligned} z(x) = \underset{u,y}{\text{minimize}} \quad & (H\omega + Zx + c)^T \omega \\ \text{subject to } & Gu + Ey \leq W + Sx \\ & u \in \mathbb{R}^n, \quad y \in \{0,1\}^p, \quad \omega = \left[u^T \; y^T\right]^T \\ & x \in X := \{x \in \mathbb{R}^q \,|CR_A x \leq CR_b\}, \end{aligned} \tag{23}$$

where $H \in \mathbb{R}^{(n+p)\times(n+p)} \succ 0$, $Z \in \mathbb{R}^{(n+p)\times q}$, $c \in \mathbb{R}^{(n+p)}$, $G \in \mathbb{R}^{m\times n}$, $E \in \mathbb{R}^{m\times p}$, $W \in \mathbb{R}^m$, $S \in \mathbb{R}^{m\times q}$ and X is compact.

The properties discussed below are also valid for multi-parametric mixed-integer linear programming (mp-MILP) problems of the form:

$$\begin{aligned} z(x) = \underset{u,y}{\text{minimize}} \quad & c^T \omega \\ \text{subject to } & Gu + Ey \leq W + Sx \\ & u \in \mathbb{R}^n, \quad y \in \{0,1\}^p, \quad \omega = \left[u^T \; y^T\right]^T \\ & x \in X := \{x \in \mathbb{R}^q \,|CR_A x \leq CR_b\}, \end{aligned} \tag{24}$$

3.1 Theoretical Properties

The properties of the solution of mp-MIQP problems of type (23) are given by the following theorem, corollary, and definitions.

Theorem 4 (Properties of mp-MIQP solution [18])**.** Consider the optimal solution of problem (23) with $H \succ 0$. Then, there exists a solution in the form

$$u_i(x) = K_i x + r_i \quad \text{if } x \in CR_i, \tag{25}$$

where CR_i, $i = 1,\ldots,M$ is a partition of the set X_f of feasible parameters x, and the closure of the sets CR_i has the following form

$$CR_i = \left\{x \in X | x^T \hat{G}_{i,j} x + \hat{h}_{i,j}^T x \leq \hat{w}_{i,j}, j = 1,\ldots,t_i\right\}, \tag{26}$$

where t_i is the number of constraints that describe CR_i.

Corollary 3 (Quadratic boundaries [18])**.** Quadratic boundaries arise from the comparison of quadratic objective functions associated with the solution of mp-QP problems for different feasible combinations of binary variables. This means that if the term $\hat{G}_{i,j}$ in Equation (26) is non-zero, it adheres to the values given by the difference of the quadratic optimal objective functions in the critical regions featuring this quadratic boundary.

The structural information known for mp-MIQP problems from Theorem 4 and Corollary 3 is clearly much more restricted than the information available for mp-QP problems from Theorems 2 and 3, Corollary 1, and Lemma 7. This is due to the fact that the presence of binary variables creates a discontinuous feasible space, which despite the convexity of H in problem (23) restricts the knowledge which can readily be inferred.

Definition 5 (Envelope of solutions [24])**.** In order to avoid the nonconvex critical regions described by Corollary 3, an envelope of solutions is created where more than one solution is associated with a critical region. The envelope is guaranteed to contain the optimal solution, and a point-wise comparison procedure among the envelope of solutions is performed online.

Definition 6 (The exact solution)**.** The exact solution of an mp-MIQP problem denotes the explicit calculation of Equations (25) and (26) for every critical region, and consequently no envelopes of solutions are present.

3.1.1 On the Notion of Exactness

The notion of the exact solution for mp-MIQP problems as the explicit calculation of Equations (25) and (26) for every critical region does not imply that solutions which feature envelopes of solutions are incorrect or approximate. As stated in Definition 5, such implicit solutions are guaranteed to describe the optimal solution. Thus, the term exactness does not indicate any difference in the evaluation of the numerical value of the solution, but a difference in the solution structure itself. The merit of an exact solution, and by extension of the algorithm presented in this chapter, is the explicit availability of the critical region description in its potentially nonconvex form given in Equation (26). This enables the assignment of one solution to each region, and consequently an assessment of the impact and meaning of each region.

This is relevant as the solution to a multi-parametric programming problem not only yields the optimal solution for any feasible parameter realization considered, but also information regarding the structure of the underlying optimization problem. For example, the consideration of when a certain binary variable is 0 or 1 may imply when a certain decision such as a valve position is decided. This enables insights and post-optimal analysis akin to sensitivity analysis. However, such an analysis is only possible if the exact solution of the problem is obtained, and not a solution featuring envelopes of solutions, as then the critical region partitioning in itself does not have any meaning.

3.2 Solution Algorithms

3.2.1 Literature Overview

Several authors have proposed strategies for the solution of mp-MILP and mp-MIQP problems. First, Acevedo and Pistikopoulos presented a branch-and-bound approach, where at the root node the binary variables are relaxed and the resulting mp-QP problem is solved [1]. For each subsequent node, a binary variable is fixed to a specific value and the resulting mp-QP problem is solved, followed by a suitable comparison procedure with a previously obtained upper bound in order to

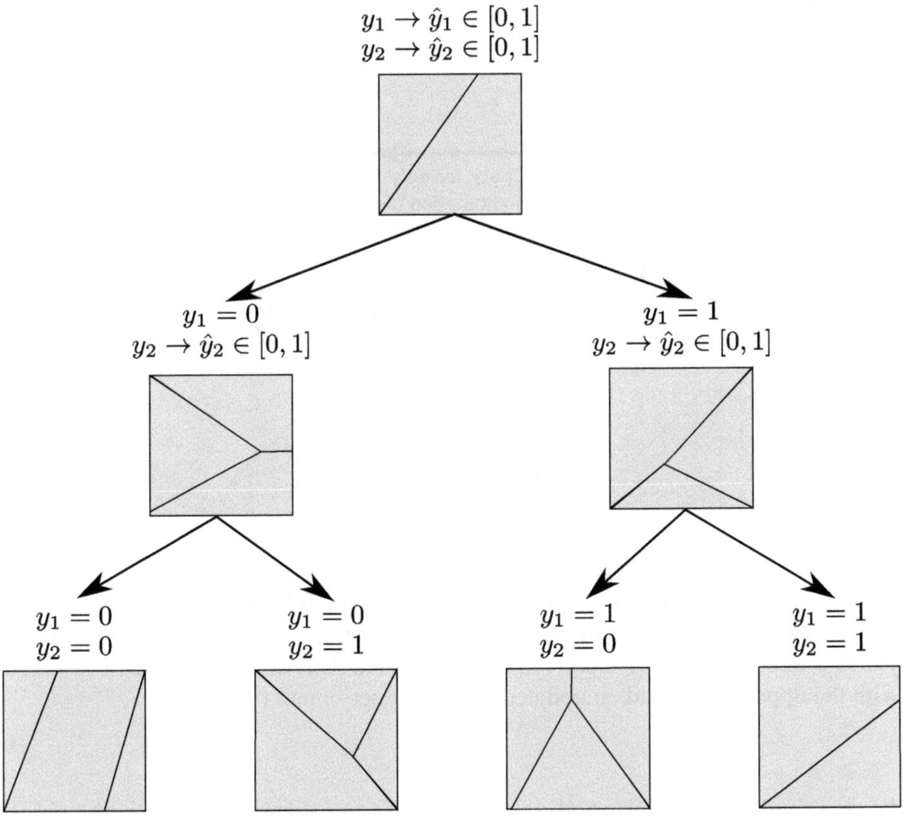

Fig. 4: A graphical representation of the branch-and-bound algorithm. The algorithm starts from the root node, where all binary variables are relaxed. Subsequently, at each node a binary variable is fixed, the resulting mp-QP problem is solved and the solution is compared to a previously established upper bound to produce an updated, tighter upper bound and to prune any part of the parameter space which is suboptimal.

produce a tighter upper bound and to prune any part of the parameter space which is guaranteed to be suboptimal. This approach was extended to mp-MIQP problems by Axehill *et al.* and Oberdieck *et al.* to consider the non-convexity of the critical regions induced by the quadratic nature of the objective function [6, 53] (Figure 4).

Later on, Dua *et al.* described a decomposition approach, where a candidate integer variable combination is found by solving a mixed-integer nonlinear programming (MINLP) problem[11] [23, 24]. After fixing this candidate solution, the resulting mp-QP problem is solved and the solution is compared to a previously obtained upper bound, which results in a new, tighter upper bound, and a new iteration begins. The introduction of suitable integer and parametric cuts to the MINLP ensures that previously considered integer combinations as well as solutions with a worse objective function value are excluded. A schematic representation of this approach is given in Figure 5.

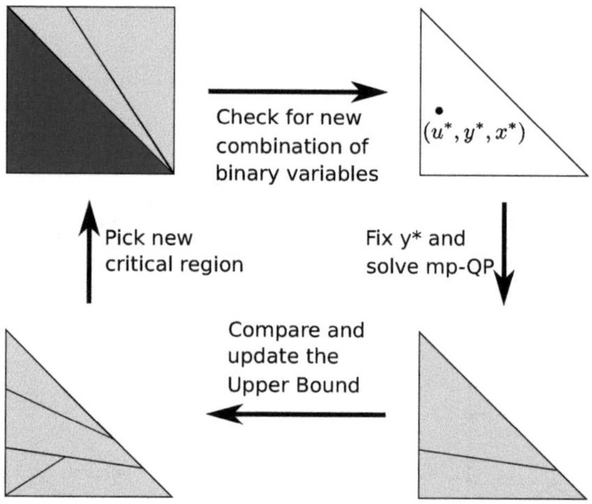

Fig. 5: A graphical representation of the decomposition algorithm. The algorithm starts with an upper bound, from where a critical region is selected. After obtaining a new candidate integer solution, the solution of the corresponding mp-QP problem yields a new solution for the given critical region. This solution is then compared with the upper bound and an updated, tighter upper bound results.

Remark 11. In addition, Borrelli *et al.* presented an exhaustive enumeration procedure of all possible combinations of binary variables [17].

As the decomposition algorithm is used in the remaining part of the chapter, it is now discussed in more detail.

[11] In the case of mp-MILP problems, the MINLP becomes a mixed-integer linear programming problem.

3.3 The Decomposition Algorithm

The decomposition algorithm consists of three parts: calculation of a new candidate integer solution *via* the solution of a MINLP problem, solving the mp-QP problem resulting from fixing the candidate integer solution in the original mp-MIQP problem, and comparing the obtained solution to a previously obtained upper bound. Note that the initial upper bound is set to ∞.

3.3.1 Calculation of a New Candidate Integer Solution

A candidate integer solution is found by solving the following global optimization problem:

$$\begin{aligned} z_{global} = \underset{u,y,x}{\text{minimize}} \quad & (H\omega + Zx + c)^T \omega \\ \text{subject to } & Gu + Ey \leq W + Sx \\ & (H\omega + Zx + c)^T \omega - \hat{z}_i(x) \leq 0 \\ & \sum_{j \in J_i} y_j - \sum_{j \in T_i} y_j \leq \text{card}(J_i) - 1 \\ & u \in \mathbb{R}^n, \quad y \in \{0,1\}^p, \quad \omega = \left[u^T \ y^T\right]^T \\ & x \in CR_i, \end{aligned} \tag{27}$$

where $i = 1, \ldots, v$ and v is the number of critical regions that constitute the upper bound, $\hat{z}_i(x)$ is the objective function value of the upper bound in the critical region CR_i considered, and J_i and T_i are the sets containing the indices of the integer variables $\hat{y}^i$ associated with the upper bound $\hat{z}_i(x)$ that attain the value 0 and 1, respectively, i.e.

$$J_i = \left\{ j | \hat{y}^i_j = 1 \right\} \tag{28a}$$

$$T_i = \left\{ j | \hat{y}^i_j = 0 \right\}. \tag{28b}$$

Remark 12. Without loss of generality, it is assumed that CR_i only features one upper bound $\hat{z}_i(x)$ in problem (27).

3.3.2 mp-QP Solution

Once a candidate integer solution has been found, it is fixed in the mp-MIQP problem, thus resulting in an mp-QP problem. This problem can be solved with any mp-QP solver described before.

3.3.3 Comparison Procedure

Within the algorithm, the solution obtained from the mp-QP problem is compared to a previously obtained current best upper bound $\hat{z}(x)$ to form a new, tighter upper bound. This can be expressed as:

$$z(x) = \min \ \{\hat{z}(x), z^*(x)\}, \tag{29}$$

where $z^*(x)$ denotes the piecewise quadratic, optimal objective function obtained by solving the mp-QP problem resulting by fixing the candidate solution of the binary variables obtained from the solution of problem (27). The solution of Equation (29) requires in turn the comparison of the corresponding objective functions in each critical region CR_i, i.e.

$$\Delta z(x) = \hat{z}(x) - z_i^*(x) = 0, \tag{30}$$

where $z_i^*(x)$ denotes the objective function within the i-th critical region of the solution of the mp-QP problem. Due to the quadratic nature of the objective functions, $\Delta z(x)$ might be nonconvex. Within the open literature, two strategies for the solution of problem (29) have been presented,

No objective function comparison: This approach, pioneered in [22] and first applied to mp-MIQP problems in [24], does not consider Equation (30) and stores both solutions, $\hat{z}(x)$ and $z_i^*(x)$, in CR_i, thus creating an envelope of solutions.

Objective function comparison over the entire CR: This approach was first presented for the solution of multi-parametric dynamic programming (mp-DP) problems [18], but has been applied to mp-MIQP problems in [6]. In this approach, Equation (30) is solved over the entire critical region CR_i, i.e. the following (possibly nonconvex) quadratic programming problem is solved:

$$\delta_{max} = \max_{x \in CR_i} \Delta z(x) \tag{31a}$$

$$\delta_{min} = \min_{x \in CR_i} \Delta z(x). \tag{31b}$$

Note that solving Equation (31) is not straightforward since it may be nonconvex. The results of solving Equation (31) allow for the following conclusions:

$$\delta_{max} \leq 0 \rightarrow \quad z_1(x) \geq z_2(x) \quad \forall x \in CR_i \tag{32a}$$

$$\delta_{min} \geq 0 \rightarrow \quad z_1(x) \leq z_2(x) \quad \forall x \in CR_i. \tag{32b}$$

If $\delta_{min} < 0$ and $\delta_{max} > 0$, then both solutions are kept and an envelope of solutions is created.

Remark 13. Without loss of generality it was assumed in Equation (30) that only one objective function is associated with each critical region, and that no envelope of solutions is present (see Definition 5).

4 Discussion and Concluding Remarks

In this chapter, we discussed the reformulation of the MPC problem in Equation (3) into the equivalent form of the multi-parametric programming problem in Equation (8) by treating the initial state vector as parameters and solving the control variables as a function thereof. In addition, alternative formulations of the control problems based on the nature of the performance index (quadratic or linear) and the existence of binary/integer decision variables have been presented. Lastly, we presented the theoretical properties and solution algorithms for mp-LP, mp-QP, mp-MILP, and mp-MIQP problems.

In the following, we qualitatively discuss the advantages and disadvantages of using multi-parametric programming for the application of MPC.

4.1 Size of Multi-Parametric Programming Problem and Offline Computational Effort

Since its conception, advances in computational hardware and algorithmic developments have significantly increased the size of problems that can be tackled. However, although multi-parametric programming has been proven to solve problems with parametric vectors with more than 40 elements [58], there is no general guideline as to what problem size can generally be solved for, as this is highly problem dependent. Note that this limitation is imposed by the computing power and available storage on a given machine, and does not indicate the inability to solve large problems from an algorithmic standpoint.

Similarly to the size of the parametric vector, a larger number of constraints and optimization variables pose a greater challenge in terms of acquiring a parametric solution. This becomes clear by considering the maximum number of critical regions as the maximum number of possible active sets, which is given by $\frac{m!}{n!(m-n)!}$ for mp-LP problems, where n is the number of optimization variables and m is the number of inequality constraints. However, for mp-QP problems this number is significantly larger as the solution is not guaranteed to lie on a vertex, and is given by $\sum_{i=0}^{n} \frac{m!}{i!(m-i)!}$.

In the case of mp-MILP and mp-MIQP problems, the offline computational effort is significantly affected by the size of binary variables. These problems require the solution of one mp-LP/mp-QP problem for every considered feasible combination of binary variables.

In order to alleviate these increasingly high computational requirements, parallelization can be employed. Parallelization inherently exploits independent aspects of an algorithm and distributes them on different machines, where these independent subproblems are computed in parallel. The disjoint nature of the critical regions thereby naturally generates independent subproblems which can be solved in parallel.

Hence it is possible to choose between continuing the current computation locally or to return the results to the main algorithm and perform a re-distribution of the problems. The resulting trade-off is between an increased overhead resulting from the information transfer between the machines and the possibility of calculating possibly suboptimal or unnecessary solutions, as the re-distribution always ensures that the algorithm performs optimally. Since at the end of the algorithm all results are combined together, the final solution is always optimal.

Consequently, the parallelization strategy can be summarized as follows:

Step 1: Formulation of the sequential solution algorithm
Step 2a: Identification of the most external iterative procedure
Step 2b: Identification of the independent elements computed at each iteration
Step 2c: Definition of ρ_{limit}[12]
Step 3: Connection to different machines and equal distribution of elements
Step 4: Execution of the current computation locally until (i) the predefined termination criteria are met or (ii) the number of iterations has reached ρ_{limit}

Details regarding the application of the parallelization algorithm in multi-parametric programming can be found in [52] and used via [54].

4.2 *Size of the Solution and Online Computational Effort*

The two major advantages of multi-parametric/explicit MPC are its ability to provide a "map of solutions" *a priori* and its effortless online applicability. In the first case, the fact that the entire control problem is solved offline provides great insight regarding the effect that the initial state vector values have on the optimal control action. This helps the control developer understand the control behavior and guarantee the optimality of the solution. Furthermore, in the presence of (measured) disturbances within the control scheme the same can be guaranteed.

The effortless online applicability is a direct result of the nature of the MPC problem. The (mixed-integer) linear or quadratic formulation of the problem guarantees that the optimal control action is linear with respect to the parameters, in this case the initial state vector values. Furthermore, the critical regions for which an optimal action remains optimal is also a polytope, except in the case of the exact solution of mp-MIQP problems.[13] The fact that the explicit expression of the optimal control action is linear results in fast MPC computations without the need of solving an op-

[12] The limiting iteration number ρ_{limit} is the maximum number of iterations performed on a single machine before the result is returned to the main algorithm.

[13] Note that in the case of the envelopes of solutions approach to the mp-MIQP problem the critical regions are polytopes but a comparison procedure between alternative solutions is necessary.

timization problem online. It is fair to say that the major burden of the optimization based MPC application has therefore been alleviated.[14]

The computationally most expensive step with respect to the application of multi-parametric/explicit MPC is the identification of the critical region within which the parameter vector lies. In essence, it is a set membership test of a series of disjoint polytopes whose union might be convex or non-convex, depending on whether the multi-parametric programming problem features binary variables. While it is possible to exhaustively enumerate all polytopes until $\theta^* \in CR_i$ has been found, such an approach becomes computationally problematic in the case of larger systems with potentially thousands of critical regions. In addition, from a practical view point, it is equally important to provide an upper bound on the worst-case scenario for the point location problem.[15] Thus, starting with works in [41, 72], several researchers have considered the design of efficient algorithms [4, 5, 8–10, 16, 19, 29, 33, 36, 40, 43, 49, 50, 57, 67, 74, 75].

4.3 Other Developments in Explicit MPC

The material outlined in this chapter covers the standard use of explicit MPC for continuous and mixed-integer systems. However, the area of explicit MPC has featured several other developments, some of the most important ones are described in the following:

Continuous-time explicit MPC: The optimal control strategies for discrete-type systems of type (1) are determined off-line by solving multi-parametric programming problems of type (9) or (23) in the case of a quadratic performance index and continuous or hybrid systems. On the other hand, for systems with continuous-time dynamics it is necessary to consider so-called multi-parametric dynamic optimization (mp-DO) problems, which lead to an infinite dimensional problem. Within the literature, two different strategies have been proposed to solve an mp-DO. One way is to transform the mp-DO problem into a finite-dimensional multi-parametric problems *via* control vector parameterization [64], while the other way is to solve the mp-DO problem directly in the infinite-dimensional form using variational approaches. While the theory presented earlier in this chapter is applicable to the finite-dimensional reformulation, for the infinite dimensional problems, it has been proposed to transform the optimization problems into a boundary value problem derived from the corresponding optimality conditions [65, 71].

These insights have led to the recent development of a unified framework, which combines the formulation of the control problem as an mp-DO with the track-

[14] Note that this claim refers mainly to alleviating the necessity of optimization hardware equipment for the application of optimization based MPC as the explicit solution enables the use of MPC-on-a-chip as described in [60].

[15] If the sampling time of a system is $1\mu s$, but the point location of the explicit MPC controller may require up to $5\mu s$, the explicit MPC controller cannot be applied in practice.

ing of the necessary conditions for optimality (NCO), namely multi-parametric NCO-tracking [71]. The aim of this method is to convert an online dynamic optimization problem into a measurement-based feedback control problem. This combination of mp-DO and NCO-tracking enables the relaxation of the fixed switching structure from an NCO-tracking perspective, as it constructs the critical regions which correspond to different optimal switching structures. This leads to a great reduction in the number of critical regions and the explicit solution of the continuous-time MPC problem.

Decentralized explicit MPC: As discussed in Section 4.1, the application of explicit MPC is often limited by the size of the problems under consideration. While in some cases it is the single system that requires a prohibitively large number of states or control variables, there are other cases where the system consists of several interconnected elements. The advantage of explicit MPC is that these elements can be solved independently, and then linked to each other by expressing the input of one element as the output of another. This has gained some recent interest in the research community, where the use of vertical and horizontal decentralization enables the use of explicit MPC, with its inherent advantages, for large and complex systems [69, 70]. This strategy was, for example, successfully applied in [58] for periodic systems and in [21] for combined heat and power systems.

References

1. Acevedo, J., Pistikopoulos, E.N.: A multiparametric programming approach for linear process engineering problems under uncertainty. Ind. Eng. Chem. Res. **36**(3), 717–728 (1997)
2. Ahmadi-Moshkenani, P., Johansen, T.A., Olaru, S.B.: On degeneracy in exploration of combinatorial tree in multi-parametric quadratic programming. In: IEEE Conference on Decision and Control (2016)
3. Ahmadi-Moshkenani, P., Olaru, S.B., Johansen, T.A.: Further results on the exploration of combinatorial tree in multi-parametric quadratic programming. In: Proceedings of the European Control Conference, pp. 116–122 (2016)
4. Airan, A., Bhartiya, S., Bhushan, M.: Linear machine: a novel approach to point location problem. In: International Symposium on Dynamics and Control of Process Systems, pp. 445–450. IFAC, Oxford (2013)
5. Airan, A., Bhushan, M., Bhartiya, S.: Linear machine solution to point location problem. IEEE Trans. Autom. Control **62**(3), 1403–1410 (2016)
6. Axehill, D., Besselmann, T., Raimondo, D.M., Morari, M.: A parametric branch and bound approach to suboptimal explicit hybrid MPC. Automatica **50**(1), 240–246 (2014)
7. Baotić, M.: An efficient algorithm for multi-parametric quadratic programming. Technical Report AUT02-04, Automatic Control Laboratory, ETH Zurich, Switzerland (February 2002)
8. Baotic, M., Borrelli, F., Bemporad, A., Morari, M.: Efficient on-line computation of constrained optimal control. SIAM J. Control Optim. **47**(5), 2470–2489 (2008)
9. Bayat, F., Johansen, T.A., Jalali, A.A.: Using hash tables to manage the time-storage complexity in a point location problem: application to explicit model predictive control. Automatica **47**(3), 571–577 (2011)

10. Bayat, F., Johansen, T.A., Jalali, A.A.: Flexible piecewise function evaluation methods based on truncated binary search trees and lattice representation in explicit MPC. IEEE Trans. Control Syst. Technol. **20**(3), 632–640 (2012)
11. Bazaraa, M.S.: Nonlinear Programming: Theory and Algorithms, 3rd edn. Wiley, Chichester (2013)
12. Bemporad, A.: A multiparametric quadratic programming algorithm with polyhedral computations based on nonnegative least squares. IEEE Trans. Autom. Control **60**(11), 2892–2903 (2015)
13. Bemporad, A., Morari, M.: Control of systems integrating logic, dynamics, and constraints. Automatica **35**(3), 407–427 (1999)
14. Bemporad, A., Borrelli, F., Morari, M.: Model predictive control based on linear programming – the explicit solution. IEEE Trans. Autom. Control **47**(12), 1974–1985 (2002)
15. Bemporad, A., Morari, M., Dua, V., Pistikopoulos, E.N.: The explicit linear quadratic regulator for constrained systems. Automatica **38**(1), 3–20 (2002)
16. Borrelli, F., Baotic, M., Bemporad, A., Morari, M.: Efficient on-line computation of constrained optimal control. In: Conference on Decision and Control, vol. 2, pp. 1187–1192 (2001)
17. Borrelli, F., Bemporad, A., Morari, M.: Geometric algorithm for multiparametric linear programming. J. Optim. Theory Appl. **118**(3), 515–540 (2003)
18. Borrelli, F., Baotić, M., Bemporad, A., Morari, M.: Dynamic programming for constrained optimal control of discrete-time linear hybrid systems. Automatica **41**(10), 1709–1721 (2005)
19. Christophersen, F.J., Kvasnica, M., Jones, C.N., Morari, M.: Efficient evaluation of piecewise control laws defined over a large number of polyhedra. In: European Control Conference (2007)
20. Columbano, S., Fukuda, K., Jones, C.N.: An output-sensitive algorithm for multi-parametric LCPs with sufficient matrices. CRM Proc. Lect. Notes **48**, 1–30 (2009)
21. Diangelakis, N.A., Avraamidou, S., Pistikopoulos, E.N.: Decentralized multiparametric model predictive control for domestic combined heat and power systems. Ind. Eng. Chem. Res. **55**(12), 3313–3326 (2016)
22. Dua, V., Pistikopoulos, E.N.: Algorithms for the solution of multiparametric mixed-integer nonlinear optimization problems. Ind. Eng. Chem. Res. **38**(10), 3976–3987 (1999)
23. Dua, V., Pistikopoulos, E.N.: An algorithm for the solution of multiparametric mixed integer linear programming problems. Ann. Oper. Res. **99**(1–4), 123–139 (2000)
24. Dua, V., Bozinis, N.A., Pistikopoulos, E.N.: A multiparametric programming approach for mixed-integer quadratic engineering problems. Comput. Chem. Eng. **26**(4–5), 715–733 (2002)
25. Feller, C., Johansen, T.A.: Explicit MPC of higher-order linear processes via combinatorial multi-parametric quadratic programming. In: 2013 European Control Conference (ECC), pp. 536–541 (2013)
26. Feller, C., Johansen, T.A., Olaru, S.B.: An improved algorithm for combinatorial multi-parametric quadratic programming. Automatica **49**(5), 1370–1376 (2013)
27. Fiacco, A.V.: Sensitivity analysis for nonlinear programming using penalty methods. Math. Program. **10**(1), 287–311 (1976)
28. Floudas, C.A.: Nonlinear and Mixed-Integer Optimization: Fundamentals and Applications. Topics in Chemical Engineering. Oxford University Press, New York (1995)
29. Fuchs, A., Axehill, D., Morari, M.: On the choice of the linear decision functions for point location in polytopic data sets – application to explicit MPC. In: Conference on Decision and Control, pp. 5283–5288 (2010)
30. Gál, T.: The historical development of parametric programming. In: Brosowski, B., Deutsch, F. (eds.) Parametric Optimization and Approximation. International Series of Numerical Mathematics/Internationale Schriftenreihe zur Numerischen Mathematik/Série internationale d'Analyse numérique, vol. 72, pp. 148–165. Birkhäuser, Basel (1985)
31. Gál, T.: Postoptimal Analyses, Parametric Programming, and Related Topics: Degeneracy, Multicriteria Decision Making, Redundancy, 2nd edn. W. de Gruyter, Berlin (1995)

32. Gál, T., Nedoma, J.: Multiparametric linear programming. Manag. Sci. **18**(7), 406–422 (1972)
33. Grancharova, A., Johansen, T.A.: Approaches to explicit nonlinear model predictive control with reduced partition complexity. In: 2009 European Control Conference (ECC), pp. 2414–2419 (2009)
34. Gupta, A., Bhartiya, S., Nataraj, P.S.V.: A novel approach to multiparametric quadratic programming. Automatica **47**(9), 2112–2117 (2011)
35. Heemels, W.P.M.H., De Schutter, B., Bemporad, A.: Equivalence of hybrid dynamical models. Automatica **37**(7), 1085–1091 (2001)
36. Herceg, M., Mariéthoz, S., Morari, M.: Evaluation of piecewise affine control law via graph traversal. In: European Control Conference, pp. 3083–3088 (2013)
37. Herceg, M., Kvasnica, M., Jones, C.N., Morari, M.: Multi-parametric toolbox 3.0. In: 2013 European Control Conference (ECC), pp. 502–510 (2013)
38. Herceg, M., Jones, C.N., Kvasnica, M., Morari, M.: Enumeration-based approach to solving parametric linear complementarity problems. Automatica **62**, 243–248 (2015)
39. Horn, R.A., Johnson, C.R.: Matrix Analysis, 2nd edn. Cambridge University Press, Cambridge (2013)
40. Jafargholi, M., Peyrl, H., Zanarini, A., Herceg, M., Mariéthoz, S.: Accelerating space traversal methods for explicit model predictive control via space partitioning trees. In: European Control Conference, pp. 103–108 (2014)
41. Johansen, T.A., Grancharova, A.: Approximate explicit constrained linear model predictive control via orthogonal search tree. IEEE Trans. Autom. Control **48**(5), 810–815 (2003)
42. Jones, C.N., Morari, M.: Multiparametric linear complementarity problems. In: 2006 45th IEEE Conference on Decision and Control, pp. 5687–5692 (2006)
43. Jones, C.N., Grieder, P., Raković, S.V.: A logarithmic-time solution to the point location problem for parametric linear programming. Automatica **42**(12), 2215–2218 (2006)
44. Jones, C.N., Kerrigan, E.C., Maciejowski, J.M.: Lexicographic perturbation for multiparametric linear programming with applications to control. Automatica **43**(10), 1808–1816 (2007)
45. Karush, W.: Minima of functions of several variables with inequalities as side constraints. Master's thesis, Dept.~of Mathematics, Univ.~of Chicago (1939)
46. Kuhn, H.W., Tucker, A.W.: Nonlinear programming. In: Proceedings of the Second Berkeley Symposium on Mathematical Statistics and Probability, pp. 481–492. University of California Press, Berkeley (1951)
47. Li, Z., Ierapetritou, M.G.: A method for solving the general parametric linear complementarity problem. Ann. Oper. Res. **181**(1), 485–501 (2010)
48. Markowitz, H.: Portfolio selection. J. Financ. **7**(1), 77–91 (1952)
49. Martinez-Rodriguez, M.C., Brox, P., Baturone, I.: Digital VLSI implementation of piecewise-affine controllers based on lattice approach. IEEE Trans. Control Syst. Technol. **23**(3), 842–854 (2015)
50. Monnigmann, M., Kastsian, M.: Fast explicit model predictive control with multiway trees. In: World Congress, IFAC proceedings volumes, pp. 1356–1361. IFAC/Elsevier, New York (2011)
51. Monnigmann, M., Jost, M.: Vertex based calculation of explicit MPC laws. In: American Control Conference (ACC), pp. 423–428 (2012)
52. Oberdieck, R., Pistikopoulos, E.N.: Parallel computing in multi-parametric programming. In: Kravanja, Z., Bogataj, M. (eds.) 26th European Symposium on Computer Aided Process Engineering, volume 38 of Computer Aided Chemical Engineering, pp. 169–174. Elsevier, Amsterdam (2016)
53. Oberdieck, R., Wittmann-Hohlbein, M., Pistikopoulos, E.N.: A branch and bound method for the solution of multiparametric mixed integer linear programming problems. J. Glob. Optim. **59**(2–3), 527–543 (2014)
54. Oberdieck, R., Diangelakis, N.A., Papathanasiou, M.M., Nascu, I., Pistikopoulos, E.N.: POP – parametric optimization toolbox. Ind. Eng. Chem. Res. **55**(33), 8979–8991 (2016)
55. Oberdieck, R., Diangelakis, N.A., Pistikopoulos, E.N.: Explicit model predictive control: a connected-graph approach. Automatica **76**, 103–112 (2017)

56. Olaru, S.B., Dumur, D.: On the continuity and complexity of control laws based on multi-parametric linear programs. In: 2006 45th IEEE Conference on Decision and Control, pp. 5465–5470 (2006)
57. Oliveri, A., Gianoglio, C., Ragusa, E., Storace, M.: Low-complexity digital architecture for solving the point location problem in explicit Model Predictive Control. J. Frankl. Inst. **352**(6), 2249–2258 (2015)
58. Papathanasiou, M.M., Avraamidou, S., Steinebach, F., Oberdieck, R., Mueller-Spaeth, T., Morbidelli, M., Mantalaris, A., Pistikopoulos, E.N.: Advanced control strategies for the multi-column countercurrent solvent gradient purification process (MCSGP). AIChE J. **62**(7), 2341–2357 (2016)
59. Patrinos, P., Sarimveis, H.: A new algorithm for solving convex parametric quadratic programs based on graphical derivatives of solution mappings. Automatica **46**(9), 1405–1418 (2010)
60. Pistikopoulos, E.N.: From multi-parametric programming theory to MPC-on-a-chip multi-scale systems applications. Comput. Chem. Eng. **47**, 57–66 (2012)
61. Rawlings, J.B., Mayne, D.Q.: Model Predictive Control: Theory and Design. Nob Hill Pub., Madison (2009)
62. Rockafellar, R.T.: Convex Analysis. Princeton Mathematical Series. Princeton University Press, Princeton (1970)
63. Rockafellar, R.T., Wets, R.J.-B.: Variational Analysis. Grundlehren der mathematischen Wissenschaften. Springer, Berlin (1998)
64. Sakizlis, V.: Design of model based controllers via parametric programming. PhD thesis, Imperial College, London (2003)
65. Sakizlis, V., Perkins, J.D., Pistikopoulos, E.N.: Explicit solutions to optimal control problems for constrained continuous-time linear systems. IEEE Proc. Control Theory Appl. **152**(4), 443–452 (2005)
66. Spjøtvold, J., Tøndel, P., Johansen, T.A.: A method for obtaining continuous solutions to multiparametric linear programs. In: World Congress, IFAC proceedings volumes, p. 902. IFAC/Elsevier, New York (2005)
67. Spjotvold, J., Rakovic, S.V., Tondel, P., Johansen, T.A.: Utilizing reachability analysis in point location problems. In: Conference on Decision and Control, pp. 4568–4569 (2006)
68. Spjøtvold, J., Kerrigan, E.C., Jones, C.N., Tøndel, P., Johansen, T.A.: On the facet-to-facet property of solutions to convex parametric quadratic programs. Automatica **42**(12), 2209–2214 (2006)
69. Spudic, V., Baotic, M.: Fast coordinated model predictive control of large-scale distributed systems with single coupling constraint. In: 2013 European Control Conference (ECC), pp. 2783–2788 (2013)
70. Spudic, V., Jelavic, M., Baotic, M.: Explicit model predictive control for reduction of wind turbine structural loads. In: 2012 IEEE 51st IEEE Conference on Decision and Control (CDC), pp. 1721–1726 (2012)
71. Sun, M., Chachuat, B., Pistikopoulos, E.N.: Design of multi-parametric NCO tracking controllers for linear dynamic systems. Comput. Chem. Eng. **92**, 64–77 (2016)
72. Tøndel, P., Johansen, T.A., Bemporad, A.: Evaluation of piecewise affine control via binary search tree. Automatica **39**(5), 945–950 (2003)
73. Tøndel, P., Johansen, T.A., Bemporad, A.: An algorithm for multi-parametric quadratic programming and explicit MPC solutions. Automatica **39**(3), 489–497 (2003)
74. Wang, Y., Jones, C., Maciejowski, J.: Efficient point location via subdivision walking with application to explicit MPC. In: European Control Conference (2007)
75. Zhang, J., Xiu, X., Xie, Z., Hu, B.: Using a two-level structure to manage the point location problem in explicit model predictive control. Asian J. Control **18**(3), 1075–1086 (2015)

Real-Time Implementation of Explicit Model Predictive Control

Michal Kvasnica, Colin N. Jones, Ivan Pejcic, Juraj Holaza, Milan Korda, and Peter Bakaráč

1 Simplification of MPC Feedback Laws

1.1 Preliminaries

We consider the control of discrete time, linear time-invariant (LTI) systems

$$x^{+} = Ax + Bu, \tag{1}$$

where $x \in \mathbb{R}^{n_x}$ is the state vector, $u \in \mathbb{R}^{n_u}$ are the control inputs, $A \in \mathbb{R}^{n_x \times n_x}$ and $B \in \mathbb{R}^{n_x \times n_u}$ are constant matrices, and the pair (A, B) is stabilizable. The state and input variables are subject to constraints $x \in \mathscr{X} \subseteq \mathbb{R}^{n_x}$, $u \in \mathscr{U} \subseteq \mathbb{R}^{n_u}$ where $\mathscr{X}$ and $\mathscr{U}$ are assumed to be polyhedral sets containing the origin in their respective interior. The constrained finite time optimal control problem for the LTI system in Equation (1) is then given by

$$\min_{U_N} x_N^\top P x_N + \sum_{k=0}^{N-1} \left(x_k^\top Q x_k + u_k^\top R u_k \right) \tag{2a}$$

$$\text{s.t. } x_{k+1} = Ax_k + Bu_k, \quad k = 0, \ldots, N-1, \tag{2b}$$

$$x_k \in \mathscr{X}, \quad k = 0, \ldots, N-1, \tag{2c}$$

$$u_k \in \mathscr{U}, \quad k = 0, \ldots, N-1, \tag{2d}$$

$$x_N \in \mathscr{X}_f, \tag{2e}$$

M. Kvasnica (✉) · J. Holaza · P. Bakaráč
Slovak University of Technology in Bratislava, Bratislava, Slovakia
e-mail: michal.kvasnica@stuba.sk

C. N. Jones · I. Pejcic · M. Korda
EPFL Lausanne, Lausanne, Switzerland
e-mail: colin.jones@epfl.ch

S. V. Raković, W. S. Levine (eds.), *Handbook of Model Predictive Control*, Control Engineering, https://doi.org/10.1007/978-3-319-77489-3_17

where $Q \in \mathbb{R}^{n_x \times n_x}$, $R \in \mathbb{R}^{n_u \times n_u}$, and $P \in \mathbb{R}^{n_x \times n_x}$ are the weighting matrices with $Q = Q^\top \succeq 0$, $R = R^\top \succ 0$, $P = P^\top \succeq 0$, N denotes the prediction horizon, x_k and u_k are the vectors of predicted states and inputs at time instant k, respectively, $U_N = [u_0^\top, \dots, u_{N-1}^\top]^\top \in \mathbb{R}^{Nn_u}$ is the sequence of optimized control actions, and $\mathscr{X}_f$ denotes the polyhedral terminal set constraint.

It is well known, see, e.g., [10, Chapter 12] that, by using the substitution $x_k = A^k x_0 + \sum_{j=0}^{k-1} A^{k-1-j} B u_j$, the MPC problem in Equation (2) can be translated into a parametric quadratic program (mp-QP) of the form

$$\min_{U_N} \, 1/2 \, U_N^\top H U_N + x_0^\top F U_N \tag{3a}$$

$$\text{s.t.} \;\; G U_N \leq w + E x_0, \tag{3b}$$

where x_0 is the initial condition of the problem in Equation (2), and $H \in \mathbb{R}^{Nn_u \times Nn_u}$, $F \in \mathbb{R}^{n_x \times Nn_u}$, $G \in \mathbb{R}^{q \times Nn_u}$, $w \in \mathbb{R}^q$, $E \in \mathbb{R}^{q \times n_x}$ are given matrices with q denoting the number of inequality constraints in Equation (3b). Moreover, with Q, R, P in Equation (2a) being positive (semi)definite symmetric matrices, the Hessian H is positive definite and thus invertible. As demonstrated, e.g., by [7], the mp-QP represented by Equation (3) admits a closed-form solution

$$U_N^\star = \kappa(x_0), \tag{4}$$

with the function $\kappa : \mathbb{R}^{n_x} \to \mathbb{R}^{Nn_u}$ mapping the initial conditions x_0 onto the sequence of optimal control inputs $U_N^\star$. Moreover, since the mp-QP Equation (3) is assumed to be strictly convex with $H \succ 0$, κ is a continuous piecewise affine (PWA) function

$$\kappa(x) = \begin{cases} L_1 x + \ell_1 & \text{if } x \in \mathscr{R}_1 \\ & \vdots \\ L_M x + \ell_M & \text{if } x \in \mathscr{R}_M \end{cases} \tag{5}$$

where

$$\mathscr{R}_i = \{x \in \mathbb{R}^{n_x} \mid Z_i x \leq z_i\}, \; i = 1, \dots, M, \tag{6}$$

are polyhedral critical regions with $Z_i \in \mathbb{R}^{c_i \times n_x}$, $z_i \in \mathbb{R}^{c_i}$ describing the half-spaces of each region and c_i denoting the number of half-spaces of the i-th region. Moreover, $L_i \in \mathbb{R}^{Nn_u \times n_x}$, $\ell_i \in \mathbb{R}^{Nn_u}$ are locally optimal gains of the i-th local feedback law. Finally, M denotes the total number of critical regions. In what follows we refer to the function κ in Equation (5) as the *explicit* solution to the MPC problem in Equation (2). Its data, i.e., the matrices Z_i, z_i, L_i, ℓ_i can be computed by solving the mp-QP in Equation (3) *off-line*, e.g., by the Multi-parametric toolbox [18], the Hybrid toolbox [5], or the POP toolbox [32].

Once the explicit MPC feedback law κ in Equation (5) is constructed, the on-line implementation of MPC, i.e., the process of obtaining $U_N^\star$ for a given initial condition $x_0 = x(t)$, boils down to a mere evaluation of $\kappa(x_0)$. Such an evaluation is fast and simple since, among other factors, it does not involve divisions. Only multiplications and additions are required to identify $U_N^\star$ for a given value of x_0. These properties foretell explicit MPC for implementation on embedded platforms with severe restrictions on available implementation resources.

1.2 Complexity of Explicit MPC

The main limitation of explicit MPC lies in its implementation complexity. Specifically, the target implementation hardware has to posses enough computational power to evaluate $\kappa(x_0)$ on-line, and requires a sufficient capacity to store the parameters Z_i, z_i, L_i, ℓ_i as in Equations (5) and (6) in the memory. We refer to the former as the *computational* complexity of explicit MPC feedback laws, while the latter will be referred to as the *space* complexity.

The computational complexity depends on the type of algorithm employed to evaluate $\kappa(x)$ for a given value of x from Equation (5). Various options are available. The simplest approach is the *sequential search* procedure where one loops through the critical regions, stopping once $x \in \mathscr{R}_{i^\star}$ is satisfied for some $i^\star \in \{1,\dots,M\}$. Subsequently, $U_N^\star$ is obtained via $U_N^\star = L_{i^\star}x + \ell_{i^\star}$. Since the critical regions are polyhedra as in Equation (6), checking the inclusion $x \in \mathscr{R}_i$ involves the validation of inequalities $Z_i x \le z_i$. Clearly, in the worst case one needs to loop through all regions. This requires a total of

$$\mathscr{C}(\kappa) = 2n_\mathrm{x}n_\mathrm{u}N + \sum_{i=1}^{M} c_i(2n_\mathrm{x}+1) \tag{7}$$

floating point operations with the first terms representing the effort of evaluating $U_N^\star = L_i x + \ell_i$ for the "active" critical region and the second one accounting for checking the inclusion $x \in \mathscr{R}_i$ for each of the M critical regions (each of them consisting of c_i half-spaces). A computationally more effective way to evaluate $\kappa(x)$ is to organize the critical regions into a binary search tree [38] or an axis-aligned search tree [11], to employ a lattice representation of the PWA function κ [39], or to resort to graph-based approaches [19]. In all cases the evaluation effort is decreased at the expense of having to construct and to store a suitable search structure.

The space complexity is determined by the amount of data required to describe the PWA function κ in Equation (5). Specifically, storing L_i, ℓ_i, Z_i, z_i in Equations (5) and (6) in the memory of the implementation hardware amounts to a total of

$$\mathscr{S}(\kappa) = MNn_\mathrm{u}(n_\mathrm{x}+1) + \sum_{i=1}^{M} c_i(n_\mathrm{x}+1), \tag{8}$$

floating point numbers. The first term in Equation (8) represents the size of local feedback gains L_i, ℓ_i, while the second term accounts for the size of the critical regions $\mathscr{R}_i$, represented by the matrices Z_i and vectors z_i. Since the prediction horizon N, as well as the state and input dimensions n_x, n_u is fixed, the main driving force behind the space complexity of explicit MPC feedback laws is twofold: the number of critical regions, and the size (i.e., the number of facets) of each critical region. It should be noted that both of these figures are entirely problem dependent.

Remark 1. In the spirit of the receding horizon implementation of MPC, only the first element of $U_N^\star$, i.e., $u_0^\star$, is applied to the plant and the remaining elements are discarded. Then the computation of $U_N^\star$ is repeated at the subsequent time instant for a fresh value of the initial condition $x_0 = x(t)$. It follows that only the first n_u rows of the matrices L_i and ℓ_i in Equation (5) are required for closed-loop implementation of explicit MPC via Equation (5), which then reduces to $u_0^\star = \kappa(x_0)$. Consequently, one can immediately decrease the computational complexity $\mathscr{C}(\kappa)$ and the space complexity $\mathscr{S}(\kappa)$ by dropping N from the first term in Equations (7) and (8), respectively.

1.3 Problem Statement and Main Results

In what follows we aim at decreasing the computational and space complexities of explicit MPC feedback laws in Equation (5), represented by Equations (7) and (8), respectively, by attacking the two main complexity drivers: the number of critical regions (M) and the amount of data required to represent each region (c_i). This task is achieved by replacing a given explicit MPC feedback law κ by a different function $\widetilde{\kappa}$ of smaller complexity. Two classes of methods are introduced. The first one aims at constructing $\widetilde{\kappa}$ that is a *performance lossless* replacement of the original function κ in the sense that $\widetilde{\kappa}(x) = \kappa(x)$ for all x in the domain of κ. In other words, the simpler feedback law $U_N^\star = \widetilde{\kappa}(x)$ preserves optimality with respect to the MPC problem in Equation (2) for all initial conditions that satisfy the constraints in Equations (2b)–(2e). The second class of methods sacrifices optimality in favor of reducing complexity, but preserves other important control-theoretical properties, such as recursive feasibility and closed-loop stability. Thus, $\widetilde{\kappa}$ will be only an *approximate* replacement of the original explicit MPC feedback law κ.

In the remainder of this chapter we review various procedures for the reduction of computational and space complexities of explicit MPC. Two classes of methods are discussed. First, in Section 2 we review three methods for obtaining an explicit MPC controller of reduced complexity while maintaining the piecewise affine nature of $\widetilde{\kappa}$ as in Equation (5). Subsequently, in Section 3 we report a procedure for obtaining approximate MPC feedback laws for control of nonlinear systems. Here, the obtained controllers involve closed-loop stability guarantees and optimize closed-loop performance with respect to a given metric.

2 Piecewise Affine Explicit MPC Controllers of Reduced Complexity

In this section we discuss the synthesis of simple explicit MPC controllers where the feedback law attains a piecewise affine structure as in Equation (5). First, in Sections 2.1 and 2.2 we derive a feedback law $\widetilde{\kappa}$ that is simpler than the optimal controller κ without sacrificing optimality, i.e., $\widetilde{\kappa}(x) = \kappa(x)$ for all x from the domain of κ. Then, in Section 2.3 we show how to construct an approximate feedback law $\widetilde{\kappa} \approx \kappa$ by trading complexity for suboptimality. In all cases, $\widetilde{\kappa}$ is a piecewise affine function as in Equation (5).

2.1 Clipping-Based Explicit MPC

The clipping-based approach to complexity reduction in explicit MPC, introduced in [25], aims at decreasing the computational and space complexity of Equations (7) and (8) by reducing the number of regions of the feedback law κ in Equation (5). It is based on the premise that, due to the presence of input constraints in Equation (2d), the optimal control action will become saturated at the boundary of the input constraint set $\mathscr{U}$ in some regions of the explicit MPC feedback law in Equation (5). A frequent case is when the input constraints are hyper-rectangular, i.e., $\mathscr{U} = \{u \mid u_{\min} \leq u \leq u_{\max}\}$. The central idea of this approach is to create a replacement feedback law $\widetilde{\kappa}$ by removing from the original feedback law κ the critical regions in which the optimal control action is saturated either at $u_{\min}$ or at $u_{\max}$, followed by covering the removed regions by geometric extensions of the remaining regions. The equivalence between κ and $\widetilde{\kappa}$ is then established by artificially clipping the value of $\widetilde{\kappa}$ at $u_{\min}$ and $u_{\max}$, respectively. As a consequence, the procedure described here allows one to replace the (complex) PWA feedback law κ by a simpler function $\widetilde{\kappa}$ without sacrificing optimality.

The idea is illustrated graphically in Figure 1. The example shows an explicit MPC feedback law $\kappa(x)$ defined over six critical regions. In three of them ($\mathscr{R}_1$, $\mathscr{R}_5$, $\mathscr{R}_6$) the optimal control action is saturated at $u_{\min} = -1$, while in $\mathscr{R}_3$ the upper saturation, i.e., $u^\star = u_{\max} = 1$ is attained. By removing the "saturated" regions $\mathscr{R}_1$, $\mathscr{R}_3$, $\mathscr{R}_5$, $\mathscr{R}_6$ and by extending the unsaturated regions $\mathscr{R}_2$ and $\mathscr{R}_4$ one obtains the function $\widetilde{\kappa}$. Consequently, $\widetilde{\kappa}(x) = \kappa(x)$ for all $x \in \mathscr{R}_2 \cup \mathscr{R}_4$. In the remainder of the domain of κ, the equivalence is attained by clipping the value of $\widetilde{\kappa}(x)$ to $u_{\min}$ and $u_{\max}$, as shown in the right-most part of Figure 1.

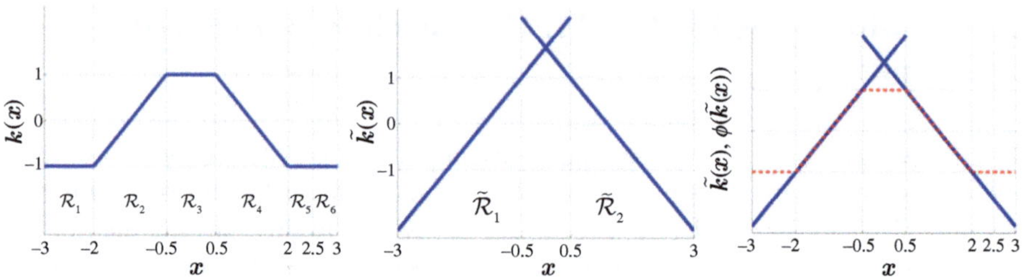

Fig. 1: From left to right: the original explicit MPC feedback law $\kappa(x)$, the augmented feedback law $\widetilde{\kappa}(x)$, the clipped version $\phi(\widetilde{\kappa}(x))$.

Technically speaking, the objective is to construct the PWA function

$$\widetilde{\kappa}(x) = \widetilde{L}_j x + \widetilde{\ell}_j \text{ if } x \in \widetilde{\mathscr{R}}_j, \tag{9}$$

where $\widetilde{\mathscr{R}}_j$, $j = 1,\ldots,\widetilde{M}$ are polyhedra, and the clipping function ϕ such that $\phi(\widetilde{\kappa}(x)) = \kappa(x)$ for all x in the domain of κ. The procedure for constructing the polyhedra $\widetilde{\mathscr{R}}_j$, along with local gains $\widetilde{L}_j$, $\widetilde{\ell}_j$, $j = 1,\ldots,\widetilde{M}$ is reported as Algorithm 1.

Algorithm 1: Clipping-based complexity reduction

INPUT: Explicit MPC feedback law κ as in Equation (5) with M critical regions.
OUTPUT: Augmented feedback law $\widetilde{\kappa}(x) = \widetilde{L}_j x + \widetilde{\ell}_j$ if $x \in \widetilde{\mathscr{R}}_j$, $j = 1,\ldots,\widetilde{M}$ with $\widetilde{M} < M$ regions.

1: Determine the index set $\mathscr{I}_{\text{unsat}} = \{i \in \{1,\ldots,M\} \mid u_{\min} < \kappa(x) < u_{\max}\}$ of unsaturated regions.
2: **for each** unsaturated region $i \in \mathscr{I}_{\text{unsat}}$ **do**
3: Identify the subset of half-space indices $\mathscr{J} \subseteq \{1,\ldots,c_i\}$ over which the neighbor of $\mathscr{R}_i$ is a saturated region.
4: Form a new polyhedron $\widetilde{\mathscr{R}}_i = \{x \,|\, \widetilde{Z}_i x \leq \widetilde{z}_i\}$ by removing from $\mathscr{R}_i$ the half-spaces indexed by $\mathscr{J}$, i.e., let $\widetilde{Z}_i = [Z_i]_{\setminus \mathscr{J}}$ and $\widetilde{z}_o = [z_i]_{\setminus \mathscr{J}}$.
5: Let $\widetilde{\mathscr{R}}_i = \widetilde{\mathscr{R}}_i \setminus \bigcup_{j \in \mathscr{I}_{\text{unsat}} \setminus i} \mathscr{R}_i$.
6: Store region(s) $\widetilde{\mathscr{R}}_i$ and matrices $\widetilde{L}_i = L_i$, $\widetilde{\ell}_i = \ell_i$.
7: **end for**

The algorithm processes all unsaturated critical regions $\mathscr{R}_i$ of the original explicit MPC feedback law κ sequentially. For each region, it first determines the half-spaces over which $\mathscr{R}_i$ has a saturated neighbor. Subsequently, these half-spaces are removed in Step 4, which leads to a new polyhedron $\widetilde{\mathscr{R}}_i$ as a geometric extension of $\mathscr{R}_i$, i.e., $\widetilde{\mathscr{R}}_i \supseteq \mathscr{R}_i$. However, when doing so, one must prevent that $\widetilde{\mathscr{R}}_i$ intersects with other unsaturated regions. This is performed by the set difference operation in Step 5. In the best case (when the set difference is empty for all unsaturated regions processed by the algorithm), the procedure constructs the augmented feedback law $\widetilde{\kappa}$ defined over $\widetilde{M} = M_{\text{unsat}}$ regions, where M_{unsat} is the number of unsaturated regions of κ. However, the set difference between a polyhedron $\widetilde{\mathscr{R}}_i$ and a (possibly non-convex) union of polyhedra $\bigcup_{j\in\mathscr{I}_{\text{unsat}}\setminus i}\mathscr{R}_i$ performed in Step 5 can lead to a subdivision of $\widetilde{\mathscr{R}}_i$ into several polyhedra, see, e.g., [4]. Software algorithms to compute such a set difference can be found, e.g., in the Multi-Parametric Toolbox.

Once the simple augmented feedback law $\widetilde{\kappa}$ as in Equation (9) is constructed, its equivalence to the original (complex) feedback κ is recovered by clipping the value of $\widetilde{\kappa}(x)$ to $u_{\min}$ and $u_{\max}$, respectively. Technically, this is achieved by evaluating the clipping function

$$\phi(\widetilde{\kappa}(x)) = \begin{cases} u_{\max} & \text{if } \widetilde{\kappa}(x) \geq u_{\max}, \\ u_{\min} & \text{if } \widetilde{\kappa}(x) \leq u_{\min}, \\ \widetilde{\kappa}(x) & \text{otherwise.} \end{cases} \tag{10}$$

Efficiency of the presented procedure, expressed as the ratio $M/\widetilde{M}$, depends on the number of unsaturated regions in Equation (5). If the feedback law κ does not contain any saturated regions, then no simplification can be achieved. As observed, e.g., in [16], the number of unsaturated regions depends mainly on two factors: tightness of the input constraint set $\mathscr{U}$ and the selection of the input penalty R in Equation (2a). The tighter the constraints and/or the lower R is, the more regions will become saturated, hence enabling the presented approach to be more efficient.

To illustrate the efficacy of the proposed clipping-based procedure to reduce complexity of explicit MPC feedback laws, consider the open-loop unstable model of an F14 fighter jet in the lateral axis [30], whose states represent the pitch and attack angles and the respective angular velocities, with the flap angle as control input:

$$\dot{x} = \begin{bmatrix} -0.015 & -60.57 & 0 & -31.170 \\ 0.0001 & -1.341 & 0.993 & 0 \\ 0.0002 & 43.25 & -0.869 & 0 \\ 0 & 0 & 1 & 0 \end{bmatrix} x + \begin{bmatrix} 13.14 \\ -0.251 \\ -1.577 \\ 0 \end{bmatrix} u. \tag{11}$$

The states are constrained by $|x_i| \leq 10$, $i = 1, \ldots, 4$, and the control command is bounded by $|u| \leq u_{\max}$. The model in Equation (11) was discretized using a sampling time of 0.01 seconds and the MPC problem in Equation (2) was formulated with $Q = P = R = 1$, and prediction horizon $N = 15$. We have investigated how tightness of input constraints (represented by $u_{\max}$) impacts the number of unsaturated regions, which is the main factor that determines efficiency of the proposed scheme. Therefore, we have computed explicit RHMPC feedback laws for $u_{\max} = \{2, 4, 6, 8, 10\}$. Each feedback law κ was then processed by Algorithm 1 to obtain the replacement function $\widetilde{\kappa}$. The results are summarized in Table 1. Columns of the table report, respectively, maximal control amplitude $u_{\max}$, complexity of the original feedback κ (in terms of the number M of critical regions, the worst-case computational complexity per Equation (7) in FLOPS, and the space complexity per Equation (8) in floating point numbers), complexity of the replacement function $\widetilde{\kappa}$, and the complexity reduction ratio. In all cases we got $\widetilde{M} = M_{\text{unsat}}$, i.e., the augmented function $\widetilde{\kappa}$ always consisted only of the unsaturated regions of κ.

As expected, both the computational complexity represented by Equation (7), and the space complexity of Equation (8) decrease proportionally to the ratio $M/\widetilde{M}$. This fraction decreases when the input constraints become less strict, as pointed out above. As a consequence, this example demonstrates that the implementation com-

Table 1: Results for the F14 example.

$u_{\max}$	Original feedback κ			Augmented feedback $\widetilde{\kappa}$			$M/\widetilde{M}$
	M	$\mathscr{C}(\kappa)$ [FLOPS]	$\mathscr{S}(\kappa)$	$\widetilde{M}$	$\mathscr{C}(\widetilde{\kappa})$ [FLOPS]	$\mathscr{S}(\widetilde{\kappa})$	
2	1,167	49,998	47,500	170	8,226	7,705	6.9
4	1,290	56,292	53,360	221	11,442	10,640	5.8
6	1,391	61,086	57,860	289	14,874	13,840	4.8
8	1,438	63,096	59,770	317	16,188	15,075	4.5
10	1,507	66,030	62,560	364	18,192	16,980	4.1

plexity of explicit MPC (both in terms of computation and space) can be improved by factors ranging from 4 to 7 just by exploiting the geometric properties of the explicit MPC feedback laws.

2.2 Regionless Explicit MPC

Another option to decrease the computational and the space complexity of explicit MPC feedback laws without sacrificing optimality is to reduce the amount of data required for storage of individual critical regions $\mathscr{R}_i$ in Equation (5), i.e., the size of the matrices $Z_i \subseteq \mathbb{R}^{c_i \times n_x}$, $z_i \subseteq \mathbb{R}^{n_x}$, $i = 1, \ldots, M$ in Equation (6). Specifically, we show that, in fact, one *does not need* to describe the critical regions using individual half-spaces captured by matrices Z_i, z_i. Instead, the critical regions are constructed *on-the-fly* using the constraints of the mp-QP problem in Equation (3). The advantage being that the underlying data (represented by matrices G, w, E in Equation (3b)) are stored just once and shared among all regions. An example is provided to illustrate that such a procedure yields considerable reduction of the space complexity of explicit MPC.

In conventional explicit MPC, the critical regions $\mathscr{R}_i$ in Equation (6) are obtained by investigating the Karush-Kuhn-Tucker (KKT) optimality conditions for the mp-QP problem in Equation (3):

$$HU^\star + F^\top x + G_{\mathscr{A}}^\top \lambda^\star = 0, \tag{12a}$$
$$GU^\star \leq w + Ex, \tag{12b}$$
$$\lambda^\star \geq 0, \tag{12c}$$
$$\lambda_i^\star (G_{\mathscr{A}_i} U^\star - w_{\mathscr{A}_i} - E_{\mathscr{A}_i} x) = 0. \tag{12d}$$

Here, $\mathscr{A} \subseteq \{1, \ldots, q\}$ is the index set of constraints that are active for some value of the parameter vector x with q denoting the total number of inequality constraints in Equation (3b), and $G_{\mathscr{A}}$ are the rows of G indexed by $\mathscr{A}$. For each fixed $\mathscr{A}$, Equations (12a) and (12d) give

$$\begin{bmatrix} H & G_{\mathscr{A}}^\top \\ G_{\mathscr{A}} & 0 \end{bmatrix} \begin{bmatrix} U^\star \\ \lambda^\star \end{bmatrix} = \begin{bmatrix} -F^\top x \\ w_{\mathscr{A}} + E_{\mathscr{A}} x \end{bmatrix}. \tag{13}$$

Recall that the Hessian H in Equation (3) is assumed to be positive definite, thus invertible, and $G_{\mathscr{A}}$ is assumed to have full row rank.[1] Then solving for $U^\star$ and $\lambda^\star$ from Equation (13) yields

$$U^\star = L_{\mathscr{A}} x + \ell_{\mathscr{A}}, \tag{14a}$$
$$\lambda^\star = S_{\mathscr{A}} x + s_{\mathscr{A}}, \tag{14b}$$

with

$$L_{\mathscr{A}} = H^{-1}\left(G_{\mathscr{A}}(G_{\mathscr{A}}H^{-1}G_{\mathscr{A}}^\top)^{-1}(E_{\mathscr{A}} + G_{\mathscr{A}}H^{-1}F^\top) - F^\top\right) \tag{15a}$$
$$\ell_{\mathscr{A}} = H^{-1}G_{\mathscr{A}}^\top(G_{\mathscr{A}}H^{-1}G_{\mathscr{A}}^\top)^{-1}w_{\mathscr{A}}, \tag{15b}$$
$$S_{\mathscr{A}} = -(G_{\mathscr{A}}H^{-1}G_{\mathscr{A}}^\top)^{-1}(E_{\mathscr{A}} + G_{\mathscr{A}}H^{-1}F^\top), \tag{15c}$$
$$s_{\mathscr{A}} = -(G_{\mathscr{A}}H^{-1}G_{\mathscr{A}}^\top)^{-1}w_{\mathscr{A}}. \tag{15d}$$

Therefore, the optimal Lagrange multipliers $\lambda^\star$ as well as the primal optimizer $U^\star$, are both affine functions of the parameter x. The region of validity of the expressions in Equations (14b) and (14a) is then obtained by plugging the respective expressions into the primal and dual feasibility constraints Equations (12b) and (12c), respectively. As both $\lambda^\star$ and $U^\star$ are affine functions of x, Equations (12b)–(12c) yield a set of linear inequalities in x:

$$\underbrace{\begin{bmatrix} GL_{\mathscr{A}} - E \\ -S_{\mathscr{A}} \end{bmatrix}}_{Z_{\mathscr{A}}} x \leq \underbrace{\begin{bmatrix} w - G\ell_{\mathscr{A}} \\ s_{\mathscr{A}} \end{bmatrix}}_{z_{\mathscr{A}}}, \tag{16}$$

that constitute the critical region $\mathscr{R}_{\mathscr{A}}$ as in Equation (6). The remaining critical regions of κ in Equation (5) are then constructed similarly by exploring all other optimal active sets $\mathscr{A}$. These can be obtained either by an exhaustive enumeration as proposed in [17], or by exploiting the geometry of existing critical regions, see, e.g., [3, 6, 9].

To sum up, by using the procedure described above, for each optimal active set $\mathscr{A}_i$ one obtains the locally optimal feedback gains $L_i := L_{\mathscr{A}_i}$, $\ell_i := \ell_{\mathscr{A}_i}$ as in Equations (5) and (14a) that define the primal optimizer $U_N^\star$, along with the associated polyhedral critical region $\mathscr{R}_i$ given per Equations (6) and (16) with $Z_i := Z_{\mathscr{A}_i}$, $z_i := z_{\mathscr{A}_i}$. In view of Equation (16), each critical region consists of q half-spaces. However, in practice, some of these half-spaces will be redundant and can be removed. Therefore each region will be defined by $c_i \leq q$ inequalities. Moreover, since the expressions in Equation (16) depend on the choice of the active set, the inequalities will, in general, be different for each region. This leads to the necessity of describing each critical region by its own set of data.

The idea behind the *regionless* approach is to abolish the necessity of explicitly storing each critical region in the memory of the implementation hardware. Instead,

[1] If $G_{\mathscr{A}}$ does not have a full row rank, it is always possible to identify a subset of $\mathscr{A}$ such that all rows of $G_{\mathscr{A}}$ are linearly independent, see, e.g., [37].

the inclusion $x \in \mathscr{R}_i$ is validated by directly checking the primal and dual feasibility conditions in Equations (12b) and (12c), respectively. To do so, one needs to store three ingredients in the memory: the primal constraints, represented by matrices G, w, E from Equation (3b), the expressions for the primal optimizer as in Equation (14a), i.e., L_i, ℓ_i, $i = 1, \ldots, M$, and the matrices S_i, s_i that describe the dual optimizer as in Equation (14b). Then the inclusion $x \in \mathscr{R}_i$, required by the sequential search procedure to evaluate the function κ for a given value of the parameter vector x is done as follows:

1. compute $U = L_i x + \ell_i$;
2. compute $\lambda = S_i x + s_i$;
3. check if $GU \leq w + Ex$ (primal feasibility) and $\lambda \geq 0$ (dual feasibility).

If all inequalities in the ultimate step are satisfied, then, by Equation (12), the pair (U, λ) is optimal for a given value of x. In practice, however, one would first compute and check the dual optimizer before validating the primal one since its cardinality is smaller. In fact, the cardinality of λ is, at most, Nn_u. On the other hand, we have q primal constraints with $q \gg Nn_u$ in practice.

The storage complexity of the regionless format is then

$$\mathscr{S}(\kappa) = 2MNn_{\mathrm{u}}(n_{\mathrm{x}} + 1) + q(Nn_{\mathrm{u}} + n_{\mathrm{x}} + 1), \tag{17}$$

where the first term accounts for the storage of the primal and the dual optimizer, i.e., the matrices L_i, ℓ_i, S_i, s_i, and the second one determines the size of the primal constraints G, w, E from Equation (3b). Comparing this figure to Equation (8) we see that the regionless format consumes less memory provided $M > N$, a condition that is often satisfied. In fact, usually $M \gg N$ in practice, since the number M of critical regions grows, in the worst case, exponentially with the prediction horizon N.

To highlight the benefits of the described approach, we have investigated the control of a 10-tray rectification column, described in [15]. The plant was modeled as an LTI system with $n_{\mathrm{x}} = 13$ states and one input. For such a system we have subsequently formulated the MPC problem per Equation (2) with the prediction horizon $N = 40$, and calculated the explicit MPC feedback law as in Equation (5) by the Multi-Parametric Toolbox. The solution consisted of $1,095$ critical regions in the 13-dimensional parametric space. The total memory footprint of such a region-based explicit controller was 40 megabytes, the majority of which was consumed by the description of the critical regions. The regionless representation of the same feedback law, on the other hand, only requires 224 kilobytes of storage capacity, a reduction of two orders of magnitude. The regionless description was subsequently exported to C-code and implemented in Simulink. Experimental results under the proposed controller are shown in Figure 2.

2.3 Piecewise Affine Approximation of Explicit MPC

The second principal approach to reducing the computational and space complexities of explicit MPC feedback laws $U_N^\star = \kappa(x)$ in Equation (5) is to replace the function κ by a different function $\widetilde{\kappa}$ such that the control sequence $U_N = \widetilde{\kappa}(x)$ satisfies the constraints in Equation (3b), but $\widetilde{\kappa}$ is allowed to be suboptimal, i.e., $\widetilde{\kappa}(x) \neq \kappa(x)$ for some (or even for all) points x. Various approaches to accomplish this task have been proposed in the literature, ranging from using barycentric interpolation [21], through the application of machine learning algorithms [12, 13, 22] and neural networks [35], up to approximating the explicit MPC control law by polynomials [26, 33, 34].

In what follows we show how to synthesize a replacement PWA function $\widetilde{\kappa}$ that, when used as a feedback law $U_N = \widetilde{\kappa}(x)$, provides recursive constraint satisfaction and is as close as possible to the original complex explicit MPC feedback law κ as in Equation (5) in some measure. The presented approach is based on [20].

Given is the explicit PWA MPC feedback law $\kappa(x)$ as in Equation (5) with a total of M polyhedral critical regions $\mathscr{R}_i$, $i = 1,\ldots,M$ as in Equation (6). Let a new set of regions $\widetilde{\mathscr{R}}_j$, $j = 1,\ldots,\widetilde{M}$ with $\widetilde{M} < M$ and $\bigcup_i \mathscr{R}_i = \bigcup_j \widetilde{\mathscr{R}}_j$ be given. We seek the parameters $\widetilde{L}_j$, $\widetilde{\ell}_j$, $j = 1,\ldots,\widetilde{M}$ of the PWA function

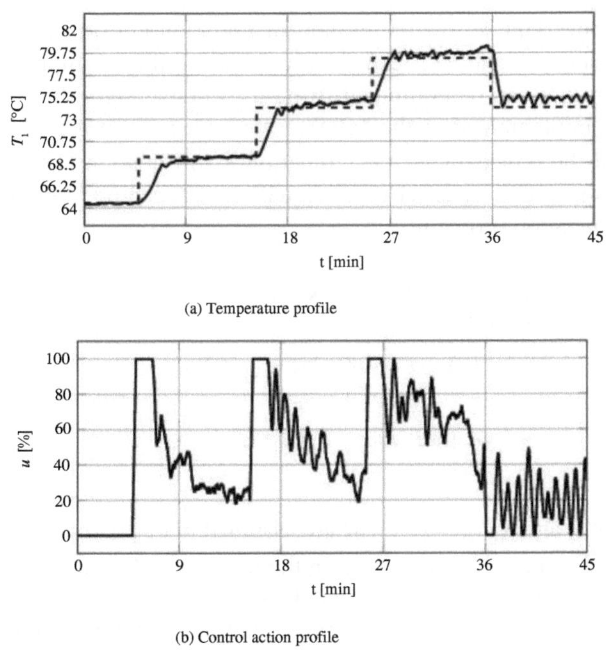

Fig. 2: Experimental results for the rectification column example. The dashed line represents the setpoint to be tracked.

$$\widetilde{\kappa}(x) = \widetilde{L}_j x + \widetilde{\ell}_j \text{ if } x \in \widetilde{\mathscr{R}}_j,\ j = 1, \dots, \widetilde{M}, \tag{18}$$

such that:

1. $U_N = \widetilde{\kappa}(x)$ satisfies the constraints in Equation (3b), i.e., U_N is always feasible;
2. the integrated squared error

$$\int_{\cup_i \mathscr{R}_i} \|\kappa(x) - \widetilde{\kappa}(x)\|_2^2 \, \mathrm{d}x \tag{19}$$

is minimized, i.e., $\widetilde{\kappa}(x)$ is as close as possible to $\kappa(x)$ with respect to the measure in Equation (19).

If $\widetilde{L}_j$, $\widetilde{\ell}_j$ satisfying these properties could be found, the replacement feedback law in Equation (18) is an approximate replacement of the original (complex) explicit MPC feedback law in Equation (5). It should be noted that various performance measures can be considered instead of Equation (19). For example, one could minimize the worst-case error between $\kappa(x)$ and $\widetilde{\kappa}(x)$, i.e., to minimize $\|\kappa(x) - \widetilde{\kappa}(x)\|_\infty$ over all $x \in \cup_i \mathscr{R}_i$. Alternatively, one could minimize the point-wise error $\sum_{i=1}^{K} \|\kappa(x_i) - \widetilde{\kappa}(x_i)\|_2^2$ over a set of points $x_1, \dots, x_K$ obtained, for instance, by gridding the domain of κ.

Let

$$\mathscr{C}_\infty = \{x \mid \exists u_k \in \mathscr{U} \text{ s.t. } Ax_k + Bu_k \in \mathscr{X},\ \forall k \geq 0\} \tag{20}$$

be the control invariant set for the system in Equation (1) subject to state and input constraint sets $\mathscr{X}$ and $\mathscr{U}$, respectively. Under mild assumptions, the set $\mathscr{C}_\infty$ is a polyhedron that can be computed using a recursive algorithm described, e.g., in [8, 14], and implemented in the MPT toolbox [18]. Then the parameters $\widetilde{L}_j$ $\widetilde{\ell}_j$, $j = 1, \dots, \widetilde{M}$ of the replacement PWA function $\widetilde{\kappa}$ in Equation (18) that minimizes the integrated squared error criterion of Equation (19) can be found by solving the following optimization problem:

$$\min_{\widetilde{L}_j, \widetilde{\ell}_j} \int_{\cup_j \widetilde{\mathscr{R}}_j} \|\kappa(x) - \widetilde{\kappa}(x)\|_2^2 \, \mathrm{d}x \tag{21a}$$

$$\text{s.t. } \widetilde{L}_j x + \widetilde{\ell}_j \in \mathscr{U},\ \forall x \in \bigcup_j \widetilde{\mathscr{R}}_j \tag{21b}$$

$$Ax + B(\widetilde{L}_j x + \widetilde{\ell}_j) \in \mathscr{C}_\infty,\ \forall x \in \bigcup_j \widetilde{\mathscr{R}}_j. \tag{21c}$$

Since the sets $\mathscr{C}_\infty$, and $\widetilde{R}_j$, $j = 1, \dots, \widetilde{M}$ are assumed to be polyhedra, the constraints in Equations (21b) and (21c) are linear in the unknowns $\widetilde{L}_j$, $\widetilde{\ell}_j$. The objective function in Equation (21a), on the other hand, is nonlinear in the decision variables. However, it can be translated into a convex quadratic function using the following result, due to [2]:

Lemma 1. *Let f be a homogeneous polynomial of degree d in n variables, and let $s_1, \dots, s_{n+1}$ be the vertices of an n-dimensional simplex Δ. Then*

$$\int_{\Delta} f(y)dy = \beta \sum_{1 \le i_1 \le \cdots \le i_d \le n+1} \sum_{\varepsilon \in \{\pm 1\}^d} \left(\left(\prod_{j=1}^{d} \varepsilon_j \right) \cdot f(\Sigma_{k=1}^{d} \varepsilon_k s_{i_k}) \right), \tag{22}$$

where

$$\beta = \frac{\mathrm{vol}(\Delta)}{2^d d! \binom{d+n}{d}}, \tag{23}$$

and $\mathrm{vol}(\Delta)$ *is the volume of the simplex.*

To apply Lemma 1 to the objective function Equation (21a) consider, for each (i, j) combination, the tessellation of the intersections $\mathscr{R}_i \cap \widetilde{\mathscr{R}}_j$ into simplices Δ_k. Moreover, let $f_{i,j,k}(x) := \|\kappa_i(x) - \widetilde{\kappa}_j(x)\|_2^2$ with $\kappa_i(x) = L_i x + \ell_i$ and $\widetilde{\kappa}_j = \widetilde{L}_j x + \widetilde{\ell}_j$ restricted to a particular simplex Δ_k. Then $f_{i,j,k}(x) := x^\top Q_{i,j} x + q_{i,j}^\top x + r_{i,j}$ with

$$Q_{i,j} = L_j^\top L_j - 2L_j \widetilde{L}_i + \widetilde{L}_i^\top \widetilde{L}_i, \tag{24a}$$

$$q_{i,j} = 2(L_j^\top \widetilde{\ell}_i + \widetilde{L}_i^\top \widetilde{\ell}_i - \widetilde{L}_i^\top \ell_j - L_j^\top \widetilde{\ell}_i), \tag{24b}$$

$$r_{i,j} = \ell_j^\top \ell_j - 2\ell_j^\top \widetilde{\ell}_i + \widetilde{\ell}_i^\top \widetilde{\ell}_i. \tag{24c}$$

It follows that the integrated squared error between κ and $\widetilde{\kappa}$ can be obtained by replacing the objective function in Equation (21a) by the evaluation of the expression in Equation (22) for all $f_{i,j,k}$ defined above. Since $f_{j,j,k}$ are convex quadratic functions of $\widetilde{L}_j$ and $\widetilde{\ell}_j$, the optimization problem in Equation (21) can thus be posed as a convex quadratic program (QP) that can be solved using off-the-shelf tools.

The complete procedure can be summarized as follows:

1. Obtain the regions $\widetilde{\mathscr{R}}_j$, $j = 1, \ldots, \widetilde{M}$ with $\widetilde{M} < M$, e.g., by solving the MPC problem in Equation (2) for a shorter prediction horizon.
2. Tessellate the intersections $\mathscr{R}_i \cap \widetilde{\mathscr{R}}_j$ into simplices Δ_k for all i, j pairs for which the intersection is non-empty.
3. Solve the QP problem in Equation (21) with the objective function in Equation (21) replaced by Equation (22) to obtain $\widetilde{L}_j$, $\widetilde{\ell}_j$.

Since the QP problem enforces input constraints via Equation (21b) and invariance via Equation (21c), the resulting PWA function $\widetilde{\kappa}$ as in Equation (18) will provide recursive constraint satisfaction. Moreover, by minimizing the integrated squared error as in Equation (19), $\widetilde{\kappa}$ will be the best possible approximation of the original explicit MPC feedback law κ. Naturally, the quality of the approximation is inversely proportional to the complexity of the function $\widetilde{\kappa}$, i.e., the higher $\widetilde{M}$ (the number of regions of $\widetilde{\kappa}$), the lower approximation error as in Equation (19) can be achieved.

To illustrate the procedure, consider an inverted pendulum mounted on a moving cart, whose linearization around the upright unstable equilibrium reads

$$\begin{bmatrix} \dot{p} \\ \ddot{p} \\ \dot{\phi} \\ \ddot{\phi} \end{bmatrix} = \begin{bmatrix} 0 & 1 & 0 & 0 \\ 0 & -0.182 & 2.673 & 0 \\ 0 & 0 & 0 & 1 \\ 0 & -0.455 & 31.182 & 0 \end{bmatrix} \begin{bmatrix} p \\ \dot{p} \\ \phi \\ \dot{\phi} \end{bmatrix} + \begin{bmatrix} 0 \\ 1.818 \\ 0 \\ 4.546 \end{bmatrix} u, \tag{25}$$

where p is the position of the cart, $\dot{p}$ is the cart's velocity, ϕ is the pendulum's angle from the upright position, and $\dot{\phi}$ denotes the angular velocity. The control input u is proportional to the force applied to the cart. The system in Equation (25) was then discretized using the sampling time of 0.1 seconds. The optimal (complex) controller $U_N^\star = \kappa(x)$ in Equation (5) was then constructed by solving the MPC problem in Equation (2) with prediction horizon $N = 8$, penalties $Q = \text{diag}(10, 1, 10, 1)$, $R = 0.1$, and constraints $|p| \leq 1$, $|\dot{p}| \leq 1.5$, $|\phi| \leq 0.35$, $|\dot{\phi}| \leq 1$, $|u| \leq 1$. Using the MPT toolbox we have obtained the explicit MPC feedback law κ defined over 943 polytopes of the 4-dimensional state-space. Subsequently, we have solved the same problem for shorter prediction horizons to obtain new sets of polyhedral regions $\widetilde{\mathscr{R}}_j$ with $\widetilde{M} = 35$ for $N = 1$, $\widetilde{M} = 117$ for $N = 2$, and $\widetilde{M} = 273$ for $N = 3$. Finally, we have searched for the parameters $\widetilde{L}_j$, $\widetilde{\ell}_j$ of the replacement PWA feedback law as in Equation (18) by solving the QP problems in Equation (21). For each investigated case, we have then evaluated the suboptimality of the replacement feedback law $\widetilde{\kappa}$ by measuring the average settling time (AST) of the pendulum, starting from a non-zero initial condition. The results are reported in Table 2. As can be seen, optimizing the parameters of $\widetilde{\kappa}$ in Equation (18) via the QP problems in Equation (21) allows one to employ fairly simple control laws at a modest loss of performance. Moreover, the complexity of the replacement feedback law is a design choice that can be used to trade performance for complexity.

Table 2: Complexity and suboptimality comparison for the pendulum example.

Prediction horizon	Number of regions	AST	Suboptimality
1	35	5.1 s	59.4 %
2	117	3.7 s	15.6 %
3	273	3.4 s	6.3 %
8	943	3.2 s	0.0 %

3 Approximation of MPC Feedback Laws for Nonlinear Systems

3.1 Problem Setup

This section describes a method for designing explicit MPC controllers for control of nonlinear systems. The obtained controllers involve closed-loop stability guarantees and closed-loop performance optimized with respect to a given metric. The method is based on the stability verification results of [23] which are in turn inspired by [36].

Figure 3 depicts the design scheme. We consider a quadratic programming (QP) based MPC controller whose coefficients defining the cost/constraint matrices and vectors are the control tuning parameters grouped into the tuning vector $\eta \in \mathbb{R}^{n_\eta}$. In particular, even though the controlled system is nonlinear, the optimization problem underlying the MPC controller is a convex QP and hence amenable to explicit MPC implementation or tailored online QP solvers.

Specifically we assume a system of the form

$$x^+ = f_x(x,u), \tag{26a}$$
$$y = f_y(x), \tag{26b}$$

with the only assumption being that f_x and f_y are multivariate polynomials. In general, the output y does not need to be the physical output of the dynamical system but, for example, the estimate of the state provided by a state estimator whose dynamics are lumped into the function f_x, or several consecutive values of the physical output whose previous values are recorded as a part of the state x. Similarly, the mapping $f_y(\cdot)$ can encode the so-called lifting of the state to a higher dimensional space, where the underlying non-linear dynamical system is well approximated by a linear dynamical system (see [24] for details of this procedure).

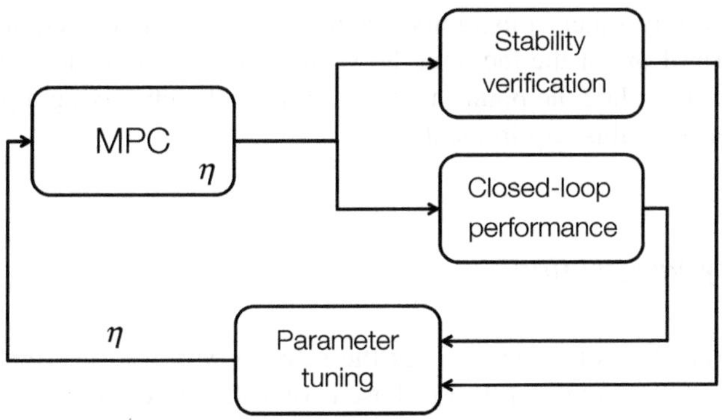

Fig. 3: Design scheme.

In the following sections we describe in detail the blocks of the control scheme depicted in Figure 3.

3.2 A QP-Based MPC Controller

We assume that at each step of the closed-loop operation the MPC controller solves a QP optimization problem parametrized by the output $y \in \mathbb{R}^{n_y}$ and the tuning parameters $\eta \in \mathbb{R}^{n_\eta}$. The QP has the form

$$\begin{aligned} &\underset{\theta\in\mathbb{R}^{n_\theta}}{\text{minimize}}\ \tfrac{1}{2}\theta^\top \mathscr{H}\theta + y^\top \mathscr{B}\theta \\ &\text{subject to } \mathscr{C}\theta \le d \end{aligned} \tag{27}$$

where $\theta \in \mathbb{R}^{n_\theta}$ is the decision vector, matrix $\mathscr{H} \in \mathbb{R}^{n_\theta \times n_\theta}$ is symmetric positive (semi)definite, $\mathscr{B} \in \mathbb{R}^{n_y \times n_\theta}$, $\mathscr{C} \in \mathbb{R}^{n_d \times n_\theta}$ and $d \in \mathbb{R}^{n_d}$. Given a solution θ to (27), the control input is obtained as

$$u = \kappa(\theta), \tag{28}$$

where κ is a given multivariate polynomial (typically just a selection of some components of θ). The tuning parameters vector η in (27) is

$$\eta = \begin{bmatrix} \mathrm{svec}(\mathscr{H}) \\ \mathrm{vec}(\mathscr{B}) \\ \mathrm{vec}(\mathscr{C}) \\ d \end{bmatrix},$$

where $\mathrm{vec}(\cdot)$ denotes a vectorization of a matrix and $\mathrm{svec}(\cdot)$ a symmetric vectorization (i.e., the elements uniquely defining a symmetric matrix arranged in a column vector). The constraints on the parameter vector η are the positive (semi)definiteness of $\mathscr{H}$ and possibly other constraints arising from application requirements, e.g., sparsity pattern of $\mathscr{H}$ or the fact that the vector θ of decision variables should not cause inputs that violate the plant input constraints (thereby fixing a part of $\mathscr{C}$ and d to constraint θ for this requirement).

3.3 Stability Verification

The tuning procedure of Figure 3 uses the stability verification from [23] for the closed-loop interconnection of the QP-based controller described in Section 3.2 and the dynamical system (26). For the purpose of stability verification, one needs to observe that the KKT system associated to the optimization problem (27) is a basic semialgebraic set, i.e., the intersection of finitely many polynomial equalities and inequalities. Indeed, the KKT conditions of (27) read

$$\mathscr{H}\theta + \mathscr{B}^\top y + \mathscr{C}^\top \lambda = 0 \tag{29a}$$

$$\lambda^\top (\mathscr{C}\theta - d) = 0 \tag{29b}$$

$$\lambda \geq 0 \tag{29c}$$

$$d - \mathscr{C}\theta \geq 0, \tag{29d}$$

where $\lambda \in \mathbb{R}^{n_d}$ is the vector of Lagrange multipliers associated with the constraint $\mathscr{C}\theta \leq d$. The KKT conditions are a system of polynomial equalities and inequalities in $(\theta, y, \eta, \lambda_a, \lambda_b)$ where the vector η contains the coefficients of the QP problem, as described in Section 3.2. By defining

$$\mathbf{T}_{y,\eta} := \{\theta \in \mathbb{R}^{n_\theta} \mid \exists \lambda \in \mathbb{R}^{n_\lambda} \text{ s.t. } g(\theta,\lambda,y,\eta) \geq 0,\ h(\theta,\lambda,y,\eta) = 0\} \tag{30}$$

where

$$h(\theta,\lambda,y,\eta)=\begin{bmatrix}\mathcal{H}\theta+\mathcal{B}^\top y+\mathcal{C}^\top\lambda\\ \lambda^\top(\mathcal{C}\theta-d)\end{bmatrix}$$

and

$$g(\theta,\lambda,y,\eta)=\begin{bmatrix}\lambda\\ d-\mathcal{C}\theta\end{bmatrix},$$

we observe that the control input generated by the MPC controller satisfies

$$u\in\kappa(\mathbf{T}_{y,\eta}), \tag{31}$$

provided that constraint qualification conditions hold such that each minimizer of (27) satisfies the KKT conditions (29).

3.3.1 Lypunov Analysis

As a result, the question of stability of the closed-loop interconnection of (26) and the QP-based MPC controller boils down to the question of stability of the difference inclusion obtained by interconnecting (26) and (31). Stability of such a difference inclusion can be analyzed using Lyapunov methods. For a given parameter value η, the closed-loop interconnection of (26) and the QP-based controller is stable if there exists a Lypunov function V_η such that

$$V_\eta(x^+,\theta^+,\lambda^+)-V_\eta(x,\theta,\lambda)\le-\|c(x)\|_2^2, \tag{32a}$$

$$V_\eta(x,\theta,\lambda)\ge 0, \tag{32b}$$

for all

$$(x,\theta,\lambda,x^+,\theta^+,\lambda^+)\in\mathbf{K}_\eta$$

where

$$\begin{aligned}\mathbf{K}_\eta=\{(x,\theta,\lambda,x^+,\theta^+,\lambda^+)\mid x^+=f_x(x,\kappa(\theta)),&\\ \hat{h}_\eta(\theta,\lambda,x)=0,\ \hat{g}_\eta(\theta,\lambda,x)\ge 0,&\\ \hat{h}_\eta(\theta^+,\lambda^+,x^+)=0,\ \hat{g}_\eta(\theta^+,\lambda^+,x^+)\ge 0\},&\end{aligned} \tag{33}$$

with

$$\hat{h}_\eta(\theta,\lambda,x):=h(\theta,\lambda,f_y(x),\eta), \tag{34a}$$

$$\hat{g}_\eta(\theta,\lambda,x):=g(\theta,\lambda,f_y(x),\eta). \tag{34b}$$

To be more precise, the conditions (32) imply only $c(x_k)\to 0$, where $c(\cdot)$ is a given vector multivariate polynomial. If stability of the full state x is desired, then one selects $c(x)=x$. However, as remarked above, the state x may not represent only the physical state of the system but also, for instance, the state of the estimator in which case one may be interested only in the stability of a certain subset of the state x which would translate to choosing $c(x) = Cx$ for some matrix C.

3.3.2 Sum-of-Squares Certificates

In general, seeking a Lyapunov function satisfying conditions (32) is hard since one searches over the class of all functions and requires the satisfaction of the inequalities (32a) and (32b) for all elements of $\mathbf{K}_\eta$. In order to obtain computable stability certificates, we restrict the class of Lyapunov functions to the set of all polynomials and replace the nonnegativity conditions (32a) and (32b) by sufficient polynomial sum-of-squares (SOS) conditions. This is possible precisely because all data is polynomial and hence the set $\mathbf{K}_\eta$ is defined by polynomial equalities and inequalities. Setting

$$\xi := (x, \theta, \lambda, x^+, \theta^+, \lambda^+),$$

the sufficient SOS conditions read

$$\begin{aligned} V(x,\theta,\lambda) - V(x^+,\theta^+,\lambda^+) - \|x\|_2^2 = \sigma_0(\xi) & \qquad (35a)\\ + \sigma_1(\xi)^\top \hat{g}(\theta,\lambda,x) + \sigma_2(\xi)^\top \hat{g}(\theta^+,\lambda^+,x^+) &\\ + p_1(\xi)^\top \hat{h}(\theta,\lambda,x) + p_2(\xi)^\top \hat{h}(\theta^+,\lambda^+,x^+) &\\ + p_3(\xi)(x^+ - f_x(x,\kappa(\theta))) &\\ V(x,\theta,\lambda) = \bar{\sigma}_0(\xi) + \bar{\sigma}_1(\xi)^\top \hat{g}(\theta,\lambda,x) + \bar{p}_1(\xi)^\top \hat{h}(\theta,\lambda,x), & \qquad (35b) \end{aligned}$$

where $\sigma_i(\xi)$ and $\bar{\sigma}_i(\xi)$ are SOS multipliers and $p_i(\xi)$ and $\bar{p}_i(\xi)$ polynomial multipliers of compatible dimensions and pre-specified degrees. The satisfaction of (35a) implies the satisfaction of (32a) and the satisfaction of (35b) implies the satisfaction of (32b) for all $\xi \in \mathbf{K}_\eta$. Therefore, a sufficient condition for stability of the closed-loop interconnection of (26) and the MPC controller is feasibility of the SOS problem:

$$\begin{array}{lll} \text{find} & V, \sigma_0, \sigma_1, \sigma_2, p_1, p_2, p_3, p_4, \bar{\sigma}_0, \bar{\sigma}_1, \bar{p}_1 & \\ \text{s.t.} & (35a),\ (35b) & \\ & \sigma_0, \sigma_1, \sigma_2, \bar{\sigma}_0, \bar{\sigma}_1 & \text{SOS polynomials} \\ & V, p_1, p_2, p_3, \bar{p}_1 & \text{arbitrary polynomials,} \end{array} \qquad (36)$$

where the decision variables are the coefficients of the polynomials

$$(V, \sigma_0, \sigma_1, \sigma_2, p_1, p_2, p_3, \bar{\sigma}_0, \bar{\sigma}_1, \bar{p}_1).$$

Since a polynomial $\sigma(\xi)$ of degree $2d$ is SOS if and only if there exists a positive semidefinite matrix W such that $\sigma(\xi) = \beta_d(\xi)^\top W \beta_d(\xi)$, where $\beta_d(\xi)$ is the vector of all monomials of total degree no more than d, the SOS problem (36) translates to a semidefinite programming (SDP) feasibility problem (see, e.g., [27] for more details on SOS programming). Importantly, the translation of the SOS problem (36) to a conic form accepted by common SDP solvers (e.g., MOSEK [38]) can be done automatically using freely-available high-level modelling tools such as Yalmip [28]. The stability of the closed-loop interconnection of a QP controller and a nonlinear dynamical system can hence be readily verified using convex optimization.

While the SOS problem (36) verifies global stability of the closed-loop system, a local stability verification over a set

$$\mathbf{X} = \{x \mid \phi_i \geq 0,\ i = 1,\dots,n_\phi\} \tag{37}$$

where ϕ_i are polynomial functions is also possible, as described in [23]. In particular, the local version is achieved by adding the inequality constraints of (37) to the set (30) and by subsequently assigning them the SOS multipliers in the conditions (35). Although the obtained Lyapunov function would involve the guaranteed decrease (35a) over the set **X**, this does not guarantee the invariance of the closed-loop system over the set **X**. A sufficient condition for invariance in **X** is that the initial state is within the largest sub-level set of the Lyapunov function which is contained in **X**. For a more detailed description of this concept and stability certification of this section, the reader is referred to [23].

3.4 Closed-Loop Performance

The goal of the design procedure is to find a stabilizing tuning parameter η that minimizes a *closed-loop* performance metric of the form

$$J_\eta(x_0) = \sum_{k=0}^{\infty} \alpha^k l(x_k, u_k) \tag{38}$$

for a range of initial states x_0 and with the discount factor $\alpha \in (0,1]$. The stage cost l can be non-convex even though the optimization problem underlying the MPC controller is a convex QP (27). In order to take into account various starting points x_0, the minimization of $J_\eta(x)$ is carried out on average over a given region of interest $\mathbf{Y} \subset \mathbb{R}^{n_x}$ by considering as the performance metric

$$P(\eta) = \int_{\mathbf{Y}} J_\eta(x) w(x)\, dx, \tag{39}$$

with w being a given nonnegative weighting function. In practice the objective function $P(\eta)$ is approximated by uniform sampling from **Y** and by truncation of the infinite sum in (38); i.e., we minimize

$$\hat{P}(\eta) = \sum_{i=1}^{M} \sum_{k=0}^{N} \alpha^k l(x_k^i, u_k^i), \tag{40}$$

where the initial conditions x_0^i, $i = 1,\dots,M$, are sampled uniformly from **Y**.

3.5 Parameter Tuning

The parameter tuning can be formulated as an optimization problem of the form

$$\begin{aligned} &\underset{\eta \in \mathbb{R}^{n_\eta}}{\text{minimize}} \; P(\eta) + \delta_{st}(\eta) \\ &\text{subject to } \eta \in \Omega, \end{aligned} \tag{41}$$

where $P(\eta)$ is the performance metric (39) possibly evaluated approximatively as (40), the set Ω is modelling some basic requirements on the tuning parameters η (e.g., symmetry and positive definiteness of the matrix $\mathscr{H}$), and $\delta_{st}(\eta)$ is a function indicating the existence of the SOS stability certificate of (36):

$$\delta_{st}(\eta) = \begin{cases} 0, & \text{for } \eta \text{ with a stability certificate (36),} \\ +\infty, & \text{otherwise.} \end{cases} \tag{42}$$

In what follows, the tuning problem (41) is addressed with a method that consists of two phases. The first phase searches for feasible solutions of (41). This is done by introducing into (35a) a SOS slack polynomial function whose presence is, as described in the sequel, minimized as much as possible by considering the slack's integral over a unit box as a cost function quantifying the slack's presence. This minimization problem can be tackled by a black-box optimization method such as Bayesian optimization which is used in the numerical example of this document. The second phase takes the tuning parameters feasible in (41) (i.e., parameters $\eta \in \Omega$ with $\delta_{st}(\eta) = 0$) obtained in the first phase and employs them as the initial conditions for minimization of the performance metric $P(\eta)$ in (41). This minimization can as well be performed by Bayesian optimization due to whose data exploitation property the region with initial stabilizing tuning parameters will be under focus for further exploration in order to minimize $P(\eta)$.

3.5.1 First Phase: Minimization of the SOS Slack

For the purpose of obtaining feasible vectors η in (41), a SOS slack polynomial function $\sigma_{slk}(\xi)$ is introduced in (35a) as

$$\begin{aligned} V(x,\theta,\lambda) - V(x^+,\theta^+,\lambda^+) - \|x\|_2^2 = {} & \sigma_0(\xi) - \sigma_{slk}(\xi) \\ & + \sigma_1(\xi)^\top \hat{g}(\theta,\lambda,x) + \sigma_2(\xi)^\top \hat{g}(\theta^+,\lambda^+,x^+) \\ & + p_1(\xi)^\top \hat{h}(\theta,\lambda,x) + p_2(\xi)^\top \hat{h}(\theta^+,\lambda^+,x^+) \\ & + p_3(\xi)(x^+ - f_x(x,\kappa(\theta))), \end{aligned} \tag{43}$$

while (35b) will be retained without modification. In the case when $\sigma_{slk}(\xi)$ is of the same degree as $\sigma_0(\xi)$, the difference of the SOS polynomials $\sigma_0(\xi) - \sigma_{slk}(\xi)$ can express any arbitrary polynomial up to that common degree [1]. Thus, provided $\sigma_0(\xi)$ and $\sigma_{slk}(\xi)$ are of a degree that is no smaller than the degrees of the other polynomial terms participating in (43), the SOS feasibility problem consisting of (43) and (35b) has a feasible solution for any fixed parameter η due to the presence of the slack $\sigma_{slk}(\xi)$. To have a solution that has as small a presence of the slack as possible for fixed η (ideally $\sigma_{slk}(\xi) = 0$ for all ξ), in addition to the constraint set consisting of (43) and (35b) we introduce a cost function that is the integral of the slack polynomial over a unit box:

$$\int_{\mathbf{B}} \sigma_{slk}(\xi) d\xi = \sum_{i=1}^{n_\beta} v_i \int_{\mathbf{B}} \beta_i(\xi) d\xi, \tag{44}$$

where $\mathbf{B}$ is the unit box and $\sigma_{slk}(\xi) = \sum_{i=1}^{n_\beta} v_i \beta_i(\xi)$ with v_i being the polynomial coefficients and $\beta_i(\xi)$ their corresponding monomials. As the integrals of the monomials $\beta_i(\xi)$ over the unit box $\mathbf{B}$ are constant values that can be precomputed in advance, the expression (44) is a weighted sum of the coefficients v_i of $\sigma_{slk}(\xi)$. Notice that due to the nonnegativity of the SOS polynomial $\sigma_{slk}(\xi)$, the integral (44) can be zero only if $\sigma_{slk}(\xi) = 0$ for all ξ which is equivalent to the case where the slack is not present.

The previous discussion summarizes to the following SOS programming problem

$$\begin{aligned} \mathrm{I}_\sigma(\eta) = \min. \quad & \int_{\mathbf{Y}} \sigma_{slk}(\xi) d\xi \\ \text{s.t.} \quad & (43),\ (35\text{b}) \\ & \sigma_{slk}, \sigma_0, \sigma_1, \sigma_2, \bar{\sigma}_0, \bar{\sigma}_1 \quad \text{SOS polynomials} \\ & V, p_1, p_2, p_3, \bar{p}_1 \quad \text{arbitrary polynomials,} \end{aligned} \tag{45}$$

where $\mathrm{I}_\sigma(\eta)$ denotes the problem's optimal value. Since the cost of (45) is a linear function of the polynomial coefficients (as can be seen in (44)), the SOS problem (45) is equivalent to an SDP and can be solved efficiently. As the control parameters η feasible in the original problem (41) are those for which the slack is identically equal to zero (i.e., they are in the set $\{\eta \mid \mathrm{I}_\sigma(\eta) = 0, \eta \in \Omega\}$), they correspond to the set of optimal solutions of the problem

$$\begin{aligned} \min. \quad & \mathrm{I}_\sigma(\eta) \\ \text{s.t.} \quad & \eta \in \Omega, \end{aligned} \tag{46}$$

where each evaluation of $\mathrm{I}_\sigma(\eta)$ is done by solving (45).

The slack minimization problem (46) can be addressed by using Bayesian optimization which is a derivative-free method for finding global optimal solutions under constraints. It is applicable to the problems of the form (41) and (46) where black-box cost functions $P(\eta)$ and $\mathrm{I}_\sigma(\eta)$ are involved. The constraint set can be specified either explicitly (like the set Ω in (41) and (46)) or as an error in the evaluation of the cost (e.g., the $+\infty$ values in (41) caused by the $\delta_{st}(\eta)$ term). The values of the cost function are allowed to be either deterministic (like with $\mathrm{I}_\sigma(\eta)$ in (46)) or stochastic (like with $P(\eta)$ in (41) in case when the approximate evaluation (40) with random samples is used). For information pertaining to the practical application of

Bayesian optimization, the reader is referred to [29]. From an algorithmic point of view, the algorithm operates in such a way that at each iteration of the Bayesian optimization the currently available cost evaluation pairs, e.g. $\{\eta_i, \mathrm{I}_\sigma(\eta_\mathrm{i})\}$, are used to build a statistical model of the cost function based on Gaussian Processes. This statistical model is then used to build a so-called acquisition function $a(\eta)$ which is such that its minimizer represents the next sampling point η that balances between exploitation of the currently known cost values $\{\eta_i, \mathrm{I}_\sigma(\eta_\mathrm{i})\}$ and exploration of the less known regions of the cost function $\mathrm{I}_\sigma(\eta)$. The feature that the next sampling point η is determined by minimizing the acquisition function $a(\eta)$ *instead* of operating with the actual cost function ($\mathrm{I}_\sigma(\eta)$ or $P(\eta)$) makes the algorithm particularly suitable for problems where the evaluation of the cost is time consuming or in some other sense expensive.

3.5.2 Second Phase: Minimization of the Performance Metric

The second phase takes a certain number of the solutions generated in the first phase (which are due to $\mathrm{I}_\sigma(\eta) = 0$ property feasible in (41)) and uses them as the initial conditions to start the Bayesian optimization algorithm on the performance optimization problem (41). Due to the exploration property of the Bayesian optimization, the solutions obtained in the first phase will be different among themselves and as the initial points in the second phase they would provide information to the Bayesian optimization algorithm about the location of a stabilizing region. As such, it would be a region of focus for further investigation due to the data exploitation property of Bayesian optimization. Addressing the problem (41) by Bayesian optimization without the stabilizing initials coming from the first phase would involve difficulties in locating a stabilizing region as the only values available to the Bayesian optimization until the first stabilizing parameter η is encountered would be $+\infty$, which (in contrast to the slack integral) are not informative about the distance to stabilizing parameters and cannot be used to give an indication where to sample next in order to reach a stabilizing region.

3.6 Numerical Example

This section demonstrates the synthesis on a system model (26) of the form

$$x_1^+ = 0.9x_1 + 0.2x_2 + 0.1x_1x_2 + 2u, \tag{47}$$

$$x_2^+ = -0.3x_1 + 0.6x_2 - 0.4x_1x_2, \tag{48}$$

$$y = x, \tag{49}$$

with input constraint

$$-u_{max} \leq u \leq u_{max} \tag{50}$$

where $u_{max} = 1$. The QP controller (27) is selected to be of size 2 (i.e., $\theta \in \mathbb{R}^2$) with the $\mathscr{C}$ and d fixed to

$$\mathscr{C} = \begin{bmatrix} 1 & 0 \\ -1 & 0 \\ 0 & 1 \\ 0 & -1 \end{bmatrix}, \quad d = \begin{bmatrix} 1 \\ 1 \\ 1 \\ 1 \end{bmatrix}, \tag{51}$$

and the elements of the symmetric positive definite matrix $\mathscr{H} \in \mathbb{R}^{2\times 2}$ and of matrix $\mathscr{F} \in \mathbb{R}^{2\times 2}$ are selected for the tuning parameters, resulting in $\eta \in \mathbb{R}^7$. The system input u is selected to be the first component of the decision vector θ resulting in (28) being $u = [1 \;\; 0]\theta$. Since the first component of θ is constrained as $-1 \leq \theta_1 \leq 1$ by the first two rows of the specified $\mathscr{G}$ and d in (51), the input constraint (50) is guaranteed to be satisfied.

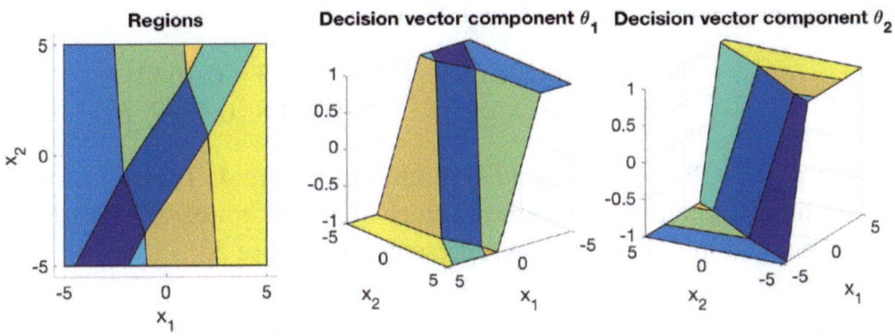

Fig. 4: The obtained QP control law. The component θ_1 of the decision vector θ corresponds to the plant input.

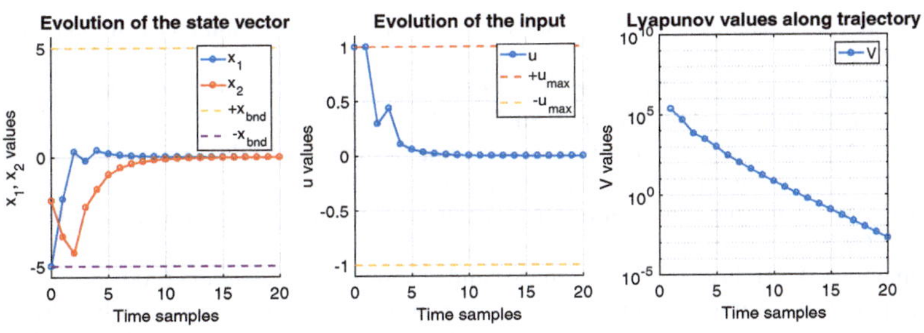

Fig. 5: Evolution of the state vector, input signal and the Lyapunov function values along the state trajectory starting from $x_0 = [-4.99, -1.97]^\top$. Without the control action (i.e., with $u = 0$), the evolution of the state vector starting from this x_0 is unstable.

The synthesis is run with a box shaped local stability set (37) of the form

$$\mathbf{X} = \{x \mid -x_{bnd} \leq x_i \leq x_{bnd},\ i = 1,2\} \tag{52}$$

where $x_{bnd} = 5$. The Lyapunov function $V(x)$ is selected to be of order four, the SOS polynomial multipliers $\sigma(\xi)$ of order two, the arbitrary polynomial multipliers $p(\xi)$ of order two, and the SOS slack polynomial $\sigma_{slk}(\xi)$ of order four. After 240 Bayesian optimization iterations applied to the phase one problem (46), 8 stabilizing controllers were obtained. The average time per Bayesian optimization iteration in the first phase was 13.68 seconds. The obtained stabilizing controllers were then used as the initial conditions for a total of 240 Bayesian optimization iterations applied to the phase two problem (41) that optimizes performance $P(\eta)$. The performance criteria $P(\eta)$ was chosen to have a stage cost $l(x,u) = x^\top x$ with discount factor $\alpha = 1$, and it was evaluated approximately by using (40) with $N = 100$ and $M = 100$. The average time per Bayesian optimization iteration in the second phase was 18.91 seconds. The obtained $\mathscr{H}$ and $\mathscr{F}$ matrices are

$$\mathscr{H} = \begin{bmatrix} 0.8957 & -0.2267 \\ -0.2267 & 1 \end{bmatrix},\ \mathscr{F} = \begin{bmatrix} 0.5193 & -0.7642 \\ 0.0388 & 0.4387 \end{bmatrix}, \tag{53}$$

which result in the QP control law (27) plotted in Figure 4. The trajectory of the state vector, input signal, and Lyapunov values along the trajectory starting from a randomly generated initial point x_0 is represented in Figure 5.

Acknowledgements M. Kvasnica, J. Holaza, and P. Bakarac gratefully acknowledge the contribution of the Slovak Research and Development Agency under the project APVV 15-0007.

References

1. Ahmadi, A.A., Hall, G.: DC decomposition of nonconvex polynomials with algebraic techniques. arxiv.org (2015)
2. Baldoni, V., Berline, N., De Loera, J.A., Köppe, M., Vergne, M.: How to integrate a polynomial over a simplex. Math. Comput. **80**(273), 297 (2010)
3. Baotić, M.: Optimal Control of Piecewise Affine Systems – a Multi-parametric Approach. Dr. sc. thesis, ETH Zurich, Zurich, Switzerland, March 2005
4. Baotić, M.: Polytopic computations in constrained optimal control. Automatica **50**(3–4), 119–134 (2009)
5. Bemporad, A.: Hybrid Toolbox - User's Guide. New Society, Gabriola (2003)
6. Bemporad, A.: A multiparametric quadratic programming algorithm with polyhedral computations based on nonnegative least squares. IEEE Trans. Autom. Control **60**(11), 2892–2903 (2015)
7. Bemporad, A., Morari, M., Dua, V., Pistikopoulos, E.: The explicit linear quadratic regulator for constrained systems. Automatica **38**(1), 3–20 (2002)
8. Blanchini, F., Miani, S.: Set-Theoretic Methods in Control. Birkhauser, Boston (2008)
9. Borrelli, F., Bemporad, A., Morari, M.: Geometric algorithm for multiparametric linear programming. J. Optim. Theory Appl. **118**(3), 515–540 (2003)

10. Borrelli, F., Bemporad, A., Morari, M.: Predictive Control for Linear and Hybrid Systems. Cambridge University Press, Cambridge (2011)
11. Christophersen, F.J., Kvasnica, M., Jones, C.N., Morari, M.: Efficient evaluation of piecewise control laws defined over a large number of polyhedra. In: Antsaklis, P.J., Tzafestas, S.G. (eds.) Proceedings of the European Control Conference ECC '07, pp. 2360–2367 (2007)
12. Domahidi, A., Zeilinger, M., Morari, M., Jones, C.: Learning a feasible and stabilizing explicit model predictive control law by robust optimization. In: 2011 50th IEEE Conference on Decision and Control and European Control Conference (CDC-ECC), pp. 513–519 (2011)
13. Domahidi, A., Ullmann, F., Morari, M., Jones, C.: Learning near-optimal decision rules for energy efficient building control. In: 2012 IEEE 51st Annual Conference on Decision and Control (CDC), pp. 7571–7576 (2012)
14. Dórea, C.E.T., Hennet, J.C.: (A, B)-invariant polyhedral sets of linear discrete-time systems. J. Optim. Theory Appl. **103**(3), 521–542 (1999)
15. Drgoňa, J., Klaučo, M., Janeček, F., Kvasnica, M.: Optimal control of a laboratory binary distillation column via regionless explicit MPC. Comput. Chem. Eng. **96**, 139–148 (2017)
16. Grieder, P., Morari, M.: Complexity reduction of receding horizon control. In: IEEE Conference on Decision and Control, Maui, December 2003, pp. 3179–3184
17. Gupta, A., Bhartiya, S., Nataraj, P.: A novel approach to multiparametric quadratic programming. Automatica **47**(9), 2112–2117 (2011)
18. Herceg, M., Kvasnica, M., Jones, C., Morari, M.: Multi-parametric toolbox 3.0. In: 2013 European Control Conference, pp. 502–510 (2013)
19. Herceg, M., Mariethoz, S., Morari, M: Evaluation of piecewise affine control law via graph traversal. In: 2013 European Control Conference (ECC), pp. 3083–3088. IEEE, Piscataway (2013)
20. Holaza, J., Takács, B., Kvasnica, M., Di Cairano, S.: Nearly optimal simple explicit MPC controllers with stability and feasibility guarantees. Optim. Control Appl. Methods **35**(6), 667–684 (2015)
21. Jones, C.N., Morari, M.: Polytopic approximation of explicit model predictive controllers. IEEE Trans. Autom. Control **55**(11), 2542–2553 (2010)
22. Klaučo, M., Drgoňa, J., Kvasnica, M., Di Cairano, S.: Building temperature control by simple MPC-like feedback laws learned from closed-loop data. In: Preprints of the 19th IFAC World Congress Cape Town (South Africa) August 24–August 29, 2014, pp. 581–586 (2014)
23. Korda, M., Jones, C.N.: Stability and performance verification of optimization-based controllers. Automatica **78**, 34–45 (2017)
24. Korda, M., Mezić, I: Linear predictors for nonlinear dynamical systems: Koopman operator meets model predictive control. arXiv preprint arXiv:1611.03537 (2017)
25. Kvasnica, M., Fikar, M.: Clipping-based complexity reduction in explicit MPC. IEEE Trans. Autom. Control **57**(7), 1878–1883 (2012)
26. Kvasnica, M., Löfberg, J., Fikar, M.: Stabilizing polynomial approximation of explicit MPC. Automatica **47**(10), 2292–2297 (2011)
27. Lasserre, J.B.: Moments, Positive Polynomials and Their Applications, 1st edn. Imperial College Press, London (2009)
28. Löfberg, J.: YALMIP: a toolbox for modeling and optimization in MATLAB. In: Proceedings of the CACSD Conference, Taipei (2004)
29. MathWorks: Statistics and machine learning toolbox: User's guide (r2016b). https://www.mathworks.com/help/pdf_doc/stats/stats.pdf, September 2016
30. Milne, G.W.: Grumman f-14 benchmark control problem solution using BLKLAB. In: IEEE Control Systems Society Workshop on Computer-Aided Control System Design, December 1989, pp. 94–101
31. MOSEK ApS: The MOSEK optimization toolbox for MATLAB manual (2016)
32. Oberdieck, R., Diangelakis, N.A., Papathanasiou, M., Nascu, I., Pistikopoulos, E.: Pop-parametric optimization toolbox. Ind. Eng. Chem. Res. **55**(33), 8979–8991 (2016)

33. Oishi, Y.: Direct design of a polynomial model predictive controlle. IFAC Proceedings Volumes **45**(13), 633–638 (2012)
34. Oishi, Y.: Simplified approaches to polynomial design of model predictive controllers. In: 2013 IEEE International Conference on Control Applications (CCA), pp. 960–965 (2013)
35. Parisini, T., Zoppoli, R.: A receding-horizon regulator for nonlinear systems and a neural approximation. Automatica **31**(10), 1443–1451 (1995)
36. Primbs, J.A.: The analysis of optimization based controllers. Automatica **37**(6), 933–938 (2001)
37. Spjøtvold, J., Tøndel, P., Johansen, T.A.: A Method for Obtaining Continuous Solutions to Multiparametric Linear Programs. In: IFAC World Congress, Prague (2005)
38. Tøndel, P., Johansen, T.A., Bemporad, A.: Evaluation of piecewise affine control via binary search tree. Automatica **39**(5), 945–950 (2003)
39. Wen, Ch., Ma, X., Ydstie, B.E.: Analytical expression of explicit MPC solution via lattice piecewise-affine function. Automatica **45**(4), 910–917 (2009)

Robust Optimization for MPC

Boris Houska and Mario E. Villanueva

1 Introduction

This chapter aims to give a concise overview of numerical methods and algorithms for implementing robust model predictive control (MPC). In contrast to nominal (certainty-equivalent) MPC, which is by now used in many industrial processes, robust MPC has—at least so far—found much fewer real-world applications. On the one hand, this is due to the fact that nominal MPC often exhibits a certain robustness—as a feedback controller it is inherently able to reject disturbances. Thus, for applications where safety is less critical, a robust MPC formulation, which explicitly models the influence of uncertainty, might simply not be needed. On the other hand, the limited deployment of robust MPC controllers in real-life may very well be due to the numerical challenges associated with their implementation. These challenges range from the need to model the uncertainty affecting the process to the intractability of general nonlinear formulations—therefore, restricting real-time implementations of robust MPC to simplified models or conservative approximations of the general nonlinear problem. It is also important to notice that the modeling decisions and the problem formulation often influence the choice of an appropriate numerical method. Thus, if one has a process for which it is necessary to explicitly take robustness aspects into account when designing an MPC controller, it is important to know about which tools and algorithms are available.

The focus of the present chapter is to discuss convex approximations of linear robust MPC as well as numerical methods for nonlinear robust MPC, leading to practical implementations. In particular, the advantages and disadvantages of various approaches are explained. Our aim is to explain and summarize highlights of the robust MPC literature from a new and somehow unifying perspective, but, as such, we do not present any new technical contributions. Moreover, different overviews

B. Houska (✉) · M. E. Villanueva
School of Information Science and Technology, ShanghaiTech University, Shanghai, China
e-mail: borish@shanghaitech.edu.cn; meduardov@shanghaitech.edu.cn

S. V. Raković, W. S. Levine (eds.), *Handbook of Model Predictive Control*,
Control Engineering, https://doi.org/10.1007/978-3-319-77489-3_18

of robust MPC can be found elsewhere in the literature. For example, the book chapter [69, Chapter 3] by J. Rawlings and D. Mayne and the plenary article [64] by S.V. Raković discuss a great variety of methods for robust MPC for linear discrete-time systems as well as the trade-off between computational complexity and accuracy. Because our aim is to give a self-consistent overview of numerical methods for robust MPC, some sections of this chapter may overlap with these existing reviews. Nevertheless, this chapter reviews these methods not only from a different, higher-level perspective but also with a much stronger focus on numerical aspects that arise when dealing with nonlinear continuous-time systems, while more theoretical properties, such as closed-loop stability of the controller, are not addressed.

2 Problem Formulation

We assume that we have a dynamic process model of the form

$$\forall t \in \mathbb{R}, \qquad \dot{x}(t) = f(x(t), u(t), w(t)),$$

which is affected by an unknown but bounded disturbance input. Here, $x : \mathbb{R} \to \mathbb{R}^{n_x}$ denotes the state trajectory, $u : \mathbb{R} \to \mathbb{R}^{n_u}$ the control input, and $w : \mathbb{R} \to \mathbb{R}^{n_w}$ the external disturbance or process noise. The function $f : \mathbb{R}^{n_x} \times \mathbb{R}^{n_u} \times \mathbb{R}^{n_w} \to \mathbb{R}^{n_x}$ is—unless otherwise stated—nonlinear, but assumed to be integrable in all its arguments and Lipschitz continuous in its first argument. It is assumed that the process model is specified together with[1]

1. a closed set $\mathbb{X} \subseteq \mathbb{R}^{n_u}$ modeling the state constraints,
2. a compact set $\mathbb{U} \in \mathbb{K}^{n_u}$ modeling the control constraints,
3. and a compact set $\mathbb{W} \in \mathbb{K}^{n_w}$ modeling the disturbance constraints.

Notice that even if the state $x_0 \in \mathbb{R}^{n_x}$ of the system at time t is known, the future state cannot be predicted accurately. This is due to the fact that the function w is unknown to us. However, one important assumption of MPC is that the system state can be measured, i.e., the controller can react to disturbances in closed-loop mode. In the following, the function $\mu : \mathbb{R} \times \mathbb{X} \to \mathbb{U}$ denotes a feedback law. For a given initial value $x_0 \in \mathbb{R}^{n_x}$, the associate closed loop system can be written in the form

$$\forall t \in [0,T], \qquad \dot{x}(t) = f(x(t), \mu(t, x(t)), w(t)), \quad \text{with} \quad x(0) = x_0 \,. \tag{1}$$

Now, a feedback law $\mu : [0,T] \times \mathbb{X} \to \mathbb{U}$ is called feasible on the time horizon $[0,T]$ and for a given initial value $x_0 \in \mathbb{R}^{n_x}$, if all solutions x of the above closed-loop system satisfy $x(t) \in \mathbb{X}$ for all $t \in [0,T]$ and for all possible uncertainties w, which satisfy $w(t) \in \mathbb{W}$ for all $t \in [0,T]$.

Robust MPC controllers search at every sampling time for feasible feedback laws μ, which are optimal among all feasible feedback laws on a given finite horizon with

[1] the set of compact and convex and compact sets in $\mathbb{R}^n$ are denoted, respectively, by $\mathbb{K}^n$ and $\mathbb{K}^n_{\mathrm{C}}$.

respect to a given control performance criterion. Putting aside theoretical and numerical difficulties associated to robust MPC, its practical implementation is mostly analogous to certainty-equivalent MPC. This means that an optimization problem is solved starting from a state measurement x_0 (whenever it is available) and shifting the time in a receding-horizon manner. The optimal control input $u(0) = \mu(0, x_0)$ is then sent to the real process. Throughout this chapter we assume that the state measurement is accurate and that the feedback is instantaneous.

The remainder of this section is devoted to different ways of formulating these problems mathematically, as well as their interconnections and particular challenges.

2.1 *Inf-Sup Feedback Model Predictive Control*

A mathematical formulation of robust MPC calls for the optimization over feasible feedback laws as defined in the previous section. Let

$$\xi(t, x_0, \mu, w) = x(t)$$

denote the solution of (1) as a function of x_0, μ, and w. An inf-sup feedback model predictive controller is given by

$$\begin{aligned} \inf_{\mu:\mathbb{R}\times\mathbb{X}\to\mathbb{U}} \quad & \sup_{w:\mathbb{R}\to\mathbb{W}} \int_0^T l(\xi(t, x_0, \mu, w))\,\mathrm{d}t + m(\xi(T, x_0, \mu, w)) \\ \text{s.t.} \quad & \sup_{w:\mathbb{R}\to\mathbb{W}} h_{\mathbb{X}}(\xi(t, x_0, \mu, w)) \leq 0 \quad \text{for all } t \in [0, T], \end{aligned} \tag{2}$$

where the function $h_{\mathbb{X}} : R^{n_x} \to \mathbb{R}$ is such that $\mathbb{X} = \{x \in \mathbb{R}^{n_x} \mid h_{\mathbb{X}}(x) \leq 0\}$. Here, $l : \mathbb{R}^{n_x} \to \mathbb{R}$ denotes the stage cost and $m : \mathbb{R}^{n_x} \to \mathbb{R}$ the terminal cost. Although the stage cost may also depend on the control, i.e., $l(\xi(t, x_0, \mu, w), \mu(t, \xi(t, x_0, \mu, w)))$, this chapter omits this dependency in favor of a more compact notation.

Min-Max MPC Variants

Mathematical theories which analyze under which assumptions the above inf-sup formulation can be replaced by a min-max formulation, can mostly be found in the field of functional analysis and optimal control theory. In particular, these can be found in the context of the analysis of viscosity solutions of the Hamilton-Jacobi-Bellman-Isaacs equation [14, 25] as well as Pontryagin's maximum principle [45, 61]. In general, however, it is difficult to check whether a Lebesgue-integrable minimizer μ of (2) exists. In the context of this paper, this question is—to a certain extent—less relevant as long as the objective is bounded from below on the feasible set, as is the case for standard tracking objectives. This is because as numerical methods for solving (2) focus on finding feasible but potentially sub-optimal (also called conservative) points of (2).

2.2 Set-Based Robust Model Predictive Control

An alternative formulation of the inf-sup feedback MPC uses the concept of reachable sets. Let[2]

$$X(t,x_0,\mu) = \left\{ x_t \in \mathbb{R}^{n_x} \;\middle|\; \begin{array}{l} \exists x \in W_{1,2}^{n_x},\ \exists w \in L_2^{n_w} : \forall \tau \in [0,t], \\ \dot{x}(\tau) = f(x(\tau),\mu(\tau,x(\tau)),w(\tau)) \\ x(0) = x_0\ ,\ x(t) = x_t \\ w(\tau) \in \mathbb{W} \end{array} \right\}$$

denote the set of all possible solutions of the closed-loop system at time t for a given feedback law μ and for a given initial value $x_0 \in \mathbb{R}^{n_x}$. Continuous-time robust MPC optimizes the future feedback policy μ by solving the optimization problem

$$\begin{aligned} \inf_{\mu:\mathbb{R}\times\mathbb{X}\to\mathbb{U}} \quad & \int_0^T \ell(X(t,x_0,\mu))\,\mathrm{d}t + \mathscr{M}(X(T,x_0,\mu)) \\ \text{s.t.} \quad & X(t,x_0,\mu) \subseteq \mathbb{X} \quad \text{for all } t \in [0,T]\,. \end{aligned} \tag{3}$$

Here, the function $\ell : \mathbb{K}^{n_x} \to \mathbb{R}$ denotes a scalar performance measure while the function $\mathscr{M} : \mathbb{K}^{n_x} \to \mathbb{R}$ denotes the scalar terminal cost.

By definition of the closed-loop reachable set $X(t,x_0,\mu)$, the feasible sets of (2) and (3) coincide. In principle, one could define the objective functions ℓ and $\mathscr{M}$ in dependence on l and m in such a way that complete equivalence of the inf-sup and the set-based formulations is achieved. However, as we will see further below a more important practical consideration is that the choice of the numerical solution method eventually depends on the particular problem formulation and, of course, also on how the objective function is modelled.

Topological Properties

From a pure mathematical perspective one might argue that problems of the form (3) are rather well-understood in the sense that the topological properties of the sets $X(\cdot,\mu,x_0)$ have been analyzed exhaustively in the context of viability theory [3] and differential inclusions [73]. This connection becomes apparent once the set

[2] We denote with L_2^n the set of n-dimensional L_2-integrable functions. Similarly, $W_{1,2}^n$ denotes the associated Sobolev space of weakly differentiable and L_2-integrable functions on $[0,T]$ with L_2-integrable weak derivatives.

$X(\cdot,\mu,x_0)$ is interpreted as the solution set of the differential inclusion

$$\dot{x}(t) \in \mathscr{F}(t,x(t),\mu) \quad \text{with} \quad x(0) = x_0$$

with set-valued right-hand side

$$\mathscr{F}(t,x(t),\mu) = \{f(x(t),\mu(t,x(t)),w(t)) \mid w(t) \in \mathbb{W}\} \ .$$

For example, it is known that if the right-hand set $\mathscr{F}(t,x(t),\mu)$ of this differential inclusion is convex and compact, the sets $X(t,\mu,x_0)$ are compact under suitable assumptions on μ and f [24]. This result can also be used as a starting point for analyzing whether minimizers of (3) exist [45, 74].

Discrete-Time Variant

Most of the methods in this chapter can also be applied for the case that the dynamic process is given in the form of a discrete-time system,

$$\forall k \in \{1,\ldots,N\}, \qquad x_{k+1} = f(x_k,u_k,w_k).$$

Here,[3] $x = [x_1,\ldots,x_N]$, $u = [u_0,\ldots,u_{N-1}]$, and $w = [w_0,\ldots,w_{N-1}]$ denote, respectively, the states, controls, and disturbances. The formulation of the associated robust MPC problem is analogous to the continuous-time case,

$$\begin{aligned} \inf_{\mu_0,\ldots,\mu_{N-1}:\mathbb{X}\to\mathbb{U}} \quad & \sum_{k=0}^{N-1} \ell\left(X_k(x_0,\mu)\right) + \mathscr{M}\left(X_N(x_0,\mu)\right) \\ \text{s.t.} \quad & X_k(x_0,\mu) \subseteq \mathbb{X} \quad \text{for all } k \in \{0,\ldots,N\} \end{aligned} \tag{4}$$

with

$$X_{k+1}(x_0,\mu) = \left\{ x_{k+1} \in \mathbb{R}^{n_x} \;\middle|\; \begin{array}{l} \exists x \in \mathbb{R}^{n_x \times N},\ \exists w \in \mathbb{R}^{n_w \times (N-1)} :\ \forall i \in \{0,\ldots,k\}, \\ x_{k+1} = f(x_k,\mu_k(x_k),w_k), \\ w_k \in \mathbb{W} \end{array} \right\} .$$

The optimization variables, $\mu_0,\ldots,\mu_{N-1} : \mathbb{X} \to \mathbb{U}$, correspond to the sequence of future feedback policies.

[3] The symbols used for continuous-time models will also be used for discrete-time. Since hybrid models are not considered in this chapter, no confusion should arise from this abuse of notation.

2.3 Numerical Challenges

So far, we have introduced two basically equivalent mathematical formulations for the robust optimal feedback control problem. Both formulations showcase different properties of the problem and also motivate different numerical solution methods. Regardless of the chosen formulation, it is clear that robust MPC problems are more difficult to solve than certainty-equivalent MPC problems. The two main reasons are that

1. the predicted vector-valued state trajectory x of certainty-equivalent MPC has to be replaced by a set-valued tube $X(\cdot, x_0, \mu)$, and
2. the optimization variable μ of the robust MPC problem (3) is a feedback law, i.e., a function of the current time t and the current state $x(t)$, rather than a single open-loop control trajectory $u : [0,T] \to \mathbb{U}$.

Looking at these points, it may appear that (3) involves the extra difficulty of computing the set-valued tube. However, (2) is a bi-level optimization problem and, as a consequence, the construction of numerical algorithms for computing rigorous solutions will require information about the solution set of the closed-loop system, too.

For anything but the simplest of systems, attempting to solve the infinite dimensional robust optimal feedback control problem with high numerical accuracy may be futile. In part, because the set-valued tube $X(\cdot, \mu, x_0)$ cannot, apart from very simple cases, be stored accurately, and, in part, because algorithms to optimize over general feedback functions do not exist. Therefore, the focus of modern numerical robust MPC algorithms is mostly on computing sub-optimal but feasible feedback laws. Thus, the remainder of this chapter is devoted to presenting numerical algorithms that either exploit particular structures in the problem, e.g. linearity of the dynamics, or construct tractable approximations of the problem.

3 Convex Approximations for Robust MPC

Because the robust MPC problem needs to be solved in real-time, it would be favorable if we could generate approximate solutions of (3) by solving a convex optimal control problem. Unfortunately, for general robust MPC problems, such convex approximations are hard to find. Nevertheless, for the case that the system dynamics is affine in the states and controls and under suitable additional conditions on the constraints and objective, it is possible to use tools from the field of convex optimization to construct tractable conservative approximations of (2) or (3). Therefore, the purpose of the following sections is to outline the main successful strategies, which lead to reasonably accurate and scalable convex approximations. Here, "scalable" means that we exclude exhaustive state-space partitioning methods as used in the field of dynamic programming or explicit robust MPC for a moment, which are, however, reviewed at a later point in this chapter.

3.1 Ellipsoidal Approximation Using LMIs

Let us consider a linear system of the form

$$f(x(t),u(t),w(t)) = A(t)x(t)+B(t)u(t) \quad \text{with} \quad w(t) = [A(t),B(t)] .$$

We assume that the sets $\mathbb{U} = \mathscr{E}(U)$ and $\mathbb{X} = \mathscr{E}(\overline{P})$ are ellipsoids[4] with given shape matrices[5] $\overline{P} \in \mathbb{S}_+^{n_x}$ and $U \in \mathbb{S}_+^{n_u}$ and that the uncertainty set[6]

$$\mathbb{W} = \mathbf{co}\left(\{[A_1,B_1],\ldots,[A_m,B_m]\}\right)$$

is a polytope with given vertices $[A_i,B_i] \in \mathbb{R}^{n_x \times (n_x+n_u)}$. Following a (conservative) linear parametrization of the feedback law, $u(t) = K(t)x(t)$, the associated closed-loop system is an uncertain linear system of the form

$$\dot{x}(t) = (A(t)+B(t)K(t))\,x(t) \quad \text{with} \quad x(0) = x_0 .$$

It can be checked easily that the reachable set of this differential equation can be overestimated by an ellipsoidal tube, $X(t,\mu,x_0) \subseteq \mathscr{E}(P(t))$, if the time-varying shape matrix P satisfies the Lyapunov differential inequality

$$\dot{P}(t) \succeq (A(t)+B(t)K(t))\,P(t) + P(t)\,(A(t)+B(t)K(t))^{\mathsf{T}} \tag{5}$$

$$P(0) \succeq x_0 x_0^{\mathsf{T}} . \tag{6}$$

This matrix inequality needs to hold for all $t \in [0,T]$ and all matrix-valued functions A, B, which satisfy $[A(t),B(t)] \in \mathbb{W}$ for all $t \in [0,T]$. The control constraint, $u(t) \in \mathbb{U}$, can be written as

$$\forall \xi' \in \mathscr{E}(P(t)), \quad K(t)\xi' \in \mathscr{E}(U) \qquad \Leftrightarrow \qquad K(t)P(t)K(t)^{\mathsf{T}} \preceq U .$$

In order to proceed, one needs to apply two convex analysis "tricks":

1. We introduce the variable substitution $Y(t) = K(t)P(t)$ to get rid of the bilinear terms in (5). As long as the search is constrained to positive functions $P(t) \succ 0$, this substitution is invertible, i.e., optimizing over the function $Y : [0,T] \in \mathbb{R}^{n_u \times n_x}$ is equivalent to optimizing over the feedback gain matrix $K(t) = Y(t)P(t)^{-1}$.

---

[4] In this chapter, $\mathscr{E}(Q) := \{Q^{\frac{1}{2}}v \mid v^{\mathsf{T}}v \leq 1\}$ denotes an n-dimensional ellipsoid with positive semidefinite shape matrix Q.

[5] The set of symmetric positive semidefinite matrices in $\mathbb{R}^{n \times n}$ is denoted by $\mathbb{S}_+^n$.

[6] We use the notation $\mathbf{co}(S)$ for the convex hull of a set S.

2. Since the right-hand expression in (5) is affine in $w(t) = [A(t), B(t)]$ it is sufficient to enforce this inequality at the vertices of the polytope $\mathbb{W}$ rather than for all points inside this polytope.

In summary, (5) holds for all A, B with $[A(t), B(t)] \in \mathbb{W}$ if

$$\begin{aligned} \dot{P}(t) &\succeq A_i P(t) + P(t) A_i^\mathsf{T} + B_i Y(t) + Y(t)^\mathsf{T} B_i^\mathsf{T}, \quad \forall t \in [0, T] \\ P(0) &\succeq x_0 x_0^\mathsf{T}, \end{aligned}$$

holds for each $i \in \{1, \ldots, m\}$, with $Y(t) = K(t)P(t)$. The state and control constraints can now be enforced through

$$P(t) \preceq \overline{P} \quad \text{and} \quad K(t)P(t)K(t)^\mathsf{T} = Y(t)P(t)^{-1}Y(t)^\mathsf{T} \preceq U\,.$$

The latter inequality is "quadratic-over-linear" in (Y, P) and thus convex. By using Schur complements, this inequality can alternatively be written in the form of the linear matrix inequality

$$\begin{pmatrix} U & Y(t) \\ Y(t)^\mathsf{T} & P(t) \end{pmatrix} \succeq 0\,,$$

which has the additional advantage that the inverse of $P(t)$ is not needed. Now, a conservative approximation of (3) is given by

$$\begin{aligned} &\inf_{P,Y} \int_0^T \ell(\mathscr{E}(P(t))\,\mathrm{d}t + \mathscr{M}\left(\mathscr{E}(P(T))\right) \\ &\text{s.t.} \begin{cases} \forall t \in [0, T],\ \forall i \in \{1, \ldots, m\}: \\ \dot{P}(t) \succeq A_i P(t) + P(t) A_i^\mathsf{T} + B_i Y(t) + Y(t)^\mathsf{T} B_i^\mathsf{T} \\ P(0) \succeq x_0 x_0^\mathsf{T} \\ 0 \preceq \begin{pmatrix} U & Y(t) \\ Y(t)^\mathsf{T} & P(t) \end{pmatrix} \\ 0 \prec P(t) \preceq \overline{P}\,. \end{cases} \end{aligned} \tag{7}$$

If one models the objective in such a way that the expression in the Lagrange and Mayer term are convex in P, for example, with

$$\ell(X') = \max_{\xi' \in X'} \|\xi'\|_2^2 \quad \Longrightarrow \quad \ell(\mathscr{E}(P(t)) = \lambda_{\max}(P(t)) \qquad \text{and} \qquad \mathscr{M}(X') = 0\,,$$

the optimization problem (7) is a convex optimal control problem. Every feasible solution (Y, P) of (7) yields a control law,

$$\mu(t, \xi') = Y(t)P(t)^{-1}\xi'\,,$$

which is a feasible point of the original robust MPC problem (3).

The above linear matrix inequality (LMI) based approximation of robust MPC has (in a very similar variant) been proposed in [39, 82]. Of course, one could think of many other variants and extensions of the convex robust MPC approximation (7). For example, one could try to extend the above analysis by working with affine rather than linear control parametrizations or by extending the formulation for other types of objectives or other uncertainty models. Some of these variants lead to convex optimization problems or at least to optimization problems, which are convex with respect to most of their optimization variables, as discussed extensively in J. Löfberg's Ph.D. thesis [47]. However, the main idea of most of these LMI relaxations is to substitute the equation

$$Y(t) = K(t)P(t)$$

(or similar variable transformations) at some point in the derivation in order to eliminate bilinear terms. For a more general overview on LMIs in systems and control, we refer to the textbook [13], because many of the methods in this book can be used as a starting point to construct LMI relaxations or other types of convex approximations of robust MPC.

Feasibility

In general, (7) is a conservative approximation of the original robust MPC problem. Thus, the semi-definite state constraint, $P(t) \preceq \overline{P}$, may lead to infeasibility—even if the original robust MPC problem was perfectly well-formulated and feasible. Nevertheless, under the additional assumption that there exists a $\bar{Y} \in \mathbb{R}^{n_u \times n_x}$ such that $\bar{P}$ satisfies

$$\forall i \in \{1,\ldots,m\}, \quad \begin{cases} 0 \succeq A_i\bar{P} + \bar{P}A_i^\mathsf{T} + B_i\bar{Y} + \bar{Y}^\mathsf{T}B_i^\mathsf{T} \\ \bar{P} \preceq \begin{pmatrix} U & \bar{Y} \\ \bar{Y}^\mathsf{T} & \bar{P} \end{pmatrix} , \end{cases} \tag{8}$$

then $\mathscr{E}(\bar{P})$ is a robust forward invariant set, i.e., (7) is feasible (and remains recursively feasible) as long as the initial value satisfies $x_0 \in \mathscr{E}(\bar{P})$.

3.2 *Affine Disturbance Feedback*

A second approach for approximating the robust feedback policy optimization problem by a convex optimization problem is based on affine disturbance feedback parameterization. This approach can be applied to linear systems of the form

$$f(x(t),u(t),w(t)) = Ax(t) + Bu(t) + w(t) \; .$$

In contrast to the model from the previous section, the matrices $A \in \mathbb{R}^{n_x \times n_x}$ and $B \in \mathbb{R}^{n_x \times n_u}$ are assumed to be given, i.e., the uncertainty w enters in the form of an additive offset only. Next, the main idea is to introduce a linear feedback parameterization *with memory*, i.e., a control law of the form

$$u(t) = \int_0^t L(t,\tau)x(\tau)\,\mathrm{d}\tau \;,$$

where the function $L : \mathbb{R} \times \mathbb{R} \to \mathbb{R}^{n_u \times n_x}$ becomes the new optimization variable. At this point, one should be clear in mind that the formulation of (3) was based on the assumption that the feedback law, $\mu(t,x(t))$, depends on the current state only, but does not have memory. This assumption is not restrictive, as the current (exact!) state measurement contains all the relevant information that is needed to predict the future evolution of the system. However, if we restrict ourselves to affine state feedback laws, it may be that the optimal feedback law depends on previous state measurements, i.e., the principle of information separation into future and past is violated. Now, the main observation is that optimizing over the set of affine state feedback laws with memory is equivalent to optimizing over the class of affine disturbance feedback laws of the form[7]

$$u(t) = \int_0^t M(t,\tau)w(\tau)\,\mathrm{d}\tau \;,$$

where $M : \mathbb{R} \times \mathbb{R} \to \mathbb{R}^{n_u \times n_x}$ is now the new optimization variable. This equivalence becomes apparent[8] by noticing that the disturbance function can be computed from $w(\tau) = \dot{x}(\tau) - Ax(\tau) - Bu(\tau)$, if we know the functions x and u on the horizon $\tau \in [0,t]$. The mathematical advantage of this change of variables is that the state at time t,

$$x(t) = H(t,x_0) \circ M$$

is an affine functional in M. Here, $H(t,x_0)$ denotes an affine operator,

$$H(t,x_0) \circ M = e^{At}x_0 + \int_0^t e^{A(t-\tau)} \left(B \int_0^\tau M(\tau,\tau')w(\tau')\,\mathrm{d}\tau' + w(\tau) \right) \mathrm{d}\tau \;,$$

mapping the function M to $x(t)$. Consequently, if the sets $\mathbb{X}$ and $\mathbb{U}$ are convex, the set of feasible disturbance feedback functions,

$$\mathscr{C}_M = \left\{ M : \mathbb{R} \times \mathbb{R} \to \mathbb{R}^{n_u \times n_x} \;\middle|\; \begin{array}{c} \forall w : [0,T] \to \mathbb{W}, \forall t \in [0,T], \\ \int_0^t M(t,\tau)w(\tau)\,\mathrm{d}\tau \in \mathbb{U} \\ H(t,x_0) \circ M \in \mathbb{X} \end{array} \right\} ,$$

[7] The idea to use affine disturbance feedback parametrization in order to approximate robust MPC can be found, in different variants, in [7, 12, 44] as well as in an early article by J. Löfberg [48].

[8] Details about this equivalence statement together with a formal proof in the discrete-time setting can be found in an article by Goulart and Kerrigan [28].

is convex. Similarly, if we have worst-case objectives of the form

$$\ell(X') = \max_{\xi' \in X'} l(\xi') \quad \text{and} \quad \mathscr{M}(X') = \max_{\xi' \in X'} m(\xi')$$

with convex functions l and m, the corresponding parametric functionals

$$\ell(X(t,\mu,x_0)) = \tilde{\ell}(t,M,x_0) = \max_{w,w(\tau)\in\mathbb{W}} l\,(H(t,x_0)\circ M)$$

$$\text{and} \quad \mathscr{M}(X(T,\mu,x_0)) = \tilde{\mathscr{M}}(M,x_0) = \max_{w,w(\tau)\in\mathbb{W}} m\,(H(t,x_0)\circ M)$$

are convex in M, since the maximum over convex functions remains convex. Consequently, a conservative approximation of (3) is given by the convex optimization problem

$$\inf_{M\in\mathscr{C}_M} \int_0^T \tilde{\ell}(t,M,x_0)\,\mathrm{d}t + \tilde{\mathscr{M}}(M,x_0)\,. \tag{9}$$

Of course, in order to solve (9) one still needs to discretize the function M—preferably without destroying convexity. For example, if one uses piecewise constant discretization of M on the 2-dimensional discrete-time grid,

$$\{t_0,t_1\ldots,t_N\}\times\{t_0,t_1\ldots,t_N\}, \quad 0=t_0<t_1<\ldots<t_N=T\ ,$$

the resulting problem is convex with $\mathbf{O}(N^2)$ matrix-valued optimization variables, as worked out in an article by Goulart and Kerrigan [28]. Thus, in general, affine disturbance feedback parametrization based robust MPC typically leads to algorithms which scale (at least) quadratically with the discrete-time prediction horizon length N. There are, however, variants which enforce additional structure of the function M, e.g., by setting $M(t,\tau)=0$ for $|t-\tau|\geq\bar{t}$ for a given constant $\bar{t}>0$. This leads to computationally less demanding but more conservative convex approximations of the original robust MPC problem. Methods for analyzing the conservatism of affine disturbance feedback parametrizations in the context of robust MPC can be found in [77].

Moreover, the above linear feedback parametrization with memory is at most as but in general less conservative than the linear feedback parametrization from Section 3.1 as used in the early article by M.V. Kothare and co-workers [39]. A discussion of more general feedback parametrization structures can also be found in the overview article by S.V. Raković [64].

4 Generic Methods for Robust MPC

There are three main classes of generic numerical algorithms for robust nonlinear MPC, namely, *(Approximate) Dynamic Programming* based approaches, *Scenario-Tree MPC* based approaches, and *Tube-MPC* based approaches.

4.1 Inf-Sup Dynamic Programming

Dynamic programming based methods for robust MPC are, at least in principle, analogous to dynamic programming methods for nominal MPC. This means that one introduces the so-called cost-to-go (or value) function

$$V(t,y) = \inf_{\mu:\mathbb{R}\times\mathbb{X}\to\mathbb{U}} \sup_{w:\mathbb{R}\to\mathbb{W}} \int_0^{T-t} l_{\mathbb{X}}(\xi(\tau,y,\mu,w))\,\mathrm{d}\tau + m(\xi(T-t,y,\mu,w))$$

for $(t,y) \in [0,T] \times \mathbb{R}^{n_x}$, where we use the shorthand

$$l_{\mathbb{X}}(\xi) = \begin{cases} l(\xi) \text{ if } \xi \in \mathbb{X} \\ \infty \quad \text{otherwise}\,. \end{cases}$$

It is known since a long time that the function V can formally be obtained as the viscosity solution of an inf-sup Hamilton-Jacobi-Bellman equation (also known under the name "Hamilton-Jacobi-Bellman-Isaacs equation"). This, is a partial differential equation (PDE) of the form

$$\begin{aligned} -\frac{\partial}{\partial t}V(t,y) &= \inf_{v\in\mathbb{U}} \sup_{\omega\in\mathbb{W}} \left\{ l_{\mathbb{X}}(y) + \nabla_y V(t,y)^\mathsf{T} f(y,v,\omega) \right\} \\ V(T,y) &= m(y) \end{aligned} \tag{10}$$

on the domain $[0,T]\times\mathbb{R}^{n_x}$. Now, one can use numerical tools from the field of partial differential equations in order to find an approximate solution. If a minimizer exists, an optimal feedback law can be picked as

$$\mu^*(t,y) \in \operatorname*{argmin}_{v\in\mathbb{U}} \sup_{\omega\in\mathbb{W}} \left\{ l_{\mathbb{X}}(y) + \nabla_y V(t,y)^\mathsf{T} f(y,v,\omega) \right\}.$$

PDE solvers for Hamilton-Jacobi-Bellman equations have been developed in [57] and there exists efficient software for solving this type of PDEs, e.g., the level-set toolbox by Mitchell and co-workers [55, 56]. An appealing feature of the above Hamilton-Jacobi-Bellman equation is that if we manage to solve this PDE, we can have access to a globally optimal feedback law. This is important for certain non-convex robust MPC scenarios, e.g., in the context of obstacle avoidance problems in robotics [27].

A common criticism of (numerically accurate) generic methods based on a direct solution of (10) is that these approaches can only be applied to problems with a small number of states. This is due to the fact that, at least in general, one has to use a grid in the state space in order to construct accurate approximations of the function V, as discussed in [31]. In practice, the corresponding adaptive grid based methods often only work well for generic problem with $n_x \leq 3$ states. Nevertheless, for the case that the differential equations have additional structure, it is possible to solve (10) for higher dimensional problems. For example, in recent articles by M. Chen, S. Bansal, and co-workers [4, 18] min-max Hamilton-Jacobi Bellman equations have been solved numerically for (simplified) quadcopter models with up to $n_x = 10$ differential states by exploiting the rather particular structures of these models.

Remark: Notice that the inf-sup Hamilton-Jacobi PDE is closely related to its associated sup-inf version, which is obtained by swapping the inf and the sup operation in (10). As discussed in [14], the solutions of the inf-sup and the sup-inf Hamilton-Jacobi PDE coincide under mild technical assumptions. This result can also be expected intuitively, as the feedback is instantaneous. To see this, consider a differential game where both us and our adverse player (nature) are choosing all control and disturbance reactions instantaneously. Clearly, if both players can choose their actions continuously in time, one cannot distinguish between who plays first and who plays last. This is in contrast to the discrete-time min-max dynamic programming recursion, which is reviewed below and for which the min and the max operation cannot be exchanged.

Discrete-Time Variant

The discrete-time analog of the Hamilton-Jacobi-Bellman PDE (10) is known under the name *dynamic programming recursion*,

$$\begin{aligned} V_k(y) &= \min_{\nu \in \mathbb{U}} \max_{\omega \in \mathbb{W}} \, l_{\mathbb{X}}(y) + V_{k+1}(f(y,\nu,\omega)) \\ V_N(y) &= m(y) \end{aligned} \tag{11}$$

for all $k \in \{0,\ldots,N-1\}$ and all $y \in \mathbb{R}^{n_x}$. Here, the function sequence $V_N, V_{N-1},\ldots,V_0$ can be found by solving the above backward recursion. Next, an optimal solution of the associated discrete-time robust MPC problem can be found as

$$\mu_k^*(y) \in \underset{\nu \in \mathbb{U}}{\operatorname{argmin}} \max_{\omega \in \mathbb{W}} \, l_{\mathbb{X}}(y) + V_{k+1}(f(y,\nu,\omega)) \, .$$

Discrete-time dynamic programming recursions are the basis for a number of existing robust MPC tools.

1. For the special case that the discrete-time system f is affine; m is piecewise quadratic; and $\mathbb{U}$, $\mathbb{X}$, and $\mathbb{W}$ are polytopic, it can be shown that the functions V_k are piecewise quadratic and can be constructed explicitly [6]. A corresponding robust MPC tool is available as part of the multi-parametric toolbox MPT [32], which can be used in combination with Yalmip [49].
2. For the special case that the discrete-time system f is affine, m is piecewise affine; and $\mathbb{U}$, $\mathbb{X}$, and $\mathbb{W}$ are polytopic, approximate robust dynamic programming methods have been developed in [9, 20]. These construct which construct

piecewise affine upper- and lower bounds on the function V_k, leading to a sub-optimal robust MPC controller with guarantees.

Other dynamic programming or approximate dynamic programming tools and methods can be found in [8, 83].

4.2 Scenario-Tree MPC

The scenario-tree approach [72], sometimes also referred to as Multi-Stage MPC, is a method for generating optimistic approximations of the discrete-time robust MPC problem (4). This means that if we have a continuous-time problem, we have to discretize this problem first. Now, the main idea is to choose a discrete inner approximation $\underline{\mathbb{W}} = \{\tilde{w}_1, \ldots, \tilde{w}_m\} \subseteq \mathbb{W}$ of the disturbance set. We also define the index set $\mathscr{I} = \{1, \ldots, m\}$. Next, an optimistic scenario based approximation of (4) can be written in the form

$$\inf_{\tilde{x},\tilde{u}} \sum_{k=0}^{N-1} \ell\left(\{\tilde{x}_{k,i_1,\ldots,i_k} \mid i_1, \ldots, i_k \in \mathscr{I}\}\right) + \mathscr{M}\left(\{\tilde{x}_{N,i_1,\ldots,i_N} \mid i_1, \ldots, i_N \in \mathscr{I}\}\right)$$

$$\text{s.t.} \begin{cases} \forall k \in \{0, \ldots, N\}, \forall i_k \in \mathscr{I}, \\ \tilde{x}_{k+1,i_1,\ldots,i_{k+1}} = f(\tilde{x}_{k,i_1,\ldots,i_k}, \tilde{u}_{k,i_1,\ldots,i_k}, \tilde{w}_{i_{k+1}}) \in \mathbb{X} \\ \tilde{u}_{k,i_1,\ldots,i_k} \in \mathbb{U} \\ \tilde{x}_0 = x_0 \,. \end{cases} \tag{12}$$

The optimization variables of this problem, $\tilde{x} \in \mathbb{R}^{n_x \cdot d_N}$ and $\tilde{u} \in \mathbb{R}^{n_u \cdot d_{N-1}}$, contain the states and associated optimal control reactions of all possible scenarios. Notice that one can pick m possible uncertanties $\tilde{w}_{i_1} \in \underline{\mathbb{W}}$ in the first step. For each of these scenarios, one can choose a control reaction $\tilde{u}_{1,i_1}$. In the second step, there are already m^2 possible scenarios; and so on. Thus, in total, there are $d_N - 1$ possible scenarios with

$$d_N = 1 + m + m^2 + \ldots + m^N = \frac{m^{N+1} - 1}{m - 1} \,.$$

A visualization of this approach for $m = 2$ can be found in Figure 1. Notice that this approach is not rigorous in the sense that it does not necessarily lead to a feasible control law. This is due to the fact that only a finite number of uncertainty scenarios is taken into account. Nevertheless, this optimistic approximation may be sufficiently accurate for practical purposes if m is large. Moreover, if a probability distribution of the disturbance inputs w_k is available, one may choose the points $\tilde{w}_i$ according to this distribution such that additional information about the probability distribution of future states can be inferred.

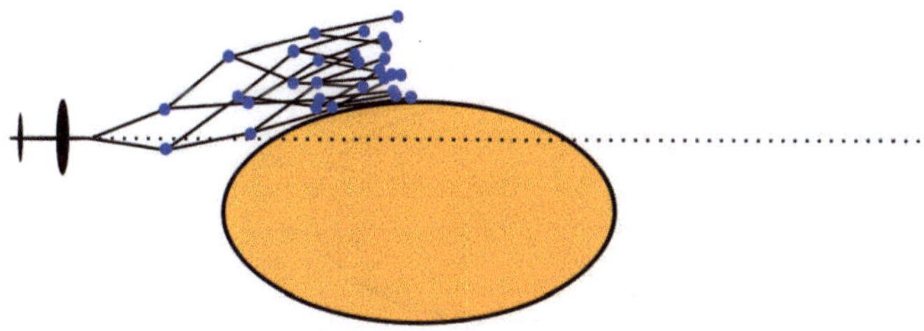

Fig. 1: Visualization of optimized discrete-time scenarios obtained for an obstacle avoidance problem and $m = 2$. After the first discrete-time step, the airplane may be at one out of 2 possible positions in the state-space. After two time steps, there are already 4 possible positions, and so on, leading to an exponentially large number of scenarios for long horizons. The corresponding feedback control inputs are optimized in such a way that the possible positions of the airplane (dots) never overlap with the obstacle.

The practical applicability of the scenario-tree approach is limited by the fact that the number of scenarios increases exponentially with the time horizon N. This means that the scenario approach can only be used for small N (in practice often $N \approx 3$). On the other hand, this approach is applicable to robust MPC problems with many states, $n_x \gg 1$, which can be seen as an advantage compared to dynamic programming or tube based approaches. Current research on scenario-tree approaches mostly focus on the development of solvers, which can exploit the particular structure of the optimization problem (12), as well as on heuristics for reducing the number of scenarios [23, 50].

4.3 Tube MPC

As mentioned in the introduction, often the problem formulation motivates the choice for an appropriate numerical method. A detailed analysis of (3) suggests that a direct method for solving this robust MPC problem can be obtained by constructing parametric outer approximations of the tube $X(\cdot,\mu,x_0)$. The main idea (sketched in Figure 2) is simple but elegant: tube-based MPC approaches, as formalized by Raković, Mayne, and collaborators [43, 63], compute an outer approximation of the set of all possible states that can be reached under all possible (continuous-time) disturbance scenarios.

Obviating for a moment that the outer optimization in (3) is still an optimal control problem over feedback laws, the development of scalable numerical algorithms for tube MPC relies on our ability to construct tractable representations of the tube. In particular, the over approximations of the closed-loop reachable sets must be constructed, stored, and propagated efficiently.

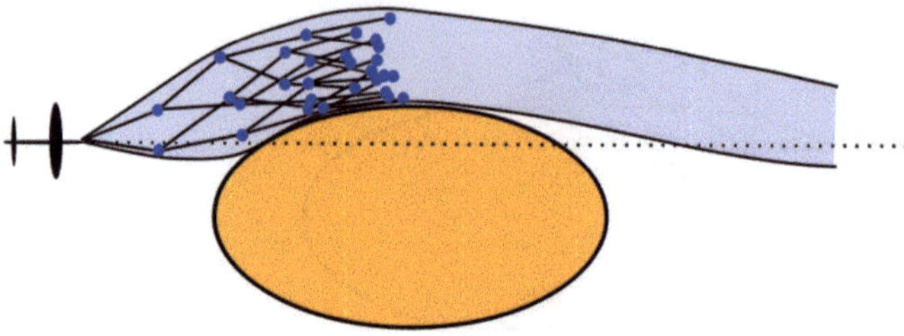

Fig. 2: In contrast to Scenario-Tree MPC from Figure 1, Tube MPC optimizes a single set-valued tube in the state space, which encloses all possible scenarios. The tube (shaded area) may not intersect with the infeasible region in the state-space, in this case an obstacle.

Due to its importance among robust MPC methods, the next section is devoted to presenting the main strategies used to construct tube MPC based control algorithms.

5 Numerical Methods for Tube MPC

In this section we present strategies to construct tube model predictive control algorithms. These strategies present some answers to the two main difficulties of tube MPC: the outer optimization over feedback laws and the practical construction of tubes for the inner optimization. Both problems can be addressed by appropriate parametrizations, first, of the feedback law, and second, of the reachable set outer approximations.

5.1 Feedback Parametrization

The most common class of feedback parametrizations for constructing approximations of (3) is that of affine feedback laws of the form

$$\tilde{\mu}[K,k](t,x) = K(t)x + k(t) \; .$$

If we substitute $\mu = \tilde{\mu}[K,k]$ in (3), the new optimization variables are the matrix-valued function $K : \mathbb{R} \to \mathbb{R}^{n_u} \times \mathbb{R}^{n_x}$ and the vector-valued additive offset function $k : \mathbb{R} \to \mathbb{R}^{n_u}$. In this case, the control constraint,

$$\forall t \in [0,T], \qquad K(t)X(t,\tilde{\mu}[K,k],x_0) + k(t) \subseteq \mathbb{U} \; ,$$

has to be added explicitly in order to ensure that the control law is feasible. Clearly, this and other feedback parameterizations lead to conservative approximations of (3). This means that if the corresponding optimization problem has a solution, we find a control law that ensures that all constraints are satisfied for all possible uncertainty scenarios. In some articles, e.g. in [84], it has been suggested to pre-compute a suitable feedback gain K and only optimize k online, which leads to a simpler optimization problem but an even more conservative approximation of (3). Similarly, one can use more expensive but less conservative feedback parametrizations, e.g., by optimizing over polynomial feedback laws or piecewise affine control laws [6].

The optimization variable μ of the robust MPC controller (3) is sometimes called an "ancillary control law" (see, e.g., [69]), because it is only needed to compute feasible set-valued tubes in the state-space. Thus, even if we substitute a linear feedback parametrization of μ, the actual robust MPC controller realizes a nonlinear control law, because μ is recomputed whenever a new measurement becomes available. However, there are two notable practical variants of robust MPC, which use the feedback law μ explicitly:

1. *Offline Robust MPC*, also called (approximate) *Explicit MPC*, solves (3) approximately offline, e.g., on a sufficiently long horizon (or with periodic boundary conditions on the tube) and over a suitable class of parametric feedback laws, and then uses the optimal feedback law to control the process.
2. *Real-Time Robust MPC* uses an optimized feedback law μ to control the process at a high sampling rate, but updates μ every now-and-then depending on how long the numerical routine needs to solve (3). For this variant, one would usually optimize over a class of feedback laws that can be evaluated efficiently in online mode.

5.2 Affine Set-Parametrizations

The formulation of (3) is based on the introduction of the set-valued tubes $X(\cdot,\mu,x_0)$. Unfortunately, it is not possible to store general sets in a computer. Consequently, many set based numerical methods for robust MPC focus on the construction of computer representable sets, which approximate the exact reachable set from outside. Examples of frequently used computer-representable sets are intervals, zonotopes, polytopes, ellipsoids, and polynomial sets. All these sets have in common that they can be written in the form

$$C * \mathbb{E} + c = \{Cx + c \mid x \in \mathbb{E}\} ,$$

where $\mathbb{E} \subseteq \mathbb{R}^m$ is a compact basis set. The matrix $C \in \mathbb{R}^{n \times m}$ and the vector $c \in \mathbb{R}^n$ are the coefficients of the set parametrization. As the coefficient matrix $[C, c]$ is finite dimensional it can be stored in a computer. For example, in order to store an interval, one would use the unit box,

$$\mathbb{E} = \{x \in \mathbb{R}^n \mid \|x\|_\infty \leq 1\} ,$$

Table 1: List of frequently used computer-representable sets with polynomial storage complexity.

Set	**Basis set $\mathbb{E}$**	**Coefficients**	**Storage complexity**
Interval 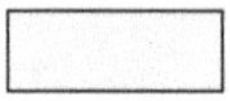	$\{x \in \mathbb{R}^n \mid \|x\|_\infty \leq 1\}$	$C \in \mathbb{R}_+^{n\times n}$, $c \in \mathbb{R}^n$ C diagonal	$\mathbf{O}(n)$
Zonotope 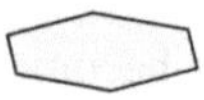	$\{x \in \mathbb{R}^m \mid \|x\|_\infty \leq 1\}$	$C \in \mathbb{R}^{n\times m}$, $c \in \mathbb{R}^n$	$\mathbf{O}(nm)$
Polytope 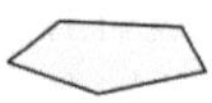	$\left\{x \in \mathbb{R}_+^m \mid \sum_i x_i = 1\right\}$	$C \in \mathbb{R}^{n\times m}$, $c \in \mathbb{R}^n$	$\mathbf{O}(nm)$
Ellipsoid 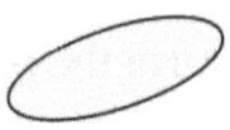	$\{x \in \mathbb{R}^n \mid \|x\|_2 \leq 1\}$	$C \in \mathbb{R}^{n\times n}$, $c \in \mathbb{R}^n$ C symmetric and p.s.d.	$\mathbf{O}(n^2)$
Polynomial set 	$\{(1, x_1, x_1x_2, \ldots, x_r^s)^\top$ with $x \in [-1,1]^r\}$	$C \in \mathbb{R}^{n\times \binom{r+s}{r}}$, $c = 0$	$\mathbf{O}(nr^s)$

as the basis set. The coefficient c can in this case be interpreted as the center of the interval box. The matrix C is then required to be non-negative and diagonal such that the diagonal entries can be interpreted as the widths of the box in different coordinate aligned directions. If one allows more general coefficient matrices C, one can also represent rotated (non-coordinate aligned) interval boxes. Table 1 lists a number of affine set parametrizations [17], which are frequently used in the context of robust MPC. Notice that some numerical set-based computing algorithms use variants of the above affine set parametrizations. For example, if we work with ellipsoids with center $c = q$ and symmetric and positive semi-definite (p.s.d.) shape matrix Q,

$$\left\{Q^{\frac{1}{2}}x + q \mid \|x\|_2 \leq 1\right\},$$

it is sometimes more convenient to store the shape matrix Q directly instead of its symmetric square-root $C = Q^{\frac{1}{2}}$.

5.3 Tube MPC Parametrization

In the context of tube MPC, the key for developing practical implementations is to construct parametric coefficient functions

$$C(\cdot,\mu,x_0):[0,T]\to\mathbb{R}^{n_x\times m} \qquad \text{and} \qquad c(\cdot,\mu,x_0):[0,T]\to\mathbb{R}^{n_x},$$

such that

$$\forall t\in[0,T], \qquad X(t,\mu,x_0)\subseteq C(t,\mu,x_0)\cdot\mathbb{E}+c(t,\mu,x_0)$$

for a suitable basis set $\mathbb{E}\subseteq\mathbb{R}^m$, $m\in\mathbb{N}$. For example, if $\mathbb{E}$ is a unit box, unit ball, or unit simplex, one obtains zonotopic, ellipsoidal, or polytopic tubes. Notice that there exist a variety of set-valued integrators from the field of set-valued computing, which can be used to construct the functions C and c systematically.

These integrators can be used in combination with a suitable feedback parameterization, e.g., the affine feedback parametrization $\tilde{\mu}[K,k]$. This leads to the conservative approximation of the original robust MPC problem

$$\begin{aligned}
\inf_{\substack{k:\mathbb{R}\to\mathbb{R}^{n_x}\\ K:\mathbb{R}\to\mathbb{R}^{n_u\times n_x}}} \quad & \int_0^T \ell(C(t,\tilde{\mu}[K,k],x_0)\cdot\mathbb{E}+c(t,\tilde{\mu}[K,k],x_0))\,\mathrm{d}t \\
& +\mathscr{M}(C(T,\tilde{\mu}[K,k],x_0)\cdot\mathbb{E}+c(T,\tilde{\mu}[K,k],x_0)) \\
\text{s.t.} \quad & \begin{cases} \forall t\in[0,T]: \\ C(t,\tilde{\mu}[K,k],x_0)\cdot\mathbb{E}+c(t,\tilde{\mu}[K,k],x_0)\subseteq\mathbb{X}, \\ K(t)\,(C(t,\tilde{\mu}[K,k],x_0)\cdot\mathbb{E}+c(t,\tilde{\mu}[K,k],x_0))\subseteq\mathbb{U} \end{cases}
\end{aligned} \tag{13}$$

Depending on the particular choice of $\mathscr{M}$, ℓ, $\mathbb{X}$, and $\mathbb{U}$, this optimization problem can be discretized and processed further in order to arrive at a standard nonlinear programming problem. One remaining challenge, however, is that the resulting optimization problem is non-convex in general. Nevertheless, in principle, the above parametrization based tube MPC approach leads to practical implementations. For example in [69] a practical implementation of this approach for robust control of an exothermic reactor can be found.

Notice that there exist a great variety of variants of the above outlined approaches for Tube MPC such as Homothetic Tube MPC [66] or Elastic Tube MPC [67]. We also refer to [64, 65] for a more general overview on parametrized Tube MPC.

5.4 Tube MPC Via Min-Max Differential Inequalities

One way to avoid parametrizing the feedback control law is to make use of a different parametrization of (3). Here, the main idea is to rewrite the problem in terms of so-called Robust Forward Invariant Tubes (RFITs). An RFIT is a set-valued function $\overline{X} : \mathbb{R} \to \mathbb{K}^{n_x}$ for which there exists a feedback control $\mu : \mathbb{R} \times \mathbb{R}^{n_x} \to \mathbb{U}$, such that

$$\overline{X}(t_2) \supseteq \bigcup_{x_1 \in \overline{X}(t_1)} X(t_2 - t_1, x_1, \mu)$$

for all $t_1, t_2 \in \mathbb{R}$ with $t_1 \leq t_2$. Let $\mathscr{X}$ denote the set of all RFITs for the dynamic system on $[0,T]$. The set-based MPC problem (3) can alternatively be written in the form

$$\begin{aligned} \inf_{\overline{X} \in \mathscr{X}} \quad & \int_0^T \ell(\overline{X}(t))\, \mathrm{d}t + \mathscr{M}(\overline{X}(T)) \\ \text{s.t.} \quad & \begin{cases} \overline{X}(t) \subseteq \mathbb{X},\ \forall t \in [0,T], \\ \overline{X}(0) = \{x_0\}\ . \end{cases} \end{aligned} \tag{14}$$

Now, we have traded optimizing over the feedback policy μ by an optimization problem over RFITs. Fortunately, if we restrict ourselves to input-affine nonlinear systems,

$$\dot{x}(t) = f(x(t), u(t), w(t)) = g(x(t), w(t)) + G(x(t))u(t)\ ,$$

and restrict the class of RFITs to those with compact and convex cross-sections $\overline{X}(t) \in \mathbb{K}^{n_x}_{\mathrm{C}}$, we can, at least in some cases, arrive at conservative but tractable approximations of (3). The construction of such approximations requires the use of the support function

$$\forall c \in \mathbb{R}^n\ , \quad \sigma[Z](c) := \max_{z \in Z} c^{\mathsf{T}} z\ ,$$

of a compact and convex set $Z \subseteq \mathbb{R}^n$. In [81] it is shown that if a set valued function $\overline{X} : \mathbb{R} \to \mathbb{K}^{n_x}_{\mathrm{C}}$ satisfies for almost all $t \in [0,T]$

$$\frac{\mathrm{d}}{\mathrm{d}t} \sigma[\overline{X}(t)](c) \geq \min_{\nu \in \mathbb{U}} \max_{\xi, \omega} \left\{ c^{\mathsf{T}} f(\xi, \nu, \omega) \;\middle|\; \begin{array}{c} c^{\mathsf{T}} \xi = \sigma[\overline{X}(t)](c) \\ \xi \in \overline{X}(t) \\ \omega \in \mathbb{W} \end{array} \right\}, \tag{15}$$

for each $c \in \mathbb{R}^{n_x}$, and the function $\sigma[\overline{X}(\cdot)](c)$ is, for all $c \in \mathbb{R}^{n_x}$, Lipschitz continuous on $[0,T]$, then it is an RFIT for the dynamic system on $[0,T]$. Thus, any solution of

$$\begin{aligned} \inf_{\overline{X}} \quad & \int_0^T \ell(\overline{X}(t))\, \mathrm{d}t + \mathscr{M}(\overline{X}(T)) \\ \text{s.t.} \quad & \begin{cases} \text{a.e. } t \in [0,T], \\ \text{Inequality (15)}, \\ \overline{X}(t) \subseteq \mathbb{X}, \\ \overline{X}(0) = \{x_0\} \end{cases} \end{aligned} \tag{16}$$

is a feasible point of (14). Problem (16) is not a standard optimal control problem, as it includes a semi-infinite differential inequality constraint. However, one can use the same set parametrization strategies as in Section 5.3 in order to reformulate (16).

This leads to a band-structured optimization problem, whose complexity scales linearly with the horizon length.

One of the key features of (16) is that it does not require a parametrization of the feedback law. Thus, the conservatism of the proposed approach depends only on the parametrization of the tube cross-sections $\overline{X}(t)$. In fact, in case the chosen parametrization has a smooth boundary with positive curvature, the feedback law inducing the tube is a nontrivial nonlinear function given by

$$\mu(t,y) = \mu_t^* \left(\mathscr{G}_{\overline{X}(t)}(y) \right) \quad \text{with} \quad \mu^*(c) := \underset{v \in \mathbb{U}}{\operatorname{argmin}} \; c^\mathsf{T} G \left(\mathscr{G}^{-1}_{\overline{X}(t)}(c) \right) v \, .$$

Here, $\mathscr{G}_{\overline{X}(t)} : \operatorname{bd} \overline{X}(t) \to \mathscr{S}^{n_x - 1}$ denotes the Gauss map of $\overline{X}(t)$ and its inverse is given by[9]

$$\mathscr{G}^{-1}_{\overline{X}(t)}(c) = \underset{\xi \in \overline{X}(t)}{\operatorname{argmax}} \, c^\mathsf{T} \xi \, .$$

As it stands, even the solution of a parametrized version of Problem (16) is nontrivial. But, as discussed in [80, 81], some parametrizations, e.g. ellipsoids, of the tube cross-sections, lead to practical implementations of robust MPC controllers, as well as explicit expressions for the feedback law μ.

Remark: The above min-max differential inequality uses properties of the boundary of robust forward invariant tubes of continuous-time systems, which have, at least in similar variants, been analyzed in earlier articles by Nagumo (see, e.g., [10] for a discussion of Nagumo's theorem) as well as in the context of viability theory [3]. As these boundary properties rely on differential analysis, the corresponding methods can, at least in the above form, only be applied to continuous-time systems.

6 Numerical Aspects: Modern Set-Valued Computing

This section gives a concise introduction to set-valued arithmetics for factorable functions and associated tools for set-valued integration. The corresponding methods can be used as a basis for the implementation of tube-based model predictive control algorithms.

6.1 Factorable Functions

A function φ is called factorable if it can be represented as a finite recursive composition of atom operations $\varphi_i \in \mathscr{L}$, with $i \in \{1, \ldots, n_\varphi\}$, from a given finite library

$$\mathscr{L} = \{+, *, \sin, \exp, \log, \ldots\} \, .$$

9 The boundary of $Z \subset \mathbb{R}^n$ is denoted by $\operatorname{bd} Z$, while $\mathscr{S}^{n-1}$ denotes the n-dimensional unit sphere.

This library typically includes binary sums, binary products, and a number of univariate atom functions such as trigonometric functions, exponentials, as well as logarithms. In practice, factorable functions over a given library $\mathscr{L}$ are represented in the form of a computational graph, which can be obtained in most object oriented programming languages by using operator overloading or source code transformation, as visualized in Figure 3. Notice that the result a_i, which is obtained after applying an atom operation φ_i, is stored temporarily. Thus, the input arguments of the atom operations φ_i are either components of the input vector x or intermediate results of previously computed operations, i.e., components of the intermediate result vector a, as shown in the example in Figure 3. In general, the recursion can be written in the form

$$\forall k \in \{1,\ldots,n_\phi\}, \quad a \leftarrow [a, \varphi_k(x,a)]^\mathsf{T} \; ,$$

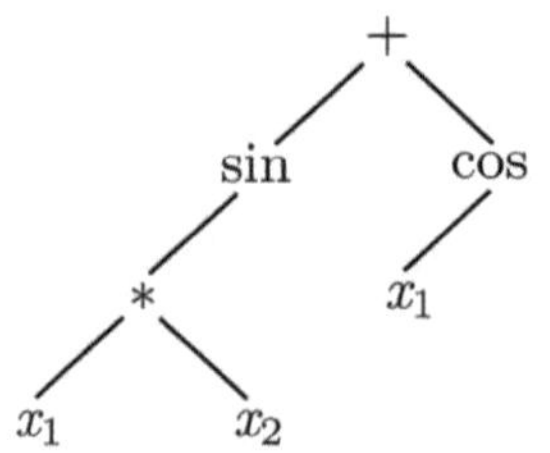

Input: x

$$\begin{aligned} a_1 &= \varphi_1(x) &&= x_1 * x_2 \\ a_2 &= \varphi_2(x,a_1) &&= \sin(a_1) \\ a_3 &= \varphi_3(x,a_1,a_2) &&= \cos(x_1) \\ a_4 &= \varphi_4(x,a_1,a_2,a_3) &&= a_2 + a_3 \end{aligned}$$

Output: $\varphi(x) = a_4$.

Fig. 3: Visualization of the computational graph of the function $\varphi(x) = \sin(x_1 * x_2) + \cos(x_1)$. The intermediate results a_1, a_2, and a_3 may or may not be deleted after a function evaluation depending on whether they are needed as part of other function evaluations.

where a is initialized with the empty vector, but then the dimension of a increases by 1 after every atom evaluation.[10] The function value $\varphi(x) = P_\varphi a$ with $P_\varphi \in \{0,1\}^{n_{\mathrm{out}} \times n_\varphi}$ corresponds to selected components of a, i.e., every row of the matrix P_φ has one entry, which is equal to 1.

Many modern computer algorithms and software explicitly exploit the structure of the computational graph of factorable functions. Examples include algorithmic differentiation [1, 2, 30], modeling environments for convex optimization [29, 49] and optimal control [34], global optimization algorithms and software [16, 54, 71], and many set arithmetic routines [15], which are reviewed in the next section.

[10] In practical implementations, the memory allocation policies for function evaluations depend on the compiler and hardware. For MPC applications on embedded hardware one often uses static memory. For example, in modern code-generation based MPC solvers for small-scale systems the memory for all components of a of all online function evaluations is pre-allocated [34].

6.2 Set Arithmetics

Let φ be a given factorable function and $\mathbb{E} \in \mathbb{K}^m$ a given basis set. The goal of a set arithmetic is to construct an enclosure function Φ of the image set map of φ with respect to the basis set $\mathbb{E}$, i.e., such that

$$\{ \varphi(x) \mid x \in C \cdot \mathbb{E} \} \subseteq \Phi(C) \cdot \mathbb{E} \tag{17}$$

for any coefficient matrix $C \in \mathbb{R}^{n \times m}$. Here, we use the same notation as in Section 5.2, but we leave away the offset parameter c, since we can always redefine

$$\mathbb{E} \leftarrow \mathbb{E} \times \{1\}$$

if we want to include such offsets. Now, the main idea is to build up the function Φ recursively by passing through the computational graph of the factorable function φ. This means that we need to construct enclosure functions Φ_i for every atom operation φ_i such that

$$\forall y \in C \cdot \mathbb{E}, \quad [y^{\mathsf{T}}, \varphi_i(y)]^{\mathsf{T}} \in \Phi_i(C) \cdot \mathbb{E} \,.$$

for all feasible coefficient matrices $C \in \mathbb{R}^{(n_{\text{in}}+i-1) \times m}$ and all $i \in \{1, \ldots, n_\varphi\}$. The enclosure function Φ is then given by the finite recursive composition

$$\Phi = P_\varphi * [\Phi_{n_\varphi} \circ \ldots \circ \Phi_2 \circ \Phi_1] \,,$$

which satisfies (17) by construction.

The key to constructing the enclosure (17) is the definition of the arithmetic rule operating on the coefficient matrix C. For example, interval arithmetics, one of the oldest and most basic set arithmetics, proceeds by defining bounding rules for binary sums and products as well as for all univariate atom operations in the given library $\mathscr{L}$. These bounding rules can, for example, be based on simple inclusions such as

$$\begin{aligned}
[\underline{c}_1, \overline{c}_1] + [\underline{c}_2, \overline{c}_2] &\subseteq [\underline{c}_1 + \underline{c}_2, \overline{c}_1 + \overline{c}_2] \,, \\
[\underline{c}_1, \overline{c}_1] * [\underline{c}_2, \overline{c}_2] &\subseteq \left[\min\{\underline{c}_1\underline{c}_2, \underline{c}_1\overline{c}_2, \overline{c}_1\underline{c}_2, \overline{c}_1\overline{c}_2\}, \max\{\underline{c}_1\underline{c}_2, \underline{c}_1\overline{c}_2, \overline{c}_1\underline{c}_2, \overline{c}_1\overline{c}_2\} \right] , \\
e^{[\underline{c}, \overline{c}]} &\subseteq \left[e^{\underline{c}}, e^{\overline{c}} \right] \,,
\end{aligned}$$

and so on—depending on which atom operations one wants to include in $\mathscr{L}$. Finally, in order to obtain the desired parametrization for the enclosure, the endpoint representation of the interval can be converted into a midpoint-radius representation, i.e.

$$[\underline{c}, \overline{c}] = \text{mid}([\underline{c}, \overline{c}]) + \text{rad}([\underline{c}, \overline{c}])[-1, 1] \,,$$

with

$$\text{mid}([\underline{c}, \overline{c}]) = \frac{\underline{c} + \overline{c}}{2} \quad \text{and} \quad \text{rad}([\underline{c}, \overline{c}]) = \frac{\overline{c} - \underline{c}}{2} \,.$$

Table 2: References about arithmetics using particular classes of basis sets (compare Table 1) and associated software tools. The second column lists the storage complexity of the used set representations, which usually coincides with the computational complexity per atom enclosure operation. However, there are some exceptions. For example, some atom operation bounding rules for ellipsoidal arithmetic may have computational complexity $\mathbf{O}(n^3)$, if dense matrix–matrix multiplications are not avoided in the implementation.

Set Representation	Complexity	Software	References
Intervals	$\mathbf{O}(n)$	`FILIB++` [44], `PROFIL` [38]	[58, 59]
Ellipsoids	$\mathbf{O}(n^2)$	`Ellipsoidal Toolbox` [42],	[40, 41]
		`MC++` [15], `CRONOS` [16]	[35, 78, 79]
Zonotopes	$\mathbf{O}(nm)$	`INTLAB` (Affine Arithmetic) [70],	[19, 26]
Polytopes	$\mathbf{O}(nm)$	`BARON` [71], `ANTIGONE` [54]	[5, 75]
		`GLOMIQO` [53], `MPT3` [32]	
Taylor models	$\mathbf{O}(nr^s)$	`COSY INFINITY` [52], `MC++`, `CRONOS`	[46, 51]
Chebychev models	$\mathbf{O}(nr^s)$	`CHEBFUN` [21], `MC++`	[22, 62, 76]

Even though intervals are easy to store and propagate, the corresponding set arithmetics can lead to large overestimation, particularly for large values of n_φ. This wrapping effect can, however, be mitigated by using more accurate set representations. For example, rules for the construction and propagation of ellipsoids, zonotopes, polytopes, and polynomial models also exist in the literature. Table 2 presents a non-exhaustive list of different set-arithmetics together with packages implementing them.

Notice that some but not all of the set arithmetic tools in Table 2 have been developed originally for applications in control. For example, the polyhedral relaxations of the software packages BARON [71, 75] and ANTIGONE [54] have been developed in the context of solving general factorable optimization problems to global optimality, although these methods can, at least in principle, be used in the context of robust control, too. Other tools, e.g., the Ellipsoidal Toolbox [42], focus on particular set operations and limited libraries $\mathscr{L}$, as ellipsoidal calculus has originally been developed in the context of reachable set computations for linear control systems [40]. For a more general overview about set theoretic methods in control (with a strong focus on linear systems) we refer to the textbook of Blanchini and Miani [10].

It is clear that none of the above affine set arithmetics yields exact enclosures, if the exact image set is not computer representable with respect to the chosen basis set. In practice, interval arithmetics often yields very conservative enclosures, and, at least, in the context of set-valued integration for nonlinear ODEs, it can be shown that ellipsoidal, zonotopic, or other affinely invariant set arithmetics lead to more stable (and, consequently more accurate) enclosures as discussed in the following section and in [36]. The accuracy of more expensive set arithmetics such as Taylor models is often analyzed approximately, i.e., for sets with a "sufficiently small" diameter [11]. The analysis of the conservatism of polynomial set arithmetics on larger domains is, however, still an active field of research [22, 62].

6.3 Set-Valued Integrators

The solution of continuous-time ODEs is typically not factorable. This implies that the set-arithmetic methods from the previous section are not directly applicable to bound the reachable set of the closed-loop system

$$\dot{x}(t) = f(x(t), \mu(t, x(t)), w(t)) \quad \text{with} \quad x(0) = x_0 \,,$$

which is needed in the context of tube based MPC. In order to slightly simplify the following discussion, we assume that this ODE can be written in the form

$$\dot{x}(t) = g(t, x(t), p) \quad \text{with} \quad x(0) = x_0$$

with a finite dimensional uncertain parameter $p \in \mathbb{P} \in \mathbb{K}^{n_p}$ by parametrizing the uncertain function w, e.g., using a polynomial parametrization

$$w(t) \approx \sum_{k=0}^{N} p_k t^k \,.$$

This parametrization can be done in a rigorous manner by over-estimating the associated parametrization error and constructing the set $\mathbb{P}$ appropriately, such that

$$X(t, \mu, x_0) \subseteq \left\{ x_t \in \mathbb{R}^{n_x} \;\middle|\; \begin{array}{l} \exists x \in W_{1,2}^{n_x},\ \exists p \in \mathbb{P} :\ \forall \tau \in [0,t], \\ \dot{x}(\tau) = g(\tau, x(\tau), p) \\ x(0) = x_0 \,,\ x(t) = x_t \end{array} \right\} ,$$

as discussed in full detail in [33, 37]. Next, a local Taylor expansion of the solution trajectory of the parametric ODE can be obtained by constructing the functions

$$\begin{aligned} \chi_0(t, x, p) &= x \\ \chi_k(t, x, p) &= \frac{1}{k} \left(\frac{\partial \chi_{k-1}(t, x, p)}{\partial x} g(t, x, p) + \frac{\partial \chi_{k-1}(t, x, p)}{\partial t} \right) \end{aligned}$$

for all $k \in \{1, \ldots, s+1\}$, where $s \in \mathbb{N}$ is the order of the expansion. If g is smooth and factorable, the functions χ_k are smooth and factorable, too, and can be generated automatically by using algorithmic differentiation [30]. Thus, any affine set arithmetic can be used to construct enclosure parameters $D_k \in \mathbb{R}^{n_x \times m}$ and $d_k \in \mathbb{R}^{n_x}$ of the factorable auxiliary functions χ_k

$$\{ \chi_k(t,x,p) \mid x \in C(t,\mu,x_0) \cdot \mathbb{E} + c(t,\mu,x_0),\ p \in \mathbb{P} \} \subseteq D_k \cdot \mathbb{E} + d_k$$

for $k \in \{0,1,\ldots,s\}$ and

$$\{ \chi_k(t,x,p) \mid t \in [0,h],\ x \in C(t,\mu,x_0) \cdot \mathbb{E} + c(t,\mu,x_0),\ p \in \mathbb{P} \} \subseteq D_{s+1} \cdot \mathbb{E} + d_{s+1}$$

for a suitable basis set $\mathbb{E} \subseteq \mathbb{K}^m$ and for a suitable step-size $h > 0$. It follows from Taylor's theorem that

$$x(t+\tau) \in C(t+\tau,\mu,x_0) \cdot \mathbb{E} + c(t+\tau,\mu,x_0)$$

with

$$C(t+\tau,\mu,x_0) = \sum_{k=0}^{s+1} D_k \tau^k \quad \text{and} \quad c(t+\tau,\mu,x_0) = \sum_{k=0}^{s+1} d_k \tau^k$$

for all $\tau \in [0,h]$ as long as $x(t) \in C(t,\mu,x_0) \cdot \mathbb{E} + c(t,\mu,x_0)$. Thus, one can construct enclosure functions C,c in a recursive way by using suitable step-sizes. This yields a continuous-time enclosure

$$X(t,\mu,x_0) \subseteq C(t,\mu,x_0) \cdot \mathbb{E} + c(t,\mu,x_0) ,$$

which is valid on the whole time horizon $t \in [0,T]$, as needed in the context of tube-based MPC having (13) in mind. Here, one remaining difficulty is how to control the step-size h during the integration. In the literature, various strategies can be found, which have been devised to deal with this problem. The first strategy proceeds by first choosing an optimistic h but then rejects the step if the over-estimation is too large. Most validated integrators, such as `COSY-INFINITY` [52], `VSPODE` [46], and `VALENCIA-IVP` [68], are based on this or variants of this strategy.[11] In [36] a reversed two-phase algorithm was introduced, which proceeds in a slightly different way by constructing a parametric enclosure that is valid on the whole time horizon. The step-size is then refined in a second phase in order to reduce the step size only if the approximation error of the Taylor expansion is large compared to the overestimation that cannot be avoided anyhow due to other set arithmetic operations.

Another practical problem with set integrators is that so-called "bound explosion phenomena" may be observed. Such bound explosions occur if one uses too conservative set arithmetics such that the associated overestimation effects are growing

[11] Early set-valued integrators, as, for example, developed by Nedialkov and Jackson [60], are not based on direct algorithmic differentiation based Taylor expansion of the solution trajectory, but more advanced Hermite-Obreschkoff integration schemes, which have the advantage that they can deal more efficiently with stiff dynamic systems. Some of the above-mentioned software packages are also using more advanced integration schemes, but the basic ideas for bounding the reachable set enclosures are, nevertheless, very similar to the easy-to-implement Taylor expansion based method, which has been outlined in this section.

over time. In the worst case, these wrapping effects can lead to unstable (or unreasonably conservative) bounds—even if the original ODE was perfectly stable. As it was shown in [36] stable set-integrators can be constructed for asymptotically stable systems and for set-parametrizations, if one uses set arithmetics, which are invariant under affine transformations. For example, an affine transformation of an ellipsoid or a zonotope is again an ellipsoid or zonotope. Thus, most implementations of ellipsoidal and zonotopic set-arithmetics can be expected to be invariant under affine transformation, while standard interval arithmetic does not have such a property. A generic implementation of stable set-valued integrators for a variety of set-parametrizations can be found as part of the CRONOS library [16].

7 Conclusions

Although exact inf-sup feedback MPC problems are intractable in general, this chapter has reviewed a large number of practical numerical methods, which can be used to construct conservative approximations of robust MPC. For the case that the system dynamics is linear, one can either rely on methods from the field of ellipsoidal approximation and linear matrix inequalities in control or use affine disturbance parametrizations to construct highly scalable implementations based on convex optimization. Moreover, this chapter has reviewed three main classes of generic methods for robust MPC, namely, Dynamic Programming, Scenario-Tree MPC, and Tube MPC. These methods have in common that they are—at least in principle—applicable to both linear and nonlinear dynamic processes, although mature software packages for general nonlinear robust MPC are not available yet and most practical implementations still focus on specialized classes of systems, often linear process models.

The second part of this chapter had a strong focus on reviewing state-of-the-art numerical methods for Tube MPC. This focus is motivated by recent developments in the field of modern set-valued computing, or, more specifically, affine set-arithmetic and set-valued integration, which can be considered as the basis for Tube MPC based methods. While traditional tools for computing set enclosures are often based on rather conservative interval arithmetic, the trend of modern set-arithmetics goes towards using polynomial set representations, often in combination with ellipsoidal or zonotopic remainder bounds, which are invariant under affine transformations and the basis for stable reach-set integration. These new tools are opening new perspectives for Tube MPC based implementations of robust MPC, although the development of mature tools for generic robust nonlinear MPC remains an important challenge for future research.

References

1. Andersson, J., Houska, B., Diehl, M.: Towards a computer algebra system with automatic differentiation for use with object-oriented modelling languages. In: 3rd International Workshop on Equation-Based Object-Oriented Modeling Languages and Tools, Oslo, Norway, October 3, pp. 99–105. Linköping University Electronic Press, Linköping (2010)
2. Andersson, J., Åkesson, J., Diehl, M.: Casadi: a symbolic package for automatic differentiation and optimal control. In: Recent Advances in Algorithmic Differentiation, pp. 297–307. Springer, Berlin (2012)
3. Aubin, J.P.: Viability theory. Systems & Control: Foundations & Applications. Birkhäuser, Boston (1991)
4. Bansal, S., Chen, M., Herbert, S., Tomlin, C.J.: Hamilton-Jacobi reachability: a brief overview and recent advances (2017). Preprint. arXiv:1709.07523
5. Bao, X., Sahinidis, N.V., Tawarmalani, M.: Multiterm polyhedral relaxations for nonconvex, quadratically constrained quadratic programs. Optim. Methods Softw. **24**(4–5), 485–504 (2009)
6. Bemporad, A., Borrelli, F., Morari, M.: Min-max control of constrained uncertain discrete-time linear systems. IEEE Trans. Autom. Control **48**(9), 1600–1606 (2003)
7. Ben-Tal, A., Goryashko, A., Guslitzer, E., Nemirovski, A.: Adjustable robust solutions of uncertain linear programs. In: Technical report, Minerva Optimization Center, Technion, Israeli Institute of Technology (2002)
8. Bertsekas, D.P.: Dynamic Programming and Optimal Control. Athena Scientific, Belmont (1995)
9. Björnberg, J., Diehl, M.: Approximate robust dynamic programming and robustly stable MPC. Automatica **42**(5), 777–782 (2006)
10. Blanchini, F., Miani, S.: Set-theoretic methods in control. Systems & Control: Foundations & Applications. Birkhäuser, Boston (2008)
11. Bompadre, A., Mitsos, A., Chachuat, B.: Convergence analysis of Taylor models and McCormick-Taylor models. J. Glob. Optim. **57**, 75–114 (2013)
12. Bosgra, O.H., van Hessem, D.H.: A conic reformulation of model predictive control including bounded and stochastic disturbances under state and input constraints. In: Proceedings of the 41st IEEE Conference on Decision and Control, pp. 4643–4648 (2002)
13. Boyd, S., El Ghaoui, L., Feron, E., Balakrishnan, V.: Linear Matrix Inequalities in System and Control Theory. Springer, London (1994)
14. Buckdahn, R., Li, J.: Stochastic differential games and viscosity solutions of Hamilton-Jacobi-Bellman-Isaacs equations. SIAM J. Control Optim. **47**(1), 444–475 (2008)
15. Chachuat, B., OMEGA Research Group: Mc++: a toolkit for construction, manipulation and bounding of factorable functions, 2006–2017. https://omega-icl.github.io/mcpp/
16. Chachuat, B., OMEGA Research Group: CRONOS: complete search optimization for nonlinear systems, 2012–2017. https://omega-icl.github.io/cronos/
17. Chachuat, B., Houska, B., Paulen, R., Peri'c, N., Rajyaguru, J., Villanueva, M.E.: Set-theoretic approaches in analysis, estimation and control of nonlinear systems. IFAC-PapersOnLine **48**(8), 981–995 (2015)
18. Chen, M., Herbert, S.L., Vashishtha, M.S., Bansal, S., Tomlin, C.J.: Decomposition of reachable sets and tubes for a class of nonlinear systems (2016). Preprint. arXiv:1611.00122
19. de Figueiredo, L.H., Stolfi, J.: Affine arithmetic: concepts and applications. Numer. Algorithms **7**(1), 147–158 (2004)
20. Diehl, M., Bjornberg, J.: Robust dynamic programming for min-max model predictive control of constrained uncertain systems. IEEE Trans. Autom. Control **49**(12), 2253–2257 (2004)
21. Driscoll, T.A., Hale, N., Trefethen, L.N.: Chebfun Guide. Pafnuty Publications, Oxford (2014). http://www.chebfun.org/docs/guide/
22. Dzetkulič, T.: Rigorous integration of non-linear ordinary differential equations in Chebyshev basis. Numer. Algorithms **69**(1), 183–205 (2015)

23. Engell, S.: Online optimizing control: the link between plant economics and process control. In: 10th International Symposium on Process Systems Engineering: Part A. Computer Aided Chemical Engineering, vol. 27, pp. 79–86. Elsevier, Amsterdam (2009)
24. Filippov, A.F.: On certain questions in the theory of optimal control. J. SIAM Control Ser. A **1**(1), 76–84 (1962)
25. Fleming, W.H., Souganidis, P.E.: On the existence of value functions of two-player, zero-sum stochastic differential games. Indiana Univ. Math. J. **38**(2), 293–314 (1989)
26. Fukuda, K.: From the zonotope construction to the Minkowski addition of convex polytopes. J. Symb. Comput. **38**(4), 1261–1272 (2004)
27. Gerdts, M., Henrion, R., Hömberg, D., Landry, C.: Path planning and collision avoidance for robots. Numer. Algebra Control Optim. **2**(3), 437–463 (2012)
28. Goulart, P.J., Kerrigan, E.C., Maciejowski, J.M.: Optimization over state feedback policies for robust control with constraints. Automatica **42**(4), 523–533 (2006)
29. Grant, M., Boyd, S.: CVX: Matlab software for disciplined convex programming, version 2.1. http://cvxr.com/cvx, March 2014
30. Griewank, A., Walther, A.: Evaluating Derivatives: Principles and Techniques of Algorithmic Differentiation. SIAM, Philadelphia (2008)
31. Grüne, L.: An adaptive grid scheme for the discrete Hamilton-Jacobi-Bellman equation. Numer. Math. **75**(3), 319–337 (1997)
32. Herceg, M., Kvasnica, M., Jones, C.N., Morari, M.: Multi-parametric toolbox 3.0. In: Proceedings of the European Control Conference, Zürich, Switzerland, 17–19 July 2013, pp. 502–510. http://control.ee.ethz.ch/~mpt
33. Houska, B., Chachuat, B.: Branch-and-lift algorithm for deterministic global optimization in nonlinear optimal control. J. Optim. Theory Appl. **162**(1), 208–248 (2014)
34. Houska, B., Ferreau, H.J., Diehl, M.: An auto-generated real-time iteration algorithm for nonlinear MPC in the microsecond range. Automatica **47**(10), 2279–2285 (2011)
35. Houska, B., Logist, F., Van Impe, J., Diehl, M.: Robust optimization of nonlinear dynamic systems with application to a jacketed tubular reactor. J. Proc. Control **22**(6), 1152–1160 (2012)
36. Houska, B., Villanueva, M.E., Chachuat, B.: Stable set-valued integration of nonlinear dynamic systems using affine set-parameterizations. SIAM J. Numer. Anal. **53**(5), 2307–2328 (2015)
37. Houska, B., Li, J.C., Chachuat, B.: Towards rigorous robust optimal control via generalized high-order moment expansion. Optimal Control Appl. Methods (2017). https://doi.org/10.1002/oca.2309
38. Keil, C.: PROFIL: Programmer's runtime optimized fast interval library, 2009–2017. http://www.ti3.tuhh.de/keil/profil/index_e.html
39. Kothare, M.V., Balakrishnan, V., Morari, M.: Robust constrained model predictive control using linear matrix inequalities. Automatica **32**(10), 1361–1379 (1996)
40. Kurzhanski, A.B., Filippova, T.F.: On the theory of trajectory tubes—a mathematical formalism for uncertain dynamics, viability and control. In: Advances in Nonlinear Dynamics and Control: A Report from Russia, volume 17 of Progress in Systems Control Theory, pp. 122–188. Birkhäuser, Boston (1993)
41. Kurzhanski, A.B., Vályi, I.: Ellipsoidal calculus for estimation and control. Systems & Control: Foundations & Applications. Birkhäuser, Boston (1997)
42. Kurzhanskiy, A.A., Varaiya, P.: Ellipsoidal toolbox (et). In: Proceedings of the 45th IEEE Conference on Decision and Control, pp. 1498–1503 (2006)
43. Langson, W., Chryssochoos, I., Raković, S.V., Mayne, D.Q.: Robust model predictive control using tubes. Automatica **40**(1), 125–133 (2004)
44. Lerch, M., Tischler, G., Gudenberg, J.W.V., Hofschuster, W., Krämer, W.: Filib++, a fast interval library supporting containment computations. ACM Trans. Math. Softw. **32**(2), 299–324 (2006). http://www2.math.uni-wuppertal.de/~xsc/software/filib.html
45. Liberzon, D.: Calculus of variations and optimal control theory: a concise introduction. Princeton University Press, Princeton (2012)

46. Lin, Y., Stadtherr, M.A.: Validated solutions of initial value problems for parametric ODEs. Appl. Numer. Math. **57**(10), 1145–1162 (2007)
47. Löfberg, J.: Minimax approaches to robust model predictive control. PhD thesis, Linköping University (2003)
48. Löfberg, J.: Approximations of closed-loop MPC. In: Proceedings of the 42nd IEEE Conference on Decision and Control, Maui, pp. 1438–1442 (2003)
49. Löfberg, J.: YALMIP: a toolbox for modeling and optimization in matlab. In: 2004 IEEE International Symposium on Computer Aided Control Systems Design, pp. 284–289. IEEE, Piscataway (2004)
50. Lucia, S., Finkler, T., Engell, S.: Multi-stage nonlinear model predictive control applied to a semi-batch polymerization reactor under uncertainty. J. Process Control **23**(9), 1306–1319 (2013)
51. Makino, K., Berz, M.: Taylor models and other validated functional inclusion methods. Int. J. Pure Appl. Math. **4**(4), 379–456 (2003)
52. Makino, K., Berz, M.: Cosy infinity version 9. Nucl. Instrum. Methods Phys. Res. Sect. A Accelerators Spectrometers Detectors Assoc. Equip. **558**(1), 346–350 (2006)
53. Misener, R., Floudas, C.A.: GloMIQO: global mixed-integer quadratic optimizer. J. Glob. Optim. **57**(1), 3–50 (2013)
54. Misener, R., Floudas, C.A.: ANTIGONE: Algorithms for coNTinuous/Integer Global Optimization of Nonlinear Equations. J. Glob. Optim. (2014). https://doi.org/10.1007/s10898-014-0166-2
55. Mitchell, I.M.: The flexible, extensible and efficient toolbox of level set methods. J. Sci. Comput. **35**, 300–329 (2008)
56. Mitchell, I.M., Templeton, J.A.: A toolbox of Hamilton-Jacobi solvers for analysis of nondeterministic continuous and hybrid systems. In: Hybrid Systems Computation and Control. Lecture Notes in Computer Science, vol. 3414, pp. 480–494 (2005)
57. Mitchell, I.M., Bayen, A.M., Tomlin, C.J.: A time-dependent Hamilton-Jacobi formulation of reachable sets for continuous dynamic games. IEEE Trans. Autom. Control **50**(7), 947–957 (2005)
58. Moore, R.E.: Interval Analysis. Prentice-Hall, Englewood Cliffs (1966)
59. Moore, R.E., Kearfott, R.B., Cloud, M.J.: Introduction to Interval Analysis. SIAM, Philadelphia (2009)
60. Nedialkov, N.S., Jackson, K.R.: An interval Hermite-Obreschkoff method for computing rigorous bounds on the solution of an initial value problem for an ordinary differential equation. In: Developments in Reliable Computing, pp. 289–310. Springer, Dordrecht (1999)
61. Pontryagin, L.S., Boltyanskii, V.G., Gamkrelidze, R.V., Mishchenko, E.F.: The Mathematical Theory of Optimal Processes. Wiley, New York (1962)
62. Rajyaguru, J., Villanueva, M.E., Houska, B., Chachuat, B.: Chebyshev model arithmetic for factorable functions. J. Glob. Optim. **68**(2), 413–438 (2017)
63. Raković, S.V.: Set theoretic methods in model predictive control. In: Nonlinear Model Predictive Control: Towards New Challenging Applications. Lecture Notes in Control and Information Sciences, vol. 384 , pp. 41–54. Springer, Heidelberg (2009)
64. Raković, S.V.: Invention of prediction structures and categorization of robust MPC syntheses. IFAC Proc. **45**(17), 245–273 (2012)
65. Rakovic, S.V., Kouvaritakis, B., Cannon, M., Panos, C., Findeisen, R.: Parameterized tube model predictive control. IEEE Trans. Autom. Control **57**(11), 2746–2761 (2012)
66. Raković, S.V., Kouvaritakis, B., Findeisen, R., Cannon, M.: Homothetic tube model predictive control. Automatica **48**(8), 1631–1638 (2012)
67. Raković, S.V., Levine, W.S., Açıkmeşe, B.: Elastic tube model predictive control. In: American Control Conference (ACC), pp. 3594–3599. IEEE, Piscataway (2016)
68. Rauh, A., Hofer, E.P., Auer, E.: VALENCIA-IVP: a comparison with other initial value problem solvers. In: 12th GAMM - IMACS International Symposium on Scientific Computing, Computer Arithmetic and Validated Numerics (SCAN 2006), pp. 36–36 (2006)

69. Rawlings, J.B., Mayne, D.Q.: Model Predictive Control: Theory and Design, 5th edn. Nob Hill Publishing, Madison (2015)
70. Rump, S.M.: INTLAB - INTerval LABoratory. In: Csendes, T. (ed.) Developments in Reliable Computing, pp. 77–104. Kluwer Academic Publishers, Dordrecht (1999). http://www.ti3.tuhh.de/rump/
71. Sahinidis, N.V.: BARON 14.3.1: Global Optimization of Mixed-Integer Nonlinear Programs, *User's Manual*, 2005–2017. http://www.minlp.com/downloads/docs/baron%20manual.pdf
72. Scokaert, P.O.M., Mayne, D.Q.: Min-max feedback model predictive control for constrained linear systems. IEEE Trans. Autom. control **43**(8), 1136–1142 (1998)
73. Smirnov, G.V.: Introduction to the Theory of Differential Inclusions. Graduate Studies in Mathematics, vol. 41, American Mathematical Society, Providence (2002)
74. Sontag, E.D.: Mathematical Control Theory: Deterministic Finite Dimensional Systems. Springer, New York (2013)
75. Tawarmalani, M., Sahinidis, N.V.: A polyhedral branch-and-cut approach to global optimization. Math. Program. **103**, 225–249 (2005)
76. Trefethen, L.N., Battles, Z.: An extension of MATLAB to continuous functions and operators. SIAM J. Sci. Comput. **25**, 1743–1770 (2004)
77. Van Parys, B.P.G., Goulart, P.J., Morari, M.: Infinite horizon performance bounds for uncertain constrained systems. IEEE Trans. Autom. Control **58**(11), 2803–2817 (2013)
78. Villanueva, M.E., Houska, B., Chachuat, B.: Unified framework for the propagation of continuous-time enclosures for parametric nonlinear ODEs. J. Glob. Optim. **62**(3), 575–613 (2015)
79. Villanueva, M.E., Rajyaguru, J., Houska, B., Chachuat, B.: Ellipsoidal arithmetic for multivariate systems. Comput. Aided Chem. Eng. **37**, 767–772 (2015)
80. Villanueva, M.E., Li, J.C., Feng, X., Chachuat, B., Houska, B.: Computing ellipsoidal robust forward invariant tubes for nonlinear MPC. In: Proceedings of the 20th IFAC World Congress, Toulouse, pp. 7436–7441 (2017)
81. Villanueva, M.E., Quirynen, R., Diehl, M., Chachuat, B., Houska, B.: Robust MPC via min-max differential inequalities. Automatica **77**, 311–321 (2017)
82. Wan, Z., Kothare, M.V.: An efficient off-line formulation of robust model predictive control using linear matrix inequalities. Automatica **39**(5), 837–846 (2003)
83. Wang, Y., O'Donoghue, B., Boyd, S.: Approximate dynamic programming via iterated bellman inequalities. Int. J. Robust Nonlinear Control **25**(10), 1472–1496 (2015)
84. Zeilinger, M.N., Raimondo, D.M., Domahidi, A., Morari, M., Jones, C.N.: On real-time robust model predictive control. Automatica **50**(3), 683–694 (2014)

Scenario Optimization for MPC

Marco C. Campi, Simone Garatti, and Maria Prandini

1 Introduction

Model Predictive Control (MPC) is a methodology to determine control actions in the presence of constraints that has proven effective in many real applications. Instead of addressing an infinite-horizon problem, which would be hard to deal with due to computational difficulties, in MPC one solves at each point in time a finite-horizon constrained problem, and implements only the first control action that has been determined; then, the procedure is repeated at the next instant of time by shifting the prediction horizon ahead of one unit of time (receding horizon).

In many control problems, disturbances are a fundamental ingredient. In MPC, disturbances have been dealt with along two different approaches, namely robust MPC and stochastic MPC. In robust MPC (e.g., [4, 5, 23, 34, 47, 50]), the control cost is optimized against the worst disturbance realization, while also guaranteeing constraints satisfaction. The drawback with this approach is that it generates conservative control actions. To overcome this drawback, an average cost with probabilistic constraints is considered in stochastic MPC where a violation of the constraints is accepted provided the probability of this to happen is kept below a given threshold (e.g., [3, 9, 18, 19, 21, 46, 49, 53]). In the stochastic optimization literature, probabilistic constraints of this type are often called "chance-constraints," see, e.g., [44, 45].

M. C. Campi (✉)
Università di Brescia - Dipartimento di Ingegneria dell'Informazione, via Branze 38, 25123, Brescia, Italy
e-mail: marco.campi@unibs.it

S. Garatti · M. Prandini
Politecnico di Milano - Dipartimento di Elettronica, Informazione e Bioingegneria, piazza L. da Vinci 32, 20133, Milan, Italy
e-mail: simone.garatti@polimi.it; maria.prandini@polimi.it

S. V. Raković, W. S. Levine (eds.), *Handbook of Model Predictive Control*,
Control Engineering, https://doi.org/10.1007/978-3-319-77489-3_19

Chance-constraints are known for being very hard to deal with. One reason is that they are highly non-convex even when the original problem is born within a convex setup where the constraints are convex for any given realization of the disturbance. As a matter of fact, solutions to MPC with chance-constrained constraints have been proposed for specific cases only such as linear systems with either bounded and i.i.d. (independent and identically distributed) [9, 18, 19] or Gaussian [21] disturbances. In this chapter, we describe an alternative scheme to deal with stochastic MPC ([43]). This scheme is grounded in some recent developments in stochastic optimization where the chance-constraints are replaced by variants obtained by sampling finitely many realizations of the disturbance (scenario approach). Considering a finite sample of realizations makes the problem computationally tractable while the link to the original chance-constrained problem is established by a rigorous theory. With this approach, one gains the important advantage that no assumptions on the disturbance, such as boundedness, independence or Gaussianity, is required.

This chapter is organized as follows. In the first section the mathematical setup of study is introduced. After a digression to summarize some relevant results of the scenario approach in Section 3, Section 4 describes how the scenario methodology can be applied to MPC. The closing Section 5 presents a simulation study for a mechanical system.

2 Stochastic MPC and the Use of the Scenario Approach

Consider a linear system whose state $x_t \in \mathbb{R}^n$ evolves according to the equation

$$x_{t+1} = Ax_t + Bu_t + Dw_t,$$

where $u_t \in \mathbb{R}^m$ is the control input, $w_t \in \mathbb{R}^l$, $l \leq n$, is a stochastic disturbance with a possibly unbounded support, and D is full-rank.

We assume that the entire state vector of the system is known at each time instant and focus on the finite-horizon optimization problem that needs to be solved at each point in time τ of a stochastic MPC scheme that implements a receding horizon strategy. Specifically, we consider the following quadratic cost

$$J = \mathbb{E}\left[\sum_{i=1}^{M} x_{\tau+i}^T Q_i x_{\tau+i} + \sum_{i=0}^{M-1} u_{\tau+i}^T R_i u_{\tau+i}\right], \tag{1}$$

where M is the horizon length and $Q_i = Q_i^T \succeq 0$ and $R_i = R_i^T \succ 0$ have appropriate dimensions, subject to the probabilistic constraint:

$$\mathbb{P}\{f(x_{\tau+1},\ldots,x_{\tau+M},u_\tau,\ldots,u_{\tau+M-1}) \leq 0\} \geq 1-\varepsilon, \tag{2}$$

where $f : \mathbb{R}^{n\times M+m\times M} \to \mathbb{R}^q$ is a q-dimensional function and the inner inequality in (2) is interpreted componentwise. In the above expressions, probability $\mathbb{P}$ refers to the stochastic nature of the disturbance and $\mathbb{E}$ is the expected value associated with it. In the probabilistic constraint (2), condition $f(x_{\tau+1},\ldots,x_{\tau+M},u_\tau,\ldots,u_{\tau+M-1}) \leq 0$ is not required to hold for all possible disturbance realizations and parameter $\varepsilon \in (0,1)$ quantifies the admissible probability of constraint violation. Allowing for an ε violation improves the control system performance and, moreover, when the disturbance has unbounded support, allowing for a small probability of constraint violation can be the only way to avoid infeasibility of the optimization problem. In practice, applications exist where violating a constraint may result in severe damages of equipments or in important malfunctionings, in which case one may not be willing to allow for an ε-violation. In many other cases, however, sporadic constraint violations are tolerable and cause little damage. For example, exceeding the load capacity in power lines for a short time does not cause any plant damages and, in a totally different field, high blood glucose is not a cause of cellular damage if it happens for short periods. Similar examples can be found in a variety of contexts. This is the frame where stochastic MPC finds application.

In many cases, function f in (2) is used to enforce input saturation constraints in addition to constraints on the allowed state values. If, for instance, f is given by

$$f(x_{\tau+1},\ldots,x_{\tau+M},u_\tau,\ldots,u_{\tau+M-1}) = \begin{bmatrix} \sup\limits_{i=0,\ldots,M-1} \|Su_{\tau+i}\|_\infty - \bar{u} \\ \sup\limits_{i=1,\ldots,M} \|Cx_{\tau+i}\|_\infty - \bar{y} \end{bmatrix}, \tag{3}$$

where S and C are matrixes in $R^{q\times m}$ and $R^{p\times n}$ respectively, then, $\bar{u}$ and $\bar{y}$ are limits on linear combinations of the inputs and of the state values. In the following, we shall consider generic but convex functions f.

Note that when the noise is Gaussian and constraints are missing, minimizing (1) gives a standard LQG control problem which admits analytical solution. Instead, in the presence of constraints, or when the noise is not Gaussian, the problem of finding the optimal solution becomes quite challenging. Hence, one can concentrate on specific structures by which the control actions are determined.

To find a suitable structure, one can think of reconstructing the noise from the state according to the equation

$$w_{\tau+i} = D^\dagger(x_{\tau+i+1} - Ax_{\tau+i} - Bu_{\tau+i}),$$

where $D^\dagger$ is pseudo-inverse and then parameterize the control action as an affine function of the disturbance

$$u_{\tau+i} = \gamma_i + \sum_{j=0}^{i-1} \theta_{i,j} w_{\tau+j}, \tag{4}$$

with $\gamma_i \in \mathbb{R}^m$ and $\theta_{i,j} \in \mathbb{R}^{m\times n}$.[1] This parametrization was indeed proposed in [30] (and, independently in [6]) where it was also shown that (4) is equivalent to considering policies that are affine in the state (i.e., to every affine in the state policy $\mu_{\tau+i}$, there correspond γ_i and $\theta_{i,j}$ such that (4) returns the same control action as $\mu_{\tau+i}$ and vice versa).

The fundamental advantage gained by adopting (4) is that the control cost and the constraints become convex in the variables γ_i and $\theta_{i,j}$ (when the control action is parameterized as an affine function of the state, this fails to be true). We shall write this control cost and the constraints explicitly in Section 4 after introducing suitable notations. There, we shall further show that by sampling the noise realizations (scenario approach) and enforcing the constraints only on the realizations that have been sampled, one obtains a standard convex problem that can be solved with conventional optimization methods. The so-found solution carries precise guarantees of satisfaction of the original chance-constrained constraint. Proving this deep result calls for the use of the scenario theory that is briefly summarized in the next section.

3 Fundamentals of Scenario Optimization

Consider the following constrained convex optimization problem

$$\begin{aligned} &\min_{x\in\mathscr{X}\subseteq\mathbb{R}^d} \ell(x) \quad \text{subject to:} \\ &x \in \mathscr{X}_\delta, \quad \delta \in \Delta, \end{aligned} \tag{5}$$

where $\ell(x)$ is a convex function, $\delta \in \Delta$ is an uncertain parameter, and $\mathscr{X}$ and $\mathscr{X}_\delta$ are convex and closed sets. In normal situations, Δ has infinite cardinality. Uncertainty in (5) can be dealt with along two distinct approaches. The first one consists in enforcing satisfaction of all constraints, that is one optimizes the cost $\ell(x)$ over the set $\bigcap_{\delta\in\Delta}\mathscr{X}_\delta$ (robust approach). Alternatively, one may want to satisfy the constraints with "high probability" (stochastic approach). Along this second approach one sees the uncertainty parameter δ as a random element with a probability $\mathbb{P}$, and seeks a solution that violates at most a fraction of the constraints that has small probability (chance-constrained solution). This second approach is often more advantageous in that it returns less conservative designs.

Notoriously, finding a solution to (5) that carries a high probability of constraints satisfaction is a very difficult task [44]. In [7, 8], the following scenario problem is introduced, where N values of δ, say $\delta^{(1)},\dots,\delta^{(N)}$, are randomly sampled from $\mathbb{P}$ one independently of the others and these N values provide the only constraints that are enforced in the optimization problem:

---

[1] Often, the total number of parameters is reduced as compared to (4) by imposing internal relation among parameters. This is further discussed in Section 4. When all $\theta_{i,j}$ are set to zero, one obtains a classical setup where optimization is directly performed on the control actions.

$$\min_{x \in \mathscr{X} \subseteq \mathbb{R}^d} \ell(x) \quad \text{subject to:} \\ x \in \mathscr{X}_{\delta^{(i)}}, \quad i \in \{1,2,\ldots,N\}. \tag{6}$$

Since (6) has a finite number of constraints, it can be solved at low computational cost. On the other hand, the obvious question to ask is whether (6) gives a chance-constrained solution. An answer is found in the following fundamental theorem that has been established in [12].[2]

Definition 1 (violation probability). The *violation probability* of a given $x \in \mathscr{X}$ is defined as $V(x) = \mathbb{P}\{\delta \in \Delta : x \notin \mathscr{X}_\delta\}$.

Theorem 1 ([12]). *Let x_N^* be the solution to* (6). *It holds that*

$$\mathbb{P}^N\{V(x_N^*) > \varepsilon\} \leq \sum_{i=0}^{d-1} \binom{N}{i} \varepsilon^i (1-\varepsilon)^{N-i}. \tag{7}$$

From (7) one obtains that $\mathbb{P}^N\{V(x_N^*) \leq \varepsilon\} \geq 1 - \sum_{i=0}^{d-1} \binom{N}{i} \varepsilon^i (1-\varepsilon)^{N-i}$, which shows that the cumulative probability distribution of $V(x_N^*)$ is bounded by a Beta distribution. This result, as all results in the scenario theory, is distribution-free, that is, it holds for all distributions $\mathbb{P}$. Moreover, it is not improvable since in [12] it is proven that the result is tight and holds with equality for a class of problems there named "fully-supported." By setting $\sum_{i=0}^{d-1} \binom{N}{i} \varepsilon^i (1-\varepsilon)^{N-i} \leq \beta$, the interpretation of Theorem 1 is that the scenario solution is, with (high) probability $1-\beta$, a feasible solution for a chance-constrained problem where one is allowed to violate an ε-fraction of the constraints.

To offer a more immediate understanding of the theorem, a pictorial representation of the result is given in Figure 1. In the figure, the N samples $\delta^{(1)}, \ldots, \delta^{(N)}$ are represented as a single multi-sample $(\delta^{(1)}, \ldots, \delta^{(N)})$ from Δ^N. In Δ^N there is a "bad set" represented in grey such that, if we extract a multi-sample in the bad set, then the theorem does not provide us with any conclusions. This, however, happens with tiny probability since β can be made very small, say 10^{-10}, without having to increase N excessively (this fact is discussed in [12] and it is also touched upon later in this chapter for the particular setup of MPC). In all other cases, the multi-sample maps into a finite convex optimization problem, the scenario problem, that we can easily solve and the corresponding solution automatically satisfies all the other unseen constraints except for a small fraction ε of them.

Scenario optimization has been introduced in [7], and has ever since attracted an increasing interest. Robustness properties have been studied in [8, 12, 20] and, under regularization and structural assumptions, further investigated in [2, 11,

[2] In [12], a mild assumption of existence and uniqueness of the solution (Assumption 1 in [12]) is made which we do not report here for conciseness of presentation. Moreover, paper [12] considers linear cost functions but the extension to generic convex functions is straightforward.

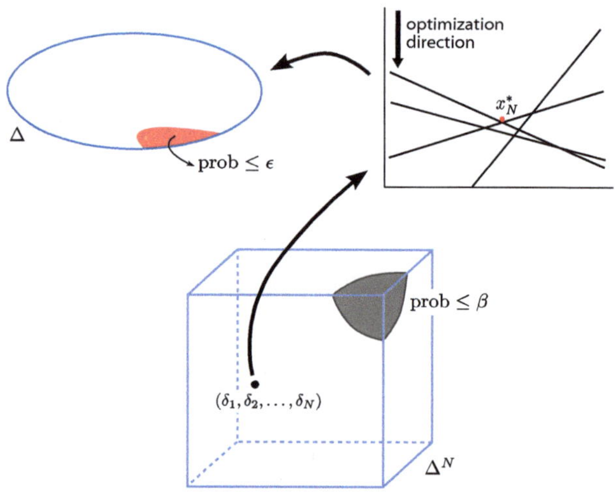

Fig. 1: Visualization of Theorem 1.

48, 55]. Papers [13, 29] consider constraints removal, and [54] examines multi-stage problems. Generalizations to a non-convex setup are proposed in [1, 28, 31]. See also [8, 16, 24, 36, 42, 51, 52] for a comparison of scenario optimization with other methods in stochastic optimization. Besides MPC, scenario optimization has found application to fields ranging from machine learning and prediction [10, 15, 22, 37] to quantitative finance [33, 39–41], from management to control design [16]. For next use in this chapter, we also recall here the main result from [29].

Theorem 2 ([29]). *Fix a value $k \leq N$; remove k constraints from problem* (6) *according to a given, arbitrary, rule; find the solution $x^*_{k,N}$ of the so-obtained problem, and assume that the rule has been designed so that the k constraints that have been removed are violated.*[3] *It holds that*

$$\mathbb{P}^N\{V(x^*_{k,N}) > \varepsilon\} \leq \binom{k+d-1}{k} \sum_{i=0}^{k+d-1} \binom{N}{i} \varepsilon^i (1-\varepsilon)^{N-i}. \qquad (8)$$

Constraints removal is important to improve the performance of the scenario program and Theorem 2 quantifies the violation when a solution that violates k constraints is considered.

[3] Violation of the removed constraints must hold almost surely with respect to the scenario realizations and for any N, see [29] for a broad discussion.

4 The Scenario Approach for Solving Stochastic MPC

By defining the following vectors of state, input, and disturbance signals

$$\mathbf{x}_+ = \begin{bmatrix} x_{\tau+1} \\ x_{\tau+2} \\ \vdots \\ x_{\tau+M} \end{bmatrix} \quad \mathbf{u} = \begin{bmatrix} u_\tau \\ u_{\tau+1} \\ \vdots \\ u_{\tau+M-1} \end{bmatrix} \quad \mathbf{w} = \begin{bmatrix} w_\tau \\ w_{\tau+1} \\ \vdots \\ w_{\tau+M-1}, \end{bmatrix}$$

one can write

$$\begin{aligned} \mathbf{x}_+ &= \mathbf{F}x_\tau + \mathbf{G}\mathbf{u} + \mathbf{H}\mathbf{w} \\ \mathbf{u} &= \Gamma + \Theta\mathbf{w}, \end{aligned} \tag{9}$$

where matrices **F**, **G**, and **H** are given by

$$\mathbf{F} = \begin{bmatrix} A \\ A^2 \\ \vdots \\ A^M \end{bmatrix} \quad \mathbf{G} = \begin{bmatrix} B & 0_{n\times m} & \cdots & 0_{n\times m} \\ AB & B & \ddots & \vdots \\ \vdots & \ddots & \ddots & 0_{n\times m} \\ A^{M-1}B & \cdots & AB & B \end{bmatrix} \quad \mathbf{H} = \begin{bmatrix} D & 0_{n\times l} & \cdots & 0_{n\times l} \\ AD & D & \ddots & \vdots \\ \vdots & \ddots & \ddots & 0_{n\times l} \\ A^{M-1}D & \cdots & AD & D \end{bmatrix},$$

and Γ and Θ contain the parameters of the control law and are given by

$$\Gamma = \begin{bmatrix} \gamma_0 \\ \gamma_1 \\ \vdots \\ \gamma_{M-1} \end{bmatrix} \quad \Theta = \begin{bmatrix} 0_{m\times l} & 0_{m\times l} & \cdots & 0_{m\times l} \\ \theta_{1,0} & 0_{m\times l} & \ddots & \vdots \\ \vdots & \ddots & \ddots & 0_{m\times l} \\ \theta_{M-1,0} & \cdots & \theta_{M-1,M-2} & 0_{m\times l} \end{bmatrix}.$$

Let us start by considering the constraints. Since the state and input vectors $\mathbf{x}_+$ and $\mathbf{u}$ are linear functions of the design parameters Γ and Θ (equation (9)), and function $f(\mathbf{x}_+,\mathbf{u})$ in (2) is convex, then $f(\mathbf{F}x_\tau + \mathbf{G}\Gamma + (\mathbf{H}+\mathbf{G}\Theta)\mathbf{w}, \Gamma + \Theta\mathbf{w})$ is a convex function of Γ and Θ.

As for the control cost (1), letting

$$\mathbf{Q} = \begin{bmatrix} Q_1 & \cdots & 0_{n\times n} \\ \vdots & \ddots & \vdots \\ 0_{n\times n} & \cdots & Q_M \end{bmatrix} \quad \mathbf{R} = \begin{bmatrix} R_0 & \cdots & 0_{m\times m} \\ \vdots & \ddots & \vdots \\ 0_{m\times m} & \cdots & R_{M-1} \end{bmatrix}$$

the cost can be expressed as follows:

$$\begin{aligned} J(\Gamma,\Theta) &= \mathbb{E}\left[\mathbf{x}_+^T\mathbf{Q}\mathbf{x}_+ + \mathbf{u}^T\mathbf{R}\mathbf{u}\right] \\ &= (\mathbf{F}x_\tau + \mathbf{G}\Gamma)^T\mathbf{Q}(\mathbf{F}x_\tau + \mathbf{G}\Gamma) + 2(\mathbf{F}x_\tau + \mathbf{G}\Gamma)^T\mathbf{Q}(\mathbf{H}+\mathbf{G}\Theta)\cdot\mathbb{E}[\mathbf{w}] \\ &\quad + \mathrm{tr}\left[(\mathbf{H}+\mathbf{G}\Theta)^T\mathbf{Q}(\mathbf{H}+\mathbf{G}\Theta)\cdot\mathbb{E}\left[\mathbf{w}\mathbf{w}^T\right]\right] + \Gamma^T\mathbf{R}\Gamma + 2\Gamma^T\mathbf{R}\Theta\cdot\mathbb{E}[\mathbf{w}] \\ &\quad + \mathrm{tr}\left[\Theta^T\mathbf{R}\Theta\cdot\mathbb{E}\left[\mathbf{w}\mathbf{w}^T\right]\right], \end{aligned}$$

which is a quadratic convex function of Γ and Θ.

Hence, the optimization problem to be solved at time τ can be written as the following chance-constrained optimization problem

$$\begin{aligned} &\min_{\Gamma,\Theta} J(\Gamma,\Theta) \quad \text{subject to:} \\ &\mathbb{P}\{f(\mathbf{F}x_\tau+\mathbf{G}\Gamma+(\mathbf{H}+\mathbf{G}\Theta)\mathbf{w},\Gamma+\Theta\mathbf{w})\leq 0\}\geq 1-\varepsilon. \end{aligned} \tag{10}$$

As it has been remarked in previous sections, the probabilistic constraint in (10) poses severe difficulties that can even lead to a conundrum in solving the problem. In the scenario approach, this difficulty is addressed by replacing the infinite amount of noise realizations with finitely many realizations sampled according to the noise distribution, as described in the following. Let $\mathbf{w}^{(i)}$, $i=1,2,\ldots,N$, be realizations of the noise vector $\mathbf{w}$ obtained by simulating the model of the noise.[4] In this context, the scenario problem is written as

$$\begin{aligned} &\min_{\Gamma,\Theta} J(\Gamma,\Theta) \quad \text{subject to:} \\ &f(\mathbf{F}x_\tau+\mathbf{G}\Gamma+(\mathbf{H}+\mathbf{G}\Theta)\mathbf{w}^{(i)},\Gamma+\Theta\mathbf{w}^{(i)})\leq 0, \quad i\in\{1,2,\ldots,N\}, \end{aligned} \tag{11}$$

which corresponds to replacing the probabilistic constraint in (10) with N deterministic constraints, one for each noise realization.

Problem (11) is a standard convex optimization problem, with a convex cost $J(\Gamma,\Theta)$ and a finite number of convex constraints. Problems of this type can be efficiently solved via standard numerical solvers like those implemented in the interfaces CVX [32] or YALMIP [35]. Moreover, by using the theory presented in the previous section, one can show that the following result holds.

Theorem 3. *Select a "confidence parameter" $\beta\in(0,1)$. Then, the solution (Γ_N^*,Θ_N^*) to the scenario problem* (11) *satisfies the relation*

$$\mathbb{P}\{f(\mathbf{F}x_\tau+\mathbf{G}\Gamma_N^*+(\mathbf{H}+\mathbf{G}\Theta_N^*)\mathbf{w},\Gamma_N^*+\Theta_N^*\mathbf{w})\leq 0\}\geq 1-\varepsilon,$$

with probability no smaller than $1-\beta$, where (d is the number of optimization variables)

$$\varepsilon=\min\left\{\frac{2}{N}\left(\ln\frac{1}{\beta}+d\right),1\right\}. \tag{12}$$

Theorem 3 states that the scenario solution is feasible for problem (10) where ε is given by the right-hand side of (12) with confidence $1-\beta$. The scenario solution cannot be guaranteed to be always feasible because of the stochastic nature of its construction. However, infeasibility is such a rare event that it can be neglected in practice. To see this, fix ε and make N explicit in (12) with respect to ε and β, so obtaining

[4] In a standard LQG setting, this would require generating M independent Gaussian noise terms for each realization. In the scenario approach, however, there is no limitation on the noise structure and the noise can, e.g., be generated by an ARMA (Auto-Regressive Moving-Average system) or by any other model.

$$N = \frac{2}{\varepsilon}\left(\ln\frac{1}{\beta} + d\right).$$

Here, N increases logarithmically with $1/\beta$, so that enforcing a very small value of β, like $\beta = 10^{-7}$ or even $\beta = 10^{-10}$, can be done without rising N to too high values.

Formula (12) provides an explicit expression for ε as a function of N and β and can be derived from Theorem 1. Precisely, equation (12) is obtained by making explicit the right-hand side of equation (7) with respect to ε; see [2] for technical details.

Further, the cost function in the scenario problem (11) can be improved by removing some of the scenario constraints. The price to pay for this is an increase in the violation probability ε. To be precise, suppose that, from the N disturbance realizations, k realizations are removed according to Algorithm 1.[5]

The so-obtained solution satisfies the remaining $N-k$ constraints and is feasible, with probability no smaller than $1-\beta$, for the chance-constrained problem (10) with a violation probability ε as given in the next theorem, which directly follows from the right-hand side of equation (8) by making it explicit with respect to ε.

Theorem 4. *Select a 'confidence parameter' $\beta \in (0,1)$. Then, the solution $(\Gamma^*_{k,N}, \Theta^*_{k,N})$ obtained by removing k of the N constraints in (11) via Algorithm 1 satisfies the relation*

$$\mathbb{P}\{f(\mathbf{F}x_\tau + \mathbf{G}\Gamma^*_{k,N} + (\mathbf{H}+\mathbf{G}\Theta^*_{k,N})\mathbf{w}, \Gamma^*_{k,N} + \Theta^*_{k,N}\mathbf{w}) \leq 0\} \geq 1-\varepsilon,$$

with probability no smaller than $1-\beta$, where (d is the number of optimization variables)

$$\varepsilon = \min\left\{\frac{k}{N} + \frac{d+h+\sqrt{h^2+2(d+k)h}}{N}, 1\right\}, \tag{13}$$

with $h = \ln\frac{1}{\beta} + d\left(1+\ln\frac{d+k}{d}\right)$.

In equation (13), k/N is the empirical violation probability and the guaranteed violation ε is obtained by adding a margin to it. Letting k be proportional to N, $k = \gamma N$, one obtains that the margin is $\mathrm{O}\left(\log N/\sqrt{N}\right)$, so that ε approaches $\gamma = k/N$ as N grows to infinity.

[5] Algorithm 1 is a greedy removal algorithm which is here introduced because it can be implemented at relatively low computational cost. Other alternatives exist, and the paper [13] offers an ample discussion on this matter. Algorithm 1 comes to termination provided that at each step an active constraint can be found whose elimination leads to a cost improvement. This is a very mild condition.

Algorithm 1: Scenario Algorithm with constraints removal

1: Solve problem (11) and store the solution.

2: Let i run over $1,2,\ldots,N$ and find the constraints violated by the stored solution (the first time that this point 2 is entered, the set of violated constraints is empty), that is, find the indexes i such that

$$f(\mathbf{F}x_\tau + \mathbf{G}\Gamma + (\mathbf{H}+\mathbf{G}\Theta)\mathbf{w}^{(i)}, \Gamma + \Theta\mathbf{w}^{(i)}) > 0.$$

Let these indexes be $j_1, j_2, \ldots, j_L$. **If** L is equal to k, then **halt** the algorithm and return the stored solution.

3: Find the active constraints for the stored solution, i.e., the indexes i such that

$$f(\mathbf{F}x_\tau + \mathbf{G}\Gamma + (\mathbf{H}+\mathbf{G}\Theta)\mathbf{w}^{(i)}, \Gamma + \Theta\mathbf{w}^{(i)}) = 0.$$

Let these indexes be $i_1, i_2, \ldots, i_q$.

4: **For** $h = 1,2,\ldots,q$

Solve problem

$$\min_{\Gamma,\Theta} J(\Gamma,\Theta) \quad \text{subject to:}$$

$$f(\mathbf{F}x_\tau + \mathbf{G}\Gamma + (\mathbf{H}+\mathbf{G}\Theta)\mathbf{w}^{(i)}, \Gamma + \Theta\mathbf{w}^{(i)}) \le 0,\ i \in \{1,2,\ldots,N\}/\{i_h, j_1, j_2, \ldots, j_L\}.$$

If the obtained cost is better than the cost of the stored solution, then delete the currently stored solution and store the last computed solution.

End For

5: **Goto 2**

As said, (13) is obtained by making equation (8) explicit with respect to ε. By instead making this same equation explicit with respect to N, one sees that the smallest N so that (8) holds scales as

$$N = \mathrm{O}\left(\frac{d + \ln\frac{1}{\beta}}{(\varepsilon-\varepsilon')^2}\right),$$

where we have put $\varepsilon' = k/N$. This relation reveals some interesting features of the computational complexity of the scenario optimization algorithm. If ε' is selected to be close to the desired violation probability ε, then N becomes large. Provably, this leads to solutions that better approximate the solution to the chance-constrained problem (10); however, this is obtained at the price of an increase of the computational burden. In a given application, the choice of a suitable ε' comes from a compromise between quality of the solution and computational tractability. In many cases the extreme choice of taking $\varepsilon' = 0$ (i.e., $k = 0$, no constraint removal) already gives acceptable results.

In closing this section, one additional word deserves to be spent on the control parametrization (4). In (4), one has $d = mM + ml\frac{(M-1)M}{2}$, where mM is the number of optimization variables in Γ and $ml\frac{(M-1)M}{2}$ is the number of those in Θ. In various applications, the quadratic dependence on the horizon length M poses a hurdle in the applicability of the scenario approach due to the linear dependence of N on d. This may suggest alternative parameterizations that keep the total number of parameters lower, and some choices are illustrated below.

1. $u_{\tau+i} = \gamma_i + \sum_{j=i-r}^{i-1} \theta_{i,j} w_{\tau+j}$, which corresponds to (blank entries are zero values):

$$\Theta = \begin{bmatrix} \theta_{1,0} & & & & \\ \vdots & \ddots & & & \\ \theta_{r,0} & \ddots & \ddots & & \\ & \ddots & \ddots & \ddots & \\ & & \theta_{M-1,M-1-r} & \cdots & \theta_{M-1,M-2} \end{bmatrix}.$$

   In this case, $d = mM + ml\left(r(M-1-r) + \frac{(r-1)r}{2}\right)$;

2. $u_{\tau+i} = \gamma_i + \sum_{j=0}^{i-1} \theta_{i-j} w_{\tau+j}$, which corresponds to:

$$\Theta = \begin{bmatrix} \theta_1 & & & \\ \theta_2 & \ddots & & \\ \vdots & \ddots & \ddots & \\ \theta_{M-1} & \cdots & \theta_2 & \theta_1 \end{bmatrix}.$$

   In this case, $d = mM + ml(M-1)$;

3. $u_{\tau+i} = \gamma_i + \sum_{j=i-r}^{i-1} \theta_{i-j} w_{\tau+j}$, which corresponds to:

$$\Theta = \begin{bmatrix} \theta_1 & & & \\ \vdots & \ddots & & \\ \theta_r & \ddots & \ddots & \\ & \ddots & \ddots & \ddots \\ & & \theta_r & \cdots & \theta_1 \end{bmatrix}.$$

   In this case, $d = mM + mlr$;

4. $u_{\tau+i} = \gamma_i$, i.e., $\Theta = 0$. In this case, $d = mM$. At times, this parametrization has been combined with a fixed linear state-feedback controller,

$$u_{\tau+i} = \gamma_i + \bar{K} x_{\tau+i}, \quad \bar{K} \text{ fixed},$$

   to improve performance [18, 19].

5 Numerical Example

We consider a numerical example inspired by [21].

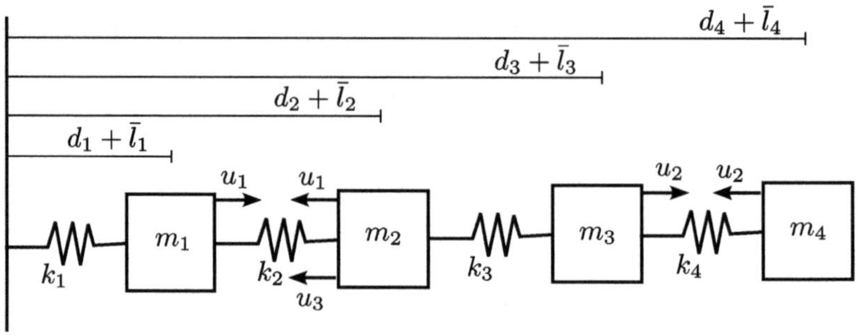

Fig. 2: Scheme of the mechanical system ($\bar{l}_1, \bar{l}_2, \bar{l}_3, \bar{l}_4$ are masses nominal positions).

The mechanical system in Figure 2 is composed by four masses and four springs. The state of the system is formed by the mass displacements, d_1, d_2, d_3, and d_4, from the nominal positions (i.e., the positions at the equilibrium when the input is zero, $\bar{l}_1$, $\bar{l}_2$, $\bar{l}_3$ and $\bar{l}_4$), and by the displacements derivatives, $\dot{d}_1$, $\dot{d}_2$, $\dot{d}_3$, and $\dot{d}_4$ (superscript dot denotes derivative). The control input is $u = [u_1, u_2, u_3]^T$, where u_1, u_2, and u_3 are forces acting on the masses as shown in Figure 2.

All masses and stiffness constants are equal to 1, i.e., $m_1 = m_2 = m_3 = m_4 = 1$ and $k_1 = k_2 = k_3 = k_4 = 1$. Assuming that the control action is held constant over the sampling period, the discrete-time model of the system is given by

$$x_{t+1} = Ax_t + Bu_t + Dw_t,$$

where

$$A = \begin{bmatrix} 0.19 & 0.35 & 0.03 & 0.00 & 0.71 & 0.14 & 0.01 & 0.00 \\ 0.35 & 0.22 & 0.35 & 0.04 & 0.14 & 0.71 & 0.14 & 0.01 \\ 0.03 & 0.35 & 0.23 & 0.39 & 0.01 & 0.14 & 0.71 & 0.14 \\ 0.00 & 0.04 & 0.39 & 0.58 & 0.00 & 0.01 & 0.14 & 0.85 \\ -1.28 & 0.44 & 0.12 & 0.01 & 0.19 & 0.35 & 0.03 & 0.00 \\ 0.44 & -1.15 & 0.45 & 0.13 & 0.35 & 0.22 & 0.35 & 0.04 \\ 0.12 & 0.45 & -1.15 & 0.57 & 0.03 & 0.35 & 0.23 & 0.39 \\ 0.01 & 0.13 & 0.57 & -0.71 & 0.00 & 0.04 & 0.39 & 0.58 \end{bmatrix},$$

$$B = \begin{bmatrix} 0.39 & 0.00 & -0.04 \\ -0.39 & 0.04 & -0.42 \\ -0.04 & 0.39 & -0.04 \\ -0.00 & -0.42 & -0.00 \\ 0.57 & 0.01 & -0.14 \\ -0.58 & 0.13 & -0.71 \\ -0.13 & 0.57 & -0.14 \\ -0.01 & -0.71 & -0.01 \end{bmatrix},$$

and w_t is an additional stochastic disturbance that affects the system. For simplicity, in this simulation section we assume that w_t is a bi-variate white Gaussian noise with zero mean and covariance matrix $I_{2\times 2}$, and that

$$D = \begin{bmatrix} 0\,0\,0\,1\,0\,0\,0\,0 \\ 0\,0\,0\,0\,0\,0\,0\,1 \end{bmatrix}^T,$$

which means that the external disturbance affects the fourth mass only.

The system is at rest at the initial time, that is, $x_\tau = 0$. The goal is to design a control action over a time-horizon $M = 5$ that gets the masses to be close to the nominal positions at the final time instant despite the presence of the noise. During operation, the springs are required to stay in their linear operation domain, a requirement which can be explicitly accounted for by imposing a constraint on the spring deformations, while the control action has also to satisfy saturation limits.

To be specific, we consider the average control cost (1) and set

$$Q_i = \begin{cases} 0_{8\times 8} & i < 5 \\ \left[\begin{array}{c|c} I_{4\times 4} & 0_{4\times 4} \\ \hline 0_{4\times 4} & 0_{4\times 4} \end{array}\right] & i = 5 \end{cases}, \quad \text{and } R_i = 10^{-6} I_{3\times 3}\ \forall i. \tag{14}$$

Moreover, in (3) we let $S = I$ and

$$C = \left[\begin{array}{cccc|c} 1 & 0 & 0 & 0 & \\ -1 & 1 & 0 & 0 & 0_{4\times 4} \\ 0 & -1 & 1 & 0 & \\ 0 & 0 & -1 & 1 & \end{array}\right],$$

so that

$$Cx_{\tau+i} = \begin{bmatrix} d_{1,\tau+i} \\ d_{2,\tau+i} - d_{1,\tau+i} \\ d_{3,\tau+i} - d_{2,\tau+i} \\ d_{4,\tau+i} - d_{3,\tau+i} \end{bmatrix}$$

represents the springs deformation at time $\tau + i$, and consider the probabilistic constraint

$$\mathbb{P}\left\{ \sup_{i=0,\dots,4} \|u_{\tau+i}\|_\infty \leq 1.8 \text{ and } \sup_{i=1,\dots,5} \|Cx_{\tau+i}\|_\infty \leq 1.8 \right\} \geq 1 - \varepsilon, \tag{15}$$

In this problem, using a probabilistic constraint finds justification because an excess of deformation can be tolerated as long as it does not happen too often, while relaxing the input saturation constraint leads to a less conservative design while hitting the saturation limits may rarely generate some deviation from the designed behavior.

For the actual implementation, the scenario approach was used. The control action was parameterized according to (4) (full parametrization), resulting in $d = 75$. $N = 1500$ realizations of the disturbance were sampled and the scenario problem in (11), with no removed constraints, was solved, which gave the solution (Γ_N^*, Θ_N^*). For the sake of comparison, the LQG solution, $(\Gamma_{LQG}^*, \Theta_{LQG}^*)$, was also computed by minimizing the cost with no input and state constraints. All numerical results were obtained by means of CVX, [32] with the solver MOSEK, [38].

The two cost values were $J(\Gamma_N^*, \Theta_N^*) = 0.709$ and $J(\Gamma_{LQG}^*, \Theta_{LQG}^*) = 0.481$, which gives a 31% improvement for the LQG cost. On the other hand, as expected, the LQG solution often violates constraints, and a Monte-Carlo simulation showed a 23% probability of constraints violation. In the scenario design, we have instead that, with high confidence $1 - 10^{-6}$, it holds that

$$\mathbb{P}\left\{ \sup_{i=0,\dots,4} \|u_{\tau+i}\|_\infty \leq 1.8 \text{ and } \sup_{i=1,\dots,5} \|Cx_{\tau+i}\|_\infty \leq 1.8 \right\} \geq 0.92^6;$$

The actual probability of constraints satisfaction, computed by means of a Monte-Carlo simulation, was found to be 97%.[7]

To better appreciate the difference between the two designs (scenario and LQG), Figure 3 displays the cumulative probability distributions of $\sup_{i=0,\dots,4} \|u_{\tau+i}\|_\infty$ and of $\sup_{i=1,\dots,5} \|Cx_{\tau+i}\|_\infty$ obtained via Monte-Carlo methods when the input and state are generated by the scenario and the LQG designs.

When N is large, resorting to the scenario approach with no constraints removal returns solutions that carry high guarantees of constraints satisfaction which, however, are also poorly performing because the design is close to the worst-case (robust) design. Here, with $N = 1500$ we already had a significant decrease of performance as compared to LQG. To improve the scenario control performance, we next resorted to constraints removal and applied Algorithm 1 with $k \neq 0$. Table 1

---

[6] The value $\varepsilon = 0.08$ was computed from (7) by bisection instead of using the explicit formula in Theorem 3.

[7] It is perhaps worth mentioning that it is possible to obtain better evaluations of constraints satisfaction by using the results of the recent contribution [14]. Specifically, from Theorem 2 of [14], it can be proven for the present setup that, with high confidence $1 - 10^{-6}$, it holds that

$$\mathbb{P}\left\{ \sup_{i=0,\dots,4} \|u_{\tau+i}\|_\infty \leq 1.8 \text{ and } \sup_{i=1,\dots,5} \|Cx_{\tau+i}\|_\infty \leq 1.8 \right\} \geq 1 - \varepsilon(s_N^*),$$

where $\varepsilon(\cdot)$ is a function defined over the integers given in the paper and s_N^* is the number of the so-called "support constraints" that have been found in the problem at hand. In other words, $\varepsilon(s_N^*)$ is not a-priori determined and it is a-posteriori tuned to the number of support constraints. The interested reader is referred to [14] for a more-in-depth discussion. In the present simulation, it turned out that the number of support constraints was 34, resulting in $1 - \varepsilon(34) = 0.949$.

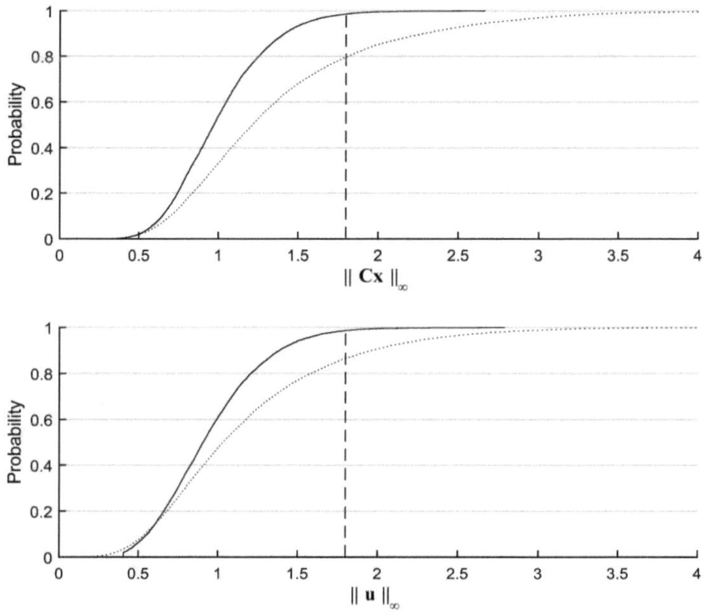

Fig. 3: Cumulative probability distributions of $\sup_{i=0,\ldots,4}\|u_{\tau+i}\|_\infty$ (lower plot) and $\sup_{i=1,\ldots,5}\|Cx_{\tau+i}\|_\infty$ (upper plot) for the scenario design (solid line) and LQG control (dotted line).

summarizes the results obtained for $k = 0, 10, 20, \ldots, 50$. In the table, "Guaranteed prob." refers to the bound on the probability of constraints satisfaction guaranteed with confidence $1 - 10^{-6}$ and obtained from (8) by means of bisection, and "Actual prob." is the actual probability computed via Monte-Carlo methods. As it appears,

Table 1: A comparison between scenario designs with different number of removed scenarios.

k	$J(\Gamma^*_{k,N},\Theta^*_{k,N})$	Guaranteed prob.	Actual prob.
0	0.709	0.92	0.971
10	0.661	0.881	0.967
20	0.629	0.871	0.959
30	0.597	0.863	0.952
40	0.575	0.854	0.947
50	0.564	0.845	0.937

constraint removal leads to a rapid improvement of the performance, while the probability of constraints satisfaction decreases more gently. This shows the ability of Algorithm 1 to remove portions of the uncertainty domain that have a strong impact on the cost function, a feature which is missing on LQG. Figure 4 shows the cumulative probability distributions of $\sup_{i=0,\dots,4} \|u_{\tau+i}\|_\infty$ and of $\sup_{i=1,\dots,5} \|Cx_{\tau+i}\|_\infty$ for $(\Gamma^*_{N,0}, \Theta^*_{N,0})$, $(\Gamma^*_{N,50}, \Theta^*_{N,50})$ and the LQG control.

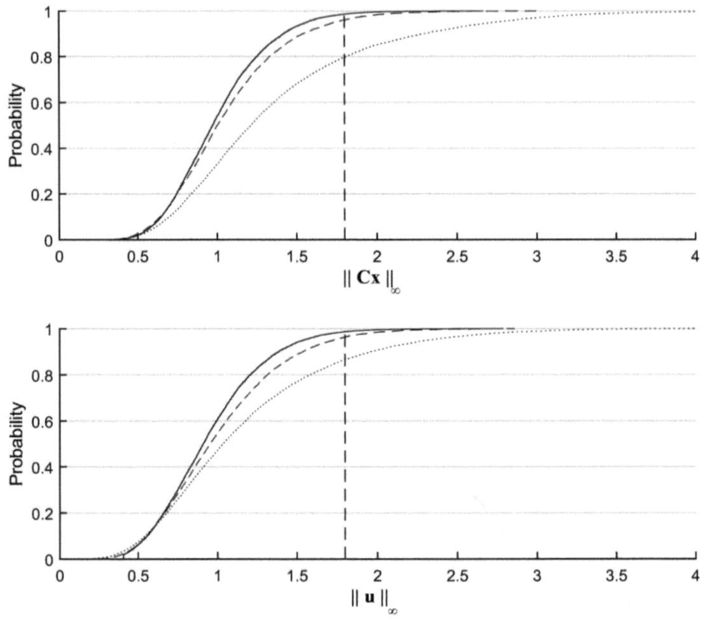

Fig. 4: Cumulative probability distributions of $\sup_{i=0,\dots,4} \|u_{\tau+i}\|_\infty$ (lower plot) and $\sup_{i=1,\dots,5} \|Cx_{\tau+i}\|_\infty$ (upper plot) for $(\Gamma^*_{0,N}, \Theta^*_{0,N})$ (solid line), $(\Gamma^*_{50,N}, \Theta^*_{50,N})$ (dashed line), and LQG control (dotted line).

6 Extensions and Future Work

In the present chapter, the main focus has been on the application of the scenario approach to the solution of the finite-horizon chance-constrained optimization problem (10). The resulting MPC scheme consists in the implementation of scenario optimization over a receding horizon; that is, at every time τ only the first control action u_τ is applied and, after that the system has moved to the new state $x_{\tau+1}$, the whole optimization process is repeated. This poses additional challenges that have been partly addressed in the literature and that are hinted at here.

A first issue concerns with the recursive feasibility of (11). It may happen that the stochastic noise pushes the state to a far distant condition such that a bounded input cannot succeed in satisfying the constraints in the next time period. This issue has been addressed in [25–27] by introducing a suitable relaxation to the scenario framework as described in this chapter.

A second issue refers to studying the constraint satisfaction in the long run. Often the constraint $f(x_{\tau+1},\ldots,x_{\tau+M},u_{\tau},\ldots,u_{\tau+M-1}) \leq 0$ comes in the form of input or state saturation limits that apply at every point in time. For this case, the paper [49] presents a study which quantifies the asymptotic proportion of times when these limits are violated over the total number of time instants passed.

A final point that deserves to be mentioned is the possibility of applying the scenario-based MPC scheme to nonlinear systems. Nonlinearity introduces an additional difficulty relating to the convexity assumption made in this chapter since convexity fails to be true for nonlinear systems even when the function f setting the constraints is convex in its arguments. More specifically, Theorems 1 and 2 of Section 3 are grounded on the fact that, in a convex setup, the solution to (6) is determined by a limited and known number of constraints (those that are called support constraints in the literature, see [7], which are no more than the number of optimization variables). In contrast, in a non-convex setup the number of support constraints cannot be a-priori bounded and it can actually be arbitrarily large. Even though the corresponding analysis has not been fully developed at the time this chapter is being written, we envisage that the wait-and-judge perspective of [14, 17] can be used in this context to circumvent this difficulty: the support constraints are determined after the solution has been found and the evaluation of constraint violation is adapted to the found number of support constraints based on the theory of [14, 17].

References

1. Alamo, T., Tempo, R., Camacho, E.F.: A randomized strategy for probabilistic solutions of uncertain feasibility and optimization problems. IEEE Trans. Autom. Control **54**(11), 2545–2559 (2009)
2. Alamo, T., Tempo, R., Luque, A., Ramirez, D.R.: Randomized methods for design of uncertain systems: sample complexity and sequential algorithms. Automatica **51**, 160–172 (2015)
3. Batina, I. Model predictive control for stochastic systems by randomized algorithms. PhD thesis, Technische Universiteit Eindhoven (2004)
4. Bemporad, A., Morari, M.: Robust model predictive control: a survey. In: Robustness in Identification and Control. Lecture Notes in Control and Information Sciences, vol. 245, pp. 207–226. Springer, Berlin (1999)
5. Bemporad, A., Borrelli, F., Morari, M.: Min-max control of constrained uncertain discrete-time linear systems. IEEE Trans. Autom. Control **48**(9), 1600–1606 (2003)
6. Ben-Tal, A., Boyd, S., Nemirovski, A.: Extending scope of robust optimization: comprehensive robust counterparts of uncertain problems. Math. Program. **107**(1–2), 63–89 (2006)
7. Calafiore, G., Campi, M.C.: Uncertain convex programs: randomized solutions and confidence levels. Math. Program. **102**(1), 25–46 (2005)
8. Calafiore, G., Campi, M.C.: The scenario approach to robust control design. IEEE Trans. Autom. Control **51**(5), 742–753 (2006)
9. Calafiore, G.C., Fagiano, L.: Robust model predictive control via scenario optimization. IEEE Trans. Autom. Control **58**(1), 219–224 (2013)
10. Campi, M.C.: Classification with guaranteed probability of error. Mach. Learn. **80**, 63–84 (2010)
11. Campi, M.C., Carè, A.: Random convex programs with L1-regularization: sparsity and generalization. SIAM J. Control Optim. **51**(5), 3532–3557 (2013)

12. Campi, M.C., Garatti, S.: The exact feasibility of randomized solutions of uncertain convex programs. SIAM J. Optim. **19**(3), 1211–1230 (2008)
13. Campi, M.C., Garatti, S.: A sampling-and-discarding approach to chance-constrained optimization: feasibility and optimality. J. Optim. Theory Appl. **148**(2), 257–280 (2011)
14. Campi, M.C., Garatti, S.: Wait-and-judge scenario optimization. Math. Program. **167**(1), 155–189 (2018)
15. Campi, M.C., Calafiore, G., Garatti, S.: Interval predictor models: identification and reliability. Automatica **45**(2), 382–392 (2009)
16. Campi, M.C., Garatti, S., Prandini, M.: The scenario approach for systems and control design. Annu. Rev. Control **33**(2), 149–157 (2009)
17. Campi, M.C., Garatti, S., Ramponi, F.: Non-convex scenario optimization with application to system identification. In: Proceedings of the 54th IEEE Conference on Decision and Control, Osaka (2015)
18. Cannon, M., Kouvaritakis, B., Wu, X.: Model predictive control for systems with stochastic multiplicative uncertainty and probabilistic constraints. Automatica **45**, 167–172 (2009)
19. Cannon, M., Kouvaritakis, B., Wu, X.: Probabilistic constrained MPC for multiplicative and additive stochastic uncertainty. IEEE Trans. Autom. Control **54**(7), 1626–1632 (2009)
20. Carè, A., Garatti, S., Campi, M.C.: Scenario min-max optimization and the risk of empirical costs. SIAM J. Optim. **25**(4), 2061–2080 (2015)
21. Cinquemani, E., Agarwal, M., Chatterjee, D., Lygeros, J.: Convexity and convex approximations of discrete-time stochastic control problems with constraints. Automatica **47**(9), 2082–2087 (2011)
22. Crespo, L.G., Giesy, D.P., Kenny, S.P.: Interval predictor models with a formal characterization of uncertainty and reliability. In: Proceedings of the 53rd IEEE Conference on Decision and Control (CDC), Los Angeles, pp. 5991–5996 (2014)
23. de la Peña, D.M., Alamo, T., Bemporad, A., Camacho, F.: A decomposition algorithm for feedback min-max model predictive control. IEEE Trans. Autom. Control **51**(10), 1688–1692 (2006)
24. de Mello, T.H., Bayraksan, G.: Monte Carlo sampling-based methods for stochastic optimization. Surv. Oper. Res. Manag. Sci. **19**(1), 56–85 (2014)
25. Deori, L., Garatti, S., Prandini, M.: Stochastic constrained control: trading performance for state constraint feasibility. In: Proceedings of the 12th European Control Conference, Zurich (2013)
26. Deori, L., Garatti, S., Prandini, M.: Stochastic control with input and state constraints: a relaxation technique to ensure feasibility. In: Proceedings of the 54th IEEE Conference on Decision and Control, Osaka (2015)
27. Deori, L., Garatti, S., Prandini, M.: Trading performance for state constraint feasibility in stochastic constrained control: a randomized approach. J. Frankl. Inst. **354**(1), 501–529 (2017)
28. Esfahani, P.M., Sutter, T., Lygeros, J.: Performance bounds for the scenario approach and an extension to a class of non-convex programs. IEEE Trans. Autom. Control **60**(1), 46–58 (2015)
29. Garatti, S., Campi, M.C.: Modulating robustness in control design: principles and algorithms. IEEE Control Syst. Mag. **33**(2), 36–51 (2013)
30. Goulart, P.J., Kerrigan, E.C., Maciejowski, J.M.: Optimization over state feedback policies for robust control with constraints. Automatica **42**(4), 523–533 (2006)
31. Grammatico, S., Zhang, X., Margellos, K., Goulart, P.J., Lygeros, J.: A scenario approach for non-convex control design. IEEE Trans. Autom. Control **61**(2), 334–345 (2016)
32. Grant, M., Boyd, S.: CVX: Matlab software for disciplined convex programming, version 1.21. http://cvxr.com/cvx, Feb 2011
33. Hong, L.J., Hu, Z., Liu, G.: Monte Carlo methods for value-at-risk and conditional value-at-risk: a review. ACM Trans. Model. Comput. Simul. **24**(4), 22:1–22:37 (2014)
34. Kothare, M., Balakrishnan, V., Morari, M.: Robust constrained model predictive control using linear matrix inequalities. Automatica **32**(10), 1361–1379 (1996)

35. Löfberg, J.: YALMIP: a toolbox for modeling and optimization in MATLAB. In: Proceedings of the CACSD Conference, Taipei (2004)
36. Margellos, K., Goulart, P.J., Lygeros, J.: On the road between robust optimization and the scenario approach for chance constrained optimization problems. IEEE Trans. Autom. Control **59**(8), 2258–2263 (2014)
37. Margellos, K., Prandini, M., Lygeros, J.: On the connection between compression learning and scenario based single-stage and cascading optimization problems. IEEE Trans. Autom. Control **60**(10), 2716–2721 (2015)
38. MOSEK ApS: The MOSEK optimization toolbox for MATLAB manual. Version 7.1 (Revision 28) (2015)
39. Pagnoncelli, B.K., Vanduffel, S.: A provisioning problem with stochastic payments. Eur. J. Oper. Res. **221**(2), 445–453 (2012)
40. Pagnoncelli, B.K., Ahmed, S., Shapiro, A.: Sample average approximation method for chance constrained programming: theory and applications. J. Optim. Theory Appl. **142**(2), 399–416 (2009)
41. Pagnoncelli, B.K., Reich, D., Campi, M.C.: Risk-return trade-off with the scenario approach in practice: a case study in portfolio selection. J. Optim. Theory Appl. **155**(2), 707–722 (2012)
42. Petersen, I.R., Tempo, R.: Robust control of uncertain systems: classical results and recent developments. Automatica **50**, 1315–1335 (2014)
43. Prandini, M., Garatti, S., Lygeros, J.: A randomized approach to stochastic model predictive control. In: 2012 IEEE 51st IEEE Conference on Decision and Control (CDC), December 2012, pp. 7315–7320
44. Prékopa, A.: Stochastic Programming. Kluwer, Boston (1995)
45. Prèkopa, A.: Probabilistic programming. In: Ruszczyǹski, A., Shapiro, A. (eds.) Stochastic Programming, volume 10 of Handbooks in Operations Research and Management Science, pp. 267–352. Elsevier, London (2003)
46. Primbs, J.A., Sung, C.H.: Stochastic receding horizon control of constrained linear systems with state and control multiplicative noise. IEEE Trans. Autom. Control **54**(2), 221–230 (2009)
47. Raimondo, D.M., Limon, D., Lazar, M., Magni, L., Camacho, E.F.: Min-max model predictive control of nonlinear systems: a unifying overview on stability. Eur. J. Control **15**, 5–21 (2009)
48. Schildbach, G., Fagiano, L., Morari, M.: Randomized solutions to convex programs with multiple chance constraints. SIAM J. Optim. **23**(4), 2479–2501 (2013)
49. Schildbach, G., Fagiano, L., Frei, C., Morari, M.: The scenario approach for stochastic model predictive control with bounds on closed-loop constraint violations. Automatica **50**(12), 3009–3018 (2014)
50. Scokaert, P.O.M., Mayne, D.Q.: Min-max feedback model predictive control for constrained linear systems. IEEE Trans. Autom. Control **43**, 1136–1142 (1998)
51. Shapiro, A., Dentcheva, D., Ruszczyński, A.: Lectures on stochastic programming: modeling and theory. MPS-SIAM, Philadelphia (2009)
52. Tempo, R., Calafiore, G., Dabbene, F.: Randomized Algorithms for Analysis and Control of Uncertain Systems, 2nd edn. Springer, London (2013)
53. van Hessem, D.H., Bosgra, O.H.: Stochastic closed-loop model predictive control of continuous nonlinear chemical processes. J. Proc. Control **16**(3), 225–241 (2006)
54. Vayanos, P., Kuhn, D., Rustem, B.: A constraint sampling approach for multistage robust optimization. Automatica **48**(3), 459–471 (2012)
55. Zhang, X., Grammatico, S., Schildbach, G., Goulart, P.J., Lygeros, J.: On the sample size of random convex programs with structured dependence on the uncertainty. Automatica **60**, 182–188 (2015)

Nonlinear Programming Formulations for Nonlinear and Economic Model Predictive Control

Mingzhao Yu, Devin W. Griffith, and Lorenz T. Biegler

1 Introduction

Model Predictive Control (MPC) is widely accepted in the process industries as a generic multivariable controller with constraint handling. More recently, MPC has been extended to Nonlinear Model Predictive Control (NMPC) in order to realize high-performance control of highly nonlinear processes. In particular, NMPC allows incorporation of detailed process models (validated by off-line analysis) and also integrates with on-line optimization strategies consistent with higher-level tasks, including scheduling and planning. NMPC for tracking and so-called "economic" stage costs, as well as associated state estimation tasks, are reviewed, formulated, and analyzed in considerable detail in [24, 28]. Due to advances described in [5, 23], fundamental stability and robustness properties of NMPC are well-known, and many of the key issues related to the applicability and relevance of NMPC are well understood.

This study expands on these key issues by examining existence and uniqueness properties of nonlinear programs (NLPs), and the stability and sensitivity of their solutions. With the former, nominal trajectories can be determined for NMPC subproblems, and robustness of the NMPC controller (through input to state stability) is guided by the latter. We also show that the formulation of the NMPC subproblem has a key impact on these NLP properties, and we emphasize how proper NLP formulations allow these stability properties to hold. Finally the existence of NLP solutions that are differentiable with respect to problem data leads to the development of sensitivity-based NMPC, which greatly reduces on-line computation and computational delay.

The remainder of this section introduces NLP-based strategies for NMPC to set the stage for the analysis. Section 2 then provides some background properties for

M. Yu · D. W. Griffith · L. T. Biegler (✉)
Chemical Engineering Department, Carnegie Mellon University, Pittsburgh, PA, USA
e-mail: mingzhay@andrew.cmu.edu; dwgriffi@andrew.cmu.edu; biegler@cmu.edu

S. V. Raković, W. S. Levine (eds.), *Handbook of Model Predictive Control*, Control Engineering, https://doi.org/10.1007/978-3-319-77489-3_20

the NLP subproblem of NMPC, including sufficient conditions for optimality and constraint qualifications, which are used to reformulate the NMPC subproblem in order to promote stability and differentiability of NLP solutions. Section 3 relates these properties to asymptotic stability, input to state stability (ISS), and input to state practical stability (ISpS). Sections 4 and 5 extend these concepts to Economic NMPC, and present three NLP formulations that lead to stability guarantees. Finally, Section 6 demonstrates these results on two case studies related to chemical process control with first principle nonlinear models, and Section 7 concludes the chapter.

1.1 NLP Strategies for NMPC

Consider the following discrete-time nonlinear dynamic model of the plant with uncertainties:

$$\begin{aligned} x_{k+1} &= \hat{f}(x_k, u_k, w_k) \\ &= f(x_k, u_k) + d(x_k, w_k) \end{aligned} \tag{1}$$

where $x_k \in \Re^{n_x}$, $u_k \in \Re^{n_u}$, and $w_k \in \Re^{n_w}$ are the plant states, controls, and disturbance signals, respectively, defined at time steps t_k with integers $k > 0$. The mapping $f : \Re^{n_x+n_u} \mapsto \Re^{n_x}$ with $f(0,0) = 0$ represents the nominal model, while the term $d : \Re^{n_x+n_u+n_w} \mapsto \Re^{n_x}$ with $d(0,0) = 0$ is used to describe modeling errors, estimation errors, and disturbances. We assume that $f(\cdot,\cdot)$ and $d(\cdot,\cdot)$ are Lipschitz continuous with respect to their arguments, and that the noise w_k is drawn from a bounded set $\mathscr{W}$.

With this model description, we compute an estimate of the current state $x(k)$ that can be used for our nonlinear model-based controller (NMPC), defined by the following nonlinear programming problem (NLP):

$$J_N(p_0) := \min_{z_l, v_l} \quad \Psi(z_N) + \sum_{l=0}^{N-1} \psi(z_l, v_l) \tag{2a}$$

$$\text{s.t. } z_{l+1} = f(z_l, v_l) \quad l = 0, \ldots N-1 \tag{2b}$$

$$z_0 = p_0 \tag{2c}$$

$$z_l \in \mathbb{X}, v_l \in \mathbb{U}, z_N \in \mathbb{X}_f. \tag{2d}$$

Here $p_0 = x_k$ is a fixed parameter in the NLP determined by the actual or estimated plant state. We assume that the states and controls are restricted to the domains $\mathbb{X}$ and $\mathbb{U}$, respectively. $\mathbb{X}_f$ is the terminal set with $\mathbb{X}_f \subset \mathbb{X}$. We also assume that N is sufficiently long and $\Psi(z_N)$ is sufficiently large. As a result, $z_N \in \mathbb{X}_f$ is always true for the solution of (2). As shown in [26] and [12], this allows $\mathbb{X}_f$ to be omitted in (2), although we do not remove terminal constraints from all formulations shown here. The set $\mathbb{U}$ is compact and contains the origin; the sets $\mathbb{X}$ and $\mathbb{X}_f$ are closed and contain the origin in their interiors. The stage cost is given by $\psi(\cdot,\cdot) : \Re^{n_x+n_u} \to \Re$, while the terminal cost is denoted by $\Psi(\cdot) : \Re^{n_x} \rightarrow \Re$; both are assumed to have Lipschitz continuous second derivatives.

Also note that we assume that the algebraic variables of dynamic and algebraic equation (DAE) systems may be rewritten as $y = \eta_y(x,u)$. A detailed treatment of DAE systems can be found in [36].

2 Properties of the NLP Subproblem

We reformulate Problem (2), with $\mathbf{x} = (z_0, \ldots, z_N, v_0, \ldots, v_{N-1})$ and $p = x_k$ as:

$$\min_{\mathbf{x}} F(\mathbf{x}, p), \; s.t. \; c(\mathbf{x}, p) = 0, \; g(\mathbf{x}, p) \leq 0. \tag{3}$$

An important characterization of the solution of (3) is the concept of a KKT point, which satisfies the Karush-Kuhn-Tucker conditions for (3):

Definition 1. (KKT, see [25]) KKT conditions for Problem (3) are given by:

$$\begin{aligned} \nabla F(\mathbf{x}^*) + \nabla c(\mathbf{x}^*)\lambda + \nabla g(\mathbf{x}^*)\nu = 0 \\ c(\mathbf{x}^*) = 0, \quad 0 \leq \nu \perp g(\mathbf{x}^*) \leq 0 \end{aligned} \tag{4}$$

for some multipliers (λ, ν), where $\mathbf{x}^*$ is a KKT point. We also define $L(\mathbf{x}, \lambda, \nu) = F(\mathbf{x}) + c(\mathbf{x})^T \lambda + g(\mathbf{x})^T \nu$ as the Lagrange function of (3).

A constraint qualification (CQ) is required so that a KKT point is a necessary condition for a local minimizer of (3) [25]. For problem (3) the following CQ is widely invoked.

Definition 2. (LICQ, [25]) The linear independence constraint qualification (LICQ) holds at $\mathbf{x}^*$ when the gradient vectors

$$\nabla c(\mathbf{x}^*, p) \text{ and } \nabla g_j(\mathbf{x}^*, p); \quad \forall j \in J \text{ where } J = \{j | g_j(\mathbf{x}^*, p) = 0\} \tag{5}$$

are linearly independent. LICQ also implies that the multipliers λ, ν are unique.

In addition, we define Strict Complementarity as follows:

Definition 3. (Strict Complementarity, [7]) At the solution $\mathbf{x}^*$ of problem (3) with multipliers (λ, ν), the strict complementarity condition (SC) holds for ν if and only if $\nu_j - g_j(\mathbf{x}^*, p) > 0$ for each $j \in J$.

The strong second order condition (SSOSC) requires positive definiteness of the Hessian in constrained directions and is given as follows.

Definition 4. (SSOSC, [29]) For the KKT point to be a strict local optimum, strong second order sufficient conditions (SSOSC) hold at $\mathbf{x}^*$ with multipliers λ and ν if

$$q^T \nabla_{\mathbf{xx}} L(\mathbf{x}^*, \lambda, \nu, p) q > 0 \quad \text{for all } q \neq 0 \tag{6}$$

such that

$$\begin{aligned} \nabla c_i(\mathbf{x}^*, p)^T q &= 0, \quad i = 1,..,n_c \\ \nabla g_j(\mathbf{x}^*, p)^T q &= 0, \quad \text{for } \nu_j > 0, j \in J. \end{aligned} \tag{7}$$

Note that if SSOSC is not satisfied at the KKT point x^*, then it is straightforward to show that it can be satisfied for a modified problem. By adding

$$\|\mathbf{x} - \mathbf{x}^*\|_Q^2 \tag{8}$$

to the objective in (3), the solution of this modified problem is still $(\mathbf{x}^*, \lambda, \nu)$, the same KKT point. Moreover, if the positive semi-definite matrix Q is chosen with sufficiently large eigenvalues for $Z^T Q Z$ (where Z is a basis of the nullspace of active constraint gradients in Definition 4), then SSOSC can always be satisfied for the related problem at $\mathbf{x}^*$. In fact, a term related to (8) is automatically added in the IPOPT solver as part of its regularization strategy, and IPOPT always solves a modified problem that satisfies SSOSC.

Finally, we state a key property related to the parametric sensitivity of (3).

Theorem 1. *(Implicit function theorem applied to (4), [7]) Let $\mathbf{x}^*(p)$ be a KKT point that satisfies* (4), *and assume that SC, LICQ, and SSOSC hold at $\mathbf{x}^*$. Further let the functions F, c, g be at least $k+1$ times differentiable in $\mathbf{x}$ and k times differentiable in p. Then*

- $\mathbf{x}^*$ *is an isolated minimizer, and the associated multipliers λ and ν are unique.*
- *for p in a neighborhood of p_0 the set of active constraints remains unchanged,*
- *for p in a neighborhood of p_0 there exists a k times differentiable function $s(p) = \left[\mathbf{x}^*(p)^T,\ \lambda(p)^T,\ \nu(p)^T\right]$ that corresponds to a locally unique minimum for* (3).

More general results on uniform continuity of the solution of (3) can be derived under the following condition.

Definition 5. (MFCQ, [25]) For Problem (3), the Mangasarian-Fromovitz constraint qualification (MFCQ) holds at the optimal point $\mathbf{x}^*(p)$ if and only if

- $\nabla c(\mathbf{x}^*, p)$ is linearly independent.
- There exists a vector q such that

$$\nabla c(\mathbf{x}^*, p)^T q = 0, \nabla g_j(\mathbf{x}^*, p)^T q < 0 \quad \forall j \in J. \tag{9}$$

MFCQ implies that the set of KKT multipliers is a closed convex polytope [9]. Another useful constraint qualification is given as:

Definition 6. (CRCQ, [16]) For Problem (3), the constant rank constraint qualification (CRCQ) holds at $(\mathbf{x}^*, p_0)$, when for all subsets $\bar{J} \subset J$, the gradients:

$$\nabla g_j(\mathbf{x}, p)\ j \in \bar{J} \text{ and } \nabla c(\mathbf{x}, p) \tag{10}$$

retain constant rank near the point $(\mathbf{x}^*, p_0)$.

Finally, if MFCQ holds at a KKT point but not LICQ, the multipliers λ, ν are no longer unique, and we need a more general second order condition.

Definition 7. (GSSOSC, [27]) The generalized strong second order sufficient condition (GSSOSC) is said to hold at $\mathbf{x}^*$ when the SSOSC holds for all KKT multipliers λ, ν.

For KKT points, MFCQ and GSSOSC are the weakest conditions under which the perturbed solution of problem (3) is locally unique [20]. Under these conditions we cannot expect $\mathbf{x}^*(p)$ to be differentiable (because active sets are nonunique). However, with these conditions and CRCQ, directional derivatives for $\mathbf{x}^*(p)$ can be calculated with a particular QP formulation [27], and this is sufficient to obtain sensitivity updates in an NMPC context. This is important for both robust reformulations and advanced-step strategies based on NLP sensitivity [34, 35, 38].

2.1 NMPC Problem Reformulation

To develop a robust problem formulation we remove $\mathbb{X}_f$ (based on arguments from [26], assuming that N is sufficiently large) and relax $\mathbb{X}$ with ℓ_1 penalty terms. We assume without loss of generality that $\mathbb{X}$ and $\mathbb{U}$ can be represented by simple upper and lower bounds on z_l and v_l, respectively, and we rewrite the bounds on $\mathbb{X}$ as inequalities $g(z_l) \leq 0$. This leads to the following reformulation of (2):

$$\begin{aligned} J_N(x(k)) = \min_{z_l, v_l, \xi_l} \quad & \Psi(z_N) + \sum_{l=0}^{N-1} (\psi(z_l, v_l) + \rho \xi_l^T \mathbf{1}) \\ \text{s.t.} \quad & z_{l+1} = f(z_l, v_l), \quad l = 0, \ldots, N-1 \\ & z_0 = x_k \\ & g(z_l) \leq \xi_l, v_l \in \mathbb{U}, \xi_l \geq 0, \quad l = 0, \ldots, N-1 \end{aligned} \tag{11}$$

where ξ_l is an auxiliary variable vector and $\mathbf{1} = [1, 1, \ldots, 1]^T$. It is easy to see that the gradients of the equality constraints contain a nonsingular basis matrix, and are linearly independent. Moreover, it is straightforward to show that the MFCQ always holds at the solution of (11) (see [17]). Under these conditions the multipliers of (11) are bounded. Moreover, when the inequalities are linear, as in most definitions of $\mathbb{X}$ and $\mathbb{U}$, CRCQ is also satisfied. Finally, with (8) GSSOSC is easy to satisfy through the addition of a sufficiently large quadratic regularization term. These terms are quite compatible with tracking stage costs as well as economic stage costs [17].

Moreover, if (2) has a solution, then selecting ρ larger than a finite threshold, $\rho > \bar{\rho}$, should drive ξ_l to zero, where $\bar{\rho}$ is the dual norm of the multipliers at the solution of problem (2). If $\xi_l = 0$, then the solution of (11) is identical to the solution of problem (2). Therefore, nominal stability properties of (11) are identical to those of (2). Since a solution with $\xi_l > 0$ for arbitrarily large values of ρ implies that problem (2) is locally infeasible, we assume that a finite $\bar{\rho}$ can be found as long as problem (2) is well-posed. This corresponds to the common assumption that there exists a feasible input sequence, which steers the system to the terminal set, i.e., the horizon N is long enough to satisfy the terminal conditions.

3 Nominal and ISS Stability of NMPC

This section reviews well-known results on nominal and robust stability for the NMPC controller $u = \kappa(x)$ [19, 23]. Here we return to the system (1) and rewrite this as:

$$x_{k+1} = \hat{f}(x_k, \kappa(x_k), w_k) \tag{12}$$

where $x \in \mathscr{X}$ is a vector of states, the set $\mathscr{X} \subset \mathbb{R}^{n_x}$ is closed and bounded, the controls $u : \mathbb{R}^{n_x} \to \mathbb{R}^{n_u}$ are a mapping of the current state (i.e., the control law), and $w \in \mathbb{W} \subseteq \mathbb{R}^{n_w}$ is a vector of disturbances.

Assumption 2 *The set $\mathscr{X} \subseteq \mathbb{R}^{n_x}$ is robustly positive invariant for $f(\cdot,\cdot,\cdot)$. That is, $\hat{f}(x,\kappa(x),w) \in \mathscr{X}$ holds for all $x \in \mathscr{X}$, $w \in \mathbb{W}$. Furthermore, $\sup_{k\in\mathbb{Z}_+} |w| = ||\mathbf{w}||$.*

Definition 8. (Comparison Functions). A function $\alpha : \mathbb{R}_+ \to \mathbb{R}_+$ is of class $\mathscr{K}$ if it is continuous, strictly increasing, and $\alpha(0) = 0$. A function $\alpha : \mathbb{R}_+ \to \mathbb{R}_+$ is of class $\mathscr{K}_\infty$ if it is a $\mathscr{K}$ function and $\lim_{s\to\infty} \alpha(s) = \infty$. A function $\beta : \mathbb{R}_+ \times \mathbb{Z}_+ \to \mathbb{R}_+$ is of class $\mathscr{KL}$ if, for each $t \geq 0$, $\beta(\cdot,t)$ is a $\mathscr{K}$ function, and, for each $s \geq 0$, $\beta(s,\cdot)$ is nonincreasing and $\lim_{t\to\infty} \beta(s,t) = 0$.

Definition 9. (Attractivity). The system (12) is attractive on $\mathscr{X}$ if $\lim_{k\to\infty} x_k = 0$ for all $x_0 \in \mathscr{X}$.

Definition 10. (Stable Equilibrium Point). The point $x = 0$ is called a stable equilibrium point of (12) if, for all $k_0 \in \mathbb{Z}_+$ and $\varepsilon_1 > 0$, there exists $\varepsilon_2 > 0$ such that $|x_{k_0}| < \varepsilon_2 \Rightarrow |x_k| < \varepsilon_1$ for all $k \geq k_0$.

Definition 11. (Asymptotic Stability). The system (12) is asymptotically stable on $\mathscr{X}$ if $\lim_{k\to\infty} x_k = 0$ for all $x_0 \in \mathscr{X}$ and $x = 0$ is a stable equilibrium point.

We highlight that asymptotic stability is only possible for disturbances w_k that converge to a constant. See Appendix B of [28] for the preceding definitions.

Assumption 3 *(Nominal Stability Assumptions of NMPC)*

- *The terminal penalty $\Psi(\cdot)$ satisfies $\Psi(z) > 0, \forall z \in \mathbb{X}_f \backslash \{0\}$,*
- *There exists a local control law $u = \kappa_f(z)$ defined on $\mathbb{X}_f$, such that $f(z,\kappa_f(z)) \in \mathbb{X}_f, \forall z \in \mathbb{X}_f$, and $\Psi(f(z,\kappa_f(z))) - \Psi(z) \leq -\psi(z,\kappa_f(z)), \forall z \in \mathbb{X}_f$.*
- *The optimal stage cost $\psi(x,u) = \psi(x,\kappa(x))$ satisfies $\alpha_p(|x|) \leq \psi(x,u) \leq \alpha_q(|x|)$ where $\alpha_p(\cdot)$ and $\alpha_q(\cdot)$ are $\mathscr{K}$ functions.*

Nominal stability of NMPC can be paraphrased by the following theorem.

Theorem 4. *(Nominal Stability of NMPC [28]) Consider the moving horizon problem (2) and associated control law $u = u^{id}$ that satisfies Assumption 3. Then, $J_N(x)$ from problem (11) is a Lyapunov function and the closed-loop system is asymptotically stable.*

For the analysis of robust stability properties of the NMPC we consider Input-to-State Stability (ISS) [18, 23].

Definition 12. (Input-to-State Stability)

- The system (1) is ISS in $\mathbb{X}$ if there exists a $\mathscr{K}\mathscr{L}$ function β, and a $\mathscr{K}$ function γ such that for all w in the bounded set $\mathscr{W}$,

$$|x(k)| \leq \beta(|x(0)|, k) + \gamma(|w|),\ \forall k \geq 0,\ \forall x(0) \in \mathbb{X} \tag{13}$$

- A function $V(\cdot)$ is called an ISS-Lyapunov function for system (1) if there exist a set $\mathbb{X}$, $\mathscr{K}$ functions $\alpha_1, \alpha_2, \alpha_3$, and σ such that $\forall x \in \mathbb{X}$ and $\forall w \in \mathscr{W}$, we have $\alpha_1(|x|) \leq V(x) \leq \alpha_2(|x|)$ and $V(\hat{f}(x,u,w)) - V(x) \leq -\alpha_3(|x|) + \sigma(|w|)$

Moreover, if $\mathbb{X}$ is a robustly invariant set for system (1) and $V(\cdot)$ is an ISS-Lyapunov function for this system, then the resulting system is ISS in $\mathbb{X}$ [5, 23]. Note that for problem (11), $\mathbb{X} = \mathfrak{R}^{n_x}$.

We note that, in the nominal case (with no disturbances), ISS reduces to asymptotic stability. The following is a useful extension of ISS.

Definition 13. (ISpS): Under Assumption 16, the system (12) is input-to-state practically stable (ISpS) on $\mathscr{X}$ if $|x_k| \leq \beta(|x_0|, k) + \gamma(||\mathbf{w}||) + c$ holds for all $x_0 \in \mathscr{X}$ and $k \geq 0$, where $\beta \in \mathscr{K}\mathscr{L}$, $\gamma \in \mathscr{K}$, and $c \in \mathbb{R}_+$.

We highlight that ISpS is more flexible than ISS as the sequence is relaxed by a non-vanishing constant c. As a result, however, there is no guarantee of asymptotic stability in the nominal case. See [22] for the preceding two definitions. Furthermore, we use a Lyapunov theorem generalized to allow for a path-dependent Lyapunov function that has $\mathbf{w}_k$, the disturbance sequence up until time k, as an argument.

Theorem 5. *If the system* (12) *admits a function $V(k, \mathbf{w}_k, x_0)$ satisfying:*

$$\alpha_1(|x_k|) \leq V(k, \mathbf{w}_k, x_0) \leq \alpha_2(|x_k|) + c_1 \tag{14a}$$

$$\begin{aligned} &V(k+1, \mathbf{w}_{k+1}, x_0) - V(k, \mathbf{w}_k, x_0) \\ &\quad \leq -\alpha_3(|x_k|) + \sigma(|w_k|) + c_2, \end{aligned} \tag{14b}$$

$$\forall\ x_0 \in \mathscr{X},\ \mathbf{w} \in \mathscr{W},\ k \in \mathbb{Z}_+$$

where $\alpha_1, \alpha_2, \alpha_3 \in \mathscr{K}_\infty$, $\sigma \in \mathscr{K}$, and $c_1, c_2 \in \mathbb{R}_+$ then $V(k, \mathbf{w}_k, x_0)$ is an ISpS Lyapunov function for (12).

Finally, we make the following assumptions and establish robust stability of the NMPC controller from the following theorem.

Assumption 6 *(Robust Stability Assumptions)*

- *The solution of problem (11), given by $s^*(p)$, satisfies MFCQ, GSSOSC, and CRCQ. From Theorem 1, the objective function and $s^*(p)$ are therefore continuous and differentiable with respect to p and the resulting feedback law, derived from $s^*(p)$, can be represented as $u = \kappa(x)$. As a result, $J_N(x)$ is Lipschitz with Lipschitz constant L_J.*
- *$d(x,w)$ is Lipschitz with respect to its arguments with Lipschitz constant L_d and $|d(x,0)| \leq \alpha_0(|x|)$, where $\alpha_0(|x|)$ is a $\mathscr{K}$ function.*

Theorem 7. *(Robust Stability of NMPC (Theorem 2 in [23], see also [18]) Under Assumptions 3 and 6 with $\alpha_0(|x|) \leq \frac{\delta}{L_J}\alpha_p(|x|)$ and $\delta \in (0,1)$ is an arbitrary real number, the cost function $J_N(x)$ obtained from the solution (11) is an ISS-Lyapunov function and the resulting closed-loop system is ISS.*

4 Economic NMPC with Objective Regularization

When an optimization-based controller such as MPC is used, a natural extension is to include the economic criterion directly into the cost function of the controller. This approach is often referred to as Economic MPC (eMPC), and has been gaining increased interest in recent years [1, 10, 12, 13, 15, 17, 32]. Moreover, a complementary chapter of this handbook [2] provides detailed development and survey of eMPC controllers and their properties.

For eMPC, we first consider the following steady state optimization problem for economic NMPC with x_s and u_s as optimal steady state solutions.

$$\begin{aligned} &\min\nolimits_{x,u} \psi^{ec}(x,u) \\ &\text{s.t.} \quad x = f(x,u) \\ &u \in \mathbb{U}, x \in \mathbb{X}. \end{aligned} \tag{15}$$

The dynamic optimization problem for economic NMPC is defined as follows:

$$\begin{aligned} V(x(k)) := &\min_{v_l, z_l} \sum_{l=0}^{N-1} \psi^{ec}(z_l, v_l) \\ \text{s.t.} \quad & z_{l+1} = f(z_l, v_l), l = 0, \dots N-1 \\ & z_0 = x_k, z_N = x_s \\ & v_l \in \mathbb{U}, z_l \in \mathbb{X}. \end{aligned} \tag{16}$$

We consider a stage cost given by $\psi^{ec}(\cdot,\cdot) : \mathfrak{R}^{n_x+n_u} \rightarrow \mathfrak{R}$, which is assumed to be Lipschitz continuous. To simplify the problem formulations, we use an NMPC formulation with terminal equality constraints that incorporate the steady state optimum x_s, instead of the origin. Therefore, in contrast to tracking NMPC, which approaches the origin, the stage cost for economic NMPC (eNMPC) $\psi^{ec}(\cdot,\cdot)$ can have arbitrary forms, which represent process economics.

Regarding the stability analysis for economic NMPC, we follow the Lyapunov stability framework and we can derive the following inequality with the standard assumptions for setpoint tracking NMPC:

$$V(x_{k+1}) - V(x_k) \leq -(\psi^{ec}(x_k, u_k) - \psi^{ec}(x_s, u_s)) \tag{17}$$

For setpoint tracking NMPC, the stage cost $\psi^{tr}(x,u)$ usually takes a quadratic form. With this inequality, the tracking objective is decreasing monotonically and thus it can be shown to be a Lyapunov function. For economic NMPC, however, the economic stage cost $\psi^{ec}(x,u)$ can have an arbitrary form that represents the economic information for process operation. For an arbitrary economic objective, the right-hand side of inequality (17) may not be always negative since the optimal solution (x_s, u_s) may not be the global minimum of $\psi^{ec}(x,u)$ for all x and u. Therefore the value function for economic NMPC may not be directly used as a Lyapunov function to demonstrate the stability of the closed-loop system.

To guarantee the stability for economic NMPC, additional properties are needed. First, as shown in [2, 3], dissipativity can be used to establish the stability for economic NMPC.

Definition 14. [3] A control system $x^+ = f(x,u)$ is dissipative with respect to a supply rate $s : \mathbb{X} \times \mathbb{U} \rightarrow \mathbb{R}$ if there exists a function $\lambda : \mathbb{X} \rightarrow \mathbb{R}$, such that

$$\lambda(f(x,u)) - \lambda(x) \leq s(x,u) \tag{18}$$

for all feasible control-input pairs. If in addition $\zeta : \mathbb{X} \rightarrow \mathbb{R}_{\geq 0}$ a positive definite function ($\zeta(x_s) = 0$ and $\zeta(x) > 0$ for all $x \neq x_s$) exists such that

$$\lambda(f(x,u)) - \lambda(x) \leq -\zeta(x) + s(x,u) \tag{19}$$

then the system is said to be strictly dissipative.

By choosing $\lambda(x) = \bar{\lambda}^T x$ for some $\bar{\lambda} \in \mathbb{R}^n$, the dissipativity assumption is equivalent to the following:

$$\min_{x,u} \quad \psi^{ec}(x,u) + \bar{\lambda}^T(x - f(x,u)) \geq \psi^{ec}(x_s, u_s) \tag{20}$$

As pointed out in [3], the dissipativity assumption can be fulfilled if the economic stage cost and dynamic model form a strongly dual problem. More importantly, this leads to the concept of rotated stage cost [3, 6] defined as follows:

$$\phi(x,u) = \psi^{ec}(x,u) + \lambda^T(x - f(x,u)) \tag{21}$$

where λ are the multipliers from the equality constraints in the steady state optimization problem (15). Moreover, it has been shown in [14, 17] that if the rotated stage cost $\phi(x,u)$ is strongly convex, then strong duality property together with the stability of the corresponding economic NMPC can be guaranteed. These results provide sufficient conditions to establish stability for economic NMPC. Note that these conditions can be satisfied by adding quadratic regularization terms to the economic stage cost, which will be discussed in the following section.

4.1 Regularization of Non-convex Economic Stage Costs

For a general economic stage cost $\psi^{ec}(x,u)$ and process dynamic model $f(x,u)$, properties like dissipativity, strong duality, or strong convexity are not fulfilled in general. To guarantee such properties, an easy remedy is to add quadratic regularization terms to the original economic stage cost. After introducing the regularization terms, the modified steady state problem and the corresponding regularized rotated stage cost are defined as follows:

$$\begin{aligned} \min_{x,u} \psi^{ec}(x,u) + \frac{1}{2}\|(x,u)-(x_s,u_s)\|_Q^2 \\ \text{s.t.} \quad x = f(x,u) \\ u \in \mathbb{U}, x \in \mathbb{X}. \end{aligned} \tag{22}$$

$$\phi_{reg}(x,u) = \psi^{ec}(x,u) + \lambda^T(x - f(x,u)) + \frac{1}{2}\|(x,u)-(x_s,u_s)\|_Q^2 \tag{23}$$

where (x_s,u_s) are the optimal solutions to the original optimization problem (15). Q is a diagonal regularization weighting matrix.

As shown in [6, 17] stability of the economic NMPC controller can be realized through the following procedure. First, we consider the *rotated* controller with rotated stage cost $\phi(x,u)$ as the objective function. With a sufficiently large regularization matrix Q, the regularized rotated stage cost $\phi_{reg}(x,u)$ becomes strongly convex. As shown in [17], a local optimal solution from problem (22) is therefore a global minimum for the regularized rotated stage cost. With this result, the value function of this rotated controller decreases monotonically based on inequality (17) and is asymptotically stable. Moreover, we know that the economic NMPC controller has the same solution as the *rotated* controller, as their objective functions only differ by a constant, which means that stability of regularized economic NMPC follows directly.

While adding regularization terms is easy, finding appropriate regularization weights that guarantee the stability of economic NMPC could be a challenging task. In [17], a systematic approach to find the sufficient regularization weights has been proposed. The key idea is to apply the Gershgorin theorem to find the sufficient regularization matrix Q, which makes the regularized rotated stage cost $\phi_{reg}(x,u)$ strongly convex. The proposed condition for Gershgorin weights is shown as follows:

$$q_i > \Sigma_{i \neq j} |h_{i,j}| - h_{i,j} \tag{24}$$

where q_i are the diagonal elements of the regularization weighting matrix Q and $h_{i,j}$ are the elements of matrix $H = \nabla^2 \phi(x,u)$, the Hessian matrix of the rotated stage cost $\phi(x,u)$ in (21). With condition (24) satisfied, the Hessian of the regularized rotated stage cost $\phi_{reg}(x,u)$ is positive definite and thus strongly convex. Based on this simple criterion, we can determine the sufficient regularization weights that guarantee stability of economic NMPC. We denote this approach, with full-state regularization, as `eNMPC-fr`.

On the other hand, it should be noted that this condition (24) must be satisfied for all $u \in \mathbb{U}, x \in \mathbb{X}$. In other words, we need to check this criterion over the entire space of (x,u) so that the regularized rotated stage cost is guaranteed to be strongly convex. In practice, we can sample a sufficient number of possible combinations of states and controls in order to check this criterion. In [4], the author divides the feasible regions of every variable, including differential states, algebraic variables and controls, into N grid points and calculates the Hessian matrix of the rotated stage cost at each grid point. Though all calculations are done offline, they can be cumbersome, especially as the required number of calculations for (24) increases exponentially with the dimension of state and control variables. With this approach, regularization may be required for most system variables (i.e., dynamic states, algebraic variables and controls), which could lead to very conservative economic performance.

To overcome this issue, we propose an economic NMPC formulation with a regularization on a reduced set of variables. The key idea is that we only focus on a subset of states, termed critical states, and determine regularization weights for these critical states only. As shown in the next subsection, such an approach leads to much easier determination of regularization weights as well as less conservative performance.

4.2 Economic NMPC with Regularization of Reduced States

With a slight notational change we restate problem (15) and denote it as `eNMPC-S`:

$$\begin{aligned} \min_{\bar{x},\hat{x},u} \ & \psi^{ec}(x,u) \\ \text{s.t.} \ & \bar{x} = f_1(\bar{x},\hat{x},u) \\ & \hat{x} = f_2(\bar{x},\hat{x},u) \\ & (\bar{x},\hat{x}) \in \mathbb{X}, u \in \mathbb{U}. \end{aligned} \tag{eNMPC-S}$$

In problem `eNMPC-S`, the system states x are divided into two subsectors $x^T = [\bar{x}^T, \hat{x}^T] \in \mathbb{X}$. Here $\bar{x}$ represent critical states of the system, which will be considered for systematic analysis and may require regularization to stabilize economic NMPC controller, while $\hat{x}$ represent noncritical system states. Critical states can be identified through structural analysis of the original optimization problem given by (15). For example, the states that are directly involved in the economic stage cost could be treated as critical states, since they directly affect the optimal solutions to the economic NMPC controller.

For the NMPC problem, we apply the robust problem formulation in [35] by relaxing $\mathbb{X}$, written as $g(z_l) \leq 0$, with ℓ_1 penalty terms as in (11). Equivalently, we can also define $g_+^{(j)}(z_l) = max(0, g^{(j)}(z_l))$, $\psi(z_l, v_l) := \psi(z_l, v_l) + \rho \|g_+(z_l)\|_1$. Note that this stage cost is no longer differentiable everywhere, but still Lipschitz continuous, which is sufficient for the stability analysis. As in (11), with constraint qualifications and second order conditions (e.g., MFCQ, CRCQ and GSSOSC) satisfied, and a sufficiently large penalty weight ρ, the optimal solution of the reformulated problem is the same as the original optimization problem and the penalty terms equal zero. Similarly, terminal equality constraints can also be removed with ℓ_1 penalty terms. For this we choose a penalty parameter ρ_t which is large enough so that $z_N = x_s$ at the optimal solution.

We define the reformulated dynamic optimization problem for the economic NMPC controller (`eNMPC`) as follows:

$$\begin{aligned} V(x(k)) &= \min_{\bar{z}_l, \hat{z}_l, v_l} \sum_{l=0}^{N-1} \psi^{ec}(z_l, v_l) + \rho_t ||z_N - x_s|| \\ \text{s.t.} \quad \bar{z}_{l+1} &= f_1(\bar{z}_l, \hat{z}_l, v_l) \qquad \text{(eNMPC)} \\ \hat{z}_{l+1} &= f_2(\bar{z}_l, \hat{z}_l, v_l), v_l \in \mathbb{U}, \qquad l = 0, \ldots N-1. \\ \bar{z}_0 &= \bar{x}_k, \hat{z}_0 = \hat{x}_k \end{aligned}$$

To partition the critical and noncritical system states for analysis, we introduce the following assumption.

Assumption 8

- *For steady state economic problem eNMPC-S, $\hat{x}$ can be uniquely determined by $(\bar{x}, u)$.*

With Assumption 8, $\hat{x}$ can be uniquely calculated via the square equation system $f_2(.,.)$ with fixed values of $\bar{x}$ and u. Under Assumption 8, the noncritical states $\hat{x}$ can be expressed as a function of critical states $\bar{x}$ and controls u, which leads to the following reformulated steady state optimization problem, which we denote as `eNMPC-SA`:

$$\begin{aligned} &\min_{\bar{x}, \hat{x}, u} \psi^{ec}(x, u) \\ \text{s.t.} \quad &\bar{x} = f_1(\bar{x}, \hat{x}, u) \qquad \text{(eNMPC-SA)} \end{aligned}$$

$$\hat{x} = \eta(\bar{x}, u)$$
$$(\bar{x}, \hat{x}) \in \mathbb{X}, u \in \mathbb{U}.$$

Note that there may not exist an explicit form for function $\eta(\cdot,\cdot)$, but we can at least determine this steady state relationship based on implicit function theorem under Assumption 8.

Next we introduce a modified DAE system, where critical states $\bar{x}$ are determined by the original dynamic model, but noncritical states $\hat{x}$ are treated as algebraic variables. We assume that this modified system is an index 1 DAE. By defining extended states $\tilde{v}_{l+1} = v_l$, we then apply the same robust reformulation and have the following economic NMPC controller eNMPC-A:

$$V(\bar{x}(k)) = \min_{z_l, \hat{z}_l, v_l} \sum_{l=0}^{N-1} \psi^{ec}(z_l, v_l) + \rho_t ||z_N - x_s||$$

$$\begin{aligned} \text{s.t.} \quad & \bar{z}_{l+1} = f_1(\bar{z}_l, \hat{z}_l, v_l), \quad l = 0, \ldots N-1 \\ & \hat{z}_l = \eta(\bar{z}_l, \tilde{v}_l), \quad l = 1, \ldots N \\ & \bar{z}_0 = \bar{x}_k \\ & \hat{z}_0 = h(\bar{x}_k, u_k) \\ & v_l, \tilde{v}_l \in \mathbb{U}. \end{aligned} \tag{eNMPC-A}$$

To simplify this process and avoid over-regularization, we consider only the critical state and control variables, and add these *reduced regularization terms* to the objective of the unregularized controller eNMPC.

To analyze the stability property with this modified regularization, we first study the stability of controller eNMPC-A, where all of the noncritical states are treated as algebraic variables. For controller eNMPC-A, a much simpler and less conservative regularization can be obtained. Then we analyze the stability of eNMPC after adding the reduced regularization obtained from eNMPC-A, by considering the effect of errors introduced by this approximation. Similar to the previous analysis, we also consider a rotated stage cost defined by the steady state problem eNMPC-SA as follows:

$$\phi(x, u) = \psi^{ec}(x, u) + \lambda^T (\bar{x} - f_1(\bar{x}, \eta(\bar{x}, u), u) \tag{25}$$

It should be noted that only a subset of model equations are rotated, and λ are the multipliers only for the equality constraints that have been rotated.

As before, using the rotated stage cost $\phi(x, u)$ as the objective has the same solution as minimizing the original economic stage cost. Moreover, using the following parametric NLP formulation pNLP(t), with parameter t, problems eNMPC and eNMPC-A can be linked by setting $t = 1$ and $t = 0$, respectively.

$$\min_{\bar{z}_l,\hat{z}_l,v_l} \sum_{l=0}^{N-1} \psi^{ec}(z_l,v_l) + \rho_t||z_N - x_s||$$

$$\text{s.t.} \quad \bar{z}_{l+1} = f_1(\bar{z}_l,\hat{z}_l,v_l), \quad l = 0,\ldots N-1 \qquad \text{(pNLP(t))}$$

$$\hat{z}_l = \eta(\bar{z}_l,\tilde{v}_l) + t(f_2(\bar{z}_{l-1},\hat{z}_{l-1},v_{l-1}) - \eta(\bar{z}_l,\tilde{v}_l)), \quad l = 1,\ldots N$$

$$\bar{z}_0 = \bar{x}_k$$

$$\hat{z}_0 = \eta(\bar{x}_k,u_k) + t(\hat{x}_k) - \eta(\bar{x}_k,u_k)$$

$$v_l,\tilde{v}_l \in \mathbb{U}.$$

Finally, for the approximation of the noncritical states, we introduce an error ("noise") vector $w(k) = [w_0 \ldots w_N]^T$ with entries defined as follows:

$$w_0 = \hat{x}(k) - \eta(\bar{x}(k),u(k-1)) \tag{26}$$

$$w_l = f_2(\bar{z}_{l-1},\hat{z}_{l-1},v_{l-1}) - \eta(\bar{z}_l,\tilde{v}_l) \quad l = 1\ldots N \tag{27}$$

This noise vector $w(k)$ represents the differences in the values of $\hat{z}_l$ given by the dynamic function and steady state relationship. For the above parametric NLP problem, when $t = 0$, we have problem eNMPC-A. On the other hand, when $t = 1$, we have problem eNMPC.

The following theorem shows that when $w(k) = 0$, the stability property can be guaranteed for controller eNMPC by adding regularization terms only for critical states $\bar{x}$ and u. In this special case, noncritical states collapse into algebraic variables and controller eNMPC is equivalent to eNMPC-A.

Theorem 9. *When $w(k) = 0$ and Assumption 8 holds, controller* eNMPC *can be made asymptotically stable by adding a sufficiently large regularization on reduced sets of states $\bar{z}$ and v.*

The proof of this theorem can be found in [36].

Then we consider the stability property for controller eNMPC for cases where $w(k) \neq 0$, and the controller eNMPC can be treated as the controller eNMPC-A corrupted with non-zero noise terms $w(k)$. The process model for controller eNMPC-A is defined as follows:

$$\overline{k+1} = f_1(\bar{x}_k,u_k,\eta(\bar{x}_k,u_{k-1})) \tag{28}$$

while the true process model is defined as follows:

$$\overline{k+1} = f_1(\bar{x}_k,u_k,\eta(\bar{x}_k,u_{k-1}) + w_0) \tag{29}$$

Here w_0 is the first element of the noise vector $w(k)$ and is defined as $\hat{k} - \eta(\bar{x}_k,u_{k-1})$, which represents the difference in the values of $\hat{x}$ at time k for eNMPC and eNMPC-A.

We now analyze the stability property for controller `eNMPC-A` when the process model is given by Equation (29), which is the nominal process model. For this case, model mismatch exists between the nominal process model and the control model (28) for controller `eNMPC-A`. We also introduce the assumption that $w(k)$ is always bounded. Along with additional standard assumptions for robust stability, we establish the Input-to-State Stability (ISS) property for controller `eNMPC-A` with Theorem 13.

Assumption 10

- *The noise vector $w(k) = [w_0 \dots w_N]^T$ is drawn from a bounded set $\mathscr{W}$ with an upper bound $\bar{w}$.*

Assumption 10 is a key assumption for the following stability analysis, which assumes bounded deviations of the dynamic noncritical states $\hat{x}$ from their algebraic approximations, i.e. dynamic states $\hat{x}$ have a similar behavior as algebraic variables. For example, this occurs with noncritical states that have very fast dynamics.

Assumption 11 *Robust stability assumptions*

- *The optimal solution to problems* `eNMPC` *and* `eNMPC-A` *is continuous with respect to x_k and w.*
- *$V(x_k)$ is Lipschitz with respect to x_k, with a positive Lipschitz constant L_v.*
- *Model equations f_1, f_2 and steady state relationship η are Lipschitz continuous with its arguments with corresponding Lipschitz constants.*

Theorem 12. *Under Assumption 8, 10, and 11, controller* `eNMPC-A`*, with a sufficiently large regularization on $\bar{z}$ and v, is ISS when the process model is given by Equation (29) and $w(k) \neq 0$.*

The proof of this theorem can be found in [36].

Because controller `eNMPC` is linked to controller `eNMPC-A` through `pNLP(t)`, we can derive the stability property for controller `eNMPC` by treating this controller as controller `eNMPC-A` corrupted with non-zero noise terms $w(k)$. This leads to the following assumption and theorem.

Theorem 13. *Under Assumption 8, 10, and 11, controller* `eNMPC` *can be made ISpS, by adding a sufficiently large regularization on $\bar{z}$ and v.*

The proof of this theorem can be found in [36].

From the above results, we can guarantee ISpS of the economic NMPC controller by regularizing critical states $\bar{x}$, under the assumption that the deviations in noncritical states $\hat{x}$ from their algebraic predictions are bounded. Unlike exogenous process disturbances, which always exist and are independent of process states, the noise vector $w(k)$ in our analysis may have some different properties that could lead to stronger stability results. With the following theorem, we can show that asymptotic stability can be established for `eNMPC` with a stronger assumption for $w(k)$.

Assumption 14 *The noise vector* $|w(k)| \leq \frac{\delta}{L_w}(|\bar{x}_k - \bar{x}_s|)$*, where* $\frac{\delta}{L_w}(|\bar{x}_k - \bar{x}_s|) \leq \bar{w}, L_w = 2L_V, \delta \in [0,1)$*, after a finite number of iterations K.*

In Assumption 14, we assume that as critical states $\bar{x}$ approach to steady state, the vector $w(k)$ is bounded by the distance of $\bar{x}$ to the optimal steady state, which is stronger than Assumption 10. However, this assumption may hold for cases where $\bar{x}_k$ and $\hat{x}_k$ are close to steady state; the deviations of dynamic states $\hat{x}_k$ and their algebraic predictions are bounded by a decaying bound and $w(k)$ will go to zero as $\bar{x}$ converges to steady state $\bar{x}_s$.

Theorem 15. *Under Assumption 14, controller* `eNMPC` *can be made asymptotically stable, by adding a sufficiently large regularization on reduced sets of states* $\bar{z}$ *and* v*.*

The proof of this theorem can be found in [36].

In this section, we have shown that, with a sufficiently large regularization on a reduced set of system states, the stability of controller `eNMPC` can still be maintained. Though controller `eNMPC-A` has a stronger stability result than controller `eNMPC`, we only use the modified process model to determine reduced regularization weight. From the modified model, we can derive a reduced Hessian in the space of critical states. By making the reduced Hessian positive definite in the reduced space, we can find sufficient regularization weights for critical states, and add these *reduced regularization terms* to the objective of the unregularized controller `eNMPC`. We denote this controller as `eNMPC-rr` and observe that it still uses the original dynamic model, which gives accurate predictions in terms of the dynamic behavior of both states $\bar{x}$ and $\hat{x}$, but with a reduced regularization for the stage cost. Additional details on calculating the reduced regularization weights and theorem proofs are given in [36].

Selection of critical states can have direct impacts on the performance of controller `eNMPC-rr`. Based on the previous stability results, we can see that dynamic states that have similar performance as their algebraic counterparts may be removed from regularization analysis by treating them as algebraic variables. These states can be located via time scale analysis of the original system. For states with very fast time scales, Assumption 10 may be satisfied implicitly and no regularization is required for these states.

Finally, selecting the appropriate critical states for regularization also requires a good understanding of the dynamic process model structure. For example, we can conduct a structural analysis of the dynamic model to see if there are inherently unstable states. If so, we need to treat these unstable states as critical states as well. In addition, the coupling of states may provide hints to remove unnecessary states for regularization analysis. For instance, if there is strong dependency of some states on the others, like the slaving relationship for algebraic variables, then these dependent states can be treated as noncritical.

5 Economic MPC with a Stabilizing Constraint

Finally, we introduce a less restrictive economic MPC formulation `eNMPC-sc` as well as related formulations to highlight advantages and disadvantages. It has been recently suggested to replace the tracking regularization terms in the objective with a stabilizing constraint [37]. The `eNMPC-sc` controller solves the NLP:

$$\min_{z_l, v_l} \quad \sum_{l=0}^{N-1} \psi^{ec}(z_i, v_i) \tag{30a}$$

$$\text{s.t. } z_{l+1} = f(z_l, v_l, 0) \quad l = 0, \ldots, N-1 \tag{30b}$$

$$z_0 = x_k \tag{30c}$$

$$z_l \in \mathbb{X},\ v_l \in \mathbb{U} \quad l = 0, \ldots, N-1 \tag{30d}$$

$$z_N = x_s \tag{30e}$$

$$\sum_{l=0}^{N-1} \psi^{tr}(z_l, v_l) - V(k-1, \mathbf{w}_{k-1}, x_0) \leq -\delta\, \psi^{tr}(x_{k-1}, u_{k-1}) \tag{30f}$$

where $\delta \in (0,1]$ is a scalar parameter. After the NLP is solved, we inject the control law $u_k = v_0$ into the system and set

$$V(k, \mathbf{w}_k, x_0) := \sum_{i=0}^{N-1} \psi^{tr}(z_i^k, v_i^k), \tag{31}$$

where (z_i^k, v_i^k) is the solution of (30) at time k. We thus note that $V(k-1, \mathbf{w}_{k-1}, x_0)$ is the value function at time $k-1$.

Once the control is injected into the system we wait for it to evolve to x_{k+1} and use the value function $V(k, \mathbf{w}_k, x_0)$ in (30f) to solve (30) at x_{k+1}, and repeat the procedure. Note that the initial value $V(0, \mathbf{w}_0, x_0)$ may simply be chosen sufficiently large to ensure that (30f) is inactive at the solution of the problem solved at time $k = 0$. The advantages of this formulation are that we do not require a solution to the tracking problem, and this formulation provides a looser constraint with the same stability properties.

To establish ISpS for the proposed economic MPC controller we must ensure uniform continuity of the value function. This property is not guaranteed for the formulation (30), but can be achieved by softening the state constraints, as is done in (11), and rewriting the problem formulation as follows:

$$\min_{z_l, v_l} \quad \sum_{l=0}^{N-1} \psi^{ec}(z_i, v_i) + \rho \left(\xi^S + \xi^{ss,L} + \xi^{ss,U} + \sum_{l=0}^{N-1} \xi_l^x \right) \tag{32a}$$

$$\text{s.t. } z_{l+1} = f(z_l, v_l, 0) \quad l = 0, \ldots, N-1 \tag{32b}$$

$$z_0 = x_k \tag{32c}$$

$$g_x(z_l) \le \xi_l^x, v_l \in \mathbb{U} \quad l = 0, \dots, N-1 \tag{32d}$$

$$-\xi^{ss,L} \le z_N - x_s \le \xi^{ss,U}, \quad \xi_l^x, \xi^S, \xi^{ss,L}, \xi^{ss,U} \ge 0 \tag{32e}$$

$$\sum_{l=0}^{N-1} \psi^{tr}(z_l, v_l) - V(k-1, \mathbf{w}_{k-1}, x_0) \le -\delta\, \psi^{tr}(x_{k-1}, u_{k-1}) + \xi^S \tag{32f}$$

where $\xi_l^x, \xi^S, \xi^{ss,L}, \xi^{ss,U}$ are auxiliary variables and $\rho \in \mathbb{R}_+$ is a penalty parameter. To analyze this controller, we consider the following assumptions and stability theorem:

Assumption 16 *(A)The set $\mathscr{X} \subseteq \mathbb{R}^{n_x}$ is robustly positive invariant for $f(\cdot,\cdot,\cdot)$. That is, $f(x, \kappa(x), w) \in \mathscr{X}$ holds for all $x \in \mathscr{X}$, $w \in \mathbb{W}$. (B) The set $\mathbb{W}$ is bounded and $||\mathbf{w}|| := \sup_{k \in \mathbb{Z}_+} |w|$. (C) f is uniformly continuous with respect to w.*

Theorem 17. *Let Assumption 16 hold. Then there exists $V(k, \mathbf{w}_k, x_0)$ such that eNMPC-sc is ISpS.*

See [11] for proof. We also note that $\delta = 1$ in (32f) corresponds to the most constrained Lyapunov function, and δ approaching zero corresponds to the least constrained. The parameter δ is thus a tuning parameter that shapes closed-loop behavior. A large δ forces a faster approach to the steady state, and small δ allows for more economic flexibility.

Moreover, by dualizing (32f) and moving it into the objective, we observe that this constraint acts as a regularization term. Its weight is determined by the optimal Lagrange multiplier of (32f) and thus changes at each time instant k (i.e., the weight is adaptive). This Lagrange multiplier can be interpreted as the *price of stability*. For more details, see [37].

6 Case Studies

6.1 Nonlinear CSTR

To compare the performance of different economic NMPC strategies proposed in the previous sections, we first consider a nonlinear continuous stirred tank reactor (CSTR), taken from [6], with a first order irreversible reaction $A \rightarrow B$. The mass balances for reactant A and product B are shown as follows:

$$\frac{dc_A}{dt} = \frac{F}{V}(c_{Af} - c_A) - kc_A \tag{33a}$$

$$\frac{dc_B}{dt} = \frac{F}{V}(-c_B) + kc_A \tag{33b}$$

where c_A and c_B denote the concentrations of components A and B, respectively, in mol/l. The manipulated input flow is F in l/min, the reactor volume is $V = 10\, l$, the rate constant is $k = 1.2\, l/(mol \cdot min)$, and $c_{Af} = 1\, mol/l$ is the feed concentration. In addition, we set variable bounds as $10 \leq q \leq 20$ and $0.45 \leq c_B \leq 1$ which are softened with auxiliary variables that are penalized in the objective function in the NLP. The economic stage cost is:

$$\psi^{ec}(c_A, c_B, F) = -F\left(3c_B - \frac{1}{2}\right) \tag{34}$$

The steady state used for the tracking objective is $c_A^* = 0.55$, $c_B^* = 0.45$, $F^* = 14.67$, so $\psi_{ss}^{ec} = -12.47$.

We compare the performance of `eNMPC-sc` with a regularized economic MPC controller. Regularization weights for the latter, $Q = diag(q_A, q_B, q_F)$ are calculated in [36]. For the full regularization strategy, regularization terms $\frac{1}{2}\|x - x_s\|_Q^2$ are required for F and $x = c_A, c_B$ with $q_A = 1.55$, $q_B = 0.44$, and $q_F = 1.98$. For reduced regularization, we chose c_A as the critical state, and find the regularization weights for the reduced system, with $q_A = 1.55$, and $q_F = 1.55$.

We discretize the CSTR model using a three-point Radau collocation with a finite element length of 0.5 min with the prediction horizon $N = 200$. The initial conditions $c_{A,0}$ = 0.2 mol/l, $c_{B,0}$ = 1 mol/l, and we apply the robust reformulation to constraints on c_B, which softens the variable bounds with ℓ_1 norm penalty in the objective function. The penalty weight ρ is 10^6. We implement the problem in AMPL [8] and solve the NLPs with IPOPT [31].

For this case study, we compare the performance of tracking NMPC and different formulations of economic NMPC, including `eNMPC-sc` with stabilizing constraint, `eNMPC-fr` with full regularization, `eNMPC-rr` with reduced regularization as well as pure economic NMPC (with $Q = 0$). A comparison of the accumulated stage costs is listed in the following table. A short simulation length of $K = 9$ is chosen so that we can focus on the dynamics that occur before the set-point is reached.

Table 1: CSTR example, comparing the accumulated cost

	$\sum_{k=0}^{K} \psi^{ec}(x_k, u_k) - \psi_{ss}^{ec}$
Tracking	−37.8078
`eNMPC-fr`	−38.5496
`eNMPC-rr`	−38.7547
`eNMPC-sc`, $\delta = 0.01$	−45.1846
Purely economic	−45.6851

From Table 1, we can see that pure economic NMPC has the best economic performance. `eNMPC-sc` has a similar cost to the pure economic case, which is better than the cases with regularization terms. Moreover, we can observe a slight improvement by reduced regularization over full regularization. We also compare the variations of economic stage cost $\psi^{ec}(x_k, u_k) - \psi_{ss}^{ec}$ for economic NMPC with different formulations in Figure 1. First we can see that, by adding full regularization, reduced regularization or imposing an inequality constraint, economic NMPC leads to fast convergence to the optimal steady state. However, pure economic NMPC leads to an oscillatory profile. We can also see that cases with reduced regularization have a similar trend as the full regularization case, and the case with a stabilizing constraint has similar behavior as the purely economic case, especially in the first few NMPC cycles.

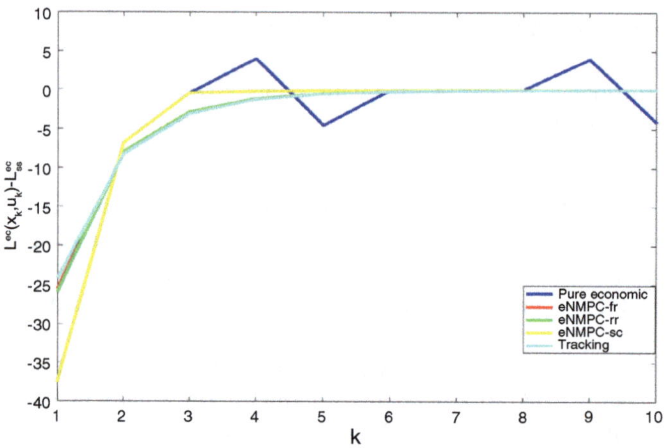

Fig. 1: Variations of stage cost for different controllers in the CSTR example.

6.2 Large-Scale Distillation System

We now consider the large-scale process system shown in Figure 2 with two distillation columns in series. The full model is omitted here for brevity, but is available in [21]. Each column is based on the model described in [30], with the main difference that we consider three components, A, B, and C. The bottom stream of the first column (with flowrate B_1) is the feed to the second column (with flowrate F_2). The distillate of the first column (with flowrate D_1) is specified at 95% A, the distillate of the second column (with flowrate D_2) is specified at 95% B, and the bottoms of the second column (with flowrate B_2) is specified at 95% C. Vapor flowrates in the reboilers of the two columns are V_{B_1} and V_{B_2}, respectively. We assume constant relative volatilities with $\alpha_A = 2, \alpha_B = 1.5$, and apply the Francis weir formula for tray hydraulics. Each column has 41 equilibrium stages including the reboiler, leading to a dynamic model with 246 state and 8 control variables.

The economic cost is the cost of feed and energy to the reboilers minus the cost of the products, that is $\psi^{ec} = p_F \cdot F_1 + p_V(V_{\bar{B},1} + V_{\bar{B},2}) - p_A \cdot D_1 - p_B \cdot D_2 - p_C \cdot B_2$, where $p_F = \$1/mol$ is the price of feed, p_i for $i = A,B,C$ is the price of compo-

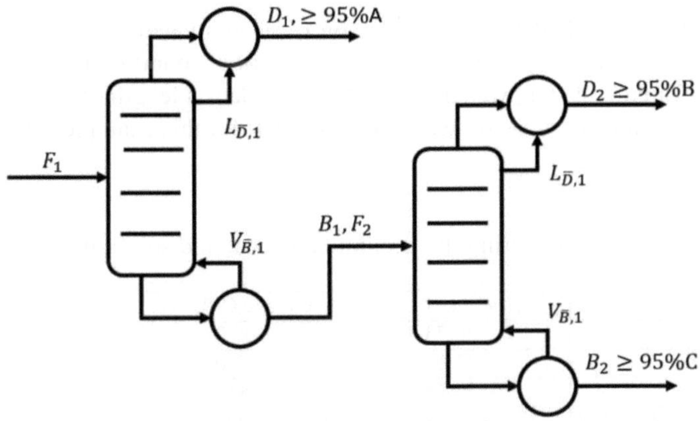

Fig. 2: Distillation Flowsheet.

nent i with $p_A = \$1/mol$, $p_B = \$2/mol$, and $p_C = \$1/mol$, $p_V = \$0.008/mol$ is the price per mole vaporized in the reboilers, and the indices represent the first or second column. The feed to the first column is saturated liquid, with composition at 40 mol% A, 20 mol% B and 40 mol% C. Product purities are implemented as inequality constraints. We discretize the DAE system using three point Radau collocation and use a finite element length of 1 min and $N = 25$. Each NLP (32) has $120,000$ variables, $108,000$ equality constraints, and $14,000$ inequality constraints. The models are implemented in AMPL and solved with IPOPT.

Finding sufficient regularization weights in the full variable space is much more cumbersome due to the size of the system. Here, the entire state space is gridded and the Hessian of the steady state problem is evaluated at each grid point. We denote the controller with full-space regularization obtained from the Gershgorin Circle theorem as `eNMPC-fr`. The Gershgorin weights for all the variables are reported in [33].

To reduce the effort to determine sufficient regularization weight, we consider the reduced regularization strategy. As critical states we choose the tray holdups at the top of the first column, and the top and bottom of the second column, as these states directly impact the distillates in the economic function. Additional regularization terms are associated with four manipulated variables, the reflux and boilup rates for the two columns. With the reduced regularization strategy, the reduced Hessian needs to be positive definite over the feasible regions of only 7 variables, rather than for all of the variables; this significantly simplifies determination of the weights. We obtain the reduced Hessian via numerical perturbations, calculate its eigenvalues $\mu_i(x,u)$ at sampling points j and find a regularization weight $Q = (q+\varepsilon)I$ where $q = max(0, max_{j,i} - \mu_i(x_j,u_j))$, $\varepsilon = 10^{-3}$.

In this example, we compare the performance of tracking NMPC and different formulations of economic NMPC, including `eNMPC-sc` with a stabilizing constraint, `eNMPC-fr` with full regularization, `eNMPC-rr` with reduced regularization as well as pure economic NMPC. A comparison of accumulated stage costs over ten time steps ($\sum_{k=0}^{K} \psi^{ec}(x_k,u_k) - \psi^{ec}_{ss}$) from the same initial condition for various cases is shown in Table 2. Again, we choose a short simulation length of $K = 9$ to emphasize dynamic performance. Also, we choose $\rho = 10^4$ for this example. The steady state cost, ψ^{ec}_{ss}, is -0.223.

Table 2: Distillation example, comparing the accumulated cost and solution times

	$\sum_{k=0}^{K}(\psi^{ec}(x_k,u_k) - \psi^{ec}_{ss})$	Average CPU sec.
Tracking	−20.7330	69.0
`eNMPC-fr`	−22.6650	72.0
`eNMPC-rr`	−26.2324	181.8
`eNMPC-sc`, $\delta = 0.01$	−28.6706	309.2
Economic	−28.6458	272.3

From the results shown in Table 2, we can see that `eNMPC-sc` provides economic benefit over regularized formulations `eNMPC-fr` and `eNMPC-rr`. Compared with full regularization strategy, `eNMPC-rr` with reduced regularization achieves better economic performance. In addition, the efforts to calculate sufficient regularization weights are greatly reduced. Lastly, all formulations of eNMPC have better economic performance over tracking NMPC.

The variations of economic stage cost $\psi^{ec}(x_k,u_k)$ for all cases are shown in the Figure 3. We can see that `eNMPC-sc` with a very small $\delta = 0.01$ has nearly the same behavior as pure economic NMPC with most significant variations. But this case gives the best economic performance. `eNMPC-fr` behaves similarly as tracking NMPC with least variations, due to conservative regularization terms on most system variables. Compared with `eNMPC-fr`, `eNMPC-rr` has more oscillations

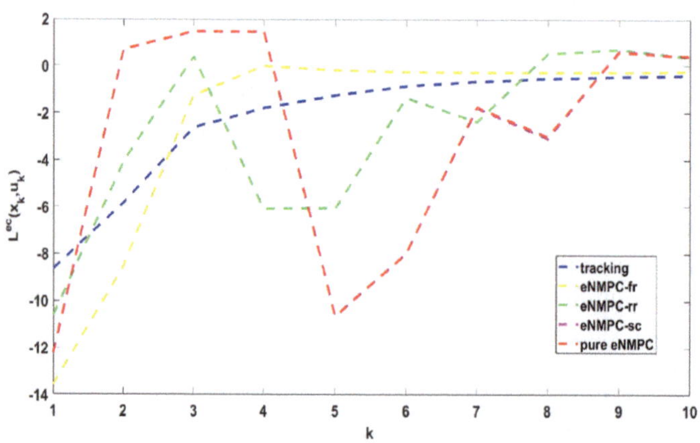

Fig. 3: Variations of stage cost for different controllers in the double distillation column example.

but improves the economic performance with smaller regularization weights and far fewer regularization terms. The computational times for these eNMPC cases are also shown in Table 2. By comparing the results from Table 2, we observe that regularization is beneficial for computational performance but sacrifices economic performance.

7 Conclusions

This work describes robustly stable NMPC formulations for both tracking and economic cases. Robustness may be obtained by softening the state variable inequality constraints with auxiliary variables that are treated as ℓ_1 penalties. We show both the nominal and robust stability of tracking NMPC reformulated in this way. For the economic case, we show multiple methods for stabilization. These include objective regularization based on the full state-space (`eNMPC-fr`), objective regularization based on a reduced set of states (`eNMPC-rr`), and the addition of a stabilizing constraint (`eNMPC-sc`). We then demonstrate these results on a small CSTR example and a large case study with two distillation column models in series. In general, we observe that tracking NMPC and `eMPC-fr` behave similarly in economic performance, while `eNMPC-sc` has economic performance closer to pure eNMPC. While `eNMPC-sc` has ISpS stability guarantees it tends to require longer solution times. Finally, `eNMPC-rr` provides a good compromise, with ISpS stability, intermediate economic performance, and relatively short solution times.

Acknowledgements

This work is partially supported by the National Science Foundation Graduate Research Fellowship Program Grant No. DGE1252522. The second author also thanks the Choctaw Nation of Oklahoma, the Pittsburgh Chapter of the ARCS foundation, the ExxonMobil Graduate Fellowship program, and the Bertucci Graduate Fellowship program for generous support.

References

1. Amrit, R., Rawlings, J.B., Biegler, L.T.: Optimizing process economics online using model predictive control. Comput. Chem. Eng. **58**, 334–343 (2013)
2. Angeli, D.: Handbook of Model Predictive Control. Chapter Economic Model Predictive Control, p. yyy. Birkhüser, Basel (2018)
3. Angeli, D., Amrit, R., Rawlings, J.: On average performance and stability of economic model predictive control. IEEE Trans. Autom. Control **57**(7), 1615–1626 (2012)
4. Biegler, L.T., Yang, X., Fischer, G.G.: Advances in sensitivity-based nonlinear model predictive control and dynamic real-time optimization. J. Process Control **30**, 104–116 (2015)

5. Chen, H., Allgöwer, F.: A quasi-infinite horizon nonlinear model predictive control scheme with guaranteed stability. Automatica **34**, 1205–1217 (1998)
6. Diehl, M., Amrit, R., Rawlings, J.: A Lyapunov function for economic optimizing model predictive control. IEEE Trans. Autom. Control **56**(3), 703–707 (2011)
7. Fiacco, A.V.: Introduction to Sensitivity and Stability Analysis in Nonlinear Programming. Academic, New York (1983)
8. Fourer, R., Gay, D.M., Kernighan, B.W.: AMPL: A Modeling Language for Mathematical Programming. Duxbury Press, Pacific Grove, Belmont (2002)
9. Gauvin, J.: A necessary and sufficient regularity condition to have bounded multipliers in nonconvex programming. Math. Program. **12**(1), 136–138 (1977)
10. Gopalakrishnan, A., Biegler, L.T.: Economic nonlinear model predictive control for the periodic optimal operation of gas pipeline networks. Comput. Chem. Eng. **52**, 90–99 (2013)
11. Griffith, D., Zavala, V., Biegler, L.: Robustly stable economic NMPC for non-dissipative stage costs. J. Process Control **57**, 116–126 (2017)
12. Grüne, L.: Economic receding horizon control without terminal constraints. Automatica **49**, 725–734 (2013)
13. Heidarinejad, M., Liu, J., Christofides, P.: Economic model predictive control of nonlinear process systems using lyapunov techniques. AIChE Journal **58**(3), 855–870 (2011)
14. Huang, R., Harinath, E., Biegler, L.T.: Lyapunov stability of economically-oriented NMPC for cyclic processes. J. Process Control **21**, 501–509 (2011)
15. Huang, R., Harinath, E., Biegler, L.T.: Economicaly-oriented nonlinear model predictive control for energy applications. J. Process Control **21**(4), 501–509 (2011)
16. Janin, R.: Directional derivative of the marginal function in nonlinear programming. In: Fiacco, A.V. (ed.) Sensitivity, Stability and Parametric Analysis. Mathematical Programming Studies, vol. 21, pp. 110–126. Springer, Berlin (1984)
17. Jäschke, J., Yang, X., Biegler, L.T.: Fast economic model predictive control based on NLP-sensitivities. J. Process Control **24**, 1260–1272 (2014)
18. Jiang, Z.P., Wang, Y.: Input-to-state stability for discrete-time nonlinear systems. Automatica **37**, 857–869 (2001)
19. Keerthi, S.S., Gilbert, E.G.: Optimal infinite-horizon feedback laws for general class of constrained discrete-time systems: stability and moving-horizon approximations. IEEE Trans. Autom. Control **57**, 265–293 (1988)
20. Kojima, M.: Strongly stable stationary solutions in nonlinear programming. In: Robinson, S.M. (ed.) Analysis and Computation of Fixed Points. Academic, New York (1980)
21. Leer, R.B.: Self-optimizing control structures for active constraint regions of a sequence of distillation columns. Master's thesis, Norweign University of Science and Technology (2012)
22. Limon, D., Alamo, T., Raimondo, D., Peña, D., Bravo, J., Ferramosca, A., Camacho, E.: Input-to-state stability: a unifying framework for robust model predictive control. In: Magni, L., Raimondo, D., Allgöwer, F. (eds.) Nonlinear Model Predictive Control: Towards New Challenging Applications. Springer, Berlin (2009)
23. Magni, L., Scattolini, R.: Robustness and robut design of mpc for nonlinear discrete-time systems. In: Findeisen, R., Allgöwer, F., Biegler, L.T. (eds.) Assessment and Future Directions of Nonlinear Model Predictive Control, pp. 239–254. Springer, Berlin (2007)
24. Mayne, D.Q., Rawlings, J.R., Rao, C.V., Scokaert, P.O.M.: Constrained model predictive control: stability and optimality. Automatica **36**, 789–814 (2000)
25. Nocedal, J., Wright, S.: Numerical Optimization. Operations Research and Financial Engineering, 2nd edn. Springer, New York (2006)
26. Pannocchia, G., Rawlings, J.B., Wright, S.J.: Conditions under which supboptimal nonlinear MPC is inherently robust. Syst. Control Lett. **60**, 747–755 (2011)
27. Ralph, D., Dempe, S.: Directional derivatives of the solution of a parametric nonlinear program. Math. Program. **70**(1–3), 159–172 (1995)
28. Rawlings, J.B., Mayne, D.Q.: Model Predictive Control: Theory and Design. Nob Hill Publishing, Madison (2009)
29. Robinson, S.M.: Strongly regular generalized equations. Math. Oper. Res. **5**, 43–62 (1980)

30. Skogestad, S.: Dynamics and control of distillation columns: a tutorial introduction. Chem. Eng. Res. Des. **75**(A), 539–562 (1997)
31. Wächter, A., Biegler, L.T.: On the implementation of a primal-dual interior point filter line search algorithm for large-scale nonlinear programming. Math. Program. **106**(1), 25–57 (2006)
32. Würth, L., Rawlings, J.B., Marquardt, W.: Economic dynamic real-time optimization and nonlinear model predictive control on infinite horizons. In: International Symposium on Advanced Control of Chemical Process, Istanbul (2009)
33. Yang, X.: Advanced-multi-step and economically oriented nonlinear model predictive control. PhD thesis, Carnegie Mellon University (2015)
34. Yang, X., Biegler, L.T.: Advanced-multi-step nonlinear model predictive control. J. Process Control **23**, 1116–1128 (2013)
35. Yang, X., Griffith, D., Biegler, L.: Nonlinear programming properties for stable and robust NMPC. In: 5th IFAC Conference on Nonlinear Model Predictive Control, IFAC-PapersOnLine **48**(23), 388–397 (2015)
36. Yu, M.: Model reduction and nonlinear model predictive control of large-scale distributed parameter systems with applications in solid sorbent-based CO_2 capture. PhD thesis, Carnegie Mellon University (2017)
37. Zavala, V.M.: A multiobjective optimization perspective on the stability of economic MPC. In: 9th International Symposium on Advanced Control of Chemical Processes, pp. 975–981 (2015)
38. Zavala, V., Biegler, L.: The advanced step NMPC controller: optimality, stability and robustness. Automatica **45**, 86–93 (2009)

Part III
Applications

Automotive Applications of Model Predictive Control

Stefano Di Cairano and Ilya V. Kolmanovsky

Model Predictive Control (MPC) has been investigated for a significant number of potential applications to automotive systems. The treatment of these applications has also stimulated several developments in MPC theory, design methods, and algorithms, in recent years. This chapter provides an overview of automotive applications for which MPC has been considered and approaches to MPC development, deployment, and implementation that have been pursued. First, a brief history of MPC applications to automotive systems and features that make MPC appealing for such applications are discussed. Then, for the main automotive control sub-domains, key first principle models and opportunities that these provide for the application of MPC are described. Next, we detail the key steps and guidelines of the MPC design process tailored to automotive systems. Finally, we discuss numerical algorithms for implementing MPC, and their suitability for automotive applications.

1 Model Predictive Control in Automotive Applications

There are very few devices that are as pervasive in our world as cars. Reports show that close to 90 million cars and light commercial vehicles were sold worldwide in 2016. Recent innovations in car mechanics, electronics, and software have been fast paced to respond to growing stringency of fuel economy, emissions, and safety regulations, as well as to market-driven pressures to provide customers with improved

S. Di Cairano (✉)
Mitsubishi Electric Research Laboratories, Cambridge, MA 02139, USA
e-mail: dicairano@ieee.org

I. V. Kolmanovsky
Department of Aerospace Engineering, University of Michigan, Ann Arbor, MI 48109, USA
e-mail: ilya@umich.edu

S. V. Raković, W. S. Levine (eds.), *Handbook of Model Predictive Control*, Control Engineering, https://doi.org/10.1007/978-3-319-77489-3_21

performance, drivability, and novel features. Advanced control methods that are capable of optimizing the vehicle operation, and can reduce the time-to-market for increasingly complex automotive systems are clearly needed.

It thus comes as no surprise that, in recent years, a significant interest in model predictive control (MPC) has been shown in the automotive industry. The research on applications of MPC to automotive systems has been steadily growing both in industry and academia to address some of the challenges of this application domain. Yet MPC is a significant step up from the classical control methods, such as PID, and its implementation in industrial practice presents challenges on its own.

The purpose of this chapter is to provide a short tutorial on the development of MPC-based solutions for automotive systems. Towards this end, we first briefly review the history of MPC applications to automotive systems, and we highlight the benefits that MPC can provide as well as the challenges faced by MPC in this domain. Then, given that MPC is a model-based control approach, for the main automotive control areas, such as powertrain control, chassis control, and energy management, we describe the key first principle models that can be used for MPC design, and the control objectives that need to be achieved. Next, we detail common steps of MPC design for automotive systems. Finally, we consider the computational aspects that are important for real-time implementation and deployment of MPC solutions on automotive computing platforms.

While this chapter represents a tutorial overview of MPC design for automotive systems based on the author's first-hand experience, due to scope and length limitations it is not able to serve as a comprehensive survey of the entire body of literature on automotive applications of MPC. A brief survey is available in [48].

1.1 A Brief History

Some of the first investigations of MPC for automotive systems can be traced back to the mid '90s, with [46] where MPC was applied to idle speed control being a notable case. In those years, the numerical algorithms for MPC were too computationally demanding for the "then-current" vehicle micro-controllers, and hence such studies were usually only simulation-based.

Two new developments in the early 2000s gave a significant boost to the investigation of MPC-based automotive control and have led to the rapid growth of related applications and literature. Firstly, the scientific community interest in hybrid dynamical systems led to the development of hybrid MPC [7], which allowed to control processes with switching dynamics. This opened up opportunities for the application of MPC to control of transmissions [2, 8, 42, 78], to traction control [15], and to control of semiactive suspensions [38]. Systems with mode-dependent objectives, such as direct injection, stratified charge engines [39], or requiring piecewise linearizations, such as camless engine actuators [23], HCCI engines [12, 68], or vehicle stability control functions [29] could now be handled. Secondly, the application of parametric programming techniques resulted in the development of explicit

MPC [6] that synthesizes the control law, and hence avoids the need to run an optimization algorithm online in the micro-controller. This led to the possibility of experimentally testing several controllers in real, production-like, vehicles including, about 12 years after the initial development, a refined MPC-based idle speed control [24], and an MPC-based diesel engine airpath control [64, 75]. From then, the applications of MPC have picked up both in powertrain control [30] and chassis (or vehicle dynamics) control [5, 29], with some industry research centers being at the forefront in developing these applications, see, e.g., [48, 61, 79].

Starting from the mid-2000s, MPC-based control has been considered for hybrid and electric vehicles, including fuel-cell vehicles. Some of the early contributions include [6, 58, 77]. The development of MPC strategies for different hybrid electric powertrain configurations has then been considered in more depth, e.g., for ERAD [70], series [28], and powersplit [14] configurations. Due to the complexity of the hybrid powertrains and the attempt to use MPC to directly optimize fuel consumption, these controllers were often rather difficult to implement in the vehicle. An interesting case is [28], where instead of optimizing directly the fuel consumption, MPC was used as an energy buffer manager to operate the engine smoothly and with slow transients, leading to a design simple enough to be implementable in a prototype production vehicle, yet still achieving significant benefits in terms of fuel economy. The resulting controller in [28] was actually implemented experimentally in such road-capable vehicle, which allowed to assess its performance in production-like computing hardware.

Currently, advanced MPC methods are being investigated both for improving existing features, and for future applications in autonomous, and connected vehicles. Some examples are Lyapunov-based MPC for network control in automotive systems [17], stochastic MPC for cooperative cruise control [74], robust and stochastic MPC for autonomous driving [18, 31, 53], and several applications exploiting V2V and V2I communications [59, 63]. Such an expansion has been also supported by the development of low complexity optimization algorithms that now allow for solving quadratic programs in automotive micro-controllers without the need to generate the explicit solutions, that have combinatorial complexity in terms of memory and computations. Still several challenges in terms of computation, estimation, and deployment remain, that will require significant investigations in the next several years, to increase the range of feasible applications. How ongoing advances in the areas of cloud computing, connectivity, large data sets, and machine learning can help tackle these challenges is also to be fully discovered.

1.2 Opportunities and Challenges

Due to regulations, competition, and customer demands, automotive control applications are driven by the need for robustness, high performance, and cost reduction all at the same time. The investigation of MPC for several automotive control problems has been mainly pursued due to MPC features that are helpful and effective in addressing such requirements and in achieving optimized operation. The key strengths of MPC are summarized in Table 1 and discussed next.

Strengths	Challenges
Simple multivariable design	High computational load
Constraint enforcement	Process models sometimes unavailable
Inherent robustness	Nonlinearities during transients
Performance optimization	Dependence on state estimate quality
Handling of time delays	Non-conventional design and tuning process
Exploiting preview information	

Table 1: Strengths and challenges for MPC in automotive applications.

A solid starting point for MPC development is that while the processes and dynamics taking place in the vehicle are interdependent and may be fairly complex, they are well studied and understood, and, for most, detailed models are available. This enables the application of model-based control methods, such as MPC.

Due to the aforementioned requirements, often driven by emissions, fuel consumption, and safety regulations, the number and complexity of actuators for influencing the vehicle operation is increasing. Some interesting examples are turbochargers, variable cam timing, electric motors, active steering, differential braking, regenerative braking. As more actuators become available, methods that can coordinate them to achieve multiple objectives, i.e., control multivariable, multiobjective systems, may achieve superior performance than control designs that are decoupled into several single-variable loops. MPC naturally handles multivariable systems without additional design complexity, thus simplifying the development of such multivariable controllers. This has been demonstrated, for instance, for spark-ignition (SI) engine speed control [26, 30, 46], vehicle-stability control by coordinated steering and braking [29, 33], and airpath control in turbocharged diesel engines [64, 75]. Furthermore, while it may still be difficult to obtain globally robust MPC designs, it is well known that often MPC provides inherent local robustness, as it can be designed to locally recover the LQR behavior, including its gain and phase margin guarantees.

Another advantage is that the tight requirements imposed by operating conditions, regulations, and interactions with other vehicle systems can often be easily formulated in terms of constraints on process variables. By enforcing constraints by design, rather than by time-consuming tuning of gains and cumbersome protection logics, MPC can reduce the development and calibration time by a significant amount [16, 29, 30, 38, 75].

The problem of ensuring high performance can often be approached through the optimization of an objective function. The ability to perform such an optimization is another key feature of MPC. In fact, this was at the root of the interest of several researchers in hybrid and electric vehicles [14, 28, 70, 81]. Even if it may be difficult to directly formulate the automotive performance measures as a cost function for MPC, it is usually possible to determine indirect objectives [26, 28] that, when optimized, imply quasi-optimal (or at least very desirable) behavior with respect to the actual performance measured.

Besides these macro-features, MPC has additional capabilities that are useful in controlling automotive processes. For instance, the capability of including time delay models, possibly of different length in different control channels, is very beneficial, as several engine processes are subject to transport delays and actuator delays. Also, new technologies and regulations in communication and connectivity, outside and inside the car, allow for obtaining preview information that MPC can exploit to achieve superior performance [32, 74]. This is even more relevant in the context of autonomous and connected vehicles [21], due to the available long-term information, for instance from mid to long range path planners, and from shared information among vehicles.

However, there are also several challenges to the large scale deployment of MPC in automotive applications [19], which are also summarized in Table 1 and discussed next.

First, MPC has larger computational load and memory footprint than classical control methods, while automotive micro-controllers are fairly limited in terms of computing power. Since the vehicle must operate in challenging environments, e.g., temperatures ranging from -40°C to $+50^{o}$C, the achievable processor and memory access frequencies are limited. The need to reduce the cost, and the shorter development and validation time often prevents introducing new processors sized for the need of a specific controller. Rather, the controller must fit in a given processor.

Second, not all the automotive processes have well-developed models. Combustion and battery charging/discharging are examples of processes that are still difficult to model precisely, and suitable models for them still remain an area under study. While some of the gaps can be closed using partially data-driven models, one has to be careful in applying MPC in this setting as such models may not generalize well.

Even for the processes that are better understood, the dynamics are intrinsically nonlinear. This third challenge is more relevant in automotive than in other fields, e.g., in aerospace, because, due to external effects, e.g., the driver, the traffic, the road, many automotive processes are continuously subject to fast transients during which the nonlinearities cannot be easily removed by linearization around a steady state.

A further complicating factor is that several variables in automotive processes are not measured, and the sensors for estimating them may be heavily quantized and noisy. Thus, a fourth challenge for MPC, which needs the state value for initializing the prediction model, is the need of state estimators, whose performance will significantly affect the overall performance of the control system. The estimator performance will depend on the sensors that in automotive applications are reduced in number and have limited capabilities, once again due to cost and harsh environment.

Fifth and final challenge, is the difference in the development process of MPC and classical controllers, e.g., PID. While the latter are mostly calibrated by gain tuning, MPC requires prediction model development and augmentation, definition of horizon and cost, and tuning of the weights of the cost function terms. As these are often not taught in basic control design courses, calibration engineers in charge of deploying and maintaining the controllers in the vehicle may find difficulties with the development of MPC. Hopefully, this entire handbook is a step towards solving this problem.

1.3 Chapter Overview

The rest of this chapter is structured based on the above discussion of strengths and challenges, with the aim of providing a guide for MPC development in automotive applications.

Due to the model-based nature of MPC, and the need for the MPC developer to acquire an understanding of the process models used for design, we first describe (Section 2) the key models to be used for MPC development in the areas of powertrain (Section 2.1), vehicle dynamics (Section 2.2), and energy management (Section 2.3). Our description of such models provides a starting point for the development of MPC solutions for these applications and enhances the understanding of the opportunities for using MPC in these applications. Then, we provide general guidelines for controller development (Section 3). Finally, we discuss the computational challenges and the key features of the algorithms used for MPC deployment in automotive applications (Section 4).

2 MPC for Powertrain Control, Vehicle Dynamics, and Energy Management

In this section we consider key automotive control areas in which the application of MPC has been considered and can have an impact. For each area we first describe the key models for model-based control development, and then, in light of these models, we briefly highlight what impact MPC may have.

2.1 Powertrain Control

Powertrain dynamics involve the generation of engine torque and transfer of such torque to the wheels to generate traction forces.

The engine model describes the effects of the operating conditions and engine actuators on the pressures, flows and temperatures in different parts of the engine, and on the torque that the engine produces. The engine actuators range from the standard throttle, fuel injectors, and spark timing to more advanced ones, such as variable ge-

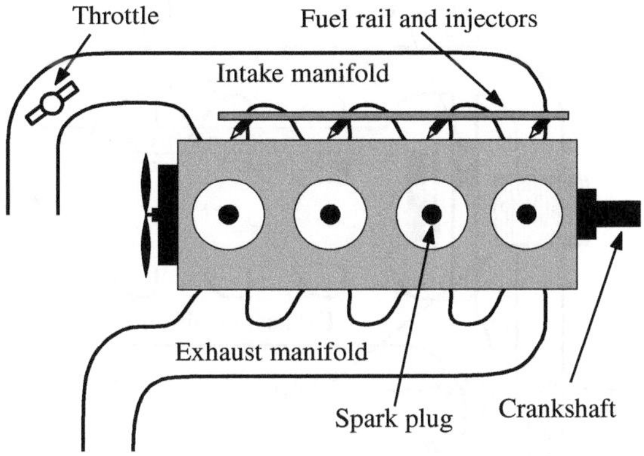

Fig. 1: Schematics of a naturally aspirated spark ignition engine, with focus on the air path.

ometry turbines (VGT), exhaust gas recirculation (EGR) valves, and variable cam timing (VCT) phasers, among others.

The engine model itself is, in general, composed of two parts, the airpath model, which describes the flow and mixing of the different gases in the engine, and the torque production model, which describes the torque generated from the combustion of the gas mixture.

For naturally aspirated spark ignition (SI), i.e., conventional gasoline, engines (see the schematic in Figure 1) the airpath model is relatively simple and represents the cycle averaged dynamics of the pressure in the intake manifold, under an isothermal assumption, and the flow from the throttle to the intake manifold and from the intake manifold into the engine cylinders,

$$\dot{p}_{\rm im} = \frac{RT_{\rm im}}{V_{\rm im}}(W_{\rm th} - W_{\rm cyl}), \tag{1a}$$

$$W_{\rm cyl} = \eta_{\rm vol}\frac{V_d p_{\rm im}}{RT_{\rm im}}\frac{N}{120} \approx \frac{\gamma_2}{\gamma_1} p_{\rm im} N + \gamma_0, \tag{1b}$$

$$W_{\rm th} = \frac{A_{\rm th}(\vartheta)}{\sqrt{RT_{\rm amb}}} p_{\rm amb}\phi\left(\frac{p_{\rm im}}{p_{\rm amb}}\right), \tag{1c}$$

where W, p, T, V denote mass flow, pressure, temperature, and volume, respectively, ϕ is a nonlinear function which represents the throttle flow dependence on the pressure ratio across the throttle [44, App.C], the subscripts *im*, *th*, *amb*, *cyl* refer to the intake manifold, the throttle, the ambient, and the cylinders, respectively, N is the engine speed, usually in revolutions per minute (RPM), V_d is the engine displacement volume, $\eta_{\rm vol}$ is the volumetric efficiency, R is the gas constant, $A_{\rm th}$ is the throttle effective flow area, which is a function of throttle angle, ϑ, and γ_i, $i \in \mathbb{Z}_{0+}$ denote engine-dependent constants, that are obtained from engine calibration data.

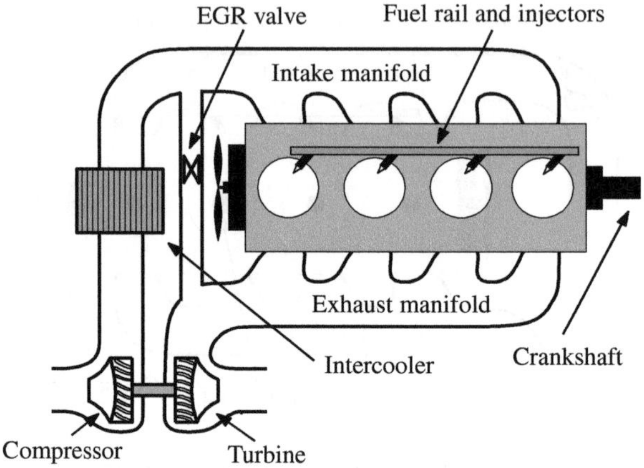

Fig. 2: Schematics of a turbocharged compression ignition engine, with focus on the air path. In comparison with the SI engine in Figure 1, note the absence of throttle and spark plugs, and the interconnected dynamics of exhaust and intake manifold, through EGR valve and turbine-compressor.

For modern compression ignition (CI), i.e., diesel, engines (see the schematic in Figure 2), the airpath model is substantially more complex, especially because these engines are usually turbocharged and exploit EGR, which renders the isothermal assumption inaccurate. Furthermore, the EGR valve and the turbocharger effectively couple the intake manifold with the exhaust manifold, which then must be included in the model. As a result, the diesel engine models include pressures, densities (ρ), and burned gas fraction (F) in both the intake (*im*) and exhaust (*em*) manifolds,

$$\dot{p}_{\text{im}} = \frac{c_p R}{c_v V_{\text{im}}}(W_{\text{com}} T_{\text{com}} - W_{\text{cyl}} T_{\text{im}} + W_{\text{egr}} T_{\text{em}}), \tag{2a}$$

$$\dot{\rho}_{\text{im}} = \frac{1}{V_{\text{im}}}(W_{\text{com}} - W_{\text{cyl}} + W_{\text{egr}}), \tag{2b}$$

$$\dot{F}_{\text{im}} = \frac{(F_{\text{em}} - F_{\text{im}})W_{\text{egr}} - F_{\text{im}} W_{\text{com}}}{\rho_{\text{im}} V_{\text{im}}}, \tag{2c}$$

$$\dot{p}_{\text{em}} = \frac{c_p R}{c_v V_{\text{em}}}(W_{\text{cyl}} T_{\text{cyl}} - W_{\text{tur}} T_{\text{em}} - W_{\text{egr}} T_{\text{em}} - \dot{Q}_{\text{em}}/c_p), \tag{2d}$$

$$\dot{\rho}_{\text{em}} = \frac{1}{V_{\text{em}}}(W_{\text{cyl}} - W_{\text{tur}} - W_{\text{egr}}), \tag{2e}$$

$$\dot{F}_{\text{em}} = \frac{(F_{\text{em}} - F_{\text{im}})W_{\text{egr}}}{\rho_{\text{em}} V_{\text{em}}}, \tag{2f}$$

where c_p, c_v are the gas specific heats at constant pressure and constant temperature, respectively, $\dot{Q}$ is the heat flow, and the subscripts *egr*, *com*, *tur* refer, respectively, to the exhaust gas being recirculated, the compressor, and the turbine.

Equations in (2a) must be coupled with the equations describing the flows. While the cylinder flow equation is the same as in (1b) for the SI engine model, and the EGR flow is controlled by a valve resulting in an equation similar to (1c), the remaining flows are determined by the turbocharger equations,

$$W_{com} = \frac{p_{\text{amb}}}{\sqrt{T_{\text{amb}}}}\phi_{\text{com}}(N_{tc}/\sqrt{T_{\text{amb}}}, p_{\text{im}}/p_{\text{amb}}), \tag{3a}$$

$$W_{tur} = \frac{p_{\text{em}}}{\sqrt{T_{\text{em}}}}\phi_{\text{tur}}(\chi_{\text{vgt}}, p_{\text{ep}}/p_{\text{em}}), \tag{3b}$$

$$\dot{N}_{tc} = \frac{\gamma_3}{J_{\text{tc}}}\frac{\eta_{\text{tur}}W_{\text{tur}}(T_{\text{em}} - T_{\text{ep}}) - \eta_{\text{com}}W_{\text{com}}(T_{\text{im}} - T_{\text{amb}})}{N_{\text{tc}}}, \tag{3c}$$

where ep refers to the exhaust pipe, χ_{vgt} is the variable geometry turbine actuator position, N_{tc}, and J_{tc} are the speed and inertia of the turbocharger, ϕ_{com} and ϕ_{tur}, η_{com}, and η_{tur}, are the flow parameters and efficiencies of turbine and compressor.

It is worth noting that in recent years downsized gasoline engines that are turbocharged have become more common. Their airpath model is a hybrid between the SI and CI models, since they have SI combustion and throttle, but also a turbocharger, although, in general, with a smaller fixed geometry turbine, and possibly a wastegate and variable valve timing instead of the EGR valve [72].

The second part of the engine model is the torque production model, which describes the net torque output generated by the engine. This model has the form,

$$M_{\text{e}} = M_{\text{ind}}(t - t_{\text{d}}) - M_{\text{fr}}(N) - M_{\text{pmp}}(p_{\text{im}}, p_{\text{em}}, N), \tag{4}$$

where M_{ind}, M_{fr}, M_{pmp} are the indicated, friction, and pumping torques, respectively. The indicated torque is the produced torque and its expression depends on the engine type. For SI engines,

$$M_{\text{ind}} \approx \kappa_{\text{spk}}(t - t_{\text{ds}})\gamma_4\frac{W_{\text{cyl}}}{N}, \tag{5a}$$

$$\kappa_{\text{spk}} \approx \left(\cos(\alpha - \alpha_{\text{MBT}})\right)^{\gamma_5}, \tag{5b}$$

where α and α_{MBT} are the ignition angle and the maximum brake torque ignition angle, and κ_{spk} is the torque ratio achieved by spark ignition timing. Since CI engines are not equipped with spark plugs, and the air-to-fuel ratio in these engines may vary over a broad range, the indicated torque equation is usually obtained from engine calibration data, e.g., as

$$M_{\text{ind}} = f_{\text{indCI}}(W_f, N, F_{\text{im}}, \delta), \tag{6}$$

where W_f is the fuel flow, and δ represents to the fuel injection parameters (e.g., start of injection).

The final component in the engine models represents the transfer of the torque from the engine to the wheels. In general, the engine speed (in RPM) is related to the engine torque M_e, inertia of the crankshaft and flywheel J_e, and load torque M_L

by

$$\dot{N} = \frac{1}{J_e}\frac{30}{\pi}(M_e - M_L). \tag{7}$$

The load torque model varies largely depending on whether the vehicle has an automatic transmission, which includes a torque converter, or a manual transmission with dry clutches. Depending on the compliance of the shafts and the actuation of the clutches, the steady state component of the torque load is

$$M_L = \frac{r_w}{g_r} F_{\text{trac}} + M_{\text{los}} + M_{\text{aux}},$$

where M_{los}, M_{aux} are the torque losses in the driveline and because of the auxiliary loads, r_w is the wheel radius and g_r is the total gear ratio between wheels and engine shaft, usually composed of final drive ratio, transmission gear ratio, and, if present, torque converter ratio.

2.1.1 MPC Opportunities in Powertrain Control

Powertrain control has likely been the first, and probably the largest, application area of MPC in automotive systems. In conventional SI engines, when the driver is pressing on the gas pedal, the vehicle is in torque control mode and there are basically no degrees of freedom available for control. Thus, the main opportunities for MPC application are in the so-called closed-pedal operation, i.e., when the gas pedal is released, and the vehicle is in a speed control mode.

An example is idle speed control [26] where the spark timing and the throttle are actuated to keep a target speed while rejecting external disturbances. The engine speed must be kept from becoming too small, otherwise the engine may stall, and the throttle and spark timing are subject to physical and operational constraints, for instance due to knocking or misfiring. Thus, the optimal control problem that defines the MPC law can be formulated as

$$\min_{\alpha,\vartheta} \quad \sum_{t=0}^{T_N} (N(t) - r_N(t))^2 + w_\vartheta \Delta\vartheta(t)^2 + w_\alpha(\alpha(t) - \alpha_r(t))^2 \tag{8a}$$

$$\text{s.t.} \quad \underline{\alpha}(t) \le \alpha(t) \le \overline{\alpha}(t), \quad \underline{\vartheta}(t) \le \vartheta(t) \le \overline{\vartheta}(t), \quad N(t) \ge \underline{N}(t) \tag{8b}$$

where w_ϑ, w_α are positive tuning weights, and r_N, α_r are references that are constant or slowly varying based on engine temperature.

During deceleration control, the engine speed is controlled to follow a reference trajectory that causes the vehicle to decelerate smoothly and energy-efficiently, and still allows for the engine to rapidly resume torque production, if acceleration is needed, see Figure 3. In this case the problem is similar to (8), except that the reference speed trajectory is time varying and a first order model for it is often available and may be used for preview.

Both idle speed control and deceleration control are multivariable control problems in which actuators are subject to constraints and the dynamics are affected by delays of different lengths in different control channels. Based on guidelines in Table 1, both idle speed control and deceleration control are clearly good application areas for MPC. On the other hand, the dynamics are clearly nonlinear in both of these problems. Since idling takes place near a setpoint, a linearized model for idling is fairly accurate. For deceleration control the system operates in a transient all the time, and hence it is often convenient to develop a low-level controller that linearizes the dynamics. In such a control architecture, MPC can exploit constraints to ensure that the interaction with the low level controller is effective. For deceleration control, a low level controller is tasked with delivering the demanded torque, thus transforming the pressure-based model into a torque-based model, where the torque response is modeled as a first order system subject to delay

$$\dot{N}(t) = \frac{1}{J_e}(\hat{\kappa}_{\text{spk}} M_{\text{air}}(t) + u_{\text{spk}}(t - t_{ds}) - M_L(t)), \tag{9a}$$

$$\dot{M}_{\text{air}}(t) = \frac{1}{\tau_{\text{air}}}(-M_{\text{air}}(t) + u_{\text{air}}(t - t_d(t)), \tag{9b}$$

$$\underline{M}_{\text{air}}(t) \leq M_{\text{air}}(t) \leq \overline{M}_{\text{air}}(t), \tag{9c}$$

$$\underline{\Delta\kappa} M_{\text{air}}(t) \leq u_{\text{spk}}(t - t_{ds}) \leq \overline{\Delta\kappa}\, M_{\text{air}}(t). \tag{9d}$$

The multiplicative relation between spark timing and torque is converted into an additive one subject to linear constraints by introducing a virtual control representing the torque modification obtained from spark actuation. This is possible with MPC due to the capability of handling constraints.

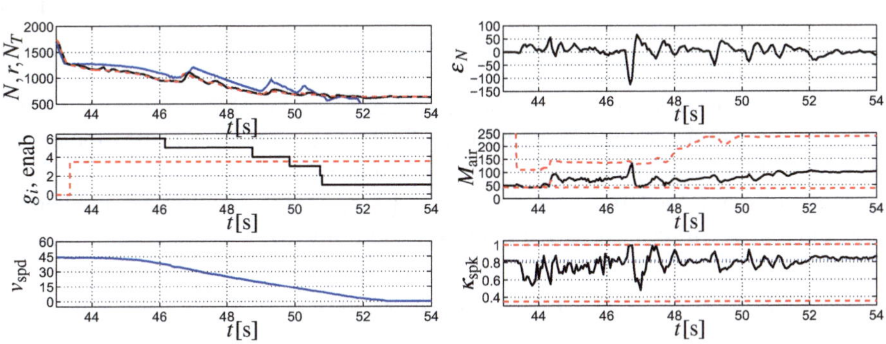

Fig. 3: Experimental test of MPC-based deceleration control from [30]. Engine speed N, reference r, and tracking error ε_N, torque converter turbine speed N_T, gear g_i, and controller enabling signal *enab*, vehicle speed v_{spd}, torque from airflow M_{air} and torque ratio from spark κ_{spk} are shown.

CI engines are far more complex and have more degrees of freedom than naturally aspirated gasoline engines, due to EGR, VGT, and multiple fuel injections that must be exploited throughout the entire operating range to achieve a suitable tradeoff between torque delivery and emissions. In general, in diesel engines the fuel flow W_f is determined based on the pedal position and current engine speed, and from that, the setpoints for other variables such as the intake manifold pressure and either mass airflow through the compressor or EGR rate are determined. Then, a feedback controller is developed that actuates the VGT, EGR valve, and possibly intake throttle, to track these setpoints. Also in this case we obtain a multivariable control problem with constraints on actuators and process variables, such as intake and exhaust manifold pressures, EGR rate, turbocharger speed, turbine temperature, compressor surge margin, etc. The MPC solution can be simplified [52] by pursuing a rate-based formulation, constraint remodeling, intermittent constraint enforcement, and by combining it with nonlinear static or dynamic inversion.

Recent publications [10, 11] have reported that MPC has been scheduled for production by General Motors in collaboration with Odys, with specific solutions disclosed for turbocharged SI engine control and control of CVT gear ratio.

2.2 Control of Vehicle Dynamics

Vehicle dynamics models are derived from the planar rigid body equations of motion. For normal driving that involves neither high performance driving nor low speed maneuvers, the single track model, also known as a bicycle model, shown in Figure 4, is common. This model is described by

$$m(\dot{v}_x - v_y\dot{\psi}) = F_{xf} + F_{xr}, \tag{10a}$$
$$m(\dot{v}_y + v_x\dot{\psi}) = F_{yf} + F_{yr}, \tag{10b}$$
$$J_z\ddot{\psi} = \ell_f F_{yf} - \ell_j F_{yr}, \tag{10c}$$

where m is the vehicle mass, ψ is the yaw rate, v_x, v_y are the components of the velocity vector in the longitudinal and lateral vehicle direction, J_z is the moment of inertia about the vertical axis, ℓ_f, ℓ_r are the distances of front and rear axles from the center of mass. In (10), F_{ij}, $i \in \{x,y\}$, $j \in \{f,r\}$ are the longitudinal and lateral, front and rear tire forces expressed in the vehicle frame [67],

$$F_{xj} = f_l(\alpha_j, \delta_j, \sigma_j, \mu, F_{zj}), \quad F_{yj} = f_c(\alpha_j, \delta_j, \sigma_j, \mu, F_{zj}), \quad F_{zj} = \frac{\ell_j}{\ell_f + \ell_r} mg, \tag{11}$$

where δ_j is the steering angle at the tires, α_j is the tire slip angle, and σ_j is the slip ratio, for front and rear tires $j \in \{f,r\}$, and μ is the friction coefficient between tires and road. The slip angles and the slip ratios relate the vehicle tractive forces with the vehicle velocity and the wheel speeds, thereby coupling the vehicle response with the powertrain response,

$$\alpha_j = \tan^{-1}\left(\frac{v_{yj}}{v_{xj}}\right), \tag{12a}$$

$$v_{lj} = v_{y_j} \sin\delta_j + v_{xj}\cos\delta_j, \quad v_{cj} = v_{y_j}\cos\delta_j - v_{xj}\sin\delta_j, \tag{12b}$$

$$\sigma_j = \begin{cases} \frac{r\omega_j}{v_{xj}} - 1 \text{ if } v_{xj} > r\omega_j, \\ 1 - \frac{r\omega_j}{v_{xj}} \text{ if } v_{xj} < r\omega_j, \end{cases} \tag{12c}$$

where v_{xj}, v_{y_j}, $j \in \{f, r\}$, are the longitudinal and lateral components of the vehicle velocity vector at the tires. In, (11), the functions f_l, f_c define the tire forces that

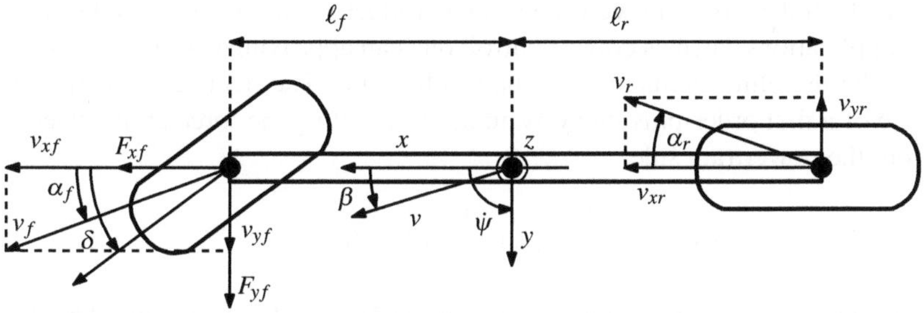

Fig. 4: Schematics of the single track model for the lateral vehicle dynamics. The most relevant vectors for describing the model are also shown.

are in general determined by data or according to a model such as Pacejka's or Lu'Gre [54].

In the above-mentioned normal driving conditions, the longitudinal and lateral dynamics are often decoupled, yielding a lateral dynamics model where v_x is constant, and, with a further linear approximation of the lateral tire forces as a function of the slip angles, resulting in

$$m\dot{v}_y = -\frac{C_f + C_r}{v_x} v_y - \left(v_x + \frac{C_f\ell_f - C_r\ell_r}{v_x}\right)\dot\psi + C_f\delta, \tag{13a}$$

$$J_z\ddot\psi = -\frac{C_f\ell_f - C_r\ell_r}{v_x} v_y - \frac{C_f\ell_f^2 + C_r\ell_r^2}{v_x}\dot\psi + \ell_f C_f\delta + M_{br}, \tag{13b}$$

where we used the relation $\alpha_f = (v_y + \ell_f\dot\psi)/v_x$, $\alpha_r = (v_y - \ell_r\dot\psi)/v_x$, and we have included a moment M_{br} that can be generated by applying non-uniform forces at different wheels, for instance by differential braking. In (13), C_f, C_r are the front and rear lateral tire stiffnesses for the single model, i.e., twice the tire stiffness of the double track model, which correspond to a linear approximation of the lateral tire forces as functions of the slip angles, $F_{yj} = C_j\alpha_j$.

Similarly, the longitudinal dynamics are also simplified by neglecting the lateral dynamics, resulting in

$$m\dot{v}_x = \sum_{j\in f,r} f_l(0, 0, \sigma_j, \mu, F_{zj}) - F_{\text{res}} \approx C_f^x\sigma_f + C_r^x\sigma_r - F_{\text{res}}, \tag{14a}$$

$$F_{\text{res}} = F_{\text{aero}} + F_{\text{roll}} + F_{\text{grade}} \approx \frac{1}{2}\rho_{\text{air}} A_f c_d v_x^2 + mgc_r\cos\theta_{\text{rd}} + mg\sin\theta_{\text{rd}}, \tag{14b}$$

where C^x_f, C^x_r are the front and rear longitudinal tire stiffnesses, that represent a linear approximation of the longitudinal tire forces as functions of the slip ratio $F_{xj} = C^x_j \sigma_j$. The slip ratio changes based on the torques exerted on the wheels by the engine and the brakes, thus relating the powertrain and braking system actuation with the vehicle motion. In (14) we have included the effects of resistance forces due to airdrag, rolling, and road grade. Here, ρ_{air} is the density of air, A_f is the vehicle frontal area, C_d is the drag coefficient, θ_{rd} is the road grade, c_r is the rolling resistance coefficient, and g is the gravity acceleration. The longitudinal vehicle dynamics can be linked to the powertrain torque production in several ways. For low bandwidth applications, such as cruise control, one can approximate $C^x_f \sigma_f + C^x_r \sigma_r \approx F_{\text{trac}}$, where the driveline shafts are assumed to be rigid. The tractive force F_{trac} is the response of a first order-plus-delay system, representing the force at the wheels applied from the powertrain side,

$$\dot{F}_{\text{trac}} = -\frac{1}{\tau_F} F_{\text{trac}} + \frac{1}{\tau_F} u_F(t - t_F).$$

If shaft compliance is considered, the tractive torque $M_{trac} = F_{\text{trac}} r_w$ is caused by the slip between the wheel half-shafts and the rigid transmission shaft, so that

$$M_{trac} = k_s(\theta_e - \theta_w g_r) + d_s(\dot{\theta}_e - \dot{\theta}_w g_r), \tag{15}$$

where k_s and d_s are the half-shafts stiffness and damping, θ_e, θ_w are the engine and wheel shaft angles, and g_r is the total gear ratio between engine and wheels.

The active control of the vertical vehicle dynamics is mainly obtained by active and semi-active suspensions. The simplest model is the quarter-car model [47], where each suspension is independent from the others. The standard quarter-car model describes the vertical vehicle dynamics as two masses, the unsprung mass M_{us} representing the car wheel, with stiffness k_{us} and damping d_{us}, and the sprung mass M_s, representing one quarter of the car body, connected by a spring-damper with stiffness and damping k_s, d_s, the passive component of the suspension, and with a force F_a acting between them. Such a force is generated by the active or semi-active suspension actuator.

The equations of motions for the sprung and unsprung mass are

$$M_s \ddot{x}_s = c_s(\dot{x}_{us} - \dot{x}_s) + k_s(x_{us} - x_s) - F_a, \tag{16a}$$
$$M_{us} \ddot{x}_{us} = c_t(\dot{r} - \dot{x}_{us}) + k_s(r - x_{us}) + c_s(\dot{x}_s - \dot{x}_{us}) + k_s(x_s - x_{us}) + F_a, \tag{16b}$$

where x_s is the position of the sprung mass, x_{us} is the position of the unsprung mass, F_a is the actuator force, and r is the local road height with respect to the average.

The objective of the suspension control is to limit tire deflections, hence ensuring that the vehicle maintains good handling, to limit suspension deflections, hence ensuring that the suspension does not run against its hard stops causing noise, vibrations and harshness (NVH) and wear, and to limit sprung mass accelerations, hence resulting in a comfortable ride. The type of actuator, e.g., hydraulic, electromagnetic, etc., and its overall capabilities, e.g., active or semi-active, may require additional models for the actuator dynamics, and possibly constraints limiting its action, such as force ranges or passivity constraints.

2.2.1 MPC Opportunities in Vehicle Dynamics

MPC of longitudinal vehicle dynamics has been applied to adaptive cruise control (ACC), see, e.g., [61]. The objective of adaptive cruise control is to track a vehicle reference speed r while ensuring a separation distance d from the preceding vehicle, related to the head-away time T_h, and comfortable ride, all of which can be formulated as

$$\begin{aligned} \min_{F_{\text{trac}}} \quad & \sum_{t=0}^{T_N} (v_x(t) - r_v(t))^2 + w_F \Delta u_F(t)^2 && \text{(17a)} \\ \text{s.t.} \quad & \underline{F}_{\text{trac}} \leq F_{\text{trac}}(t) \leq \overline{F}_{\text{trac}}, && \text{(17b)} \\ & d(t) \geq T_h v_x(t), && \text{(17c)} \end{aligned}$$

where w_F is a positive tuning weight. For ACC, interesting opportunities are opened when a stochastic description of the velocity of the traffic ahead is available or can be estimated [13], or, in the context of V2V and V2I, when there is perfect preview through communication [74]. Also, using an economic cost can help reduce fuel consumption, with minimal impact on travel time [65]. Additional potential applications in longitudinal vehicle dynamics still to be investigated in depth are launch control and gear shifting. More recent applications involve braking control for collision avoidance systems, see, e.g., [59] possibly by using again V2X to exploit preview information.

The interest in MPC for lateral dynamics spans multiple applications, especially lateral stability control and lane keeping, up to autonomous driving. A challenging case [29] is the coordination of differential braking moment and steering to enforce cornering, i.e., yaw rate reference r_ψ tracking, and vehicle stability, i.e., avoiding that the slip angles become so large that the vehicle spins out of control. Such a problem is challenging due to its constrained multivariable nature and to the need to consider nonlinear tire models. A viable approach has been to consider piecewise linear tire models, resulting in the optimal control problem

$$\begin{aligned} \min_{\Delta\delta, M_{\text{br}}} \quad & \sum_{t=0}^{T_N} (\dot{\psi}(t) - r_\psi(t))^2 + w_\delta \Delta\delta(t)^2 + w_{br} M_{\text{br}}(t)^2 && \text{(18a)} \\ \text{s.t.} \quad & |M_{\text{br}}(t)| \leq \overline{M}_{\text{br}}, \quad |\delta(t) - \delta(t-1)| \leq \overline{\Delta\delta}, \quad |\delta(t)| \leq \overline{\delta}, && \text{(18b)} \\ & f_{c\,j}(\alpha_j) = \begin{cases} -d_j\alpha_j + e_j & \text{if} \quad \alpha_j > p_j, \\ C_j\alpha_j & \text{if} \quad |\alpha_j| \leq p_j, \\ d_j\alpha_j - e_j & \text{if} \quad \alpha_j < -p_j, \end{cases} && \text{(18c)} \\ & |\alpha_j| \leq \overline{\alpha}_j, && \text{(18d)} \end{aligned}$$

where w_δ, w_{br} are positive tuning weights, and then using either a hybrid MPC or a switched MPC approach, where the current linear model is applied for prediction during the entire horizon.

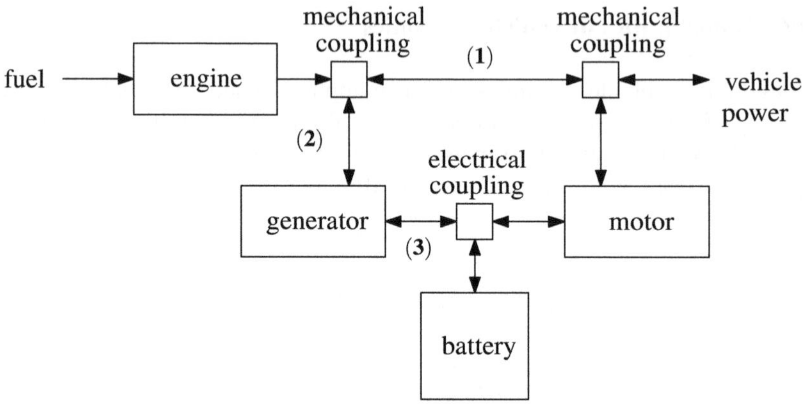

Fig. 5: Schematic of a powersplit HEV architecture. The arrows indicate the allowed power flow directions. The series HEV architecture is obtained by removing the link (1), thus the mechanical couplings are simply mechanical connections. The parallel HEV architecture is obtained by removing the generator and hence links (2) and (3).

As for the vertical dynamics, MPC offers interesting possibilities for active suspension control when preview of the road is available [32, 60], for instance obtained from a forward looking camera. MPC may also be beneficial in semi-active suspension control, since the passivity condition

$$F_a(\dot{x}_s - \dot{x}_{us}) \geq 0, \tag{19}$$

can be enforced in MPC as a constraint, which ensures that only commands that are realizable by a semi-active actuator are actually issued. Constraint (19) is nonlinear, but can be enforced by mixed-logical constraints

$$[\delta_v = 1] \leftrightarrow [\dot{x}_s - \dot{x}_{us} \geq 0], \tag{20a}$$
$$[\delta_F = 1] \leftrightarrow [F_a \geq 0], \tag{20b}$$
$$[\delta_v = 1] \leftrightarrow [\delta_F = 1], \tag{20c}$$

where δ_v, δ_F are auxiliary binary variables, thus resulting in a hybrid MPC [38].

Finally, for more advanced systems that aim at coordinating all the suspensions in the vehicle [41], the multivariable nature of MPC can be very effective.

2.3 Energy Management in Hybrid Vehicles

The novel element in hybrid powertrains is the presence of multiple power generation devices, e.g., engine, motor, generator, and energy storage devices, e.g., fuel tank, battery, flywheel. The most common hybrid vehicles are hybrid electric vehicles (HEV) where the internal combustion engine is augmented with electric motors and generators, and batteries for energy storage. For HEV there are a several component topologies that determine the configurations of the power coupling, the most common being series, parallel, powersplit, and electric rear axle drive (ERAD).

The presence of multiple power generation devices requires modeling the power balance. A general model for the mechanical power balance that ultimately describes the amount of power delivered to the wheels is

$$P_{\text{veh}} = P_{\text{eng}} - P_{\text{gen}} + P_{\text{mot}} - P_{\text{mec}}^{\text{los}}, \tag{21}$$

where P_{veh} is the vehicle power for traction, P_{eng} is the engine power, P_{gen} is the mechanical power used to generate electrical energy to be stored in the battery, P_{mot} is the electrical power used for traction, and $P_{\text{mec}}^{\text{los}}$ accounts for the mechanical power losses. In advanced HEV architectures, such as the powersplit architecture in Figure 5, motoring and electric energy generation can be accomplished by multiple components, since, despite the names that indicate their preferred usage, both the motor and the generator can convert mechanical energy into electrical energy, and the other way around.

The electrical power balance, that is used to determine the power delivered to/from the battery, is often modeled as

$$P_{\text{bat}} = P_{\text{mot}} - P_{\text{gen}} + P_{\text{gen}}^{\text{los}} + P_{\text{mot}}^{\text{los}}, \tag{22}$$

where P_{bat} is the power flowing from the battery and $P_{\text{gen}}^{\text{los}}$, $P_{\text{mot}}^{\text{los}}$ are the losses in electric energy generation and in the electric motoring, respectively.

As opposed to the fuel tank, the battery power flow is bi-directional and the stored energy is usually quite limited, thus the energy stored in the battery should be tracked and it is, in fact, the main state of the HEV energy model. The energy stored in the battery is related to the stored charge, which is normalized with respect to the maximum to obtain the battery state of charge (SoC) $SoC = \frac{Q_{\text{bat}}}{Q_{\text{max}}}$. The battery power, voltage, and current are related by

$$P_{\text{bat}} = (V_{\text{bat}}^{\text{oc}} - I_{\text{bat}} R_{\text{bat}}) I_{\text{bat}},$$

where $V_{\text{bat}}^{\text{oc}}$ is the open circuit battery voltage, I_{bat} is the battery current, and R_{bat} is the battery internal resistance. This results in the state of charge dynamics

$$S\dot{o}C = -\frac{V_{\text{bat}}^{\text{oc}} - \sqrt{V_{\text{bat}}^{\text{oc}\,2} - 4R_{\text{bat}}P_{\text{bat}}}}{2R_{\text{bat}}Q_{\text{max}}}.$$

Considering a power coupling that is kept under voltage control and representing the battery as a large capacitor, i.e., ignoring internal resistance, we obtain a simpler representation

$$S\dot{o}C = -\eta_{\text{bat}}(P_{\text{bat}}, SoC)\frac{P_{\text{bat}}}{V_{\text{bat}}^{\text{cc}} Q_{\text{max}}},$$

where η_{bat} is the battery efficiency, that, for feedback control, can be well approximated by one or two constants [28], the latter case modeling different efficiencies in the charging and discharging process.

The main control problem specific to HEV powertrains is the management of the energy in order to minimize the consumed fuel subject to charge sustaining constraints

$$\min \int_{t_i}^{t_f} W_f(t)dt \tag{23a}$$

$$\text{s.t. } SoC(t_i) = SoC(t_f), \tag{23b}$$

where the fuel flow W_f is related to the engine power by a function that depends on the engine operating point, $W_f = f_f(P_{\text{eng}}, N)$. As opposed to conventional powertrains, in most HEV configurations, even for a given engine power and wheel speed, there are degrees of freedom in selecting the engine operating point, i.e., engine speed and engine torque, that can be leveraged by the energy management strategy.

2.3.1 MPC Opportunities in Hybrid Vehicles

Due to the focus on optimizing the energy consumption subject to constraints on power flows and battery state of charge, HEV energy management has been an extensively studied target for MPC application.

The key idea is to construct a finite horizon approximation of the fuel consumption cost function (23a), augmented with a term penalizing large differences of SoC at the end of the horizon, which can be interpreted as the augmented Lagrangian form of the charge sustaining constraint (23b). The cost function can also include additional terms such as SoC reference tracking. Furthermore, constraints on various power flows and battery SoC can also be included, resulting in the following optimal control problem,

$$\min_{P_{\text{bat}},\dots} F_N(SoC(T_N)) + \sum_{t=0}^{T_N-1} W_f(t) + w_{soc}(SoC(t) - r_{SoC}(t))^2 \tag{24a}$$

$$\text{s.t. } P_{\text{eng}}(t) - P_{\text{gen}}(t) + P_{\text{mot}}(t) - P_{\text{mec}}^{\text{los}}(t) = P_{\text{drv}}(t), \tag{24b}$$

$$\underline{SoC} \leq SoC(t) \leq \overline{SoC}, \tag{24c}$$

$$|P_{\text{bat}}(t)| \leq \overline{P}_{\text{bat}}. \tag{24d}$$

The mechanical power equation (21) is enforced as a constraint in (24) to ensure that the vehicle power is equal to the driver-requested power P_{drv}. The actual number of degrees of freedom in (24) varies with the HEV architecture. For a powersplit architecture, shown in Figure 5, in its entirety, it is equal to two. For a parallel architecture, where there is no generator, and for a series architecture, where there is no pure mechanical connection between engine and wheels, it is equal to one.

Exploiting the simplicity of the latter, an MPC was developed in [28] which was deployed on a prototype production series HEV that was fully road drivable.

It is interesting to note that the cost function in HEV energy management is of economic type and, in retrospect, HEV energy management was probably the first real-world application of economic MPC, showing in fact the possibility of steady state limit cycles or offsets [28]. In recent years, multiple advanced MPC methods have been applied to HEV energy management, including stochastic MPC in [71] where the driver-requested power is predicted using statistical models, possibly learned from data during vehicle operation.

2.4 Other Applications

Given that the field of automotive control is large, it is impossible to provide a comprehensive account for all automotive applications of MPC in a single chapter. In this chapter we focused on the three areas described above, which have been very actively researched over the last few years. However, there are several other applications that could be noted, including, among others, emission control in SI, e.g., [73, 76], and CI engines, e.g., [49, 50] engines, transmission control, e.g., [2, 8, 42, 78], control of gasoline turbocharged engines, e.g., [1, 72], control of homogeneous combustion compression ignition (HCCI) engines, e.g., [12, 68], and energy management of fuel-cell vehicles, e.g., [3, 4, 77].

3 MPC Design Process in Automotive Applications

This section aims at providing guidelines for a design process for MPC in automotive applications. While not a standard, it has been applied by the authors in multiple designs that were eventually tested in vehicles, and it has been proved useful and effective in those applications. We focus on linear MPC, because, as discussed later, this has been so far the main method used in automotive applications, primarily due to computational and implementation requirements. However, the design process extends almost directly to nonlinear MPC.

The MPC design amounts to properly constructing components of the finite horizon optimal control problem to achieve the desired specifications and control-oriented properties. We consider the finite horizon optimal control problem

$$\min_{U(t)} x'_{t+N|t}Px_{t+N|t} + \sum_{k=0}^{N-1} z'_{t+k|t}Qz_{t+k|t} + u'_{t+k|t}Ru_{t+k|t} \tag{25a}$$

$$x_{t+k+1|t} = Ax_{t+k|t} + Bu_{t+k|t}, \tag{25b}$$

$$y_{t+k|t} = Cx_{t+k|t} + Du_{t+k|t}, \tag{25c}$$

$$z_{t+k|t} = Ex_{t+k|t}, \tag{25d}$$

$$u_{t+k|t} = \kappa_f x_{t+k|t}, \quad k = N_u, \ldots, N-1, \tag{25e}$$
$$x_{t|t} = x(t), \tag{25f}$$
$$\underline{y} \leq y_{t+k|t} \leq \overline{y}, \quad k = N_i, \ldots N_{cy}, \tag{25g}$$
$$\underline{u} \leq u_{t+k|t} \leq \overline{u}, \quad k = 0, \ldots N_{cu} - 1, \tag{25h}$$
$$H_N x_{t+N|t} \leq K_N, \tag{25i}$$

where the notation $t+k|t$ denotes the k-step prediction from measurements at time t, $U(t) = \{u_{t+0|t} \ldots u_{t+N-1|t}\}$ in the control input sequence to be optimized, x, u, y, z are the prediction model state, input, constrained outputs, and performance output vectors, $\underline{u}, \overline{u}, \underline{y}, \overline{y}$ are lower and upper bounds on input and constrained output vectors, P, Q, R are weighting matrices, N, N_{cu}, N_{cy}, N_u are non-negative integers defining the horizons, κ_f is the terminal controller, and H_N, K_N describe the terminal set. Next, we discuss the role of each of these components in achieving the specifications that are common in automotive applications.

3.1 Prediction Model

Several dynamical processes occurring in automotive applications are well studied and have readily available physics-based models, some of which have been described in Sections 2.1–2.3. In MPC design it is desirable to start from such physics-based models. However, many of them may be of unnecessarily high order, may be nonlinear, and may have several parameters to be estimated for different vehicles. Hence, the first step in MPC design is usually to refine the physics based model by:

- simplifying the model to capture the relevant dynamics based on the specific application and controller requirements, e.g., the sampling period, by linearization, model order reduction, etc.
- estimating the unknown parameters by gray-box system identification methods, e.g., linear/nonlinear regression, step response analysis, etc.;
- time-discretizing the dynamics to obtain a discrete-time prediction model.

Even for the relatively simple case of idle speed control, in [26] due to computational requirements, the powertrain model (1), (4) is linearized around the nominal idle operating point. The model structure is known from physics, and it consists of two transfer functions, from throttle and spark to engine speed, each of second order and subject to delays, where the latter also has a stable zero. The model parameters are identified from the step responses, with the models for the delays removed during identification, to be added again later, see Figure 6.

The result of the first model construction step is a linear, discrete-time constrained model for the physical process,

$$x_m(t+1) = A_m x_m(t) + B_m u_m(t), \tag{26a}$$
$$y_m(t) = C_m x_m(t) + D_m u_m(t), \tag{26b}$$

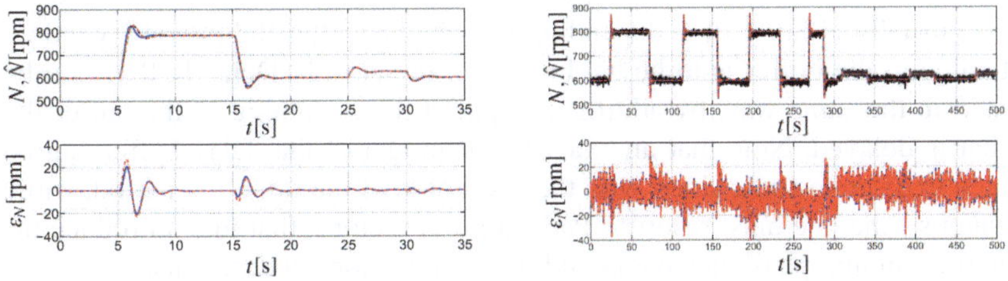

Fig. 6: Validation of identification of engine model for idle speed control from nonlinear model data (left) and experimental data (right) in throttle and spark up-down steps from [26]. Upper plot: data (solid), continuous-time linear model (dash), discrete-time linear model (dash-dot). Lower plot: continuous-time model error (solid) discrete-time model error (dash-dot).

$$z_m(t) = E_m x_m(t), \tag{26c}$$

where $x_m \in \mathbb{R}^{n_m}$ is the state vector, $u_m \in \mathbb{R}^{m_m}$ is the input vector, $y_m \in \mathbb{R}^{p_m}$ is the constrained output vector, and $z_m \in \mathbb{R}^{q_m}$ is the performance output vector.

The process model (26) usually needs to be augmented with additional states and artificial dynamics in order to achieve the problem specifications, such as tracking of references, non-zero, and possibly unknown, steady state input values, rejection of certain classes of disturbances, e.g., constant, sinusoidal, etc. Further modifications may be made to account for additional information available in the system, such as preview on disturbances or references, or known models for those. Typical augmentations are the incremental input formulation

$$u(t+1) = u(t) + \Delta u(t),$$

the inclusion of integral action to track constant references and reject constant unmeasured disturbances,

$$\iota(t+1) = \iota(t) + T_s C_\iota z(t),$$

and the inclusion of disturbance models

$$\begin{aligned} \eta(t+1) &= A_d \eta(t), \\ d(t) &= C_d \eta(t), \end{aligned}$$

where the disturbance model state η is measured in the case of measured disturbances, while in the case of unmeasured disturbance it is estimated by disturbance observers. A case of particular interest is the inclusion of buffers, which allow to account for preview on disturbances and references,

$$\begin{aligned} \xi(t+1) &= \begin{bmatrix} 0 & I \\ 0 & c \end{bmatrix} \xi(t), \\ \chi(t) &= \begin{bmatrix} 1 & 0 & \dots & 0 \end{bmatrix} \xi(t), \end{aligned}$$

where c is usually either 1 or 0 depending on whether, after the preview window, the last value in the buffer is to be held constant or set to 0. Delay buffers can be formulated in the same way, by adding as input at the beginning of the buffer the signal that is delayed. Note that an exact linear model of the delay buffer can be formulated in discrete-time, albeit with the resolution of the sampling period, while in continuous-time one must resort to Padé approximations, that may introduce fictitious non-minimum phase behaviors and mislead the control decisions.

Due to its intrinsic feedforward-plus-feedback nature, MPC is often applied for reference tracking. However, the application of MPC to these problems is not as simple as for linear controllers, because the constraints usually prevent from simply "shifting the origin" to translate the tracking problem into a regulation problem. In automotive applications, it is also difficult to compute the equilibrium associated with a certain reference value r, due to the uncertainties in the model and the unmeasured disturbances. If one wants to avoid adding a disturbance observer, it may be effective to apply the velocity (or rate-based) form of the model, where both the state and input are differentiated, and the tracking error e_m is included as an additional state

$$\Delta x_m(t+1) = A_m \Delta x_m(t) + B_m \Delta u_m(t), \tag{27a}$$
$$e_m(t) = e_m(t-1) + E_m \Delta x_m(t) + \Delta r(t), \tag{27b}$$
$$y_m(t) = y_m(t-1) + C_m \Delta x_m(t) + D_m \Delta u_m(t), \tag{27c}$$

where Δr is the change in reference signal. For MPC applications one need to add integrators (27c) to reformulate y_m in terms of the state and input changes, Δx_m, Δu_m, except for the cases where the constraints are originally in differential form.

The more common augmentations in relations to specifications commonly found in automotive control applications are summarized in Table 2. Applying all the augmentations results in a higher order prediction model,

Specification	**Model Augmentation**
Piecewise constant reference or measured disturbance	Incremental input
Measured non-predictable disturbance	Constant disturbance model
Previewed reference/disturbance	Preview reference/disturbance buffer
Known time delay	Delay buffer
Unmeasured constant disturbance	Output/tracking error integral action Output disturbance and observer
Reference tracking	Reference model and tracking error, velocity form

Table 2: List of common specifications and related augmentations to the process model to handle them.

$$x_{t+k+1|t} = Ax_{t+k|t} + Bu_{t+k|t}, \quad x = \begin{bmatrix} x_m \\ x_p \end{bmatrix} \in \mathbb{R}^n, \; u = \begin{bmatrix} u_m \\ u_p \end{bmatrix} \in \mathbb{R}^m, \tag{28a}$$

$$y_{t+k|t} = Cx_{t+k|t} + Du_{t+k|t}, \quad y = \begin{bmatrix} y_m \\ y_p \end{bmatrix} \in \mathbb{R}^p, \tag{28b}$$

$$z_{t+k|t} = Ex_{t+k|t}, \quad z = \begin{bmatrix} z_m \\ z_p \end{bmatrix} \in \mathbb{R}^q, \tag{28c}$$

where x, u, y, z are the prediction model state, input, constrained outputs, and performance output vectors, x_p, u_p, y_p, z_p are the augmented state, input, constrained outputs, and performance output vectors.

3.2 Horizon and Constraints

The constraints are usually enforced on the constrained output vector, and on the input. While enforcing the constraints directly on the states is certainly possible, it is more convenient to introduce the vector y specifically for this use, which allows to enforce state, mixed state-inputs, and, possibly even pure input constraints through a single vector. Thus, the constraints are formulated as

$$y_{t+k|t} \in \mathcal{Y}_m, \quad u_{t+k|t} \in \mathcal{U}_m, \tag{29}$$

where $\mathcal{Y}_m$ and $\mathcal{U}_m$ are the admissible sets for constrained output and input vectors, respectively. Enforcing constraints on the prediction model y and u, which include augmentation, often allows to formulate (29) as simple bounds

$$\underline{y} \leq y_{t+k|t} \leq \overline{y}, \quad \underline{u} \leq u_{t+k|t} \leq \overline{u}, \tag{30}$$

which are easier to specify and to handle in an optimization problem.

In automotive applications, the sampling period T_s is often equal to the period of the control cycle for the function being developed. However, for the prediction model to be accurate, it is expected that the sampling period T_s is small enough to allow for 3-10 steps in the settling of the fastest dynamics, following a reference step change. If this is not the case, upsampling or downsampling may be advised, resulting in the prediction model sampling period and the control loop period to be different, and appropriate strategies, such as move blocking or interpolation, are applied to bridge such a difference.

The choice of the prediction horizon N is related to the prediction model dynamics. In general, N should be slightly larger, e.g., $1.5\times$-$3\times$, than the number of steps for the settling of the slowest (stable) prediction model dynamics, following a reference step change. This requirement relates to the choice of the sampling period so that the total amount of prediction steps is expected to be 5-30 times the ratio between the slowest and the fastest (stable) system dynamics. To be more correct, since the controller usually alters the response of the open-loop system, the relevant settling time for the choice of the prediction horizon is that of the closed-loop system, which leads to an iterative selection procedure.

For adjusting the computational requirements in solving the MPC problem, other horizons can be defined. The control horizon N_u determines the number of control steps left as free decision variables to the controller, where for $k \geq N_u$ the input is not an optimization variable but rather assigned by a pre-defined terminal controller,

$$u_{t+k|t} = \kappa_f x_{t+k|t}, \qquad k = N_u, \ldots, N-1. \tag{31}$$

The constraint horizons, for outputs and inputs, N_{cy}, N_{cu}, respectively, determine for how many steps the constraints are enforced,

$$\underline{y} \leq y_{t+k|t} \leq \overline{y}, \quad k = N_i, \ldots, N_{cy}, \qquad \underline{u} \leq u_{t+k|t} \leq \overline{u}, \quad k = 0, \ldots, N_{cu}-1, \tag{32}$$

where $N_i \in \{0,1\}$ depending on whether output constraints are enforced or not at the initial step. Enforcing output constraints at the initial step is reasonable only if the input directly affects the constrained outputs. By choosing N_u, one determines the number of optimization variables, $n_v = N_u m$ and by choosing N_{cy}, N_{cu}, one determines the number of constraints, $n_c = 2(p(N_{cy}+N_i) + m(N_{cu}))$. This defines the size of the optimization problem.

3.3 Cost Function, Terminal Set and Soft Constraints

The cost function encodes the objectives of MPC and their priority. For classical, i.e., not economic, MPC, the control specifications are formulated as variables that are to be controlled to 0. The variables are either a part of the process model (26) or are included in the prediction model (28) by the augmentations discussed in Section 3.1. In general the MPC cost function is

$$J_t = x'_{t+N|t} P x_{t+N|t} + \sum_{k=0}^{N-1} z'_{t+k|t} Q z_{t+k|t} + u'_{t+k|t} R u_{t+k|t}, \tag{33}$$

where the performance outputs of the prediction model $z = Ex$ can model objectives such as tracking, e.g., by $z = Cx - C_r x_r$, where x_r is the reference model state, and $r = C_r x_r$ is the current reference. In (33), $Q \geq 0$, $R > 0$ are the matrix weights that determine the importance of the different objectives: for a diagonal weight matrix Q, the larger the ith diagonal component, the faster the ith performance output will be regulated to 0. It is important to remember that weights determine relative priorities between objectives. Hence, increasing the jth performance output weight may slow down the regulation of the ith performance output.

Very often, the control specifications are given in terms of "domain quantities," such as comfort, NVH (noise, vibration, harshness), consistency, i.e., repeatability of the behavior, and it is not immediately clear how to map them to corresponding weights in the cost function (33). While the mappings tend to be application dependent, some of the common mappings are reported in Table 3, where we stress that the outputs and inputs, and their derivatives, refer to the plant outputs, which may be a part of the performance outputs or inputs, depending on the performed model augmentations.

Specification	Corresponding weights
Regulation/Tracking error	weight on plant output error
Energy	weight on plant input
Noise, vibration, and harshness (NVH)	weight on plant output acceleration, and plant input rate of change
Consistency	weight on plant output velocity, and plant input rate of change
Comfort	weight on plant output acceleration and jerk, and model input rate of change

Table 3: List of "domain terms" specifications and weights that often affect them.

In (33), $P \geq 0$ is the terminal cost, which is used to guarantee at least local stability. There are multiple ways to design P. The most straightforward and more commonly used approach in automotive applications is to choose P to be the solution of the Riccati equation, constructed from A, B in the prediction model (28), and Q, R in the cost function (33),

$$P = A'PA + Q - A'PB(B'PB + R)^{-1}B'PA.$$

This method can be used, after some modifications, also for output tracking [25]. Alternative approaches are based on choosing P to be the solution of a Lyapunov equation for systems that are asymptotically stable, or on choosing P as the closed-loop matrix of the system stabilized by a controller, which, for linear plants and controllers, can be computed via LMIs. The terminal controller $u = \kappa_f x$ used after the end of the control horizon is then chosen accordingly, being either the LQR controller, constantly 0, or the stabilizing controller, respectively.

The use of terminal set constraint $H_N x_{t+N|t} \leq K_N$ to guarantee recursive feasibility and stability in the feasible domain of the MPC optimization problem has seen a fairly limited use in automotive applications. This is primarily due to the many disturbances and modeling errors acting on the prediction model, which may cause infeasibility of such a constraint, and to the need to keep the horizon short because of computational requirements. In practice, for ensuring that the optimization problem admits a solution, constraint softening is often applied. A quadratic program (QP) with soft constraints can be formulated as

$$\min_{v,s} \frac{1}{2} v'Hv + \rho s^2 \tag{34a}$$

$$\text{s.t. } Hv \leq K + Ms, \tag{34b}$$

where s is the slack variable for softening the constraints, M is a vector of 0 and 1 that determines which constraints are actually softened, and ρ is the soft penalty, that usually satisfies $\rho I \gg Q$. In general, only output constraints are softened, because input constraints should always be feasible for a well-formulated problem. More advanced formulations with multiple slack variables giving different priorities to different constraints are also possible, as well as different weighting functions for the constraint violation, such as using the absolute value of s.

4 Computations and Numerical Algorithms

As mentioned in Section 1, a key challenge for implementing MPC in automotive applications is accommodating its significantly larger computational footprint when compared to standard automotive controllers, i.e., PID. As MPC is based on solving a finite time optimal control problem, the MPC code is significantly more complex, and it may involve iterations, checking of termination conditions, and sub-routines, as opposed to the integral and derivative updates and the "one-shot" computation of the three terms in the PID feedback law.

The embedded software engineers that are ultimately responsible for controller deployment need to face this additional complexity, and need to shift from a controller that evaluates a function to a controller that executes an algorithm. For the technology transfer of MPC from research to production to have some chances of success, one must, at least initially, reduce such a gap as much as possible by proposing simple MPC implementations, and then gradually move towards more advanced implementations when confidence in MPC builds up.

Furthermore, cost is a key driver in the automotive development. Automotive engineers often consider advanced control methods as a pathway to reducing the cost of sensors and actuators, while still achieving robustness and efficiency through software. If the control algorithms are so complex that they require the development of new and more expensive computational platforms, their appeal is significantly reduced. Hence, the control developers should always strive to fit the controller in the existing computing hardware, rather than assume that computing hardware that is able to execute it will become available.

	Clock	**Instructions**	**Instr./s**	**RAM**	**ROM**
dCPU	1000s MHz	CISC	100s GIPS	10s GB	1000s GB
aMCU	100s MHz	RISC	1000s MIPS	1000s kB	10s MB

Table 4: A comparison of current characteristic ranges for desktop processors (dCPU) and automotive micro-controllers (aMCU).

The common practice found in many research papers of extrapolating the real-time behavior of MPC in an automotive micro-controller unit (aMCU) from the one that is seen in a desktop CPU (dCPU) is potentially misleading as aMCUs and dCPUs have significantly different capabilities, see Table 4. First, one needs to consider that powerful aMCUs usually run more than ten feedback loops, and probably an even larger number of monitoring loops, and hence the actual computation power available for a single controller is only a fraction of what is available in the entire aMCU. The difference between instruction sets (RISC vs CISC) and the simpler structure of the aMCU memory architecture results in a significantly different numbers of instructions per seconds (IPS) for the aMCUs with respect to dCPUs. Most of the differences are due to the need for the aMCU to work in extreme environments, e.g., ambient temperature ranges between -40°C and $+50^{o}$C, and even higher near the engine, in which a dCPU is not required to operate and may be even prevented from starting. This is also the cause of the major differences in the size of execution memory, which is normally DRAM in dCPU, but is in general permanent (e.g., SRAM, EPROM, or Flash) in aMCU, with higher cost and physical size, and hence lower quantities and speeds. In fact, for several embedded platforms, memory access may actually be the bottleneck [83].

Because of this, and since also processor-specific code optimization, by engineers or custom compilers, may play a very significant role, the evaluation of the computational load of an MPC controller in an aMCU can only be extrapolated using the following approximations (in decreasing order of reliability)

- computing the worst case number of operations per type, the number of instruction per operation type, and hence the total number of instructions per controller execution,
- executing the controller on the specific platform, computing the cost per iteration, and estimating the allowed number of iterations per sampling period,
- evaluating the execution time in a dCPU, estimating the ratio of IPS between dCPU and aMCU, and using that to estimate the execution time in aMCU.

However, according to Table 4, what is often restricting is the memory, both for execution and data. Hence the memory occupancy of the controller is something to be very mindful of. Indeed, PIDs need a minimal amount of memory, 3 gains, 2 extra variables for integral and derivative error, and very few instructions. MPC requires significantly more program and data memory than PID, and hence a careful choice of the numerical algorithm is often critical to the success of the application. Based on the previous discussions, algorithms with limited memory usage and relatively simple code may be preferred, at least initially.

4.1 Explicit MPC

Explicit MPC has had a significant impact on the implementation of the first MPC solutions in experimental vehicles. Some examples, tested in fully functional and road-drivable (and in several cases road-driven) vehicles are in [15, 26, 28–30, 61, 64, 75, 78, 84].

In explicit MPC, the optimizer of the MPC problem is obtained by evaluating a pre-computed function of the current prediction model state, possibly including references and other auxiliary states,

$$u_{\mathrm{mpc}}(x) = \begin{cases} F_1 x + G_1 & \text{if } H_1 x \leq K_1, \\ \vdots & \\ F_s x + G_s & \text{if } H_s x \leq K_s, \end{cases} \tag{35}$$

where s is the total number of regions, see Figure 7 for some examples.

The main advantages of explicit MPC and the reasons it has been the first effective method of MPC deployment in experimental automotive applications are:

- Simple execution code: explicit MPC is a lookup table of affine controllers, selected by evaluating linear inequalities.
- Basic operations: the controller implementation requires only sums, multiplications, and comparisons.
- Predictability: the worst case number of operations is easily computed.

Additional benefits include the possibility of computing explicit MPC, with few modifications, also for hybrid and switched systems, which allowed for the application of switched and hybrid MPC to automotive control problems [15, 29, 61, 64,

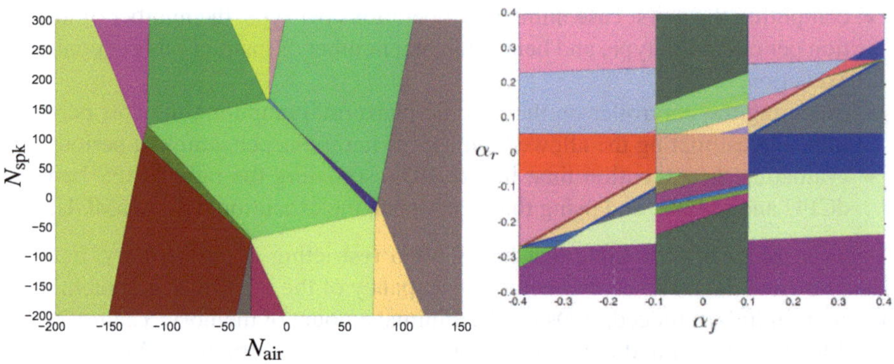

Fig. 7: Section of the partitions of the explicit multivariable linear MPC for idle speed control from [26] (left) and for the switched linear MPC steering controller from [29].

75, 78], the possibility of building explicitly the closed-loop system and hence studying its local and global stability, and the possibility of using only the regions that are most often visited, while utilizing a backup controller otherwise, thus reducing memory requirements.

On the other hand, the explicit MPC data memory occupancy and worst case number of operations grows proportionally to the number of active-sets of constraints n_{as}, and hence exponentially with the number of constraints/variables according to

$$n_{as} \leq \sum_{h=0}^{\min\{n_c, n_v\}} \binom{n_c}{h}, \tag{36}$$

where n_v is the number of variables, and n_c is the number of constraints, both of which are proportional to the length of the horizon(s). Because of (36), explicit MPC is limited to applications with relatively short horizon, few control inputs, and few constraints. Another limitation is that the algorithm to construct the explicit law (35) is too complex to be implemented in the vehicle, and hence explicit MPC cannot be easily tuned or adjusted after deployment.

4.2 Online MPC

In problems with many control inputs, long prediction horizons and many constraints, explicit MPC may be too complex to store for real-time usage, or even to compute. Also, if the prediction model changes over time, it is very difficult to adjust accordingly the explicit solution, while it is relatively simple to update the data of the optimal control problem. In these cases, the online solution of the MPC optimal control problem may be preferable, and hence online MPC has been applied to automotive problems with the above features.

Among such problems, in [33], a quadratic programming solver and a nonlinear programming solver were used online for controlling an autonomous vehicle, which enabled using a long horizon to take maximum advantage of the known reference trajectory. The nonlinear programming solver was based on sequential quadratic programming, and both the nonlinear program and quadratic program solvers used active-set methods. In [34] an online active-set solver was used to solve quadratic programs for controlling MAP and MAF in a diesel engine using EGR and VGT. The online solver was used due to the many constraints imposed by the problem, and the need to update the matrices of the model depending on the current operating point, i.e., using in fact a linear-parameter varying model of the diesel engine. In [5], due to the need for using a relatively long horizon for vehicle dynamics cornering performance and stability control at the limit of performance, an interior point solver was used online. While the solvers have been tested in real vehicles, the computing platforms were dedicated rapid prototyping units and custom micro-controllers that may be more capable than production aMCU, in particular, because they are dedicated to the controller being developed, while aMCUs run multiple controllers.

Active-set and interior-point methods have fast convergence rate, but they often use linear algebra libraries for solving systems of linear equations at each iteration of the optimization. Using these libraries may require a fairly large amount of memory, for both data and code storage, and for execution, and achieving portability of these libraries to certain micro-controllers may not be straightforward. Alternatively, first-order methods often have slower convergence rate, but have much simpler code and hence require less memory, and they are independent of third parties libraries.

First order methods are essentially based on updating the current solution $z_s^{(h)}$ in the direction determined by a function h_s of the gradient of the cost function J with a stepsize α_s, followed by the projection onto the feasible set $\mathscr{F}$

$$\hat{z}_s^{(h+1)} = \hat{z}_s^{(h)} - \alpha_s(z_s^{(h)}) \cdot h_s \left(\frac{d}{dz_s} J(z_s) \Big|_{z_s^{(h)}} \right), \tag{37a}$$

$$z_s^{(h+1)} = \text{proj}_{\mathscr{F}} \left(\hat{z}_s^{(h+1)} \right), \tag{37b}$$

where z_s may contain additional variables other than those of the original optimization problem, and the choice of the stepsize α_s, of the function h_s, and of the additional variables, differentiate the methods.

In recent years, several low complexity first-order methods for MPC have been proposed, based on Nesterov's fast gradient algorithm [69], Lagrangian methods [57], nonnegative least squares [27], and alternating direction method of multipliers (ADMM) [37, 40, 66]. For instance, in [31] the method in [27] was used for vehicle trajectory tracking.

4.3 Nonlinear MPC

Most of the previous discussion focused on linear MPC because, at least until very recently, that was the only class of MPC that realistically could have been implemented in automotive systems. Nonlinear MPC is significantly more complex, which is to a large extent due to the algorithm for solving the nonlinear programming problem. For instance, in [33] the authors after implementing a nonlinear method chose to implement a linear time-varying method based on local linearization and quadratic programming, to reduce the computational load. As another example, the paper [43] concerned with diesel engine air path control states that "currently it is not possible to implement NMPC in real time due to the limited computational power available." However, this is starting to change in more recent years, in part thanks to the research aimed at tailoring nonlinear solvers to MPC problems.

Some applications of nonlinear MPC to automotive systems [36, 55, 82] are based on the C/GMRES method reported in [62]. This method appears quite effective if the objective is to solve a nonlinear optimal control problem with equality constraints, only few inequality constraints, and few changes of the active-set. This is due to the changes to the active sets possibly causing discontinuities in the dual variables, and sometimes also in the primal variables, see, e.g., [56], which is in conflict with the smooth update rule of the continuation methods. Various inequality constraint enforcement techniques for diesel engines in the context of the method in [62] are considered and compared in [51].

More recently [1, 35], some applications are being investigated based on the so-called real-time iteration (RTI) scheme [45], which is based on combining effective integrators for multiple-shooting methods with sequential quadratic programming, where usually only one step of optimization is actually performed.

Expanding the reliability and reducing the computational cost of nonlinear MPC methods will probably be a key effort in the upcoming years to allow for significant use in automotive applications.

5 Conclusions and Future Perspectives

MPC has been extensively employed for research in automotive control, and is maturing for product deployment. The main opportunities are in using MPC for optimal multivariable constrained control, for exploiting preview information, and for handling systems subject to time delays. As a consequence, MPC has been investigated in several applications in powertrain, lateral and longitudinal vehicle dynamics, and energy management of HEV. In the upcoming years the number of applications of MPC to autonomous vehicles [33, 53, 59] is expected to grow, where MPC may need to be integrated with higher level planning methods [21, 31], decision logics [80], and connected cooperative driving [63, 74], possibly within some kind of distributed architecture.

While many challenges presented by MPC deployment have been overcome by research on specific applications, research efforts are still necessary to address them in general ways, and some challenges for product deployment are still not entirely solved. These include the construction of models, and effective approximations thereof, the numerical algorithms for linear [27, 40, 66, 69] and nonlinear MPC [45, 62], the calibration and reconfiguration methods [20, 22], and the design process, to which, hopefully, this chapter has contributed.

References

1. Albin, T., Ritter, D., Abel, D., Liberda, N., Quirynen, R., Diehl, M.: Nonlinear MPC for a two-stage turbocharged gasoline engine airpath. In: Proceedings of 54th IEEE Conference on Decision and Control, pp. 849–856 (2015)
2. Amari, R., Alamir, M., Tona, P.: Unified MPC strategy for idle-speed control, vehicle start-up and gearing applied to an Automated Manual Transmission. In: Proceedings of IFAC World Congress, Seoul (2008)
3. Arce, A., Alejandro, J., Bordons, C., Ramirez, D.R.: Real-time implementation of a constrained MPC for efficient airflow control in a pem fuel cell. IEEE Trans. Ind. Electron. **57**(6), 1892–1905 (2010)
4. Bambang, R.T., Rohman, A.S., Dronkers, C.J., Ortega, R., Sasongko, A., et al.: Energy management of fuel cell/battery/supercapacitor hybrid power sources using model predictive control. IEEE Trans. Ind. Inf. **10**(4), 1992–2002 (2014)
5. Beal, C.E., Gerdes, J.C.: Model predictive control for vehicle stabilization at the limits of handling. IEEE Trans. Control Syst. Technol. **21**(4), 1258–1269 (2013)
6. Beck, R., Richert, F., Bollig, A., Abel, D., Saenger, S., Neil, K., Scholt, T., Noreikat, K.E.: Model predictive control of a parallel hybrid vehicle drivetrain. In: Proceedings of 44th IEEE Conference on Decision and Control, pp. 2670–2675 (2005)
7. Bemporad, A., Morari, M.: Control of systems integrating logic, dynamics, and constraints. Automatica **35**(3), 407–427 (1999)
8. Bemporad, A., Borrelli, F., Glielmo, L., Vasca, F.: Optimal piecewise-linear control of dry clutch engagement. In: 3rd IFAC Workshop on Advances in Automotive Control, Karlsruhe, pp. 33–38 (2001)
9. Bemporad, A., Morari, M., Dua, V., Pistikopoulos, E.: The explicit linear quadratic regulator for constrained systems. Automatica **38**(1), 3–20 (2002)

10. Bemporad, A., Bernardini, D., Long, R., Verdejo, J.: Model predictive control of turbocharged gasoline engines for mass production. In: WCX: SAE World Congress Experience (2018)
11. Bemporad, A., Bernardini, D., Livshiz, M., Pattipati, B.: Supervisory model predictive control of a powertrain with a continuously variable transmission. SAE Technical Paper, No. 2018-01-0860 (2018)
12. Bengtsson, J., Strandh, P., Johansson, R., Tunestal, P., Johansson, B.: Model predictive control of homogeneous charge compression ignition (hcci) engine dynamics. In: Proceedings of IEEE International Conference on Control Applications, Munich, pp. 1675–1680 (2006)
13. Bichi, M., Ripaccioli, G., Di Cairano, S., Bernardini, D., Bemporad, A., Kolmanovsky, I.: Stochastic model predictive control with driver behavior learning for improved powertrain control. In: Proceedings of 49th IEEE Conference on Decision and Control, Atlanta, GA, pp. 6077–6082 (2010)
14. Borhan, H., Vahidi, A., Phillips, A.M., Kuang, M.L., Kolmanovsky, I.V., Di Cairano, S.: MPC-based energy management of a power-split hybrid electric vehicle. IEEE Trans. Control Syst. Technol. **20**(3), 593–603 (2012)
15. Borrelli, F., Bemporad, A., Fodor, M., Hrovat, D.: An MPC/hybrid system approach to traction control. IEEE Trans. Control Syst. Technol. **14**(3), 541–552 (2006)
16. Canale, M., Milanese, M., Novara, C.: Semi-active suspension control using fast model-predictive techniques. IEEE Trans. Control Syst. Technol. **14**(6), 1034–1046 (2006)
17. Caruntu, C.F., Lazar, M., Gielen, R.H., van den Bosch, P., Di Cairano, S.: Lyapunov based predictive control of vehicle drivetrains over can. Control Eng. Pract. **21**(12), 1884–1898 (2013)
18. Carvalho, A., Gao, Y., Lefevre, S., Borrelli, F.: Stochastic predictive control of autonomous vehicles in uncertain environments. In: 12th International Symposium on Advanced Vehicle Control (2014)
19. Di Cairano, S.: An industry perspective on MPC in large volumes applications: potential benefits and open challenges. In: 4th IFAC Symposium on Nonlinear Model Predictive Control, pp. 52–59 (2012)
20. Di Cairano, S.: Model adjustable predictive control with stability guarantees. In: Proceedings of the American Control Conference, pp. 226–231 (2015)
21. Di Cairano, S.: Control and optimization of autonomous vehicles. In: IEEE-VTS Connected and Autonomous Vehicles Summer School. http://resourcecenter.vts.ieee.org/vts/product/events/VTSEVTWPI003 (2016)
22. Di Cairano, S., Bemporad, A.: Model predictive control tuning by controller matching. IEEE Trans. Autom. Control **55**(1), 185–190 (2010)
23. Di Cairano, S., Bemporad, A., Kolmanovsky, I., Hrovat, D.: Model predictive control of magnetically actuated mass spring dampers for automotive applications. Int. J. Control **80**(11), 1701–1716 (2007)
24. Di Cairano, S., Yanakiev, D., Bemporad, A., Kolmanovsky, I., Hrovat, D.: An MPC design flow for automotive control and applications to idle speed regulation. In: Proceedings of 48th IEEE Conference on Decision and Control, pp. 5686–5691 (2008)
25. Di Cairano, S., Pascucci, C.A., Bemporad, A.: The rendezvous dynamics under linear quadratic optimal control. In: Proceedings of 51st IEEE Conference on Decision and Control, pp. 6554–6559 (2012)
26. Di Cairano, S., Yanakiev, D., Bemporad, A., Kolmanovsky, I.V., Hrovat, D.: Model predictive idle speed control: design, analysis, and experimental evaluation. IEEE Trans. Control Syst. Technol. **20**(1), 84–97 (2012)
27. Di Cairano, S., Brand, M., Bortoff, S.A.: Projection-free parallel quadratic programming for linear model predictive control. Int. J. Control **86**(8), 1367–1385 (2013)
28. Di Cairano, S., Liang, W., Kolmanovsky, I.V., Kuang, M.L., Phillips, A.M.: Power smoothing energy management and its application to a series hybrid powertrain. IEEE Trans. Control Syst. Technol. **21**(6), 2091–2103 (2013)

29. Di Cairano, S., Tseng, H., Bernardini, D., Bemporad, A.: Vehicle yaw stability control by coordinated active front steering and differential braking in the tire sideslip angles domain. IEEE Trans. Control Syst. Technol. **21**(4), 1236–1248 (2013)
30. Di Cairano, S., Doering, J., Kolmanovsky, I.V., Hrovat, D.: Model predictive control of engine speed during vehicle deceleration. IEEE Trans. Control Syst. Technol. **22**(6), 2205–2217 (2014)
31. Di Cairano, S., Kalabić, U., Berntorp, K.: Vehicle tracking control on piecewise-clothoidal trajectories by MPC with guaranteed error bounds. In: Proceedings of 55th IEEE Conference on Decision and Control, pp. 709–714 (2016)
32. Donahue, M.D., Hedrick, J.K.: Implementation of an active suspension preview controller for improved ride comfort. In: Johansson, R., Rantzer, A. (eds.) Nonlinear and Hybrid Systems in Automotive Control, vol. 146, pp. 1–22. Springer, London (2003)
33. Falcone, P., Borrelli, F., Asgari, J., Tseng, H., Hrovat, D.: Predictive active steering control for autonomous vehicle systems. IEEE Trans. Control Syst. Technol. **15**(3), 566–580 (2007)
34. Ferreau, H.J., Ortner, P., Langthaler, P., Del Re, L., Diehl, M.: Predictive control of a real-world diesel engine using an extended online active set strategy. Annu. Rev. Control **31**(2), 293–301 (2007)
35. Frasch, J.V., Gray, A., Zanon, M., Ferreau, H.J., Sager, S., Borrelli, F., Diehl, M.: An auto-generated nonlinear MPC algorithm for real-time obstacle avoidance of ground vehicles. In: Proceedings of European Control Conference, pp. 4136–4141 (2013)
36. Gagliardi, D., Othsuka, T., del Re, L.: Direct C/GMRES control of the air path of a diesel engine. In: Proceedings of 19th IFAC World Congress, pp. 3000–3005 (2014)
37. Ghadimi, E., Teixeira, A., Shames, I., Johansson, M.: Optimal parameter selection for the alternating direction method of multipliers (ADMM): quadratic problems. IEEE Trans. Autom. Control **60**(3), 644–658 (2015)
38. Giorgetti, N., Bemporad, A., Tseng, E.H., Hrovat, D.: Hybrid model predictive control application towards optimal semi-active suspension. Int. J. Control **79**(5), 521–533 (2006)
39. Giorgetti, N., Ripaccioli, G., Bemporad, A., Kolmanovsky, I., Hrovat, D.: Hybrid model predictive control of direct injection stratified charge engines. IEEE/ASME Trans. Mechatron. **11**(5), 499–506 (2006)
40. Giselsson, P., Boyd, S.: Linear convergence and metric selection for Douglas-Rachford splitting and ADMM. IEEE Trans. Control Syst. Technol. **62**(2), 532–544 (2017)
41. Gohrle, C., Schindler, A., Wagner, A., Sawodny, O.: Design and vehicle implementation of preview active suspension controllers. IEEE Trans. Control Syst. Technol. **22**(3), 1135–1142 (2014)
42. Hatanaka, T., Yamada, T., Fujita, M., Morimoto, S., Okamoto, M.: Explicit receding horizon control of automobiles with continuously variable transmissions. In: Nonlinear Model Predictive Control. Lecture Notes in Computer Science, vol. 384, pp. 561–569. Springer, Berlin (2009)
43. Herceg, M., Raff, T., Findeisen, R., Allgower, F.: Nonlinear model predictive control of a turbocharged diesel engine. In: Proceedings of IEEE International Conference on Control Applications, pp. 2766–2771 (2006)
44. Heywood, J.: Internal Combustion Engine Fundamentals. McGraw-Hill, New York (1988)
45. Houska, B., Ferreau, H.J., Diehl, M.: An auto-generated real-time iteration algorithm for nonlinear MPC in the microsecond range. Automatica **47**(10), 2279–2285 (2011)
46. Hrovat, D.: MPC-based idle speed control for IC engine. In: Proceedings of FISITA Conference (1996)
47. Hrovat, D.: Survey of advanced suspension developments and related optimal control applications. Automatica **33**(10), 1781–1817 (1997)
48. Hrovat, D., Di Cairano, S., Tseng, H., Kolmanovsky, I.: The development of model predictive control in automotive industry: a survey. In: Proceedings of IEEE International Conference on Control Applications, Dubrovnik, pp. 295–302 (2012)

49. Hsieh, M.F., Wang, J.: Diesel engine selective catalytic reduction (scr) ammonia surface coverage control using a computationally-efficient model predictive control assisted method. In: ASME Dynamic Systems Control Conference, pp. 865–872 (2009)
50. Hsieh, M.F., Wang, J., Canova, M.: Two-level nonlinear model predictive control for lean NOx trap regenerations. J. Dyn. Syst. Meas. Control **132**(4), 041001 (2010)
51. Huang, M., Nakada, H., Butts, K., Kolmanovsky, I.: Nonlinear model predictive control of a diesel engine air path: a comparison of constraint handling and computational strategies. In: 5th IFAC Symposium on Nonlinear Model Predictive Control, pp. 372–379 (2015)
52. Huang, M., Zaseck, K., Butts, K., Kolmanovsky, I.: Rate-based model predictive controller for diesel engine air path: design and experimental evaluation. IEEE Trans. Control Syst. Technol. **24**(6), 1922–1935 (2016)
53. Jalalmaab, M., Fidan, B., Jeon, S., Falcone, P.: Model predictive path planning with time-varying safety constraints for highway autonomous driving. In: International Conference on Advanced Robotics, pp. 213–217 (2015)
54. Jazar, R.N.: Vehicle Dynamics: Theory and Application. Springer, New York (2013)
55. Kaijiang, Y., Mukai, M., Kawabe, T.: Performance of an eco-driving nonlinear MPC system for a power-split hev during car following. SICE J. Control Meas. Syst. Integr. **7**(1), 55–62 (2014)
56. Kalabić, U.V., Gupta, R., Di Cairano, S., Bloch, A.M., Kolmanovsky, I.V.: MPC on manifolds with an application to the control of spacecraft attitude on SO(3).Automatica, **76**, 293–300 (2017)
57. Kögel, M., Findeisen, R.: Fast predictive control of linear systems combining Nesterov's gradient method and the method of multipliers. In: Proceedings of 50th IEEE Conference on Decision and Control, pp. 501–506 (2011)
58. Koot, M., Kessels, J.T., De Jager, B., Heemels, W., Van den Bosch, P., Steinbuch, M.: Energy management strategies for vehicular electric power systems. IEEE Trans. Veh. Technol. **54**(3), 771–782 (2005)
59. Makarem, L., Gillet, D.: Model predictive coordination of autonomous vehicles crossing intersections. In: 16th International IEEE Conference on Intelligent Transportation Systems, pp. 1799–1804 (2013)
60. Mehra, R.K., Amin, J.N., Hedrick, K.J., Osorio, C., Gopalasamy, S.: Active suspension using preview information and model predictive control. In: Proceedings of IEEE International Conference on Control Applications, pp. 860–865. IEEE, Hartford (1997)
61. Naus, G., Ploeg, J., Van de Molengraft, M., Heemels, W., Steinbuch, M.: Design and implementation of parameterized adaptive cruise control: an explicit model predictive control approach. Control Eng. Pract. **18**(8), 882–892 (2010)
62. Ohtsuka, T.: A continuation/GMRES method for fast computation of nonlinear receding horizon control. Automatica **40**(4), 563–574 (2004)
63. Ong, H.Y., Gerdes, J.C.: Cooperative collision avoidance via proximal message passing. In: Proceedings of American Control Conference, pp. 4124–4130 (2015)
64. Ortner, P., del Re, L.: Predictive control of a diesel engine air path. IEEE Trans. Control Syst. Technol. **15**(3), 449–456 (2007)
65. Ozatay, E., Onori, S., Wollaeger, J., Ozguner, U., Rizzoni, G., Filev, D., Michelini, J., Di Cairano, S.: Cloud-based velocity profile optimization for everyday driving: a dynamic-programming-based solution. IEEE Trans. Intell. Transp. Syst. **15**(6), 2491–2505 (2014)
66. Raghunathan, A.U., Di Cairano, S.: Infeasibility detection in alternating direction method of multipliers for convex quadratic programs. In: Proceedings of 53rd IEEE Conference on Decision and Control, pp. 5819–5824 (2014)
67. Rajamani, R.: Vehicle Dynamics and Control. Springer, New York (2011)
68. Ravi, N., Liao, H.H., Jungkunz, A.F., Widd, A., Gerdes, J.C.: Model predictive control of hcci using variable valve actuation and fuel injection. Control Eng. Pract. **20**(4), 421–430 (2012)
69. Richter, S., Jones, C.N., Morari, M.: Computational complexity certification for real-time MPC with input constraints based on the fast gradient method. IEEE Trans. Autom. Control **57**(6), 1391–1403 (2012)

70. Ripaccioli, G., Bemporad, A., Assadian, F., Dextreit, C., Di Cairano, S., Kolmanovsky, I.: Hybrid modeling, identification, and predictive control: an application to hybrid electric vehicle energy management. In: Hybrid Systems: Computation and Control. Lecture Notes in Computer Science, vol. 5469, pp. 321–335. Springer, Berlin (2009)
71. Ripaccioli, G., Bernardini, D., Di Cairano, S., Bemporad, A., Kolmanovsky, I.: A stochastic model predictive control approach for series hybrid electric vehicle power management. In: Proceedings of American Control Conference, Baltimore, MD, pp. 5844–5849 (2010)
72. Santillo, M., Karnik, A.: Model predictive controller design for throttle and wastegate control of a turbocharged engine. In: Proceedings of American Control Conference, pp. 2183–2188 (2013)
73. Schallock, R., Muske, K., Peyton Jones, J.: Model predictive functional control for an automotive three-way catalyst. SAE Int. J. Fuels Lubr. **2**(1), 242–249 (2009)
74. Stanger, T., del Re, L.: A model predictive cooperative adaptive cruise control approach. In: Proceedings of American Control Conference, pp. 1374–1379 (2013)
75. Stewart, G., Borrelli, F.: A model predictive control framework for industrial turbodiesel engine control. In: Proceedings of 48th IEEE Conference on Decision and Control, Cancun, pp. 5704–5711 (2008)
76. Trimboli, S., Di Cairano, S., Bemporad, A., Kolmanovsky, I.: Model predictive control for automotive time-delay processes: an application to air-to-fuel ratio. In: Proceedings of 8th IFAC Workshop on Time-delay Systems, pp. 1–6 (2009)
77. Vahidi, A., Stefanopoulou, A.G., Peng, H.: Current management in a hybrid fuel cell power system: a model-predictive control approach. IEEE Trans. Control Syst. Technol. **14**(6), 1047–1057 (2006)
78. Van Der Heijden, A.C., Serrarens, A.F.A., Camlibel, M.K., Nijmeijer, H.: Hybrid optimal control of dry clutch engagement. Int. J. Control **80**(11), 1717–1728 (2007)
79. Vermillion, C., Butts, K., Reidy, K.: Model predictive engine torque control with real-time driver-in-the-loop simulation results. In: Proceedings of American Control Conference, pp. 1459–1464 (2010)
80. Wongpiromsarn, T., Topcu, U., Murray, R.M.: Receding horizon temporal logic planning. IEEE Trans. Autom. Control **57**(11), 2817–2830 (2012)
81. Yan, F., Wang, J., Huang, K.: Hybrid electric vehicle model predictive control torque-split strategy incorporating engine transient characteristics. IEEE Trans. Veh. Technol. **61**(6), 2458–2467 (2012)
82. Yu, K., Mukai, M., Kawabe, T.: A battery management system using nonlinear model predictive control for a hybrid electric vehicle. In: IFAC Symposium on Advances in Automotive Control, pp. 301–306 (2013)
83. Yu, L., Goldsmith, A., Di Cairano, S.: Efficient convex optimization on gpus for embedded model predictive control. In: Proceedings of ACM General Purpose GPUs, pp. 12–21 (2017)
84. Zhao, D., Liu, C., Stobart, R., Deng, J., Winward, E., Dong, G.: An explicit model predictive control framework for turbocharged diesel engines. IEEE Trans. Ind. Electron. **61**(7), 3540–3552 (2014)

Applications of MPC in the Area of Health Care

G. C. Goodwin, A. M. Medioli, K. Murray, R. Sykes, and C. Stephen

1 Introduction

Health care represents one of the largest expenditures of GDP throughout the world. Control problems are ubiquitous in health care systems. Indeed, any situation in which measurements are used to modify an action can be viewed as a feedback control problem. At this level of abstraction, almost all functions in the health care system can be viewed as feedback control problems.
Examples include:

- Staff rostering with real-time adjustment to account for illness or varied workload.
- Delivery of drugs to a hospital in response to the outbreak of an infectious disease.
- Allocation of patient priority (triage) in emergency departments to ensure appropriate and timely management.
- Ambulance scheduling to minimise arrival times and account for recent ambulance movements, traffic conditions & weather.
- Adjustment of a long-term treatment plan for a patient based on, say, a monthly visit by the patient to the treating physician.
- Real-time adjustment of drug delivery based on minute-by-minute observations of the patient's response, e.g., in an intensive care situation.

G. C. Goodwin (✉) · A. M. Medioli · R. Sykes · C. Stephen
The University of Newcastle, Newcastle, NSW, Australia
e-mail: graham.goodwin@newcastle.edu.au; adrian.medioli@newcastle.edu.au; rebecca.sykes@uon.newcastle.edu.au; carly.stephen@newcastle.edu.au

K. Murray
University of Sydney, Sydney, NSW, Australia
e-mail: kr.murray@outlook.com

S. V. Raković, W. S. Levine (eds.), *Handbook of Model Predictive Control*, Control Engineering, https://doi.org/10.1007/978-3-319-77489-3_22

The goal of the current chapter is to examine the role that Model Predictive Control (MPC) can play in developing improved control strategies in the area of health care. The aim is to illustrate the broad applicability of MPC rather than to give a comprehensive survey.

2 Is MPC Relevant to Health Problems?

One of the core advantages of MPC is that it provides a rigorous framework within which control problems can be articulated. Thus MPC establishes a set of core questions that underly feedback control problems. Examples of these questions are:

- What defines the "state" of the system?
- How does the state evolve with time?
- How do manipulated inputs and external disturbances influence the evolution of the state?
- Are there hard constraints on input, output and/or state variables?
- What are the performance goals?
- Are the model parameters fixed or do they change with time?
- What direct measurements are available?
- What indirect data is available which may help define the "state"?
- Is an observer necessary to combine model information with past data to estimate the state?

MPC directly addresses these questions through the associated problem formulation.

Some of the attributes most commonly claimed for MPC are:

- its capacity to handle multi-variable systems,
- its capacity to include input/output constraints,
- its capacity to treat nonlinear systems, and
- its ability to facilitate precise articulation of goals via the associated cost function.

These attributes are germane to health care problems making the MPC framework a natural way to formulate the associated feedback design problem.

3 Special Characteristics of Control Problems in the Area of Health

There are several aspects of health problems that uniquely impact the application of MPC in this area. Some of these are discussed below:

3.1 Safety

Patient safety, in health applications, is understandably paramount. Hence, any automation system must satisfy a set of regulations before it can be applied to, or tested on, patients. Some of the mechanisms that ensure patient safety are:

- Formulation of research that clearly defines the need, efficacy, safety and its ethical appropriateness.
- Approval from a Government agency (FDA in the USA, or TGA in Australia) before any new drug, device or technology is released to the public.

Due to these safety regulations, the application of complex control theory (including MPC) is necessarily more conservative in health care than in other areas. New ideas usually go through an exhaustive set of checks and balances before they appear in real-world products. To quote just one area with which the authors are familiar, the reader may be surprised to learn that PID (with some embellishments) aimed at insulin delivery for Type 1 Diabetes [7] has just reached the market place. MPC has been contemplated for this problem but a large-scale free-living trial is only now about to begin. Moreover, it is not immediately obvious whether MPC (or indeed, even PID) offers real benefits over a simple injection of insulin proportional to the carbohydrate content of meals (with some simple adjustments for current blood glucose level and upcoming exercise).

3.2 Background Knowledge

Current medical practice is based on many years of past clinical experience. Thus there are no "quick fixes" in the area of health. Indeed, anyone contemplating applying MPC to a health problem will first need to become very familiar with the biology and clinical practice relevant to the problem. Thus, in common with all real-world problems, a control engineer working in the area of health is faced with the substantial task of learning about the problem. This is necessary to achieve a "common language" when discussing health problems with medical professionals and to gain an appreciation of the true complexity of the problem. Only through this collaboration can we hope to define the actual need and then be in a position to select the control approach that will appropriately meet that need.

3.3 Models

Models in the health area have distinctive characteristics and terminology. For example, in drug therapy, it is usual to distinguish:

- Pharmacokinetics—the dynamics of the transport of a drug through the body, and

- Pharmacodynamics—the dynamics of the drug on the patient's state once the drug arrives at an action site

The above distinction is useful when developing the structure of models but may not be quite so relevant if one focuses on the observed input–output behaviour. Biological insights can also play an important role in suggesting a model structure. The level of biological insight used in models leads to questions of black box, grey box or white box modelling.

3.4 Population Versus Personalised Models

Many models used in the area of health are based on averages across a particular population of patients. However, when treating an individual, a personalised model may be more appropriate. The distinction between these classes of models can be very important when developing MPC solutions for health problems, see [63].

4 Specific Examples Where MPC Has Been Used in the Area of Health

As discussed above, MPC has the potential to be used in many areas of health care. To illustrate the broad range of applications achieved to date, we will briefly discuss the following problems:

- Ambulance Scheduling.
- Joint Movement
- Management of Type 1 Diabetes
- Anaesthesia
- HIV
- Cancer
- Chronic Inflammation

The first of these examples is an operational issue in the health area. The second example relates to mechanicals aids. The remaining five examples refer to drug delivery systems. Further details of these applications are given in the sequel.

4.1 Ambulance Scheduling

The application of stochastic MPC to ambulance scheduling has been described in [30]. The basic idea of the ambulance scheduling problem is that when an ambulance becomes free (e.g. by being released at a hospital) or an emergency incident occurs, then a real-time control decision arises, that is, where should the ambulance be sent. Possible control decisions could be

- stay at the hospital,
- go to the home station for the specific ambulance,
- go to another station,
- do a lower priority job (e.g. patient transfer), or
- take a rest break.

The best decision depends on a host of factors, including

- the state of the system (e.g. where all ambulances currently are, what the ambulances are doing, the current emergency incidents and their status),
- the time of day, day of week, week of year,
- the probable location and time of future incidents,
- current emergency department loads, and whether or not they are able to take a patient and treat them appropriately, and
- current environmental conditions (e.g., traffic, weather, etc.).

The goal of the associated feedback control problem is to minimise the (average) time to reach high-priority incidents. Indeed, most countries have goals of, say, reaching 50% of 'priority-one' emergencies within a specified time period, e.g. 12 minutes. Even small improvements can be important since ambulance scheduling is associated with life and death issues. For example, research has shown that every minute of delay in reaching a cardiac emergency reduces the probability of survival by 10% [70]. In view of the huge impact on society, it is not surprising that emergency services have been the subject of considerable research effort. Work on this problem can be found in [48, 70] and [80]. Also, there exists commercial software that addresses related problems [38]. A common theme in past work has been to utilise approximate dynamic programming. This has been shown to lead to significant improvements in operational performance [80].

There are six key requirements necessary to formulating this problem in the context of Stochastic MPC:

1. Choice of a model that describes the movement of emergency vehicles in response to dispatch instructions. The dispatch orders are the 'control actions' and are discrete in nature, i.e., send vehicle A to location B.
2. Selecting a cost function that measures the desired performance.
3. Formulation of a suitable system state for the system which will typically contain both integer (e.g., vehicle A has a patient on-board) and real variables (e.g., vehicle A is at location B).
4. Articulation of the environment in which the system operates including disturbances (e.g., where and when future incidents may occur). These are random variables with an (associated) probability distribution.
5. Choice of a simulation tool. The model of the system is complex and not typically based on a set of differential equations. Hence, calculating the response to a specified decision and a given set of future disturbance scenarios is achieved through simulations.

6. Choice of a sampling strategy. In [30] an event based strategy was chosen rather than time based one.

Details of the application of MPC to this problem are available in [30].

4.2 Joint Movement

As an illustration of the broad spectrum of applications of MPC in the health area, we next outline the application of MPC to joint movement control.

Joint movement has been extensively researched in biomedical engineering with the aim of better understanding joint control and developing prosthetics and exoskeletons that can replace and/or guide a damaged or absent extremity. Joint movement involves planning and execution of muscle contractions across a joint in a controlled manner so as to produce a desired force on, or movement of, the bones to which the muscle is attached. This controlled movement is achieved through the pairing of muscles: one agonist which generates movement in the desired direction and the other an antagonist which provides a counter force correcting over-contraction and providing a steadying force.

MPC has been proposed by many researchers as a method of accurately controlling the desired movement of prosthetics or exoskeletons [6, 41, 79] and for humanoid robots [43]. Some conditions that this method of control aims to address include foot drop [6], full or partial paralysis, and providing aid to patients who are amputees.

There are several physiological aspects that need to be modelled for MPC to work effectively for joint movement. Some of these are:

- Reflexive muscle characteristics including stiffness and torque
- Joint dynamics
- Activation dynamics
- Antagonist activation
- Stretch reflex (joint angle vs fibre length)

The use of MPC in this area is appealing since it can use nonlinear models. These models are widely accepted. One example is an inverted pendulum model to describe heel and toe link and bipedal posture. This has been used [6, 76, 77] as a model for the motion of walking. MPC has also been used to simulate the neural plasticity of the brain. Reference [76] suggests a related learning algorithm. Reference [41] uses an exoskeleton to map the dynamics of a finger joint to generate a more accurate model via equivalent equations.

MPC of joints is often considered a single input, single output system, requiring strong input and output constraints to generate the desired effect. The input is the torque (or force) needed to reach the desired equilibrium point. This must be consistent with the force capable of being produced by the joint [79]. The output is typically either the angle of one bone with respect to another in the joint, or the position of the limb in space. Whatever the output, it must abide by the constraints relevant to human muscles, ligaments, tendons and bones [6, 41, 76, 77, 79]. These constraints are aimed at ensuring that the prosthetic has similar mechanics to the human joint and therefore that no harm will come to the joint. When an equilibrium is required to be reached, then the output constraints need to be altered to satisfy this equilibrium [43].

Both implicit and explicit MPC control laws have been considered for this area. The associated time constant is of the order of seconds. Endpoint MPC has been proposed in [79] for an exoskeleton controlling gait. Reference, [6] used offline MPC, with no environmental feedback, to derive the best motion based on constraints, which conserves energy for correction of foot drop. Although accurate models have been well developed in the field of joint movement, further progress in modelling is likely to enhance the effectiveness of MPC in this area.

4.3 Type 1 Diabetes Treatment

Type 1 diabetes is a disease which affects the cells in the pancreas that are responsible for producing insulin [2, 65]. Insulin is required to transport glucose into the cells for energy and into muscles and liver for storage. Further, if insufficient insulin is available then this can result in the acute and possibly life-threatening condition called diabetic ketoacidosis (DKA).

As a result, people with type 1 diabetes are unable to regulate their blood glucose level (BGL). If the BGL rises above 140mg/dL (7.8 mM), the patient is said to be hyperglycaemic. Extended periods of hyperglycaemia result in long-term complications including cardiovascular disease, kidney damage, blindness, and nerve damage [2]. Similarly, serious short-term effects result from hypoglycaemia, the state in which the BGL drops below 70mg/dL (3.9 mM). These short-term complications include anxiety, tremor, pallor, poor concentration, seizures, coma, neurological damage and, in extreme cases, death [2, 29].

With an autoimmune attack having compromised the body's ability to regulate blood glucose levels, type 1 diabetes patients use the ingestion of carbohydrate to raise the BGL and the injection of insulin analogues to lower the BGL. It is common practice that the required insulin dosing is determined by a clinician based on the patient's insulin requirement per gram of carbohydrate ingested. This allows the patient to calculate a single injection of insulin for each meal [78]. Additional insulin may also be given if the patient's BGL is too high. This is based on the patient's historical trends.

There are currently multiple models used in the area of type 1 diabetes. These are used for both simulation and control purposes. Some of the models used are the Bergman Minimal Model [9], Hovorka model [37], and the Dalla Man model [20]. Modified versions of these are also used.

In type 1 diabetes there are several effects and delays that need to be captured in the model. These factors include:

- The rate of subcutaneous insulin delivery and its associated rate of absorption.
- Subcutaneous and plasma insulin concentration and rate of change of concentration.
- The rate of insulin clearance (how long it takes for the insulin to leave the body).
- Blood glucose concentration and the effect of insulin on plasma glucose.
- The food absorption dynamics including the rate of absorption of glucose from meals.
- The patient's sensitivity to insulin.
- The rate of endogenous glucose production.
- The increase in glucose uptake, and decrease in endogenous glucose production in response to the presence of glucose.
- Sensor dynamics and the relationship between the measured interstitial fluid glucose concentration (typically measured by a continuous glucose monitor (CGM)) and the blood glucose concentration.

Several types of MPC have been suggested for the management of type 1 diabetes. Some of the controllers suggested are linear [61, 72], non-linear [69, 82], implicit [1, 47, 61, 82], explicit [62], and stochastic [34]. The majority of controllers used are one-sided i.e., only account for the positive flow of insulin. This is because insulin can be injected but not removed from the body. Some very recent research has simulated the efficacy of bi-hormonal controllers where the input can be either insulin or glucagon. The hormone glucagon acts to raise the BGL in the case of a hypoglycaemic event [5].

Typically, the BGL is the only available measurement but there can be ten or more states. Thus various observers have been proposed. One of the main observers that has been implemented for type 1 diabetes is the Kalman filter [5, 12, 61, 72].

Feedforward controllers have been suggested in diabetes control to mitigate the impact of delays experienced in the body [1]. The use of feedforward is consistent with the well-known fact that the optimal BGL response occurs when insulin is delivered before a meal is ingested [33]. A major difficulty associated with applying MPC to blood glucose regulation is related to the inability to predict future disturbances such as exercise, stress and food consumption.

For this system, a hard constraint is usually placed on the input, since negative insulin flows are impossible. Also, it is important to constrain the output (BGL) to a range of 70–140mg/dL (3.9–7.8 mM) [34]. The lower limit is particularly important since hypoglycaemia has the potential for short-term catastrophic consequences.

MPC is under intense scrutiny for diabetes treatment. Many issues need to be addressed for this application including:

- Model type—linear or nonlinear [6, 8, 34].
- How to calibrate the model for an individual [42].
- Is an observer necessary [46]?

- How to deal with past disturbances. For example, should measurements of past insulin delivery, food and exercise be made available to the observer when estimating the patient's current state?
- Cost function—should one regulate to a set-point or does it suffice to regulate to a range [44]?
- Input constraints—insulin flow must be positive.
- Output constraints—long-term health issues arise if BGL exceeds an upper level, short-term health issues arise if BGL falls below a lower level.
- Prediction horizon—the time constants are of the order of hours. Thus actions taken now may not manifest themselves for many hours. This suggests that prediction horizons of the order of many hours are necessary.
- Upcoming disturbances such as food, exercise and stress are important and these events can be associated with a high degree of uncertainty [34].
- The existence of fundamental limitations due to the nature of one side control action [33].
- Does the available model apply to a population or has it been calibrated for an individual [63]?

4.4 Anaesthesia

Anaesthesia plays an important role in surgical operations. The three main aspects of anaesthesia are hypnosis (level of consciousness and amnesia), muscle relaxation and analgesia. Specific drugs for each aspect are used in combination to induce, and maintain, unconsciousness, prevent unwanted movement, and block pain from being felt by the patient during medical procedures [53]. The typical drugs used include inhaled drugs such as Sevoflurane (hypnosis) and intravenously administered drugs such as Propofol (hypnosis), Pancuronium bromide (muscle relaxant) and Remifentanil (analgesia) [4, 15, 28, 39, 40, 53]. Hence, an automatic feedback controller for anaesthesia would be useful due to its potential of increasing patient safety throughout their procedure [54].

The patient's level of hypnosis is calculated using the Bispectral Index (BIS), which is measured using an electroencephalogram (EEG) [40] sometimes in conjunction with pulse rate [28]. A scale from 0–100 is used where 100 denotes the patient being fully awake [53]. If the BIS level is too low, this can result in excessive sedation with associated side effects including nausea and vomiting [28, 68], and a longer recovery time, whilst a BIS level that is too high can result in a patient becoming aware of their surroundings during surgery [68]. Therefore, it is desirable that the controller be informed of relevant output constraints including a BIS level between approximately 40–60, corresponding to a moderate level of hypnosis [40]. Also, there are input constraints: the amount of drug used must be non-negative and there is an upper limit on the amount of drug that can safely be administered [68].

The models used in this area are typically compartmental and include:

- Type of drug administration (intravenous or inhalation), and rate of infusion
- Drug concentration and rate of distribution between the central (circulatory) and peripheral (muscle and fat) compartments
- Time delay between the drug delivery, and its response on the body
- The age, weight, height, gender, and body mass of the patient
- The observed effect of the drug on the patient

The associated pharmacokinetic model is typically linear whilst the pharmacodynamic model is generally nonlinear. A common choice uses the Hill Curve [4, 15]. Related to the models used in anaesthesia are those described by Schnider, Minto, Krieger, and Schüttler and Ihmsen [45, 53, 68].

The time constants for anaesthesia are relatively small (order of minutes) when compared to other medical applications such as diabetes or HIV. Both implicit [28, 67, 68] and explicit [39, 52, 54] MPC have been suggested. The controller needs to be able to cope with disturbances. However, these usually cannot be suitably predicted. Therefore, feedforward control has not been used to date. This can be a disadvantage, as the controller is unaware of the progress of the surgery, and cannot raise the desired BIS level in response to the surgery coming to an end. This can prolong the recovery time of the patient [68].

An observer is required for the controller to estimate the current "state" of the patient and to maintain the desired depth of anaesthesia during the procedure. The observers that have been proposed in the context of anaesthesia include the Kalman filter [45], the Extended Kalman filter [67], and multi-parametric Moving Horizon Estimation [52–54].

4.5 HIV

Human Immunodeficiency Virus, commonly known as HIV, is responsible for the gradual decline of helper T cells in the body [81]. Helper T cells are an essential part of the human adaptive immune system. Therefore, a significant decline in their concentration can result in patients becoming highly susceptible to infection [84]. When the level of helper T cells drops too low, the patient will develop Acquired Immune Deficiency Syndrome (AIDS). The current treatment for HIV is Highly Active Anti-Retroviral Therapy (HAART). This treatment utilises reverse transcriptase inhibitors and protease inhibitors which act in combination to prevent or slow viral reproduction and delay the progression of the disease [49]. HAART is an ongoing treatment since the virus is unable to be entirely eliminated from the body. The harsh side effects and the high cost of the treatment make an optimal control strategy for drug delivery desirable.

The models used for HIV control are typically nonlinear and may include some of the following aspects:

- The concentration and rate of production and destruction of healthy helper T cells.
- The rate of viral infection.
- Concentration and depletion rate of infected helper T cells.
- Concentration, rate of virus production by infected cells, and rate of progress of the virus.
- Development of immune memory, and the immune system's effect on killing the virus.
- The effectiveness of the drugs in use.

These interactions are nonlinear, and current models such as the Wordarz-Nowak model reflect this [60, 84, 85]. The current models for HIV still have many shortcomings. This is generally believed to be a major limitation when designing an MPC controller for HIV.

MPC is being investigated for HIV treatment in multiple ways. The main applications of MPC in HIV treatment are Structured Treatment Interruptions (STI) [4, 60, 81, 84] and development of an optimal administration schedule to reduce the concentration of free virus particles within a specified period of time [64]. The purpose of both of these strategies is to minimise the amount of drug administered, and consequently minimise the adverse side effects and overall cost of the treatment. The added purpose of the STI is to encourage the immune system to control the virus [84, 85].

Due to the large time constant and sampling period (typically weeks to months) the main type of MPC proposed to date has been implicit, since the computational time of the controller is not a key limiting factor. The controllers are entirely based on feedback as there is usually no information available about future disturbances.

The measured variables are typically the concentration of the virus, and sometimes the number of healthy helper T-cells [60]. Since there are other variables, of relevance, that are not directly measured, this means that an observer is required. Some of the observers that have been proposed include the Extended Kalman filter and nonlinear multistate estimation [64], a nonlinear observer [84] and a deadbeat observer [60].

There are also constraints placed on the amount of drug that should be administered. This is important to prevent the virus from becoming drug resistant, which could occur if the dose is too low. Also, this accounts for the toxicity of the drug, should the dose be too large. For STI, the input constraints are typically between 0 and 1, where 0 denotes no treatment and 1 denotes full treatment. This constraint has been implemented on the input either as a continuous input or a discrete input [81]. Output constraints can include an upper bound on the viral concentration, and a lower bound on the concentration of healthy helper T cells.

4.6 Cancer

Around the world, 39% of the population will be diagnosed with cancer at some point in their life [56]. In the developed world, cancer is the leading cause of death in people aged under 85 [25]. The increasing threat of cancer has prompted the investigation of new and better ways of administering treatment [19]. Developing treatments for cancer is difficult since the nature of the condition stems from the replication cycle of normal cells in the body becoming mutated, resulting in uncontrollable cell growth and proliferation. These cells can spread throughout the body undetected since they are not foreign and therefore do not cause an immune reaction. Death often results from these metastases disrupting body tissues and organ function [14].

Current treatment options include:

1. surgery to remove tumours,
2. radiotherapy which uses concentrated doses of radiation to kill cancer cells,
3. chemotherapy which stops, or slows, the rate of growth of cancer. This treatment can shrink tumours (although it also attacks healthy cells),
4. immunotherapy which helps the immune system to recognise and kill cancer cells,
5. targeted therapy which treats cancer by targeting the changes in cancer cells that help them grow, divide, and spread,
6. hormone therapy that inhibits the production of, or blocks, the action of hormones produced by the body that are known to promote the growth of some cancers, e.g., oestrogen promotes breast cancer growth,
7. stem cell transplants which can treat cancer in blood cells, and
8. precision medicine which helps doctors select treatments that are most likely to help patients based on a genetic understanding of their disease [55].

The responsiveness of the patient to particular chemotherapies, hormone and targeted molecular therapies and the associated toxicity of these treatments is dependent on the genetic profile of both the tumour and of the individual. Thus the action of treatments needs to be monitored and appropriate remedial action taken. Hence, cancer treatment is again a quintessential feedback control problem.

The physiological aspects of cancer and its treatment (chemotherapy, hormone therapy and radiotherapy) are typically modelled as follows:

- Tumour growth modelling. Modelling of tumour growth rate and cell phase distribution via a saturating rate cell cycle model or Gompertz model has been described in many published papers [13, 19, 24, 25]. It is usually represented by multiple non-linear differential equations. In the case of leukaemia, a pluripotential stem cell model has been used which predicts the number of red blood cells compared to white blood cells [57]. For androgen treatment of prostate cancer a three parameter on-treatment, off-treatment model has been used to describe the non-linear exponential growth of cells. A relapse factor has also been included to optimise the treatment protocol [36].

- Pharmacokinetics. This is usually represented as a linear compartmental oral dosing structure.
- Pharmacodynamics. A bilinear kill term in the differential equation is usually added for rate of cell volume growth [24]. This has been used to convey pharmacodynamic effects in the tumour growth model. In the case of leukaemia, a linear relationship of drug to nucleated cell count has been used.
- Lymphocyte levels in the body. This is used to monitor the effect of chemotherapy on the immune system as well as antibody production aimed at increasing the tumour destruction rate.
- In the case of ultrasound hyperthermia treatment, typical models include a tissue-tumour model. This represents a one dimensional view of the tumour as well as surrounding tissue. Different properties of the tumour and surrounding tissue are used in the Pennes' bioheat transfer equation [10] which would then be used to simulate tissue temperature response. For small tumours, or for more concentrated therapy, a single point model has been used which simplifies the Pennes' bioheat transfer equation to an ordinary differential equation [3].
- In the case of real-time motion compensation for radiotherapy treatment, the models used include the patient support system dynamics [59].

MPC has been investigated for cancer treatment. MPC has been proposed for optimising drug delivery and other interventions for chemotherapy, radiotherapy, immunotherapy, hormone therapy and other anti-cancer agents. It has also been proposed as a way of optimally delivering ultrasound hyperthermia treatment methods [3] and for motion compensation in adaptive radiotherapy [59].

Both implicit and explicit MPC have been considered. When the states cannot be directly measured, an observer is used to estimate the tumour size and characteristics. A Kalman filter has been considered for states that are linear. For states that are nonlinear, an Extended Kalman filter, unscented Kalman filter or particle filter has been proposed [19]. The associated time constants are of the order of days. Ultrasound therapy has used implicit MPC and a Kalman filter to estimate the temperature distribution in the tissue, derived from sensors at discrete points in the tumour. The time constant for this application is of the order of seconds [3]. Tumour motion compensation utilises sensors on all 3 axes to predict the motion of the tumour associated with the body's external movement. A full order state observer has also been used to estimate velocities from position measurement. An algorithm has also been used to predict the patient support system feedback. The time constant for this particular task is of the order of seconds [59].

All cancer treatment options involve constraints which further suggest MPC as a desirable strategy. Constraints limit the maximum amount of drug that can be delivered to the patient. In the case of leukaemia the constraint is on the concentration of the drug metabolite found in the blood [57]. In the case of androgen suppression for prostate cancer treatment, each variable must be non-negative and changes are limited to 20% per day [36]. Lymphocyte levels are also monitored and used to constrain the upper limit of both dose and period of time for which the body is exposed to the treatment. As the lymphocyte level depletes, treatment should be lowered or stopped [13, 19]. For the case of ultrasound hyperthermia therapy, a constraint on the maximum temperature used is necessary to minimise the impact of increased temperature on surrounding tissue. This is monitored via discomfort and blood flow rates of surrounding vessels [3]. Output constraints are also applied to motion compensation as dictated by the limits of the patient support system dimensions [59].

4.7 Inflammation

Inflammation is a complex biological response to harmful stimuli in tissues often described as the body's immunovascular response. It usually presents as oedema or swelling, redness of the skin if localised, amplified nociception and heat of the affected area or fever. Too little inflammation can result in progressive tissue destruction due to the harmful stimuli, too much inflammation can be crippling and can lead to patient death from cell damage in various body tissues [66].

Inflammation can be classified as acute or chronic. Acute inflammation is in general a result of an event such as microbial invasion or physical injury. This is an innate form of defence which the body employs to treat injury by creating an environment that allows rapid access by cells and mediators in response to the injury. For example, in the case of microbial infection, the body creates an adverse environment for the foreign pathogens and allows phagocytes to gain access to the infected area. This cause of inflammation is usually treated with antibiotics.

Chronic inflammation, for example in arthritis, is not a part of the body's natural healing process and can cause major discomfort and an inability to use the part of the body that is affected due to excessive pain and swelling [21]. This calls for effective management using anti-inflammatory medication, typically Non-Steroidal Anti Inflammatory Drugs (NSAIDs). Antihistamine drugs can also be used to combat the oedema, redness and heat [66]. Immunomodulation is employed for chronic and acute inflammation treatment [21]. The use of pro-inflammatory drugs has been proposed for combating the excessive effects of anti-inflammatory medication, though it is not commonly used clinically.

All of the presented applications of MPC are in the treatment of chronic inflammation. MPC has been proposed by many authors as an effective scheme to regulate anti and/or pro-inflammatory drugs to produce the best outcome for the patient [21–23, 83].

All proposed schemes use a model comprising non-linear differential equations which consider:

- Number of pathogen cells in the body.
- Number of phagocytic cells (pro-inflammatory/anti-pathogen agents) .
- Tissue damage.
- Anti-inflammatory mediators (for example, cortisol and interleukin-10).

In [21, 22, 83] the model was based on data collected from the patient regarding the number of activated phagocytes and number of anti-inflammation mediators. Stochastic differential equations have also been used to account for uncertainty.

Regarding constraints, the control input, or amount of drug delivered, is constrained to be non-negative. Also, an upper bound is typically used to avoid aseptic death. This is achieved by calculating the difference between the actual number of anti-inflammatory/pro-inflammatory mediators in the body and the maximum allowable number [23].

The only type of MPC that has been used to date for inflammation control is implicit. The associated time constant is of the order of hours. Many different observers have been considered for state estimation. For example, [23] proposed an "ad hoc" observer, while [83] uses a particle filter after determining that a Kalman filter would be inadequate to estimate pathogen levels and tissue damage.

5 Appraisal

Whilst MPC has clear potential to make a "game-changing" contribution to many problems in the area of health, it is not a panacea.

Major improvements in performance are not always a result of utilising a more sophisticated control law. On the contrary, significant improvements often arise from simply understanding the problem better, being able to act upon the system in a more authoritative fashion, or by embellishing the control system architecture e.g., by adding feed-forward.

Thus, before jumping into the application of MPC to any problem, including those in the area of health, there are a set of questions that need to be asked. These questions include:

- Does the system already meet the performance objectives? If not, then what are the key factors limiting the achievable performance?
- If the performance is (predominantly) limited by model accuracy, how can one improve the model?
- If the performance is (predominantly) limited by sensors, how can the sensors be improved or what additional sensors would be helpful?
- If the performance is (predominantly) limited by the actuators, how can one enhance the existing actuators or what additional manipulated variables would be helpful?

A difficulty with MPC is that it delivers the "best possible" performance under the limitations inherited from the available model, sensors and actuators. It does not explicitly point to the issues that limit performance. To achieve the latter one may need to temporarily step aside from the MPC framework. In this context, a useful mechanism may be to ask the following question, "If we were to remove all of the MPC infrastructure (e.g., constraints, nonlinear behaviours, etc.) what would classical control ideas tell us?" Issues that naturally arise in this context are:

- *Sum of sensitivity and complementary sensitivity is one.*
  A feedback system cannot simultaneously deliver low sensitivity to measurement errors and disturbances having similar characteristics [11, 74].

- *Bode sensitivity integrals and "sensitivity dirt".*
  Reducing the sensitivity of a feedback loop in some (frequency) range necessarily leads to an increase in sensitivity elsewhere [11, 17, 18, 35, 58, 73].
- *Robustness versus performance trade-offs.*
  High performance is inevitably accompanied by greater sensitivity to measurement and model errors [32, Chapter 3], [51, 71].
- *Bandwidth limitations arising from model uncertainty.*
  The achievable closed-loop bandwidth is below the frequency at which the magnitude of the relative model error approaches one [32, Section 8.5].
- *Impact of delays and/or large lags.*
  Delays and large lags inevitably inhibit high closed loop performance since the information provided by the measurements is "out-of-date" [32, Section 8.6.2], [27].
- *Sensor limitations (e.g. measurement accuracy).*
  Poor sensors always result in poor performance since they deliver misleading information about the system [75].
- *Unmeasured and/or unpredictable disturbances.*
  Unmeasured disturbances negatively impact the performance of observers and make accurate future predictions problematic [16, 31, 71].
- *Inverse response (non-minimum phase behaviour).*
  Inverse response means that achieving a fast response is associated with undershoot [26, 50].
- *Actuation imperfections such as slip-stick friction, nonlinearities, response times etc.*
  If actuators do not respond as required then high performance control is problematic since the desired control actions are not delivered in an uncorrupted fashion to the system [32, Section 8.8.2].

Within this framework, MPC is undeniably a valuable tool. It provides a framework to set the associated control design question and to provide a rigorous solution that ensures stability in the presence of known model imperfections, input and output constraints and certain classes of disturbances.

6 Conclusion

Health is a major component of national expenditure in all developed countries. Many problems in the area of health are quintessential feedback control problems since treatment (i.e., adjustment of manipulated variables) is usually a function of observations (i.e., measured response variables.). Typical control problems in the area of health involve multi-variable interactions, are nonlinear and have hard constraints. Thus MPC arises as a natural tool to achieve improved management and treatment strategies.

This chapter has described a number of health related problems to which MPC has already been applied. The results achieved to date are extremely encouraging. Beyond the current applications, there exist many other health problems which could similarly benefit from the application of advanced control tools including MPC. We are only limited by our vision and courage.

Acknowledgements The authors gratefully acknowledge input into this chapter from co-workers in the areas of (i) Ambulance Scheduling (Dr. Paul Middleton, Dr. Katie O'Donnel, Dr. Rosemary Carney and Mr. John Dent); (ii) Diabetes management (Dr. Bruce King, Dr. Carmel Smart, Mrs. Megan Patterson, Ms. Tenele Smith, Mr. Jordan Rafferty, Dr. Diego Carrasco and Mr. Hieu Phan) and (iii) in the broader health area (Dr. Corrine Balit).

References

1. Abu-Rmileh, A.: A gain scheduling model predictive controller for blood glucose control in type 1 diabetes. IEEE Trans. Biomed. Eng. **57**(10), 2478–2484 (2009)
2. Aronoff, S.L., Berkowitz, K., Shreiner, B., Want, L.: Glucose metabolism and regulation: beyond insulin and glucagon. Diabetes Spectr. **17**(3), 183–190 (2004). https://doi.org/10.2337/diaspect.17.3.183. http://spectrum.diabetesjournals.org/cgi/doi/10.2337/diaspect.17.3.183
3. Arora, D., Skliar, M., Roemer, R.B.: Model predictive control of ultrasound hyperthermia treatments of cancer. In: Proceedings of the 2002 American Control Conference (IEEE Cat. No.CH37301), vol. 4, pp. 2897–2902 (2002). https://doi.org/10.1109/ACC.2002.1025229
4. Bamdadian, A., Towhidkhah, F., Moradi, M.H.: Generalized predictive control of depth of anesthesia by using a pharmocokinetic-pharmacodynamic model of the patient. In: 2nd International Conference on Bioinformatics and Biomedical Engineering, iCBBE 2008, pp. 1276–1279 (2008). https://doi.org/10.1109/ICBBE.2008.647
5. Batora, V., Tarnik, M., Murgas, J., Schmidt, S., Nørgaard, K., Poulsen, N.K., Madsen, H., Jørgensen, J.B.: Bihormonal model predictive control of blood glucose in people with type 1 diabetes. In: IEEE Multi-conference on Systems and Control, pp. 1693–1698 (2014)
6. Benoussaad, M., Mombaur, K., Azevedo-Coste, C.: Nonlinear model predictive control of joint ankle by electrical stimulation for drop foot correction. In: IEEE International Conference on Intelligent Robots and Systems, pp. 983–989 (2013). https://doi.org/10.1109/IROS.2013.6696470
7. Bergenstal, R.M., Garg, S., Weinzimer, S.A., Buckingham, B.A., Bode, B.W., Tamborlane, W.V., Kaufman, F.R.: Safety of a hybrid closed-loop insulin delivery system in patients with type 1 diabetes. JAMA **316**(13), 1407 (2016). https://doi.org/10.1001/jama.2016.11708. http://jama.jamanetwork.com/article.aspx?doi=10.1001/jama.2016.11708
8. Bergman, R., Ider, Y., Bowden, C.R., Cobelli, C.: Quantitative estimation of insulin sensitivity. Am. J. Physiol. Endocrinol. Metab. **6**(236), 667–677 (1979). http://ajpendo.physiology.org/content/236/6/E667.short
9. Bergman, R.N., Phillips, L.S., Cobelli, C.: Physiologic evaluation of factors controlling glucose tolerance in man: measurement of insulin sensitivity and beta-cell glucose sensitivity from the response to intravenous glucose. J. Clin. Investig. **68**(6), 1456–67 (1981). https://doi.org/10.1172/JCI110398
10. Bergman, T., Incropera, F., DeWitt, D., Lavine, A.: Fundamentals of Heat and Mass Transfer. Wiley, New York (2011). https://books.google.com.au/books?id=vvyIoXEywMoC
11. Bode, H.: Network Analysis and Feedback Amplifier Design. Bell Telephone Laboratories Series. Van Nostrand, Princeton (1945). https://books.google.com.au/books?id=kSYhAAAAMAAJ

12. Boiroux, D., Duun-Henriksen, A.K., Schmidt, S., Nørgaard, K., Poulsen, N.K., Madsen, H., Jørgensen, J.B.: Adaptive control in an artificial pancreas for people with type 1 diabetes. Control Eng. Pract. **58**, 2115–2120 (2016). https://doi.org/10.1016/j.conengprac.2016.01.003. http://www.sciencedirect.com/science/article/pii/S096706611630003X
13. Bumroongsri, P., Kheawhom, S.: Optimal dosing of breast cancer chemotherapy using robust MPC based on linear matrix inequalities. Eng. J. **19**(1), 97–106 (2015). https://doi.org/10.4186/ej.2015.19.1.97
14. Cancer Council Australia: What Is Cancer? http://www.cancer.org.au/about-cancer/what-is-cancer (2016)
15. Cardoso, N., Lemos, J.M.: Model predictive control of depth of anaesthesia: guidelines for controller configuration. In: 30th Annual International IEEE EMBS Conference, vol. 2008, pp. 5822–5825 (2008). https://doi.org/10.1109/IEMBS.2008.4650538. http://www.ncbi.nlm.nih.gov/pubmed/19164041
16. Carrasco, D.S., Goodwin, G.C.: Connecting filtering and control sensitivity functions. Automatica **50**(12), 3319–3322 (2014). https://doi.org/10.1016/j.automatica.2014.10.042. http://dx.doi.org/10.1016/j.automatica.2014.10.042
17. Chen, J.: Sensitivity integral relations and design trade-offs in linear multivariable feedback systems. IEEE Trans. Autom. Control **40**(10), 1700–1716 (1995). https://doi.org/10.1109/9.467680
18. Chen, J., Nett, C.N.: Sensitivity integrals for multivariable discrete-time systems. Automatica **31**(8), 1113–1124 (1995). https://doi.org/10.1016/0005-1098(95)00032-R
19. Chen, T., Kirkby, N.F., Jena, R.: Optimal dosing of cancer chemotherapy using model predictive control and moving horizon state/parameter estimation. Comput. Methods Programs Biomed. **108**(3), 973–983 (2012). https://doi.org/10.1016/j.cmpb.2012.05.011.
20. Dalla Man, C., Rizza, R.A., Cobelli, C.: Meal simulation model of the glucose-insulin system. IEEE Trans. Biomed. Eng. **54**(10), 1740–1749 (2007). https://doi.org/10.1109/TBME.2007.893506
21. Day, J.D., Rubin, J., Florian, J., Parker, R.P., Clermont, G.: Modulating inflammation using nonlinear model predictive control. J. Crit. Care **21**(4), 349–350 (2006). https://doi.org/10.1016/j.jcrc.2006.10.012
22. Day, J., Rubin, J., Vodovotz, Y., Chow, C., Clermont, G.: Using nonlinear model predictive control to optimize inflammation-modulating therapy. J. Crit. Care **22**(4), 350–351 (2007). https://doi.org/10.3934/mbe.20xx.xx.xx
23. Day, J., Rubin, J., Clermont, G.: Using nonlinear model predictive control to find optimal therapeutic strategies to modulate inflammation. Math. Biosci. Eng. **7**(4), 739–763 (2010)
24. Florian, J.A., Eiseman, J.J.L., Parker, R.S.: A nonlinear model predictive control algorithm for breast cancer treatment. In: Proceedings of the Dycops 7 (2004)
25. Florian, J.A., Eiseman, J.L., Parker, R.S.: Nonlinear model predictive control for dosing daily anticancer agents using a novel saturating-rate cell-cycle model. Comput. Biol. Med. **38**(3), 339–347 (2008). https://doi.org/10.1016/j.compbiomed.2007.12.003
26. Freudenberg, J.S., Looze, D.P.: Right half plane poles and zeros and design tradeoffs in feedback systems. IEEE Trans. Autom. Control **30**(6), 555–565 (1985). https://doi.org/10.1109/TAC.1985.1104004
27. Freudenberg, J.S., Looze, D.P.: A sensitivity tradeoff for plants with time delay. IEEE Trans. Autom. Control **32**(2), 99–104 (1987). https://doi.org/10.1109/TAC.1987.1104547
28. Furutani, E., Tsuruoka, K., Kusudo, S., Shirakami, G., Fukuda, K.: A hypnosis and analgesia control system using a model predictive controller in total intravenous anesthesia during day-case surgery. In: Proceedings of the SICE Annual Conference, pp. 223–226 (2010). http://www.scopus.com/inward/record.url?eid=2-s2.0-78649270395&partnerID=tZOtx3y1
29. Gerich, J.E.: Control of glycaemia. Bailliere Clin. Endocrinol. Metab. **7**(3), 551–586 (1993). https://doi.org/10.1016/S0950-351X(05)80207-1
30. Goodwin, G.C., Medioli, A.M.: Scenario-based, closed-loop model predictive control with application to emergency vehicle scheduling. Int. J. Control **86**(8), 1338–1348

(2013). https://doi.org/10.1080/00207179.2013.788215. http://www.tandfonline.com/doi/abs/10.1080/00207179.2013.788215
31. Goodwin, G.C., Mayne, D.Q., Shim, J.: Trade-offs in linear filter design. Automatica **31**(10), 1367–1376 (1995)
32. Goodwin, G.C., Graebe, S.F., Salgado, M.E.: Control System Design, 1st edn. Prentice Hall PTR, Upper Saddle River (2001)
33. Goodwin, G.C., Medioli, A.M., Carrasco, D.S., King, B.R., Fu, Y.: A fundamental control limitation for linear positive systems with application to type 1 diabetes treatment. Automatica **55**, 73–77 (2015). https://doi.org/10.1016/j.automatica.2015.02.041. http://linkinghub.elsevier.com/retrieve/pii/S000510981500103X
34. Goodwin, G.C., Medioli, A.M., Phan, H.V., King, B.R., Matthews, A.D.: Application of MPC incorporating stochastic programming to type 1 diabetes treatment. In: 2016 American Control Conference, vol. 2016, pp. 907–912 (2016). https://doi.org/10.1109/ACC.2016.7525030
35. Hara, S., Sung, H.k.: Constraints on sensitivity characteristics in linear multivariable discrete-time control systems. Linear Algebra Appl. **122–124**(C), 889–919 (1989). https://doi.org/10.1016/0024-3795(89)90679-4
36. Hirata, Y., Azuma, S.i., Aihara, K.: Model predictive control for optimally scheduling intermittent androgen suppression of prostate cancer. Methods **67**(3), 278–281 (2014). https://doi.org/10.1016/j.ymeth.2014.03.018.
37. Hovorka, R., Canonico, V., Chassin, L.J., Haueter, U., Massi-Benedetti, M., Orsini Federici, M., Pieber, T.R., Schaller, H.C., Schaupp, L., Vering, T., Wilinska, M.E., Howorka, R., Federici, O.M.: Nonlinear model predictive control of glucose concentration in subjects with type 1 diabetes. Physiol. Meas. **25**(4), 905–20 (2004). https://doi.org/10.1088/0967-3334/25/4/010
38. Indermedix: Optima live dispatch^{TM} (2011)
39. Ingole, D., Holaza, J., Takács, B., Kvasnica, M.: FPGA-based explicit model predictive control for closed-loop control of intravenous anesthesia. In: 2015 International Conference on Process Control, pp. 42–47 (2015)
40. Ionescu, C.M., Keyser, R.D., Struys, M.M.R.F.: Evaluation of a propofol and remifentanil interaction model for predictive control of anesthesia induction. In: IEEE Conference on Decision and Control and European Control Conference, pp. 7374–7379 (2011)
41. Kakoty, N.M., Hazarika, S.M., Koul, M.H., Saha, S.K.: Model predictive control for finger joint trajectory of TU Biomimetic hand. In: 2014 IEEE International Conference on Mechatronics and Automation, IEEE ICMA 2014, pp. 1225–1230 (2014). https://doi.org/10.1109/ICMA.2014.6885874
42. Kanderian, S.S., Weinzimer, S., Voskanyan, G., Steil, G.M.: Identification of intraday metabolic profiles during closed-loop glucose control in individuals with type 1 diabetes. J. Diabetes Sci. Technol. **3**(5), 1047–1057 (2009)
43. Kouchaki, E., Sadigh, M.J.: Constrained-optimal balance regulation of a biped with toe-joint using model predictive control. In: International Conference on Robotics and Mechatronics, ICRoM 2013 (1), pp. 248–253 (2013). https://doi.org/10.1109/ICRoM.2013.6510113
44. Kovatchev, B., Patek, S., Dassau, E., Doyle, F.J., Magni, L., De Nicolao, G., Cobelli, C.: Control to range for diabetes: functionality and modular architecture. J. Diabetes Sci. Technol. **3**(5), 1058–65 (2009)
45. Krieger, A., Pistikopoulos, E.N.: Model predictive control of anesthesia under uncertainty. Comput. Chem. Eng. **71**, 699–707 (2014). https://doi.org/10.1016/j.compchemeng.2014.07.025
46. Magni, L., Forgione, M., Toffanin, C., Dalla Man, C., Kovatchev, B., De Nicolao, G., Cobelli, C.: Run-to-run tuning of model predictive control for type 1 diabetes subjects: in silico trial. J. Diabetes Sci. Technol. **3**(5), 1091–8 (2009)
47. Markakis, M.G., Mitsis, G.D., Papavassilopoulos, G.P., Marmarelis, V.Z.: Model predictive control of blood glucose in type 1 diabetes: the principal dynamic modes approach. In: 30th Annual International IEEE EMBS Conference, 20–24 August 2008, pp. 5466–5469. https://doi.org/10.1109/IEMBS.2008.4650451

48. Maxwell, M.S., Restrepo, M., Henderson, S.G., Topaloglu, H.: Approximate dynamic programming for ambulance redeployment. INFORMS J. Comput. **22**(2), 266–281 (2009). https://doi.org/10.1287/ijoc.1090.0345. http://joc.journal.informs.org/cgi/doi/10.1287/ijoc.1090.0345
49. Mhawej, M., Moog, C., Biafore, F., Brunet-François, C.: Control of the HIV infection and drug dosage. Biomed. Signal Process. Control **5**(1), 45–52 (2010). https://doi.org/10.1016/j.bspc.2009.05.001
50. Middleton, R.H.: Trade-offs in linear control system design. Automatica **27**(2), 281–292 (1991). https://doi.org/10.1016/0005-1098(91)90077-F
51. Morari, M., Zafiriou, E.: Robust Process Control. Prentice-Hall, Englewood Cliffs (1989)
52. Nascu, I., Pistikopoulos, E.N.: A multiparametric model-based optimization and control approach to anaesthesia. Can. J. Chem. Eng. **94**(11), 2125–2137 (2016). https://doi.org/10.1002/cjce.22634
53. Nascu, I., Lambert, R.S.C., Pistikopoulos, E.N.: A combined estimation and multi-parametric model predictive control approach for intravenous anaesthesia. In: IEEE International Conference on Systems, Man, and Cybernetics, pp. 2458–2463 (2014)
54. Nascu, I., Krieger, A., Ionescu, C.M., Pistikopoulos, E.N.: Advanced model-based control studies for the induction and maintenance of intravenous anaesthesia. IEEE Trans. Biomed. Eng. **62**(3), 832–841 (2015). https://doi.org/10.1109/TBME.2014.2365726. http://www.scopus.com/inward/record.url?eid=2-s2.0-84923855365&partnerID=tZOtx3y1
55. National Cancer Institute: Types of Treatment. https://www.cancer.gov/about-cancer/treatment/types (2015)
56. National Cancer Institute: Cancer Stat Facts: Cancer of Any Site. https://seer.cancer.gov/statfacts/html/all.html (2016)
57. Noble, S.L., Sherer, E., Hannemann, R.E., Ramkrishna, D., Vik, T., Rundell, A.E.: Using adaptive model predictive control to customize maintenance therapy chemotherapeutic dosing for childhood acute lymphoblastic leukemia. J. Theor. Biol. **264**(3), 990–1002 (2010). https://doi.org/10.1016/j.jtbi.2010.01.031.
58. O'Young, S.D., Francis, B.A.: Sensitivity tradeoffs for multivariable plants. IEEE Trans. Autom. Control **30**(7), 625–632 (1985). https://doi.org/10.1109/TAC.1985.1104019
59. Paluszczyszyn, D., Skworcow, P., Haas, O., Burnham, K.J., Mills, J.A.: Model predictive control for real-time tumor motion compensation in adaptive radiotherapy. IEEE Trans. Control Syst. Technol. **22**(2), 635–651 (2014). https://doi.org/10.1109/Tcst.2013.2257774
60. Pannocchia, G., Laurino, M., Landi, A.: A model predictive control strategy toward optimal structured treatment interruptions in anti-HIV therapy. IEEE Trans. Biomed. Eng. **57**(5), 1040–1050 (2010). https://doi.org/10.1109/TBME.2009.2039571
61. Parker, R.S., Doyle, F.J., Harting, J.E., Peppas, N.A.: Model predictive control for infusion pump insulin delivery. In: 18th Annual International Conference of the IEEE Engineering in Medicine and Biology Society, pp. 1822–1823 (1996)
62. Percival, M.W., Wang, Y., Grosman, B., Dassau, E., Zisser, H., Jovanovič, L., Doyle, F.J.: Development of a multi-parametric model predictive control algorithm for insulin delivery in type 1 diabetes mellitus using clinical parameters. J. Process Control **21**(3), 391–404 (2011). https://doi.org/10.1016/j.jprocont.2010.10.003
63. Phan, H., Goodwin, G., Carrasco, D., Medioli, A.M., Stephen, C., Feuer, A.: Individualization of stochastic models from population statistics for blood glucose regulation in type one diabetes patients. In: Australian Control Conference, Newcastle (2016)
64. Pinheiro, V., Lemos, M.: Multi-drug therapy design for HIV-1 infection using nonlinear model predictive control. In: 19th Mediterranean Conference on Control and Automation, pp. 485–490 (2011)
65. Poretsky, L.: Principles of Diabetes Mellitus, Springer, New York (2010). https://doi.org/10.1007/978-0-387-09841-8
66. Reynolds, A., Rubin, J., Clermont, G., Day, J., Vodovotz, Y., Bard Ermentrout, G.: A reduced mathematical model of the acute inflammatory response: I. Derivation of model and analysis

of anti-inflammation. J. Theor. Biol. **242**(1), 220–236 (2006). https://doi.org/10.1016/j.jtbi.2006.02.016
67. Rezvanian, S., Towhidkhah, F., Ghahramani, N.: Controlling the depth of anesthesia using model predictive controller and Extended Kalman Filter. In: 2011 1st Middle East Conference on Biomedical Engineering, pp. 213–216 (2011). https://doi.org/10.1109/MECBME.2011.5752103
68. Sawaguchi, Y., Furutani, E., Shirakami, G., Araki, M., Fukuda, K.: A model-predictive hypnosis control system under total intravenous anesthesia. IEEE Trans. Biomed. Eng. **55**(3), 874–887 (2008). https://doi.org/10.1109/TBME.2008.915670
69. Schlotthauer, G., Nicolini, G.A., Gamero, L.G., Torres, M.E.: Type I diabetes: modeling, identification and non-linear model predictive control. In: Second Joint EMBS/BMES Conference, vol. 1, pp. 226–227 (2002)
70. Schmid, V.: Solving the dynamic ambulance relocation and dispatching problem using approximate dynamic programming. Eur. J. Oper. Res. **219**(3), 611–621 (2012). https://doi.org/10.1016/j.ejor.2011.10.043. http://linkinghub.elsevier.com/retrieve/pii/S0377221711009830
71. Seron, M.M., Braslavsky, J.H., Goodwin, G.C.: Fundamental Limitations in Filtering and Control. Springer, Berlin (1997)
72. Soru, P., De Nicolao, G., Toffanin, C., Dalla Man, C., Cobelli, C., Magni, L.: MPC based artificial pancreas: strategies for individualization and meal compensation. Annu. Rev. Control **36**(1), 118–128 (2012). https://doi.org/10.1016/j.arcontrol.2012.03.009.
73. Stein, G.: Respect the unstable. IEEE Control Syst. Mag. **23**(4), 12–25 (2003). https://doi.org/10.1109/MCS.2003.1213600
74. Sung, H.k., Hara, S.: Properties of sensitivity and complementary sensitivity functions in single-input single-output digital control systems. Int. J. Control **48**(6), 2429–2439 (1988). https://doi.org/10.1080/00207178808906338.
75. Tarbouriech, S., Garcia, G., Glattfelder, A.H. (eds.): Advanced Strategies in Control Systems with Input and Output Constraints. Lecture Notes in Control and Information Sciences, vol. 346. Springer, Berlin/Heidelberg (2007). https://doi.org/10.1007/978-3-540-37010-9. http://link.springer.com/10.1007/978-3-540-37010-9
76. Towhidkhah, F.: Model predictive impedance control: a model for joint movement control. Ph.D. thesis, University of Saskatchewan (1996). https://doi.org/10.16953/deusbed.74839. http://hdl.handle.net/10388/etd-10212004-000718
77. Towhidkhah, F., Gander, R.E., Wood, H.C.: Model predictive impedance control: a model for joint movement. J. Mot. Behav. **29**(3), 209–222 (1997). https://doi.org/10.1080/00222899709600836. http://www.ncbi.nlm.nih.gov/pubmed/12453780
78. Wang, Y., Dassau, E., Doyle, F.J.: Closed-loop control of artificial pancreatic beta-cell in type 1 diabetes mellitus using model predictive iterative learning control. IEEE Trans. Biomed. Eng. **57**(2), 211–219 (2010). https://doi.org/10.1109/TBME.2009.2024409
79. Wang, L., Van Asseldonk, E.H.F., Van Der Kooij, H.: Model predictive control-based gait pattern generation for wearable exoskeletons. In: IEEE International Conference on Rehabilitation Robotics (2011). https://doi.org/10.1109/ICORR.2011.5975442
80. Yue, Y., Marla, L., Krishnan, R., Heinz, H.J., College, I.I.I.: An efficient simulation-based approach to ambulance fleet allocation and dynamic redeployment. In: Twenty-Sixth AAAI Conference on Artificial Intelligence, pp. 398–405 (2012). http://www.aaai.org/ocs/index.php/AAAI/AAAI12/paper/viewFile/5148/5169
81. Zarei, H., Kamyad, A.V., Effati, S.: Model predictive control for optimal anti-HIV drug administration. Adv. Model. Optim. **13**(3), 403–417 (2011)
82. Zarkogianni, K., Mougiakakou, S.G., Prountzou, A., Vazeou, A., Bartsocas, C.S., Nikita, K.S.: An insulin infusion advisory system for type 1 diabetes patients based on non-linear model predictive control methods. In: 29th Annual International Conference of the IEEE EMBS, vol. 2007, pp. 5972–5975 (2007). https://doi.org/10.1109/IEMBS.2007.4353708

83. Zitelli, G., Djouadi, S.M., Day, J.D.: Combining robust state estimation with nonlinear model predictive control to regulate the acute inflammatory response to pathogen. Math. Biosci. Eng. **12**(5), 1127–1139 (2015). https://doi.org/10.3934/mbe.2015.12.1127
84. Zurakowski, R.: An output-feedback MPC-based scheduling method for enhancing immune response to HIV. In: 2006 American Control Conference, vol. 1–12, pp. 4800–4805 (2006)
85. Zurakowski, R., Teel, A.R.: A model predictive control based scheduling method for HIV therapy. J. Theor. Biol. **238**(2), 368–382 (2006). https://doi.org/10.1016/j.jtbi.2005.05.004

Model Predictive Control for Power Electronics Applications

Daniel E. Quevedo, Ricardo P. Aguilera, and Tobias Geyer

1 Introduction

Advances in the field of power electronics allow engineers to manipulate electrical power and to control its flow efficiently with power levels ranging from milliwatt to gigawatt. The utilization of power electronics has increased considerably in recent years. In 2015, the overall market size was USD 36 billion [42]. The power electronics market can be divided into industrial applications, utility-scale power electronics [13], automotive [14], consumer electronics, aerospace and defense, and information and communication technology. Notable examples of industrial applications include renewable energy systems [8], rail traction and motor drives [15]. Power converters have been constantly advanced regarding their semiconductors, packaging, passive materials, topologies and control techniques [30].

From a control systems perspective, power electronic systems give rise to intrinsically challenging design problems. Specifically, three major challenges can be identified:

D. E. Quevedo (✉)
Chair for Automatic Control (EIM-E), Paderborn University, 33098 Paderborn, Germany
e-mail: dquevedo@ieee.org

R. P. Aguilera
School of Electrical and Data Engineering, University of Technology Sydney, Sydney, NSW 2007, Australia
e-mail: raguilera@ieee.org

T. Geyer
ABB Corporate Research, ABB Switzerland Ltd, Power Electronic Systems, Segelhofstrasse 1 K, 5405 Baden-Dättwil, Switzerland
e-mail: t.geyer@ieee.org

S. V. Raković, W. S. Levine (eds.), *Handbook of Model Predictive Control*,
Control Engineering, https://doi.org/10.1007/978-3-319-77489-3_23

1. Switched dynamics. The main building blocks of power electronic systems are linear circuit elements, such as inductors, capacitors and resistors, which are complemented by semiconductor switches. The latter are either actively controlled or (passive) diodes. As a result, when controlling currents, fluxes and voltages and manipulating the switch positions, power electronic systems constitute switched *linear* systems, provided that saturation effects of magnetic material, delays and safety constraints can be neglected [23, 53].
   In general, however, power electronic systems represent switched *nonlinear* systems. Nonlinearities arise, for example, when machine variables such as the electromagnetic torque or stator flux magnitude are directly controlled; both quantities are nonlinear functions of currents or flux linkages. For grid-connected converters, the real and reactive power is nonlinear in terms of the currents and voltages. Saturation effects in inductors and current constraints lead to additional nonlinearities.
2. MIMO systems. Three-phase power converters have at least three manipulated variables, i.e., one switch position per phase. In the simplest case, the current of an inductive load needs to be controlled. When the star point of the load floats, two linearly independent currents arise, resulting in a system with two controlled variables and three manipulated variables. For more complicated systems, such as converters with *LC* filters and inductive loads, six controlled variables result. Dc-ac modular multilevel converters (MMC) [36] with n modules per arm are significantly more complex with up to $6n$ manipulated variables and up to $6n+6$ controlled variables.
3. Short computation times. The third challenge results from the short sampling intervals of 1 ms and less that are typically used in power electronic systems. These short sampling intervals limit the time available to compute the control actions. To reduce the cost of power electronic converters sold in high volumes, cheap computational hardware is usually deployed as the control platform. Replacing existing control loops with only low computational requirements by new and computationally more demanding methods exasperates the challenge of short sampling intervals. This is particularly the case for direct control methods that avoid the use of a modulator. These methods typically require very short sampling in the range of 25 μs.

To address these challenges, various embodiments of model predictive control (MPC) principles have emerged as a promising control alternative for power conversion applications [9, 20, 31, 50, 52, 54]. As we shall see in this chapter, this popularity of MPC is due to the fact that predictive control algorithms present several advantages that make them suitable for the control of power electronic systems:

1. The concepts are intuitive and easy to understand;
2. MPC can handle converters with multiple switches and states, e.g., current, voltage, power, torque, etc.;

3. constraints and nonlinearities can be easily included; and
4. the resulting controller is, in general, easy to implement.

2 Basic Concepts

Various MPC methods have been proposed for controlling power electronic systems. Here, one can distinguish between formulations that use system models governed by linear time-invariant dynamics, and those that incorporate nonlinearities. Most MPC strategies are formulated in a discrete-time setting with a fixed sampling interval, say $h > 0$. System inputs are restricted to change their values only at the discrete sampling instants, i.e., at times $t = kh$, where $k \in \mathbb{N} \triangleq \{0,1,2,\dots\}$ denotes the sampling instants.

Since power electronics applications are often governed by nonlinear dynamic relations, it is convenient to represent the system to be controlled in discrete-time state space form via:

$$x(k+1) = f(x(k),u(k)), \quad k \in \mathbb{N}, \tag{1}$$

where $x(k) \in \mathbb{R}^n$ denotes the state value at time k and $u(k) \in \mathbb{R}^m$ is the plant input. Depending on the application at hand, the system state is a vector, which may contain capacitor voltages, inductor and load currents, and fluxes.

2.1 System Constraints

An interesting feature of the MPC framework is that it allows one to incorporate state and input constraints, say:

$$\begin{aligned} x(k) &\in \mathbb{X} \subseteq \mathbb{R}^n, & k &\in \{0,1,2,\dots\}, \\ u(k) &\in \mathbb{U} \subseteq \mathbb{R}^m, & k &\in \{0,1,2,\dots\}. \end{aligned} \tag{2}$$

State constraints can, for example, correspond to constraints on capacitor voltages in flying capacitor converters or neutral point clamped converters. Constraints on load currents can also be modeled as state constraints. Throughout this chapter we will focus on input constraints, since they naturally arise when controlling power converters.

Input constraints, $u(k) \in \mathbb{U}$, are related to the switch positions during the interval $(kh,(k+1)h]$. If a modulator is used, then $u(k)$ will be constrained to belong to a bounded continuous set. For example, the components of $u(k)$ could correspond to duty cycles, $d(k)$, or PWM reference signals. In this case, the control input is constrained by

$$u(k) = d(k) \in \mathbb{U} \triangleq [-1,1]^m \subset \mathbb{R}^m, \qquad k \in \{0,1,2,\dots\}, \tag{3}$$

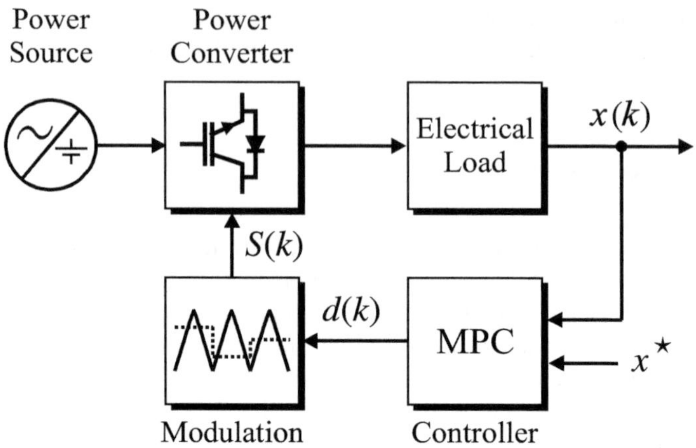

Fig. 1: MPC with continuous control set.

where m denotes the number of phases, see Figure 1. Clearly, the above model can only approximate switching effects, see also [35]. Nevertheless, as we will see, several interesting and powerful controllers for power converters have been developed by using this simple setting.

On the other hand, in the so-called *direct control* applications, where no modulator is used, $u(k)$ is constrained to belong to a finite set describing the available switch combinations. Such approaches have attracted significant attention in the power electronics community, often under term finite control set MPC (FCS-MPC) [52]. The main advantage of this predictive control strategy comes from the fact that switching actions, say $S(k)$, are directly taken into account in the optimization procedure as constraints on the system inputs, see Figure 2. Thus, the control input is restricted to belong to a finite set represented by

$$u(k) = S(k) \in \mathbb{U} \subset \mathbb{R}^m, \qquad k \in \{0,1,2,\ldots\}, \tag{4}$$

where $\mathbb{U}$ is an integer set obtained by combining the m switch values. For the control of multilevel topologies, it is often convenient to consider the resultant phase voltage level as the control input rather than the switch position of each semiconductor switch. For example, for a five-level inverter, $\mathbb{U} = \{-2,-1,0,1,2\}^m$.

2.2 Cost Function

A distinguishing element of MPC, when compared to other control algorithms, is that at each time instant k and for a given (measured or estimated) plant state $x(k)$,

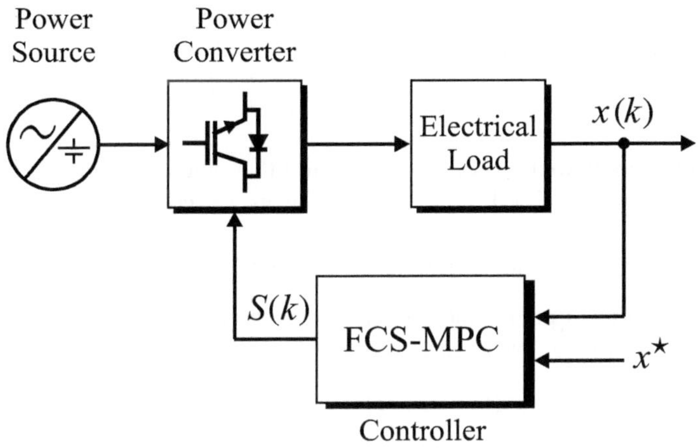

Fig. 2: MPC with finite control set (direct control).

a cost function over a finite horizon of length N is minimized. The following choice encompasses many alternatives documented in the literature:

$$V(x(k), \mathbf{u}'(k)) \triangleq F(x'(k+N)) + \sum_{\ell=k}^{k+N-1} L(x'(\ell), u'(\ell)). \tag{5}$$

Here, $L(\cdot,\cdot)$ and $F(\cdot)$ are weighting functions, which serve to penalize predicted system behaviour, e.g., differences between references for voltages and currents and their predicted values, see Section 2.4.

For example, for a two-level three-phase inverter in orthogonal $\alpha\beta$ coordinates, one can use (see [51])

$$L(x'(\ell), u'(\ell)) = \lambda_1 (i_\alpha(\ell) - i_\alpha^\star)^2 + \lambda_2 (i_\beta(\ell) - i_\beta^\star)^2.$$

For a one-phase three-cell flying capacitor converter (FCC) one can choose (see, e.g., [37])

$$L(x'(\ell), u'(\ell)) = \lambda_1 (i_a(\ell) - i_a^\star)^2 + \lambda_2 (v_{c1}(\ell) - v_{c1}^\star)^2 + \lambda_3 (v_{c2}(\ell) - v_{c2}^\star)^2.$$

In (5), predicted plant state values, $x'(\ell)$, are formed using the system model (1):

$$x'(\ell+1) = f(x'(\ell), u'(\ell)), \quad \ell \in \{k, k+1, \ldots, k+N-1\} \tag{6}$$

where

$$u'(\ell) \in \mathbb{U}, \quad \ell \in \{k, k+1, \ldots, k+N-1\}$$

refers to tentative plant inputs (to be decided). The recursion (6) is initialized with the current plant state measurement (or estimate), i.e.:

$$x'(k) \longleftarrow x(k). \tag{7}$$

Thus, (6) refers to predictions of the plant states that would result if the plant inputs at the update times $\{k, k+1, \ldots, k+N-1\}$ were set equal to the corresponding values in

$$\mathbf{u}'(k) \triangleq [u'^T(k)\ u'^T(k+1)\ \ldots\ u'^T(k+N-1)]^T. \tag{8}$$

Both, the predicted plant state trajectory and the plant inputs are constrained in accordance with (2), i.e., we have:

$$\begin{aligned} u'(\ell) &\in \mathbb{U}, \quad \forall \ell \in \{k, k+1, \ldots, k+N-1\} \\ x'(\ell) &\in \mathbb{X}, \quad \forall \ell \in \{k+1, k+2, \ldots, k+N\}. \end{aligned}$$

Constrained minimization of $V(\cdot,\cdot)$ in (5) gives the optimizing control sequence at time k and for state $x(k)$:

$$\mathbf{u}^{\text{opt}}(k) \triangleq [(u^{\text{opt}}(k))^T\ (u^{\text{opt}}(k+1;k))^T\ \ldots\ (u^{\text{opt}}(k+N-1;k))^T]^T. \tag{9}$$

It is worth emphasizing here that, in general, plant state predictions, $x'(\ell)$, will differ from actual plant state trajectories, $x(\ell)$. This is a consequence of possible model inaccuracies and the moving horizon optimization paradigm described next.

2.3 Moving Horizon Optimization

Despite the fact that the optimizer $\mathbf{u}^{\text{opt}}(k)$ in (9) contains feasible plant inputs over the entire horizon, $(kh, (k+N-1)h]$, in most MPC approaches, only the first element is used, i.e., the system input in (1) is set to

$$u(k) \longleftarrow u^{\text{opt}}(k).$$

At the next sampling step, i.e., at discrete-time $k+1$, the system state $x(k+1)$ is measured (or estimated), the horizon is shifted by one step, and another optimization is carried out. This yields $\mathbf{u}^{\text{opt}}(k+1)$ and its first element provides $u(k+1) = u^{\text{opt}}(k+1)$, etc. As illustrated in Figure 3 for a horizon length $N = 3$, the horizon taken into account in the minimization of the cost function V slides forward as k increases.

The design of observers for the system state is beyond the scope of this chapter. The interested reader is referred to [2, 17, 25], which illustrate the use of Kalman filters for MPC formulations in power electronics.

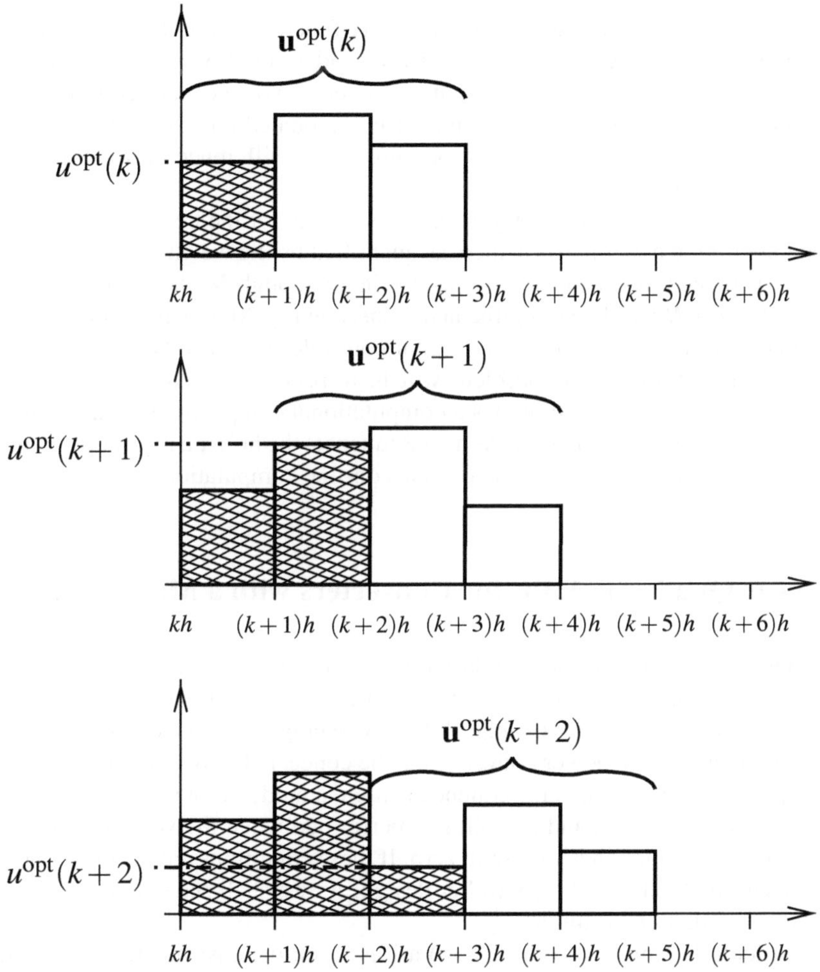

Fig. 3: Moving horizon principle with horizon length $N = 3$.

2.4 Design Parameters

As seen above, MPC allows one to treat multi-variable nonlinear systems in an, at least conceptually, simple way. In addition to choosing the sampling interval h (which, amongst other things, determines the system model (1)), MPC design essentially amounts to selecting the cost function, i.e., the weighting functions $F(\cdot)$ and $L(\cdot,\cdot)$, and the horizon length N.

As we shall see, the design of the weighting functions should take into account the actual control objectives and may also consider stability issues [3, 43]).[1] For example, tracking of the desired output and internal voltages and currents (which are assumed to be given, cf., [48]) can be accommodated into the MPC framework by choosing weights that penalize a measure of the difference between predicted and reference values.

For a given sampling frequency $1/h$, larger values for the horizon length N will in general provide better performance, as quantified by the weighting functions $F(\cdot)$ and $L(\cdot,\cdot)$. Indeed, one can expect that, for large enough N, the effect of $u(k)$ on $x'(\ell)$ for $\ell > k+N$ will be negligible and, consequently, MPC will approximate the performance of an infinite horizon optimal controller [27, 45]. On the other hand, the constrained optimization problem which, in principle, needs to be solved on-line to find the controller output, has a computational complexity which, in general, increases with the horizon length. As a consequence, the horizon parameter N allows the designer to trade-off performance versus on-line computational effort.

3 Linear Quadratic MPC for Converters with a Modulator

Most power converters use a modulation stage to synthesize the switching signals. To simplify the design of control strategies, it is common practice to separate control and modulation issues, see, e.g., [29]. By averaging the switching signal, the switching nature of the power converter can be concealed, provided that the switching frequency per fundamental frequency is high, a modulation method with a fixed modulation cycle is used and sampling is performed when the voltage and current ripples due to modulation are close to zero. If these conditions are fulfilled, one may use standard methods for the controller design. As we shall see below, for the case of MPC, the situation is similar.

A particularly simple case of (5)–(6) arises when the cost function is quadratic and the system model is linear and time-invariant, i.e.:

$$
\begin{aligned}
V(x(k),\mathbf{u}'(k)) &= x'^T(k+N)Px'(k+N) + \sum_{\ell=k}^{k+N-1} \left\{ x'^T(\ell)Qx'(\ell) + u'^T(\ell)Ru'(\ell) \right\}, \\
x'(\ell+1) &= Ax'(\ell) + Bu'(\ell), \\
& x'(\ell) \in \mathbb{X} \subseteq \mathbb{R}^n,\; u'(\ell) \in \mathbb{U} \subseteq \mathbb{R}^m, \quad \ell \in \{k, k+1, \dots k+N-1\},
\end{aligned}
\tag{10}
$$

where A and B denote the state-update and input matrices, and P, Q and R are positive semi-definite matrices of appropriate dimensions. The constraint sets $\mathbb{X}$ and $\mathbb{U}$ are polyhedra.

[1] Note that the weighting functions should be chosen such that $V(\cdot,\cdot)$ depends on the decision variables contained in $\mathbf{u}'(k)$, see (8).

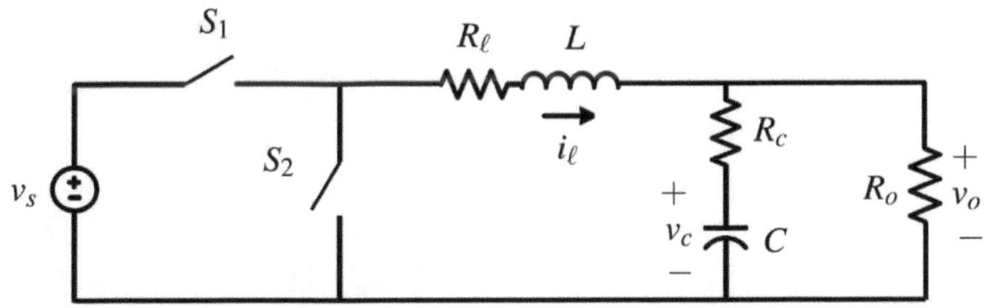

Fig. 4: Topology of the dc-dc buck converter.

Despite the ever growing computational power available and recent advances in implementing quadratic programming (QP) solvers on embedded system architectures, solving the QP in real-time for power electronics applications poses a highly challenging problem. When using sampling intervals in the μs range, the computation times needed to solve the QP typically exceed the sampling interval—often by one or two orders of magnitude. Rather than solving the mathematical optimization problem in real-time for the given state vector at the current time-step, the optimization problem can be solved offline for *all possible* states. Specifically, the so-called (explicit) *state-feedback control laws* as presented in previous parts of this book can be computed for all states $x(k) \in \mathbb{X}$ [6]. Explicit control laws are characterized via a polyhedral partition of the state space which can be stored in a look-up table. The optimal control input can thus be read from the look-up table in a computationally efficient manner.

Example 1. To further illustrate the derivation and properties of the explicit state-feedback control law of MPC, consider a dc-dc step-down synchronous converter. The latter is commonly referred to as a buck converter, and it is shown in Figure 4. Using the classic technique of averaging between the on and off modes of the circuit, the discrete-time system model

$$x(k+1) = Ax(k) + Bv_s d(k) \tag{11}$$

can be obtained, where v_s denotes the unregulated input voltage and $d(k)$ the duty cycle. The state vector contains the inductor current i_ℓ and the output voltage v_o, i.e. $x = [i_\ell \; v_o]^T$. From Figure 4, the continuous-time system matrices are

$$F = \begin{bmatrix} -R_\ell/L & -1/L \\ \frac{R_o}{R_o+R_c}\frac{L-R_cR_\ell C}{LC} & -\frac{1}{R_o+R_c}\frac{L+R_cR_oC}{LC} \end{bmatrix}, \quad G = \begin{bmatrix} 1/L \\ \frac{R_o}{R_o+R_c}\frac{R_c}{L} \end{bmatrix}, \tag{12}$$

whereas their discrete-time representations in (11) are given by

$$A = e^{Fh}, \quad B = \int_0^h e^{F\tau} G \, d\tau. \tag{13}$$

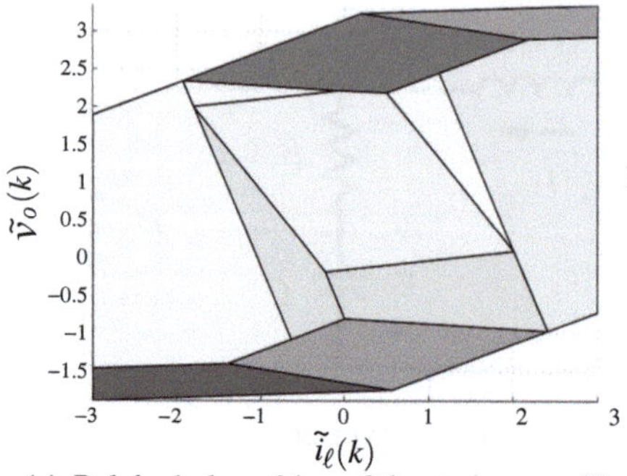

(a) Polyhedral partition of the state-space $\mathbb{X}$

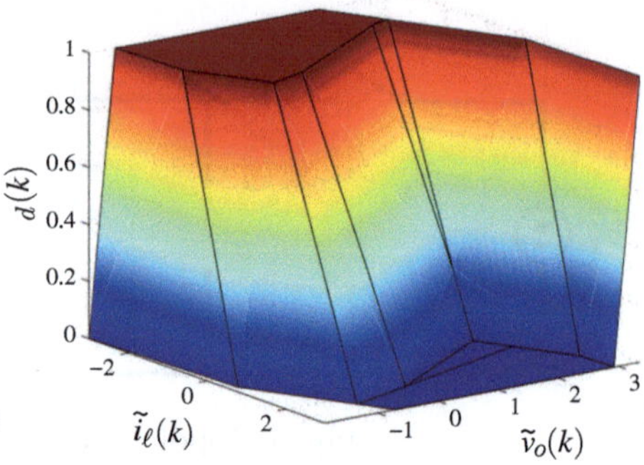

(b) Control law $u^{\text{opt}}(k)$

Fig. 5: Explicit state-feedback control law for the dc-dc buck converter over the state-space $\mathbb{X}$ spanned by the scaled inductor current $\tilde{i}_\ell(k)$ and the scaled output voltage $\tilde{v}_o(k)$

Adopting the per unit (pu) system, the parameters in (12) are here taken as the inductor $L = 3$ pu, capacitor $C = 20$ pu and output resistor $R_o = 1$ pu. The internal resistor of the inductor is set to $R_\ell = 0.05$ pu and the equivalent series resistance of the capacitor is $R_c = 0.005$ pu. The nominal input voltage is assumed to be $v_s = 1.8$ pu.

To allow for variations in the input voltage, it is convenient to scale the system equations by v_s, as proposed in [25]. To this end, we define $\tilde{i}_\ell = i_\ell / v_s$, $\tilde{v}_o = v_o / v_s$ and $\tilde{x} = [\tilde{i}_\ell \; \tilde{v}_o]^T$, and rewrite (11) as

$$\tilde{x}(k+1) = A\tilde{x}(k) + Bd(k). \tag{14}$$

Note that, unlike (11), (14) is linear in the state vector and the duty cycle.

The control objective is to regulate the output voltage to its reference $v_o^\star$ and to maintain the inductor current below its maximal allowed limit $i_{\ell,\max}$ by manipulating the duty cycle. The latter is bounded between zero and one. This control problem can be captured by the optimization problem (cf., (10))

$$\begin{aligned} V(\tilde{x}(k), \mathbf{u}(k)) &= \sum_{\ell=k}^{k+N-1} \left\{ (\tilde{x}'(\ell) - \tilde{x}^\star)^T Q (\tilde{x}'(\ell) - \tilde{x}^\star) + R(u'(\ell))^2 \right\}, \\ \tilde{x}'(\ell+1) &= A\tilde{x}'(\ell) + Bu'(\ell), \\ \tilde{x}'(\ell) &\in \mathbb{X}, \; u'(\ell) \in \mathbb{U}, \quad \ell \in \{k, k+1, \dots k+N-1\}, \end{aligned} \tag{15}$$

where we set $Q = \text{diag}(0,1)$, $R = 0.1$, $\mathbb{X} = [-\tilde{i}_{\ell,\max}, \tilde{i}_{\ell,\max}] \times [-10, 10]$ and $\mathbb{U} = [0,1]$. Note that $\tilde{i}_{\ell,\max} = i_{\ell,\max}/v_s$ and $u = d$. To facilitate the regulation of the output voltage to a non-zero reference, we define $\tilde{x}^\star = [0 \; \tilde{v}_o^\star]^T$ with $\tilde{v}_o^\star = v_o^\star / v_s$. We assume $\tilde{v}_o^\star = 0.5$ and choose the horizon $N = 3$.

The explicit control law can be computed using the MPT toolbox [33]. The two-dimensional state-space is partitioned into 20 polyhedra. Using optimal complexity reduction [24], an equivalent control law with 11 polyhedra can be derived, as shown in Figure 5(a). The corresponding state-feedback controller providing $u(k) = u^{\text{opt}}(k)$ is shown in Figure 5(b). Note that the duty cycle is limited by zero and one as a result of the design procedure. An additional patch, such as an anti-windup scheme, is not required, see also [12]. ■

A similar MPC scheme was proposed in [40]. This rather basic controller can be enhanced in various ways. In the context of dc-dc converters, it is usually preferred to penalize the *change* in the duty cycle rather than the duty cycle as such, by introducing $\Delta u(k) = u(k) - u(k-1)$ and penalizing $R(\Delta u(\ell))^2$ rather than $R(u(\ell))^2$ in (15). To enhance the voltage regulation at steady-state by removing any dc offset, an integrator state can be added [40]. Load variations can be addressed by a Kalman filter, see [25].

In the context of power electronics and drives applications, such MPC formulations have been studied extensively. One of the earliest references is [38], which proposes an explicit MPC controller in a field-oriented controller setting for an electrical drive. These initial results are extended in [41]. In [7], the speed and current control problem of a permanent-magnet synchronous machine is solved using MPC. Drives with flexible shafts are considered in [11], whereas [39] focuses on active rectifier units with LC filters.

4 Linear Quadratic Finite Control Set MPC

Controlling power converters without a modulator has received significant interest in recent years, leading to *direct control* methods. These methods combine the inner control loop, which typically controls the load currents, and the modulator in one computational stage. In doing so, the intrinsic delay of the modulator is avoided and the switching nature of the power converter can be directly addressed.

One of the most popular predictive control strategy for power electronic systems is FCS-MPC [9, 52]. This predictive control strategy explicitly models the switch positions by means of a finite control set. This implies that the input constraint set has a finite number of elements, as, for example, in (4).

In general, large prediction horizons N are preferable when using MPC. However, finding the optimal input sequence in case of FCS-MPC typically requires one to solve a combinatorial optimization problem [47]. Interestingly, for some topologies, one-step horizon MPC provides already good closed-loop performance [31, 50].

4.1 Closed-Form Solution

Consider again a quadratic cost function and a linear time-invariant system model. Unlike as in (10), however, the input constraint set $\mathbb{U}$ is now a finite control set. Hereafter, we revisit the closed-form expression for the solution to this linear quadratic FCS-MPC problem, as presented in [26, 47].

Firstly, we define the predicted state sequence

$$\mathbf{x}'_{[1:N]}(k) \triangleq [x'^T(k+1)\; x'^T(k+2)\; \dots\; x'^T(k+N)]^T. \tag{16}$$

The subscript $_{[1:N]}$ indicates that, unlike in (8), the state sequence is shifted by one time step.

Considering an initial system state $x'(k) = x(k)$, see also (7), we obtain

$$\mathbf{x}'_{[1:N]}(k) = \Phi \mathbf{u}'(k) + \Lambda x'(k),$$

where

$$\Phi \triangleq \begin{bmatrix} B & 0 & \cdots & 0 & 0 \\ AB & B & \cdots & 0 & 0 \\ \vdots & \vdots & \ddots & \vdots & \vdots \\ A^{N-1}B & A^{N-2} & \cdots & AB & B \end{bmatrix}, \quad \Lambda \triangleq \begin{bmatrix} A \\ A^2 \\ \vdots \\ A^N \end{bmatrix}.$$

In the following, we drop the time dependence of the state and input sequences in order to simplify the notation. The cost function (10) can then be re-written as

$$V(x, \mathbf{u}') = v(x) + \mathbf{u}'^T W \mathbf{u}' + 2\mathbf{u}'^T F x, \tag{17}$$

where $x = x(k)$, $\mathbf{u}' = \mathbf{u}'(k)$ and the term $v(x)$ is independent of $\mathbf{u}'$. In (17),

$$W \triangleq \Phi^T \mathscr{Q} \Phi + \mathscr{R} \in \mathbb{R}^{Nm \times Nm},$$
$$F \triangleq \Phi^T \mathscr{Q} \Lambda \in \mathbb{R}^{Nm \times n},$$

with

$$\mathscr{Q} \triangleq diag\{Q, \dots, Q, P\} \in \mathbb{R}^{Nn \times Nn},$$
$$\mathscr{R} \triangleq diag\{R, \dots, R\} \in \mathbb{R}^{Nm \times Nm}.$$

Notice that, if Q and R are positive definite, so is W.

Remark 1 (Unconstrained Solution) *If system constraints are not taken into account, i.e.* $\mathbb{U} \triangleq \mathbb{R}^m$ *and* $\mathbb{X} \triangleq \mathbb{R}^n$, *then* $V(x, \mathbf{u}')$ *is minimized when*

$$\mathbf{u}^{\text{opt}}_{uc}(x) \triangleq \arg\left\{ \min_{\mathbf{u}' \in \mathbb{R}^{Nm}} V(x, \mathbf{u}') \right\} \triangleq -W^{-1} F x. \tag{18}$$

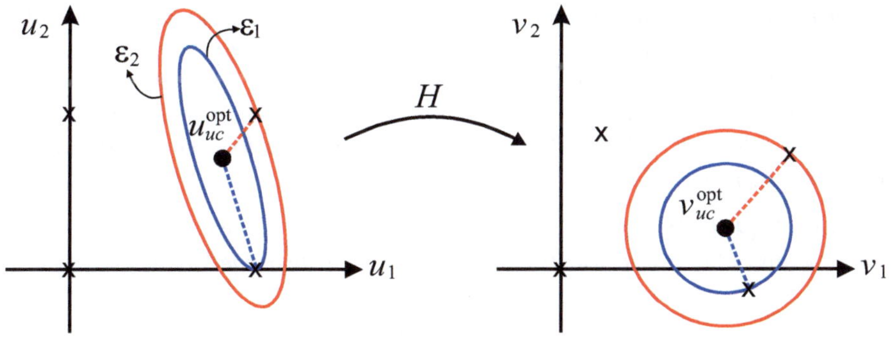

Fig. 6: Geometrical representation of the optimal solution for FCS-MPC with $u_1, u_2 \in \{0,1\}$ and the horizon $N = 1$.

Based on the unconstrained optimum, it is convenient to rewrite the cost function (17) as:

$$V(x,\mathbf{u}') = (\mathbf{u}' - \mathbf{u}_{uc}^{\text{opt}}(x))^T W(\mathbf{u}' - \mathbf{u}_{uc}^{\text{opt}}(x)) + g(x), \tag{19}$$

where the term $g(x)$ is independent of $\mathbf{u}'$.

To obtain the optimal finite set constrained solution one must find the control input which minimizes $V(x,\mathbf{u}')$. From (19), it follows that level sets of the cost function are ellipsoids, where the eigenvectors of W define the principal directions of the ellipsoid. Thus, the constrained optimizer $\mathbf{u}^{\text{opt}}(x)$ does not necessarily correspond to the nearest neighbour of $\mathbf{u}_{uc}^{\text{opt}}(x)$ within the constraint set $\mathbb{U}^N$.

Example 2. Consider the case where a power converter, modeled as a linear time-invariant model, has two semiconductor switches, which can take two values, i.e., $u_1, u_2 \in \{0,1\}$. Thus, the control input belongs to the following finite set:

$$u \in \mathbb{U} \triangleq \left\{ \begin{bmatrix}0\\0\end{bmatrix}, \begin{bmatrix}0\\1\end{bmatrix}, \begin{bmatrix}1\\0\end{bmatrix}, \begin{bmatrix}1\\1\end{bmatrix} \right\} \subset \mathbb{R}^2. \tag{20}$$

A geometrical representation of the situation for the case when the horizon is $N = 1$ is depicted in Figure 6 (left). Here, the ellipses, ε_i centred in u_{uc}^{opt}, represent all the points that lead to the same cost. Formally, if $a, b \in \varepsilon_i$ then, $V(x,a) = V(x,b)$.

As we move away from the centre, the ellipses become larger, increasing the cost function value, i.e., if $a \in \varepsilon_1$ and $b \in \varepsilon_2$ then, $V(x,a) < V(x,b)$. Thus, in this example, the optimal solution, which produces the minimum cost function value is $u^{\text{opt}} = [1\ 0]^T$, despite the nearest vector to the unconstrained solution being $u = [1\ 1]^T$. Clearly, the optimal solution is, in general, not the nearest neighbour to the unconstrained solution. ■

Based on the above observations, one can derive a closed-form solution to the finite-set constrained optimization problem at hand.

Definition 1 (Vector Quantizer (see, e.g., [18])) *Consider a set $\mathscr{A} \subseteq \mathbb{R}^n$ and a countable (not necessarily finite) set $\mathscr{B} \triangleq \{b_i\} \subset \mathbb{R}^n$, $i \in \mathscr{I} \subseteq \mathbb{N}$ which satisfies that $\exists \varepsilon > 0 : | b_i - b_j | \geq \varepsilon, \forall i, j \in \mathscr{I}$. A function $q_{\mathscr{B}}(\cdot) : \mathscr{A} \to \mathscr{B}$ is an Euclidean vector quantizer if $q_{\mathscr{B}}(a) = b_i \in \mathscr{B}$ if and only if b_i satisfies that $| a - b_i | \leq | a - b_j |$, for all $b_j \neq b_i$, where $b_j \in \mathscr{B}$. The associated quantization error is defined as $\bar{\eta}_{\mathscr{B}}(a) \triangleq q_{\mathscr{B}}(a) - a$.*

Theorem 1 ([47]) *Denote the elements of $\mathbb{U}^N \triangleq \mathbb{U} \times \cdots \times \mathbb{U}$ via $\{\mu_1, \ldots, \mu_r\}$. Consider a matrix H that satisfies $H^T H = W$. Then, the constrained optimizer*

$$\mathbf{u}^{opt}(x) \triangleq \arg\left\{ \min_{\mathbf{u}' \in \mathbb{U}^N} V(x, \mathbf{u}') \right\} \tag{21}$$

is given by

$$\mathbf{u}^{opt}(x) = H^{-1} q_{\mathbb{V}}\left(H^{-1} \mathbf{u}_{uc}^{opt}(x)\right) = H q_{\mathbb{V}}\left(-H^{-T} F x\right), \tag{22}$$

where the vector quantizer $q_{\mathbb{V}}$ maps $\mathbb{R}^{Nm}$ to $\mathbb{V}$. The latter set is defined via $\mathbb{V} \triangleq \{v_1, \ldots, v_r\} \subset \mathbb{R}^{Nm}$, in which $v_i = H\mu_i$ for all $\mu_i \in \mathbb{U}^N$.

Proof. To obtain the optimal solution, we define $\mathbf{v}' = H\mathbf{u}'$. Now, the cost function (19) can be expressed as:

$$V(x, \mathbf{v}') \triangleq (\mathbf{v}' - \mathbf{v}_{uc}^{\text{opt}}(x))^T (\mathbf{v}' - \mathbf{v}_{uc}^{\text{opt}}(x)) + g(x), \tag{23}$$

where

$$\mathbf{v}_{uc}^{\text{opt}}(k) \triangleq H\mathbf{u}_{uc}^{\text{opt}}(x).$$

Thus, in terms of $\mathbf{v}'$, the level sets of the cost function describe spheres centred at $\mathbf{v}_{uc}^{\text{opt}}$, as depicted in Figure 6 (right). Therefore, in terms of these transformed variables, the nearest vector to the unconstrained solution $\mathbf{v}_{uc}^{\text{opt}}(x)$ is indeed the (constrained) optimal solution. ■

Notice that if W is symmetric and positive definite, then it is always possible to obtain a matrix H that satisfies $H^T H = W$, e.g., $H = W^{1/2}$, as chosen in [47].

4.2 Design for Stability and Performance

We will next investigate stabilizing properties of FCS-MPC. For that purpose, we will include additional terminal constraints in the problem formulation of Section 2.2. This will allow us to adapt robust control concepts to suit the problem at hand.

For our subsequent analysis, we shall assume that the pair (A, B) is stabilizable and that the matrices Q and R are positive definite. A widely used idea to establishing stability of MPC is based on finding a known control policy, say $\kappa_f(x)$, which stabilizes the system model within a given terminal region $\mathbb{X}_f$, see [49]. In particular, for a disturbance-free LTI system with convex constraints, say

$$x(k+1) = Ax(k) + B\bar{u}(k), \tag{24}$$

using quadratic MPC, one can use a fixed state feedback gain as a stabilizing controller for the terminal region $\mathbb{X}_f$ (see Section 2.5 in [49]). To adapt this idea to systems with finite control inputs, we first introduce an associated convex set via:

$$\bar{\mathbb{U}} \triangleq \{\bar{u} \in \mathbb{R}^m : |\bar{u}| \leq \bar{u}_{\max}\},$$

where $\bar{u}_{\max} \in (0,\infty)$ is a design parameter. Since $\bar{\mathbb{U}}$ is bounded, so is the quantization effect, i.e.,

$$\Delta_q \triangleq \max_{\bar{u}\in\bar{\mathbb{U}}} |q_{\mathbb{U}}(\bar{u}) - \bar{u}| < \infty. \tag{25}$$

Note that Δ_q depends upon $\bar{u}_{\max}$.

Based on this, stability of FCS-MPC can be examined by investigating properties of a local controller $\kappa_f(x)$ corresponding to the optimal solution presented in (22) with prediction horizon $N = 1$. In this case, one has $F = B^T PA$ and $W = B^T PB + R$, so that

$$\mathbf{u}_{uc}^{\text{opt}}(x) = Kx, \quad K = -W^{-1}F. \tag{26}$$

The above motivates one to impose that the terminal state in the optimization lies inside a terminal region: $x(k+N) \in \mathbb{X}_f$, with

$$\mathbb{X}_f \triangleq \{x \in \mathbb{R}^n : |x| \leq b\}, \quad b \triangleq \frac{\bar{u}_{\max}}{|K|}. \tag{27}$$

Within this region the local controller satisfies

$$\kappa_f(x) = Kx + H^{-1}\eta_{\mathbb{V}}(x), \quad x \in \mathbb{X}_f, \tag{28}$$

where $\eta_{\mathbb{V}}(x) \triangleq \bar{\eta}_{\mathbb{V}}(W^{-1/2}Kx)$. Clearly,

$$\begin{aligned} |\eta_{\mathbb{V}}(x)| &\leq |q_{\mathbb{V}}(HKx) - HKx| \leq |Hq_{\mathbb{U}}(Kx) - HKx| \\ &\leq |H||q_{\mathbb{U}}(Kx) - Kx| \leq |H|\Delta_q, \end{aligned} \tag{29}$$

where we have used (25).

Consequently, system (24) with the proposed local controller $\kappa_f(x)$ in (28) can be expressed via:

$$x(k+1) = A_K x(k) + w_f(x(k)), \quad \forall x(k) \in \mathbb{X}_f, \tag{30}$$

where $A_K = A + BK$, and $w_f(x(k)) = BH^{-1}\eta_v(x(k))$ represents the effect of the quantization on the "nominal system", $x(k+1) = A_K x(k)$.

Notice that, in (30), $w_f(x)$ is not an external disturbance but a known discontinuity produced by the quantization, which makes (30) a nonlinear system. The key point here is that $w_f(x)$ is bounded on $\mathbb{X}_f$. Therefore, the local controller can be

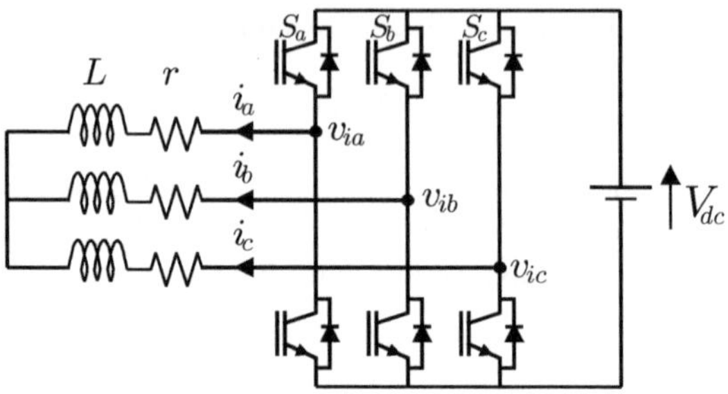

Fig. 7: Two-level inverter topology.

shown to be stabilizing if it is robust to bounded input disturbances. As shown in [3], it is convenient to choose the matrix P in (10) as the (unique) solution to the algebraic Riccati equation

$$A_K^T P A_K + Q + K^T R K - P = 0. \tag{31}$$

With this choice, $\kappa_f(x)$ in (28) can be used to guarantee closed-loop stability of FCS-MPC. Theorem 2, given below, establishes that for all $x(0)$ that belong to the feasible set X_N, the system will be steered by the multi-step predictive controller towards the terminal region $\mathbb{X}_f \subseteq X_N$ and then (with the same controller) into an ultimately bounded set $\mathscr{D}_{\delta_N} \subset \mathbb{X}_f$.

Theorem 2 ([3]) *Let $\mathscr{D}_{\delta_N} \triangleq \{x \in \mathbb{X}_f : |x| \leq \delta_N\}$ be a neighbourhood of the origin, where*

$$\delta_N^2 \triangleq \gamma_N \Delta_q^2, \quad \gamma_N \triangleq \left(\frac{1 + (1-\rho)N}{\lambda_{\min}(Q)(1-\rho)} \right) |W|. \tag{32}$$

Suppose that $x(0) \in X_N$ and the matrix P in (10) satisfies (31). If Δ_q in (25) is bounded by

$$\Delta_q^2 < \frac{b^2}{\gamma_N}, \tag{33}$$

then $\limsup_{k\to\infty} |x(k)| \leq \delta_N$. Furthermore, there exists a finite instant $t > 0$, such that after that instant, the system state $x(k)$ converges at an exponential rate, i.e., there exists $c > 0$ and $\rho \in [0,1)$, such that

$$|x(k)|^2 \leq c\rho^{k-t}|x(t)|^2 + \gamma_N \Delta_q^2, \forall k \geq t, \tag{34}$$

where $c = \lambda_{\max}(P)/\lambda_{\min}(Q)$ and $\rho = 1 - 1/c$, with $\lambda_{\min}(Q) \leq \lambda_{\max}(P)$.

4.3 Example: Reference Tracking

The topology of a two-level inverter is presented in Figure 7. The associated continuous-time dynamic model for the three-phase output current, $i_{abc} \triangleq [i_a\ i_b\ i_c]^T$, is

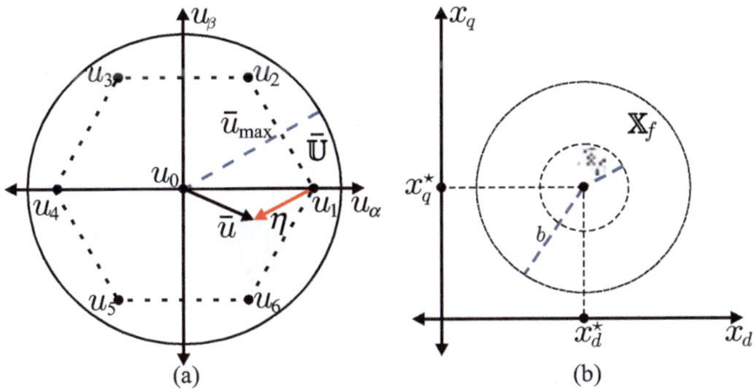

Fig. 8: Sets involved in the cost function design; (a) finite control set $\mathbb{U}$ and nominal input set $\bar{\mathbb{U}}$; (b) terminal region $\mathbb{X}_f$ and bounded set $\mathscr{D}_\delta$.

$$\frac{di_{abc}(t)}{dt} = -\frac{r}{L}i_{abc}(t) + \frac{1}{L}(V_{dc}S_{abc}(t) - v_o(t)I_{3\times 1}), \tag{35}$$

where V_{dc} denotes the dc-link voltage and v_o stands for the common-mode voltage. The latter is defined as $v_o = \frac{1}{3}(v_a + v_b + v_c)$, where v_a, v_b and v_c are the voltages at the inverter terminals, see Figure 7. The switch positions, $S_{abc} \triangleq [S_a\ S_b\ S_c]^T$, belong to the following finite set

$$\mathbb{S} = \left\{ \begin{bmatrix}0\\0\\0\end{bmatrix}, \begin{bmatrix}0\\0\\1\end{bmatrix}, \begin{bmatrix}0\\1\\0\end{bmatrix}, \begin{bmatrix}0\\1\\1\end{bmatrix}, \begin{bmatrix}1\\0\\0\end{bmatrix} \begin{bmatrix}1\\0\\1\end{bmatrix}, \begin{bmatrix}1\\1\\0\end{bmatrix}, \begin{bmatrix}1\\1\\1\end{bmatrix} \right\}. \tag{36}$$

For this converter, the control target is to track three-phase sinusoidal references of the form:

$$i^\star_{abc}(t) = a^\star \left[\sin(\omega t)\ \sin(\omega t - 2\pi/3)\ \sin(\omega t + 2\pi/3)\right]^T \tag{37}$$

We will next illustrate how the preceding ideas can be applied to this situation. For that purpose, we first note that sinusoidal quantities in a three-phase system can be transformed into a rotating orthogonal dq reference frame using the so-called Park transformation. More specifically, the three-phase current i_{abc} in (35) is transformed into the dq frame by the transformation

$$i_{dq}(t) = \Gamma(t) i_{abc}(t), \tag{38}$$

where:

$$\Gamma(t) \triangleq \frac{2}{3}\begin{bmatrix}\sin(\omega t) & \sin(\omega t - \frac{2\pi}{3}) & \sin(\omega t + \frac{2\pi}{3})\\ \cos(\omega t) & \cos(\omega t - \frac{2\pi}{3}) & \cos(\omega t + \frac{2\pi}{3})\end{bmatrix}, \tag{39}$$

and $i_{dq} \triangleq [i_d\ i_q]^T$.

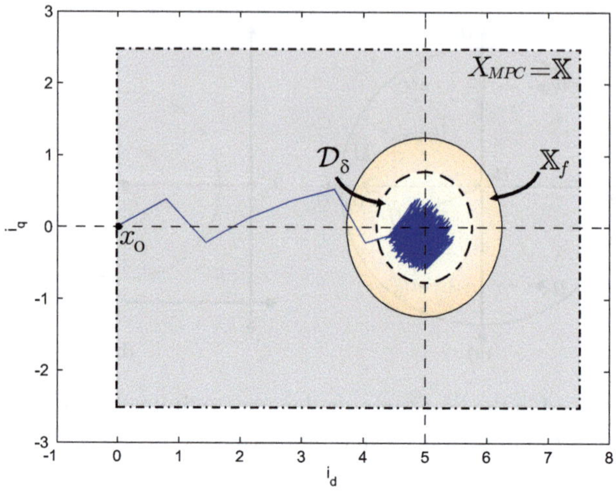

Fig. 9: Convergence of the two-level inverter for $R = 0.0001 I_{2\times 2}$.

Figure 8 (left) shows the typical output voltages of a two-level inverter in a vectorial representation in the stationary orthogonal $\alpha\beta$ coordinate system. The finite input set, $\mathbb{U}$, contains the 7 inverter vectors, which are contained by the nominal input set, $\bar{\mathbb{U}}$, i.e.,

$$\mathbb{U} = \{u_0, \ldots, u_6\} \subset \bar{\mathbb{U}} \subset \mathbb{R}^2. \tag{40}$$

In this case, the quantization of the nominal input $\bar{u} \in \bar{\mathbb{U}}$ is given by $q_{\mathbb{U}}(\bar{u}) = u_1$, thus $\eta_{\mathbb{U}}(\bar{u}) = u_1 - \bar{u}$. Notice that the inverter voltage vectors rotate in the dq reference frame. However, they always will be contained by the nominal input set, $\bar{\mathbb{U}}$, producing the same maximum quantization error Δ_q as in the $\alpha\beta$ coordinate system.

Thus, considering $x = i_{dq}$ and $u = S_{dq}$, the discrete-time model of the two-level inverter in the dq frame is

$$x(k+1) = Ax(k) + Bu(k), \quad u(k) \in \mathbb{U}(k), \tag{41}$$

$$A = \begin{bmatrix} 1 - hr/L & \omega h \\ -\omega h & 1 - hr/L \end{bmatrix}, \quad B = (h/L) V_{dc} I_{2\times 2}, \tag{42}$$

where h is the sampling period and

$$\mathbb{U}(k) = \Gamma(kh)\mathbb{S}. \tag{43}$$

In this case, with a current reference of constant amplitude $a^\star$, the reference

$$x^\star = i^\star_{dq} = [a^\star \; 0]^T \tag{44}$$

directly follows. The input required to keep this state value is given by

$$u^\star = S^\star_{dq} = [ra^\star/V_{dc} \quad \omega La^\star/V_{dc}]^T. \tag{45}$$

Here, experimental results of the performance of FCS-MPC when applied to a three-phase two-level inverter are presented. The inverter prototype was built using discrete insulated-gate bipolar transistors (IGBTs) IRG4PC30KD. The electrical parameters of the converter-load system are $V_{dc} = 200\ V$, $r = 5\ \Omega$ and $L = 17\ mH$, see Figure 7. The predictive strategy was implemented in a standard TMS320C6713 DSP considering a sampling period of $h = 100\ \mu s$. The desired amplitude for the output current is $a^\star = 5\ A$ with an angular frequency of $\omega = 2\pi 50$ rad/s.

Following the result in Theorem 1, one obtains for the weighting matrices $Q = I_{2x2}$ and $R = 2I_{2x2}$ that

$$P = 1.7455 I_{2\times 2}, \quad K = \begin{bmatrix} -0.4514 & -0.0146 \\ 0.0146 & -0.4514 \end{bmatrix}. \tag{46}$$

A key observation is that the time-varying constraint set $\mathbb{U}$ in (43) can be bounded by a fixed nominal set $\bar{\mathbb{U}}$. In Figure 8, one can see that when the nominal input $\bar{u}$ is inside the hexagon-shaped boundary, the maximum quantization error, Δ_q, is given by the centroid of the equilateral triangle formed by the adjacent inverter vectors. Therefore, the maximum quantization error is given by $\Delta_q = 2\frac{\sqrt{3}}{9}$. The associated nominal input set can be chosen as:

$$\bar{\mathbb{U}} \triangleq \{\bar{u} \in \mathbb{R} : |\bar{u}| \leq 2\Delta_q\},$$

while the terminal region can be characterized via (see [4] for details):

$$\mathbb{X}_f \triangleq \left\{ x \in \mathbb{R}^n : |x - x^\star| \leq \frac{u_{\max} - |u^\star|}{|K|} = 1.3 \right\}$$

which provides that

$$|\eta_{\mathbb{U}}(\bar{u})| \leq \Delta_q = 2\frac{\sqrt{3}}{9}, \quad \forall x \in \mathbb{X}_f.$$

Thus, one can anticipate that the system state will be led by the predictive controller to the ultimately invariant set:

$$\mathscr{D}_{\delta_N} \triangleq \{x \in \mathbb{R}^n : |x - x^\star| \leq \delta = 0.8088\}. \tag{47}$$

The evolution of the two-level inverter using FCS-MPC with $N = 1$ and starting from $i_d = i_q = 0$ is depicted in Figure 9. Here, one can see that the predictive controller leads the system state to the terminal region, $\mathbb{X}_f$, and then to $\mathscr{D}_{\delta_N}$. As expected for this kind of controller, the inverter voltage spectrum is spread, as can be observed in Figure 10. If this is undesired, then one can use noise shaping techniques, as described in [10, 46].

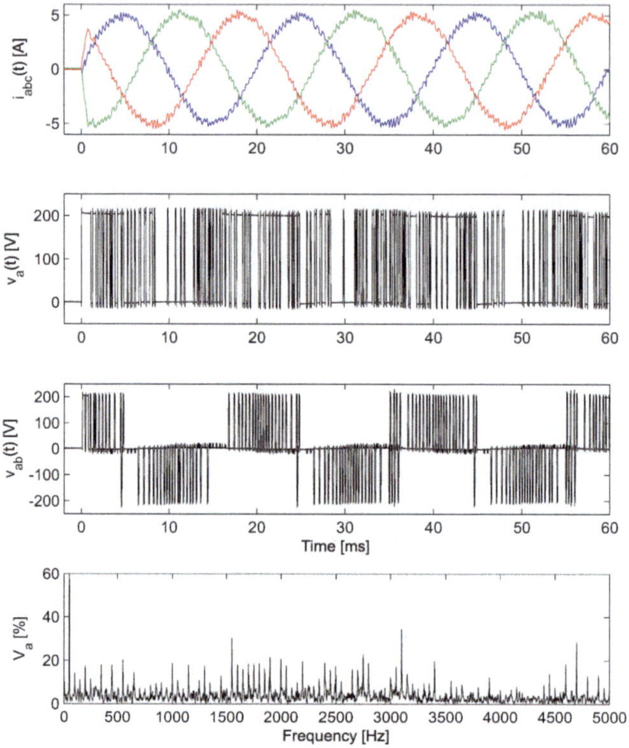

Fig. 10: System state and input trajectories, and inverter voltage spectrum.

5 An Efficient Algorithm for Finite-Control Set MPC

In this section we consider the cost function

$$V(x(k),\mathbf{u}(k)) = \sum_{\ell=k}^{k+N-1} (y^*(\ell+1)-y'(\ell+1))^T(y^*(\ell+1)-y'(\ell+1)) + \lambda_u(\Delta u'(\ell))^T\Delta u'(\ell), \tag{48}$$

which penalizes the predicted output errors and the control effort

$$\Delta u'(\ell) \triangleq u'(\ell)-u'(\ell-1).$$

The latter is weighted by the non-negative scalar weighting factor λ_u. The cost function (48) is minimized subject to

$$\begin{aligned} &\mathbf{u}(k) \in \mathbb{U}^N \\ &\|\Delta u'(\ell)\|_\infty \leq 1, \quad \forall \ell \in \{k, k+1, \ldots, k+N-1\}, \end{aligned} \tag{49}$$

where the first constraint restricts the sequence of manipulated variables to the set of feasible switch positions of the converter. In many converters the second constraint is required to avoid switching in a phase by more than one step up or down.

Owing to the discrete nature of the decision variable $\mathbf{u}(k)$, minimizing (48) subject to (49) is difficult, except for short horizons. In fact, as the prediction horizon is enlarged and the number of decision variables is increased, the (worst-case) computational complexity grows exponentially, thus, cannot be bounded by a polynomial, see also [47]. The difficulties associated with minimizing V become apparent when using exhaustive search. With this method, the set of admissible switching sequences $\mathbf{u}(k)$ is enumerated and the cost function evaluated for each such sequence. The switching sequence with the smallest cost is (by definition) the optimal one and its first element is chosen as the control input.

It is easy to see that exhaustive search is computationally feasible only for very small horizons N, such as one or two. In fact, for $N = 5$, assuming a three-level converter, the number of switching sequences amounts to $1.4 \cdot 10^7$.

Techniques from vector quantization [18] and from mathematical programming, such as branch and bound [19, 34, 44], can be used to reduce the computational burden. However, none of the general methods take advantage of the particular structure of (48) and the fact that in MPC the solution is implemented in a moving horizon manner.

To address computational issues, we will exploit the geometrical structure of the underlying MPC optimization problem and present a practical optimization algorithm. The algorithm uses elements of sphere decoding [28] to provide optimal switching sequences, requiring only little computational resources, thus, enabling the use of longer prediction horizons in practical applications [5, 21, 22].

We will illustrate the ideas on a variable speed drive application consisting of a three-level neutral point clamped voltage source inverter driving an induction machine. The methods proposed and results obtained are directly applicable to both the machine-side inverter in an ac drive setting and to grid-side converters. The ideas can also be used for other converter topologies and are particularly promising for topologies with a high number of voltage levels.

5.1 Modified Sphere Decoding Algorithm

Using algebraic manipulations akin to those mentioned in Section 4, it is easy to show that the minimization of (48) amounts to finding

$$\mathbf{u}^{\text{opt}}(k) = \arg\min_{\mathbf{u}} (\mathbf{z} - H\mathbf{u})^T (\mathbf{z} - H\mathbf{u}), \quad \text{subject to (49)}, \tag{50}$$

where H is an invertible lower-triangular matrix. In (50), we use

$$\mathbf{z} = H\mathbf{u}^{\text{uc}},$$

where $\mathbf{u}^{\text{uc}}$ is the sequence obtained from optimizing (48) *without constraints*, i.e., with $\mathbb{U} = \mathbb{R}^3$. Thus, we have rewritten the MPC optimization problem as a (truncated) *integer least-squares* problem. Interestingly, various efficient solution algorithms for (50) subject to finite-set constraints have been developed in recent years; see, e.g.,[1] and the references therein. We will next show how to adapt the sphere decoding algorithm [16, 28] to find the optimal switching sequence $\mathbf{u}^{\text{opt}}(k)$.

The basic idea of the algorithm is to iteratively consider candidate sequences, say $\mathbf{u} \in \mathbb{U}^N$, which belong to a sphere of radius $\rho(k) > 0$ centred in $\mathbf{z}$,

$$(\mathbf{z} - H\mathbf{u})^T(\mathbf{z} - H\mathbf{u}) \leq \rho(k). \tag{51}$$

Especially in the case of multilevel converters (where $\mathbb{U}$ has many elements; see, e.g., [37]), the set of candidate sequences satisfying the above conditions is much smaller than the original constraint set $\mathbb{U}^N$. Not surprisingly, computation times can be drastically reduced compared to exhaustive search.

A key property used in sphere decoding is that, since H is triangular, for a given radius, identifying candidate sequences which satisfy (51) is very simple. In particular, for the present case, H is lower triangular, thus (51) can be rewritten as

$$\rho^2(k) \geq (z_1 - H_{(1,1)}u_1)^2 + (z_2 - H_{(2,1)}u_1 - H_{(2,2)}u_2)^2 + \dots \tag{52}$$

where z_i denotes the i-th element of $\mathbf{z}$, u_i is the i-th element of $\mathbf{u}$, and $H_{(i,j)}$ refers to the (i, j)-th entry of H. Therefore, the solution set of (51) can be found by proceeding in a sequential manner akin to Gaussian elimination, in the sense that at each step only a one-dimension problem needs to be solved; for details, see [28].

The algorithm requires an initial value for the radius used at time k to determine $\mathbf{u}$. On the one hand, the radius $\rho(k)$ should be as small as possible, enabling us to remove as many candidate solutions *a priori* as possible. On the other hand, $\rho(k)$ must not be too small, to ensure that the solution set is non-empty. As shown in [21], it is convenient to choose the initial radius by using the following *educated guess* for the optimal solution:

$$\mathbf{u}^{\text{sub}}(k) = \begin{bmatrix} 0 & I & 0 & \dots & 0 \\ 0 & 0 & I & \ddots & \vdots \\ \vdots & & \ddots & \ddots & 0 \\ 0 & \dots & \dots & 0 & I \\ 0 & \dots & \dots & 0 & I \end{bmatrix} \mathbf{u}^{\text{opt}}(k-1), \tag{53}$$

which is obtained by shifting the previous solution by one time-step and repeating the last switch position. This is in accordance with the moving horizon optimization paradigm. Since the optimal solution at the previous time-step satisfies the constraint, $\mathbf{u}^{\text{sub}}(k)$ is a feasible solution candidate of (48). Given (53), the initial value of $\rho(k)$ is then set to:

$$\rho(k) = (\mathbf{z} - H\,\mathbf{u}^{\text{sub}}(k))^T(\mathbf{z} - H\,\mathbf{u}^{\text{sub}}(k)). \tag{54}$$

Algorithm 1: Modified sphere decoding algorithm

function $\mathbf{u}^{\text{opt}}(k)$ = MSphDec $\mathbf{u}$, d^2, i, ρ^2, $\mathbf{z}$
  for each $u \in \{-1,0,1\}$ **do**
    $u_i \leftarrow u$
    $d'^2 \leftarrow (z_i - H_{(i,1:i)}\mathbf{u}_{1:i})^T(z_i - H_{(i,1:i)}\mathbf{u}_{1:i}) + d^2$
    if $d'^2 \leq \rho^2$ **then**
      if $i < 3N$ **then**
        $\text{MSPHDEC}(\mathbf{u}, d'^2, i+1, \rho^2, \mathbf{z})$
      else
        if u meets (49) **then**
          $\mathbf{u}^{\text{opt}} \leftarrow \mathbf{u}$
          $\rho^2 \leftarrow d'^2$
        end if
      end if
    end if
  end for
end function

At each time-step k, the controller first uses the current system state $\mathbf{x}(k)$, the future reference values, the previous switch position $\mathbf{u}(k-1)$ and the previous optimizer $\mathbf{u}^{\text{opt}}(k-1)$ to calculate $\mathbf{u}^{\text{sub}}(k)$, $\rho(k)$ and $\mathbf{z}$. The optimal switching sequence $\mathbf{u}^{\text{opt}}(k)$ is then obtained by invoking Algorithm 1 (see [21]):

$$\mathbf{u}^{\text{opt}}(k) = \text{MSPHDEC}(\emptyset, 0, 1, \rho^2(k), \mathbf{z}), \tag{55}$$

where $\emptyset$ is the empty set[2].

As can be seen in Algorithm 1, this modification to sphere decoding operates in a recursive manner. Starting with the first component, the switching sequence $\mathbf{u}$ is built component by component, by considering the admissible single-phase switch positions in the constraint set $\{-1,0,1\}$. If the associated squared distance is smaller than the current value of ρ^2, then one proceeds to the next component. If the last component,i.e., u_{3N}, has been reached, meaning that $\mathbf{u}$ is of full dimension $3N$, then $\mathbf{u}$ is a candidate solution. If $\mathbf{u}$ meets the switching constraint (49) and if the distance is smaller than the current optimum, then one updates the incumbent optimal solution $\mathbf{u}^{\text{opt}}$ and also the radius ρ.

The computational advantages of this algorithm stem from adopting the notion of branch and bound [34, 44]. Branching is done over the set of single-phase switch positions $\{-1,0,1\}$; bounding is achieved by considering solutions only within the sphere of current radius. If the distance d' exceeds the radius, a certificate has been found that the branch (and all its associated switching sequences) provides only

[2] The notation $H_{(i,1:i)}$ refers to the first i entries of the i-th row of H; similarly, $\mathbf{u}_{1:i}$ are the first i elements of the vector $\mathbf{u}$. Note that the matrix H is time-invariant and does not change when running the algorithm. Therefore, H can be computed once offline before the execution of the algorithm.

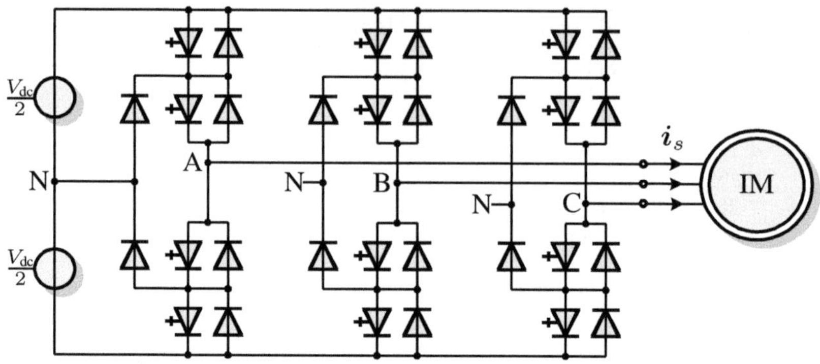

Fig. 11: Three-level three-phase neutral point clamped voltage source inverter driving an induction motor with a fixed neutral point potential.

solutions worse than the incumbent optimum. Therefore, this branch can be pruned, i.e., removed from further consideration without exploring it. During the optimization procedure, whenever a better incumbent solution is found, the radius is reduced and the sphere thus tightened, so that the set of candidate sequences is as small as possible, but non-empty. The majority of the computational burden relates to the computation of d' via evaluating the terms $H_{(i,1:i)}\mathbf{u}_{1:i}$. Thanks to (52), d' can be computed sequentially, by computing only the squared addition due to the ith component of $\mathbf{u}$. In particular, the sum of squares in d, accumulated over the layers 1 to $i-1$, does not need to be recomputed.

5.2 Simulation Study of FCS-MPC

As an illustrative example of a power electronics system, we consider a medium-voltage variable speed drive system consisting of a neutral point clamped (NPC) voltage source inverter (VSI) and a squirrel-cage induction machine (IM). This setup is shown in Figure 11. The inverter can synthesize three output voltage levels at each of its three phase terminals. The total dc-link voltage V_{dc} is assumed constant and the neutral point potential N is fixed.

System Model

Let the integer variables u_a, u_b, $u_c \in \{-1, 0, 1\}$ denote the switch positions in the three phase legs. The voltage vector applied to the machine terminals in the stationary orthogonal $\alpha\beta$ coordinate system is

$$v_{s,\alpha\beta} = \begin{bmatrix} v_{s\alpha} \\ v_{s\beta} \end{bmatrix} = \frac{1}{2} V_{\text{dc}} \mathscr{P} u \tag{56}$$

with

$$\mathscr{P} \triangleq \frac{2}{3}\begin{bmatrix} 1 & -\frac{1}{2} & -\frac{1}{2} \\ 0 & \frac{\sqrt{3}}{2} & -\frac{\sqrt{3}}{2} \end{bmatrix}, \quad u \triangleq \begin{bmatrix} u_a \\ u_b \\ u_b \end{bmatrix} \in \mathbb{U} \text{ and } \mathbb{U} \triangleq \{-1,0,1\}^3. \tag{57}$$

For the state-space model of an induction machine in the stationary coordinate system, we choose the stator currents $i_{s\alpha}$ and $i_{s\beta}$ and the rotor flux linkages $\psi_{r\alpha}$ and $\psi_{r\beta}$ as state vector

$$x \triangleq \begin{bmatrix} i_{s\alpha} & i_{s\beta} & \psi_{r\alpha} & \psi_{r\beta} \end{bmatrix}^T.$$

The model input are the stator voltages $v_{s\alpha}$ and $v_{s\beta}$ as defined in (56). The model parameters are the stator and rotor resistances R_s and R_r, and the stator, rotor and mutual reactances X_{ls}, X_{lr} and X_m, respectively. Assuming operation at a constant speed, the angular velocity of the rotor, ω_r, is also a parameter. The continuous-time state-space equations of the squirrel-cage induction machine are then (see [32])

$$\frac{\mathrm{d}i_{s,\alpha\beta}}{\mathrm{d}t} = -\frac{1}{\tau_s} i_{s,\alpha\beta} + \left(\frac{1}{\tau_r} - \omega_r \begin{bmatrix} 0 & -1 \\ 1 & 0 \end{bmatrix} \right) \frac{X_m}{D} \psi_{r,\alpha\beta} + \frac{X_r}{D} v_{s,\alpha\beta} \tag{58a}$$

$$\frac{\mathrm{d}\psi_{r,\alpha\beta}}{\mathrm{d}t} = \frac{X_m}{\tau_r} i_{s,\alpha\beta} - \frac{1}{\tau_r} \psi_{r,\alpha\beta} + \omega_r \begin{bmatrix} 0 & -1 \\ 1 & 0 \end{bmatrix} \psi_{r,\alpha\beta}, \tag{58b}$$

where we have used

$$X_s \triangleq X_{ls} + X_m, \; X_r \triangleq X_{lr} + X_m, \; D \triangleq X_s X_r - X_m^2, \; \tau_s \triangleq \frac{X_r D}{R_s X_r^2 + R_r X_m^2} \text{ and } \tau_r \triangleq \frac{X_r}{R_r}.$$

The objective of the current controller is to manipulate the three-phase switch position u such that the stator current vector $i_{s,\alpha\beta}$ closely tracks its reference. To this end, we define the system output vector $y \triangleq i_{s,\alpha\beta}$ and its reference $y^* \triangleq i^*_{s,\alpha\beta}$. The second control objective is to minimize the switching effort, i.e., the switching frequency or the switching losses.

Performance Evaluation

As an example of a typical medium-voltage induction machine, consider a 3.3 kV and 50 Hz squirrel-cage induction machine rated at 2 MVA with a total leakage inductance of 0.25 pu. The dc-link voltage is $V_{\mathrm{dc}} = 5.2\,\mathrm{kV}$ and assumed to be constant. The parameters of the drive system are provided in [22]. We consider operation at the fundamental frequency 50 Hz and full torque. The controller uses the sampling interval $h = 25\,\mu\mathrm{s}$.

During steady-state operation, the key control performance criteria are the device switching frequency f_{sw} and the total harmonic distortions (THD) of the current I_{THD}. We will also investigate the empirical *closed-loop* cost, V_{cl}, which—in

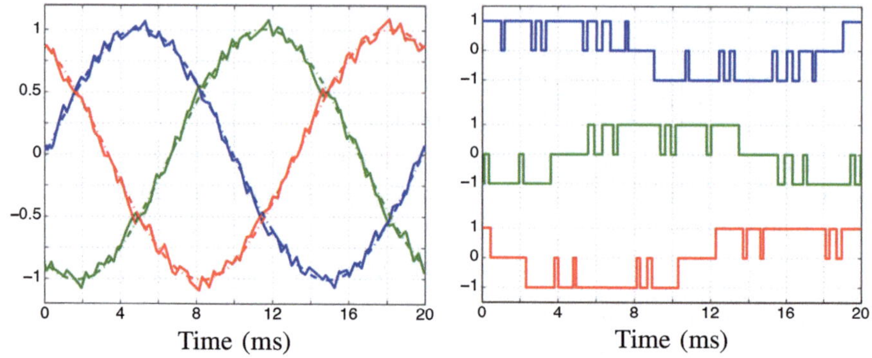

(a) Stator currents i_s (b) Switch positions u

Fig. 12: Simulated waveforms for MPC with horizon $N = 10$ and weight $\lambda_u = 0.103$.

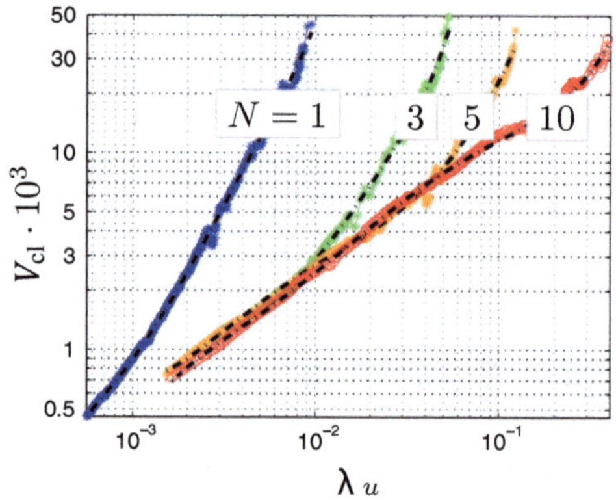

Fig. 13: The closed-loop cost is shown as a function of the tuning parameter λ_u for different prediction horizons. The individual simulations are indicated using dots, their overall trend is approximated using dash-dotted polynomials.

accordance with (48)—captures the squared RMS current error plus the weighted averaged and squared switching effort.

We start by investigating the steady-state performance of MPC with prediction horizon $N = 10$ and weighting factor $\lambda_u = 0.103$. An average device switching frequency of $f_{\text{sw}} = 300\,\text{Hz}$ results, which is typical for medium-voltage applications, and a current THD of $I_{\text{THD}} = 5.03\%$. Figure 12(a) illustrates three-phase stator current waveforms along with their (dash-dotted) references over one fundamental period. The three-phase switch positions are shown in Figure 12(b).

The influence of λ_u on the empirical closed-loop is investigated next. Steady-state simulations were run for each of the horizons $N = 1$, 3, 5 and 10 and for more

Table 1: Average and maximal number of switching sequences that need to be considered by the sphere decoding and exhaustive search algorithms to obtain the optimal result, depending on the length of the prediction horizon

Prediction	Sphere decoding		Exhaustive search	
Horizon N	Avg.	Max.	Avg.	Max.
1	1.18	5	11.8	18
2	1.39	8	171	343
3	1.72	14	2350	4910
5	2.54	35	467,000	970,000
10	8.10	220		

than 1000 different values of λ_u, ranging between 0 and 0.5. Focusing on switching frequencies between 100 Hz and 1 kHz, and current THDs below 20%, the results are shown in Figure 13, using a double logarithmic scale. The cost is significantly reduced as the prediction horizon is increased, suggesting the use of $N > 1$.

Computational Burden

Last, we investigate the computational burden of the modified sphere decoder for different prediction horizons. The switching frequency is held constant at 300 Hz for all prediction horizons by tuning the weight λ_u accordingly. We use the number of switching sequences that are investigated by the algorithm at each time-step as a measure of the computational burden. The average and the maximal number of switching sequences is monitored over multiple fundamental periods. Table 1 shows that the computational burden of the algorithm grows modestly as the prediction horizon is increased, despite being exponential in the worst case. In contrast to that, exhaustive search becomes computationally intractable for prediction horizons exceeding three.

6 Conclusions

In this chapter, basic aspects and methods underlying model predictive control for power electronics applications have been presented. Algorithms and system theoretic properties depend on whether the discrete switch positions are directly manipulated, or a modulator is used. Special attention has been paid on (practical) stability and computational issues.

Our presentation has been kept at a basic system-theoretic level and was illustrated on simple converter topologies, which can be described via LTI dynamics. Some configurations like active front end converters [48] and modular multilevel converters [55] require a more careful consideration of both control theoretic tools and also physical system knowledge for the design of high-performance model predictive controllers.

References

1. Agrell, E., Eriksson, T., Vardy, A., Zeger, K.: Closest point search in lattices. IEEE Trans. Inf. Theory **48**(8), 2201–2214 (2002)
2. Aguilera, R.P., Quevedo, D.E.: Capacitor voltage estimation for predictive control algorithm of flying capacitor converters. In: IEEE International Conference on Industrial Technology, Melbourne (2009)
3. Aguilera, R.P., Quevedo, D.E.: Stability analysis of quadratic MPC with a discrete input alphabet. IEEE Trans. Autom. Control **58**(12), 3190–3196 (2013)
4. Aguilera, R.P., Quevedo, D.E.: Predictive control of power converters: designs with guaranteed performance. IEEE Trans. Ind. Inf. **11**(1), 53–63 (2015)
5. Baidya, R., Aguilera, R.P., Acuna, P., Vazquez, S., Mouton, H.D.: Multistep model predictive control for cascaded h-bridge inverters: formulation and analysis. IEEE Trans. Power Electron. **PP**(99), 1–1 (2017). https://doi.org/10.1109/TPEL.2017.2670567
6. Bemporad, A., Morari, M., Dua, V., Pistikopoulos, E.N.: The explicit linear quadratic regulator for constrained systems. Automatica **38**(1), 3–20 (2002)
7. Bolognani, S., Bolognani, S., Peretti, L., Zigliotto, M.: Design and implementation of model predictive control for electrical motor drives. IEEE Trans. Ind. Electron. **56**(6), 1925–1936 (2009)
8. Carrasco, J.M., Franquelo, L.G., Bialasiewicz, J.T., Galván, E., Guisado, R.C.P., Prats, A.M., León, J.I., Moreno-Alfonso, N.: Power-electronic systems for the grid integration of renewable energy sources: a survey. IEEE Trans. Ind. Electron. **53**(4), 1002–1016 (2006)
9. Cortés, P., Kazmierkowski, M.P., Kennel, R.M., Quevedo, D.E., Rodríguez, J.: Predictive control in power electronics and drives. IEEE Trans. Ind. Electron. **55**(12), 4312–4324 (2008)
10. Cortés, P., Rodríguez, J., Quevedo, D.E., Silva, C.: Predictive current control strategy with imposed load current spectrum. IEEE Trans. Power Electron. **23**(2), 612–618 (2008)
11. Cychowski, M., Szabat, K., Orlowska-Kowalska, T.: Constrained model predictive control of the drive system with mechanical elasticity. IEEE Trans. Ind. Electron. **56**(6), 1963–1973 (2009)
12. De Doná, J.A., Goodwin, G.C., Serón, M.M.: Anti-windup and model predictive control: reflections and connections. Eur. J. Control **6**(5), 467–477 (2000)
13. De Doncker, R.W., Meyer, C., Lenke, R.U., Mura, F.: Power electronics for future utility applications. In: Proceedings of IEEE International Conference on Power Electronics and Drive Systems, Bangkok (2007)
14. Emadi, A., Lee, Y.J., Rajashekara, K.: Power electronics and motor drives in electric, hybrid electric, and plug-in hybrid electric vehicles. IEEE Trans. Ind. Electron. **55**(6), 2237–2245 (2008)
15. Finch, J.W., Giaouris, D.: Controlled AC electrical drives. IEEE Trans. Ind. Electron. **55**(2), 481–491 (2008)
16. Fincke, U., Pohst, M.: Improved methods for calculating vectors of short length in a lattice, including a complexity analysis. Math. Comput. **44**(170), 463–471 (1985)
17. Fuentes, E.J., Silva, C.A., Yuz, J.I.: Predictive speed control of a two-mass system driven by a permanent magnet synchronous motor. IEEE Trans. Ind. Electron. **59**(7), 2840–2848 (2012)
18. Gersho, A., Gray, R.M.: Vector Quantization and Signal Compression. Kluwer Academic, Boston (1992)
19. Geyer, T.: Computationally efficient model predictive direct torque control. IEEE Trans. Power Electron. **26**(10), 2804–2816 (2011)
20. Geyer, T.: Model Predictive Control of High Power Converters and Industrial Drives. Wiley, London (2016)
21. Geyer, T., Quevedo, D.E.: Multistep finite control set model predictive control for power electronics. IEEE Trans. Power Electron. **29**(12), 6836–6846 (2014)
22. Geyer, T., Quevedo, D.E.: Performance of multistep finite control set model predictive control for power performance. IEEE Trans. Power Electron. **30**(3), 1633–1644 (2015)

23. Geyer, T., Papafotiou, G., Morari, M.: Model predictive control in power electronics: a hybrid systems approach. In: Proceedings of IEEE Conference on Decision and Control, Seville (2005)
24. Geyer, T., Torrisi, F., Morari, M.: Optimal complexity reduction of polyhedral piecewise affine systems. Automatica **44**(7), 1728–1740 (2008)
25. Geyer, T., Papafotiou, G., Morari, M.: Hybrid model predictive control of the step-down DC-DC converter. IEEE Trans. Control Syst. Technol. **16**(6), 1112–1124 (2008)
26. Goodwin, G.C., Mayne, D.Q., Chen, K., Coates, C., Mirzaeva, G., Quevedo, D.E.: An introduction to the control of switching electronic systems. Annu. Rev. Control. **34**(2), 209–220 (2010)
27. Grüne, L., Rantzer, A.: On the infinite horizon performance of receding horizon controllers. IEEE Trans. Autom. Control **53**(9), 2100–2111 (2008)
28. Hassibi, B., Vikalo, H.: On the sphere-decoding algorithm I. Expected complexity. IEEE Trans. Signal Process. **53**(8), 2806–2818 (2005)
29. Holmes, D.G., Lipo, T.A.: Pulse width Modulation for Power Converters: Principles and Practice. IEEE Press, New York (2003)
30. Holtz, J.: Power electronics—a continuing challenge. IEEE Ind. Electron. Mag. **5**(2), 6–15 (2011)
31. Kouro, S., Cortés, P., Vargas, R., Ammann, U., Rodríguez, J.: Model predictive control—a simple and powerful method to control power converters. IEEE Trans. Ind. Electron. **56**(6), 1826–1838 (2009)
32. Krause, P.C., Wasynczuk, O., Sudhoff, S.D.: Analysis of Electric Machinery and Drive Systems, 2nd edn. Wiley, London (2002)
33. Kvasnica, M., Grieder, P., Baotić, M., Morari, M.: Multi parametric toolbox (MPT). In: Alur, R., Pappas, G. (eds.) Hybrid Systems: Computation and Control. Lecture Notes in Computer Science, vol. 2993, pp. 448–462. Springer, Philadelphia (2004). http://control.ee.ethz.ch/~mpt
34. Lawler, E.L., Wood, D.E.: Branch and bound methods: a survey. Oper. Res. **14**(4), 699–719 (1966)
35. Lehman, B., Bass, R.M.: Extensions of averaging theory for power electronic systems. IEEE Trans. Power Electron. **11**(4), 542–553 (1996)
36. Lesnicar, A., Marquardt, R.: An innovative modular multilevel converter topology suitable for a wide power range. In: Proceedings of IEEE Power Tech Conference, Bologna (2003)
37. Lezana, P., Aguilera, R.P., Quevedo, D.E.: Model predictive control of an asymmetric flying capacitor converter. IEEE Trans. Ind. Electron. **56**(6), 1839–1846 (2009)
38. Linder, A., Kennel, R.: Model predictive control for electrical drives. In: Proceedings of IEEE Power Electronics Specialists Conference (PESC), Recife, Brazil, pp. 1793–1799 (2005)
39. Mariéthoz, S., Morari, M.: Explicit model predictive control of a PWM inverter with an *LCL* filter. IEEE Trans. Ind. Electron. **56**(2), 389–399 (2009)
40. Mariethoz, S., Beccuti, A., Papafotiou, G., Morari, M.: Sensorless explicit model predictive control of the DC-DC buck converter with inductor current limitation. In: Applied Power Electronics Conference and Exposition, pp. 1710–1715 (2008)
41. Mariéthoz, S., Domahidi, A., Morari, M.: High-bandwidth explicit model predictive control of electrical drives. IEEE Trans. Ind. Appl. **48**(6), 1980–1992 (2012)
42. Markets and markets: power electronics market—global forecast to 2022. Tech. rep. (2017)
43. Mayne, D.Q., Rawlings, J.B., Rao, C.V., Scokaert, P.O.M.: Constrained model predictive control: optimality and stability. Automatica **36**(6), 789–814 (2000)
44. Mitten, L.G.: Branch–and–bound methods: general formulation and properties. Oper. Res. **18**(1), 24–34 (1970)
45. Müller, C., Quevedo, D.E., Goodwin, G.C.: How good is quantized model predictive control with horizon one? IEEE Trans. Autom. Control **56**(11), 2623–2638 (2011)
46. Quevedo, D.E., Goodwin, G.C.: Control of EMI from switch-mode power supplies via multi-step optimization. In: Proceedings of American Control Conference, Boston, MA, vol. 1, pp. 390–395 (2004)

47. Quevedo, D.E., Goodwin, G.C., De Doná, J.A.: Finite constraint set receding horizon quadratic control. Int. J. Robust Nonlinear Control **14**(4), 355–377 (2004)
48. Quevedo, D.E., Aguilera, R.P., Pérez, M.A., Cortés, P., Lizana, R.: Model predictive control of an AFE rectifier with dynamic references. IEEE Trans. Power Electron. **27**(7), 3128–3136 (2012)
49. Rawlings, J., Mayne, D.: Model Predictive Control: Theory and Design. Nob Hill Publishing, Madison (2009)
50. Rodríguez, J., Cortés, P.: Predictive Control of Power Converters and Electrical Drives, 1st edn. Wiley-IEEE Press, Chichester (2012)
51. Rodríguez, J., Pontt, J., Silva, C., Correa, P., Lezana, P., Cortés, P., Ammann, U.: Predictive current control of a voltage source inverter. IEEE Trans. Ind. Electron. **54**(1), 495–503 (2007)
52. Rodríguez, J., Kazmierkowski, M.P., Espinoza, J., Zanchetta, P., Abu-Rub, H., Young, H.A., Rojas, C.A.: State of the art of finite control set model predictive control in power electronics. IEEE Trans. Ind. Inf. **9**(2), 1003–1016 (2013)
53. Senesky, M., Eirea, G., Koo, T.J.: Hybrid modelling and control of power electronics. In: Pnueli, A., Maler, O. (eds.) Hybrid Systems: Computation and Control. Lecture Notes in Computer Science, vol. 2623, pp. 450–465. Springer, Berlin (2003)
54. Vazquez, S., Rodríguez, J., Rivera, M., Franquelo, L.G., Norambuena, M.: Model predictive control for power converters and drives: advances and trends. IEEE Trans. Ind. Electron. **PP**(99), 1 (2016)
55. López, A.M., Quevedo, D.E., Aguilera, R.P., Geyer, T., Oikonomou, N.: Limitations and accuracy of a continuous reduced-order model for modular multilevel converters. IEEE Trans. Power Electron. **33**, 6292–6303 (2018)

Learning-Based Fast Nonlinear Model Predictive Control for Custom-Made 3D Printed Ground and Aerial Robots

Mohit Mehndiratta, Erkan Kayacan, Siddharth Patel, Erdal Kayacan, and Girish Chowdhary

1 Introduction

In almost all robotic applications, there are always time-varying system dynamics and/or environmental variations throughout the operation. For instance, off-road agricultural robots, including fruit picking robots, driverless tractors, and sheep shearing robots, must be operated on varying soil conditions. Furthermore, there are always topological challenges, such as bumps and hollows in a field. All these challenges bring additional uncertainties to the system which can be modeled as longitudinal and lateral slip variations [17]. Since the performance of a model-based controller is guaranteed for an accurate mathematical model of the system, any plant-model mismatch results in suboptimal performance. Therefore, for a guaranteed performance from a model-based controller, the aforementioned variations must be learnt over time, and the controller must adapt itself to the changing conditions autonomously. Another example is mass variations in package delivery problems of aerial robots. When the total mass of a multi-rotor unmanned aerial vehicle (UAV) is considered, the payload changes may result in massive variations in its

M. Mehndiratta · S. Patel
Nanyang Technological University, 50 Nanyang Avenue, Singapore 639798
e-mail: mohit005@e.ntu.edu.sg; PATE0006@e.ntu.edu.sg

E. Kayacan (✉)
Massachusetts Institute of Technology, Cambridge, MA 02139 USA
e-mail: erkank@mit.edu

E. Kayacan
Aarhus University, Department of Engineering, DK-8000 Aarhus C, Denmark
e-mail: erdal@eng.au.dk

G. Chowdhary
University of Illinois at Urbana–Champaign, Champaign, IL, USA
girishc@illinois.edu

S. V. Raković, W. S. Levine (eds.), *Handbook of Model Predictive Control*, Control Engineering, https://doi.org/10.1007/978-3-319-77489-3_24

dynamic model which will also result in suboptimal performance for a model-based controller. Motivated by the challenges listed above, our goal is to use an online learning-based nonlinear model predictive control (NMPC) for systems with uncertain and/or time-varying dynamic models.

As a solution to modeling mismatch problem between the plant to be controlled and its corresponding mathematical model, adaptation of either controller parameters or deployed mathematical model parameters is not a novel idea. In adaptive control, controller adapts itself systematically to compensate lack of modeling due to uncertain and/or time-varying parameters. This feature, apparently, exterminates the effect of parameter uncertainties on the closed-loop system's performance [7]. A well-utilized strategy in this area is the adaptive-optimal control, which comprises of the use of an adaptive controller for stability during the learning phase, followed by the switch to the main model-based optimal controller that eventually optimizes the performance. An online switching metric is developed that initiates the switching to model predictive control (MPC) after gaining enough confidence in the parameter estimates, as realized in [5, 6]. On the other hand, an alternative learning approach could be to combine the control with some optimization-based estimation scheme including predictive filtering and moving horizon estimation (MHE) [1], which is also the case in this work. These estimators are model-based estimators, which can incorporate parameter variations along with the state estimation, to learn the uncertain system parameters online. This learning-based NMPC has been utilized for numerous robotic applications including constrained path tracking of a mobile robot in [26], control of a 3 degree of freedom helicopter in [23], control of lateral dynamics of a fixed-wing UAV in [30], control of a quadrotor in [4], teleoperation of an underwater vehicle in [11], and robust obstacle avoidance in [9, 21].

In addition to MHE, extended Kalman filter (EKF) can also be utilized for online learning. However, EKF is based on the linearization of the nonlinear system at the current estimate and is only suitable for unconstrained problems. In other words, EKF might give irrational estimation results, e.g. less than zero or larger than one for slip parameters [13, 15, 16, 18]. On the contrary, MHE strategy exploits the past measurements available over a window and solves an optimization problem to estimate the system's states and unknown parameters [22]. Additionally, MHE is a powerful nonlinear estimator that is not only suitable for non-Gaussian disturbance but is also competent in handling constraints explicitly [28]. This implies, MHE will never give irrational estimation results for the aforementioned slip parameters [29].

In this work, the efficacy of the learning-based NMPC is elaborated for the trajectory tracking of two custom-made 3D printed robotic platforms: an off-road agricultural ground vehicle and an aerial robot for package delivery problem. In the first application, NMPC is utilized for controlling a field robot in an off-road terrain. Since the ground conditions, including surface quality (loose soil, grass) and terrain topography (uphill and downhill), may change over the time, modeling errors are induced [14]. As an artifice, nonlinear MHE (NMHE) is employed to learn the changing operational conditions, so that a better performing NMPC can be realized. Secondly, an in-flight payload dropping application of a tilt-rotor tricopter UAV is addressed. With each drop of payload, the total UAV mass varies and this results in a plant-model mismatch. Therefore, in order to eliminate this mismatch and hence, achieve a superior tracking accuracy from NMPC, NMHE is utilized to learn the UAV mass online. For both the applications, fast NMPC and NMHE solution methods are incorporated and the test-results are obtained from real-time experiments.

The remaining part of this study is organized as follows: Section 2 illustrates the receding horizon control and estimation methods in terms of NMPC and NMHE problem formulations. In Section 3, the leaning-based NMPC-NMHE framework is demonstrated for the tracking problems of two robotic systems. Finally, the drawn conclusions are presented in Section 4.

2 Receding Horizon Control and Estimation Methods

In this section, we briefly discuss the optimal control problems (OCPs) of NMPC and NMHE. For both the OCPs, the considered nonlinear system is modelled as:

$$\dot{\mathtt{x}}(t) = \mathtt{f}\big(\mathtt{x}(t),\mathtt{u}(t),\mathtt{p}\big), \tag{1}$$

where $\mathtt{x}(t) \in \mathbb{R}^{n_\mathtt{x}}$, $\mathtt{u}(t) \in \mathbb{R}^{n_\mathtt{u}}$ and $\mathtt{p}(t) \in \mathbb{R}^{n_\mathtt{p}}$ are the state, input, and system parameter vectors, respectively, at time t; $\mathtt{f}(\cdot,\cdot,\cdot) : \mathbb{R}^{n_\mathtt{x}+n_\mathtt{u}+n_\mathtt{p}} \longrightarrow \mathbb{R}^{n_\mathtt{x}}$ is the continuously differentiable state update function and $\mathtt{f}(0,0,\mathtt{p}) = 0\ \forall t$. The derivative of $\mathtt{x}$ with respect to t is denoted by $\dot{\mathtt{x}} \in \mathbb{R}^{n_\mathtt{x}}$.

Similarly, a nonlinear measurement model denoted as $\mathtt{y}(t)$ can be described with the following equation:

$$\mathtt{y}(t) = \mathtt{h}\big(\mathtt{x}(t),\mathtt{u}(t),\mathtt{p}\big), \tag{2}$$

where $\mathtt{h}(\cdot,\cdot,\cdot) : \mathbb{R}^{n_\mathtt{x}+n_\mathtt{u}+n_\mathtt{p}} \longrightarrow \mathbb{R}^{n_\mathtt{x}}$ is the measurement function which describes the relation between the variables of the system model and the measured outputs of the real-time system.

2.1 *Nonlinear Model Predictive Control*

NMPC is an advanced, dynamic optimization-based strategy for feedback control that solely relies on the accuracy of the mathematical model for its optimum performance. In NMPC strategy, a parametric OCP is formulated in the form of a least square function, in order to penalize deviations of predicted system's trajectory (including states and control inputs) from the specified reference. The parametric nature of the OCP is due to its dependence on the current state (measured or estimated). In addition, to keep the computational burden realizable for a real-time application (especially for fast robotic systems), the optimization problem is solved over a finite window, commonly known as *prediction horizon* (N_c). It may be worth noting that NMPC typically leads to non-convex optimization problems, in contrast to linear MPC in which nearly all formulations use convex cost and constraint functions [24].

In NMPC, the dynamic optimization problem is recursively solved for the optimal control inputs, over the given prediction horizon ($t_j \leq t \leq t_{j+N_c}$) at each sampling instant. We formulate the following least-square type cost function in discrete time, which is commonly utilized for tracking applications:

$$\min_{\mathtt{x}_k,\mathtt{u}_k} \quad \frac{1}{2}\left\{\sum_{k=j}^{j+N_c-1}\left(\left\|\mathtt{x}_k-\mathtt{x}_k^{\mathrm{ref}}\right\|_{W_\mathtt{x}}^2+\left\|\mathtt{u}_k-\mathtt{u}_k^{\mathrm{ref}}\right\|_{W_\mathtt{u}}^2\right)+\left\|\mathtt{x}_{N_c}-\mathtt{x}_{N_c}^{\mathrm{ref}}\right\|_{W_{N_c}}^2\right\} \tag{3a}$$

$$\text{s.t.} \quad \mathtt{x}_j = \hat{\mathtt{x}}_j, \tag{3b}$$

$$\mathtt{x}_{k+1} = \mathtt{f_d}(\mathtt{x}_k, \mathtt{u}_k, \mathtt{p}), \quad k = j, \cdots, j+N_c-1, \tag{3c}$$

$$\mathtt{x}_{k,\min} \leq \mathtt{x}_k \leq \mathtt{x}_{k,\max}, \quad k = j, \cdots, j+N_c, \tag{3d}$$

$$\mathtt{u}_{k,\min} \leq \mathtt{u}_k \leq \mathtt{u}_{k,\max}, \quad k = j, \cdots, j+N_c-1, \tag{3e}$$

where $\mathtt{x}_k \in \mathbb{R}^{n_\mathtt{x}}$ is the differential state, $\mathtt{u}_k \in \mathbb{R}^{n_\mathtt{u}}$ is the control input and $\hat{\mathtt{x}}_j \in \mathbb{R}^{n_\mathtt{x}}$ is the current state estimate; time-varying state and control references are denoted by $\mathtt{x}_k^{\mathrm{ref}}$ and $\mathtt{u}_k^{\mathrm{ref}}$, respectively; the terminal state reference is denoted by $\mathtt{x}_{N_c}^{\mathrm{ref}}$; the discrete time dynamical model is represented by $\mathtt{f_d}(\cdot,\cdot,\cdot)$; $W_\mathtt{x} \in \mathbb{R}^{n_\mathtt{x}\times n_\mathtt{x}}$, $W_\mathtt{u} \in \mathbb{R}^{n_\mathtt{u}\times n_\mathtt{u}}$ and $W_{N_c} \in \mathbb{R}^{n_\mathtt{x}\times n_\mathtt{x}}$ are the corresponding weight matrices, which are assumed constant for simplicity, however, their time-varying formulation can also be included in a similar manner. Furthermore, $\mathtt{x}_{k,\min} \leq \mathtt{x}_{k,\max} \in \mathbb{R}^{n_\mathtt{x}}$ and $\mathtt{u}_{k,\min} \leq \mathtt{u}_{k,\max} \in \mathbb{R}^{n_\mathtt{u}}$ specify the lower and upper bounds on the states and control inputs, respectively.

Once the solution to the OCP (3) at t_j is available, the first computed control input ($\mathtt{u}_j$) is applied to the system for a short time period, that typically coincides with the sampling time [19]. This sampling time has to be kept short enough with respect to the system's dynamics, while sufficiently long at the same time to facilitate timely computation of the optimized solution. Subsequently, a new optimization problem is solved for the prediction window $[t_{j+1}, t_{j+N_c+1}]$, which itself is moving forward with time. Due to this shifting property of the prediction window, the NMPC is also known as *receding horizon* control technique.

The last expression in (3a) represents the final cost incurred due to the finite prediction horizon and is generally referred to as the *terminal penalty* term. This term is often included in the problem formulation for stability reasons [19]. In addition, some other stability results include a problem formulation with sufficiently long horizon [10], an additional prediction horizon and a locally stabilizing control law [25].

2.2 Nonlinear Moving Horizon Estimation

Typically, MHE is considered as a *dual problem* of MPC as they exploit the similar optimization problem structure; despite the fact that MPC predicts the future of the system, while MHE utilizes the past measurements over an *estimation horizon* for state estimation [20, 31]. Moreover, the two main differences of optimization problem formulation of MHE from MPC are: (i) there is no initial state constraint like in (3b), and (ii) the optimization variables are the states and unknown system parameters, excluding the control inputs as they are already given to the system in the past.

In a similar manner to NMPC, the NMHE scheme is also formulated using a least square function to penalize the deviation of estimated outputs $\big(\mathtt{h}(\cdot,\cdot,\cdot)\big)$ from measurements ($\mathtt{z}$). The performance of NMHE also relies on the availability of an accurate system model, while a mismatch in the form of process noise between the system model and the real plant may deteriorate the optimal estimation solution, which eventually may lead to an unstable closed-loop. To address this issue, a suitable component (*arrival cost*) is included in the final optimization problem formulation of NMHE, as done in [20]. The NMHE formulation includes an estimation horizon containing M measurements $(\mathtt{z}_\mathtt{S},\cdots,\mathtt{z}_j,)$ taken at time $t_\mathtt{S} < \cdots < t_j$, where the length of the horizon is given by $T_E = t_j - t_\mathtt{S}$, and $j - M + 1 \stackrel{\text{def}}{=} \mathtt{S}$ is taken for notational convenience. Finally, the discrete time dynamic optimization problem to estimate the constrained states ($\hat{\mathtt{x}}$) as well as the unknown parameter ($\hat{\mathtt{p}}$) at time t_j using the process model $\mathtt{f}(\cdot,\cdot,\cdot)$, measurement model $\mathtt{h}(\cdot,\cdot,\cdot)$ and available measurements within the horizon, is of the form [20]:

$$\min_{\hat{\mathtt{x}}_k,\hat{\mathtt{p}}} \quad \left\{ \left\| \begin{matrix} \hat{\mathtt{x}}_\mathtt{S} - \bar{\mathtt{x}}_\mathtt{S} \\ \hat{\mathtt{p}} - \bar{\mathtt{p}}_L \end{matrix} \right\|^2_{P_\mathtt{S}} + \sum_{k=\mathtt{S}}^{j} \|\mathtt{z}_k - \mathtt{h}(\hat{\mathtt{x}}_k,\mathtt{u}_k,\mathtt{p})\|^2_V + \sum_{k=\mathtt{S}}^{j-1} \|w_k\|^2_W \right\} \tag{4a}$$

$$\text{s.t.} \quad \hat{\mathtt{x}}_{k+1} = \mathtt{f}_\mathtt{d}(\hat{\mathtt{x}}_k,\mathtt{u}_k,\mathtt{p}) + w_k, \qquad k = \mathtt{S},\cdots,j-1, \tag{4b}$$

$$\hat{\mathtt{x}}_{k,\min} \le \hat{\mathtt{x}}_k \le \hat{\mathtt{x}}_{k,\max}, \qquad k = \mathtt{S},\cdots,j, \tag{4c}$$

$$\hat{\mathtt{p}}_{\min} \le \hat{\mathtt{p}} \le \hat{\mathtt{p}}_{\max}, \tag{4d}$$

where w_k represents the added process noise; $\hat{\mathtt{x}}_{k,\min} \le \hat{\mathtt{x}}_{k,\max}$ and $\hat{\mathtt{p}}_{\min} \le \hat{\mathtt{p}}_{\max}$ specify the lower and upper bounds on the estimated state and parameter vectors, respectively; $\bar{\mathtt{x}}_\mathtt{S}$ and $\bar{\mathtt{p}}_\mathtt{S}$ denote the estimated state and parameter values (arrival cost data) at the start of estimation horizon, i.e., at $t_\mathtt{S}$. The weight matrices $P_\mathtt{S}$, V, and W are interpreted as the inverse of the covariance matrices and are evaluated as:

$$P_\mathtt{S} = Q_0^{-\frac{1}{2}} = \begin{bmatrix} Q_0^\mathtt{x} & 0 \\ 0 & Q_0^\mathtt{p} \end{bmatrix}^{-\frac{1}{2}}, \quad V = R^{-\frac{1}{2}}, \quad W = Q^{-\frac{1}{2}} = \begin{bmatrix} Q^\mathtt{x} & 0 \\ 0 & Q^\mathtt{p} \end{bmatrix}^{-\frac{1}{2}}, \tag{5}$$

where Q_0 is the initial covariance matrix (incorporating state and parameter, both), R is the measurement noise covariance matrix and Q is the process noise covariance matrix. With the above choice of weight matrices, it is assured that the NMHE scheme results in a maximum-likelihood estimate for the very likely trajectories [31].

The first term in (4a) is generally referred to as the arrival cost. It is incorporated into the objective function in order to accommodate the effect of past measurements (before the beginning of estimation horizon), in the current state and parameter estimates. This can be interpreted as analogous to terminal penalty term of NMPC which summarizes the response of the system after the prediction horizon. EKF is often utilized to update the arrival cost for practical implementation, as also done in [20].

Another parameter that affects the performance of NMHE is the choice of estimation window length M, which in general is problem-specific. It basically represents a trade-off between computational liability and estimation accuracy that simultaneously grow with M. In the case of small but fast robotic systems, like ground robots and UAVs, we cannot indefinitely increase M as limited computation power is available on-board. Moreover, it is not necessarily true that the estimation accuracy always increases with M, as the plant-model mismatch degrades the significance of model prediction which adversely affects the estimation performance [19]. That is, the selection of a too high value of M for the system in which the unknown parameter (to be estimated) is radically changing, plant-model anomalies may arise that eventually may result in deteriorated overall estimation quality.

3 Real-Time Applications

In this section, two real-time robotic applications will be presented to show how we have addressed the two main problems encountered in NMPC application, which are lack of modeling and online solution of the nonlinear optimization problem. The applications include the trajectory tracking problems of the ground and aerial robotic systems with time-varying dynamic model parameters which are estimated using NMHE. Owing to the similarities between the optimization problems of NMPC and NMHE defined in (3) and (4), respectively, we solve them utilizing the direct multiple shooting method and real-time iteration approach, which is incorporated in ACADO toolkit [2]. In ACADO toolkit, firstly the optimization problem, in terms of system equations and constraints, is defined in a C++ environment and then, the self-contained C codes are obtained using its code generation package [2]. Finally, these generated C codes can be utilized to run on C/C++ or MATLAB/Simulink based software platforms.

3.1 Ultra-Compact Field Robot

Firstly, we illustrate NMPC for the trajectory tracking problem of a 3D printed field robot, operating in an off-road terrain. Since the soil conditions and terrain topography may vary over the operation, modeling uncertainties would arise. In order to tackle these operational uncertainties and hence, achieve optimum control performance, NMHE is utilized to estimate two slip (or traction) parameters, namely, (α, κ), in addition to performing the state estimation task.

3.1.1 System Description

The 3D printed field robot as shown in Figure 1 has been built utilizing practical, hands-on experience with various sensors and actuators. A real-time kinematic (RTK) differential global navigation satellite system (GNSS), i.e., a Septentrio Altus

(a) Field robot in corn plots (b) CAD drawing of the field robot

Fig. 1: Ulta-compact 3D printed field robot.

APS-NR2 GNNS receiver (Septentrio Satellite Navigation NV, Belgium), is used to obtain highly accurate positional information, which has a specified position accuracy of 0.03 m at a 5-Hz measurements rate. The Trimble network supplies real-time kinematic correction signals via 4G internet. A gyroscope (PmodGYRO with an ST L3G4200D, Digilent Inc., USA) is mounted on the body of the robot to measure the yaw rate of the 3D printed field robot at a rate of 5-Hz with a resolution of 1°. Four powerful 12V brushed DC motors with $131:1$ metal gearboxes (Pololu Corporation, USA) are used as actuators, and four integrated quadrature encoders for brushed DC motors (Pololu Corporation, USA) are used to measure the speed of the wheels of the field robot with an accuracy of 0.05 m/s.

The real-time NMHE and NMPC are implemented and executed on an on-board computer, i.e., Raspberry Pi 3, which is equipped with Quad Core 1.2 GHz Broadcom BCM2837 64bit CPU and 1 GB of RAM. The inputs of NMHE are the position, speed and yaw rate, while the outputs are full state and parameter vectors that are fed to the NMPC. In addition to the full state and parameter information, NMPC receives the reference trajectory throughout the prediction horizon and then generates a control signal, i.e., the desired yaw rate, and sends it to the low-level controller, i.e., Kangaroo x2 motion controller (Dimension Engineering, USA). Apart from the desired yaw rate, the low-level controller receives the measured speed information from encoders and generates voltage values which are sent to the motor driver (Sabertooth dual 12A motor driver, Dimension Engineering, USA) to control the speeds of the DC motors. The low-level controller is executed at a rate of 50-Hz, which is 10 times more than the high-level controller.

3.1.2 System Model

In this section, we represent the nonlinear system and measurement models of the field robot according to (1) and (2), respectively. Instead of using the traditional kinematic model of a mobile robot, an adaptive nonlinear kinematic model, which is an extension of the traditional model, is derived as the system model of the field robot in this study. Two traction parameters (α, κ) are added to minimize deviations between the real-time system and system model. These parameters, i.e., α and κ, represent the effective speed and steering of the field robot, respectively. It is noted that they must be between zero and one, and it is inherently arduous to measure them. The field robot's model can be formulated with the following equations:

$$\dot{x} = \alpha v \cos\psi, \tag{6a}$$
$$\dot{y} = \alpha v \sin\psi, \tag{6b}$$
$$\dot{\psi} = \kappa r, \tag{6c}$$

where x and y denote the position of the field robot, ψ denotes the yaw angle, v denotes the speed and r denotes the yaw rate. The state, parameter, input and measurement vectors are, respectively, denoted as follows:

$$\mathrm{x} = \left[x\ y\ \psi\right]^T, \tag{7}$$
$$\mathrm{p} = \left[v\ \alpha\ \kappa\right]^T, \tag{8}$$
$$\mathrm{u} = r, \tag{9}$$
$$\mathrm{z} = \left[x\ y\ v\ r\right]^T. \tag{10}$$

3.1.3 Control Scheme

The control objective is to design NMPC in order to track a predefined trajectory. The optimized solution as the desired set point is forwarded to the low-level controller, which is a proportional-integral-derivative (PID) controller. The response of this low-level PID controller is finally given to the motors of the field robot.

3.1.4 Implementation of NMHE

The inputs of NMHE are the position, speed and yaw rate of the field robot as defined in (10). The outputs of NMHE, the position, yaw angle, speed and traction parameters, are the full state and parameter vectors (7)-(8). The NMPC requires full state and parameter as input to generate the desired yaw rate applied to the field robot; therefore, the full estimated state and parameter values by NMHE are fed to NMPC.

The NMHE formulation is solved at each sampling instant with the following constraints on the traction parameters:

$$0 \leq \alpha \leq 1, \tag{11a}$$
$$0 \leq \kappa \leq 1. \tag{11b}$$

The standard deviations of the measurements are set to $\sigma_x = \sigma_y = 0.03\,\text{m}$, $\sigma_v = 0.05\,\text{m/s}$, $\sigma_r = 0.0175\,\text{rad/s}$, based on the experimental analysis. Therefore, the following weighting matrices V, P_S and W are used in NMHE design:

$$\begin{aligned} V &= \text{diag}(\sigma_x^2, \sigma_y^2, \sigma_v^2, \sigma_r^2)^{-1/2}, \\ &= \text{diag}(0.03^2, 0.03^2, 0.5^2, 0.0175^2)^{-1/2}, \end{aligned} \tag{12a}$$

$$\begin{aligned} P_S = W &= \text{diag}(x^2, y^2, \psi^2, v^2, \alpha^2, \kappa^2)^{-1/2}, \\ &= \text{diag}(10.0^2, 10.0^2, 0.1^2, 1.0^2, 0.25^{2/2}, 0.25^2)^{-1/2}. \end{aligned} \tag{12b}$$

3.1.5 Implementation of NMPC

The NMPC formulation is solved at every sampling instant with the following constraints on the input:

$$-0.1(\text{rad/s}) \leq r \leq 0.1(\text{rad/s}). \tag{13}$$

The state and input references for the field robot are changed online and defined as follows:

$$\mathrm{x}_r = [x_r, y_r, \psi_r]^T \quad \text{and} \quad \mathrm{u}_r = r_r, \tag{14}$$

where x_r and y_r are the position references, r_r is the yaw rate reference, and the yaw angle reference is calculated from the position references as:

$$\psi_r = \text{atan2}\,(\dot{y}_r, \dot{x}_r) + \lambda\pi, \tag{15}$$

where λ describes the desired direction of the field robot ($\lambda = 0$ for forward and $\lambda = 1$ for backward). If the yaw rate reference, calculated from the reference trajectory, is used as the input reference, steady state error might occur in case of a mismatch between the system model and the real system. Therefore, the measured yaw rate is used as the input reference to penalize the input rate in the objective function.

The weighting matrices W_{x}, W_{u} and W_{N_c} are selected as follows:

$$W_{\mathrm{x}} = \text{diag}(1,1,1), \quad W_{\mathrm{u}} = 10 \quad \text{and} \quad W_{N_c} = 10 \times W_{\mathrm{x}}. \tag{16}$$

The weighting matrix for the input W_{u} is set larger than the weighting matrix for the states W_{x}, in order to ensure a well-damped closed-loop system behaviour. In addition, the weighting matrix for the terminal penalty W_{N_c} is set 10 times larger than the weighting matrix for the states W_{x}. This implies that the last deviations between the predicted states and their references in the prediction horizon are minimized in the objective function 10 times more than the previous points in the prediction horizon. The reason for doing that is the error at the end of the prediction horizon plays a critical role in terms of the stability of the control algorithm.

If the prediction horizon is large, the computation burden for NMPC increases unreasonably, such that solving a non-convex optimization problem online will be infeasible. Moreover, if the prediction horizon is selected to be too small, NMPC cannot stabilize the system. Therefore, the prediction horizon of the NMPC has to be large enough in reference to the velocity of the vehicle, in order to obtain a stable control performance. Since the field robot is a quite slow system, it is not required to select a very large value for the prediction horizon. Thus, it is set to 3 seconds.

3.1.6 Results

Throughout the real-time experiments, a reference trajectory consisting of straight and curved lines is tracked by the 3D printed robot, which is controlled employing the NMPC-NMHE framework. Thus, the performance of the framework can be investigated for different path geometries. The system has a constant speed, and yaw rate is the input to the system. The closest point on the reference trajectory to the 3D printed robot is calculated and then, the next 15 points are fed to the NMPC as reference trajectory due to the fact that the length of the prediction horizon (N_c) is set to 15.

The control performance of the 3D printed robot is shown in Figure 2. As can be seen in Figure 2a, the robot is capable of staying on-track throughout the experiment and tracking the target trajectory accurately. The variation of Euclidean error with time is shown in Figure 2b and its mean value is approximately 0.0459 m, which is within the tolerance for an agricultural application.

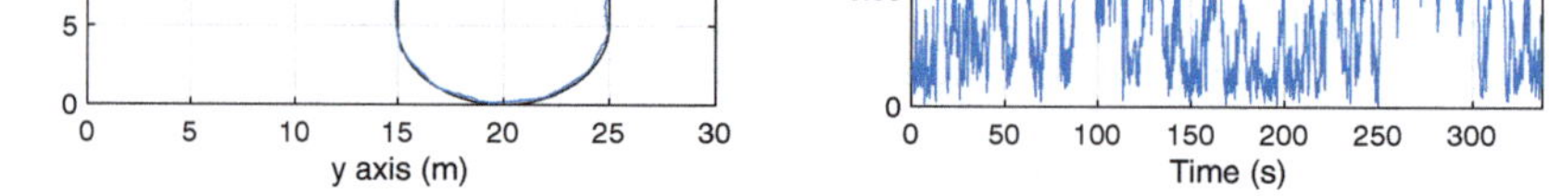

(a) Reference and actual trajectories (b) Euclidean error

Fig. 2: Trajectory tracking control performance.

The performance of NMHE in estimating the yaw angle and traction parameters is shown in Figure 3. NMPC needs full state information to generate a control signal. The position in x- and y-axes is measured; however, the yaw angle cannot be measured in practice. Therefore, NMHE estimates the yaw angle, which plays a very important role in the trajectory tracking performance. As seen in Figure 3a, the yaw angle has been controlled very accurately. Moreover, the traction parameters are immeasurable and the constraints on these parameters are defined in (11). It is

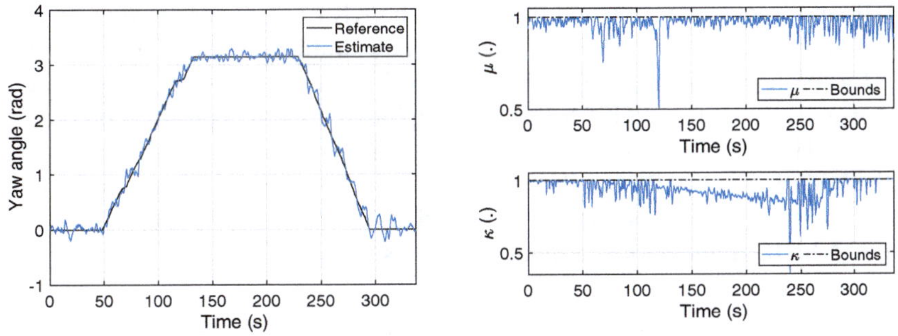

(a) Reference and estimated yaw angle

(b) Traction parameters

Fig. 3: Estimation performance of NMHE.

important to estimate the traction parameters, because soil conditions can change over the time. Therefore, the online estimation of the parameters is required to learn soil conditions and thus, adapt NMPC to the changing working conditions. As can be seen in Figure 3b, the estimated values are within the bounds. Moreover, it is observed that the traction parameter estimates stabilize at certain values, so that a stable trajectory tracking performance is ensured.

The measured and estimated speed of the 3D printed robot is shown in Figure 4a. NMHE is capable of filtering noisy measurements. Additionally, the control signal,

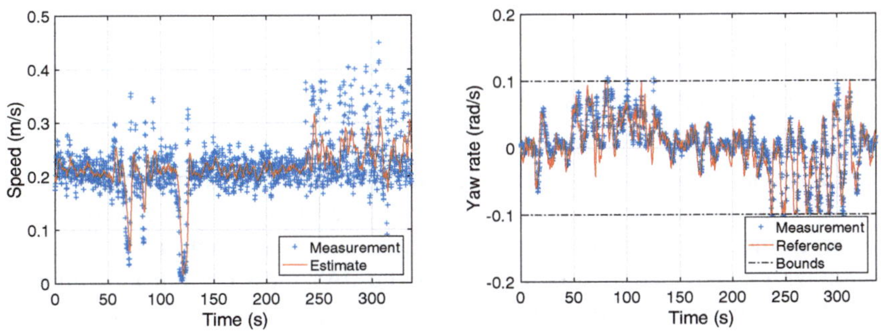

(a) Measured and estimated speed

(b) Reference and actual yaw rate

Fig. 4: Speed and yaw rate.

i.e., yaw rate reference, generated by the NMPC is shown in Figure 4b. It is observed that the NMPC is capable of dealing with the input constraints and the low-level controller shows a good control performance.

It is necessary to check the optimality of the NMPC-NMHE framework, because a single quadratic programming iteration at each sampling time instant may result in a suboptimal solution. Therefore, the Karush-Kuhn-Tucker (KKT) tolerances for

NMHE and NMPC are shown in Figure 5a. The KKT tolerances are very small, but they are not equal to zero. The reason is that a quadratic program is solved precisely only for the linear systems, such that the KKT tolerance becomes zero. Moreover, the low and non-drifting KKT tolerances emphasize that the optimization problems in the NMHE and NMPC are well defined and properly scaled. The execution times for the NMHE and NMPC are shown in Figure 5b. Their mean values are 0.2101 ms and 0.3813 ms, respectively, which implies that the overall computation time for the NMPC-NMHE framework is around 0.5914 ms.

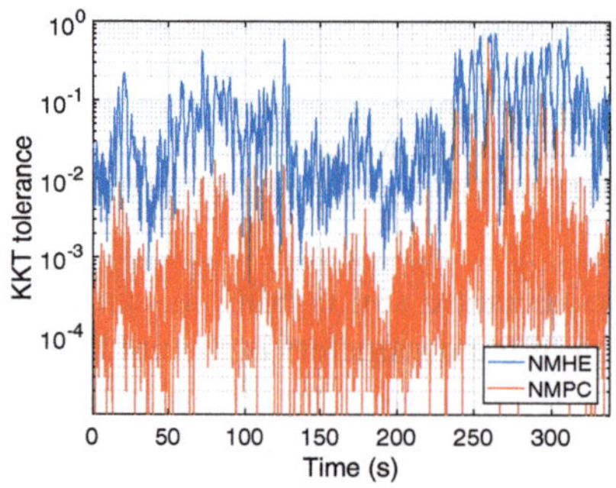

(a) KKT Tolerances

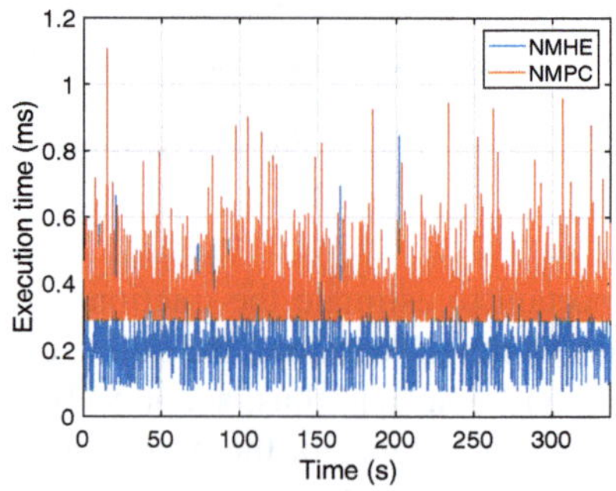

(b) Execution time

Fig. 5: KKT tolerances and execution times of NMHE and NMPC.

3.2 Tilt-Rotor Tricopter UAV

In this application, we tackle a real-life package delivery problem, where a UAV takes off with the full payload, tracks a predefined trajectory in 3D, drops each package to the time-based designated location, and finally, returns to its starting location with no payload. In this application, the UAV mass is 1.608 kg without any payload. The dropping mechanism is designed to drop four payloads in the sequence 55 g, 75 g, 77 g, and 86 g, respectively, which makes the total takeoff mass to be 1.901 kg. Considering the total payload of 293 g, it is almost 18% of the total mass of the UAV. This means a massive change in the model parameters which has to be handled during the control of the system. In this application, we learn the variations in the mass online and feed the estimated mass value to the model which is used by the NMPC.

3.2.1 System Description

The aerial robot used in this application is a 3D printed tilt-rotor tricopter, as shown in Figure 6a. It is a custom-made system, which is developed based on the other Talon tricopter frames available in the market. The frame is customized, such that it provides flexibility to accommodate all the electronics as needed. The Pixhawk flight controller is used as the low-level stabilization controller. In addition, the tricopter also houses the on-board computer, i.e., Raspberry Pi 3, which serves two vital functions. One is wireless communication with the ground-station computer, and the other is controlling the servomotor for the payload drop mechanism.

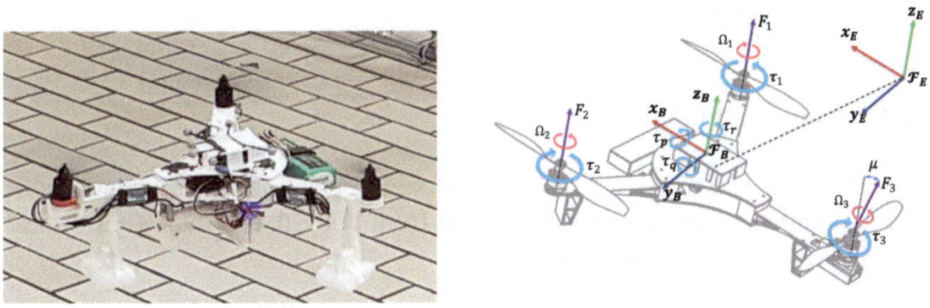

(a) Actual setup (b) Coordinate frame and sign conventions

Fig. 6: 3D printed tilt-rotor tricopter UAV.

The mechanism used to hold and drop the payload throughout the flight has two plates, which are supported at the base of the UAV. Amongst them, one houses the servomotor, while the other holds the payload blocks to be dropped. A circular gear mounted on the servomotor drives a linear gear, shown in Figure 7a, that results in a linear motion. This linear motion pulls the rod attached to the linear gear and thus,

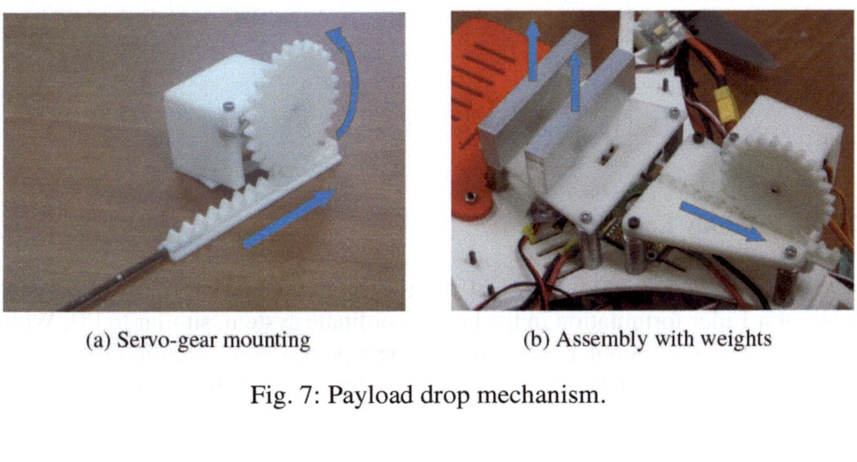

(a) Servo-gear mounting (b) Assembly with weights

Fig. 7: Payload drop mechanism.

drops the blocks one-by-one in the process. The dropping mechanism is shown in Figure 7b with an inverted view.

3.2.2 System Model

The tilt-rotor tricopter is considered as a rigid-body having two stationary rotors and one non-stationary (or tilting) rotor, as shown in Figure 6b. In our configuration, the two stationary rotors – RR (right rotor) rotating clockwise and LR (left rotor) rotating counter-clockwise – are placed in the front of the body (CG - centre of gravity), while the tilting rotor – BR (back rotor) rotating counter-clockwise – is mounted at the rear part of the body.

3.2.3 Kinematic Equations

The translational and rotational motion, describing position and orientation of the UAV, are obtained using the transformation from body-fixed frame ($\mathcal{F}_B$) to Earth-fixed frame ($\mathcal{F}_E$). They are written as:

$$\begin{bmatrix} \dot{x} \\ \dot{y} \\ \dot{z} \end{bmatrix} = R_{EB} \begin{bmatrix} u \\ v \\ w \end{bmatrix}, \quad \begin{bmatrix} \dot{\phi} \\ \dot{\theta} \\ \dot{\psi} \end{bmatrix} = T_{EB} \begin{bmatrix} p \\ q \\ r \end{bmatrix} \tag{17}$$

where x, y, z and ϕ, θ, ψ are the translational position and rotational attitude, respectively, which are defined in $\mathcal{F}_E$; u, v, w and p, q, r are the translational and rotational velocities that are defined in $\mathcal{F}_B$; R_{EB} is the translation transformation matrix between frames $\mathcal{F}_E$ and $\mathcal{F}_B$, while T_{EB} maps the rotational velocity component from $\mathcal{F}_B$ to $\mathcal{F}_E$. The matrices R_{EB} and T_{EB} are given as ($c : cos$, $s : sin$, $t : tan$):

$$R_{EB} = \begin{bmatrix} c\theta c\psi & s\phi s\theta c\psi - s\psi c\phi & c\phi s\theta c\psi + s\phi s\psi \\ c\theta s\psi & s\phi s\theta s\psi + c\psi c\phi & c\phi s\theta s\psi - s\phi c\psi \\ -s\theta & s\phi c\theta & c\phi c\theta \end{bmatrix}, \tag{18a}$$

$$T_{EB} = \begin{bmatrix} 1 & s\phi t\theta & c\phi t\theta \\ 0 & c\phi & -s\phi \\ 0 & \frac{s\phi}{c\theta} & \frac{c\phi}{c\theta} \end{bmatrix}. \tag{18b}$$

3.2.4 Rigid-Body Equations

The rigid-body dynamic equations of the tilt-rotor tricopter are derived based on the Newton-Euler formulation in the body coordinate system, similar to [3]. Within these equations, the tricopter is assumed to be a point mass, wherein all the forces and moments act at the CG. The corresponding force and moment equations can be written as:

Force Equations

$$\dot{u} = rv - qw + g\sin(\theta) + \frac{1}{m}F_x, \tag{19a}$$

$$\dot{v} = pw - ru - g\sin(\phi)\cos(\theta) + \frac{1}{m}F_y, \tag{19b}$$

$$\dot{w} = qu - pv - g\cos(\phi)\cos(\theta) + \frac{1}{m}F_z, \tag{19c}$$

Moment Equations

$$\dot{p} = \left(\frac{1}{I_{xx}I_{zz} - I_{xz}^2}\right)\Big[\{-pq(I_{xz}) + qr(I_{yy} - I_{zz})\}I_{zz} - \{qr(I_{xz}) + pq(I_{xx} - I_{yy})\}I_{xz} + \tau_x(I_{zz}) - \tau_z(I_{xz})\Big], \tag{20a}$$

$$\dot{q} = pr\left(\frac{I_{zz} - I_{xx}}{I_{yy}}\right) - (r^2 - p^2)\left(\frac{I_{xz}}{I_{yy}}\right) + \tau_y\left(\frac{1}{I_{yy}}\right), \tag{20b}$$

$$\dot{r} = \left(\frac{1}{I_{xx}I_{zz} - I_{xz}^2}\right)\Big[\{qr(I_{xz}) + pq(I_{xx} - I_{yy})\}I_{xx} - \{-pq(I_{xz}) + qr(I_{yy} - I_{zz})\}I_{xx} + \tau_z(I_{xx}) - \tau_x(I_{xz})\Big], \tag{20c}$$

where F_x, F_y, F_z, are the total external forces and τ_x, τ_y, τ_z, are the total external moments acting on the tricopter body in frame $\mathcal{F}_B$. In addition, I_{xx}, I_{yy}, I_{zz} and I_{xz} represent the moments of inertia of the whole tricopter along axes $\mathcal{F}_{B_x}$, $\mathcal{F}_{B_y}$, $\mathcal{F}_{B_z}$ and $\mathcal{F}_{B_{xz}}$, respectively. One may note that unlike a quadrotor UAV, the tilt-rotor tricopter only has a single plane of symmetry, i.e., along $\mathcal{F}_{B_{xz}}$ plane. Therefore, the effect of asymmetric moment I_{xz} is explicitly considered in (20), in contrast to what is done in [3].

3.2.5 External Forces and Moments

The external forces and moments generated by the rotors rotating at a certain angular velocity Ω are modelled as:

$$F_i = K_F\Omega_i^2 \quad \text{and} \quad \tau_i = K_\tau\Omega_i^2, \tag{21}$$

where F_i and τ_i are the external force and drag-moment generated, respectively. Also, K_F and K_τ are positive intrinsic parameters of the rotor and are commonly known as the *force* and *drag-moment* coefficients, respectively. According to the tilt-rotor tricopter configuration shown in Figure 6b, the expression for total external force acting on the tricopter body in $\mathcal{F}_B$ frame is written as:

$$F_{ext} = \begin{bmatrix} F_x \\ F_y \\ F_z \end{bmatrix} = \begin{bmatrix} 0 \\ -F_3\sin(\mu) \\ F_1 + F_2 + F_3\cos(\mu) \end{bmatrix}, \tag{22}$$

where μ is the tilting angle of the back rotor. On the other hand, the total external moment acting on the tricopter platform is the summation of moment due to propeller's rotation τ_{prop}, and the moment due to change in orientation of propeller's rotation plane τ_{gyro}. The latter is commonly known as gyroscopic moment and can be written as:

$$\tau_{gyro} = \sum_{n=1}^{3} J_P(\mathrm{x}_{rate} \times r_n)\Omega_n, \tag{23}$$

where J_P is propeller's moment of inertia and r_n is the unit reaction vector along the rotational axis of nth rotor and x_{rate} is the angular velocity vector. Finally, the expression for total external moment is:

$$\tau_{ext} = \tau_{prop} + \tau_{gyro}, \tag{24}$$

where

$$\tau_{prop} = \begin{bmatrix} (F_2 - F_1)l_2 \\ (F_3 cos(\mu))l_1 - (F_1 + F_2)l_3 + \tau_3 sin(\mu) \\ \tau_1 - \tau_2 - \tau_3 cos(\mu) + (F_3 sin(\mu))l_1 \end{bmatrix}, \tag{25}$$

$$\tau_{gyro} = \begin{bmatrix} J_P\{q(\Omega_1 - \Omega_2) - \Omega_3(\cos(\mu)q + \sin(\mu)r)\} \\ J_P\{p(\Omega_2 - \Omega_1) + \cos(\mu)\Omega_3\} \\ J_P\{-p\sin(\mu)\Omega_3\} \end{bmatrix}. \tag{26}$$

Furthermore, the constant intrinsic parameters for the considered tilt-rotor tricopter UAV are listed in Table 1. These parameters are either obtained by experiments or by any of the system identification method. In this application, we physically measured the mass (m) of the UAV (without the payload mass) and the moment arm lengths (l_1, l_2, l_3). However, for the evaluation of the moment of inertias ($I_{(..)}$) and thrust (K_f) as well as drag-moment (K_τ) coefficients, simple experiments are performed; details of which can be referred from [8, 12].

Table 1: Tilt-rotor tricopter intrinsic parameters

Parameter	Description	Value
m	Mass of tricopter UAV	1.608 kg
l_1	Moment arm	0.284 m
l_2	Moment arm	0.212 m
l_3	Moment arm	0.092 m
I_{xx}	Moment of Inertia about $\mathcal{F}_{B_x}$	0.016053 kg-m^2
I_{yy}	Moment of Inertia about $\mathcal{F}_{B_y}$	0.028158 kg-m^2
I_{zz}	Moment of Inertia about $\mathcal{F}_{B_z}$	0.032752 kg-m^2
I_{xz}	Moment of Inertia about $\mathcal{F}_{B_{xz}}$	0.029763 kg-m^2
K_f	Aerodynamic force coefficient	3.76×10^{-5} N-s^2
K_τ	Aerodynamic drag-moment coefficient	2.56×10^{-6} Nm-s^2

3.2.6 Control Scheme

In contrast to what is done in [27], NMPC in this implementation is designed to be responsible for tracking a given position trajectory. Based upon the current feedback of the other states, optimized solutions for the control inputs in terms of the total thrust and attitude angles are computed. These optimized solutions are then passed to the low-level controller as their desired setpoints. Moreover, the low-level attitude controller is selected as a PID controller (implemented in Pixhawk), which is designed individually for each axis.

3.2.7 Implementation of NMPC

The state, parameter, control and measurement vectors for the high-level NMPC are considered to be composed of:

$$\mathbf{x}_{\mathrm{NMPC}} = [x, y, z, u, v, w]^T, \tag{27}$$

$$\mathbf{p}_{\mathrm{NMPC}} = m, \tag{28}$$

$$\mathbf{u}_{\mathrm{NMPC}} = [F_z, \phi, \theta, \psi]^T, \tag{29}$$

$$\mathbf{z}_{\mathrm{NMPC}} = [x, y, z, u, v, w]^T. \tag{30}$$

Additionally, the final nonlinear programming (NLP) formulation for high-level NMPC also requires the parametrization of the nonlinear model (in translation) with respect to the three rotational rates namely, p, q and r. Therefore, to obtain the solution of the formulated NLP, the three rotational rates are fed to the NMPC along with the other states at each sampling instant. Furthermore, the following state and control nominal values are selected for the parametrization of the state and control trajectories:

$$\mathbf{x}^{\mathrm{ref}} = \mathbf{x}_{N_c}^{\mathrm{ref}} = [x_r, y_r, z_r, 0, 0, 0]^T, \quad \text{and} \quad \mathbf{u}^{\mathrm{ref}} = [\mathrm{mg}, -0.0414, 0, 0]^T, \tag{31}$$

where m and g are the UAV mass and gravitational constant, respectively.

Some constraints are introduced in the definition of NMPC due to the restrictions put up by the real setup. Typically, these are the input constraints that are imposed in order to achieve a stable behaviour from the low-level controller:

$$0.5mg\ (\mathrm{N}) \leq F_z \leq 1.5mg\ (\mathrm{N}), \tag{32a}$$

$$-15\ (^\circ) \leq \phi \leq 15\ (^\circ), \tag{32b}$$

$$-15\ (^\circ) \leq \theta \leq 15\ (^\circ). \tag{32c}$$

Also, the following weight matrices are selected by trial-and-error:

$$W_{\mathbf{x}} = \mathrm{diag}(25, 26, 32, 1.0, 1.0, 1.1), \tag{33a}$$

$$W_{\mathbf{u}} = \mathrm{diag}(0.024, 22, 25, 80), \tag{33b}$$

$$W_{N_c} = \mathrm{diag}(40, 40, 40, 1, 1, 1). \tag{33c}$$

Furthermore, the prediction window $N_c = 30$ is selected to facilitate the real-time applicability of the control framework. One may note that for defining the constraints and obtaining NMPC weights, the UAV mass is selected to be the maximum takeoff mass, i.e., $m = 1.901$ kg.

3.2.8 Implementation of NMHE

In this application, the main task of NMHE is to estimate the UAV mass (m) online, which is made time-varying by a sequential drops of payload. The overall state, parameter, control and measurement vectors for NMHE design are considered to be composed of:

$$\mathrm{x}_{\mathrm{NMHE}} = [u, v, w]^T, \tag{34}$$

$$\mathrm{p}_{\mathrm{NMHE}} = m, \tag{35}$$

$$\mathrm{u}_{\mathrm{NMHE}} = [F_z, \phi, \theta]^T, \tag{36}$$

$$\mathrm{z}_{\mathrm{NMHE}} = [u, v, w, F_z, \phi, \theta]^T. \tag{37}$$

One may note that the state vector for NMHE in (34) is different than the state vector for NMPC in (27). This is because m only appears in the force equations of (19). Moreover, the three rotational rates are included in the measurements along with states and inputs in order to solve the underlying NLP, as also done in NMPC.

For the selected tricopter model, the weight matrices P_{S}, V and W are chosen to be:

$$P_{\mathrm{S}} = \mathrm{diag}(5.4772^2, 5.4772^2, 5.4772^2, 8.9443^2)^{-1/2}, \tag{38a}$$

$$V = \mathrm{diag}(0.0447^2, 0.0447^2, 0.0447^2, 0.2236^2, 0.1^2, 0.1^2)^{-1/2}, \tag{38b}$$

$$\mathrm{W} = \mathrm{diag}(0.01^2, 0.0316^2, 0.0316^2, 0.0316^2)^{-1/2}. \tag{38c}$$

The above values of the weight matrices are decided based upon experience, incorporating the definitions in (5). Additionally, in order to achieve a constrained estimation of the UAV mass, the initial knowledge about the maximum takeoff mass and the minimum assembly mass is exploited and hence, the following constraints are imposed:

$$1.5\ (\mathrm{kg}) \leq m \leq 2.0\ (\mathrm{kg}). \tag{39}$$

Furthermore, the estimation window length M is selected to be equal to 70, which is more than the prediction horizon length of 30 for NMPC. This is purposely kept in order to realize a slower learning from NMHE.

3.2.9 Results

In this section, we present the results of the implementation of NMPC for the high-level position tracking of a tilt-rotor tricopter UAV. In addition to tracking, we also analyse its robustness for the time-varying dynamics of the system by conducting experiments for two scenarios: *NMPC without learning*, and *NMPC with learning* (also referred to as NMPC-NMHE framework).

The real-time implementation of the entire process is summarized in Figure 8. The NMPC and NMHE, which are running at 50-Hz and 30-Hz, respectively, are designed using 's-functions' in Simulink, which are generated via the ACADO toolkit. The OptiTrack motion capture system, consisting of eight cameras running at 240 frames/sec, is used to get the feedback during experiments. The controller commands along with the feedback (position and orientation) are sent to the UAV via Raspberry Pi over a wireless network. The low-level controller is running on the Pixhawk module, which takes the NMPC commands from Raspberry Pi 3 and further computes the actuator commands that are finally given to the tricopter motors. The communication between the UAV and ground-station is achieved through ROS running on a hardware consisting of Intel® Core™i7-4710MQ CPU@2.50GHz processor with 15.6 GB of memory on a 64-bit Ubuntu Linux system.

The initial state of the UAV is $\mathrm{x}(0) = [0,0,1.5,0,0,0]^T$. In addition to the discretization of the nonlinear tricopter model utilizing multiple shooting method (implemented in ACADO toolkit), a fixed integration process consisting of $2N_c$ steps is also performed based on *Runge-Kutta* 4th order method. Moreover, a time-based circular reference trajectory ($[x_r, y_r, 1.5]$) of radius 1 m is selected for these experiments. In both the scenarios, first a complete circle is performed with full takeoff mass of the UAV and then, the payload in terms of four blocks are sequentially dropped (55g $\rightarrow$ 75g $\rightarrow$ 77g $\rightarrow$ 86g) at fixed time-intervals during the trajectory.

Remark I: In the two scenarios – NMPC without learning and NMPC with learning – analysed here, NMHE in the latter case is only utilized to perform the parameter estimation, i.e., only the estimation of mass (m) is fed to NMPC, not the state values. This is done in order to achieve a consistent comparison between the two cases.

3.2.10 Circular Reference Tracking

The overall position tracking performance of the UAV for both scenarios can be seen in Figure 9a and b, where the vertical magenta lines represent the instants of payload drop. In Figure 9b, one may notice a negative offset along z since the beginning for the case of NMPC without leaning. The reason for this is the slight mismatch that exists between the model and the system. Nevertheless, the offset is around 5 cm, which is reasonable compared to the size of the UAV. Moreover, it is illustrated from Figure 9a and b that the position tracking of the NMPC-NMHE framework is better than the NMPC without learning case. The controller's performance is stable and especially, the tracking along z is more accurate. This is because NMHE is able

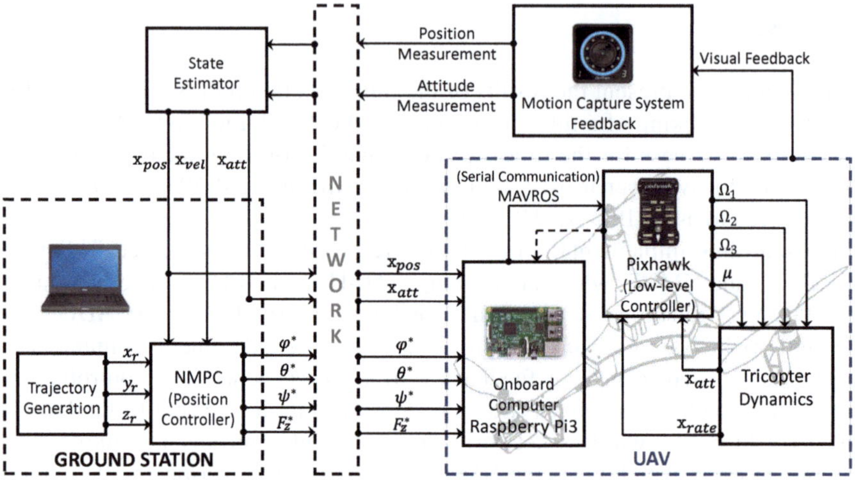

Fig. 8: Schematic diagram for real-time implementation.

to learn the change in UAV mass and thus, the offset created due to mass change diminishes with time. Also, the Euclidean errors for position tracking in the two cases are shown in Figure 9c, where their mean value over the entire run time for NMPC without and with learning are 0.2064 m and 0.1618 m, respectively.

The performance of NMPC can be appreciated by observing its F_z command throughout the trajectory for both without and with learning cases, as shown in Figure 10a, where the vertical magenta lines again represent the instants of payload drop. It is implied that for both the scenarios, the F_z command of NMPC never crosses the bounds specified in (32a), but gets adjusted at every instant of a payload drop. Additionally, the attitude angles commanded by NMPC for without and with learning are given in Figure 10b and c, respectively. It can be seen that ϕ and θ angles of the UAV are well within the specified bounds defined in (32b) and (32c), respectively, while ψ response has some irregularities following the heading command of NMPC. This behaviour is due to the PID tuning of the Pixhawk's low-level controller, but the system is stable enough to maintain the heading.

The performance of NMHE in estimating the mass is shown in Figure 10d, along with its true value, wherein the estimation is observed to stay within the specified bounds defined in (39). As NMPC is a model-based controller, a good estimate of the time-varying model parameters is crucial for its optimum performance. Moreover, in package delivery applications of the UAV that we are considering, the total mass changes with time. Therefore, it is important to estimate it, so that NMPC can adapt itself to the changing working conditions. It is worth noting that although m is a physical parameter in the system model, when it is estimated by NMHE, it also accommodates the effects of modeling uncertainties that are injected during operation. Overall, it can be interpreted as an adaptive parameter that facilitates NMPC to achieve an offset free tracking along z.

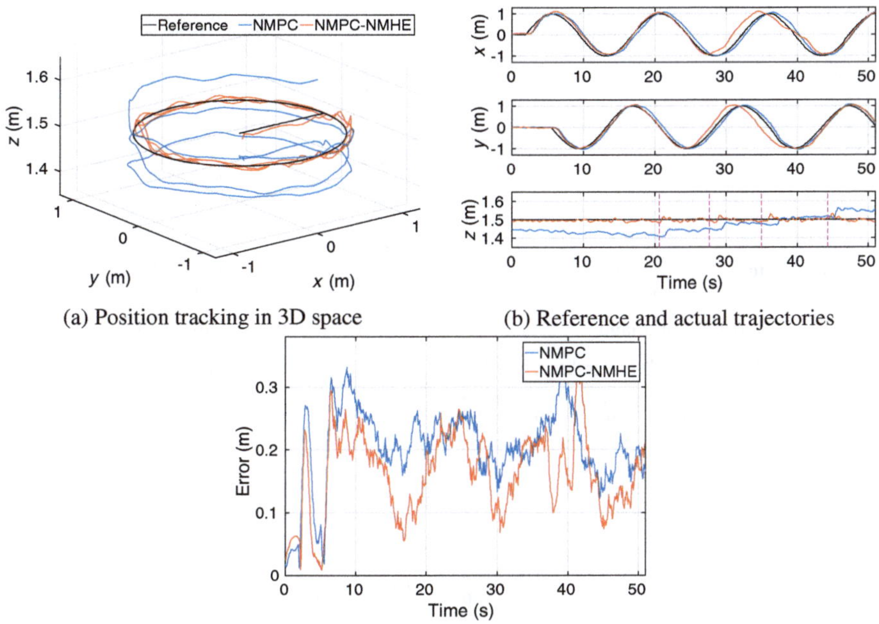

(a) Position tracking in 3D space
(b) Reference and actual trajectories
(c) Euclidean error

Fig. 9: Trajectory tracking performance.

In order to check the optimality of NMPC and NMPC-NMHE framework, their KKT tolerances are obtained and plotted in Figure 11a and b, respectively. As mentioned in the previous application, it is necessary to check the optimality of the solution because a single quadratic programming iteration at each sampling time instant (performed in ACADO) may result in a suboptimal solution. As visualized in Figure 11a and b, the KKT tolerances for NMPC and NMHE are small, but not zero. This is happening due to the nonlinearities in the system dynamics. It is to be noted that the tilt-rotor tricopter UAV is far more nonlinear and inherently unstable in comparison to other multirotor UAVs including quadrotors and hexacopters. This is mainly due to the odd number of rotors, which results in an unbalanced yaw moment. In addition, one may point out the difference in KKT values for NMPC and NMHE, and the reason lies in the selection of their prediction and estimation horizons, respectively. Nonetheless, their low and non-drifting magnitudes still represent the well-defined and properly scaled optimization problems of both NMPC without learning and the NMPC-NMHE framework.

Finally, the execution times for each entity in NMPC without learning and NMPC-NMHE framework are displayed in Figure 12a and b, respectively. In addition, the combined mean execution times for both without and with learning cases are 0.7763 ms and 3.7 ms, respectively. These values are less compared to the selected sampling time of 20 ms (position controller) and hence, support the implementation of both on a cheaper embedded hardware including Raspberry Pi 3.

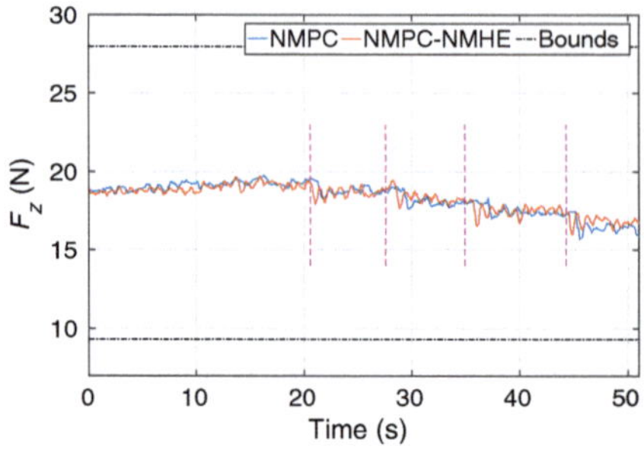

(a) F_z command

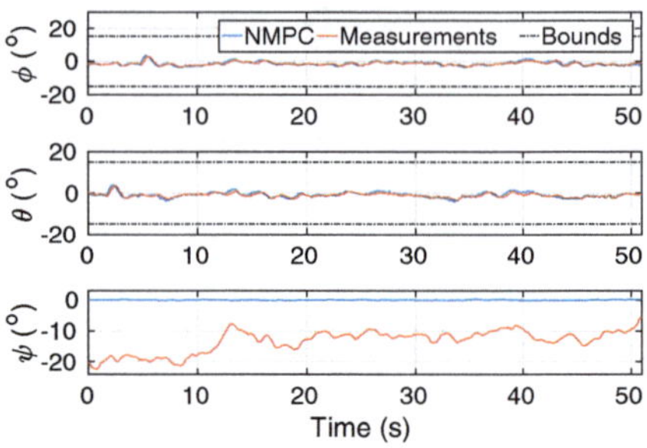

(b) Attitude command of NMPC

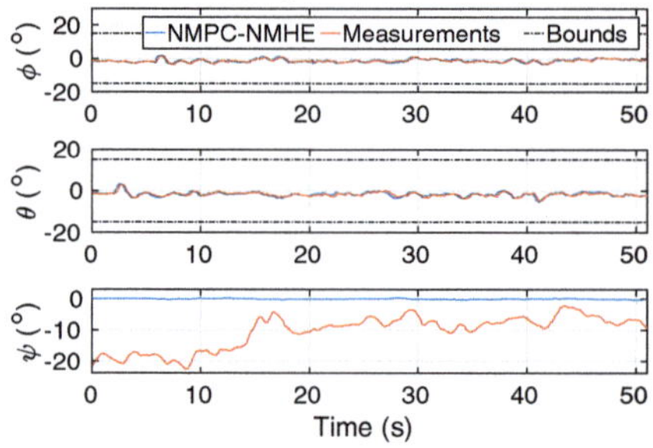

(c) Attitude command of NMPC with learning

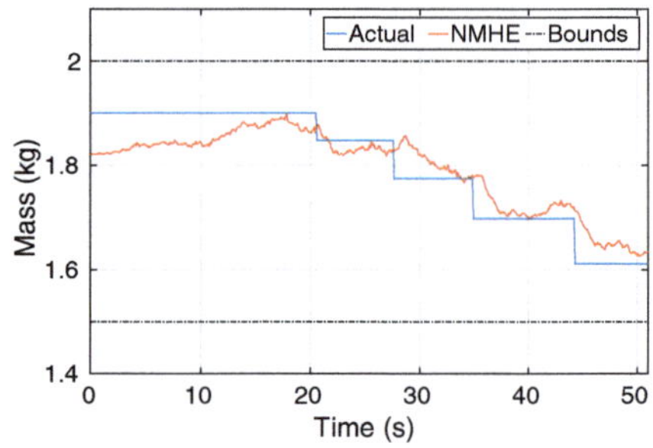

(d) Estimation performance of NMHE

Fig. 10: Controller and estimator outputs.

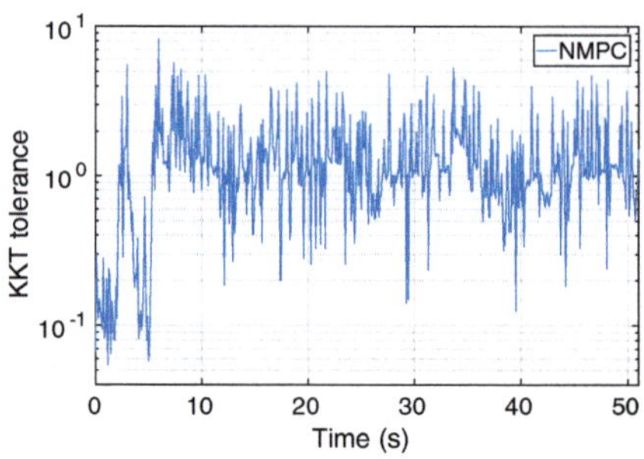

(a) NMPC without learning

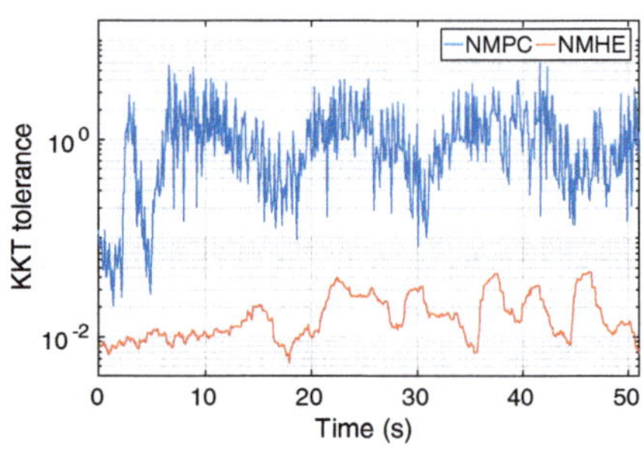

(b) NMPC-NMHE framework

Fig. 11: KKT tolerance.

Remark II: The tuning of NMPC weights is a problem that is generally encountered while performing the experiments. Any minor change in the system including discharging of the battery leads to a change in the controller weights mainly along F_z. Moreover, the best way to tune the combined NMPC-NMHE framework is to first tune the NMPC separately, and then utilize those weights as a reference for the combined framework.

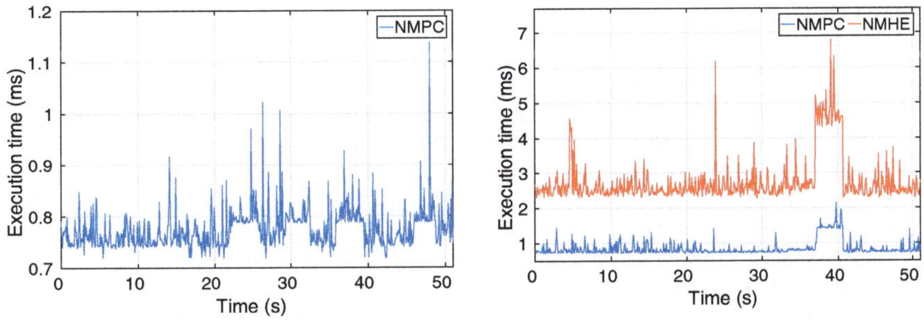

(a) NMPC without learning

(b) NMPC-NMHE framework

Fig. 12: Execution time.

Remark III: Selecting an appropriate estimation horizon for NMHE is problem specific as it directly affects the rate of learning. That is, for a shorter horizon, the mass learning is fast which eventually makes NMPC to be aggressive towards the change. On the other hand, for longer horizon, the NMHE gradually learns the mass parameter and hence, a smooth response is obtained from NMPC.

4 Conclusion

We have incorporated NMPC-NMHE framework for the system with uncertainties including slip variations in the off-road ground robotic vehicle due to the change in soil conditions and mass variations for the considered package delivery problem of the aerial robot. Thanks to its learning capability, the accuracy of the NMPC is enhanced by the estimation of parameters by NMHE in both the applications. The outcome of the first application, in which we have estimated the soil condition variations, is that the Euclidean error for the NMPC is about 0.0459 m, which is satisfactory for any agricultural application. In the presented second application, where we have estimated the mass of the UAV for a package delivery task, the learning-based NMPC gives better tracking performance than that of the NMPC without learning. The average Euclidean error for the learning-based NMPC (0.1618 m) is less than that of NMPC without learning (0.2064 m). Since the true values of the mass are known throughout the trajectory, we have also presented the estimation results versus their true values. Overall, the obtained results from both the applications imply that the online leaning-based NMPC substantially improves the tracking performance for the presented robotic applications.

Acknowledgements This research is supported by the National Research Foundation, Prime Minister's Office, Singapore under its Medium-Sized Centre funding scheme. The information, data, or work presented herein was funded in part by the Advanced Research Projects Agency-Energy (ARPA-E), U.S. Department of Energy, under Award Number DE-AR0000598.

References

1. Allgöwer, F., Badgwell, T.A., Qin, J.S., Rawlings, J.B., Wright, S.J.: Nonlinear Predictive Control and Moving Horizon Estimation — An Introductory Overview. Springer, London (1999). https://doi.org/10.1007/978-1-4471-0853-5_19
2. Ariens, D., Houska, B., Ferreau, H., Logist, F.: ACADO: toolkit for automatic control and dynamic optimization. Optimization in Engineering Center (OPTEC), 1.0beta edn. (2010). http://www.acadotoolkit.org/
3. Bouabdallah, S.: Design and Control of Quadrotors with Application to Autonomous Flying, p. 155. EPFL, Lausanne (2007)
4. Bouffard, P., Aswani, A., Tomlin, C.: Learning-based model predictive control on a quadrotor: onboard implementation and experimental results. In: 2012 IEEE International Conference on Robotics and Automation (ICRA), pp. 279–284 (2012). https://doi.org/10.1109/ICRA.2012.6225035
5. Chowdhary, G., Mühlegg, M., How, J.P., Holzapfel, F.: Concurrent Learning Adaptive Model Predictive Control, pp. 29–47. Springer, Berlin (2013). https://doi.org/10.1007/978-3-642-38253-6_3
6. Chowdhary, G., Mühlegg, M., How, J.P., Holzapfel, F.: A concurrent learning adaptive-optimal control architecture for nonlinear systems. In: 52nd IEEE Conference on Decision and Control, pp. 868–873 (2013). https://doi.org/10.1109/CDC.2013.6759991
7. Eren, U., Prach, A., Koçer, B.B., Raković, S.V., Kayacan, E., Açıkmeşe, B.: Model predictive control in aerospace systems: current state and opportunities. J. Guid. Control. Dyn. **40**(7), 1541–1566 (2017). https://doi.org/10.2514/1.G002507
8. Fum, W.Z.: Implementation of simulink controller design on Iris+ quadrotor. Ph.D. thesis, Monterey, California, Naval Postgraduate School (2015)
9. Garimella, G., Sheckells, M., Kobilarov, M.: Robust obstacle avoidance for aerial platforms using adaptive model predictive control. In: 2017 IEEE International Conference on Robotics and Automation (ICRA), pp. 5876–5882 (2017). https://doi.org/10.1109/ICRA.2017.7989692
10. Grúne, L.: NMPC without terminal constraints. IFAC Proc. Vol. **45**(17), 1–13 (2012)
11. Havoutis, I., Calinon, S.: Supervisory teleoperation with online learning and optimal control. In: 2017 IEEE International Conference on Robotics and Automation (ICRA), pp. 1534–1540 (2017)
12. Hoffmann, G.M., Huang, H., Waslander, S.L., Tomlin, C.J.: Quadrotor helicopter flight dynamics and control: theory and experiment. In: Proceedings of the AIAA Guidance, Navigation, and Control Conference, vol. 2, p. 4 (2007)
13. Kayacan, E., Kayacan, E., Ramon, H., Saeys, W.: Distributed nonlinear model predictive control of an autonomous tractor-trailer system. Mechatronics **24**(8), 926–933 (2014)
14. Kayacan, E., Kayacan, E., Ramon, H., Saeys, W.: Nonlinear modeling and identification of an autonomous tractor-trailer system. Comput. Electron. Agric. **106**, 1–10 (2014). https://doi.org/10.1016/j.compag.2014.05.002
15. Kayacan, E., Kayacan, E., Ramon, H., Saeys, W.: Learning in centralized nonlinear model predictive control: application to an autonomous tractor-trailer system. IEEE Trans. Control Syst. Technol. **23**(1), 197–205 (2015)
16. Kayacan, E., Kayacan, E., Ramon, H., Saeys, W.: Robust tube-based decentralized nonlinear model predictive control of an autonomous tractor-trailer system. IEEE/ASME Trans. Mechatron. **20**(1), 447–456 (2015)
17. Kayacan, E., Kayacan, E., Ramon, H., Saeys, W.: Towards agrobots: Identification of the yaw dynamics and trajectory tracking of an autonomous tractor. Comput. Electron. Agric. **115**, 78–87 (2015)
18. Kayacan, E., Peschel, J.M., Kayacan, E.: Centralized, decentralized and distributed nonlinear model predictive control of a tractor-trailer system: a comparative study. In: 2016 Ameri-

can Control Conference (ACC), pp. 4403–4408 (2016). https://doi.org/10.1109/ACC.2016.7525615
19. Kraus, T., Ferreau, H., Kayacan, E., Ramon, H., Baerdemaeker, J.D., Diehl, M., Saeys, W.: Moving horizon estimation and nonlinear model predictive control for autonomous agricultural vehicles. Comput. Electron. Agric. **98**, 25–33 (2013)
20. Kühl, P., Diehl, M., Kraus, T., Schlöder, J.P., Bock, H.G.: A real-time algorithm for moving horizon state and parameter estimation. Comput. Chem. Eng. **35**(1), 71–83 (2011)
21. Liu, Y., Rajappa, S., Montenbruck, J.M., Stegagno, P., Bülthoff, H., Allgöwer, F., Zell, A.: Robust nonlinear control approach to nontrivial maneuvers and obstacle avoidance for quadrotor UAV under disturbances. Robot. Auton. Syst. **98**, 317–332 (2017). https://doi.org/10.1016/j.robot.2017.08.011
22. López-Negrete, R., Biegler, L.T.: A moving horizon estimator for processes with multi-rate measurements: a nonlinear programming sensitivity approach. J. Process Control **22**(4), 677–688 (2012)
23. Mehndiratta, M., Kayacan, E.: Receding horizon control of a 3 DOF helicopter using online estimation of aerodynamic parameters. Proc. Inst. Mech. Eng. Part G J. Aerosp. Eng. (2017). https://doi.org/10.1177/0954410017703414
24. Morari, M., Lee, J.H.: Model predictive control: past, present and future. Comput. Chem. Eng. **23**(4), 667–682 (1999). https://doi.org/10.1016/S0098-1354(98)00301-9
25. Nicolao, G.D., Magni, L., Scattolini, R.: Stabilizing receding-horizon control of nonlinear time-varying systems. IEEE Trans. Autom. Control **43**(7), 1030–1036 (1998)
26. Ostafew, C.J., Schoellig, A.P., Barfoot, T.D.: Robust constrained learning-based NMPC enabling reliable mobile robot path tracking. Int. J. Robot. Res. **35**(13), 1547–1563 (2016). https://doi.org/10.1177/0278364916645661
27. Prach, A., Kayacan, E.: An MPC-based position controller for a tilt-rotor tricopter VTOL UAV. Optim. Control Appl. Methods https://doi.org/10.1002/oca.2350
28. Rao, C.V., Rawlings, J.B., Mayne, D.Q.: Constrained state estimation for nonlinear discrete-time systems: stability and moving horizon approximations. IEEE Trans. Autom. Control **48**(2), 246–258 (2003). https://doi.org/10.1109/TAC.2002.808470
29. Robertson, D.G., Lee, J.H., Rawlings, J.B.: A moving horizon-based approach for least-squares estimation. AIChE J. **42**(8), 2209–2224 (1996)
30. Shin, J., Kim, H.J., Park, S., Kim, Y.: Model predictive flight control using adaptive support vector regression. Neurocomputing **73**(4), 1031–1037 (2010). https://doi.org/10.1016/j.neucom.2009.10.002
31. Vukov, M., Gros, S., Horn, G., Frison, G., Geebelen, K., Jørgensen, J., Swevers, J., Diehl, M.: Real-time nonlinear MPC and MHE for a large-scale mechatronic application. Control Eng. Pract. **45**, 64–78 (2015)

Applications of MPC to Building HVAC Systems

Nishith R. Patel and James B. Rawlings

1 Introduction to Building HVAC Systems

Heating, ventilation, and air conditioning (HVAC) systems are an integral part of most buildings. They are responsible for temperature regulation by heating or cooling as needed to keep occupants in the building comfortable, as shown in Figure 1. Heating and cooling against temperature gradients requires a large amount of energy: cooling consumes electricity and heating consumes fuel. In fact, commercial buildings are responsible for 20% of the total U.S. energy consumption, and total expenditures exceed $200 billion every year [7]. Even small percentage savings in this industry can have a significant impact due to the magnitude of these numbers.

A key motivation for improving control of HVAC systems stems from the pricing structures utilized by power companies. These pricing structures are motivated by the nature of loads. Figure 2 shows a typical week-long profile for the ambient temperature of the southern U.S. in the summer. The temperature is highest in early afternoon and lowest overnight into the early morning hours, which also roughly represents the cooling load placed on the HVAC system. This load on the HVAC system results in a similar power load placed on power companies. Since a constant load profile allows power plants to operate more efficiently, power companies enforce pricing structures to incentivize customers to purchase less electricity during busier hours and more during other hours so they have a flatter load profile.

Power companies charge customers according to time-varying electricity prices as well as a peak *demand* charge on the peak power usage over a given month. An example of one time-varying electricity price profile is given in Figure 3. The key

N. R. Patel · J. B. Rawlings (✉)
University of Wisconsin–Madison, 1415 Engineering Dr., Madison, WI 53706, USA
e-mail: nishith.patel@wisc.edu; james.rawlings@wisc.edu

S. V. Raković, W. S. Levine (eds.), *Handbook of Model Predictive Control*, Control Engineering, https://doi.org/10.1007/978-3-319-77489-3_25

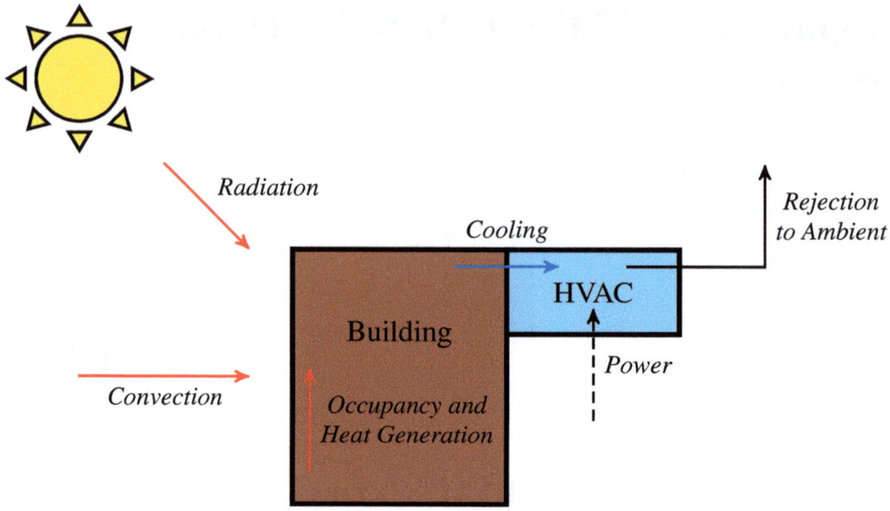

Fig. 1: Heat transfer during the summer. The HVAC system uses electricity to cool the building by rejecting all loads due to radiation, convection, occupancy, and internal heat generation back to the ambient.

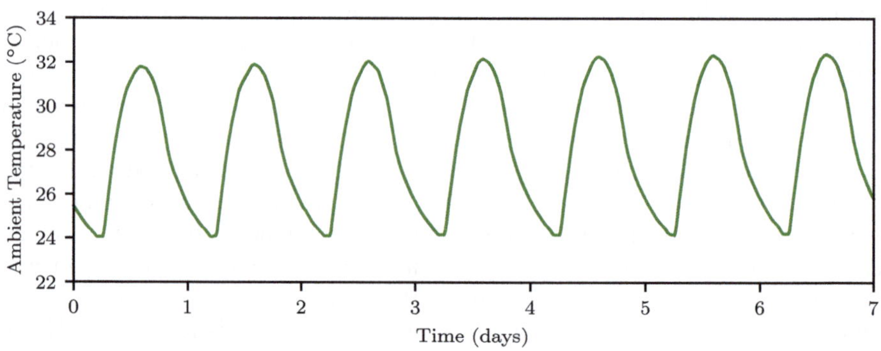

Fig. 2: Representative ambient temperature data over a 7-day period in the summer [26]. In this plot, zero corresponds to midnight.

feature to note is that typically the cost is higher during afternoon hours (referred to as peak hours) and lower overnight (off-peak hours). As a result, HVAC customers pay more for using electricity during the peak hours than off-peak hours. If they are able to do more cooling at night and less cooling during the day, operating costs will significantly decrease. Thermal energy storage (TES) is required to store this "cooling" energy from overproduction. Two forms of TES include passive TES (e.g., mass of the buildings) and active TES (e.g., chilled water and ice tanks). The shifting of the power load from peak to off-peak hours using TES can provide significant cost savings.

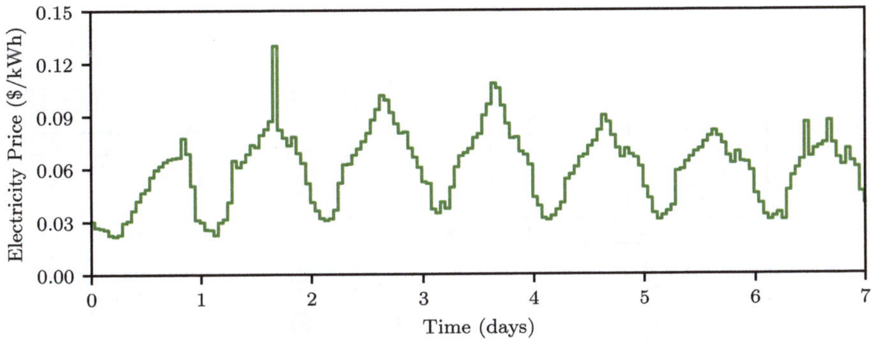

Fig. 3: Representative electricity pricing data over a 7-day period in the summer. In this plot, zero corresponds to midnight. Data provided by Johnson Controls, Inc. [26].

The prospect of achieving such cost savings has made commercial HVAC systems an attractive area for academic research, particularly in the area of control. Most buildings currently use on/off, logic-based, and PID controllers [1]. These existing control systems are not well-suited for load shifting and, thus, cannot provide significant cost savings. In this paper, we explore model predictive control as a control architecture to achieve load shifting and realize these benefits. In Section 2, the HVAC control problem is defined. In Section 3, several challenges and opportunities are discussed. In Section 4, a hierarchical decomposition control architecture is presented. In Section 5, the decomposition is performed and solved for an illustrative example system. In Section 6, a real-world implementation of some of these ideas at Stanford University is outlined. In Section 7, a few potential next steps for this technology in the HVAC industry are listed.

2 Problem Statement

The main parts of a large-scale commercial HVAC system are depicted in Figure 4. The commercial HVAC system is typically divided into two subsystems: *airside* and *waterside*. As shown in Figure 4, the airside subsystem consists of the buildings and air handling units (AHUs) that are responsible for temperature regulation. Typically, independent regulatory controllers exist in each of the zones to track a given temperature setpoint. The waterside subsystem consists of all equipment, such as chillers, heat recovery chillers, boilers, and cooling towers as well as storage tanks, that is used to meet the load from the airside subsystem.

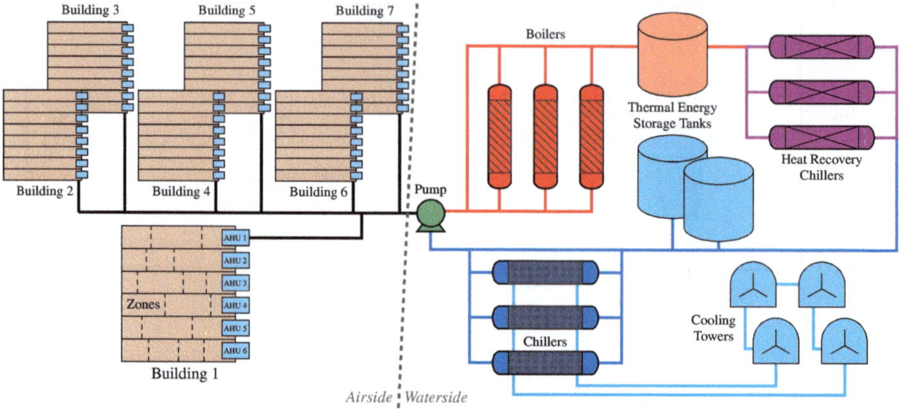

Fig. 4: Diagram of a typical large-scale commercial application with the airside system (buildings) on the left and waterside system (central plant) on the right [20].

The supervisory control system must make the following decisions:

- What are the zone temperature setpoints?
- What is the total load generated by the airside system?
- What is the equipment operation schedule required to meet that total load?
- When is storage charging or discharging and at what rate?

When making these decisions, the goal is to minimize the energy costs while respecting comfort bounds to keep occupants satisfied and capacity constraints on equipment. For simplicity, only cooling in the summer is discussed in this paper due to the pricing structures associated with electricity markets, but heating in the winter can be treated in a similar manner.

2.1 MPC

Model predictive control (MPC) is well-suited for this application due to its ability to treat large-scale multivariable interactive processes, to make predictions, to respect constraints, and to optimize performance. An optimization problem can be solved at each timestep to determine the schedule for waterside equipment operation and the setpoints that are sent to the airside regulatory controllers through the building automation system (BAS), as shown in Figure 5. Models of the buildings can be used to determine what the loads are, models of the waterside system can be used to determine the corresponding power consumption, and models of the regulatory controllers can be used to incorporate the relevant setpoint dynamics.

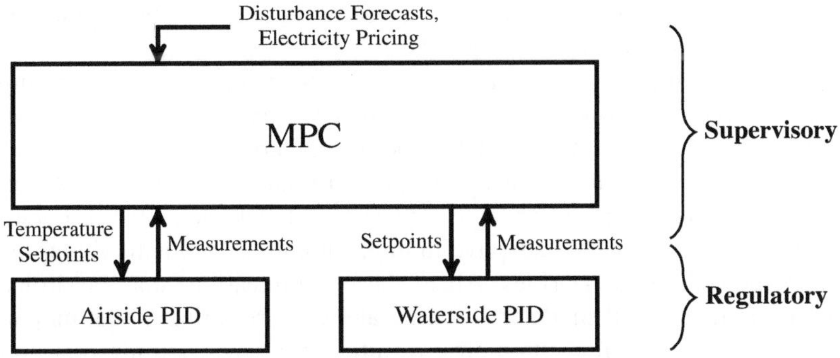

Fig. 5: Supervisory control system (MPC) sends setpoints to the waterside system and the airside building automation system (BAS) which consist of regulatory PID controllers.

Since the objective is to minimize energy costs in the buildings application, an economic cost function is used in the MPC optimization, hence it is an economic MPC framework. Many academic researchers have proposed using economic MPC for control of HVAC systems. See [11, 14, 16, 19, 31, 32] as a few examples. These and many other works show the significant savings that are made possible by load shifting, which occurs naturally as a result of the MPC optimization. Although these works have shown that significant benefits that can be realized with MPC, wide-scale implementation of MPC-based systems has not yet taken place.

3 Challenges and Opportunities

There are many challenges and opportunities when applying MPC to HVAC systems. In this section, we outline a few of them, including modeling, forecasting loads, making discrete decisions, scaling formulation to handle large system, and addressing demand charges. Many of these topics remain active areas of academic research. There are many other areas (e.g., MPC performance monitoring) that are equally important, but not covered here due to space limitations.

3.1 Modeling

Since MPC relies on making predictions, a dynamic model of the system is necessary to relate control actions to measured outputs. For HVAC systems, the MPC decisions are usually zone temperature setpoints and equipment operation, while the measured outputs are zone temperatures and energy consumption. Hence, the relevant models include power consumptions curves for the various waterside equipment, temperature dynamics for each zone, and control laws for the regulatory temperature controllers receiving the setpoints.

Modeling buildings can be challenging due to their highly nonlinear nature and presence of large disturbances [12, 30]. Equipment curves for waterside are rarely linear. In energy balances, heat transfer coefficients are functions of temperature; bilinear terms arise from the product of temperatures and flow rates [15]. Regulatory control laws in buildings are often logic- or rule-based. Additionally, PID controllers may include features such as adaptive tuning, saturation, and anti-windup, which further increase the nonlinearities. All of these nonlinearities may be difficult to capture accurately, and if modeled too accurately, the resulting model may not be suitable for real-time optimization. The resulting nonlinear optimization problem may have multiple local solutions. Hence, a tradeoff exists between accuracy and simplicity for optimization.

Since a large engineering effort may be required to model each building from first principles, models can be instead identified from operating data. However, slow time-varying disturbances and poor sensors (single thermostat measurements may not represent the entire "zone" mass well) can complicate system identification. To reduce data requirements, structure can be added to the identification procedure and grey-box models can be used in place of black-box models [20].

Since modeling buildings can be difficult, the MPC architecture relies on feedback to correct for these inevitable model errors. Dynamic linear models of lower order can be used for optimization as long as the mismatch is addressed appropriately.

3.2 Load Forecasting

In order to make predictions with the model, forecasts of future loads are necessary. Weather data can be used to estimate radiation and convection loads. Historical data can be used to estimate typical loads due to occupancy. Regression-based techniques, including artificial neural networks, may be used for this purpose [22]. Alternatively, stochastic techniques can be used to generate the full distribution of load profiles to characterize uncertainty [15]. Typically, the farther the forecast is into the future, the greater the uncertainty is in the prediction.

As in the case before, prediction errors can be corrected through feedback. However, performance loss may occur if there are significant differences between predictions and actual loads since suboptimal use of TES may take place. One key challenge in this area is how to use current measurements of loads to update the entire future forecast for MPC. While several viable choices may exist, it is not clear which results in the best control performance (i.e., minimal cost).

3.3 Discrete Decisions

Since the supervisory control system must decide when to turn equipment on and off in order to dispatch the equipment operation schedule, discrete decisions must be made during the optimization. Traditionally, these discrete decisions have been made using heuristics or manually by human operators, leaving considerable savings behind. Additionally, some heuristic-based control strategies may actually lead to increased operating costs when an active TES tank is included [3]. However, with advances in hardware and algorithms for mixed-integer optimization, modern computers are able to solve such problems in reasonable times. Since the use of mixed-integer optimization for online control is in its infancy, there may be challenges associated with implementation. However, there are also opportunities for using state-of-the-art tools since MPC is new in the HVAC industry; there are fewer restrictions on what systems can be implemented compared to other industries where MPC has been widely used for several decades and preexisting MPC systems would have to be displaced.

Such applications have also motivated efforts to understand the theory of MPC with discrete actuators. Conventional MPC theory treats continuous decision variables. However, if appropriate assumptions are made about the constraint sets, the stability theory naturally extends to handle the discrete actuator case without additional restrictions [24]. This emerging field opens up a wide array of other rich applications of MPC using discrete actuators. Lessons from these applications can be applied to HVAC systems.

3.4 Large-Scale Applications

In many MPC applications, a single optimization problem is typically formulated and solved. However, a key obstacle in implementing a single monolithic MPC formulation for HVAC is that many large-scale applications (e.g., university campuses) can have hundreds of buildings and thousands of zones. Solving a single optimization with both continuous and discrete decisions for such applications is not practical in real-time, and such control systems are more difficult to maintain. Decomposing the centralized problem into smaller subproblems alleviates these issues.

Distributed MPC solves smaller optimization problems, as discussed in [23, Chapter 6] and [5, 25]. Iterative methods have been proposed for the buildings problem [4, 10, 13, 28]. Due to the limitations placed on information exchanges by existing communication protocols in HVAC systems [18], the drawback is that they may involve many exchanges and iterations before converging to the solution. Addressing the full complexity of HVAC applications including both airside and waterside systems for these large systems using a decomposition is still an open research question. Several viable decompositions have been proposed, but a consensus on which may become the gold standard has not been reached.

3.5 Demand Charges

In addition to the time-varying electricity prices shown in Figure 3, power companies also levy a peak demand charge based on the peak power usage over a fixed time period (e.g., a month) [6, 14]. This peak demand charge couples all of the buildings across the campus. Since it comprises a significant portion of the total energy cost, it cannot be neglected. Completely decentralized MPC for each building is insufficient since all buildings may precool simultaneously incurring a large peak. Hence, a system-wide optimization must take place to manage when the various regions consume power. This feature motivates the need for a coordination layer to manage the total load. However, there is no consensus of handling the demand charge in a truly distributed setting without iteration.

4 Decomposition

To solve the MPC optimization problem with discrete variables online for large-scale applications, a decomposition of the centralized problem is necessary. There are several viable ways to decompose the problem. In this section, a hierarchical control system is presented and some of its advantages are discussed.

The hierarchical decomposition is shown in Figure 6. The MPC problem is divided into two layers: high-level and low-level layers. The low-level is further broken up into airside and waterside subsystems and these low-level problems can be solved in parallel to reduce computation time.

4.1 High-Level

In this decomposition, the high-level problem serves as a coordinator to manage energy usage across the entire campus. It performs an economic optimization, considering both time-varying electricity prices and demand charges, to compute the loads for each subsystem. The inclusion of demand charges in the high-level problem serves to manage the total peak usage across the entire campus. For computational reasons, the full detailed models of the airside and waterside subsystems are not used by the high-level problem. Instead, lower order aggregate models are utilized. The high-level models the aggregate performance of the central plant as well as any active TES tanks for the waterside model and uses average subsystem (e.g., building) temperature models for the airside. The decision variable for the high-level problem is the load profile for each airside subsystem as well as the central plant production and storage schedule. These computed profiles are then passed down to the low-level airside and waterside problems.

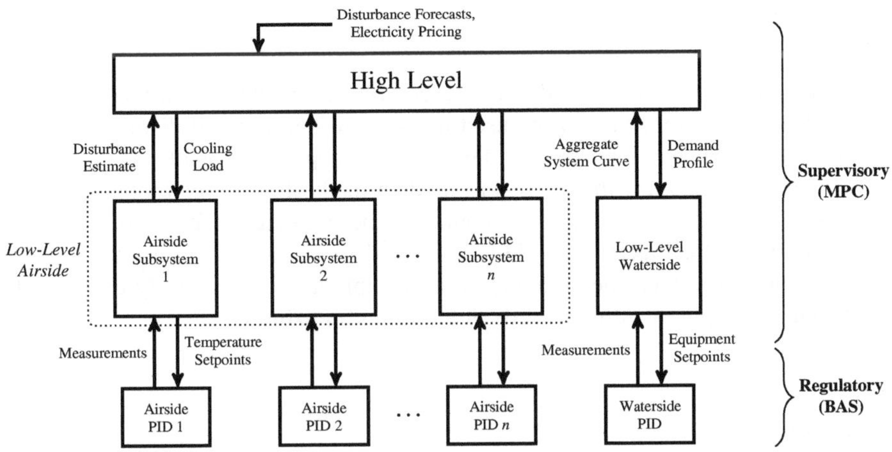

Fig. 6: Hierarchical Decomposition. At each timestep, the high-level optimization is solved first, followed by the low-level airside and waterside problems (solved in parallel). Feedback occurs on both at fast time-scale with disturbance estimates and on a slow time-scale with model adjustments.

4.2 Low-Level Airside

To reduce computational complexity, the entire airside system is divided into various subsystems. One way to decompose into subsystem is by building, so that the subsystems are noninteracting as zones in different buildings do not interact. This choice eliminates the need for iterations between the distributed controllers. Each low-level airside subsystem is allocated a certain load from the solution of the high-level problem. The objective is to minimize energy usage while not exceeding the allocation from the high-level problem. Each subsystem computes the temperature setpoints for all zones in that particular subsystem while satisfying comfort constraints. For details about the mathematical formulation of the high-level and low-level airside subproblems, see [21].

4.3 Low-Level Waterside

The total campus load computed from the high-level optimization is sent to the low-level waterside problem. The waterside problem computes the equipment on/off schedule as well as their loads to minimize cost while meeting the load from high-level layer. Detailed equipment models are used as well as storage models, hence the production schedule from the high-level problem can be refined. Discrete decisions are handled via mixed-integer optimization. The computed solution is dispatched to the central plant. For more details about the mathematical formulation of the waterside subproblem, see [27].

4.4 Feedback

Measurements are fed back to the low-level MPC controllers from the regulatory layer. These measurements can be used to estimate unmeasured disturbances and update any disturbance forecasts for the next time step. Additionally, as the central plant operation changes, the aggregate models in the high-level problem can be updated on a slower time scale to account for these changes. To demonstrate this architecture, an illustrative example is presented in the next section for a modest-sized campus. However, the architecture is scalable and can be easily extended to handle large campuses with hundreds of buildings and zones.

5 Example

In this example, we consider a campus that has four buildings each with five zones and a large central cooling plant. Dynamic linear models are used to represent the heat transfer and regulatory temperature controller in each of the 20 zones. The comfort zone for each zone is between 20.5 °C and 22.5 °C. The ambient temperature and electricity pricing data used are shown in Figures 2 and 3. The horizon for the MPC problems can range from 24 h to 1 week and timesteps can range from 15 min to 1 h.

The central cooling plant contains eight conventional chillers, six pumps, ten cooling towers, and a small active TES tank. The minimum and maximum capacities of each chiller are 2.5 MW and 12.5 MW, respectively. The maximum cooling capacity of the storage tank is 100 MWh. Piecewise-linear power consumption models are used for the chillers. The chilled water supply temperature is held at 5.5 °C.

With the system now defined, the hierarchical decomposition from the previous section can be executed. The results of the high-level optimization are shown in Figure 7. The top plot shows the production schedule generated using the aggregate models of the airside and waterside systems. For brevity, only the total loads are shown in the plot, but each building's load is computed in the optimization. During periods of low prices, storage is charged and during periods of high prices, direct production from chillers is reduced. Note that the overall production profile is flat due to the peak demand charge on the maximum rate of power usage. The bottom plot shows the average building temperatures. A similar trend is also observed there. The buildings are precooled during morning hours when prices are low and allowed to naturally heat back up the upper comfort zone limit in the early afternoon hours when prices are high. The corresponding loads are sent to the two low-level problems.

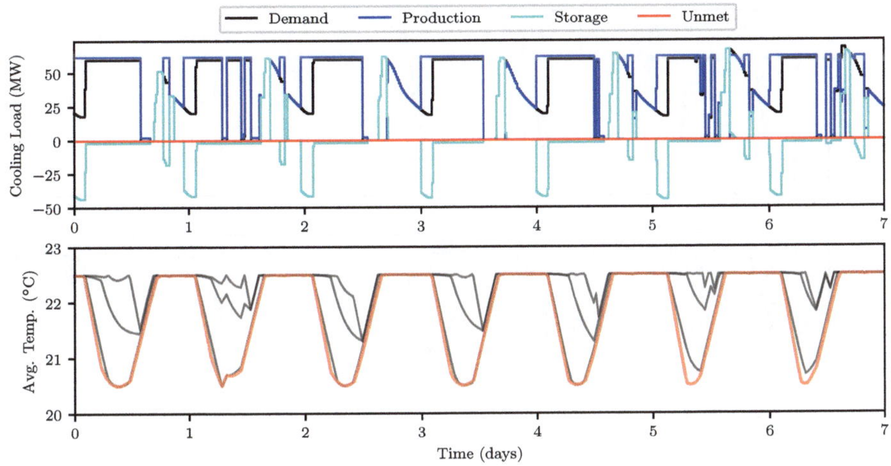

Fig. 7: High-Level Results. Optimal production schedule and average building temperatures computed from solving the high-level problem. In upper plot, negative values of storage denote charging of tank, while positive values indicate discharging of tank. In lower plot, red line highlights one particular building, with all others shown in black.

Each of the four buildings has its own low-level airside MPC controller. They receive an allocation from the high-level problem and determine the setpoints for the zones in that building. Figure 8 shows the temperature and setpoint trajectories for all 20 zones computed from these problems. The individual zones follow a similar profile as the building average. Note that the setpoints drop sharply in advance of precooling period. This phenomenon is due to the fact that since the regulatory controller dynamics are modeled in the MPC problem, the setpoints are adjusted accordingly to counteract sluggish responses which are common in buildings.

The initial production schedule estimated by the high-level problem is sent to the low-level waterside problem for refinement using more accurate equipment models. The resulting production schedule and associated Gantt chart detailing the equipment operation schedule from the waterside problem is presented in Figure 9. The production profiles are similar, but the one in Figure 9 shows significant jumps, which are due to the discrete nature of turning pieces of equipment on and off.

In this example, the high-level and low-level airside problems are formulated as linear programming problems. The low-level waterside problem is formulated as a mixed-integer linear program. Hence, widely available commercial or open-source solvers can be used to solve these problems efficiently, in a matter of seconds.

To evaluate the benefits of this approach, comparisons are made to a baseline strategy. One baseline is to avoid optimization and precooling altogether by employing the minimum energy usage control strategy: stay at the upper bound of the comfort region at all times. While this baseline minimizes energy usage, the economic MPC framework presented can achieve significant cost savings. Typical energy cost

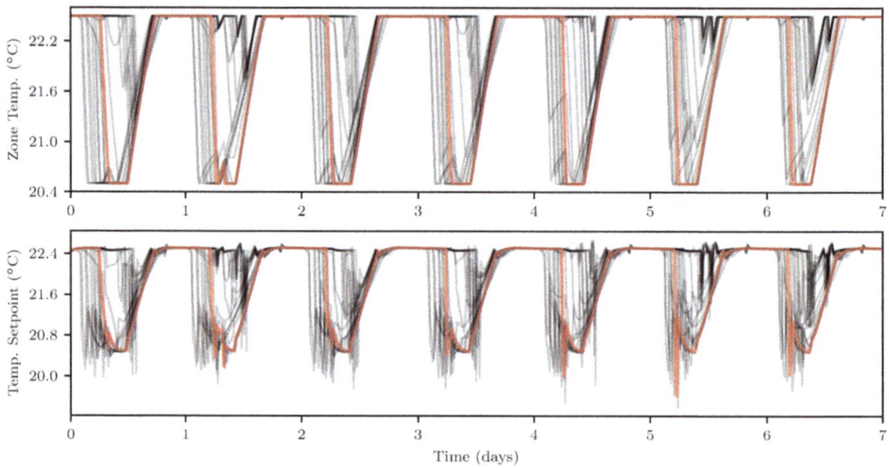

Fig. 8: Low-Level Airside Results. Optimal zone temperatures and setpoints computed from solving the low-level airside problems. Red line denotes one particular zone as an example, with all others shown in black.

savings can vary from 15% to 40% depending on the amount of the TES available and incentives for load shifting, namely the difference in peak and off-peak electricity prices.

While the decomposition presented provides significant cost savings, detailed analysis of the loss in performance due to decomposing when compared to the centralized problem has not yet been conducted. Small-scale systems for which the centralized problem can be solved must be used for this comparison, which is the subject of future work along with benchmarks against other decomposition methods. The concepts illustrated by the simulation can be applied in practice as exemplified in the next section.

6 Stanford University Campus

6.1 SESI Project

Stanford University recently overhauled their campus-wide HVAC system. They embarked on the $485-million Stanford Energy System Innovations (SESI) project to replace an existing 50-MW natural-gas-fired cogeneration plant with a central chiller plant to service the cooling and heating needs of the campus. The central plant includes conventional chillers, boilers, and thermal energy storage tanks. In addition, there are three 3.5-MW heat-recovery chillers that reject waste heat from the return chilled water stream to the supply hot water stream rather than rejecting

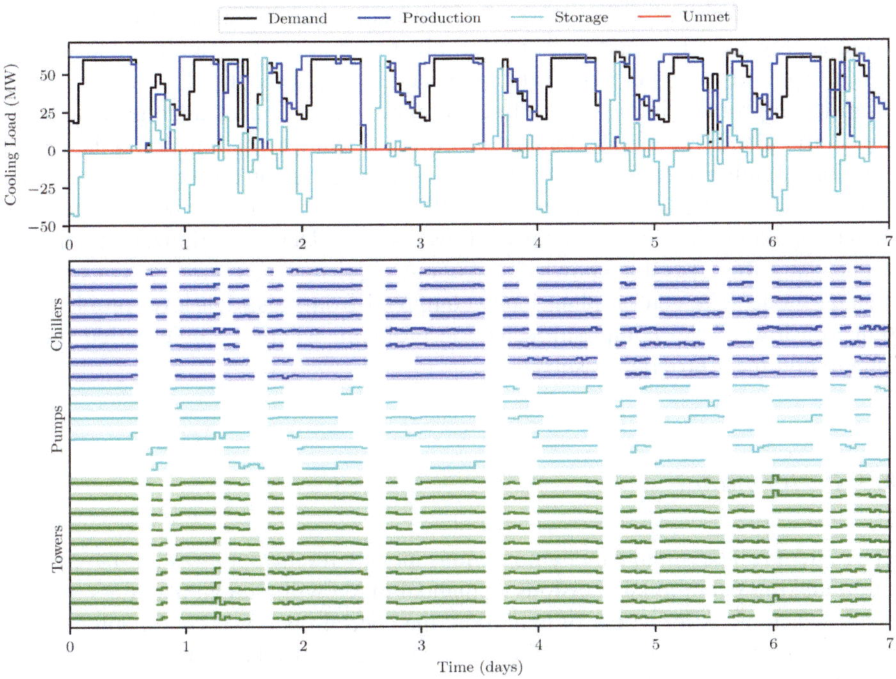

Fig. 9: Low-Level Waterside Results. Optimal production plot and Gantt chart for central plant equipment computed from solving the low-level waterside problem. In lower plot, boxes show on/off state of the equipment with dark line inside showing the loading.

that heat back to the ambient via cooling towers [2]. These heat-recovery chillers improve the efficiency of the overall plant. Without cogeneration, a new 80-megavolt-ampere electrical substation is used to bring electricity from the grid to power the equipment. As part of this project, they also converted 155 campus buildings from steam to hot-water utility, including the installation of a 22-mile network of new hot-water piping.

6.2 Control System

Johnson Controls, Inc. designed and deployed a control system for this new HVAC system. The implemented control architecture resembles one half of the decomposition in Figure 6, namely the high-level and low-level waterside problems. The operation of waterside equipment is optimized to minimize energy costs. Airside optimization was outside of the scope of this project.

In order to minimize operating costs, an economic optimization is performed using aggregate representation of central plant equipment to determine the storage tank usage and loads per period for the waterside equipment [32]. This economic (high-level) problem is solved as a linear program using convex aggregate model of the central plant. The airside demand is estimated using a combination of historical data and weather forecasts. The regression-based procedure used for forecasting heating and cooling loads is discussed in [9]. The predictions can be overridden by human operators to handle unexpected events [33]. Hence, optimization can still occur using this operator knowledge without requiring the entire system to be in manual mode. The central plant loads computed from this high-level problem are sent to a lower-level equipment selection problem, where a single-period mixed-integer nonlinear program is solved.

6.3 Performance

The system can be run in either autonomous mode with the optimization-based algorithm making operational decisions or manual mode with human operators making these decisions. After a year of operation, the central plant was run in autonomous mode 90% of the time, which includes maintenance periods when it was taken offline. The optimization-based system achieved about 10–15% additional cost savings compared to the performance of the best team of trained humans [29]. This large-scale implementation of MPC in the HVAC industry clearly demonstrates that significant benefits are attainable. Additional opportunities still exist to drive cost down even further and improve energy efficiency of current operations.

Since the airside load is not a control decision in this particular application, additional savings can be achieved by modeling and optimizing the airside system alongside the waterside system to utilize passive TES in load shifting. Optimization can also reduce unnecessary overproduction of resources. Several commercial offerings for airside optimization technology have already been deployed on commercial buildings, including qCoefficient and BuildingIQ. Early projects have shown the significant savings are also achievable via airside optimization. Given these promising results, industrial projects involving simultaneous optimization of both airside and waterside systems are already underway.

7 Outlook

While significant progress has been made in this field to identify opportunities for cost savings, wide-scale implementation of these ideas has not yet taken place. Early projects, such as SESI, have demonstrated that such benefits are attainable. Some obstacles for implementation of this technology include the formulation and identification of models that are both accurate and suitable for optimization, forecasting of future loads based on historical data and weather predictions, choice of decomposition for large-scale applications, and software for solving mixed-integer optimization quickly and robustly enough for online use. The biggest obstacle may be the natural opposition for replacing the existing control technology that has been around for decades. However, with the HVAC industry being willing to innovate and the absence of a pre-existing MPC technology in the field, tremendous opportunities exist for capitalizing on ideas from recent areas such as economic MPC and MPC with discrete actuators. MPC certainly seems to be a promising part of the future of HVAC control.

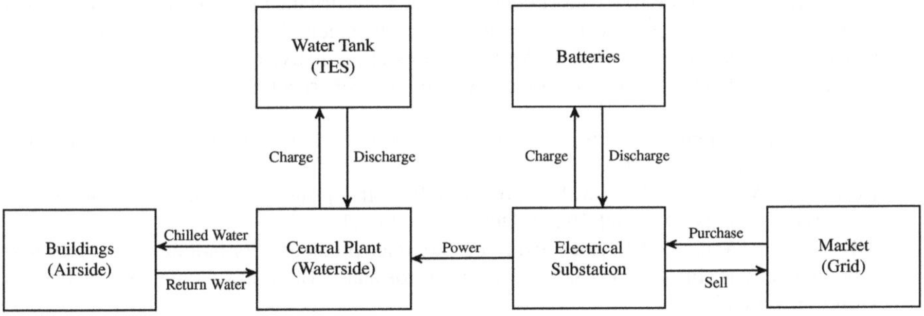

Fig. 10: Flow of resources in network with both thermal energy and electricity storage units.

The potential benefits do not end with load shifting using only thermal energy storage. Batteries are emerging as an economically viable option for efficient storage of electricity. Large-scale batteries may be used for central plant equipment or the batteries may be embedded directly into any piece of equipment that draws powers, such as fans, pumps, air-handler units, root-top units, and variable refrigerant flow units. As shown in Figure 10, not only do batteries provide an alternative path for load shifting, but they permit buildings to participate in real-time electricity markets for revenue generation [8]. Grid-scale integration of such large consumers of energy moves power plants to more efficient operation since they can operate closer to design specifications with constant loads. Hence, both the supplier and the consumer can benefit in this arrangement.

To manage these integrated resources operating on multiple time-scales, an optimization-based system such as MPC is necessary. Hence, additional opportunities exist for designing MPC architecture for these applications. As the electricity market is uncertain and highly volatile at times, ideas from stochastic MPC can be used to treat the random variables directly [17]. HVAC systems in buildings is a rich application that involves many disciplines: traditional process control (airside system), planning and scheduling (waterside system), and stochastic variables (pricing and forecasting). A multidisciplinary approach is required to fully optimize such systems.

References

1. Afram, A., Janabi-Sharifi, F.: Theory and applications of HVAC control systems—a review of model predictive control (MPC). Build. Environ. **72**, 343–355 (2014)
2. Blair, S.: Editors' choice and best energy/industrial: stanford energy system innovations. Engineering News-Record (2016). http://www.enr.com/articles/39005-editors-choice-best-energyindustrial-stanford-energy-system-innovations
3. Braun, J.E.: A near-optimal control strategy for cool storage systems with dynamic electric rates. HVAC R Res. **13**(4), 557–580 (2007)
4. Cai, J., Kim, D., Jaramillo, R., Braun, J., Hu, J.: A general multi-agent control approach for building energy system optimization. Energ. Build. **127**, 337–351 (2016)
5. Christofides, P., Scattolini, R., de la Peña, D., Liu, J.: Distributed model predictive control: a tutorial review and future research directions. Comput. Chem. Eng. **51**, 21–41 (2013)
6. Cole, W.J., Edgar, T.F., Novoselac, A.: Use of model predictive control to enhance the flexibility of thermal energy storage cooling systems. In: American Control Conference (ACC), pp. 2788–2793 (2012)
7. Department of energy: 2011 buildings energy data book. http://buildingsdatabook.eren.doe.gov/ChapterIntro3.aspx (2012)
8. Dowling, A.W., Kumar, R., Zavala, V.M.: A multi-scale optimization framework for electricity market participation. Appl. Energ. **190**, 147–164 (2017)
9. ElBsat, M.N., Wenzel, M.J.: Load and electricity rates prediction for building wide optimization applications. In: 4th International High Performance Buildings Conference at Purdue, West Lafayette (2016)
10. Elliott, M., Rasmussen, B.: Neighbor-communication model predictive control and HVAC systems. In: American Control Conference, Montreal, pp. 3020–3025 (2012)
11. Henze, G.P.: Energy and cost minimal control of active and passive building thermal storage inventory. J. Solar Ener. Eng. **127**(3), 343–351 (2005)
12. Killian, M., Kozek, M.: Ten questions concerning model predictive control for energy efficient buildings. Build. Environ. **105**, 403–412 (2016)
13. Lamoudi, M.Y., Alamir, M., Béguery, P.: Distributed constrained model predictive control based on bundle method for building energy management. In: 50th IEEE Conference on Decision and Control and European Control Conference, Orlando, pp. 8118–8124 (2011)
14. Ma, J., Qin, J., Salsbury, T., Xu, P.: Demand reduction in building energy systems based on economic model predictive control. Chem. Eng. Sci. **67**(1), 92–100 (2012)
15. Ma, Y., Matuško, J., Borrelli, F.: Stochastic model predictive control for building HVAC systems: complexity and conservatism. IEEE Trans. Control Syst. Technol. **23**(1), 101–116 (2015)
16. Mendoza-Serrano, D.I., Chmielewski, D.J.: HVAC control using infinite-horizon economic MPC. In: 2012 IEEE 51st Annual Conference on Decision and Control (CDC), pp. 6963–6968 (2012)
17. Mesbah, A.: Stochastic model predictive control. IEEE Control Syst. Mag. **36**, 30–44 (2016)
18. Moroşan, P.D., Bourdais, R., Dumur, D., Buisson, J.: Building temperature regulation using a distributed model predictive control. Energ. Build. **42**, 1445–1452 (2010)
19. Oldewurtel, F., Parisio, A., Jones, C.N., Gyalistras, D., Gwerder, M., Stauch, V., Lehmann, B., Morari, M.: Use of model predictive control and weather forecasts for energy efficient building climate control. Energ. Build. **45**, 15–27 (2012)
20. Patel, N.R., Rawlings, J.B., Wenzel, M.J., Turney, R.D.: Design and application of distributed economic model predictive control for large-scale building temperature regulation. In: 4th International High Performance Buildings Conference at Purdue, West Lafayette (2016)
21. Patel, N.R., Risbeck, M.J., Rawlings, J.B., Wenzel, M.J., Turney, R.D.: Distributed economic model predictive control for large-scale building temperature regulation. In: American Control Conference, Boston, pp. 895–900 (2016)

22. Powell, K.M., Sriprasad, A., Cole, W.J., Edgar, T.F.: Heating, cooling, and electrical load forecasting for a large-scale district energy system. Energy **74**, 877–885 (2014)
23. Rawlings, J.B., Mayne, D.Q.: Model Predictive Control: Theory and Design, 576 pp. Nob Hill Publishing, Madison (2009). ISBN 978-0-9759377-0-9
24. Rawlings, J.B., Risbeck, M.J.: Model predictive control with discrete actuators: theory and application. Automatica **78**, 258–265 (2017)
25. Rawlings, J.B., Stewart, B.T.: Coordinating multiple optimization-based controllers: new opportunities and challenges. J. Process Control **18**, 839–845 (2008)
26. Rawlings, J.B., Patel, N.R., Risbeck, M.J., Maravelias, C.T., Wenzel, M.J., Turney, R.D.: Economic MPC and real-time decision making with application to large-scale HVAC energy systems. Comput. Chem. Eng. **114**, 89–98 (2018)
27. Risbeck, M.J., Maravelias, C.T., Rawlings, J.B., Turney, R.D.: Cost optimization of combined building heating/cooling equipment via mixed-integer linear programming. In: American Control Conference, Chicago, pp. 1689–1694 (2015)
28. Scherer, H., Pasamontes, M., Guzmán, J., Álvarez, J., Camponogara, E., Normey-Rico, J.: Efficient building energy management using distributed model predictive control. J. Process Control **24**(6), 740–749 (2014)
29. Stagner, J.: Enterprise optimization solution (EOS) cost savings vs. manual plant dispatching. Report on Central Energy Facility, Stanford Energy System Innovations (2016)
30. Sturzenegger, D., Gyalistras, D., Morari, M., Smith, R.S.: Model predictive climate control of a Swiss office building: implementation, results, and cost-benefit analysis. IEEE Trans. Control Syst. Technol. **24**(1), 1–12 (2016)
31. Touretzky, C.R., Baldea, M.: A hierarchical scheduling and control strategy for thermal energy storage systems. Energ. Build. **110**, 94–107 (2016)
32. Wenzel, M.J., Turney, R.D., Drees, K.H.: Model predictive control for central plant optimization with thermal energy storage. In: 3rd International High Performance Buildings Conference at Purdue, West Lafayette (2014)
33. Wenzel, M.J., Turney, R.D., Drees, K.H.: Autonomous optimization and control for central plants with energy storage. In: 4th International High Performance Buildings Conference at Purdue, West Lafayette (2016)

Toward Multi-Layered MPC for Complex Electric Energy Systems

Marija Ilic, Rupamathi Jaddivada, Xia Miao, and Nipun Popli

1 Introduction

Model predictive control (MPC) is a well-established control technique and has been successfully implemented, mainly in chemical plants and oil refineries [15]. Centralized and distributed MPC have been widely studied for industrial applications, as summarized in [39, 42]. In addition, theoretical conditions regarding MPC stability, robustness and convergence have been studied [7, 15, 35]. This chapter is primarily dedicated to the domain applications in electric power systems. We start this chapter by first briefly summarizing in Section 2 the main sources of temporal uncertainties introduced by intermittent resources and electricity industry restructuring.

The complexity of temporal and spatial uncertainties brought about by fundamental changes in electric load characteristics are discussed in Section 3. Load modeling needs more detailed representation because of customers' participation in electricity markets. In Section 4, we briefly review today's organization of electric power systems. We emphasize that these systems can be modeled in the time domain using dynamical systems theory. Somewhat specific to the MPC formulation is the need to characterize temporal uncertainties in the context of an on-line control implementation. To understand this and its implications on the MPC problem formulations and solutions described in this chapter, we conceptualize the classification of system disturbances and other exogenous inputs into their predictable and hard-to-predict components. It is explained in Section 4 how this classification is implicitly used in today's hierarchical control of electric power systems. The concepts throughout this chapter are discussed w.l.o.g using a two-area small electric energy system.

The objective often entails decomposition of an otherwise computationally unsolvable problem into several layers of control design, in particular, feed-forward scheduling for predictable exogenous inputs/disturbances and feedback control for

M. Ilić (✉) · R. Jaddivada · X. Miao · N. Popli
Massachusetts Institute of Technology, Cambridge, MA, USA
e-mail: ilic@mit.edu; rjaddiva@mit.edu; xmiao@mit.edu; npopli@mit.edu

S. V. Raković, W. S. Levine (eds.), *Handbook of Model Predictive Control*, Control Engineering, https://doi.org/10.1007/978-3-319-77489-3_26

compensating hard-to-predict disturbances and model uncertainties. This is discussed in Section 5.

In order to decompose the algorithms for controlling complexity described in Section 3, a novel formulation of multi-layered MPC algorithms is introduced. Instead of having fully separable decision-making, it is essential to have interactive MPC formulations for multi-temporal and multi-spatial management so that the benefits of these interactions are accounted for. In Sections 6 and 7, we propose a novel formulation for nested temporal and nested spatial MPC in complex multi-temporal and multi-spatial dynamical network systems, such as electric energy systems. While the lifting idea for MPC has its deep roots in numerical methods for real-time optimal control [5], these ideas have only recently begun to penetrate into the type of applications as described in this chapter [44]. Based on these nested temporal and spatial multi-layered MPC formulations, we propose a novel MPC formulation for the changing electric energy systems.

In Section 8, the discrete time formulation of the multi-layered MPC is summarized. In Section 9, we claim that our earlier proposed DyMonDS concept, which promotes the idea of abstraction and a minimal information exchange framework, is analogous to the nested MPC formulations. Notably, depending on which variables are exchanged between the layers (physical or Lagrange multipliers), economic and technical signals may be aligned, or not. Finally, in Section 10, we return to the two-area example used throughout in this chapter to identify open questions for MPC algorithms in future electric energy systems. The use of DyMonDS tools is illustrated for efficient integration of temporally diverse generation and demand response. It is shown that this can be done while ensuring stable operation with minimal fast expensive storage.

2 Temporal and Spatial Complexities in the Changing Electric Power Industry

The fundamental objective of power system operation is uninterrupted service to the customers. This necessitates balancing supply and demand in real time. Currently, numerous changes have occurred in the field, such as the development of sensing and communication infrastructure, large-scale integration of renewable energy resources, smart buildings, electric vehicles, and numerous others. These technological innovations are expected to enhance the sustainability, flexibility, and reliability of future electric energy systems, but there is a limited operating experience and understanding of these new technologies. In particular, the industry lacks well-established modeling, analysis and/or decision-making paradigms to support the deployment of these new technologies. Furthermore, it is imperative to facilitate new functionalities for power system operations to accommodate multi-temporal uncertainties introduced by the renewable energy resources and smart dynamic loads as an enabler of demand response [18, 24]. For instance, Figure 1 depicts the renewable electricity production in the grid operated by California Independent System Operator (CAISO). The electricity production from the wind farms, depicted by the blue curve, drops by 700 MW which is about 4% of the net electricity demand during the early morning hours when the actual load is increasing. The utilities frequently experience difficulty in scheduling the resources owing to such uncertainties [37] over multiple hours. In addition, Figure 2 depicts the disturbances created by a large photovoltaic installation evolving at the time scale of few minutes.

Such temporal uncertainties are even more pronounced in the electric power industry under restructuring. Instead of having single system-level smooth input, it is often necessary to have more granular spatial information about market participants and their characteristics. This calls for spatial lifting in order to account for costs/benefits incurred by different industry participants.

To summarize, multi-temporal complexity is the unique feature rooted in power systems, which is in sharp contrast to many problems in other domains such as chemical processes. The problem of real-time supply-demand balancing introduced above can be posed as a set of MPC problems over multiple time horizons driven by multi-temporal disturbances. Nevertheless, objectives at different time scales may have conflicting performance metrics or cost functions. In addition, the multi-spatial complexity of typical market-managed electric power systems requires further for-

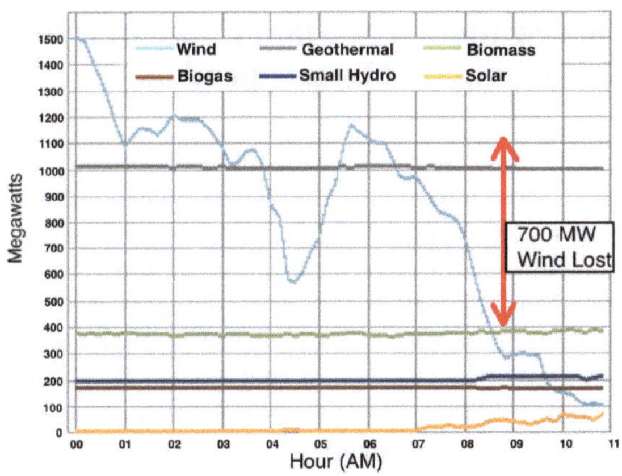

Fig. 1: CAISO renewable generation profile [6].

malization of MPC through nested spatial lifting. Therefore, application of MPC in electric power systems poses theoretical challenges because of the necessity to ensure system stability, despite temporally composite disturbances.

Next, multi-spatial complexity typical of market-managed dynamical network systems requires formalization of MPC through nested spatial lifting. A vast amount of literature seeking solutions to this problem exists. A very general formulation that can accommodate most of these approaches is presented first to familiarize readers non-conversant with the area of power systems. Specifically, the content is presented in the language of systems theory. Hence, people with any level of expertise in

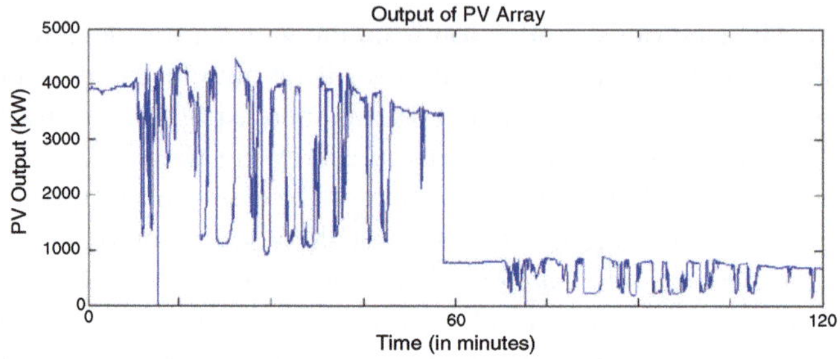

Fig. 2: High frequency disturbance [41].

the application area of power systems can appreciate the problem and its ingrained complexity, particularly due to the unique multi-temporal features that serve as roadblocks to system level analysis. However, this complexity can be taken advantage of by suitably designing the control. We propose a multilayer nested framework based on the functionalities of electric power systems.

3 Load Characterization: The Main Cause of Inter-Temporal Dependencies and Spatial Interdependencies

The networked electric power systems are driven by disturbances varying over vastly different time scales. Most importantly, the power consumptions and injections of the unpredictable devices are referred to as a major component of the disturbance. In addition, disturbance may also be experienced due to imperfect control action, malfunctioning of controllers (as well as loss of equipment). Apart from that, there is always noise in the system, which is often modeled as white noise with a probability distribution of zero mean. The variance of such noise gets lower over coarser time scales and spatial scales. Shown in Figure 3 is the disturbance resulting from heterogeneous electricity consumers (end-devices) over a time scale of an hour. It can be seen that there is a large variability exhibited by these end devices and are often unpredictable. However, a group of these heterogeneous end-devices when aggregated can be seen to be much smoother and predictable as shown in Figure 4a. This aggregated demand further when zoomed-in shows larger variance over a smaller time window. The zoomed-in profile of the first time interval is shown in Figure 4b. To summarize, these figures emphasize spatial and temporal variability in the load or the disturbance profile. Such features are to be accounted for the design of model predictive controllers for networked electric power systems. Towards this goal, we categorize the temporally composite load profiles or the multi-rate disturbances as:

- Slow and predictable component $\mathbf{m}[kT_t]$
- Fast and hard-to-predict $\mathbf{m}[nT_s]$
- Fast and unpredictable $\mathbf{m}[pT_p]$

where T_t, T_s, and T_p also denote the sampling time for tertiary, secondary, and primary control of the system, respectively, as will be explained in Section 4.

Note that even for the slow and predictable component, the accumulated error in prediction may grow as time progresses. For instance, consider Figure 5 from [38], depicting the actual or the real-time wind farm output, as well as predicted and observed average values of the wind power output over time intervals of constant length T_t. The predictions tend to be less accurate farther in time. In the context of MPC on the other hand, as time progresses, the predictions are made iteratively. Consequently, the predictions keep getting better for use in control design and its implementation at that instant of time.

In order to balance the predictable load, over time T_t, feed-forward scheduling (MPC) needs to be designed. However, the feedback controllers at each component are expected to stabilize any fast hard-to-predict deviations from the feed-forward schedule. Correspondingly, these problems necessitate multi-rate dispatch algorithms and robust nonlinear controllers at the component level. Primarily, we focus on the feed-forward control. It should also be noted that an alternative to fast nonlinear feedback controllers is fast MPC at the component. Furthermore, one of the objectives of emerging power grid operations is to allow integration of controllable demand at value. This makes feed-forward design more complicated due to the consumer behavior, policy and market constraints involved. The temporal decomposition of loads and modeling of the load thus plays a crucial role in feed-forward scheduling of the resources. In this section, we introduce the categorization of the loads into controllable and uncontrollable ones and further explain one method that

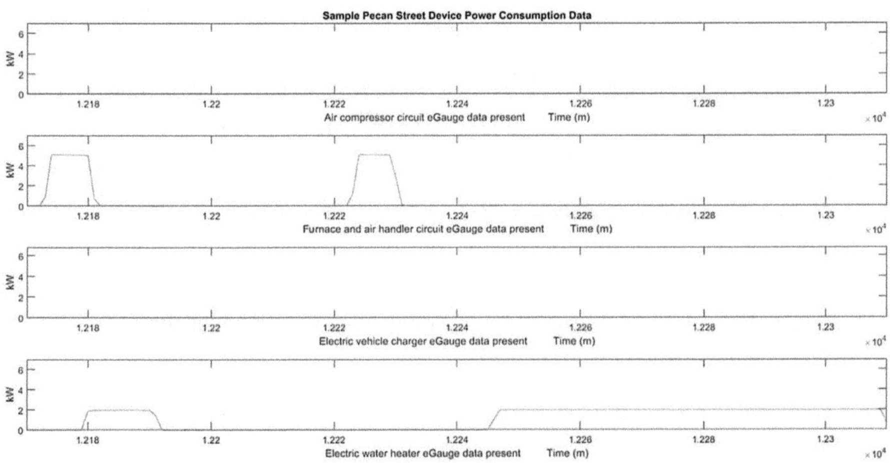

Fig. 3: Consumption patterns.

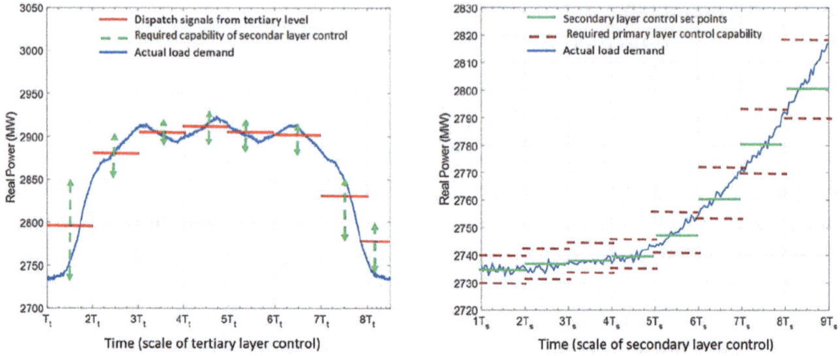

(a) Over a slower time scale

(b) Over a faster time scale

Fig. 4: Aggregate load demand.

is utilized by industry to model the inflexible demand. The model is utilized to forecast electricity consumption that will be used in determining the feed-forward schedule. However, the accuracy of such model defines the accuracy of the feed-forward schedule found by the MPC.

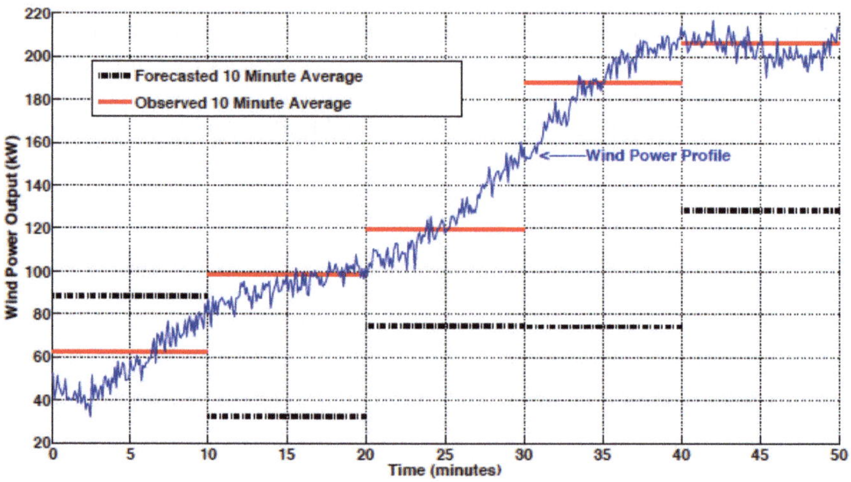

Fig. 5: Decreasing accuracy of wind power predictions over time [38].

3.1 Multi-Temporal Load Decomposition

Consumer devices can be broadly classified as being controllable (flexible demand) or uncontrollable (inflexible demand). Figure 6 shows a typical load profile at any arbitrary bus j (dark line in the top) which is split into controllable (dash-dotted line in the bottom) and uncontrollable (dashed line in the middle) components. The uncontrollable load at bus j can further be decomposed into two components: Firstly, a predictable component P_j evolving slowly over T_t timescale, secondly, a fast fluctuating unpredictable component ΔP_j evolving at T_s time scale, modeled as the deviation from the predictable component over the slow time horizon T_t. In Figure 6, $T_t = 10$ minutes and $T_s = 2$ minutes.

$$\begin{aligned} P_j[nT_s] &= P_j[kT_t] + \Delta P_j[nT_s]; \\ |\Delta P_j[nT_s]| &\leq \Delta P_j^{max}[kT_t] \qquad T_s << T_t \end{aligned} \tag{1}$$

In (1), k and n denote the sampling numbers for the aforementioned coarser and finer granularity time scales, respectively. Accordingly, the controllable resources can be decomposed into a slowly varying component $P_{Dj}[kT_t]$ and a fast varying component $\Delta P_{Dj}[nT_s]$, the bounds of which are given by $B_{Dj}[kT_t]$ as shown in (2).

$$\begin{aligned} P_{Dj}[nT_s] &= P_{Dj}[kT_t] + \Delta P_{Dj}[nT_s]; \\ |\Delta P_{Dj}[nT_s]| &\leq B_{Dj}[kT_t] \qquad T_s << T_t \end{aligned} \tag{2}$$

3.2 Inflexible Load Modeling

The modeling of inflexible load P_j plays a significant role in market operations. The electric utilities, operating the networked power systems, use stochastic models. These models are generated by processing historical data to determine the uncertainty in load or electricity demand. Here, we briefly outline a simple technique based on a statistical time-series approach. The model thus obtained can be used to forecast the load over a future time horizon, which is specifically useful for designing control in a model predictive way. The historical data can be used to fit the model parameters for predicting the hourly total seasonal system mean [14]. The hourly seasonal mean values are interpolated to find the values within an hour given by $L[kT_t]$. The load at each bus j is then modeled using the sum of scaled mean and a stochastic process of correlated noise $w_j[t]$. This is modeled using a Seasonal Auto Regressive Moving Average model (SARMA) in order to correlate the noise at the previous time instant and the innovation 24 hours prior. The predicted load at each location j is then given by

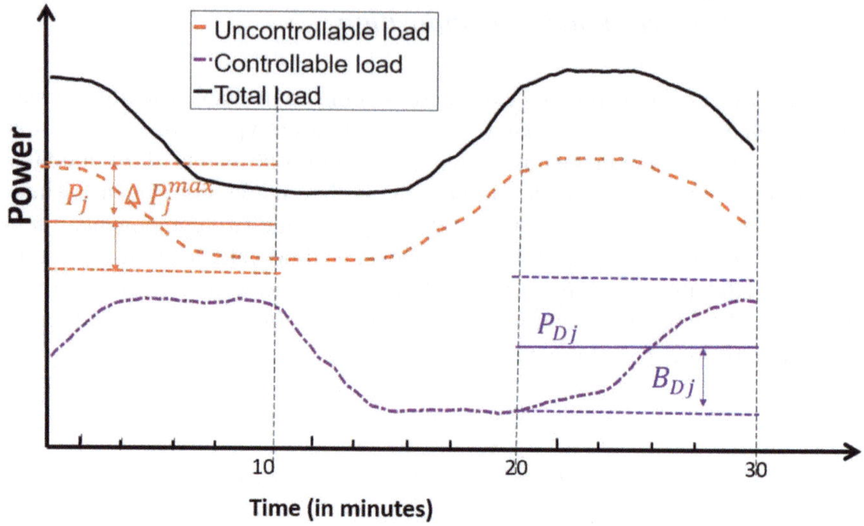

Fig. 6: Aggregate load profile and its decomposition into controllable and bounds on uncontrollable parts.

$$\begin{aligned} P_j[kT_t] &= s_j L[kT_t] + x_j[nT_t], \\ x_j[nT_t] &= \phi x_j[nT_t - 1] + \Phi(w_j[nT_t - D] - w_j[nT_t - D - 1]) + \varepsilon_j. \end{aligned} \tag{3}$$

In (3), D is the number of time steps of length T_t in a day and s_j represents the fraction of total system load incident on bus j. Historic data is then used to fit the model to find the parameters used for this model. The noise at each bus ε_j is modeled as a multivariate Gaussian distribution with zero mean and a time-invariant covariance matrix relating the deviations at each bus. The choice of the covariance matrix significantly affects the market decision algorithm. For example, the high correlation between the loads at different buses may have high standby reserve requirements to account for uncertainties.

More accurately, the above description of load should, however, be appended with the fast unpredictable components $P_{Dj}[pT_p]$ and $P_j[pT_p]$ for controllable and inflexible loads, respectively. However, in the state-of-the-art operations of the grid, zero mean value of load at T_p timescale is assumed. Hence the feed-forward scheduling is done only for the time scales of T_t and to some extent at T_s. The rationale behind doing so is explained in detail in the next section to simultaneously understand the missing signals for efficient and reliable operation for power grids.

4 Hierarchical Control in Today's Electric Power Systems

To start with, consider a small electric power system as shown in Figure 7. An electrically interconnected system is horizontally organized into two control areas which are in today's industry responsible for balancing their own supply and demand during normal operations. Each area first schedules its own generation at the T_t time scale in anticipation of its area loads. For example, the schedule of Area I is $L^I = L_1 + L_2 + F_{I,II}$, where $F_{I,II}$ denotes the pre-agreed power exchange with Area II. Then, at the rate T_s, hard-to-predict slow deviations are managed by each area using so-called automatic generation control (AGC) by balancing other area power imbalances and cancelling out the effects of unpredictable disturbance in $F_{I,II}$. Finally, controllable components, generators, in particular, have very fast primary controllers which respond to the set points computed by scheduling and AGC. Current state-of-the-art of these controllers is embedded PID control gains, typically without using an advanced control such as MPC.

These areas have been historically interconnected to share resources during large contingencies. If, for example, in area I, suppose generator G_1 goes out of service. The remaining generator G_3 may not have enough generation to supply its total load which is the sum of power consumed by loads L_1 and L_2. During such extreme conditions, area II has a typical memorandum of understanding (MOU) to start generating as much as it can beyond its need to supply its own loads L_4 and L_5. Roughly speaking, generators G_4 and G_5 are expected to increase their production to the capacity of the generator G_1 lost in Area I. Throughout this process, it is assumed that loads remain the same (no load shedding) and must be balanced up to 30 minutes during these extreme events. In other words, planning as well as operations do not rely on proactive demand adjustments, i.e., the system load is assumed to be inflexible. In the following section, we first introduce the general mathematical formulation of the main objectives of hierarchical control in today's electric power systems. Then, a unified modeling framework for electric power systems is proposed which forms the basis for control design and is used throughout the chapter. We close the section with brief discussions of the assumptions and limitations of existing hierarchical control.

4.1 Main Objectives of Hierarchical Control

Hierarchical power system control can be interpreted as a composite control comprising a primary stabilizing control $\mathbf{u}[pT_p]$, a secondary regulation control $\mathbf{u}[nT_s]$ and a tertiary feed-forward control $\mathbf{u}[kT_t]$ in response to disturbances $m(t) = \mathbf{m}[pT_p] + \mathbf{m}[nT_s] + \mathbf{m}[kT_t]$, the decomposition of which was described in Section 3. A general form of such control is given as follows:

$$\mathbf{u}^*(t) = \mathbf{u}[pT_p] + \mathbf{u}[nT_s] + \mathbf{u}[kT_t] \tag{4}$$

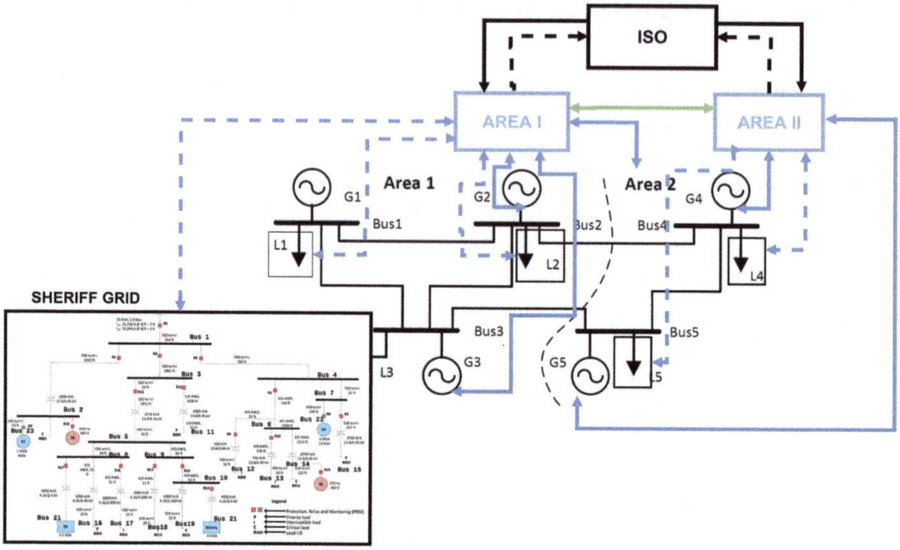

Fig. 7: BPS control hierarchy of today.

Here, T_p, T_s, and T_t denote the sampling time with respective sample numbers p, n and k used to solve the primary ($\mathbf{u}[pT_p]$), secondary ($\mathbf{u}[nT_s]$), and tertiary ($\mathbf{u}[kT_t]$) control problems, respectively.

The industry practice mainly focuses on scheduling resources $\mathbf{u}(t)$, with an objective of optimizing the cost of power production $\mathbf{z}(t)$ subject to the balancing of disturbances $\mathbf{m}(t)$. This is the major objective of the tertiary level feed-forward control, which is conducted by the independent system operators (ISOs) at system level. The scheduling is often done at several different time horizons so that supply and demand imbalance is managed as accurately as possible. In large systems, several of these time horizons are in turn managed by hierarchically organized entities at different spatial granularity. These entities are called area coordinators, which further manage the supply demand imbalance more accurately in response to area level disturbances. It should be noted that the component level disturbances are unobservable at the system level and hence ISO rather schedules resources for an aggregate of several of these devices. Such aggregation results in smoothening of disturbance, thus requiring another intermediate entity closer to end devices, to further coordinate the unscheduled disturbances at area level.

In the context of MPC for hierarchical control, the control signal can be decomposed into three components. The slowest component $\mathbf{u}[kT_t]$ is obtained at the system level by assuming that the load (disturbance $\mathbf{m}[kT_t]$) is inflexible and predictable with high accuracy. This layer further assumes that the unpredictable component has zero mean deviations of disturbance within the scheduling interval. The actual deviation of the disturbances from the predictable component $\mathbf{m}[nT_s]$, which are hard-to-predict, is instead scheduled at area level at finer time granularity by the control component $\mathbf{u}[nT_s]$. This component essentially adjusts the output set points of controllable components within the area as $\mathbf{y}[nT_s] = \mathbf{y}[kT_t] + \Delta\mathbf{y}[nT_s]$ to offset the hard-to-predict disturbances $\mathbf{m}[nT_s]$. Here, $y[kT_t]$ represents the outputs over slower T_t time scale corresponding to the distrubance $m[kT_t]$ and control $u[kT_t]$ under steady-state assumption. $\Delta y[nT_s]$ is the output set-point adjustment over a faster T_s timescale. Finally, fast primary layer control is embedded within components $\mathbf{u}[pT_p]$, ensure instantaneous output $\mathbf{y}(t)$ reaches the desired output $y[nT_s]$ (so-called Quality of Service).

4.2 General Formulation of Main Objectives

A general electric power system can be regarded as a networked system with $\mathcal{N}$ nodes or junctions, and $\mathcal{E}$ edges denoting the set of nodes on which components are incident and the lines connecting these nodes, respectively. The components are heterogeneous in nature with different rates of evolution defined by their states $\mathbf{x}^i(t)$ $i \in \mathcal{N}$ vectorized as $\mathbf{x}(t)$ varying as a function $\mathbf{f}$ of control input $\mathbf{u}(t)$ and disturbance $\mathbf{m}(t)$.

With the above definitions, basic functional objectives of a general electric energy system can be formulated as a flat complex nonlinear dynamic optimization problem [19]. The objective is to optimize the cost subject to system dynamics and control constraints [36]. The general performance objective is expressed in terms of output variables and control cost as shown in (5):

$$\min_{\mathbf{u}(t)} J = \min_{\mathbf{u}(t)} \int_{t=0}^{\infty} c(\mathbf{z}(t), \mathbf{u}(t))dt \tag{5a}$$

subject to

$$\frac{d\mathbf{x}(t)}{dt} = \mathbf{f}(\mathbf{x}(t), \mathbf{u}(t), \mathbf{m}(t));\ \mathbf{x}(0) = \mathbf{x}_0 \tag{5b}$$

$$\mathbf{y}(t) = \mathbf{g}(\mathbf{x}(t), \mathbf{u}(t)) \tag{5c}$$

$$\mathbf{z}(t) = \mathbf{h}(\mathbf{x}(t), \mathbf{y}(t)) \tag{5d}$$

$$\mathbf{y}^{min} \preceq \mathbf{y}(t) \preceq \mathbf{y}^{max} \tag{5e}$$

$$\mathbf{m}^{min}(t) \preceq \mathbf{m}(t) \preceq \mathbf{m}^{max}(t) \tag{5f}$$

$$\mathbf{u}^{min}(t) \preceq \mathbf{u}(t) \preceq \mathbf{u}^{max}(t) \tag{5g}$$

Here $\mathbf{z}(t), \mathbf{u}(t), \mathbf{m}(t)$ are the vectors of interactions with the grid, the control input and disturbances, respectively. These variables affect the cost function c for efficient grid operations and are in general assumed to be smooth and continuous. It should be noted that Equations (5b) to (5g) are invoked for all times t in $[0, \infty)$.

The functions $\mathbf{g}$ and $\mathbf{h}$, respectively, denote the mapping of states to output variables $\mathbf{y}(t)$ and interaction variables $\mathbf{z}(t)$ which define the cost function of the system.

Notice that the control $\mathbf{u}(t)$ affects the interaction variable $\mathbf{z}(t)$ indirectly through the mapping of $\mathbf{y}(t)$. Most power system operations are based on controller performance assumptions leading to a simplification of the above problem subject to the constraints in (5f), (5g) alone and establishing limits on z through quasi-stationary or empirical relations obtained by utilizing the steady state assumptions on (5b), (5c), (5d), and (5e).

The above optimization problem needs to be solved for all time t to get the optimal control input $\mathbf{u}(t) = \mathbf{u}^*(t)$. However, in reality, it is numerically difficult to directly solve this nonlinear problem in a single level centralized way. Firstly, the disturbances for a long period are not known exactly over a fine time granularity. These are generally learned as the time progresses. Secondly, the controllable sources are heterogeneous with vastly different response times, making each of the technologies suitable for balancing power at different time scales. Furthermore, the fact that power grids are spread over a large geographical area makes it impossible to have observability of all the end-devices due to the limited installment of sensors. Even with more sensors, it becomes difficult to coordinate all the end-devices due to their often contradicting sub-objectives, leading to a failure in optimum decision making. Hence, we introduce the notion of spatial lifting to handle the problem at different levels of hierarchy in Section 7. Combination of the lifting techniques based on the physical system and cyber design can then be used for partitioning that facilitates control design of a large-scale power system.

It should be noted that consistent information exchange can be made possible only through a unified modeling framework that is comprehensible by an agent irrespective of the spatial and temporal granularity that it belongs to. We thus briefly introduce the proposed unified modeling framework and the resulting optimal control formulation. Details of the importance and derivation of such models are further highlighted in [19].

4.3 Unified Modeling Framework

In [19, 21, 26, 36], a transformed state space is introduced to unify the modeling of heterogeneous components by defining the internal and interaction dynamics. In the proposed unified modeling framework, the state variables of a general component n are defined as:

$$\mathbf{x}^n = [\mathbf{x}^n_{int}, \mathbf{z}^n_{out}]^T$$

where x^n_{int} and z^n_{out} denote the internal and interaction state variables, respectively. Then, the component dynamics can be written in the following form:

$$\frac{d\mathbf{x}^n_{int}(t)}{dt} = \mathbf{f}_x(\mathbf{x}^n_{int}(t), \mathbf{z}^n_{out}(t), \mathbf{u}^n(t), \mathbf{m}^n(t)); \qquad \mathbf{x}^n_{int}(0) = \mathbf{x}^n_0 \qquad (6a)$$

$$\frac{d\mathbf{z}^n_{out}(t)}{dt} = \mathbf{f}_z(\mathbf{x}^n_{int}(t), \mathbf{z}^n_{out}(t), \mathbf{z}^m_{out}(t), \dot{\mathbf{z}}^m_{out}(t), \mathbf{m}^n(t), \dot{\mathbf{m}}^n(t)); \qquad \mathbf{z}^n_{out}(0) = \mathbf{z}^n_0 \qquad (6b)$$

$$\forall n \in \mathcal{N} \qquad \forall m \in \mathcal{C}_n$$

where $\mathscr{N}$ denotes the number of nodes in the system. $\mathscr{C}_n$ denotes the indexes of neighboring components connected directly to component n.

Applying the relations above and vectorizing the variables of all components, the problem in (5) can be re-written as

$$\min_{\mathbf{u}(t)} J = \min_{\mathbf{u}(t)} \int_{t=0}^{\infty} c(\mathbf{z}_{out}(t), \mathbf{u}(t))dt \tag{7a}$$

subject to

$$\frac{d\mathbf{x}(t)}{dt} = \mathbf{f}(\mathbf{x}(t), \mathbf{u}(t), \mathbf{m}(t));\ \mathbf{x}(0) = \mathbf{x}_0 \tag{7b}$$

$$\mathbf{y}(t) = \mathbf{g}(\mathbf{x}(t), \mathbf{u}(t)) \tag{7c}$$

$$\mathbf{y}^{min} \preceq \mathbf{y}(t) \preceq \mathbf{y}^{max} \tag{7d}$$

$$\mathbf{m}^{min}(t) \preceq \mathbf{m}(t) \preceq \mathbf{m}^{max}(t) \tag{7e}$$

$$\mathbf{u}^{min} \preceq \mathbf{u}(t) \preceq \mathbf{u}^{max} \tag{7f}$$

Note that the vector $\mathbf{z_{out}}$ used in the cost function is an element of the vector $\mathbf{x}$. Furthermore, the objective function and constraints formulated above are invoked for all time t in $[0, \infty)$.

4.4 Assumptions and Limitations Rooted in Today's Hierarchical Control

4.4.1 Tertiary Level Control

Tertiary level control is to feed-forward schedule controllable resources at the slowest time T_t. This includes power flow, energy management, unit commitment, and economic dispatch.

The key assumption made at the tertiary level is that the system frequency is nominal and there are no instabilities, which is equivalent to saying that $\frac{d\mathbf{x}(t)}{dt} = 0$. Note that this assumption requires that the secondary and primary control meet their objectives as will be discussed later. Therefore, one consequence of the assumption is that both the system model and the objective function given in Equations (7b) and (7a) can be simplified accordingly. The obtained static model or the linearized small signal model (often decoupled) is widely used [4, 10, 17]. Recently, [45] introduced a dynamic dispatch model based control to capture inter-temporal dependencies for ramp rate limited dispatch.

However, all these models used by existing control methods are approximations themselves. No provable performance can be ensured due to the uncertainties in the secondary and primary level control. Validation of the above assumption becomes even more critical when a large portion of hard-to-predict heterogeneous components are integrated.

4.4.2 Secondary Level Control

Secondary level control is designed to compensate frequency (voltage) deviation caused by area-level total power mismatch. In operation, the secondary level control provides the set points to the primary controllers embedded in the components participating in frequency (voltage) regulation.

One assumption made here is that the primary level control has stabilized the system. Thus, most of the existing secondary level controls are designed based on quasi-static models [11, 29, 40]. To relax these assumptions and consider the effects of time-varying disturbances, some progress has been made in [32, 33] by utilizing the interaction variable concept [20].

4.4.3 Primary Level Control

Primary level control is designed to stabilize the fast local deviations in both frequency and voltage caused by the fluctuations in supply-demand imbalances and external disturbances. Primary controllers normally are physical control hardware, such as governors and exciters on generators, switches in power electronics devices, etc. The most common design approach is first to simplify the model by linearizing Equation (7b) around the operating point. Then, the control logic is designed based on the obtained linearized model. It should be noted that such a linearized model is only valid for a small region around the operating point. In [2, 9, 13, 34], some nonlinear controllers are proposed to avoid linearization. However, one critical issue of nonlinear control methods is that high gain may be required when a large disturbance occurs, which may cause additional problems due to saturation.

To summarize, it should be emphasized that the assumptions stated for hierarchical control mostly rely on stationary disturbances. Hence, these assumptions may not be valid for the emerging electric energy systems which are driven by disturbances evolving at multiple rates as explained in Section 2. It becomes important to thus pose the control design problem of such a hierarchically organized grid driven by multi-rate disturbances as an interactive MPC problem.

5 Need for Interactive Multi-Layered MPC in Changing Industry

The changing electric energy system has varying needs for energy over different spatial, temporal, and functional granularities. Fundamentally, this is only achievable by a cyber implementation that has just-in-time (JIT) and just-in-place (JIP) functionalities needed to manage all of these misalignments. We treat this problem as two interrelated problems and separate it by entities that have an objective and by time horizons. In the first part, we recognize the time-varying uncertainty over different horizons and introduce the need for interactive MPC. Next, we emphasize the need for coordinating the distributed control of multiple agents for efficient operation through the interactive spatial MPC.

5.1 Temporal Aspect

The unpredictability of loads as discussed in Section 3 calls for control design at different time scales as the disturbance predictions get better and finally the unpredictable component is supposed to be taken care of by a robust feedback controller. There has been a lot of work promoting the decentralized control of components by reacting to the disturbance sensed locally, however, this may not be realizable owing to limitations on the control such as rate of change of control and its saturation. Hence, control design at different time scales is critical and one needs to have a planning horizon chosen based on the respective physical model to ensure deliverability of the required control. Also, end-user preferences are not known to the user itself a-priori, in addition to other uncertainties due to renewable energy sources. This information used for decision making in shorter time intervals needs to be discovered incrementally as the uncertainty gradually decreases. At the same time, it must be noted that the time available to make such decisions also diminishes. For example, in real-time markets, decisions may have to be taken within 5 minutes and furthermore, decisions for frequency regulation may have to be taken within a minute.

In the past, the load was predictable to a very good accuracy. Hence, the intra-prediction time scale deviations from the predictable component were assumed to be of zero mean and were further assumed to be bounded. Hence, composite control composed of the feed-forward component $\mathbf{u}[kT_t]$ found at each prediction time step T_t and the primary stabilizing feedback control $\mathbf{u}[pT_p]$ was sufficient for proper grid functioning. However, in changing electric energy systems, consider, for instance, the fast disturbances due to solar radiation as shown in Figure 2. Managing such fast disturbances requires MPC problem formulation using small time steps. On the other hand, managing rather slowly evolving disturbances caused by inflexible demand as shown in Figure 4 can't be accommodated in the same MPC problem. This is because of rate of response limits on slow controllable resources, that results in scheduling of fast expensive resources. This leads to sub-optimal operation of the grid. Thus, there arises a need to consider interactive sub-problems formulated using different discretization time steps to handle such multi-rate disturbances.

5.2 Spatial Aspect

Power systems are composed of different types of entities some of which are physical while others are cyber. Traditional operations of power systems do not explicitly consider the model of the demand and as a result, fail to take into account its economic preferences. Historical data is often used to model the load at different time periods by scheduling the generation sources and at times by curtailing the demand when there is a shortage of supply and/or when the curtailment of the demand is seen to be more efficient compared to the increase in generation. However, this is being taken into consideration at an aggregate level through non-utility owned cyber entities called Load Serving Entities. In doing so, the end-user preferences are ignored. Presently, it is not clear how to design the cyber layer to coordinate millions of end-user devices. In addition, the short-term scheduling operations cannot be done with numerous end-devices within a short period of time owing to the complication of resolving numerous constraints with often contradicting sub-objectives. Furthermore, the internal dynamics of each end-user may not be communicated to the system coordinator due to privacy concerns. Thus, there arises a need to coordinate these devices through nested spatial hierarchies of control to align temporal and economic signals. In Section 10.2, this is described and illustrated in more detail.

In order to resolve the issues emphasized here in a systematic manner, we pose the time lifted and spatially lifted problems of the centralized problem formulation in (5) to better understand the structure of the resulting MPC formulations at different granularities and the significance of interactive information exchange between these sub-problems.

6 Temporal Lifting for Decision Making with Multi-Rate Disturbances

As outlined in Section 5, one of the major problems causing the complexity in power grid analysis is the unpredictability in the disturbances seen by the components. This calls for the need to first formulate the problem as a time-lifted model predictive control problem to find the optimum control policy as the disturbances are learned. Hence, we lift the long-horizon problem by partitioning into multiple stages. In the context of model predictive control, we can apply it as shown in (8). These set of equations defining the objective functions and the constraints are referred to as a problem $\mathscr{P}_{N_1,T_1}(t)$ which is solved using time windows of length T_1 for N_1 prediction horizons starting at time t'.

$$\mathscr{P}_{N_1,T_1}(t'): \min_{\mathbf{u}_k(\tau)} \sum_{k=1}^{N_1} \left[\int_{\tau=0}^{T_1} c(\mathbf{z}_{k,T_1}(\tau), \mathbf{u}_{k,T_1}(\tau)) d\tau \right] + Q^{term}_{N_1,T_1}(t') \tag{8a}$$

subject to

$$\frac{d\mathbf{x}_{k,T_1(\tau)}}{d\tau} = \mathbf{f}(\mathbf{x}_{k,T_1}(\tau), \mathbf{u}_{k,T_1}(\tau), \mathbf{m}_{k,T_1}(\tau)) \tag{8b}$$

$$\mathbf{x}_{k,T_1}(0) = \mathbf{x}_{k-1,T_1}(T_1) \qquad (\Gamma_{k,T1}(t')) \tag{8c}$$

$$\mathbf{y}_{k,T_1}(\tau) = \mathbf{g}(\mathbf{x}_{k,T_1}(\tau), \mathbf{u}_{k,T_1}(\tau)) \tag{8d}$$

$$\mathbf{y}^{min} \preceq \mathbf{y}_{k,T_1}(\tau) \preceq \mathbf{y}^{max} \tag{8e}$$

$$\mathbf{m}^{min} \preceq \mathbf{m}_{k,T_1}(\tau) \preceq \mathbf{m}^{max} \tag{8f}$$

$$\mathbf{u}^{min} \preceq \mathbf{u}_{k,T_1}(\tau) \preceq \mathbf{u}^{max} \tag{8g}$$

$$\forall \tau \in [0, T_1) \qquad \forall k = 1, 2, ..N_1$$

In the above set of equations, all the variables are appended with a subscript 'k, T_1' to denote the trajectory evolution in the kth time window, where the windows are created using a time step T_1. Precisely, the notation for a state variable $\mathbf{x}$ used in the problem $\mathscr{P}_{N_1,T_1}(t')$ is given by (9)

$$\mathbf{x}_{k,T_1}(\tau) = \mathbf{x}(T_1(k-1) + \tau + t') \tag{9}$$

(8c) is the time coupling constraint that binds the otherwise decoupled problem in different time windows. This has an associated Lagrange multiplier given by $\Gamma_{k,T1}$ for the kth time window initial condition, where the time windows are created every T_1 time. Note that the infinite horizon problem in (5) has been terminated after a time of $N_1 T_1$ and hence the cost function is appended with the terminal cost $Q^{term}_{N_1,T_1}(t')$. This term captures the cost incurred due to the intervals beyond the planning horizon used, i.e. after time $t' + N_1 T_1$. For a given MPC problem where the planning horizon starts from t', this quantity is a function of terminal variable values $\mathbf{z_{m,T_1}}(\tau) \forall m \geq N_1$. As the planning horizon start time moves from t' to $t' + T_1$, the terminal cost accordingly changes as being a function of $\mathbf{z_{m,T_1}}(\tau) \forall m \geq (N_1 + 1)$.

In this formulation, it is assumed that the system has knowledge of disturbance evolution every T_1 time step starting at t' until $t' + N_1 T_1$. However, it is not always possible to have the knowledge of disturbance evolution at the finest possible granularity. Since the disturbance can be predicted only as time progresses, there arises a need to also consider a nested temporally lifted problem and further analyze how the longer time optimal control policy affects the shorter term optimal control policy found in a receding horizon manner.

6.1 Nested Temporal Lifting

We have seen in Section 3 that there are hard-to-predict non-zero-mean deviations injected into the grid at vastly different time scales. For instance, wind energy integration may cause deviations on a sub-hourly basis while solar may result in fluctuations over a time scale of minutes. In addition, the demand response programs that are being encouraged by grid operators are leading to consumption patterns with high variability. To analyze the implications of factors like these on the system level problem, the time lifting needs to be done at multiple time scales and the effect of decision making at one timescale over the others needs to be explicitly considered in quantifiable terms. In the following section, the mathematical formulation and the information exchange framework facilitating such analysis is proposed.

First, we will establish a recurrence relation on the terminal costs by following the derivation in [44]. Throughout this derivation, we consider only the time coupling constraint which is of importance here and the rest of the constraints are ignored. It should however be noted that similar relations can be derived by considering the other constraints as well. The recurrence relation will then be used to establish relation between the terminal costs of time lifted problems $\mathscr{P}_{N_1,T_1}(t')$ and $\mathscr{P}_{N_1,T_2}(t')$ evolving at time scales T_1 and T_2, respectively.

Expanding the problems $\mathscr{P}_{N_1,T_1}(t')$ and $\mathscr{P}_{N_1+1,T_1}(t')$ with respective terminal costs lets us write the following equation.

$$Q^{term}_{N_1,T_1}(t') = \min_{\mathbf{u}_{N_1+1,T_1}(\tau)} \int_0^{T_1} c(\mathbf{z}_{N1+1,T1}(\tau), \mathbf{u}_{N1+1,T1}(\tau))d\tau + Q^{term}_{N_1+1,T_1}(t') \tag{10}$$

This is subject to the additional time coupling constraint with an associated lagrange multiplier $\Gamma_{N_1+1,T_1}(t')$.

$$\mathbf{x}_{N_1+1,T_1}(0) = \mathbf{x}_{N_1,T_1}(T_1) \qquad (\Gamma_{N_1+1,T_1}(t')) \tag{11}$$

The Lagrangian can then be written as

$$\begin{aligned} Q^{term}_{N_1,T_1}(t') = & \min_{\mathbf{u}_{N_1+1,T_1}(\tau)} \int_0^{T_1} c(\mathbf{z}_{N_1+1,T_1}(\tau), \mathbf{u}_{N_1+1,T_1})dt + \\ & Q^{term}_{N_1+1,T_1} + \Gamma_{N_1+1,T_1}(t')\Big(\mathbf{x}_{N_1,T_1}(T_1) - \mathbf{x}_{N_1+1,T_1}(0)\Big) \end{aligned} \tag{12}$$

Since the integral above is not directly sensitive to the decision variables in $\mathscr{P}_{N_1,T_1}(t')$, the terminal cost sensitivity with respect to variables in problem (8) can be written as

$$\partial Q^{term}_{N_1,T_1}(t') = \Gamma_{N_1+1,T_1}(t')\Big(\mathbf{x}_{N_1,T_1}(T_1) - \mathbf{x}_{N_1+1,T_1}(0)\Big) + \partial Q^{term}_{N_1+1,T_1} \tag{13}$$

In the above equation, the last two terms are constant for the problem within the time window of N_1T_1. Hence, the problem in (8a) can now be rewritten as

$$\mathscr{P}_{N_1,T_1}(t') : \min_{\mathbf{u}_k(\tau)} \sum_{k=1}^{N_1} \Big[\int_{t=0}^{T_1} c(\mathbf{z}_{k,T_1}(\tau), \mathbf{u}_{k,T_1}(\tau))d\tau\Big] + \Gamma_{N_1+1,T_1}(t')\mathbf{x}_{N_1,T_1}(T_1) \tag{14}$$

The problem in (8) can be re-formulated for a time scale of T_2 by breaking down the time space into intervals of T_2 length. Suppose the timescale $T_2 \leq T_1$ is such that $N_2T_2 = T_1$, we can then obtain the terminal cost function for T_2 time scale problem as a function of the lagrange multiple of T_1 time scale problem. The objective function can then be written as

$$\mathscr{P}_{N_2,T_2}(t') : \min_{\mathbf{u}_k(\tau)} \sum_{k=1}^{N_2} \Big[\int_{t=0}^{T_2} c(\mathbf{z}_{k,T_2}(\tau), \mathbf{u}_{k,T_2}(\tau))d\tau\Big] + \Gamma_{N_2+1,T_2}(t')\mathbf{x}_{N_2,T_2}(T_2) \tag{15}$$

The intuition behind the last term is to account for the cost incurred if one considers the time horizon beyond N_2T_2. However, if the disturbances for finer time granularity of T_2 time scale are available only till the time of N_2T_2, the terminal cost is better approximated using the adjoint variable $\Gamma_{N_1+1,T_1}(t')$ for the time coupling constraint found in the problem $\mathscr{P}_{N1,T1}(t')$ when it finds the optimal control policy in the first time window at T_1 time scale. This exchange of information is shown in Figure 8.

As an example, the problem in (15) after problem $\mathscr{P}_{N_1,T1}(t')$ communicates the value of $\Gamma_{1,T_1}(t')$ can be written as

$$\mathscr{P}_{N_2,T_2}(t') : \min_{\mathbf{u}_k(\tau)} \sum_{k=1}^{N_2} \left[\int_{t=0}^{T_2} c(\mathbf{z}_{k,T_2}, \mathbf{u}_{k,T_2}) d\tau \right] + \Gamma_{1,T_1}(t') \mathbf{x}_{N_2,T_2}(T_2) \tag{16}$$

This information exchange between the different MPC problems is shown schematically in Figure 8. In this figure, the upper window (in dashed lines) shows the prediction horizon over which the T_1 time scale problem is being solved. The optimal control is found for each T_1 given the disturbance $m[kT_1]$ for all k in the prediction horizon. However, only the control in the first window is applied. Further, the hard-to-predict disturbances observed at T_2 time scale granularity are available only till the time T_1 starting at $t = 0$. This information is used to further adjust the control input found at T_1 time scale at faster rates with the ones found at T_2 time scale $u[pT_2]$ every p^{th} sampling time in a receding horizon manner. Here, the smaller time scale problems $\mathscr{P}_{N_2,T_2}(t')$ are shown in bottom windows in dotted lines, and the optimal control adjustments are shown in dark lines evolving over T_2 timescale.

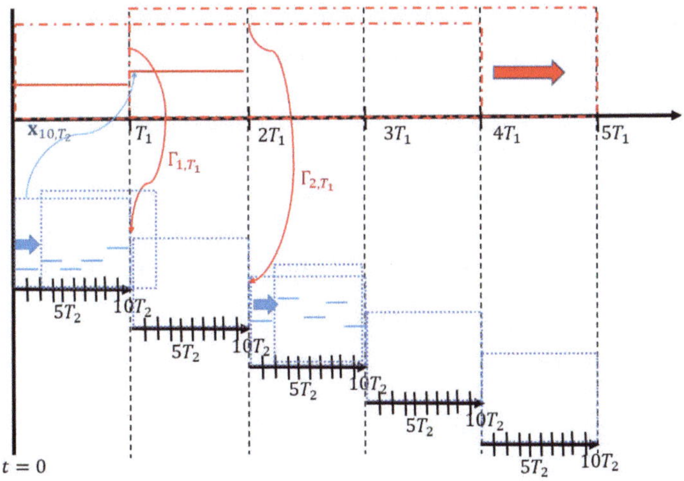

Fig. 8: Information exchange between temporally lifted problems.

Once the problem $\mathscr{P}_{N_2,T_2}(t')$ has been completely solved for an interval of $t' + T_2 = t' + N_1 T_1$, the present value of the state gets communicated back to the problem $\mathscr{P}_{N_2,T_2}(t')$ which uses the communicated value to find the optimal strategy by moving the prediction horizon by one time window as shown in Figure 8.

Since these smaller time scale problems are also solved in a receding horizon manner, the terminal cost information to be sent to the problem $\mathscr{P}_{N_2,T_2}$ is not obvious and a choice has to be made between sending the previous prediction horizon time coupling Lagrange multiplier or the next one or a combination of both. In the figure in the first time window, we show this information to be constant until the faster time scale problem reaches the time T_1. However, to maintain consistency of terminal cost consideration for the time windows, the problem $\mathscr{P}_{N_2,T_2}(t')$ may also be solved using decreasing window length as shown in the third time window. The problems at T_2 time scale can thus be solved repeatedly as the information of the finer granularity disturbances gets available. This can further be formulated as multiple layers of problems defined with different granularities. The T_i for each layer i is chosen based on the decomposition of time scales of the disturbances that enter into the grid, thus giving rise to nested time lifting of the problem. The case for two layers is shown in Figure 8.

7 Spatial Lifting for Multi-Agent Decision Making

The future power system operations would rely on the participation of numerous heterogeneous end devices with vastly different time scales. Mathematically, this means that the cardinality of vectors considered above may become so large that they cannot be solved by one single operator. In order to better show the complexity involved, the spatially lifted problem is shown below and is denoted as $\mathscr{P}^c$ for partitioning the problem in (7) component-wise.

$$\mathscr{P}^c : \min_{\mathbf{u}^n(t)} \int_{t=0}^{\infty} \sum_{n\in\mathscr{N}} c^n(\mathbf{z}^n_{out}(t), \mathbf{u}^n(t))dt \tag{17a}$$

subject to system dynamics, disturbance, and control constraints:

$$\frac{d\mathbf{x}^n(t)}{dt} = \mathbf{f}^n(\mathbf{x}^n(t), \mathbf{u}^n(t), \mathbf{m}^n(t), \mathbf{z}^n_{in}(t));\ \mathbf{x}^n(0) = \mathbf{x}^n_0 \tag{17b}$$

$$\mathbf{y}^n(t) = \mathbf{g}^n(\mathbf{x}^n(t), \mathbf{u}^n(t)) \tag{17c}$$

$$\mathbf{z}^n_{in}(t) = \sum_{m\in\mathscr{C}_n} \mathbf{z}^m_{out}(t) + \mathbf{m}^n(t) = 0 \tag{17d}$$

$$-\sum_{n\in\mathscr{N}} \mathbf{z}^n_{out}(t) + \mathbf{m}^S(t) = 0 \tag{17e}$$

$$\mathbf{y}^{n,min} \preceq \mathbf{y}^n(t) \preceq \mathbf{y}^{n,max} \tag{17f}$$

$$\mathbf{m}^{n,min}(t) \preceq \mathbf{m}^n(t) \preceq \mathbf{m}^{n,max}(t) \tag{17g}$$

$$\mathbf{u}^{n,min}(t) \preceq \mathbf{u}^n(t) \preceq \mathbf{u}^{n,max}(t) \tag{17h}$$

In the equations above, all the symbols are appended with superscript n to denote the vectors and functions corresponding to the nth component in the index set of components in the system $\mathscr{N}$. The constraints (17b) - (17d) and (17f) - (17h) are written for all components in the system identified by the index n and all the constraints are invoked for (almost) all time $t \in [0,\infty)$. Note that (17e) has been changed for component level analysis, where the output interaction variables of component in steady-state sum up to zero if there is no external disturbance $\mathbf{m}^S$ injected into the system. This has a physical interpretation of conservations of energy, the details of which can be referred to in [17]. This equality constraint has an associated system level Lagrange multiplier denoted as $\lambda^S(t)$, commonly referred to as market clearing price. Furthermore, the coupling term at the component level is given by (17d) which also takes into account disturbance $\mathbf{m}^n$ seen at the component level. Note that the output interaction variable $\mathbf{z}^n_{out}$ is one of the entities of the vector $\mathbf{x}^n$.

The purpose of such spatial lifting is to create partitions in space dimension that make the system level problem computationally scalable in addition to satisfying privacy requirements of components. However, the coupling constraint in (17e) hinders such partitioning. This is thus dealt with by most present-day utilities, by applying a dual decomposition algorithm, where the coupling constraint is relaxed in individual problems. The coordination of solutions found by individual sub-problems is then done by the system level coordinating entity ensuring that the coupling constraint is satisfied. Furthermore, at each component n, the constraint in (17d) requires just the local measurements of the output interaction variables of component m directly connected. It should also be noted that if the privacy is not much of a concern, the coupling constraint relaxation can also be done in conventional state space by applying primal decomposition techniques, where the physical variables $\mathbf{x}^m$ of neighboring components are duplicated instead [1]. This methodology is adopted in the partitioning of most electromagnetic transient programs today for analysis at faster time scales [43]. However, this approach requires fast communication between the cores to ensure time synchronization, and is thus often done on one central computer with many high-performance GPUs (Graphics Processing Units).

7.1 Nested Spatial Lifting

The problem has now been decomposed into different parts from a spatial perspective. In a large power system, the number of these end devices cannot be kept track of by a single system operator. There thus arises a need to group together some of them into groups which are coordinated by an entity. A group of such entities are further coordinated by a system operator, thus forming layers of hierarchical control. These coordinating entities are then better defined by interaction variable dynamics $\mathbf{z}$ alone. The Nth area interaction variable can be defined as the sum of all its constituents' interaction variables. Similarly, the disturbance $\mathbf{m}^N$ seen by the area and control capability of an area $\mathbf{u}^N$ is the sum of disturbances seen and the sum of the control inputs, respectively, of constituent members. If the set of areas coordinated by the system is grouped in the set $\mathscr{N}_a$, the problem can now be posed as

$$\mathscr{P}^a : \min_{\mathbf{u}^N(t)} \int_{t=0}^{\infty} \sum_{N\in\mathscr{N}_a} c^N(\mathbf{z}^N_{out}(t), \mathbf{u}^N(t))dt \tag{18a}$$

subject to system dynamics, disturbance, and control constraints:

$$\frac{d\mathbf{z}_{out}^{N}(t)}{dt} = \mathbf{f}^{N}(\mathbf{z}^{N}(t),\mathbf{u}^{N}(t),\mathbf{m}^{N}(t),\mathbf{z}_{in}^{N}(t),\dot{\mathbf{z}}_{in}^{N});\ \mathbf{z}^{N}(0) = \mathbf{z}_{0}^{N} \tag{18b}$$

$$\mathbf{z}_{in}^{N}(t) = \sum_{M\in\mathscr{C}_N} \mathbf{z}_{out}^{M}(t) + \mathbf{m}^{N}(t) \tag{18c}$$

$$-\sum_{N\in\mathscr{N}_a} \mathbf{z}_{out}^{N}(t) + \mathbf{m}^{N}(t) = 0 \qquad (\lambda^{S}(t)) \tag{18d}$$

$$\mathbf{y}^{N,min} \preceq \mathbf{y}^{N}(t) \preceq \mathbf{y}^{N,max} \tag{18e}$$

$$\mathbf{m}^{N,min}(t) \preceq \mathbf{m}^{N}(t) \preceq \mathbf{m}^{N,max}(t) \tag{18f}$$

$$\mathbf{u}^{N,min} \preceq \mathbf{u}^{N}(t) \preceq \mathbf{u}^{N,max} \tag{18g}$$

All the symbols are marked with superscript N to denote the variables and functions corresponding to the Nth aggregate component. Constraints (18b) and (18e) - (18g) are invoked for all $N \in \mathscr{N}_a$ and all the constraints are written for (almost) all time $t \in [0,\infty)$. Equation (18c) establishes the coupling relation for the Nth area while (18d) models the fact that the total interaction variables of all the areas sum up to zero in the absence of disturbance seen by the system, which is again associated with an associated system level lagrange multiplier $\lambda^{S}(t)$.

Each area-level problem can further be solved to obtain component level decision variables. The problem written for coordinating all the components at once in (17) can instead be split into several hierarchies, where system-level communicates Lagrange multiplier $\lambda^{S}(t)$ to respective areas. Relaxation of (18d) can then let us write formulation $\mathscr{P}^{N}$ of the Nth area coordinating its entities as

$$\mathscr{P}^{N} : \min_{\mathbf{u}^{n}(t)} \int_{t=0}^{\infty} \sum_{n\in N} c^{n}(\mathbf{z}_{out}^{n}(t),\mathbf{u}^{n}(t))dt - \lambda^{S}(t)\mathbf{z}_{out}^{N}(t) \tag{19a}$$

subject to system dynamics, disturbance, and control constraints:

$$\frac{d\mathbf{z}_{out}^{n}(t)}{dt} = \mathbf{f}^{n}(\mathbf{x}^{n}(t),\mathbf{u}^{n}(t),\mathbf{m}^{n}(t),\mathbf{z}_{in}^{n}(t),\dot{\mathbf{z}}_{in}^{n});\ \mathbf{z}^{n}(0) = \mathbf{z}_{0}^{n} \tag{19b}$$

$$\mathbf{z}_{in}^{n}(t) = \sum_{m\in\mathscr{C}_n} \mathbf{z}_{out}^{m}(t) + \mathbf{m}^{n}(t) \tag{19c}$$

$$-\sum_{n\in N} \mathbf{z}_{out}^{n}(t) + \mathbf{z}^{N}(t) = 0 \qquad (\lambda^{N}(t)) \tag{19d}$$

$$\mathbf{y}^{n,min} \preceq \mathbf{y}^{n}(t) \preceq \mathbf{y}^{n,max} \tag{19e}$$

$$\mathbf{m}^{n,min} \preceq \mathbf{m}^{n}(t) \preceq \mathbf{m}^{n,max} \tag{19f}$$

$$\mathbf{u}^{n,min} \preceq \mathbf{u}^{n}(t) \preceq \mathbf{u}^{n,max} \tag{19g}$$

In the above, the constraints (19b) - (19c) and (19e) - (19g) are written for every component n in area N and all the constraints are invoked for (almost) all time $t \in [0,\infty)$. The area level coordinator thus solves the problem at area level to give the control signal u^{n} to the components belonging to the cluster N by just utilizing the interaction variable dynamics of its constituent components. The interaction of this area with the neighboring areas is considered through the coupling constraint in (19c). Next, the system level problem communicates the cleared value of area level interaction variable $\mathbf{z}^N$ as shown in (19d), resulting in an associated Lagrange multiplier λ_N for denoting the sensitivity factor for imbalance in area N.

This kind of partitioning finally thus enables the component i to do its own decision making with the internal dynamics involved to perform fast MPC or it can otherwise react using feedback control. The formulation for the former is shown in the following lines, which utilizes corresponding area level Lagrange multiplier $\lambda^N(t)$ such that $i \in N$, as follows:

$$\mathscr{P}^i : \min_{\mathbf{u}^i(t)} \int_{t=0}^{\infty} c^i(\mathbf{z}^i_{out}(t), \mathbf{u}^i(t))dt - \lambda^N(\mathbf{z}^i_{out}) \tag{20a}$$

subject to system dynamics, disturbance, and control constraints:

$$\frac{d\mathbf{x}^i(t)}{dt} = \mathbf{f}^i(\mathbf{x}^i(t), \mathbf{u}^i(t), \mathbf{m}^i(t), \mathbf{z}^i_{in}(t)); \; \mathbf{x}^i(0) = \mathbf{x}^i_0 \tag{20b}$$

$$\mathbf{z}^i_{in}(t) = \sum_{j \in \mathscr{C}_n} \mathbf{z}^j_{out}(t) + \mathbf{m}^i(t) = 0 \tag{20c}$$

$$\mathbf{z}^i_{out}(t) = \mathbf{z}^{i,high}(t) \tag{20d}$$

$$\mathbf{y}^{i,min} \preceq \mathbf{y}^i(t) \preceq \mathbf{y}^{i,max} \tag{20e}$$

$$\mathbf{m}^{i,min}(t) \preceq \mathbf{m}^i(t) \preceq \mathbf{m}^{i,max}(t) \tag{20f}$$

$$\mathbf{u}^{i,min}(t) \preceq \mathbf{u}^i(t) \preceq \mathbf{u}^{i,max}(t) \tag{20g}$$

The above constraints are invoked for (almost) all time $t \in [0, \infty)$. Equation (20c) represents the coupling with the neighboring components while (20d) requires the component i to track the set point $z^{i,high}_{out}$ sent by its coordinating entity, which was found by solving (19). It should further be noted that the control objective here at the component level is thus to ensure tracking these set points while also ensuring the control inputs and the outputs of interest are within pre-specified limits.

7.1.1 Functional Bids

It may seem from the above formulations that the coordinating entity at any layer needs to know the inner details of the components. However, abstraction of inner details is possible if there is a bottom-up communication from lower layer to higher layer as well through functional bids instead of point-wise bids, which are more prevalent in the literature [16, 20, 30]. The methodology for creating functional bids is to solve the optimization problem of the agents in response to the Lagrange multiplier sent by the higher layer for small perturbations. The sensitivity is thus captured by interpolating the points created through this methodology, which is also illustrated in Figure 9. The slope and the intercept in the resulting marginal cost/benefit curve along with the minimum and maximum limits are communicated to the coordinating entity, which then solves its own problem. It further adopts a similar approach to simultaneously create bid functions capturing the sensitivity of aggregate interaction variable: for perturbation in the Lagrange multiplier sent by its coordinating entity.

The problems in power systems most generally are designed using quadratic cost functions for ease of finding a solution, and the coupling constraint in most problems of interest is the power balance equation, which is linear. Hence, the sensitivity of the interaction variable with respect to the Lagrangian multiplier corresponding to the power balance equation captures the descent direction of the entire problem. Utilizing this sensitivity, each sub-problem can thus reach the global optimum of the system with enough accuracy. In other words, functional bids provides a mechanism to communicate the slope which dictates the adjustment around the present operating point in response to price. This approach leads to faster convergence as opposed to point-wise bids that need to be cleared iteratively.

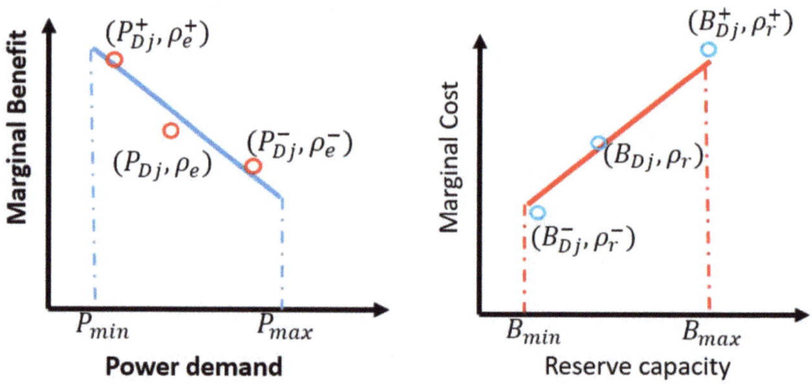

(a) (b)

Fig. 9: Creation of bids.

8 Digital Implementation

From the formulations of spatial and temporal lifting, non-causality of information flow can clearly be seen. In the case of sequential decision making by agents, the first agent assumes a value of the Lagrange multipliers and other information from neighboring modules and passes on the needed information to neighboring agents. This proceeds in a sequential way until the pre-specified stopping criterion is met [44]. This can also be done concurrently through Jacobi methods [12]. In real-time applications where time is a critical factor, trade-off analysis is done between the optimality in control and time available for computation. Accordingly, control inputs which may not necessarily be optimal are applied [44]. This can be supported by the claim that the error obtained by stopping after one Gauss Siedel/ Jacobi iteration is negligible compared to the one obtained due to inaccuracies in measurements and parameter uncertainties.

Furthermore, the continuous time integral in each of the problems and the coupled ODEs need to be discretized for digital implementation, which is done through a method called transcription by choosing right time grid points. The differential equations can be discretized using one of the many existing schemes [8] shown here is the basic technique called forward Euler method for the differential equations of the time-lifted problem in (8).

$$\mathscr{P}_{N_1,T_1}[n'] : \min_{\mathbf{u}_{k,j,T1}} \left[\sum_{k=1}^{N_1} \sum_{j=1}^{M_1} c(\mathbf{z}_{k,j,T_1}, \mathbf{u}_{k,j,T_1}) \right] + Q^{term}_{N_1,T_1}[n'] \tag{21a}$$

subject to

$$\begin{aligned}
\mathbf{x}_{k,j+1,T_1} &= \mathbf{x}_{k,j,T_1} + (T_1/(M_1-1))\mathbf{f}(\mathbf{x}_{k,j,T_1}, \mathbf{u}_{k,j,T_1}, \mathbf{m}_{k,j,T_1}) && \text{(21b)}\\
\mathbf{x}_{k,1,T_1} &= \mathbf{x}_{k-1,M_1,T_1} \qquad (\Gamma_{k,T1}[n']) && \text{(21c)}\\
\mathbf{y}_{k,j,T_1} &= \mathbf{g}(\mathbf{x}_{k,j,T_1}, \mathbf{u}_{k,j,T_1}) && \text{(21d)}\\
\mathbf{y}^{min} &\le \mathbf{y}_{k,j,T_1} \le \mathbf{y}^{max} && \text{(21e)}\\
\mathbf{m}^{min} &\le \mathbf{m}_{k,j,T_1} \le \mathbf{m}^{max} && \text{(21f)}\\
\mathbf{u}^{min} &\le \mathbf{u}_{k,j,T_1} \le \mathbf{u}^{max} && \text{(21g)}\\
&\forall j \in 1,2,\ldots M_1 \qquad \forall k = 1,2,..N_1
\end{aligned}$$

In the above formulation, the problem $\mathscr{P}_{N_1,T_1}[n']$ and the terminal cost $Q^{term}_{N_1,T_1}[n']$ need to be considered at discrete steps rather than continuous time t' which results in the Lagrangian multiplier corresponding to the time coupling in (21c) evolving in discrete time given by the notation $\Gamma_{k,T1}[n']$ as well. Furthermore, the continuous time variables within the T_1 time window are being described using M_1 points, each of which is indexed by j. The notation of $\mathbf{x}_{k,j,T_1}$ refers to the jth discrete point within the kth time window, where these windows are created using the T_1 time step. Several other methods, such as collocation methods [3] and multiple shooting methods [5], have been introduced for model predictive control in process plants, but have not been applied to power grids yet. These techniques coupled with spatial and temporal lifting seem to be promising given the computational advancements that the world has seen recently.

These mathematical formulations of the system partitions seem to be promising for handling the spatial and temporal complexities. However, implementation of such techniques needs a framework for exchanging the right set of information at the right time steps for provable performance.

9 Framework for Implementing Interactive Multi-Spatial Multi-Temporal MPC: DyMonDS

It has been discussed briefly in Sections 6 and 7 how the complexity of the problem formulated in (5) can be reduced by handling objectives of different entities and time scales in a distributed way through a minimum exchange of information for coordination. In order to facilitate such multi-temporal and multi-layer interactive information exchange simultaneously, we need to have an information exchange protocol that enables seamless integration of end devices in a scalable way while also ensuring efficient operation of power grids with provable performance. Dynamic Monitoring and Decision Systems (DyMonDS) [18], introduced some time ago is one such framework that facilitates intertwining of the physical power network in support of JIT, JIP, and JIC functionalities. This architecture is completely in alignment with the partitions created by spatial and temporally lifted sub-problems. Based on nested temporal and nested spatial lifting, one can define interactive exchange as shown in Figure 10.

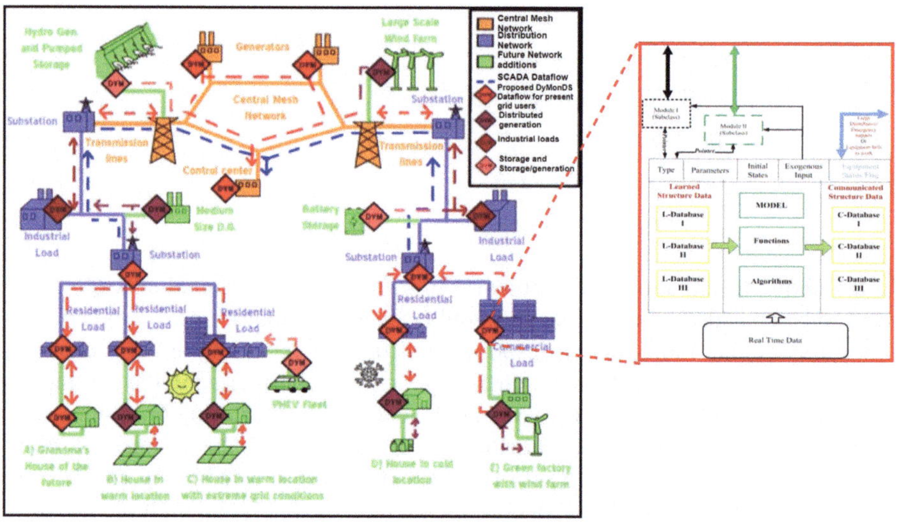

Fig. 10: DyMonDS interactive information exchange.

This framework supports next-generation SCADA, by having local sensors and control embedded into diverse components (DyMonDS). We call them local DyMonDS. The main research question has been new multi-layered modular modeling, which requires only minimal information exchange between local DyMonDS. Next generation SCADA effectively becomes a Cyber Physical System (CPS) with well-defined communication signals. The distinction to note here is that the red lines in the figure go all the way to end-users, which are absent in the present electric energy systems.

The unified modeling framework introduced in Section 4.3 supports consistent information exchange with any other partition or module. Each local DyMonDS has the same internal structure as shown in Figure 10 and is called an actor. Each of these actors can be embedded with the respective partition of the spatially and temporally nested MPC problem. The combination of DyMonDS framework coupled with the unified modeling framework and the software shell that facilitates such interactions can be used to make large-scale power system design and analysis scalable [25].

Shown in Figure 11a is the proposed DyMonDS information exchange framework within a single geographical area of the power grid. Each box in orange represents the end-device controllers which can be interacting with a power producer, power consumer, or a power transmission device. These components communicate with the area-level coordinator at the required timescale which is dictated by the internal logic embedded. Similarly, a bunch of these area level modules interacting with the system coordinator compose the entire power grid which is shown in Figure 11b. The intra-layer communication is done at vastly different time scales simultaneously. For example, the Power Producer module in a single area communicates the bids to the system operator at T_t time scale while it may also communicate the deviations in power generation and local output variable to the area-level controller at T_s timescale for frequency regulation. Similarly, the information exchange using economic signals is shown in Figure 12.

Notice in this figure that each DyMonDS module, with the embedded partition of the spatially nested MPC, exchanges only the economic signals λ, μ with its coordinating DyMonDS modules and its neighbors. Only the bottom up information exchange is either in term of physical variables or the functional bids reacting to price signals as explained in Section 7.1.1. In these figures, we have shown only the information exchange between several spatially nested MPC formulations. However, it must be noted that each spatial entity has temporally nested MPC formulations embedded as well. The MPC problem relevant to the particular disturbance time scale gets activated as and when the entity senses relative changes in the operating

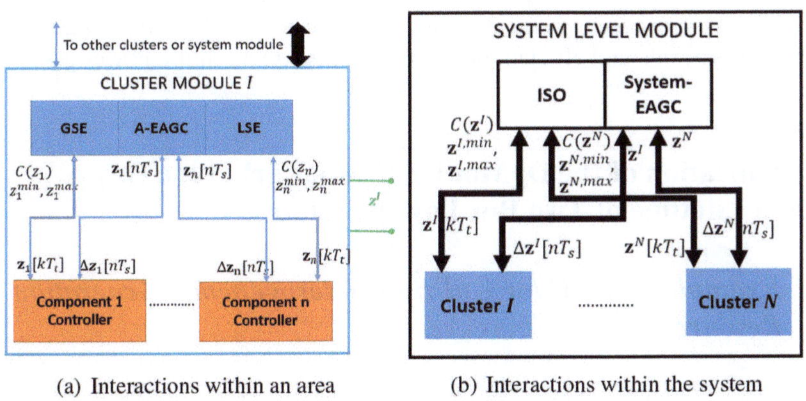

(a) Interactions within an area (b) Interactions within the system

Fig. 11: DyMonDS information exchange based on physical signals between spatially nested layers.

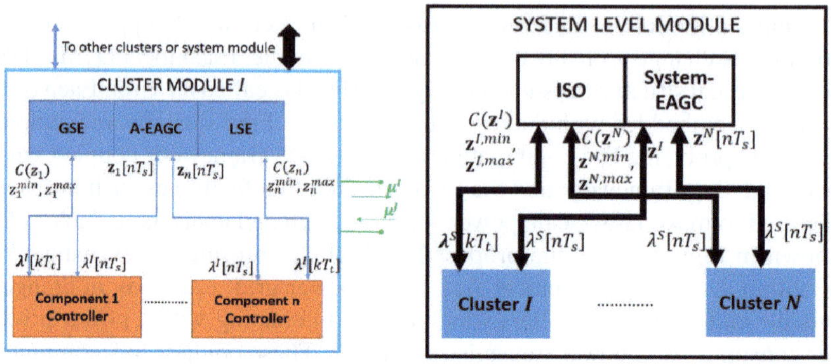

(a) Interactions within an area (b) Interactions within the system

Fig. 12: DyMonDS information exchange based on economic signals between spatially nested layers.

point, relevant for analysis at respective time scales. For instance, the disturbance at aggregate level generally evolves slower relative to its constituent components, i.e. $\Delta\mathbf{m}^N[k_1T_1] << \Delta\mathbf{m}^i[k_1T_1]\ \forall i \in N$. Area N solves the problem $\mathscr{P}^N_{N_2,T_2}(t')$ resulting in area level cleared price of $\lambda^N[kT_2]$. As discussed in Section 7.1, all its constituent members $i \in N$ utilize this signal for solving the problem $\mathscr{P}^i_{N_1,T_1}(t')$. Note that the discretization used by area level and component level problems are generally different. This is because component may be exposed to fast varying disturbances requiring a smaller time step $T_1 < T_2$. Note also the fact that the cleared price computed over coarser time granularity by the area level problem serves as a better estimate of terminal cost from temporal perspective as described in Section 6.1.

Relations obtained by such nesting of temporal and spatial hierarchies have existed in power systems operations for a long time. These are the basis on which the hierarchical operation exists today, although not in a provable manner. Rethinking of this notion in the context of spatially and temporally nested MPC problems begins to quantify several missing signals in today's industry, while also letting us design control for provable and efficient operation of present and future power grids.

10 Application of the DyMonDS Framework: One Day in a Lifetime of Two Bus Power System

10.1 Example 1: MPC for Utilizing Heterogeneous Generation Resources

Now, we begin with one of the most fundamental example to illustrate the effects of temporal lifting-based MPC. In this example, we consider the two-area 5-bus bulk power system depicted in Figure 7. Here, the net system load is forecasted over a prediction time horizon of $T = 30$ minutes. Now, fundamental to efficient scheduling of generator power is the fact that the static optimization, performed over prediction time horizon T, is significantly less effective than the optimal solution obtained by performing MPC at each time $T_1 = 5$ minutes. As an illustrative representation, shown in Figure 13a are progressive predictions of system load at each discrete-time interval of constant length $T = 5$ minutes. It can be seen that, closer to the real-time, the system load predictions are more accurate. If the receding-horizon approach is implemented based on temporal lifting at each $T_1 = 5$ minutes, then the generation schedule tracks the system load than when just static optimization is performed over $T = 30$ minutes. Consider the three plots in Figure 13b: 1) The first plot, with legend "**System Load**," represents the observed 5-minutes average for the net disturbance. 2) The second plot with legend "**Schedule with MPC**" represents the feedforward output, or electrical power, scheduled in response to the forward-shifting predictions in Figure 13a. It is based on temporal lifting each $T_1 = 5$ minutes. 3) The third plot with legend "**Schedule without MPC**" represents a static or a single snapshot optimization over the horizon $T = 5$. It can be observed that the MPC-based feedforward schedule ("**Schedule with MPC**") is much closer to the net system load ("**System Load**") than the static approach ("**Schedule without MPC**"). Now, the larger the difference between the system load and the feedforward schedule, the higher will be the operating cost in terms of feedback control for after-the-fact regulation. Specifically, a static optimization approach to the feedforward schedule will entail an excessive overall cost of feedback action. It must be noted that in this example, only the improved knowledge of disturbance temporal lifting is updated. Our ongoing work concerns nested temporal lifting for scheduling of balancing resources. The nested approach is critical to compute multi-rate feedforward schedules, particularly in anticipation of disturbances forecasted over temporally composite rates.

10.2 Example 2: MPC Spatial and Temporal Lifting in Microgrids to Support Efficient Participation of Flexible Demand

This example is to illustrate the benefits obtained by letting controllable consumer end devices participate in the decision making. We consider the microgrid connected to bus 3 in Figure 7. The fleets of Electric Vehicles (EVs), marked in triangles, are the controllable loads. Each of the electric vehicle has its own requirement for battery charging. In addition, the objective of two generators of capacities 1 MVA and 4 MVA is to minimize their respective generation costs. The goal is to coordinate fast reacting EVs and slow reacting generators to offset the disturbances caused by the inflexible load. The multi-temporal disturbances that the microgrid sees are shown in Figure 4.

We formulate the problem for coordinating these devices every T_t for meeting their sub-objectives while ensuring the supply-demand balance. Furthermore, we assume that the bounds on the load deviations from the predictable components can

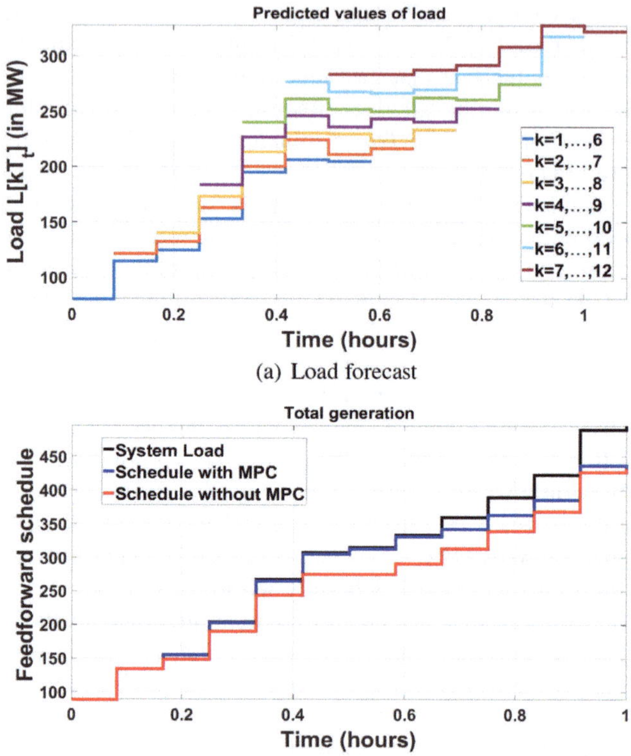

(b) Net scheduled generation with and without MPC

Fig. 13: MPC for feed-forward scheduling of generation resources.

be predicted every T_t. We thus pose it as a co-optimization problem to ensure supply-demand balance and to ensure regulation reserves capacity is scheduled to meet the maximum deviation of the inflexible load. We assume that we have the knowledge on the bounds of deviations every T_t $\Delta\hat{P}[kT_t]$ as shown in Figure 4a.

In relation to the problem formulation in (17), the time T_1 considered is equal to T_t and since we assume the disturbance evolution to be slow in the interval T_t, this lets us derive quasi-stationary input–output relations. The device constraints need to be satisfied by 500 electric vehicles. Efficient coordination of these, by a single entity, is not realizable for real-time operations. Hence, nested spatial lifting needs to be applied to form different layers of coordination. We introduce two Electric Vehicle Load Serving Entities (EVLSEs) as coordinating entities to monitor 200 and 300 electric vehicles respectively, which are further coordinated by a system operator along with the generators. The DyMonDS information exchange framework for solving such nested spatially lifted MPC problem is shown in Figure 14. Detailed formulations of the spatially lifted problems can be found in [27].

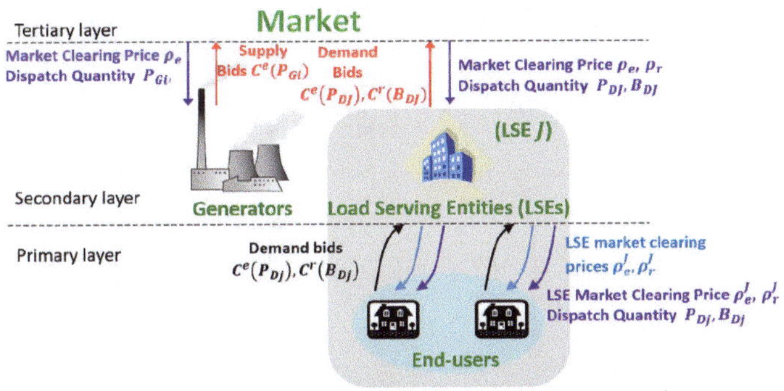

Fig. 14: Information exchange framework for embedding nested spatial lifted MPC [20].

The spatially nested MPC problem relieves the system level coordinator of the need to keep track of 500 communication channels. These algorithms coupled with the DyMonDS information exchange framework would reduce the system operator communication requirement to just 4 channels (2 each for the generators and the EVLSEs). The microgrid system under study has been simulated with the above information exchange framework for two hours when the responsive demand participates in both energy and reserve markets and when it does not participate. The total EV dispatch for each of these cases is shown in Figure 15a and the system energy price variation over the time is shown in Figure 15b

The effect of participation of EVs in both energy and reserve markets is seen to be advantageous since the system prices remain smaller compared to the other two cases for most time intervals.

10.3 Example 3: The Role of MPC in Reducing the Need for Fast Storage While Enabling Stable Feedback Response

Currently, ensuring system stability at the fastest time scale relies on the primary control. Although many control methods, including nonlinear control, have been proposed, their performance remains questionable, due to some issues such as model validation, control saturation, insufficiency of measurements, etc. An alternative solution proposed by researchers is to install fast storage. However, even with storage installed, there is no guarantee that the interconnected system will be always stable. Thus, from both economic and performance perspectives, one fundamental question should be asked first: have we fully utilized the existing devices? If the answer is no, the next question to ask is what would be the minimum storage capacity required to ensure the stability. The existing challenges and the aforementioned two questions indeed make MPC an appealing solution.

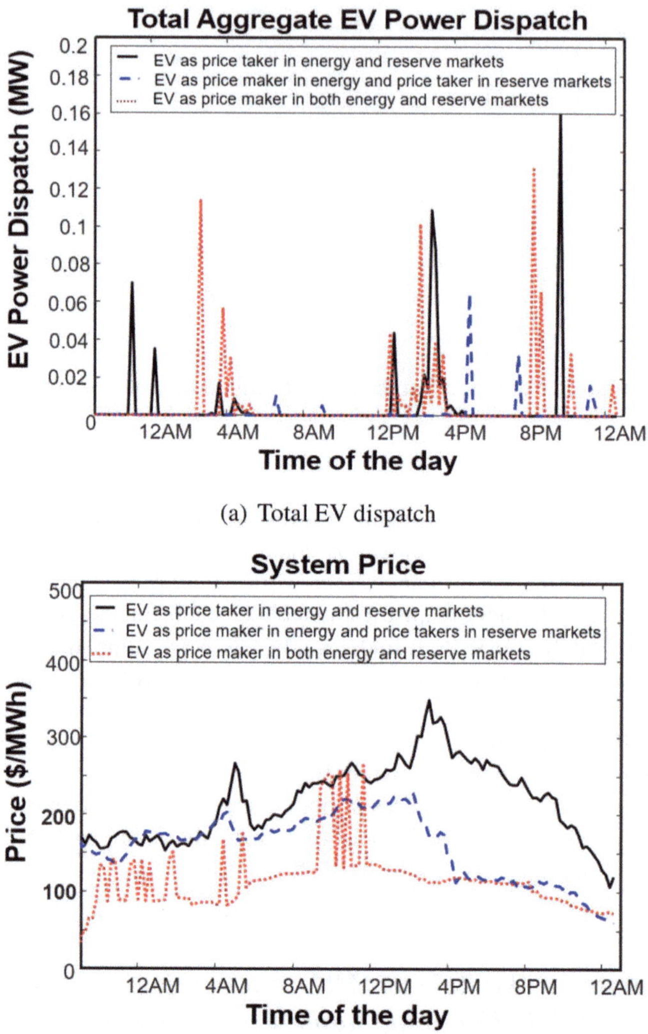

(a) Total EV dispatch

(b) System Price

Fig. 15: Participation of flexible demand in a microgrid.

At the fastest time scale, a singular perturbation argument can be made. That is, one can think of the stabilization problem as though the slow components are stationary. Therefore, without loss of generality, the 5 bus system can be simplified as a two-area system with two synchronous machines, a solar PV, and a load. Suppose that two synchronous machines are the only controllable components. Then the primary control design can be posed as an MPC problem with the objective to minimize the deviation of state variables from their reference set-points within the prediction horizon (T_s).

This example is given to illustrate the potential benefits of MPC at the fastest time scale, where the disturbance is not predictable but the performance metrics must be met at the T_s time scale. This is an extremely difficult problem because of model uncertainties and nonlinearities. Here, we consider two scenarios that can arise in the two-area system:

- There are unpredictable fluctuations in PV profile, as shown in Figure 2
- A sudden contingency in which we lose one synchronous machine (Two areas are disconnected)

In the simulation setup, PV fluctuations are modeled as perturbations of the PV state variables while the disconnection is modeled as a 10% deviation from the nominal grid frequency. The performance of the proposed MPC is compared with FBLC [9].

It should be noted that the prediction model used in the proposed MPC is derived using the unified modeling framework proposed in Section 6. The detailed derivation is omitted. Interested readers are referred to [36]. The objective is to track $x(T_s)$ by predicting the evolution of $x[pT_p]$ within T_s interval.

Simulation results of scenario 1 are shown in Figure 16. Both FBLC and MPC can stabilize the system. However, a significant reduction in voltage overshoot is achieved by the proposed MPC. This comparison supports the claim that MPC can be used to reduce the need for fast storage.

For scenario 2, voltage responses of FBLC and MPC are shown in Figure 17. As shown in Figure 17a, the terminal voltage collapses. The singularity issue of FBLC is the main cause of this instability. Due to the large frequency perturbation, phase angle changes dramatically, which leads the system to a singularity point, where feedback linearization is no longer valid. In contrast, under the same disturbance, the proposed MPC is able to stabilize the system. Furthermore, the terminal voltage,

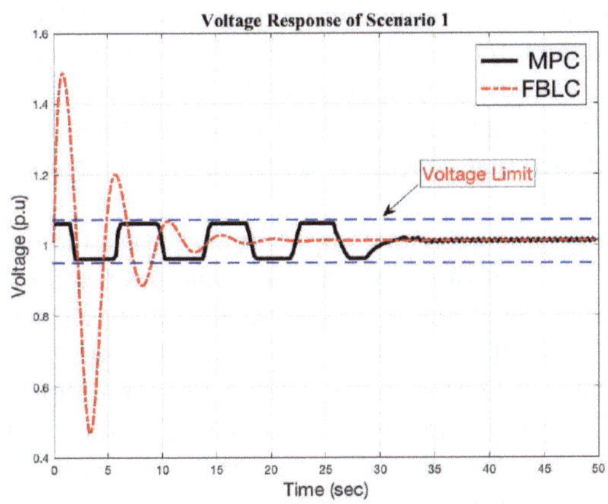

Fig. 16: Scenario 1: Voltage Response.

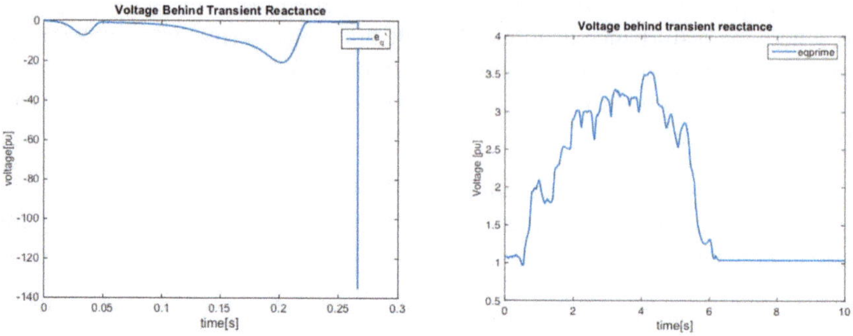

(a) Over a slower time scale (b) Voltage Response of MPC

Fig. 17: Scenario 2: Voltage Response.

as shown in Figure 17b, is limited within the feasible region in the steady state. Since no voltage collapse occurs, expensive storage is not required. Thus, we can conclude that the proposed MPC greatly improves the system performance. By solving real-time optimization, the need for expensive storage will also be minimized.

10.4 Example 4: The Role of MPC Spatial Lifting in Normal Operation Automatic Generation Control (AGC)

Control of the small-signal dynamics (SSD) is one of the major tasks in electric power system operation. Recently, as renewable energy resources (RESs) are increasingly integrated and the size of electric power systems is growing, this task becomes even more challenging than in the past. First, the RESs create persistent disturbances around the forecast, which could cause system frequency variation and further lead to poor Quality of Response (QoR). Moreover, today's primary speed controllers and voltage controllers are tuned locally without considering the rest of the system. It is assumed that generators under control remain stable when they are interconnected.

However, as the system size and complexity increases, the interactions between different machines and controllers could deteriorate or even destabilize the system dynamics. In addition, the classic secondary control is based on area control error (ACE), which represents the conserved net power imbalance. But this concept is defined at the control-area level with the steady state assumption. Due to the integration of hard-to-predict renewable resources, it may be hard for future power systems to reach the steady state. More precisely, the steady-state assumption will no longer be valid in the future. Therefore, an enhanced secondary layer control is needed to solve above problems. In the following, the same 5-bus system shown in Figure 7 is used to illustrate the proposed MPC-based secondary control.

In [21, 26, 28, 31], the authors proposed a new approach which generalizes the conventionally used steady-state concepts such as the ACE and the inadvertent energy exchange (IEE). For instance, the interaction variable (IntV) used in the proposed secondary control represents the accumulated net energy imbalance which can be interpreted as the dynamic IEE. In this example, we further relax the small signal assumption and extend the Enhanced-AGC by using the proposed framework, which leads to the multi-layered MPC with unified communication (via IntVs).

The problem formulation exactly follows the procedure discussed in Section 7. So we skip the derivation but explain the intuition behind the proposed E-AGC instead. The frequency dynamics is assessed by investigating the variations of IntVs. At the control area level, constant IntV implies the non-existence of inadvertent power exchange with other areas. Furthermore, at the lower component level, constant IntV implies a zero net power imbalance of the component, which gives rise to a satisfactory frequency quality. Higher level MPC will coordinate resources in the system which ensures economic performance.

It should be noted that only IntVs and control signals are exchanged between different layers. Not all information is exposed to every module, thus, improving the cyber security of the given electrical power system. From an implementation point of view, this is also critical because less complicated communication channels are needed as only two types of information are exchanged.

The simulation results for change in reference set points for two snapshots with and without E-AGC are shown in Figure 18.

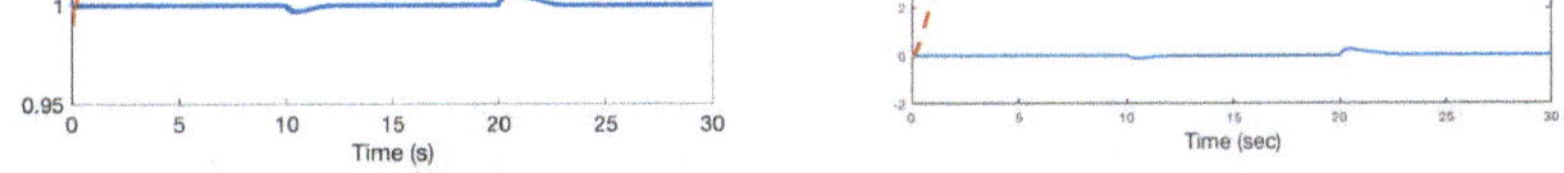

(a) frequency response (b) IntV response

Fig. 18: Simulation Result of Area I.

Without E-AGC, the system gradually becomes unstable as set points are changing. In comparison, the proposed multi-layered E-AGC is able to regulate the frequency back to the nominal value. The same conclusion can be drawn from the interaction variable response as well.

11 Conclusions

In this chapter, we put forward the hypothesis that multi-layered modeling and decision making will play a fundamental role in the changing electric energy industry. Today's industry is based on the assumptions that: (1) the system load is highly predictable and inflexible; and (2) the generation is fully controllable at pre-specified rates of producing required power. Also, most of the software tools used in industrial control centers and those embedded into system equipment are designed under the assumptions of spatial and temporal decoupling. It is described in this paper that emerging uncertainties in the changing industry create a highly uncertain environment. To plan and operate complex electric power systems, it is essential to rely on predictions over multiple time horizons so that long-term performance is achieved. We make the case that this can be done by formalizing decision making in the changing industry as a multi-temporal and multi-spatial MPC network problem. Temporal and spatial lifting is used to decompose a long-term decision-making problem under complex uncertainties into a family of interdependent MPC sub-problems. The information exchange required to relate these sub-problems is defined and it is claimed and illustrated that such information protocols are critical to both efficient and physically near-optimal operation. An earlier proposed DyMonDS framework for enhanced operation and planning in the changing industry was used as a motivation for the methods proposed. As future work, we are formalizing the use of MPC multi-layered approach and its DyMonDS implementation for enhanced planning and operation methods. In particular, we are working on formalizing multi-temporal market designs in support of interdependent capacity, energy and regulation markets. Capturing the interdependencies under large uncertainties is the key to efficient and sustainable utilization of energy services. In parallel work in progress is underway towards standardizing technical performance requirements for industry participants so that system-level technical objectives such as stability, frequency, and voltage regulation are achieved. Further work is needed for formalizing these requirements using unified modeling.

Acknowledgements This material is based upon work supported by the Department of Energy under Air Force Contract No. FA8721-05-C-0002 and/or FA8702-15-D-0001. Any opinions, findings, conclusions, or recommendations expressed in this material are those of the authors and do not necessarily reflect the views of the Department of Energy. It is also supported in part by the National Institute of Standards and Technology (NIST) under the project of "Smart Grid in a Room Simulator" and PSERC project S-64 entitled "Monitoring and maintaining limits of area transfers with PMUs."

References

1. Bachovchin, K.D., Ilić, M.D.: Automated modeling of power system dynamics using the lagrangian formulation. Int. Trans. Electr. Energy Syst. **25**, 2087–2108 (2015)

2. Bachovchin, K.D., Ilić, M.D.: Transient stabilization of power grids using passivity-based control with flywheel energy storage systems. In: Power & Energy Society General Meeting, 2015 IEEE, pp. 1–5. IEEE, New York (2015).
3. Biegler, L.T.: Solution of dynamic optimization problems by successive quadratic programming and orthogonal collocation. Comput. Chem. Eng. **8**(3–4), 243–247 (1984)
4. Biskas, P.N., Bakirtzis, A.G., Macheras, N.I., Pasialis, N.K.: A decentralized implementation of DC optimal power flow on a network of computers. IEEE Trans. Power Syst. **20**(1), 25–33 (2005)
5. Bock, H.G., Plitt, K.J.: A multiple shooting algorithm for direct solution of optimal control problems. IFAC Proc. **17**(2), 1603–1608 (1984)
6. California ISO: What the duck curve tells us about managing a green grid, California ISO, Shaping a Renewed Futur, pp. 1–4
7. Camacho, E.F., Alba, C.B.: Model Predictive Control. Springer, London (2013)
8. Cellier, F.E., Kofman, E.: Continuous System Simulation. Springer, New York (2006)
9. Chapman, J.W., Ilić, M.D., King, C.A., Eng, L., Kaufman, H.: Stabilizing a multimachine power system via decentralized feedback linearizing excitation control. IEEE Trans. Power Syst. **8**(3), 830–839 (1993)
10. Chowdhury, B.H., Rahman, S.: A review of recent advances in economic dispatch. IEEE Trans. Power Syst. **5**(4), 1248–1259 (1990)
11. Cohn, N.: Control of Generation and Power Flow on Interconnected Systems. Wiley, New York (1961)
12. Crow, M.L., Ilic, M.: The parallel implementation of the waveform relaxation method for transient stability simulations. IEEE Trans. Power Syst. **5**(3), 922–32 (1990)
13. Cvetković, M., Ilić, M.D.: Ectropy-based nonlinear control of facts for transient stabilization. IEEE Trans. Power Syst. **29**(6), 3012–3020 (2014)
14. Donadee, J.: Operation and valuation of multi-function battery energy storage under uncertainty. Ph.D. dissertation, Department of Electrical Engineering, Carnegie Mellon University, Pittsburgh, PA (2015)
15. Garcia, C.E., Prett, D.M., Morari, M.: Model predictive control: theory and practice a survey. Automatica **25**(3), 335–348 (1989)
16. Hortacsu, A., Puller, S.L.: Understanding strategic bidding in multi-unit auctions: a case study of the Texas electricity spot market. RAND J. Econ. **39**(1), 86–114 (2008)
17. Huneault, M., Galiana, F.D.: A survey of the optimal power flow literature. IEEE Trans. Power Syst. **6**(2), 762–770 (1991)
18. Ilic, M.D. Dynamic monitoring and decision systems for enabling sustainable energy services. Proc. IEEE **99**(1), 58–79 (2011)
19. Ilic, M.D.: Toward a unified modeling and control for sustainable and resilient electric energy systems. Found. Trends Electr. Energy Syst. **1**(1–2), 1–141 (2016)

20. Ilic, M.D., Liu, X.: A simple structural approach to modeling and analysis of the inter-area dynamics of the large electric power systems: Part I: linearized models of frequency dynamics. In: Proceedings of the 1993 North American Power Symposium, Washington, pp. 560–569 (1993)
21. Ilić, M.D., Liu, S.X.: Hierarchical Power Systems Control: Its Value in a Changing Electric Power Industry. Advances in Industrial Control. Springer, New York (1996)
22. Ilic, M.D., Liu, Q.: Toward sensing, communications and control architectures for frequency regulation in systems with highly variable resources. In: Chapter 1, Control and Optimization Methods for Electric Smart Grids. Springer, New York (2012)
23. Ilic, M.D., Zaborszky, J.: Dynamics and Control of Large Electric Power Systems. Wiley, New York (2000)
24. Ilić, M., et al.: Physics-based foundations for cyber and market design in complex electric energy systems. In: 2014 IEEE 53rd Annual Conference on Decision and Control (CDC). IEEE, Piscataway (2014)
25. Ilić, M., Jaddivada, R., Miao, X.: Scalable electric power system simulator. In: Innovative Smart Grid Technologies-Europe (ISGT Europe), 2018 IEEE (2018, accepted for publication)
26. Ilić, M., Liu, S.X., Eidson, B.D., Vialas, C., Athans, M.: BA new structure-based approach to the modeling and control of electric power systems in response to slow reactive power/voltage load variations. Automatica **33**, 515–531 (1997)
27. Jaddivada, R., Ilić, M.D.: A distribution management system for synthetic regulation reserve. In: 2017 IEEE 49th North American Power Symposium. IEEE, Piscataway (2017)
28. Joo, J.-Y.: Adaptive load management: multi-layered and multi-temporal optimization of the demand aide. Ph.D. Dissertation, Department of Electrical Engineering, Carnegie Mellon University, Pittsburgh (2013)
29. Jaleeli, N., VanSlyck, L.S., Ewart, D., Fink, L., Hoffmann, A.: Understanding automatic generation control. IEEE Trans. Power Syst. **7**(3), 1106–1122 (1992)
30. Wu, F.F., Varaiya, P.: Coordinated multilateral trades for electric power networks: theory and implementation. Int. J. Electr. Power Energy Syst. **21**(2), 75–102 (1999)
31. Liu, Q.: A large-scale systems framework for coordinated frequency control of electric power systems. Ph.D. dissertation, Carnegie Mellon University (2013)
32. Liu, Q., Ilic, M.D.: Enhanced automatic generation control (e-agc) for future electric energy systems. In: 2012 IEEE Power and Energy Society General Meeting, pp. 1–8. IEEE, Piscataway (2012)
33. Liu, Q., Miao, X., Ilic, M.D.: Enhanced analysis and control of frequency dynamics in large-scale electric power systems (2017, to be submitted)
34. Loukianov, A.G., et al.: Discontinuous controller for power systems: sliding-mode block control approach. IEEE Trans. Ind. Electron. **51**(2), 340–353 (2004)
35. Mayne, D.Q., et al.: Constrained model predictive control: stability and optimality. Automatica **36**(6), 789–814 (2000)
36. Miao, X., Ilić, M.D.: Distributed model predictive control of synchronous machines for stabilizing microgrids. In: 49th North American Power Symposium, 2017 IEEE. IEEE, Piscataway (2017)
37. Navid, N., Rosenwald, G.: Ramp capability for load following in miso markets 2011 [Online]. Available: https://www.misoenergy.org
38. Popli, N., Ilić, M.D.: Modeling and control framework to ensure intra dispatch regulation reserves. In: Chapter-14, Engineering IT-Enabled Sustainable Electricity Services: The Tale of Two Low-Cost Green Azores Islands, vol. 30. Springer, Boston (2013)
39. Qin, S.J., Badgwell, T.A.: An overview of nonlinear model predictive control applications. In: Nonlinear Model Predictive Control, pp. 369–392. Birkhäuser, Basel (2000)
40. Rebours, Y.G., Kirschen, D.S., Trotignon, M., Rossignol, S.: A survey of frequency and voltage control ancillary services Part I: technical features. IEEE Trans. Power Syst. **22**(1), 350–357 (2007)
41. Salcedo, R.O., Nowocin, J.K., Smith, C.L., Rekha, R.P., Corbett, E.G., Limpaecher, E.R., LaPenta, J.M.: Development of a real-time hardware-in-the-loop power systems simulation

platform to evaluate commercial microgrid controllers, No. TR-1203. Massachusetts Institute of Technology, Lexington Lincoln Lab (2016)
42. Scattolini, R.: Architectures for distributed and hierarchical model predictive control: a review. J. Process Control **19**(5), 723–731 (2009)
43. Snider, L., et al.: Today's power system simulation challenge: high-performance, scalable, upgradable and affordable COTS-based real-time digital simulators. In: Joint International Conference on Power Electronics, Drives and Energy Systems and 2010 Power India, pp. 1–10 (2010)
44. Victor, M.Z.: New architectures for hierarchical predictive control. IFAC-PapersOnLine **49**(7), 43–48 (2016). ISSN 2405–8963
45. Xie, L., Carvalho, P.M.S., Ferreira, L.A.F.M., Liu, J., Krogh, B.H., Popli, N., Ilic, M.D.: Wind integration in power systems: operational challenges and possible solutions. Proc. IEEE **99**(1), 214–232 (2011)

[illegible]

Applications of MPC to Finance

James A. Primbs

1 Introduction

In recent years, a new research direction in MPC has been its application to stochastic control problems that arise in the field of finance. The most prominent of these are the *portfolio optimization* and *dynamic option hedging* problems, where MPC is providing improved solutions that are able to successfully incorporate realistic price dynamics, market constraints, transaction costs, and other features. The purpose of this chapter is to introduce the key problems of portfolio optimization and dynamic option hedging from a control perspective, and provide an overview of the MPC formulations and methods that are being successfully used to address them.

1.1 Portfolio Optimization

The portfolio optimization problem is a central problem in modern finance theory, and, at its most basic level, involves the question of how to best trade a portfolio of stocks[1] in order to maximize some measure of one's future wealth. This basic question is faced by many market participants, from an individual self-managing a retirement account, to a multi-billion dollar university endowment seeking to ensure long-term financial stability for an academic institution.

J. A. Primbs (✉)
Department of Finance, Mihaylo College of Business and Economics, California State University Fullerton, Fullerton, CA 92831, USA
e-mail: jprimbs@fullerton.edu

[1] In this chapter we will limit our discussion to the trading of stocks, but the formulations presented directly translate to the trading of other securities as well.

S. V. Raković, W. S. Levine (eds.), *Handbook of Model Predictive Control*, Control Engineering, https://doi.org/10.1007/978-3-319-77489-3_27

The quantitative approach to portfolio optimization began with the pioneering work of Markowitz [17] in the 1950s, who mathematically modeled the random returns of stocks and showed that a quadratic program could be solved to determine the portfolio with the best trade-off of risk (as measured by the portfolio's variance) and return (as measured by the portfolio's expected return) in a single period setting.

The single period formulation of Markowitz soon gave way to dynamic formulations of the portfolio optimization problem in which stock price movement was modeled in a stochastic process framework, and traders were allowed to dynamically adjust their portfolio holdings over time. Such a dynamic formulation is naturally cast as a stochastic control problem, where the control variables that the trader can affect are the number of shares held of each stock, and the objective is to maximize some measure of the future value of the trader's wealth.

Some of the most prominent early stochastic control approaches to the dynamic portfolio optimization problem date back to the work of Samuelson [30] in the late 1960s and more notably Merton [21] in the early 1970s. In their work, the portfolio optimization problem is explicitly formulated and solved using stochastic control methods and dynamic programming. In particular, in Merton's seminal paper on dynamic portfolio optimization [21], closed form solutions are obtained for an economically important class of objective functions under the assumption that trading takes place in an *idealized frictionless market*. The key features of such a market are that trading is unconstrained, can take place in continuous time and does not influence prices, there are no transaction costs or broker's fees, and there are no collateral or so-called margin requirements associated with borrowing.

In the ensuing years from Merton's seminal paper to the present, much of the work in dynamic portfolio optimization has focused on solving the problem without many of the idealized market assumptions. For example, actual trading may be subject to a number of constraints, such as no short selling, or margin requirements. Moreover, when a trader purchases or sells a share of stock, the so-called bid-ask spread comes into play and a commission fee is charged by the broker. These are examples of so-called transaction costs.

It is in addressing such issues that MPC is now finding great application to finance problems. This chapter will provide an introduction to some of the basic Stochastic MPC formulations for portfolio optimization, and highlight how they are being used to incorporate such realistic and important market features.

1.2 Dynamic Option Hedging

The second class of problems that will be covered in this chapter, and where Stochastic MPC has also found application, is *dynamic option hedging*, otherwise known as *option replication*. This problem is intimately related to the pricing of options, and more generally, so-called *derivative securities*. The basics of options in the context of the hedging problem will be covered later in the chapter, but readers seeking more in-depth coverage are referred to the books of Luenberger [16], Hull [14], and Primbs [28].

To provide a brief example that will be used to illustrate the dynamic hedging problem later, a *European call option* is a contract between a buyer and seller that gives the buyer the right (but not the obligation) to purchase from the seller a share of a specified stock (called the *underlying* stock) at a specified price (called the *strike price*) and specified time (called the *expiration date*). Market participants who sell options are exposed to the liability that the buyer may want to "exercise" the option at the expiration time and purchase shares from them at the specified strike price. However, if the seller can in turn trade a portfolio whose value replicates the liability of the sold option, then that traded portfolio serves to "hedge" the seller's liability. For this reason, option replication is commonly referred to as *dynamic hedging*. Selling of options is performed by many financial market participants, especially investment banks and market makers, and thus the ability to offset their risks by option replication is of great value.

The dynamic option hedging problem can be cast as a stochastic control problem, but with the objective of having the trader's account value match the payoff value of an option at its expiration time. In the context of control, this is the problem of matching a target random variable as closely as possible at a specified future time. Under specific stock price dynamics and idealized frictionless markets, the seminal work of Black and Scholes [4] and Merton [22] provided a strategy to exactly match the target payoff of a European call option.

Once again, Stochastic MPC formulations are of value when realistic market environments are considered, and features such as transaction costs are encountered. The world of derivative securities, of which European call options are the classical example, has also greatly expanded since the original work of Black and Scholes, and now many derivative securities exist where exact replication, even in idealized frictionless markets, is not possible.

While sharing much in common with the dynamic portfolio optimization problem, the unique characteristics of the dynamic option hedging problem have led to distinct Stochastic MPC formulations. The basic approaches to this problem from an MPC perspective will be covered in Section 4.

1.3 Organization of Chapter

The rest of this chapter is devoted to explaining how MPC is being used to address the problems of portfolio optimization and dynamic option hedging. In the following section, we begin by modeling the account value dynamics of a trader in a control context. This provides the basic system dynamics for both of the finance problems that we consider. In Section 3, we address the portfolio optimization problem. We first develop a simple control formulation of the problem, and then highlight the ways in which MPC has been used to provide improved solutions by incorporating relevant market features and constraints. We then turn our attention to the dynamic option hedging problem in Section 4, where again we begin with a basic control formulation of the problem. We then tackle the issues involved with applying MPC to such a problem, and highlight some approaches that have been taken. Finally, Section 5 concludes with a brief discussion of the value MPC is bringing to the field of finance, and opportunities for the future.

2 Modeling of Account Value Dynamics

Central to both classes of problems that we will consider is a trader who is buying and selling stocks dynamically over time. Thus, our first order of business is to model such a trader, and in particular, the dynamic evolution of the trader's account value.

To this end, we consider a market containing n stocks evolving in discrete time that are available to buy and sell at the times $k = 0, 1, \ldots$. The price per share at time k of each stock is denoted by $S_i(k)$ for $i = 1, \ldots, n$. For example, these could represent the daily closing prices of the stocks in the S&P 500, where we would then have $n = 500$.

Our goal is to model a trader with initial account value $V(0) = V_0$, who uses this wealth to buy and sell shares of the stocks at each time instant. Specifically, let $u_i(k)$ denote the number of shares of stock $S_i(k)$ held by the trader at time k. This collection of shares held, i.e., $u_i(k)$ for $i = 1, \ldots, n$, is referred to as the trader's *portfolio*. If $u_i(k) > 0$, then this means that the trader owns shares of stock i at time k, and is commonly referred to as being "long." For example, $u_1(k) = 10$ means that the trader's account holds 10 shares of stock 1 at time k. On the other hand, if $u_i(k) < 0$, this means that the trader has borrowed shares and sold them, and they represent a liability. This is known as being "short," and is commonly allowed in financial markets.

In a control context, the shares of each stock held by the trader $u_i(k)$ for $i = 1, \ldots, n$ represent the control variables. That is, the trader is able to choose values for $u_i(k)$ (subject to possible constraints) at each time instant k. It is also possible that the trader holds some funds purely in cash. This cash balance is assumed to earn a guaranteed interest rate of r_f per period, which is termed the *risk-free* rate of return. Let $u_0(k)$ denote this amount held in cash at time k. With this notation, the

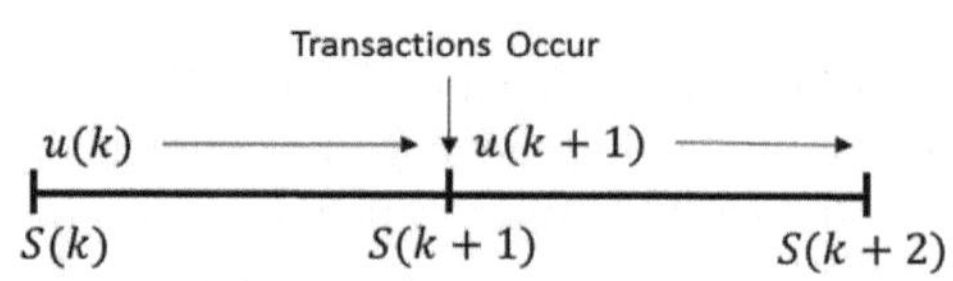

Fig. 1: Timing of stock movement and transactions in trading.

total account value of the trader at time k is given by the sum of the values of the positions in stocks, plus the amount held in cash,

$$V(k) = u_0(k) + \sum_{i=1}^{n} u_i(k)S_i(k). \tag{1}$$

To obtain dynamics for the evolution of the account value $V(k)$ over time, we consider the timing of events, as shown in Figure 1, as we move from time k to $k+1$. First, the value of each stock changes from $S_i(k)$ to $S_i(k+1)$, and the risk-free rate r_f is earned on the cash balance $u_0(k)$. Thus, at time $k+1$ *prior to any trades being made*, the account value is

$$V(k+1) = u_0(k)(1+r_f) + \sum_{i=1}^{n} u_i(k)S_i(k+1). \tag{2}$$

Next, at time $k+1$ we are allowed to trade and change our holdings in the stocks from $u_i(k)$ to $u_i(k+1)$. We will assume that this is done in a so-called *self-financing* manner, which means that no "outside" money is allowed to be added to or removed from the account. In this case, our account value immediately preceding the trades at time $k+1$, as given in (2), must be exactly equal to our account value following the trades. That is, we must also have

$$V(k+1) = u_0(k+1) + \sum_{i=1}^{n} u_i(k+1)S_i(k+1), \tag{3}$$

which is the $k+1$ counterpart of Equation (1).

These steps can be combined into a single difference equation for the account value by solving for $u_0(k)$ in Equation (1) and substituting into Equation (2). This leads to

$$V(k+1) = V(k)(1+r_f) + \sum_{i=1}^{n} u_i(k)(S_i(k+1) - (1+r_f)S_i(k)). \tag{4}$$

One way to view this difference equation is that each input $u_i(k)$ is multiplied by a random disturbance term generated by the movement of the corresponding stock price S_i. These are then summed and contribute to the new account value at the end of the period, $V(k+1)$. This input/output block structure is shown in Figure 2.

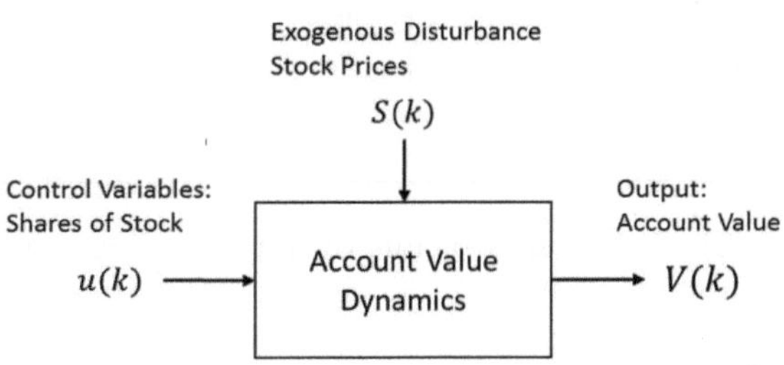

Fig. 2: Account value dynamics block with inputs and outputs.

2.1 Stock Price Dynamics

As depicted in Figure 2, the stock prices enter the account value dynamics as an exogenous disturbance.[2] An important aspect of any market model is an accurate description of the movement of these stock prices.

A common starting point is to model the return of each stock as

$$\frac{S_i(k+1)-S_i(k)}{S_i(k)} = r_i(k)$$

where $r_i(k)$ is a random variable. When $r_i(k)$ is assumed to be Gaussian, with mean $\mu\Delta t$ and variance $\sigma^2\Delta t$, where Δt represents the discrete time increment measured in years, the model of the stock can be written as

$$\frac{S_i(k+1)-S_i(k)}{S_i(k)} = \mu_i\Delta t + \sigma_i z_i(k) \tag{5}$$

where $z_i(k) \sim N(0,1)$ is a standard normal random variable. Moreover, returns are often assumed to be independent over each time period; i.e., $\mathbb{E}[z_i(k)z_j(l)] = 0$ for all $i, j = 1\ldots n$ and $k \neq l$. The continuous time limit of this model is the so-called Geometric Brownian Motion (GBM) stock model

$$\frac{dS_i}{S_i} = \mu_i dt + \sigma_i dZ \tag{6}$$

where $Z(t)$ is a standard Brownian motion [23]. This model pervades much of the classical financial literature; see, for example, [4, 21].

In a similar manner, when dealing with n stocks, their returns are often represented as a Gaussian random vector

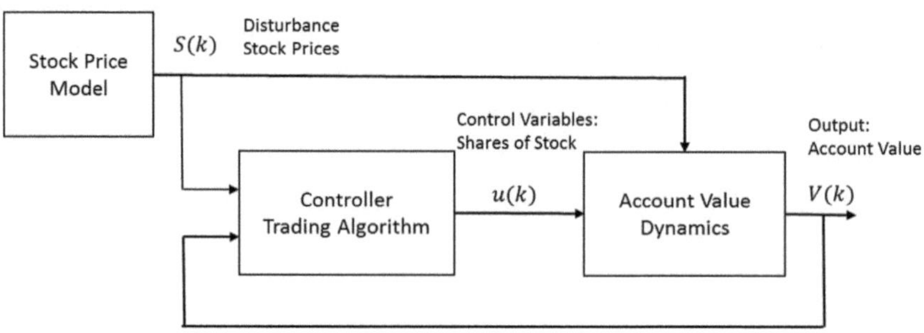

Fig. 3: Block diagram structure of simple trading strategy.

[2] We assume that the trader's actions do not affect stock prices, which is referred to as being a *price taker*, otherwise we would require a model for stock prices that depends on the buying and selling of the trader.

$$r(k) = \begin{bmatrix} r_1(k) \\ r_2(k) \\ \vdots \\ r_n(k) \end{bmatrix}$$

with corresponding mean and covariance structure given by

$$\mathbb{E}[r(k)] = \mu \Delta t,\ \mathbb{E}\left[(r(k) - \mathbb{E}[r(k)])\,(r(k) - \mathbb{E}[r(k)])^T\right] = \Sigma \Delta t.$$

This vector GBM model is frequently used as a starting point for stock price movement, but numerous other models exist. In fact, the range of possible models for stock price movement is so large that no attempt will be made to survey the possibilities here. Moreover, for many problems in finance, it is important to be able to allow significant flexibility in the modeling of stock prices, given that the range of phenomena possible in finance markets is vast. For example, one may want to allow prices generated via an algorithm or drawn from historical data rather than from an analytical model. Practically speaking, this means that in some cases the system dynamics will have structure that can be taken advantage of in MPC formulations, but in many others it won't.

2.2 Control Structure of Trading Algorithms

To provide a more complete diagram of the information flow in trading, we note that in its simplest form, a trading strategy is a mapping from stock prices $S(k)$ and account value $V(k)$ to the number of shares held of each stock, $u(k)$. This trading strategy, along with the stock price dynamics, then drives the account value dynamics. These interconnections are shown in block diagram form in Figure 3.

More generally, a trading algorithm may use additional information about the past and present state of stock prices in the market, and perhaps outside exogenous information (such as news sources, financial statements, economic conditions, etc.) to determine the control variables $u(k)$.

With this basic control description of the dynamics involved in trading as background, in the following sections we turn our attention to the problems of portfolio optimization and dynamic option hedging.

3 Portfolio Optimization Problems

As stated previously, the goal of portfolio optimization is to trade a portfolio of stocks in order to maximize some measure of wealth (or account value) at a future time. That measure of future wealth defines the objective function for the portfolio optimization problem, and typically involves a tradeoff between achieving a high expected value of wealth, but at the same time attempting to minimize risk. That is, ideally one would like to make a lot of money in the least risky manner possible.

With these considerations in mind, there are a number of objective functions that one can use in a portfolio optimization problem. For the purpose of illustration, in this chapter we will consider one of the most standard and widely used objective functions that simply trades off the expected value of future wealth with its variance. That is, consider maximizing an objective function of the form

$$\mathbb{J}_0(V(T)) = \mathbb{E}_0[V(T)] - \frac{\gamma}{2}\mathbb{V}_0[V(T)] \tag{7}$$

where T is some future time of interest, and $\mathbb{E}_0[\cdot]$ and $\mathbb{V}_0[\cdot]$ denote the expectation and variance, respectively, conditioned upon the information at time 0. The parameter γ, specified by the trader, is a *risk aversion coefficient* that influences the trade-off between expected return and risk. The higher the value of γ, the more risk averse the trader. For example, traders with a longer horizon for generating wealth, such as a university endowment, will likely have a lower value of γ, whereas an individual who is nearing retirement and more risk averse will tend to choose a higher value.

Given such an objective function, the basic portfolio optimization problem is given as

$$\max_{u(\cdot)} \; \mathbb{J}_0(V(T)) \tag{8}$$

$$\text{s.t.} \quad V(0) = V_0, \tag{9}$$

$$V(k+1) = V(k)(1+r_f) + \sum_{i=1}^{n} u_i(k)(S_i(k+1) - (1+r_f)S_i(k)), \tag{10}$$

$$k = 0, \ldots, T-1 \tag{11}$$

where the control actions $u(\cdot)$ are taken over some non-anticipating admissible set of functions. Note that we have omitted a model for the stock price dynamics, which must also be included and will depend on the details of the problem.

There are various instances in which this basic portfolio optimization problem can be solved in closed form. For example, when stock price movement in (5) is combined with the mean-variance objective of (7), the problem dynamics can be transformed to a linear system with multiplicative noise, which is then amenable to dynamic programming methods [32]. Moreover, the continuous time version of this problem with the GBM stock price dynamics of (6) and the hyperbolic absolute risk aversion class of utility functions as the objective was solved explicitly by Merton [21].

However, when realistic market features are added to the problem, such as constraints and transaction costs, closed form solutions often fail to exist. This is where MPC methods, as described next, have been used with great success.

3.1 MPC Formulations

The standard MPC implementation repeatedly solves an on-line, *open-loop* version of the portfolio optimization problem at each time step k, and implements only the first step of the optimal control sequence in typical receding horizon fashion.

That is, consider the open-loop optimization problem conditioned on the information at time k with MPC horizon N,

$$\begin{aligned}
&\max_{u(\cdot|k)} \mathbb{J}_k^{(N)}(V(N|k)) \\
\text{s.t. } & V(0|k) = V(k), \\
& V(j+1|k) = V(j|k)(1+r_f) + \sum_{i=1}^{n} u_i(j|k)(S_i(j+1|k+1) - (1+r_f)S_i(j|k)), \\
& j = 0,\ldots,N-1,
\end{aligned}$$

where, based on some specified stock price model $S_i(j|k)$ for $j = 0,\ldots,N$, this problem predicts N steps into the future from the current time k to obtain $V(N|k)$. The objective function $\mathbb{J}_k^{(N)}(\cdot)$ is used to capture the risk-return characteristics of the trader over the horizon N, and the problem is solved open-loop to obtain the optimal control actions $u^*(j|k)$, for $j = 0,\ldots,N-1$. The MPC control strategy then uses $u^*(0|k)$ at time k, resolving this optimization problem at each new time step. Figure 4 provides a pictorial representation of this implementation.

The advantage of the MPC approach is that many of the features encountered in real financial markets can be incorporated into the MPC on-line optimization, ultimately leading to greatly improved trading strategies. To provide a feel for the range of features that have already been addressed in this manner, next we provide an overview of some of the literature on MPC for portfolio optimization. This is followed by a detailed treatment of the issue of transaction costs, and the specification of some commonly encountered portfolio optimization constraints.

3.1.1 Overview of MPC Literature

The exact form of the on-line optimization is highly dependent on the assumed model for stock price dynamics and the objective function. Here, we briefly mention some of the models that have been used in the context of MPC formulations.

When the basic stock price model of Equation (5) is used, and a mean-variance objective function is considered, the on-line optimization can often be solved as a convex quadratic program. Tractable on-line optimizations are also possible for more advanced models of stock price movement and objectives. For example, in Herzog et al. [12, 13], a linear factor model is used to drive stock returns, and an approximation to the log-wealth of the account value is considered. Dombrovskii and coauthors have worked extensively on random parameter systems with multiplicative noise [8, 10, 11] and have included features such as Markovian jumps [7].

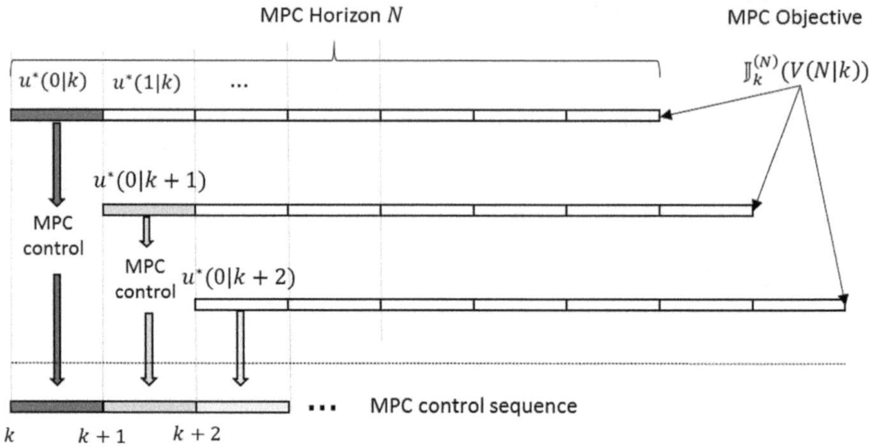

Fig. 4: Standard MPC implementation applied to the portfolio optimization problem.

In Sridharan et al. [31], a mini-max objective is used in a wealth tracking context, whereas in Yamada and Primbs [33], MPC is applied to the trading of pairs of stocks, where the relative value between stocks is modeled as mean-reverting.

Additional formulations appearing in the literature include Lee [15], who uses MPC to solve a portfolio optimization problem involving the dynamics of illiquid asset classes, Meindl [18] who develops an MPC approach in the context of bond portfolios, and Piccoli and Marigo [24] and Primbs and Sung [29] who consider portfolio tracking related problems.

While most of the MPC formulations given above use an open-loop on-line optimization, others have used some form of feedback control, such as Calafiore [5, 6], Primbs and Sung [29], and Primbs [25].

As evidenced from this literature, the range of possible formulations is extremely varied and often depends on the exact market and trading objective under consideration. For example, here we have not mentioned the stock price models that appear in dynamic option hedging that will be considered in Section 4. An important takeaway is that the stock price model, objective function, and constraints can greatly affect the MPC formulation and its corresponding on-line solution methodologies, providing a nearly endless supply of challenges and opportunities for the MPC community.

The rest of this section is used to provide specific examples of commonly encountered market features and how they are addressed in MPC formulations. We begin with the issue of transaction costs, followed by various forms of portfolio constraints.

3.1.2 Transaction Costs

One of the most successful applications of MPC has been in its ability to incorporate the effects of transaction costs. They have been explicitly considered in [6, 9] in a portfolio optimization context, and in [2, 20, 26, 27] in the dynamic option hedging context of Section 4.

Transaction costs are the costs that are incurred when trades are made (buying or selling). They come from two main sources. The first is a commission that is paid to the broker, and is often a fixed amount per trade. These days, the commission per trade ranges from $15 or so down, and is generally not a significant cost for large traders.

The second cost associated with buying and selling is related to the so-called *bid-ask spread*. The *bid* is the price that one can sell a stock for immediately, while the *ask* is the price that one can buy for immediately. When a trader uses so-called *market* orders to buy and sell, they buy at the ask and sell at the bid, in effect losing the difference between the bid and ask (the so-called *bid-ask spread*) on each round trip transaction.[3] This can be considered a cost of transacting. Note that this cost scales in proportion to the number of shares transacted, and is often modeled as a fixed percentage of the dollar amount of each transaction.

The bid-ask spread ranges from a penny on up, depending on the liquidity of the stock. While it may appear to be a very small amount, and often represents a fraction of a percent of the dollar amount transacted (for example, $0.01 on a $10 stock is just 0.1% of the dollar amount transacted) it can have a significant effect on the profitability of high-turnover strategies.

MPC has been extremely successful at taking into account the effects of transaction costs. Next, we illustrate how transaction costs may be incorporated into MPC formulations via altering the account value dynamics.

Modeling of Transaction Costs in Account Value Dynamics

Assume that at time k, a transaction takes place in which the shares held in stock i change from $u_i(k-1)$ to $u_i(k)$. The total dollar amount of this transaction is the price of stock i at time k, $S_i(k)$, multiplied by the number of shares bought or sold $u_i(k)-u_i(k-1)$, which totals $S_i(k)(u_i(k)-u_i(k-1))$.

If we model the transaction cost associated with this trade as a fixed percentage χ_i of the dollar amount, then the cost of this single transaction will be $\chi_i S_i(k)|u_i(k)-u_i(k-1)|$. Therefore, the cost of all transactions corresponding to the n stocks is just the sum of the individual costs,

$$\mathscr{T}(k) = \sum_{i=1}^{n} \chi_i S_i(k)|u_i(k)-u_i(k-1)|. \tag{12}$$

[3] Due to a lack of liquidity, large market orders will sometimes transact at prices that are worse than the existing bid or ask. This represents a further cost.

Now, if the account value *immediately prior* to the transactions at time k was $V(k^-)$, then the account value immediately after the transactions (assuming the self-financing constraint is in effect) is simply the previous account value minus the transaction cost,

$$V(k^+) = V(k^-) - \mathscr{T}(k). \tag{13}$$

The account value dynamics over the next time step is given by

$$V(k+1^-) = V(k^+)(1+r_f) + \sum_{i=1}^{n} u_i(k)(S_i(k+1) - (1+r_f)S_i(k)). \tag{14}$$

If desired, one can combine the above two equations to obtain

$$V(k+1^-) = (V(k^-) - \mathscr{T}(k))(1+r_f) + \sum_{i=1}^{n} u_i(k)(S_i(k+1) - (1+r_f)S_i(k)). \tag{15}$$

To incorporate this into the on-line MPC optimization, one simply replaces the account value dynamics over the horizon with Equations (12) and (15). Moreover, such simple proportional transaction costs are often quite tractable when added to the on-line optimizations.

If a different form for the transaction cost is desired, one replaces the definition of $\mathscr{T}(k)$ in (12) by the appropriate model. For example, for combined fixed and proportional costs, see the model used by Bemporad et al. [2] in the context of dynamic option hedging.

3.1.3 Constraints in Portfolio Optimization

To complete this section, we outline a number of constraints that are frequently encountered in financial markets and trading. Some arise out of practical considerations, while others are imposed via regulation. Most of these constraints can be directly incorporated into open-loop versions of the on-line MPC optimization; e.g., see [7, 9, 29].

No Short Selling

Short selling is when a trader borrows shares of a stock and sells them in the market. In our model, this corresponds to holding a negative number of shares $u(k) < 0$. Short selling is extremely common in financial markets, but in some cases is restricted.[4] In the context of our portfolio optimization formulation, short selling corresponds to the constraint that the decision variables must be non-negative, $u(k) \geq 0$.

[4] For example, short selling is not possible in Individual Retirement Accounts (IRAs) in the US.

Limits on Percent Invested

When trading, it may be desirable to hold a portfolio that is diversified and not overly concentrated in a single stock or industry. In the case of a single stock, a constraint of the form

$$|u_i(k)S_i(k)| \leq \beta V(k)$$

with β some fraction, such as 5%, will ensure that stock i cannot comprise more than β of the portfolio. A similar constraint can be imposed over all stocks in a specific industry $\mathscr{I}$ as

$$\sum_{i \in \mathscr{I}} |u_i(k)S_i(k)| \leq \beta V(k).$$

Turnover Constraints

In order to avoid excessive trading, one may seek to impose a so-called *turnover* constraint. This is a constraint that limits that dollar value of the trading that takes place, and can be imposed as

$$\sum_{i=1}^{n} |u_i(k) - u_i(k-1)| S_i(k) \leq \beta V(k)$$

where β specifies the maximum turnover allowed, expressed as a percent of the total portfolio value.

The above considered constraints are some of the most commonly encountered in portfolio optimization, however others may appear depending on problem specifics. In such cases, a great advantage of MPC is its ability to incorporate such constraints into the on-line optimizations. Next, we turn our attention to the dynamic option hedging problem.

4 MPC in Dynamic Option Hedging

The second main class of problems we consider in this chapter is dynamic option hedging. As mentioned in the introduction, the dynamic option hedging problem involves trading to replicate, as closely as possible, the payoff of an option. The classical dynamic hedging problem is that of replicating the payoff of a *European call option* on a single stock. We will use this problem to illustrate the application of MPC methods. Thus, we begin this section with a brief description of a European call option and its associated hedging problem.

4.1 European Call Option Hedging

A European call option is a contract that gives the owner the right, but not the obligation, to purchase a share of a specified stock, S, called the *underlying stock*, at a fixed price K, called the *strike price*, at a fixed time T, called the *expiration time*.

For example, a European call option on AAPL with strike price \$100 and expiration in 2 months gives the holder of the option the right to purchase AAPL in 2 months for a price of \$100. If, in 2 months, AAPL is selling for \$120, the holder of the option will "exercise" it and purchase AAPL for \$100, thus saving $S(T) - K = \$120 - \$100 = \$20$. The seller of the option will have to deliver the option holder the share of AAPL in exchange for the \$100, and thus will lose \$20. On the other hand, if AAPL is selling for \$90 in 2 months, the holder of the option will not "exercise" it, thus letting it expire worthless. In this case, the seller of the option has no remaining obligation to the option holder.

Using this reasoning, a European Call Option on a stock S can be thought of as a security with a payoff value of $c(T) = \max\{S(T) - K, 0\}$ at the expiration time T. That is, if the stock price at expiration T is greater than the strike price K, the holder of the option will exercise it and purchase the stock for K. This represents a savings of $S(T) - K$ over the market price of the stock. On the other hand, if the stock price at expiration T is below the strike price, the option will not be exercised and will instead expire worthless. Combining these two scenarios indicates that the value of an option at expiration is given by the function $c(T) = \max\{S(T) - K, 0\}$. A plot of this payoff function is provided in Figure 5.

The option replication problem involves trading the underlying stock over time so that the value of the trader's account at expiration time T matches the payoff value of the option $c(T) = \max\{S(T) - K, 0\}$. The ability to create such a trading strategy serves two purposes. First, if provides a notion of a "fair" price for the option.[5] The reasoning is simple. Since the portfolio replicates the payoff of the option, a trader should be indifferent between holding the option or the replicating portfolio. Thus, the cost of creating the replicating portfolio should be equal to the price of the option since they provide equivalent payoffs.

Second, if one sells an option, then they are responsible for the payoff $c(T) = \max\{S(T) - K, 0\}$. With a replicating strategy, the seller can take the proceeds of the option sale and use it to "hedge" their liability by replicating the required $\max\{S(T) - K, 0\}$ payoff. For this reason, option replication is often called *dynamic hedging*. This hedging is particularly important for investment banks and option market makers that routinely sell options.

[5] More specifically, it leads to the concept of the absence of arbitrage price for the option. See [16] or [28].

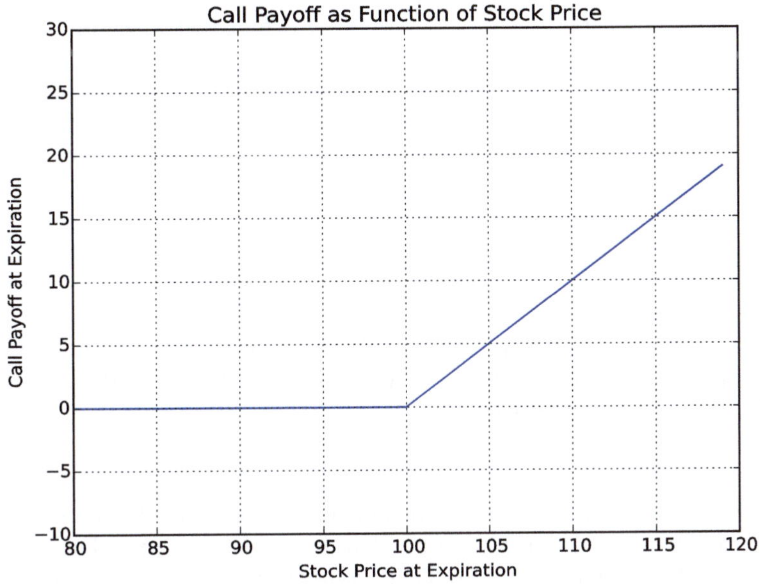

Fig. 5: The payoff function of a European call option with strike price $K = \$100$.

4.2 Option Replication as a Control Problem

As explained previously, the option replication problem involves trading the underlying stock in order to replicate the payoff of the option $c(T)$ as closely as possible. As a control problem, this may be formulated as follows. Let $\rho(V(T), c(T))$ represent some measure of the error between the option payoff $c(T) = \max\{S(T) - K, 0\}$ and the trader's account value $V(T)$. For example, a common choice is the mean-squared error $\rho(V(T), c(T)) = \mathbb{E}_0[(V(T) - c(T))^2]$, but many other objectives have been used as well. Then, the dynamic option hedging control problem can be stated as

$$\begin{aligned} &\min_{u(\cdot)} \rho(V(T), c(T)) \\ \text{s.t.}\quad &V(k+1) = V(k)(1+r_f) + u(k)(S(k+1) - (1+r_f)S(k)), \\ &V(0) = V_0, \end{aligned}$$

where again the optimization is taken over admissible non-anticipative trading strategies $u(\cdot)$. A pictorial representation of this control problem is given in Figure 6.

In some cases, the trader's initial wealth V_0 may also be a decision variable. That is, the trader is able to choose the best initial value of the account in order to replicate the payoff. This best initial account value serves as a notion of the "fair" price for the option, since it represents the initial capital needed to best replicate the option.

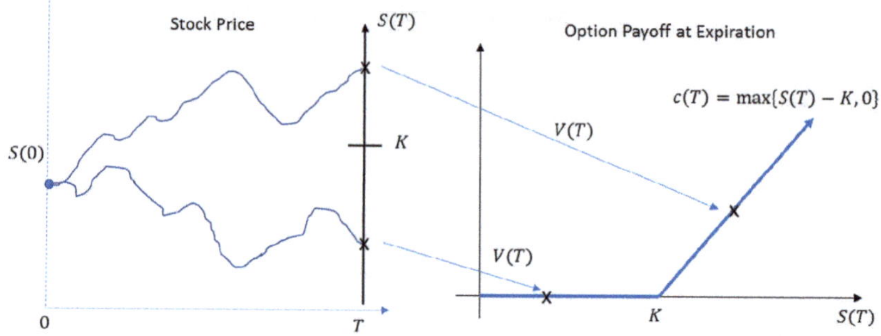

Fig. 6: The figure on the left shows two paths of the stock. The right plot shows the desired replication of the payoff function $c(T) = \max\{S(T) - K, 0\}$ by the account value $V(T)$ at expiration time T.

In idealized frictionless markets when the underlying stock S follows the GBM price dynamics of Equation (6), the problem has a closed form solution that was first given by Black and Scholes [4]. On the other hand, when markets are not ideal, such as when transaction costs exist, no closed form solution is possible. This has led to the development of MPC methods, as discussed next.

4.3 MPC Option Hedging Formulations

The basic MPC approach to option replication involves using a model of the stock price movement to predict the account value N steps into the future from the current time, $V(N|k)$, so as to minimize the hedging replication error. That is, the basic structure of the on-line optimization is typically of the form,

$$\begin{aligned} &\min_{u(\cdot|k)} \rho_k^{(N)}(V(N|k)) \\ \text{s.t. } &V(j+1|k) = V(j|k)(1+r_f) + u(j|k)(S(j+1|k) - (1+r_f)S(j|k)) \\ &V(0|k) = V(k), \end{aligned}$$

where $\rho_k^{(N)}(\cdot)$ captures the desired replication objective, but suitably transformed to the end of the horizon $k+N$.

There are two important ways in which this on-line optimization tends to differ from those used in the portfolio optimization problem. The first is that the optimization is often *not* solved open-loop. That is, rather than assuming that the control variables $u(\cdot|k)$ depend only on time, stochastic programming formulations that allow stock price dependence are used. For example, in the work of Meindl and Primbs [19, 20] and Bemporad et al. [1–3] scenario-based stochastic programming formulations that allow dependence on the stock price value are developed. In a similar vein, in Primbs [26], $u(\cdot|k)$ is taken to be a linear combination of basis functions that also depend on the stock price.

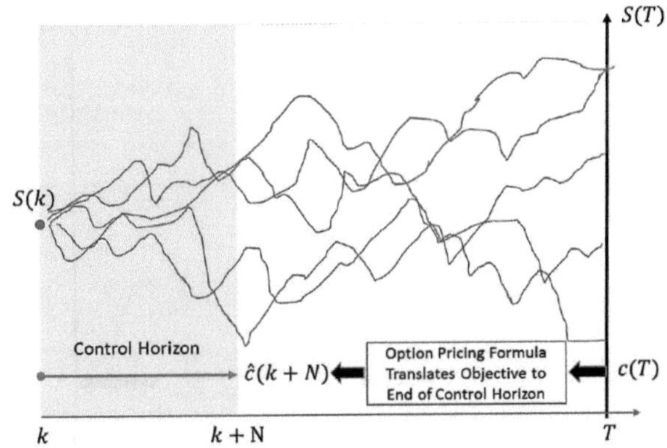

Fig. 7: An option pricing formula is used to translate the objective of replicating the option payoff at expiration T to an objective of replicating the option pricing formula used at the end of the MPC horizon $k+N$.

Second, different approaches have been taken over how to assign a reasonable objective function $\rho_k^{(N)}(\cdot)$ to reflect the goal of achieving replication at the expiration time T. The issue arises because typically the horizon used in MPC does not come close to reaching the expiration time T of the option. For example, the expiration of the option may occur in 2 months, but the MPC horizon may only extend 1 week into the future. We explain two different approaches to this objective function issue next.

4.3.1 Using an Option Pricing Model

One approach to assigning the objective for the on-line MPC optimization is to use an existing option pricing formula as the replication target at the end of the MPC horizon, $k+N$. That is, while the payoff of the option $c(T)$ is only known at expiration T, a pre-existing and often simplified option pricing formula, call it $\hat{c}(k+N)$, is used to compute a theoretical value for the option at the end of the MPC horizon $k+N$. The on-line objective function in the MPC problem, $\rho_k^{(N)}(V(N|k))$, then becomes to replicate the option pricing formula $\hat{c}(k+N)$. Moreover, the market features that were ignored in order to compute $\hat{c}(k+N)$ are then included in the MPC optimization. This approach is depicted in Figure 7, and has been successfully employed in [1–3].

4.3.2 Predicting to Expiration

A second approach is to predict from the end of the MPC control horizon $k+N$ all the way to expiration T using a predefined hedging strategy. The objective is then

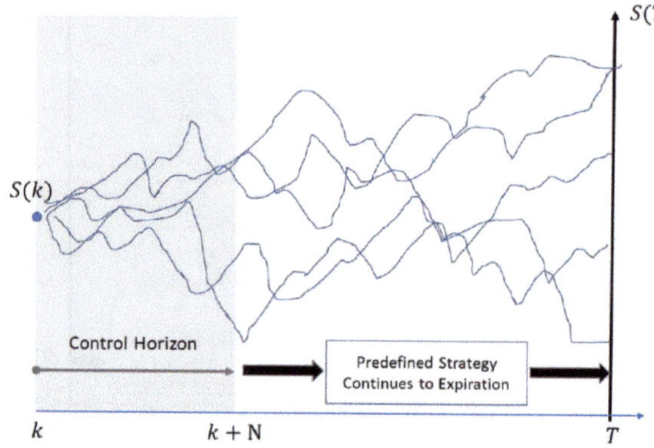

Fig. 8: A predefined strategy is used to continue to expiration where the objective is to replicate the payoff value of the option.

the replication error at expiration, and the MPC implementation takes on a shrinking horizon approach. This is depicted in Figure 8. The hedging strategy used to predict to expiration is usually quite simple and makes little attempt to incorporate detailed market features that are included over the MPC control horizon. Variations of this approach were employed in [18–20, 27].

4.4 Additional Considerations in Option Hedging

Because option replication involves dynamically trading a stock, some of the considerations that appear in the portfolio optimization problem are also relevant in this case. For example, since option replication strategies involve frequent trading, transaction costs are a highly significant issue, and were key motivations behind the MPC formulations in [2, 18–20, 27]. On the other hand, most participants in options markets, especially those that seek to hedge option sales, will not be restricted from short selling. Thus, such constraints are often not relevant to the hedging problem. Moreover, option pricing and hedging strategies are highly dependent on the volatility of the underlying stock. Therefore, using models that accurately capture market features that are relevant to dynamic hedging is important.

Another issue is the choice of a measure of replication. That is, many reasonable choices can be made for the replication objective function $\rho(\cdot)$. For example, as mentioned previously, a popular choice is the mean-squared error between the account value $V(T)$ and the option payoff $c(T)$; i.e., $\mathbb{E}[(V(T)-c(T))^2]$. This can lead to simplified calculations for the on-line optimization; see, for example, [27]. On the other hand, in the context of a seller hedging an option, the seller will lose money if $V(T) < c(T)$, but is actually rewarded when $V(T) > c(T)$. This has motivated other replication or outperformance criteria, including mean-variance [27], expected absolute error [20], and conditional Value-at-Risk and mini-max [2, 3]. Thus, there are many different objective functions that can be reasonably used for the dynamic hedging problem, each leading to various trade-offs in computation and performance characteristics.

Overall, dynamic option hedging is an important application area in the field of finance, and MPC approaches are able to effectively and elegantly address many of the critical features of this problem.

5 Conclusions

This chapter provided an overview of the application of MPC to the field of finance. The two classical finance problems of portfolio optimization and dynamic option hedging were considered. Both of these are naturally formulated as stochastic control problems and thus are amenable to Stochastic MPC techniques.

MPC is particularly relevant to problems involving trading because financial markets are naturally a dynamic and constrained environment where the inclusion of realistic features, such as transaction costs, has a significant effect on the quality of solutions. Moreover, the modeling of stock price movement, which plays a key role in determining the stochastic dynamics, and trading objectives are extremely varied. Some formulations lend themselves to problem structure that facilitates efficient on-line optimization, while others are better addressed via more general stochastic programming methods. All of these facts play to the strengths of MPC, which is flexible enough to locally incorporate important features in its on-line optimization, while globally provide solutions that outperform existing methods.

Overall, finance is an exciting application area where MPC is being used with great success. As new Stochastic MPC methods are developed, they will continue to find fertile ground for application in financial markets and their associated stochastic control problems.

References

1. Bemporad, A., Gabbriellini, T., Puglia, L., Bellucci, L.: Scenario based stochastic model predictive control for dynamic option hedging. In: Proceedings of 49th IEEE Conference on Decision and Control, Atlanta (2010)
2. Bemporad, A., Puglia, L., Gabbriellini, T.: A stochastic model predictive control approach to dynamic option hedging with transaction costs. In: Proceedings of American Control Conference, San Francisco (2011)
3. Bemporad, A., Bellucci, L., Gabbriellini, T.: Dynamic option hedging via stochastic model predictive control based on scenario simulation. Quant. Finan. **14**(10), 1739–1751 (2014)
4. Black, F., Scholes, M.: Pricing of options and corporate liabilities. Eur. J. Polit. Econ. **81**(3), 637–654 (1973)
5. Calafiore, G.C.: Multi-period portfolio optimization with linear control policies. Automatica **44**, 2463–2473 (2008)

6. Calafiore, G.C.: An affine control method for optimal dynamic asset allocation with transaction costs. SIAM J. Control Optim. **48**(4), 2254–2274 (2009)
7. Dombrovskii, V.V., Obyedko, T.Y.: Predictive control of systems with Markovian jumps under constraints and its application to the investment portfolio optimization. Autom. Remote Control **72**(5), 989–1003 (2011)
8. Dombrovskii, V.V., Odyedko, T.Y.: Model predictive control for constrained systems with serially correlated stochastic parameters and portfolio optimization. Automatica **54**, 325–331 (2015)
9. Dombrovskii, V.V., Dombrovskii, D.V., Lyashenko, E.A.: Investment portfolio optimisation with transaction costs and constraints using model predictive control. In: Proceedings of the 8th Russian-Korean International Symposium on Science and Technology, pp. 202–205 (2004)
10. Dombrovskii, V.V., Dombrovskii, D.V., Lyashenko, E.A.: Predictive control of random-parameter systems with multiplicative noise: application to investment portfolio optimization. Autom. Remote Control **66**, 583–595 (2005)
11. Dombrovskii, V.V., Dombrovskii, D.V., Lyashenko, E.A.: Model predictive control of systems random dependent parameters under constraints and its application to the investment portfolio optimization. Autom. Remote Control **67**, 1927–1939 (2006)
12. Herzog, F., Keel, S., Dondi, G., Schumann, L.M., Geering, H.P.: Model predictive control for portfolio selection. In: Proceedings of the 2006 American Control Conference, Minneapolis, pp. 1252–1259 (2006)
13. Herzog, F., Dondi, G., Geering, H.P.: Stochastic model predictive control and portfolio optimization. Int. J. Theor. Appl. Finance **10**(2), 203–233 (2007)
14. Hull, J.C.: Options, Futures, and Other Derivatives, 9th edn. Pearson, New York (2014)
15. Lee, J.H.: Dynamic portfolio management with private equity funds. PhD thesis, Stanford University (2012)
16. Luenberger, D.G.: Investment Science, 2nd edn. Oxford University Press, Oxford (2013)
17. Markowitz, H.: Portfolio selection. J. Finance **7**(1), 77–91 (1952)
18. Meindl, P.: Portfolio optimization and dynamic hedging with receding horizon control, stochastic programming, and Monte Carlo simulation. PhD thesis, Stanford University (2006)
19. Meindl, P., Primbs, J.: Dynamic hedging with stochastic volatility using receding horizon control. In: Proceedings of Financial Engineering Applications, Cambridge, MA, pp. 142–147 (2004)
20. Meindl, P., Primbs, J.: Dynamic hedging of single and multi-dimensional options with transaction costs: a generalized utility maximization approach. Quant. Finan. **8**(3), 299–312 (2008)
21. Merton, R.C.: Optimum consumption and portfolio rules in a continuous-time model. J. Econ. Theory **3**(4), 373–413 (1971)
22. Merton, R.C.: Theory of rational option pricing. Bell J. Econ. Manag. Sci. **4**, 142–183 (1973)
23. Oksendal, B.: Stochastic Differential Equations: An Introduction with Applications, 6th edn. Springer, Berlin (2010)
24. Piccoli, B., Marigo, A.: Model predictive control for portfolio optimization. In: Proceedings of 2nd IFAC Symposium on System, Structure and Control (2004)
25. Primbs, J.: Portfolio optimization applications of stochastic receding horizon control. In: Proceedings of American Control Conference, pp. 1811–1816 (2007)
26. Primbs, J.: Dynamic hedging of basket options under proportional transaction costs. Int. J. Control **82**(10), 1841–1855 (2009)
27. Primbs, J.: LQR and receding horizon approaches to multi-dimensional option hedging under transaction costs. In: Proceedings of American Control Conference, pp. 6891–6896 (2010)
28. Primbs, J.A.: A Factor Model Approach to Derivative Pricing. CRC Press, West Palm Beach (2016)
29. Primbs, J., Sung, C.H.: A stochastic receding horizon control approach to constrained index tracking. Asia-Pacific Finan. Markets **15**(1), 3–24 (2008)

30. Samuelson, P.A.: Lifetime portfolio selection by dynamic stochastic programming. Rev. Econ. Stat. **51**(3), 239–246 (1969)
31. Sridharan, S., Chitturi, D., Rodriguez, A.A.: A receding horizon control approach to portfolio optimization using a risk-minimax objective for wealth tracking. In: Proceedings of IEEE Conference on Control Applications, Denver, pp. 1282–1287 (2011)
32. Wonham, W.M.: Optimal stationary control of a linear system with state-dependent noise. SIAM J. Control Optim. **5**, 486–500 (1967)
33. Yamada, Y., Primbs, J.: Model predictive control for optimal portfolios with cointegrated pairs of stocks. In: Proceedings of the 51st IEEE Conference on Decision and Control, pp. 5705–5710 (2012)

Index

S. V. Raković, W. S. Levine (eds.), *Handbook of Model Predictive Control*,
Control Engineering, https://doi.org/10.1007/978-3-319-77489-3

MIX
Papier aus verantwortungsvollen Quellen
Paper from responsible sources
FSC® C105338
FSC
www.fsc.org

If you have any concerns about our products,
you can contact us on
ProductSafety@springernature.com

In case Publisher is established outside the EU,
the EU authorized representative is:
Springer Nature Customer Service Center GmbH
Europaplatz 3, 69115 Heidelberg, Germany

Printed by Libri Plureos GmbH
in Hamburg, Germany